Environmental Science and Engineering

Environmental Science

Series Editors

Rod Allan
Ulrich Förstner
Wim Salomons

For further volumes:
http://www.springer.com/series/3234

Jacek Kozak · Katarzyna Ostapowicz
Andrzej Bytnerowicz · Bartłomiej Wyżga
Editors

The Carpathians: Integrating Nature and Society Towards Sustainability

Springer

Editors
Jacek Kozak
Katarzyna Ostapowicz
Institute of Geography and Spatial Management
Jagiellonian University
Kraków
Poland

Bartłomiej Wyżga
Institute of Nature Conservation
Polish Academy of Sciences
Kraków
Poland

Andrzej Bytnerowicz
USDA Forest Service
Pacific Southwest Research Station
Riverside
USA

ISSN 1431-6250
ISBN 978-3-642-12724-3 ISBN 978-3-642-12725-0 (eBook)
DOI 10.1007/978-3-642-12725-0
Springer Heidelberg New York Dordrecht London

Library of Congress Control Number: 2013932852

Printed on acid-free paper

Springer is part of Springer Science+Business Media (www.springer.com)

Preface

The Carpathian Mountains exhibit the same set of conditions that are typical of other mountain regions inhabited by humans for centuries: rich resources, high biodiversity, landscape values, complex history, diverse political and social settings, and opportunities offered by tourism that are confronted with peripheral economic locations and a demanding environment. These issues were presented and discussed during the 1st Forum Carpaticum held in Kraków, Poland, in September 2010.

It was the first of a series of conferences planned in the Carpathian region and initiated by the Science for the Carpathians (S4C): a network of scientists that has been active since 2008, attempting to merge various fields of expertise in Carpathian research and to contribute to the sustainable development of the region. The 1st Forum Carpaticum gathered more than 200 scientists and stakeholders from the whole region, initiating a discussion on vital scientific and practical issues. The 1st Forum was only the beginning, as already in 2012 the 2nd Forum took place at Stará Lesná in Slovakia, and further events are planned as well, including the 3rd Forum Carpaticum 2014 in Ukraine. The large number of scientists interested in participating in these events and those subscribing to the S4C newsletter are a testimony to the growing momentum of Carpathian science. This increase has been caused by a worldwide and European recognition of the value of the Carpathian region, resulting in the amount of research efforts concentrated in this mountain region. Equally important, since the political changes in 1989, scientists from the Carpathian countries have been able to communicate with the global mountain community in a much more open and active way than ever before. In this process, they increasingly take advantage of the long history of science in the region and the vast knowledge that has been gathered there for centuries.

This book is a direct outcome of many excellent presentations and stimulating discussions during the 1st Forum Carpaticum. Our gratitude is extended first to the participants of the event, especially those who submitted their research papers for publication. We would also like to thank all our colleagues who voluntarily reviewed chapters. Invaluable editorial help of Laurie Dunne and Joanna Zawiejska is greatly appreciated.

The 1st Forum Carpaticum was a joint effort of many, especially those involved in the S4C activities. We would like to express our gratitude to all who have contributed to various steps of the S4C development, in particular the former and present members of the S4C Scientific Steering Committee, members of the Scientific and Organizational Committees of the 1st Forum Carpaticum, and the supporting institutions, including the Mountain Research Initiative—Europe; Institut für Gebirgsforschung: Mensch und Umwelt; and the United Nations Environment Programme Vienna—Interim Secretariat of the Carpathian Convention.

We would like to thank the authorities of the Jagiellonian University and its Institute of Geography and Spatial Management that hosted the 1st Forum Carpaticum. We also thank the Faculty of Forestry, University of Agriculture in Kraków, who prepared the special session on mountain forestry. We would like to acknowledge the support granted by the Director General of the State Forests National Holding in Poland, Regional Directorate of the State Forests in Kraków, the International Visegrad Fund, and the Committee on Geographical Sciences of the Polish Academy of Sciences.

Jacek Kozak
Katarzyna Ostapowicz
Andrzej Bytnerowicz
Bartłomiej Wyżga

Contents

Part III Human Dimension

The Carpathian Mountains: Challenges for the Central and Eastern European Landmark

Jacek Kozak, Katarzyna Ostapowicz, Andrzej Bytnerowicz and Bartłomiej Wyżga

Abstract The Carpathians are a very distinct mountain chain and a major landscape landmark in the core of Central and Eastern Europe (CEE). The Carpathians are characterized by valuable biological resources and a rich cultural heritage that is of high importance for the Carpathian countries (Austria, Czech Republic, Hungary, Poland, Romania, Serbia, Slovakia and Ukraine). Historically the Carpathian Region has experienced many political changes and conflicts including more than 40 years of communism that restricted development of an efficient pan-Carpathian scientific co-operation. In contrast, the last 20 years allowed remarkable increase in the awareness of the value of the Carpathians for the European society and opened new opportunities for most Carpathian countries after their accession to the European Union. The 1st Forum Carpaticum held in 2010 in Kraków, Poland, is a perfect example of this trend. This book presents a subset of research problems related to the sustainable development of the Carpathian region discussed at the event. Four sections of the book deal with various issues related to abiotic environment, forests and biodiversity, human activities in the region and research methods allowing a better understanding of the past, present and future of the Carpathians.

J. Kozak (✉) · K. Ostapowicz
Institute of Geography and Spatial Management, Jagiellonian University, Gronostajowa 7
30-387 Kraków, Poland
e-mail: jkozak@gis.geo.uj.edu.pl

K. Ostapowicz
e-mail: kostapowicz@gis.geo.uj.edu.pl

A. Bytnerowicz
Pacific Southwest Research Station, USDA Forest Service, 4955 Canyon Crest Drive,
Riverside, CA 92507, USA
e-mail: abytnerowicz@fs.fed.us

B. Wyżga
Institute of Nature Conservation, Polish Academy of Sciences, al. Mickiewicza 33
31-120 Kraków, Poland
e-mail: wyzga@iop.krakow.pl

J. Kozak et al. (eds.), *The Carpathians: Integrating Nature and Society Towards Sustainability*, Environmental Science and Engineering,
DOI: 10.1007/978-3-642-12725-0_1, © Springer-Verlag Berlin Heidelberg 2013

1 Geographical Background

Mountainous areas have attracted the attention of scientists for many years, due to their exceptional environmental value, beauty, and a range of services they provide to the communities living in and near them. Mountains have been referred to as water towers of the world and recognized as important hotspots of biodiversity. In addition, these areas are frequently considered as regions that are highly sensitive to global change and thus can be treated as early warning systems to humanity. Currently, mountains worldwide are the target of various national or international programs and policies oriented toward sustainable development. Since the 1990s, these initiatives have increased especially after the United Nations Conference on Environment and Development in Rio de Janeiro in 1992 and the International Year of the Mountains in 2002 (European Environment Agency 2010; Messerli 2012).

Compared to other mountainous areas, the Carpathians are perhaps one of the lowest in elevation, yet a very distinct mountain chain. The Carpathian Mountains occupy the core of Central and Eastern Europe (CEE) and are its most eminent landmark. Hardly exceeding 2,000 m in altitude, they also form the easternmost mountain rampart in Europe before the vast lowlands extending to the Ural and the Caucasus Mountains. There are remarkable differences among regional divisions of this part of Europe: Kondracki (1989) includes in the Carpathians a hilly region in northern Austria and the uplands of Transylvania and excludes the mountain ranges in northeastern Serbia, contrary to the division that is used by the Carpathian Convention (United Nations Environment Programme 2007). Various regional divisions result in area estimates of the Carpathians that differ significantly; for instance, the extent of the Carpathians as estimated by the Carpathian Convention is slightly above 160,000 km^2 (United Nations Environment Programme 2007), while the area of the Carpathians as delimited by Kondracki (1989) exceeds 200,000 km^2. There is a general agreement, however, that the backbone of the mountains forms an arc-shaped chain stretching from the Danube River near Vienna to the northeast, then east, southeast, south, and west again to the Danube at the Iron Gate, and further into Serbia (Fig. 1).

The Carpathians are divided among eight CEE countries: Austria, the Czech Republic, Hungary, Poland, Romania, Serbia, Slovakia, and Ukraine, and most of that territory is now within the European Union. The Carpathians occupy a significant part of Slovakia and Romania (71 and 29 %, respectively) and up to 10 % of the territories of other countries. In addition, in the Czech Republic, Hungary, Poland, and Ukraine the Carpathians have a peripheral position; as for Austria and Serbia, various regional divisions may exclude either of these two countries from the Carpathian region.

The Carpathian arc exceeds the altitude of 2,000 m in its northern part, the Tatra Mountains, the highest range of the Carpathians (Gerlachovský štít, Gerlach, 2,655 m), and in the Lower Tatra Mountains (Ďumbier, 2,043 m). Further eastward elevations of the main range decrease below 1,000 m, only to rise again in the eastern section of the Carpathian chain to above 2,000 m, with the culmination in the Rodna

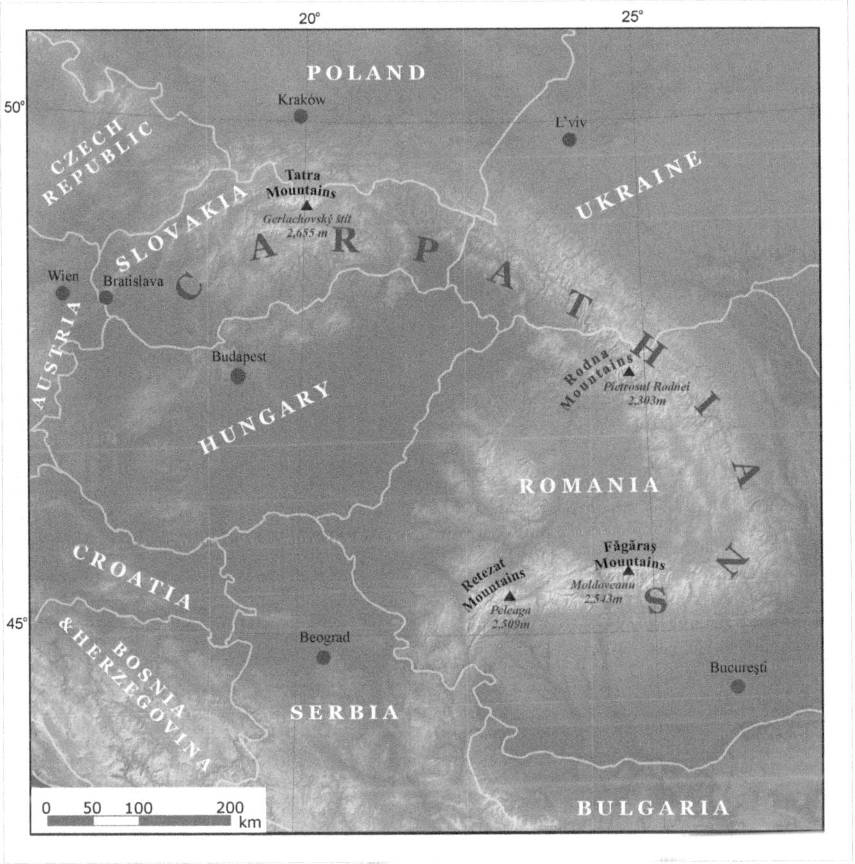

Fig. 1 The Carpathians

Mountains (Pietrosul Rodnei 2,303 m). The Făgăraş and Retezat Mountains in the Southern Carpathians (Moldoveanu 2,543 m, Peleaga 2,509 m) are lower than the Tatra Mountains, yet the area above 2,000 m in the southern part of the Carpathians is 405 km^2, significantly larger than in its northern and eastern parts (55 and 10 km^2, respectively). The mountain arc forms an important European watershed divide between the Danube in the south and east, and the Oder, the Vistula, and the Dniester Rivers in the north. The Danube drains most of the Carpathian area: all territory of the Hungarian, Romanian, Austrian, and Serbian Carpathians; almost all territory of the Slovak Carpathians; significant parts of the Czech and Ukrainian Carpathians; and small sections of the Polish Carpathians belong to its drainage basin. The Czech Carpathians are also partially drained by the Oder, the Polish Carpathians mostly by the Vistula, and the Ukrainian Carpathians by the Dniester. In this way, the Carpathians secure water for a number of regions far beyond their boundaries, being an important water tower for most countries of the region.

Fig. 2 Rodna Mountains in Romania (photo credit: Dominik Kaim)

Although several mountain ranges of the Carpathians reach high above the current timberline (1,400–1,800 m), their relatively low elevation does not allow the persistence of glaciers even in the highest locations (Fig. 2). Forests below the timberline (Fig. 3) constitute the major resource of the Carpathians, with the dominance of beech, spruce, and fir. Estimates of the forest cover proportion of the Carpathians vary among various reports, depending on regional delimitations and mountain boundaries. For instance, Kozak et al. (2008) provide a value of 48 % using regional boundaries as found in Kondracki (1989), while the Carpathian Environment Outlook results (United Nations Environment Programme 2007) show a cover of nearly 60 %. Higher up, yet below the timberline, the mountains are almost fully forested: between 1,000–1,500 m a.s.l. the forest cover exceeds 80 % (Kozak 2009).

The Carpathian landscape not only includes a vast "wilderness," which is typically under the most strict nature protection regime, but also historical, man-made features: a legacy of the rich history of the mountainous area, especially of the advances in agriculture, forestry and social systems in the late eighteenth and the nineteenth century. In this period, the Carpathians constituted the northern, eastern, and southeastern margins of a single political entity: the Austrian and later on the Austro-Hungarian Empire. Hence, the eighteenth and nineteenth centuries were marked by a relative homogeneity of the political and land use system in the

Fig. 3 The Carpathian forest (photo credit: Jacek Kozak)

Carpathian area. The formation of the settlement system and the expansion of agriculture were at its end, and due to social and land reforms, the feudal land systems were becoming a market-oriented economy. This also meant that forests started to be managed as a land resource not directly linked to agricultural activities, leading to major changes in forest composition throughout the area, such as widespread introduction of spruce in the northern and western parts of the mountains. The nineteenth century saw the emergence of a specific land use and land cover patterns in the Carpathians: a mosaic of arable lands, grasslands, and forests, with different proportions of these three components at various elevations. This mosaic was partially erased by collectivization of agriculture in the postwar period in several CEE countries.

Currently, after the privatization of land carried out in the 1990s and 2000s, two major processes threaten the Carpathian cultural landscapes: land abandonment and forest succession on the one hand and urbanization and large-scale investments in the tourism industry on the other (Kozak 2009). While the former increases the naturalness of the mountain region, and to some extent may benefit nature conservation (Fig. 4), the latter can cause numerous societal conflicts and disputes on the allowable and desirable use of the mountain environment by humans (Fig. 5). Regardless of the current trends of land use and land cover change, the diverse Carpathian landscape with a dominance of forests harbors a significant part of European biological resources. In particular, large mammals that have vanished from most of Western Europe are still abundant in the mountains, due to their vast forest areas, and include the brown bear, European bison, lynx, and wolf (United Nations Environment Programme 2007).

Fig. 4 Land abandonment that results in forest succession in the Polish Carpathians (photo credit: Bartłomiej Wyżga)

Fig. 5 Are there any mountains left? A scene from Zakopane, Poland (photo credit: Jacek Kozak)

2 Toward Regional Cooperation

In the late nineteenth and throughout the twentieth century, numerous political conflicts including two world wars reshaped the political boundaries in this part of Europe in three distinct phases: first the disintegration of the Austro-Hungarian Empire into several independent states, then the shift of the western boundary of the Soviet Union, and finally the complete change in the political status of the region after the collapse of communism in 1989. The 1990s witnessed the breakdown of the former Soviet Union and Yugoslavia, resulting in the emergence of Ukraine and Serbia, and the division of Czechoslovakia into the Czech Republic and Slovakia. The past political turmoil and numerous conflicts among the countries of the region, together with political barriers that did not allow for full cooperation during the communist period, impeded the emergence of the pan-Carpathian identity.

The frequent and major changes of political boundaries caused a continuous redefinition of the core-periphery relationships in the region, breaking old links of settlement and transportation systems and establishing new ones. For example, the Zakarpattia region was a periphery of Czechoslovakia in the interwar period, and later it became the westernmost periphery of the Soviet Ukraine, separated from the rest of the country by high Carpathian ridges.

Only recently several attempts of regional integration have been promoted by the national authorities of the respective Carpathian countries. In 2003 this process led to the adoption of the Framework Convention on the Protection and Sustainable Development of the Carpathians by the Carpathian countries (The Carpathian Convention 2012), resembling to some extent the forms of regional integration in the Alps. Recently, active cooperation of Carpathian scientists and stakeholders has been initiated within the Science for the Carpathians (S4C) network (Björnsen Gurung 2013).

There are several obstacles, however, that constrain the development of the truly pan-Carpathian cooperation in a manner resembling that of the Alps. First, the economic potential and welfare of the Carpathian countries is much lower than that of the Alpine ones, with a more pronounced regional diversification (United Nations Environment Programme 2007). The economic gap was significantly deepened by the 40 years of an inefficient, environmentally disastrous economy during communism. Also, in the last 20 years after the fall of communism, the paths of growth differed among various countries of the region and practically all of them suffered a deep economic crisis. Currently, the Carpathians show all levels of economic development: from regions experiencing depopulation or high unemployment to the quickly expanding regional centers of tourism, providing jobs and high income. The accession of Poland, Slovakia, Czech Republic, Hungary, and later Romania to the European Union (EU) opened a variety of new possibilities for this relatively poor part of Europe. With Ukraine and Serbia remaining outside the EU, a new partitioning of the Carpathian region took place and led to another significant redefinition of the core-periphery relations in the Carpathians.

Regional cooperation based on joint multilateral projects is also difficult due to the geography of the mountains and their unique, arched shape. The distance between, for example, the Czech Republic or Poland and Romania or Serbia is relatively large, and the shortest paths linking these countries do not lead through mountain valleys or passes, but across the plains of Hungary. Several countries, like Poland, the Czech Republic, Austria, Romania, and Serbia are neighbors of less than half of the other Carpathian countries, typically three out of seven or even two in the case of Serbia. This distinguishes the Carpathians from the Alpine region where all countries except Slovenia are neighbors of more than half of the

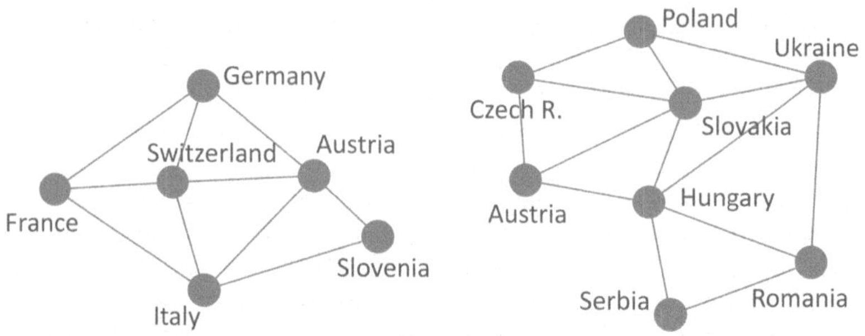

Fig. 6 Neighborhoods of the Alpine (*left*) and Carpathian countries (*right*). For simplicity, Liechtenstein is omitted

other Alpine countries (Fig. 6). Hence, the geographic and political situation of the Carpathians does not stimulate integration of the Carpathian countries. However, the Carpathian countries have shown a remarkably active cooperation on several issues such as nature conservation of the Danube watershed area (WWF 2012) or the Black Sea catchment (enviroGRIDS 2012).

3 About this book

The most important aspects of Carpathian sustainability are dealt with in 44 chapters of this book, written by more than 100 authors from 10 countries who have considerable expertise in various subjects and a deep knowledge of the Carpathian region.

All contributions to this volume are a diverse outcome of 10 thematic sessions of the 1st Forum Carpaticum, divided into 4 parts: The Abiotic Environment (edited by Bartłomiej Wyżga), Forests (edited by Andrzej Bytnerowicz), The Human Dimension (edited by Jacek Kozak), and Methods (edited by Katarzyna Ostapowicz). This classification clearly illustrates the major Carpathian challenges, both for researchers and for the stakeholder community.

The first part of the book, the Abiotic Environment, focuses on climate change and various natural hazards induced by heavy rainfall, such as floods, landslides, and soil erosion. These factors may soon exert important impacts on Central European societies, even those far outside the Carpathians. This is well illustrated by a recent sequence of events in Poland: in May and June 2010 enormous amounts of Carpathian waters carried by the Vistula River flooded several villages in central Poland, even 400 km away from the mountains, while in 2011 a long autumn drought in southern Poland caused significant problems with water supply in many areas of the Polish Carpathians (Fig. 7).

The next part (Forests) highlights challenges faced by forest management that needs to be truly multifunctional to support various ecosystem services, including

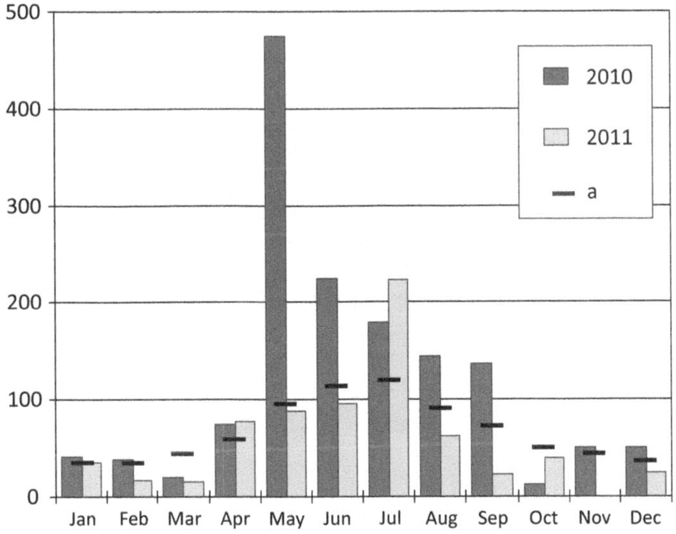

Fig. 7 Monthly rainfall in 2010 and 2011 at Gaik-Brzezowa, the Carpathian Foothills in Poland; a = mean monthly rainfall for 1988–2011 (data provided by Anita Bokwa, Jagiellonian University)

protection of water resources and mountain slopes, maintenance of forest biodiversity and habitats for endangered species, supply of timber and non-timber forest products, and finally a provision of recreational values for the growing number of tourists. This complex set of tasks is tackled in various ways and models in Carpathian forestry management, which for many years has maintained close international cooperation. Spruce decline (Fig. 8), a legacy of the past planting of the ecologically unfit provenances of this species and high air pollution in the past several decades, has been identified as one of the major problems in the Carpathian region. At present and in the future the role of forests in carbon sequestration is of extremely high importance for climate change mitigation.

A variety of other human activities range from policies to preserve the Carpathian landscape and its cultural heritage, through the potential use (or abuse) of regional values by the tourist industry, to the ability of local communities to undergo sustainable development based on industry or services. These topics are discussed in the third part of the book (The Human Dimension). This complex and diverse set of issues is based on various facets of the sustainable development idea. At its core, the Carpathian landscape can be understood as congruent with the definition adopted at the European Landscape Convention (Council of Europe 2000): "An area, as perceived by people, whose character is the result of the action and interaction of natural and/or human factors." That is a major asset of Carpathian societies, and proper governance is needed to find rational solutions to a number of conflicting ideas and pathways to local prosperity. This is exemplified for instance, by many discussions about future major ski infrastructure investments

Fig. 8 Spruce decline in the Silesian Beskid Mountains, Poland (photo credit: Jacek Kozak)

in the Tatra Mountains, both in Poland and in Slovakia. In Poland, these discussions have continued with changing intensity since the 1930s, involving local governments, business associations, politicians, nature conservation bodies, scientists, and various nongovernmental organizations (NGOs). They illustrate tremendous difficulties in accepting a coherent vision of local development that would not lead to long-lasting harmful effects on the environment, yet would provide new opportunities for the mountain communities.

Widening the knowledge about different phenomena and processes in the Carpathians is possible only with the availability and use of accurate data, advanced technologies and methods. Examples of different data types and sources (historical maps, satellite imagery, airborne laser scanning) and their various uses (from simple map overlay methods to more sophisticated point-cloud modeling or 3D visualization) is the focus of the last part of the book, Methods. In addition, it was one of the clear conclusions of the 1st Forum Carpaticum that a complete description of the Carpathian environment and its changes, in the pan-Carpathian scale, is only possible with a wide use of new tools for data acquisition and analysis and a coherent environmental information system using all benefits of the geospatial technologies (Kozak et al. 2011). The first attempts to build a pan-Carpathian database, e.g., the Carpathian geoportal developed under the INTERREG CADSES framework (Carpathian Project 2009), were important steps toward this goal, yet they are not sufficient for the scientific community and various stakeholders. However, considering the progress of the implementation of INSPIRE Directive (European Commission 2012) and huge inflow of remotely sensed data, a substantially more elaborate information system for the Carpathians seems to be at hand. The full use of these current and future capabilities, by researchers and practitioners, requires better access to and knowledge about the existing or new data sets. This is a key for the efficiency of the data-knowledge-action cycle that was the main topic of the 2nd Forum Carpaticum held in 2012 in Slovakia.

References

Björnsen Gurung A (2013) Science for the Carpathians: using regional capacity to cope with global change. In: Kozak J, Ostapowicz K, Bytnerowicz A, Wyżga B (eds) The Carpathians: Integrating Nature and Society Towards Sustainability. Springer-Verlag Berlin-Heidelberg

Carpathian Project (2009) VASICA Visions and Strategies in the Carpathian Area. Carpathian Project, Vienna. http://www.carpathianproject.eu/portal/. Accessed 30 April 2012

Council of Europe (2000) European Landscape Convention. http://conventions.coe.int/Treaty/en/Treaties/Html/176.htm. Accessed 30 April 2012

enviroGRIDS (2012) Building capacity for a Black Sea Catchment Observation and Assessment System supporting sustainable development. http://envirogrids.net. Accessed 30 April 2012

European Environment Agency (2010) Europe's ecological backbone: recognising the true value of our mountains. EEA Report 6/2010, Copenhagen

European Commission (2012) INSPIRE Infrastructure for Spatial Information in the European Community. http://inspire.jrc.ec.europa.eu. Accessed 17 May 2012

Kondracki J (1989) Karpaty, 2nd edn. Wydawnictwa Szkolne i Pedagogiczne, Warszawa

Kozak J, Estreguil C, Ostapowicz K (2008) European forest cover mapping with high resolution satellite data: the Carpathians case study. Int J Appl Earth Obs Geoinf 10:44–55

Kozak J (2009) Land use change in the Northern Carpathians. In: Jandl R, Borsdorf A, van Miegroet H, Lackner R, Psenner R (eds) Global change and sustainable development in mountain regions. Alpine space—man and environment 7, Innsbruck University Press, Innsbruck

Kozak J, Björnsen Gurung A, Ostapowicz K (2011) Research agenda for the Carpathians 2010–2015. Integrating nature and society towards sustainability. Institute of Geography and Spatial Management, Jagiellonian University, Kraków

Messerli B (2012) Global change and the World's mountains. Mt Res Dev 32:S55–S63

The Carpathian Convention (2012) History of the Convention. http://www.carpathianconvention.org/history.htm. Accessed 17 May 2012

United Nations Environment Programme (2007) Carpathian Environment Outlook. UNEP, Geneva

WWF (2012) WWF's work in the Danube-Carpathian region. http://wwf.panda.org/what_we_do/where_we_work/black_sea_basin/danube_carpathian/. Accessed 17 May 2012

Science for the Carpathians: Using Regional Capacity to Cope with Global Change

Astrid Björnsen Gurung

Abstract Emphasizing the global and regional importance of mountain ecosystem services and referring to the anticipated future environmental changes affecting the provision of these services, this chapter takes a closer look at the Carpathian Mountains. In addition to climate change and general effects of globalization, rapid socioeconomic transformations after the fall of the Iron Curtain pose an extra challenge to the sustainable development of the region. Describing the early efforts of organizing mountain science through programs such as UNESCO MAB and UNEP at the global scale, this chapter focuses on the recent history of research coordination for the European mountains, in particular on the activities of the Carpathian Convention and the European Program of the Mountain Research Initiative, which were among main driving factors for the initiation of the Science for the Carpathians (S4C) network. This regional mountain research network was established in 2008 to foster scientific collaboration and communication and to promote applied research and capacity building, which in turn would support sustainable development in the Carpathian Mountains. Forum Carpaticum, a biennial open science conference, has become a central activity of the S4C network counting more than 400 members today.

A. Björnsen Gurung (✉)
Mountain Research Initiative—Europe, University of Berne,
Erlachstr. 9aTrakt 3 3012 Berne, Switzerland
e-mail: astrid.bjoernsen@giub.unibe.ch

A. Björnsen Gurung
Institute of Interdisciplinary Mountain Research, Austrian Academy of Sciences,
Technikerstraße 21a, Otto Hittmair-Platz 1, ICT 6020 Innsbruck, Austria

J. Kozak et al. (eds.), *The Carpathians: Integrating Nature and Society Towards Sustainability*, Environmental Science and Engineering,
DOI: 10.1007/978-3-642-12725-0_2,

1 Global Change Research in Mountains

Mountains are complex and fragile ecosystems characterized by verticality, highly differentiated climatic conditions and often by an abundance of water and rich biodiversity. At the same time, life in mountain areas can be threatened by avalanches, landslides, glacial lake outbursts, strong winds, or earthquakes. Remoteness and difficult access hamper economic and social development.

Despite these constraints, mountains offer significant opportunities. Mountain dwellers have adapted to life under harsh conditions and developed techniques to optimize farming, water use and forestry. Mountains provide a wide range of ecosystem services. Provisioning services come from agricultural and forestry systems, natural ecosystems and rivers. Regulating services relate particularly to climate, air quality, water flow, and the minimization of natural hazards. Supporting services to nutrient dispersal and cycling and seed dispersal, and cultural services are associated with tourism, recreation, aesthetics, protected areas and locations of religious importance (EEA 2010). People living in lowland areas or in big cities benefit from these various services. Global change may threaten, or at least alter, the capacity of mountain ecosystems to provide those goods and services necessary for both highland and lowland people (Huber et al. 2005).

The awareness of the global and regional importance of mountains, the fragility of their resources and difficult living conditions of many mountain people has increased significantly over the last decades. This insight, together with the challenge to manage and preserve mountain ecosystems under global change, created a need for a better understanding of the mountain ecosystems functioning. Even though mountain research has a long history, going back at least 200 years (Messerli 2001), it lacked a comprehensive knowledge, as the activities were scattered and conducted by individual scientists. As a global concern, and with an understanding that mountain research requires cooperation between science and policy, the topic gained attention only after 1970.

Two factors provided the impetus for organizing mountain science at the global and regional scales. In 1971, the launch of the Man and the Biosphere Programme of UNESCO provided a basis for more integrated mountain research. The 1st UN Conference on "Human Environment" in Stockholm in 1972 stirred and influenced an entire series of mountain conferences to follow. It declared, that "research should be promoted, assisted, coordinated, or undertaken by the Man and the Biosphere Programme" (UN Conference on the Human Environment 1972a) and further stated that international cooperation "is essential to effectively control, prevent, reduce and eliminate adverse environmental effects" (UN Conference on the Human Environment 1972b). It asked for cooperation at the international level in order to prevent environmental problems going beyond national borders.

Secondly, the vision and spirit of a group of scientists from various countries[1] promoted the idea of interdisciplinary research cooperation to address the challenges posed to mountain ecosystems at the global level. Their personal commitment contributed to the growing recognition of mountains in the global arena. In the favorable international political environment it facilitated the inclusion of Chap. 13, the "Mountain Chapter", in Agenda 21, the plan of action endorsed at the UN Conference on Environment and Development in Rio de Janeiro in 1992. Placing mountains in the context of sustainable development, that event was crucial in instigating numerous mountain publications and follow-up activities.

The next milestone in the mountain research history was the first UN resolution on the theme of mountains in 1998, which included the designation of the year 2002 as the International Year of the Mountains (IYM). The IYM reinforced the implementation of Chap. 13 of Agenda 21 placing mountains on an equal footing with climate change, tropical deforestation and desertification. Among other objectives, the IYM was supposed to initiate new mountain research programmes. A solid knowledge base about mountain ecosystems and their responses to global change was identified as a precondition for the successful follow-up of the IYM, the implementation of Chap. 13, the development of national strategies for sustainable development and the formulation of mountain-specific policies (Hofer 2005). During the year 2002, the World Summit on Sustainable Development adopted a Plan of Implementation in which paragraph 42 was specifically devoted to mountains. In the wake of these events a number of international research programmes were launched, with the Mountain Research Initiative as one of the most prominent ones.

2 Mountain Research in Europe

The Mountain Research Initiative (MRI) is a scientific organization addressing global change issues in mountains around the world. As such, it was one of many factors stimulating the emergence of the Science for the Carpathians (S4C) initiative. The origins of the MRI are found in the 1990s, coincident with the preparations for the IYM in 2002. The International Geosphere Biosphere Programme (IGBP), together with the International Human Dimensions Programme (IHDP) and the Global Terrestrial Observation System (GTOS) proposed a joint initiative to "achieve an integrated approach for observing, modeling and investigating Global Change phenomena and processes in mountain regions, including the impacts of these changes and of human activities on mountain ecosystems" (UN General Assembly 2000).

---

[1] Lawrence Hamilton, Ruedi Högger, Jack Ives, Bruno Messerli, Jayanta Bandyopadhyay, Peter Stones, Maurice Strong, and others.

A small group of devoted scientists translated this goal into an integrated interdisciplinary approach (Becker and Bugmann 2001) spanning a range of activities: monitoring, process studies, modeling, as well as guidance to policy and management. Based on this methodological approach, the two-year Global Change and Mountain Regions (GLOCHAMORE) project, a joint 6th Framework Programme initiative of the MRI, the UNESCO Man and the Biosphere Programme and the University of Vienna, formulated an integrated state-of-the-art research strategy for mountain regions (Björnsen Gurung 2006), which drew on the expertise of more than 250 scientists and managers of mountain biosphere reserves worldwide.

To move the GLOCHAMORE Research Strategy towards implementation, the MRI started in 2006 initiating and supporting regional networks in North and South America, Africa, Asia and Europe, with regional initiatives in the Carpathians and Southeastern Europe.

Since its establishment in 2007, the European Program of the MRI (MRI-Europe) has put great emphasis on the establishment of active communities of mountain researchers driving forward the identification and prioritization of research needs at the regional scale, while taking the GLOCHAMORE Research Strategy as a departure point. This vision proved to be of importance for the future collaboration of MRI with scientists working in the Carpathians.

3 Mountain Research in the Carpathians

Until recently, the significance of the Carpathian mountain research to the global mountain research community has not been adequately reflected. Even though this eco-region offers perfect conditions for a "natural laboratory" for global change research and has a rich mountain research history of its own with a wealth of local expertise and data, publishing in international journals was relatively low (Körner 2009). Especially during the communist time research results were only exceptionally published in international journals. After the fall of the Iron Curtain, the research collaboration and visibility has started to improve (e.g., the special issue of Mountain Research and Development dedicated to Environmental Transformation and Human Impact in the Polish Tatra Mountains, Ives 1992), and several international projects have been successfully launched, for instance, in the field of air pollution and forest health (Szaro et al. 2002; Bytnerowicz et al. 2002, 2003). Still, insufficient networking among Carpathian researchers and between Carpathian scientists and those from other countries has been identified as one of the major obstacles for the development of Carpathian research (Ostapowicz and Sitko 2009). Research collaboration is of particular significance here, as the environmental and socioeconomic problems observed in the Carpathians today cannot be adequately addressed by individual disciplines but require interdisciplinary research that focuses on elucidating the region in systemic terms.

The global mountain research community could greatly benefit from the experiences and knowledge of researchers in the Carpathian region. This mountain range not only hosts great natural diversity, exceptional at the European scale, but is also of high interest to human geographers and social scientists, as 20 years after the fall of the Iron Curtain, rapid socioeconomic transformation poses numerous challenges to sustainable development. On top of that, the region is affected by direct and indirect impacts of climate change and globalization (Kozak et al. 2011). Therefore, information exchange with the global mountain research communities needs to be intensified both in terms of stronger involvement in EU-funded research projects, peer-reviewed publications and participation in international conferences, and in terms of interdisciplinary research efforts. To make research useful to society, a dialogue and information transfer between research, policy and practice needs to be established.

4 The Science for the Carpathians Initiative

The idea of the Science for the Carpathians initiative goes back to the negotiation process of the Framework Convention on the Protection and Sustainable Development of the Carpathians (Carpathian Convention) in 2001. The Ukrainian Government requested the United Nations Environment Programme—Regional Office for Europe (UNEP/ROE) to facilitate an intergovernmental consultation process among the Carpathian countries with the aim of drafting an international convention on the Carpathian mountains to be adopted at the Fifth Ministerial Conference "Environment for Europe" in 2003. At that time, the Alpine Convention was the only legally binding regional agreement specifically relating to mountain areas. The UNEP forwarded the request to the Italian Ministry for the Environment and Protection of Territory, as the Italian Government demonstrated a keen interest in promoting sustainable development in mountain areas after the Johannesburg Summit 2002.

The first significant step took place with the formation of the Alpine-Carpathian Partnership, launched during the International Year of the Mountain (2002) and supported by the Presidency of the Alpine Convention, which in that period was held by Italy. An initial informal meeting was hosted by the Ukrainian government in Kiev in November 2001, where participants agreed to a list of areas of cooperation. The formal negotiations between the Carpathian countries facilitated by UNEP/ROE took place during five preparatory meetings (Bolzano, Vaduz, Geneva, Vienna, Bolzano) with several international organizations, academic institutions and NGOs providing support.

After acceptance of the Carpathian Convention in Kiev in 2003, European Academy (EURAC) played a central role in the framing of scientific issues related to the Convention. The first Ad Hoc Expert Meeting of the Carpathian Convention in Bolzano in May 2004 welcomed the scientific support activities offered by EURAC and recommended stronger cooperation between the Carpathian

Convention's Interim Secretariat at UNEP in Vienna and EURAC. In October 2004, the two parties signed a Memorandum of Cooperation, including scientific, logistical and communication support.

Apart from the political environment, the INTERREG CADSES "Carpathian project" stimulated the creation of a Carpathian science initiative, as it clearly showed the necessity of organizing science in the region (Borsa et al. 2009). Initiated by the International Scientific Committee on Research in the Alps (ISCAR) and the Interim Secretariat of the Carpathian Convention (ISCC), a small group of scientists and stakeholders met at the Forum Alpinum 2007 in Engelberg, Switzerland, to develop first ideas for a science network in the Carpathians similar to the one existing in the Alps. These ideas were further elaborated during a consecutive meeting in Bolzano in October 2007. During the informal S4C meeting at the COST Conference on "Global Change and Sustainable Development in Mountains" (Innsbruck, April 2008), the S4C task force joined hands with the Carpathian Forestry Group. This resulted not only in the inclusion of active US scientists in the S4C information loop but had also a determinant impact on the quality and content of the 1st Forum Carpaticum.

The Institute of Geography and Spatial Management, Jagiellonian University, Kraków, Poland, EURAC, Joanneum Research, The University of Applied Sciences Eberswalde, Humboldt Universität zu Berlin, UNEP, and the MRI organized an S4C launching event in May 2008, which aimed at people from the Carpathian region. A crucial prerequisite for this event was the generation of a "Who's Who in Carpathian science" database established by the MRI including scientists having an interest in shaping future research activities in the region. Seventy people attended the first workshop and initiated the development of a research strategy for the Carpathian Mountains (Björnsen Gurung et al. 2009; Ostapowicz and Sitko 2009).

In June 2008, two weeks after its official launch, the vision and preliminary goals of the S4C initiative were presented to the 2nd Ministerial Conference of Parties (COP2) of the Carpathian Convention in Bucharest, Romania (Björnsen Gurung 2008). The Meeting adopted the Decision COP2/9, which welcomes the establishment of S4C as a platform of researchers for and in the Carpathians for the implementation of Article 12 (Environmental assessment and information system, monitoring and early warning) of the Carpathian Convention (ISCC 2008).

The partnership between S4C and MRI-Europe, starting in early 2008, turned out to be of mutual benefit for both, Carpathian researchers and the global MRI community. Similarly, the linkage between the science network and the ISCC not only has been an important stimulus to the strategic development of the initiative, but also a motivation factor for scientists to make their expertise available to policy makers. In 2008, S4C activities received an additional impetus when the MRI and the Institute of Interdisciplinary Mountain Research, at the Austrian Academy of Sciences started a joint implementation program with a focus on Central and Southeastern Europe.

In June 2009, the Institute of Landscape Ecology of the Slovak Academy of Sciences organized the first official planning meeting of the S4C initiative in

Bratislava, in cooperation with MRI-Europe, the Institute of Geography and Spatial Management, Jagiellonian University, ISCAR, the Swiss and Austrian Academies of Sciences, with the support of IGF, EURAC and the ISCC-UNEP. The meeting addressed the institutional setup and planning of the S4C network and nominated 14 members for the S4C Interim Scientific Steering Committee. A Memorandum describes tasks and responsibilities of the Steering Committee and outlines the objectives, as they are, for instance, the promotion of applied research, capacity building, improved networking and communication, and the support of sustainable development in the Carpathian region. At the same occasion, the 13 representatives of the Carpathian Academies of Sciences (Czech Republic, Hungary, Poland, Romania, Ukraine and Slovakia) who had accepted the invitation of the presidents of the Swiss and Austrian Academies discussed future directions of Carpathian research. They expressed high interest in the S4C activities and defined fields in which the Academies could contribute to foster research cooperation in the Carpathian region.

Since its launch in 2008, the S4C network has grown to 400 members by 2011 and is still growing. Its functioning is largely dependent from voluntary contributions of individual scientists. Despite national borders, a relatively weak tradition of scientific collaboration in the region and insufficient funding, the S4C network has an extraordinary sense of community and strong will to organize science at a pan-Carpathian scale. The Forum Carpaticum has already become a core activity of the network, complemented with communication tools such as the bi-monthly newsflash and the S4C website.

5 The Forum Carpaticum

The Bratislava meeting was crucial to the development of the Forum Carpaticum: an open scientific conference devoted to the Carpathian region. Preparations to the 1st Forum gave a major impetus to the network development in the months to follow.

S4C benefited from the experience and expertise of ISCAR whose Office Manager was invited to Bratislava to provide insights on this Alpine research coordination effort with potential use for S4C. ISCAR promotes international cooperation in Alpine research. In 2000, the Alpine Conference recognized ISCAR as an official observer. In this function ISCAR provides research and scientific expertise to the official bodies of the Convention. Since 1994, ISCAR has organized the biennial Forum Alpinum with national partners. The Forum Alpinum is primarily a scientific conference aiming to promote international research cooperation on topics of relevance across the entire Alpine area. It has also been designed as an interface between the scientific community and the general public, providing opportunities for dialogue among various groups.

The 1st Forum Carpaticum 2010 was organized by the Jagiellonian University, Institute of Geography and Spatial Management, Kraków (15–18 September 2010) and the S4C. In the style of the Forum Alpinum, it combined elements of a scientific conference, as a key to network development in the Carpathians, while offering workshops to stimulate the dialogue between research and practice. As the first mountain conference of this kind in the Carpathian region, the Forum's topic "Integrating nature and society towards sustainability" provided a wide framework within which representatives from a broad variety of disciplines could engage in the discussion on future research directions. 200 participants from 13 countries, mostly scientists but also practitioners and stakeholders (e.g., from the forestry sector), discussed the current status and emerging themes for future research (Ostapowicz 2010; Ostapowicz et al. 2010). The results were compiled and refined in the "Research Agenda for the Carpathians" (Kozak et al. 2011) and presented during the 3rd Meeting of the Conference of Parties to the Framework Convention on the Protection and Sustainable Development of the Carpathians in May 2011 in Bratislava, Slovakia.

6 Outlook

The Carpathian countries have the potential to develop a community of science, policy and practice based on the presumption that these forces can work together to tackle mountain issues at the Carpathian level. Building such community offers the opportunity to situate Carpathian mountain research and development at the European level in terms of recognition, resource use, as well as funding.

Together, the S4C network and the 1st Forum Carpaticum not only contributed to the visibility of the Carpathians in the global change research agendas for mountains but also induced new research ideas and activities with an attempt to link research, policy and practice. With the publication of the "Research Agenda for the Carpathians" (Kozak et al. 2011) the phase of developing research priorities draws to a close and attention shifts toward freeing capacity for the implementation of the identified priorities. As a next step, forces need to be mobilized and appropriate funding raised for research activities at the pan-Carpathian scale, which is both, a scientific and a political challenge.

References

Becker A, Bugmann H (eds) (2001) Global change and Mountain Regions: the Mountain Research Initiative. IGBP Report Series 49; GTOS 28; IHDP Report Series 13. Stockholm

Björnsen Gurung A (ed) (2006) Global Change and Mountain Regions Research Strategy. MRI/UNESCO Publication, ADAG Copy AG, Zürich

Björnsen Gurung A (ed) (2008) S4C Bucharest Statement. http://mri.scnatweb.ch/archive/s4c-presentationcop2. Accessed Feb 24 2012

Björnsen Gurung A, Bokwa A, Chełmicki W, Elbakidze M, Hirschmugl M, Hostert P, Ibisch P, Kozak J, Kuemmerle T, Matei E, Ostapowicz K, Pociask-Karteczka J, Schmidt L, van der

Linden S, Zebisch M (2009) Global change research in the Carpathian mountain region. Mt Res Dev 28(3):282–288

Borsa M, Chifelea C, Egerer H, Gal Z, Glowacki W, Halas M, Hopfgartner V, Illes I, Niewiadomski Z, Ptacek P, Wiederwald D (2009) VASICA: Visions and Strategies in the Carpathian Area. The Carpathian Project, UNEP, Vienna

Bytnerowicz A, Godzik B, Frączek W, Grodzińska K, Krywult M, Badea O, Barancok P, Blum O, Cerny M, Godzik S, Mankovska B, Manning W, Moravčik P, Musselman R, Oszlanyi J, Postelnicu D, Szdzuj J, Varšavova M, Zota M (2002) Distribution of ozone and other air pollutants in forests of the Carpathian Mountains in central Europe. Environ Pollut 116:3–25

Bytnerowicz A, Badea O, Barbu I, Fleischer P, Frączek W, Gancz V, Godzik B, Grodzińska K, Grodzki W, Karnosky D, Koren M, Krywult M, Krzan Z, Longauer R, Mankovska B, Manning WJ, McManus M, Musselman RC, Novotny J, Popescu F, Postelnicu D, Prus-Głowacki W, Skawiński P, Skiba S, Szaro R, Tamas S, Vasile C (2003) New international long-term ecological research on air pollution effects on the Carpathian Mountain forests, Central Europe. Environ Int 29:367–376

European Environment Agency (2010) Europe's ecological backbone: recognizing the true value of our mountains. EEA, Copenhagen

Huber UM, Bugmann H, Reasoner M (eds) (2005) Global change and mountain regions. An overview of current knowledge. Springer, Dordrecht

Hofer Th (2005) The international year of mountains: challenge and opportunity for mountain research. In: Huber UM, Bugmann H, Reasoner M (eds) Global change and mountain regions. Springer, Dordrecht

Interim Secretariat of the Carpathian Convention (2008) Conference of Parties to the Framework Convention on the Protection and Sustainable Development of the Carpathians. Report 2nd Meeting, Bucharest, Romania, pp 17–19 June 2008

Ives J (1992) Editorial (to the Special Issue on Environmental Transformation and Human Impact in the Polish Tatra Mountains). Mt Res Dev 12:103–103

Körner Ch (2009) Global Statistics of "Mountain" and "Alpine" Research. Mt Res Dev 29(1):97–102

Kozak J, Björnsen Gurung A, Ostapowicz K (eds) (2011) Research agenda for the Carpathians 2010–2015: Integrating Nature and Society towards Sustainability. Science for the Carpathians, Kraków

Messerli B (2001) The International Year of Mountains, the Mountain Research Initiative and PAGES. PAGES Newslett 1:2

Ostapowicz K (2010) Integrating Nature and Society Towards Sustainability. Newslett Mt Res Initiative 5:32–34

Ostapowicz K, Björnsen Gurung A, Bytnerowicz A, Cudlin P, Elbakıdze M, Halada L, Hirschmugl M, Hostert P, Kozak J, Kruhlov I, Matei E, Mázsa K, Nijnik M, Prots B, Zebisch M (2010) Integrating nature and society towards sustainability—current status of global change research in the Carpathian Mountains. Parallel Session 6.7, Global Change and the World's Mountains, Perth, Scotland, UK, 27–30 Sept 2010, Abstracts CD

Ostapowicz K, Sitko I (eds) (2009) Science for the Carpathians Strategy Development and Networking Workshop, 27–28 May 2008. Report. Bern, Switzerland

Szaro RC, Bytnerowicz A, Oszlanyi J (eds) (2002) Effects of air pollution on forest health and biodiversity in forests of the Carpathian Mountains. NATO Science Series, Series Life and Behavioral Sciences, IOS Press, Amsterdam, p 345

UN Conference on the Human Environment (1972a) Recommendation 24. http://www.unep.org/Documents.Multilingual/default.asp?DocumentID=97&ArticleID=1506&l=en. Accessed 15 Dec 2011

UN Conference on the Human Environment (1972b) Declaration U.N. Doc. A/Conf.48/14/Rev. 1 (1973); 11 I.L.M. 1416 (1972). http://www.unep.org/Documents.Multilingual/Default.asp?documentid=97&articleid=1503. Accessed 15 Dec 2011

UN General Assembly (2000) Status of preparations for the International Year of Mountains, 2002. Report of the Secretary General A/55/218. New York

Part I
Abiotic Environment

Progress in Understanding the Dynamics of Carpathian Abiotic Environment

Bartłomiej Wyżga

Characterized by complex lithology, ranging in altitude from 300–500 m in the marginal, foothill areas to 2,655 m in the Tatra massif, and running in progressively changing direction on the length of about 1,300 km, the Carpathians exhibit a complex mosaic of natural conditions. Moreover, due to the considerable spatial extent of the mountains, their different parts are differently affected by global environmental impacts such as a climate change. For instance, the Southern Carpathians are situated within the area of a forecast reduction in precipitation totals, whereas northern parts of the mountain chain may exhibit no change in precipitation totals over the twenty first century as they fall within the boundary area between the zone of predicted decreased precipitation in Southern Europe and increased precipitation in Northern Europe (European Commission 2005). During the 1st Forum Carpaticum, analyses of various components of the natural environment of the Carpathians, modifications of these components under the influence of global changes and their impacts on the management of the mountain areas were the subject of four sessions dealing with *Climate Change*, *Chemical Environment*, *Water Resources and Management*, and *Natural Hazards and Risks* (Kozak et al. 2011). This book section comprises ten contributions from the total of 28 oral presentations and 24 posters offered during these sessions: three from Poland, three from Slovakia, two from Romania, one from the Czech Republic and one from Hungary.

Landslides are common forms of relief in mountains and their formation frequently poses a serious risk for settlements, infrastructure and land use practices in mountain areas (Fig. 1). Evaluation of the landslide susceptibility of particular areas is thus crucial for the safe and effective management of mountain regions. Grozavu et al. present a methodology of landslide susceptibility assessment and apply it to the Putna River drainage basin in the Romanian Carpathians characterized

B. Wyżga (✉)
Institute of Nature Conservation, Polish Academy of Sciences, al. Mickiewicza 33, 31-120, Kraków, Poland
e-mail: wyzga@iop.krakow.pl

J. Kozak et al. (eds.), *The Carpathians: Integrating Nature and Society Towards Sustainability*, Environmental Science and Engineering,
DOI: 10.1007/978-3-642-12725-0_3, © Springer-Verlag Berlin Heidelberg 2013

Fig. 1 Landslide activated by a heavy rainfall in May 2010 destroyed houses and agricultural land on the slopes of the Wieliczka Foothills, Polish Carpathians, at Winiary

by a high structural diversity and lithologic heterogeneity. Outcomes from this analysis indicated that areas exhibiting high and very high susceptibility to landsliding cover three times greater area than the mapped landslide area and represent over 30% of the total area of the analyzed region.

Three chapters deal with threats to soils and their quality. Due to the inclination of cultivated surfaces and increased precipitation, mountain areas are vulnerable to soil erosion. Soil erosion models have greatly advanced since their early formulation in the 1960s, due to both theoretical progress and the use of GIS procedures in the analysis of complex terrains. Šir et al. present a methodology of the prediction of erosion and deposition phenomena with the use of empirically based USPED model (Mitasova et al. 1996) and apply it to the analysis of processes caused by a convective rainstorm in a third-order, montane to foreland catchment in NE Czech Republic. While the results of the modeling cannot be viewed as informative about absolute amounts of erosion and deposition at particular locations, they nevertheless indicate the spatial distribution of erosion-sedimentation phenomena within the catchment with complex topography, soil types and land use patterns. In turn, Kenderessy used a physically based EROSION 3D model to simulate soil loss on cultivated land in southern Slovakia. His chapter presents results of the simulation for a number of scenarios reflecting the variability of crops, soil properties and erosion-mitigation measures at three stages of the vegetation period. Also this model yields results that should be treated in relative rather than absolute terms. However, it satisfactorily indicates the erosion risk at particular locations and provides information allowing the optimal localization and scaling of site-specific soil conservation measures.

The rates of production and decay of leaf litter and wood debris in forests may be altered as a result of climate changes, thus affecting the availability of nutrients in the soil and soil pH. Toth et al. simulated such effects through experimental manipulation of litter quality and quantity in an oak forest in northern Hungary. The results of 8-year-long litter manipulation indicated that a reduction in litter production caused by climate change will tend to decrease C, total N, Ca^{2+} and Mg^{2+} content in the soil and soil pH, with detrimental effects on the soil organic matter and fertility. A positive relationship between soil pH and Ca^{2+} and Mg^{2+} concentrations in the soil was demonstrated, suggesting that decreased litter production may diminish the content of these elements in the soil, with the resultant reduction of the buffering capacity of soil leading to its acidification. These authors' conclusions should be completed by the notion that alterations in the amount of litter in forest soil may result not only from ongoing and future climate changes but also from changes in forest management. A large proportion of fallen leaves in the Carpathian forests was historically taken as bedding for livestock and wood debris was used for fuel (Bürgi and Gimmi 2007); currently, abandonment of these practices will thus tend to increase the amount of litter on the forest bottom, with beneficial effects of the increase on the nutrient content of soil and soil pH.

Climate changes are among the most apparent, presently occurring global environmental disturbances and three chapters deal with climate trends in Polish and Slovak parts of the Carpathians. Bokwa et al. analyze the variability and extremes of air temperature and precipitation that typified different vertical zones of the Polish Carpathians during the last 50 years, and compare them with those recorded in the Eastern Alps. The western part of the Polish Carpathians experienced some warming during the study period, especially pronounced at lower locations, whereas in the Eastern Alps the temperature increase was greater and spatially more uniform. At the same time, no trend in precipitation was observed in either mountain region. Pribullová et al. examine the variability in annual and monthly air temperature in the High Tatra Mountains over a similar study period using two independent data sources: measurements from a set of ground meteorological stations and records from a nearby aerological station. Both datasets indicated a consistent increase in the temperature over the whole vertical range of the High Tatras, the most rapid in the summer and winter months. An aspect of the study especially interesting for mountain research is the documented increase in the altitude of annual and monthly isotherms, a causative factor of possible subsequent upward shift of vegetation belts in this high-mountain massif. Melo et al. analyze changes in air temperature and precipitation in the Slovak Carpathians during the twentieth century and show that the whole area experienced warming over the examined period, coupled with a reduction in precipitation totals in the south and increased precipitation in the north. These trends resulted in northwards and upwards shifts of the climatic regions and subregions distinguished in the area according to the Köppen and Konĉek climate classifications. The chapter finishes with presentation of the climate change scenarios for the Slovak Carpathians for the twenty first century that are based on downscaling of the outputs from the General Circulation Models.

Regions with flysch lithology constitute a large proportion of the total area of the Carpathians and water conditions are among the environmental characteristics that make such regions distinct from other mountain areas. Two chapters describe hydrochemical features of small streams draining flysch catchments in the Polish Carpathians. Janusz Siwek et al. analyze spring water chemistry in an unmanaged catchment in the High Bieszczady Mts., aiming to identify natural factors responsible for spatial differences in the chemical composition of the spring water.

The lithology of aquifer, elevation of springs controlling climate conditions and plant communities in the alimentation area and the varied role of shallow water circulation in springs alimentation were indicated as the principal factors differentiating the content of main and biogenic ions in the spring water. Joanna Siwek et al. describe how water chemistry in the streams draining catchments of differing land use in the Carpathian Foothills area changes between baseflow conditions and floods of diverse types and attribute the changes to the operation of natural and anthropogenic factors. They demonstrate that processes operating in the agricultural and woodland catchments change the concentration of some main and biogenic ions in flood waters in the opposite directions, depending on the degree of pollution of pre-event waters and the intensity of element uptake by vegetation. Finally, a short note of Tudose et al. presents an analysis of damages of torrent-control hydrotechnical structures in the Cârcinov River catchment in the Romanian Carpathians.

As this book section covers only a small proportion of the contributions to the four abiotic sessions of the 1st Forum Carpaticum, some conclusions from the other contributions and the entire sessions are briefly outlined below. Studies on climate change in the Polish, Slovak and Romanian Carpathians consistently demonstrate an increase in temperature during the second half of the twentieth century, although the timing and rate of the increase vary among regions and altitudinal belts. By contrast, no consistent trends in the annual sums of precipitation were found within the Carpathians, with precipitation totals decreased in the Romanian Carpathians and showing no temporal trend in Poland and Slovakia. While these observations well fit the directions of changes in climate parameters forecast for particular regions of Europe (European Commission 2005), there is an urgent need for better international collaboration and common protocols in future climate research to make comparisons between particular Carpathian countries reliable (Kozak et al. 2011).

Studies on the air pollution by ozone as well as sulfur and nitrogen air pollution and deposition constituted a large proportion of the session on *Chemical Environment*. High concentrations of tropospheric ozone, detrimental for sensitive vegetation, were indicated for southern Slovakia and mountain ranges of the Czech Republic. Moreover, the studies identified plant species useful as bioindicators of the ozone pollution. While considerable reductions in S deposition have been recorded in the Carpathian region for about 20 years, N deposition, especially in the Western Carpathians, still remains one of the highest in Europe. Current levels of N deposition facilitate leaching of base cations and inhibit plant growth in acid soils; they also result in increased mobilization and hence toxicity of heavy metals in soil. Future efforts should aim at organizing a network of Carpathian field stations and developing integrated long-term monitoring systems for air, water and soil pollution. Understanding the spatial distribution of pollution and mechanisms of its impact on ecosystems requires ecosystem-level research and large-scale studies across Carpathian catchments (Kozak et al. 2011).

An important part of the 1st Forum Carpaticum session on *Water Resources and Management*, not represented in the book, was a set of contributions dealing with

Fig. 2 Following straightening, considerable narrowing and partitioning by a number of concrete weirs, the contemporary Czarny Dunajec River in its channelized reach exhibits poor hydromorphological quality (Wyżga et al. 2010), reflected in degraded fish and benthic invertebrate communities. As many previously channelized reaches of Polish Carpathian rivers run through riparian forest, far from settlements and infrastructure, further maintenance of the channelization schemes is unjustified and there are no constraints on restoration of these reaches

hydromorphological quality of rivers. According to the Water Framework Directive of the European Union, rivers in the member countries should attain good ecological state by 2015 (European Commission 2000), and hydromorphological modifications are one of important causes of the ecological degradation of rivers. Recognition of the patterns in hydromorphological quality of watercourses related to various human impacts such as channelization and river partitioning by transversal hydrotechnical structures (Fig. 2), in-channel gravel mining and flow regulation, as well as their impacts on riverine and riparian biota is thus crucial for the attainment of river recovery.

Natural hazards are a growing concern to mountain communities and research on them could allow the development of early-warning systems and informed decisions about the management practices mitigating the risk for the communities. Apart from the hazards considered in the book, the session also comprised presentations on snow avalanches and forest fires. The former represent a risk for tourists in the Tatra and Făgăraş Mountains. The latter are relatively rare in the Carpathians but their risk increased recently as the ongoing climate change leads to more frequent occurrence of prolonged droughts.

Studies that increase knowledge of the functioning of Carpathian abiotic environment and its changes are important not only for the scientific community in the region. Equally valuable is their practical significance, allowing to improve management of natural resources by the replacement of former practices, frequently costly and detrimental to the environment, by more sustainable practices, to increase the awareness of ongoing environmental alterations and to help the societies adapt to them. A selection of studies presented in this volume is expected to assist with these tasks.

References

Bürgi M, Gimmi U (2007) Three objectives of historical ecology: the case of litter collecting in Central European forests. Landsc Ecol 22:77–87

European Commission (2000) Directive 2000/60/EC of the European Parliament and of the Council of 23 October 2000 establishing a framework for Community action in the field of water policy. Off J Eur Comm L 327(43):1–72

European Commission (2005) Climate change and the European water dimension. EU Report No. 21553, European Commission—Joint Research Centre, Ispra

Kozak J, Björnsen-Gurung A, Ostapowicz K (2011) Research agenda for the Carpathians 2010–2015. Integrating nature and society towards sustainability. Institute of Geography and Spatial Management, Jagiellonian University, Kraków

Mitasova H, Hofierka J, Zlocha M, Iverson LR (1996) Modeling topographic potential for erosion and deposition using GIS. Int J GIS 10:629–641

Wyżga B, Zawiejska J, Radecki-Pawlik A, Amirowicz A (2010) A method for the assessment of hydromorphological river quality and its application to the Czarny Dunajec, Polish Carpathians. In: Radecki-Pawlik A, Hernik J (eds) Cultural landscapes of river valleys. Agricultural University in Kraków, Kraków

Landslide Susceptibility Assessment: GIS Application to a Complex Mountainous Environment

Adrian Grozavu, Sergiu Pleşcan, Cristian Valeriu Patriche, Mihai Ciprian Mărgărint and Bogdan Roşca

Abstract This study attempts to quantify landslide susceptibility in the upper Putna River basin in the Romanian Carpathians Bend using GIS techniques and logistic regression. First, a detailed landslide inventory was carried out and a GIS database was built, comprising potential predictors of landslide occurrence. The GIS database included 11 quantitative predictors, mostly geomorphometric parameters, and 4 qualitative predictors which were transformed into quantitative variables using landslide density approach. The logistic regression analysis, combined with a stepwise selection of the predictors, showed that landslide occurrence is best explained by slope inclination class, altitude, soil class, distance to drainage network and surface geology. The results show that the potentially unstable terrains, displaying high and very high landslide susceptibility values, cover an area about 3 times greater than the mapped landslide area.

1 Introduction

Landslides are a very common geomorphic hazard with considerable economic and ecological consequences. In Romania, significant landslide areas occur in hilly and mountain regions, especially those underlain by molasse and flysch formations. The studies carried out to date have tried to explain manifestation, typology and evolution of landslides as well as the relations between geology and landslide

A. Grozavu (✉) · S. Pleşcan · C. V. Patriche · M. C. Mărgărint · B. Roşca
Geography Department, Alexandru Ioan Cuza University of Iaşi, Carol I Blvd., no. 20A
700505 Iasi, Romania
e-mail: adriangrozavu@yahoo.com

C. V. Patriche · B. Roşca
Iaşi Branch, Geography Team, Romanian Academy, Carol I blvd., no. 8
700505 Iasi, Romania

J. Kozak et al. (eds.), *The Carpathians: Integrating Nature and Society Towards Sustainability*, Environmental Science and Engineering,
DOI: 10.1007/978-3-642-12725-0_4, © Springer-Verlag Berlin Heidelberg 2013

distribution. Recent studies attempt to apply new research methods and techniques, such as landslide susceptibility assessment and appropriate mapping (Micu and Bălteanu 2009; Bălteanu et al. 2010; Grozavu et al. 2010).

At international level, landslide susceptibility assessment has recently been a subject of numerous studies; however, the application of this knowledge utilizes various conceptual and methodological approaches. Several authors provide good reviews of the recent methodology and evaluations of the subsequent approaches (Mantovani et al. 1996; Soeters and van Westen 1996; Aleotti and Chowdhury 1999; Guzzetti et al. 1999; Castellanos Abella and Van Westen 2008; Corominas and Moya 2008; Grozavu et al. 2010).

Generally, landslide susceptibility is defined as a quantitative and qualitative assessment of the classification, volume (or area) and spatial distribution of landslides which exist or potentially may occur in an area (Fell et al. 2008). Therefore, the purpose of landslide susceptibility mapping is to highlight the regional distribution of potentially unstable slopes based on a detailed study of the factors responsible for landsliding. Thus, the focus is on the recognition of landslide-prone areas achieved by mapping susceptibility.

One of the problems related to the accountability of landslide susceptibility maps is the lack of standardization in analytical methods (Ayalew et al. 2005) and consequently, the need for a common language and standard procedures in landslide risk zoning (Fell et al. 2008).

This chapter focuses on the evaluation of landslide susceptibility by applying a logistic regression analysis, to a typical region in the Romanian Carpathians Bend area. Here, the extension of built-up areas as a consequence of a clear intensification of touristic activities during the last two decades has complex, systemic implications at the local level. Our main goal is to identify, weigh and integrate the different parameters determining landslide susceptibility and to achieve an adequate spatial model for the study area.

2 Study Area

The studied region is located in the Romanian Carpathians Bend, in the upper Putna River basin (Fig. 1). The area of the region is 210.15 km^2 and the altitude ranges from 460 to 1,588 m with an average of around 900 m.

Geologically, the study region belongs to the outer flysch, known by its active tectonics, structural diversity and lithological heterogeneity (Dumitrescu et al. 1970). The geological layers are of Cretaceous (Senonian), Paleogene (Eocene and Oligocene) and lower Miocene (Aquitanian and Burdigalian) age. They form two major structural units separated by Tarcău fault: Vrancea Nappe, which appears in a tectonic window, south from Tarcău fault, and Tarcău Nappe, to the north of the fault (Fig. 2).

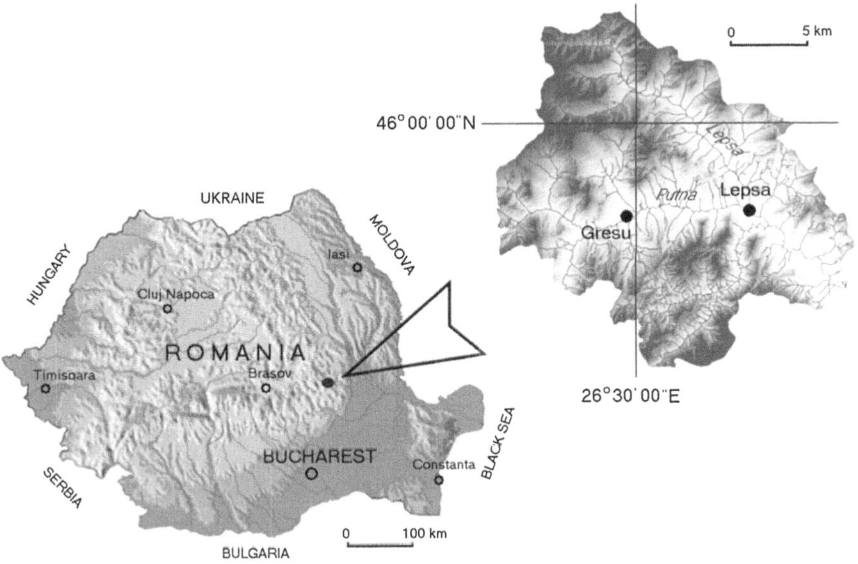

Fig. 1 Location of the study area

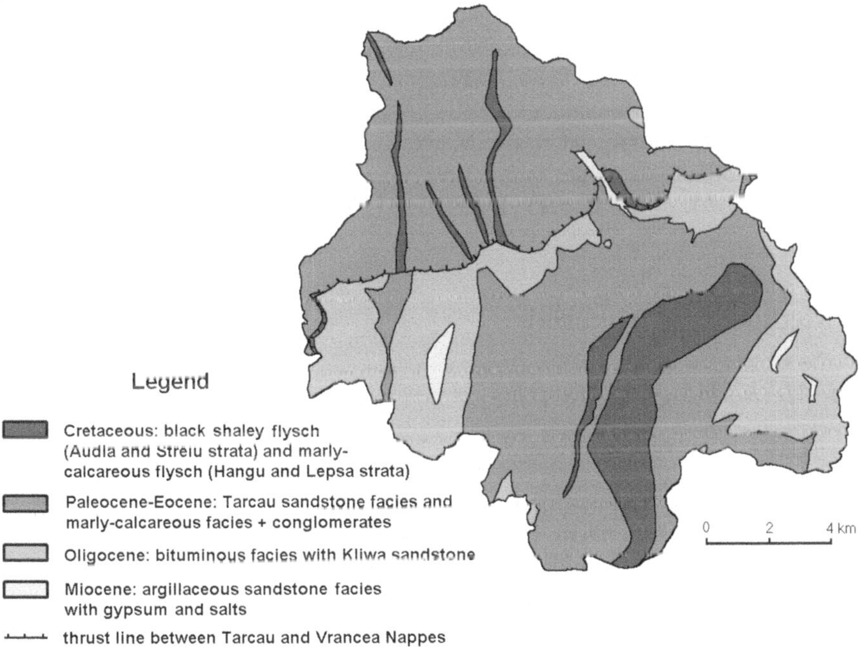

Fig. 2 Schematic lithological and structural map of the study area

The following main lithofacies occur in the study area (Ichim et al. 1998):

- black shaley flysch (Streiu strata) of Cretaceous age, occurring only in the Vrancea Nappe, composed by bituminous schistous shales with conglomerates, marls and limestones,
- marly limestone facies, including Cretaceous Hangu and Lepşa strata and Paleocene-Eocene lateral variation of Tarcău Sandstone deposits,
- Tarcău Sandstone facies (Paleocene-Eocene), consisting of sandstones with mica, forming massive beds with thin intercalations of marls,
- bituminous facies with Kliwa sandstone of Oligocene age, composed of thick-bedded, white quartzouse sandstones with intercalations of conglomerates, menilites, bituminous marls and disodilic shales,
- shaley sandstone facies with gypsum and salt of Miocene age.

Morphologically, the region is dominated by steep slopes, frequently affected by intense denudation processes, while flat areas (terraces and floodplains) occupy minor surfaces (Tufescu 1966; Ichim et al. 1996).

The area is drained by the east-flowing Putna River and its main tributaries: Lepşa, Greşu and Tişiţa.

The region is characterized by a high variety of hydrogeological conditions, due to the diverse lithology, with springs occurring on valley bottoms and slopes (Ichim et al. 1998).

Geological conditions, geomorphometric parameters, climate characteristics and human activities, especially construction on slopes, favor slope processes such as landslides. Large-scale mass movements can also be triggered by powerful earthquakes and the Romanian Carpathians Bend area is well-known for its high seismicity. For example, after the March 1977 earthquake, the volume of the material mobilized on slopes was 20–50 times greater than a multi-year average (Bălteanu 1979).

3 Methods and Materials

3.1 Methodological Review

Various methods for landslide susceptibility assessment can be encountered in the scientific literature. Qualitative methods, such as ranking and rating (Anbalagan 1992) or analytical hierarchy process, AHP (Barredo et al. 2000; Ayalew et al. 2005; Komac 2006), are simple and rely on subjective assessment. Quantitative methods, such as bivariate statistical analyses, BSA (Yin and Yan 1988; Binaghi et al. 1998) or multivariate statistical analyses, MSA (Carrara et al. 1991), are based on complex mathematical concepts. These initial efforts were followed by many multivariate statistical studies based on the application of multiple regression and discriminant analysis. Among these, logistic regression is a frequently used method, considered particularly suitable as it reduces the subjectivity in the landslide susceptibility analysis (e.g. Aleotti and Chowdhury 1999; Malczewski 1999; Süzen and Doyuran 2004a, b; Van Westen et al. 2006; Thiery et al. 2007; Nefeslioglu et al. 2008; Van Den Eeckhaut et al. 2010).

A third category of methods providing good results in landslide susceptibility analysis is represented by hybrid methods, including index-based methods, such as BSA + AHP (Ayalew et al. 2004) and training-based methods, such as BSA + Neural Networks (Lee et al. 2004; Borgogno Mondino et al. 2009).

3.2 Materials, Database and Methodological Approach

The primary input data consisted of ortophotos, 1:5,000 topographic maps, 1:100,000 geological map and 1:100,000 soil map. Based on this input, a landslide causative factors database was built in GIS environment, including the following information layers:

- geology, soil, land use (as qualitative variables),
- Digital Elevation Model (DEM) derived from 1:5,000 topographic maps at a resolution of 5 × 5 m,
- geomorphometric parameters (slope angle, slope height, slope aspect, mean curvature, plan curvature, profile curvature, wetness index, modified catchment area),
- distance to drainage network, distance to roads.

The data on geology, soil and land use were digitized from the respective maps and terrains with particular parameters were grouped into 5 susceptibility classes (very low, low, medium, high and very high) according to their susceptibility for landsliding. Next, these classes were intersected with landslide polygons and landslide density for each class was computed (Bai et al. 2010). In this manner, qualitative variables were transformed into quantitative variables and further used as predictors in the logistic regression approach.

We also tested the use of landslide density for slope inclination classes, taking into account that the relation between landslide distribution and slope inclination is not linear, with most of the landslides occurring on moderately steep slopes (7–20°).

The other potential predictors (DEM, slope aspect, mean curvature, plan curvature, profile curvature, wetness index, distance to drainage network and roads) were used as continuous variables.

The susceptibility mapping started with the preparation of a landslide inventory map. We identified 198 landslides with the total area of 2,326.06 ha covering 11.07 % of the study area. The inventory was performed by means of large-scale mapping, using topographic maps (1:25,000, 1:5,000), orthophotos and field surveys for validation of landslide areas. Landslides were classified according to their activity into active landslides, semi-active landslides (the most frequent ones) and stabilized landslides (recently forested).

Useful information regarding the typology, stage of landslide evolution and the relations between geology and landslide distribution in the region is provided by Ichim et al. (1996) and Ursu (2006). The first category distinguished by these authors comprises the old and large landslides with a dominant translational character, which affect the in situ geological structures, reaching depths up to 80–90 m. The second category includes shallow landslides with a rotational character, which affect surface deposits. Most of the landslides belong to this category, but they have a small extent and represent approximately 20 % of the total landslide area. Our analysis also takes into account the sections affected by rocky landslides and soil landslides, encountered along steep slopes.

Descriptive statistics for landslide area and the potential predictors are given in Table 1.

Spatial analysis was performed using TNTmips 6.9 software and ArcGIS 9.3 software, while statistical analysis was carried out using Excel 2003 and XLSTAT 2010 software.

From the wide variety of methods potentially useful for quantifying landslide susceptibility, we chose a multivariate statistical approach based on the application of the logistic regression model. This method links the presence/absence of a phenomenon to a set of quantitative or qualitative variables, generating a continuous spatial probability model:

$$P = \frac{1}{1 + e^{-z}} \tag{1}$$

Table 1 Descriptive statistics for landslide area and potential predictors

Variables	Minimum	Maximum	Mean	Standard deviation
Landslide area (ha)	0.014	524.555	11.748	56.872
DEM (m)	459.22	1588.09	899.39	165.76
Slope inclination (degree)	0	68.483	19.974	10.143
Slope height (m)	0.39	377.93	40.21	42.74
Slope aspect (degree)	0.15	359.99	176.36	98.57
Modified catchment area (ha)	0.0025	227.329	0.701	5.073
Curvature (rad/m)	− 0.053	0.045	0.00	0.008
Plan curvature (rad/m)	− 0.036	0.039	0.00	0.005
Profile curvature (rad/m)	− 0.047	0.034	0.00	0.005
Wetness index	4.184	17.465	8.187	1.766
Distance to roads (m)	0	3881	617	619
Distance to drainage (m)	0.03	1064.3	172.8	160.1
Geology class (landslide density)	0.547	1.259	1.027	0.249
Soil class (landslide density)	0.116	1.754	1.057	0.315
Land use class (landslide density)	0.651	2.040	1.058	0.436
Slope class (landslide density)	0.154	1.976	1.025	0.694

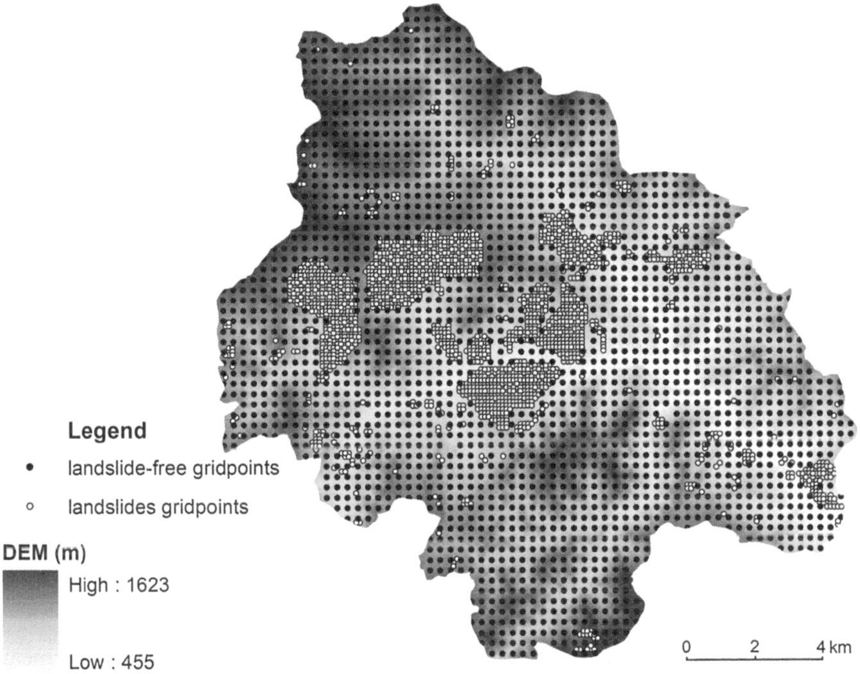

Fig. 3 Locations of grid points inside and outside the landslide area

where P is the probability of an event (landslide) to occur, which varies from 0 to 1 on an S-shaped curve, computed on the basis of a linear combination (z) of predictors ($x_1 \ldots x_n$):

$$z = b_0 + \sum_{i=1}^{n} b_i \cdot x_i \tag{2}$$

where b_0 is the intercept of the model and b_i are the regression coefficients.

In order to extract predictors' values from a raster layer, a total number of 3,999 equally distanced grid points were generated for the landslide and landslide-free areas (Fig. 3). For the reason of preserving the relative equality of the two point samples, required by the nature of the statistical analysis, the density of points inside landslide area is markedly higher than in the landslide-free area.

4 Results and Discussion

A comparison of the logistic regression models analyzing landslide densities in either continuous slope inclination values or slope inclination classes indicated that the latter model explains greater proportion of the variance. The stepwise selection

Table 2 Standard regression coefficients of the logistic regression model using slope classes instead of slope continuous values

Predictors	Standardized regression coefficients	Standard error	Wald chi square	Pr > chi square
DEM	−0.348	0.025	189.572	< 0.0001
Slope height[a]	*0.000*	*0.000*		
Slope aspect	0.038	0.021	3.519	0.061
Modified catchment area[a]	*0.000*	*0.000*		
Curvature[a]	*0.000*	*0.000*		
Plan curvature[a]	*0.000*	*0.000*		
Profile curvature	0.098	0.022	19.431	< 0.0001
Wetness	0.053	0.024	4.821	0.028
Distance to roads	−0.048	0.027	3.222	0.073
Distance to drainage	0.163	0.021	59.189	< 0.0001
Geology	0.152	0.021	50.701	< 0.0001
Soil	0.221	0.024	85.279	< 0.0001
Land use	0.114	0.022	26.149	< 0.0001
Slope inclination classes	0.406	0.025	255.324	< 0.0001

[a] Variables excluded from the model by the stepwise selection procedure are shown in italics

of the predictors eliminated slope height, modified catchment area, plan and mean terrain curvature. According to the standardized regression coefficients (Table 2), landslides occurrence is best explained by slope inclination class, altitude, soils (soil class), distance to drainage network, and geology (Fig. 4), with land use, profile curvature, wetness index, distance to roads and terrain aspect being less significant predictors.

The logistic regression analysis showed that the spatial distribution of landslide occurrence probability is determined by the following combination of linear relationships:

$$\begin{aligned} Z = &-1.659 - 3.813\text{E}{-03} \cdot \mathit{DEM} + 7.069\text{E}{-04} \cdot \mathit{ASPECT} + 37.714 \cdot \mathit{PROFILEC} \\ &+ 5.502\text{E}{-02} \cdot \mathit{WETNESS} - 1.406\text{E}{-04} \cdot \mathit{DIST\_ROADS} + 1.846\text{E}{-03} \cdot \mathit{DIST\_DRAINAGE} \\ &+ 1.107 \cdot \mathit{GEOLOGY} + 1.287 \cdot \mathit{SOIL} + 0.478 \cdot \mathit{LAND\_USE} + 1.058 \cdot \mathit{SLOPE\_CLASS} \end{aligned}$$

The outcome of the application of the logistic regression equation in GIS environment is displayed in Fig. 5. Considering the high complexity of the analyzed mountainous area, the quality regression parameters indicate a fairly good model using a cutoff value of 0.5 (Tables 3 and 4). The percentage of correctly classified points is 74.93 % in the landslide area and 70.74 % in the landslide-free area, while the overall prediction accuracy of the model is 72.84 %, with an area under the Receiver Operating Characteristic (ROC) curve of 0.802.

Landslide susceptibility was classified into five classes (very low, low, medium, high and very high susceptibility), using the natural breaks (Jenks) method (Fig. 6). The method identifies significant changes in the histogram distribution

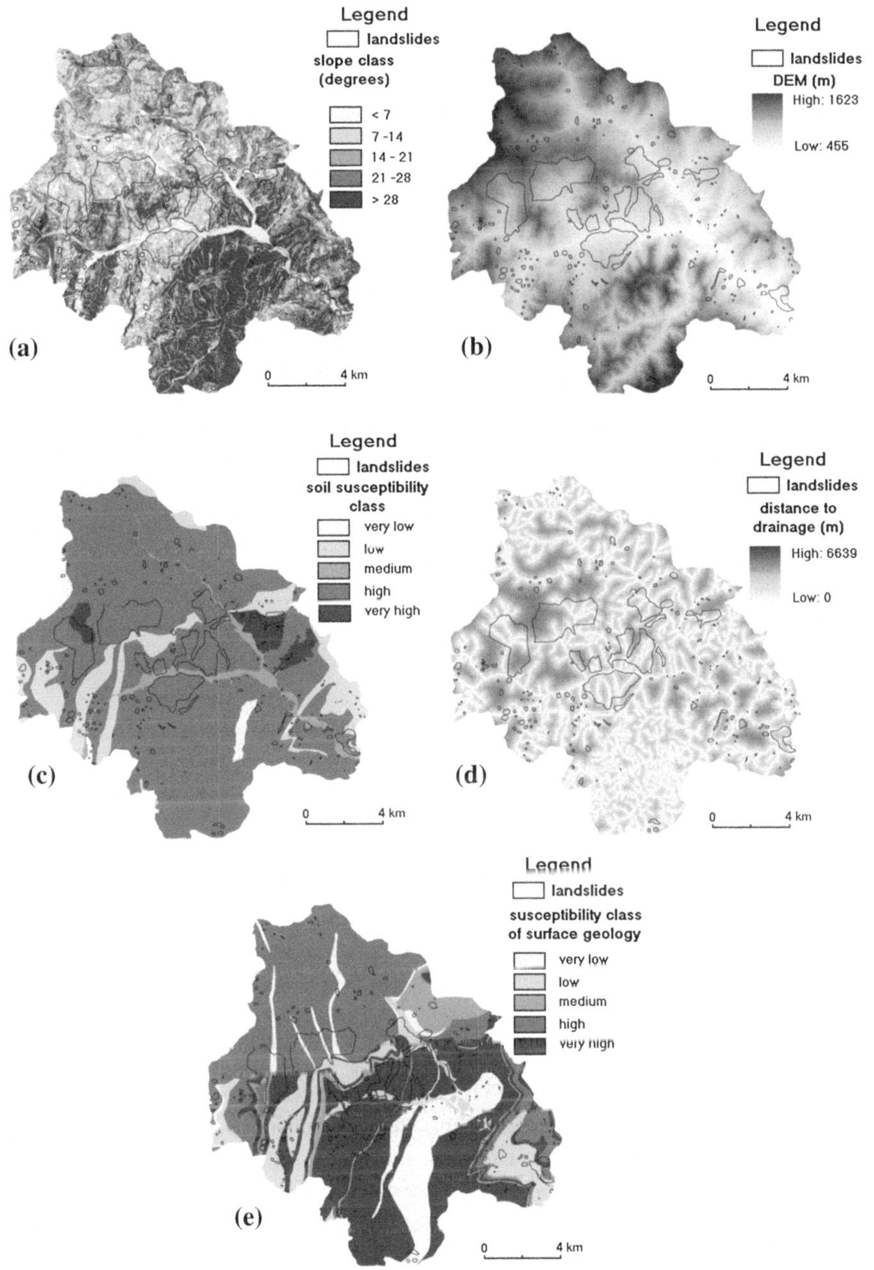

Fig. 4 Distribution in the study area of the main predictors used for deriving landslide susceptibility index: **a** slope classes, **b** DEM, **c** soil classes, **d** distance to drainage network, **e** surface geology

Fig. 5 Distribution of the landslide susceptibility index in the study area

Table 3 Quality parameters of the logistic regression model

Quality parameters	Values
-2 Log(Likelihood)	4,337.806
R^2 (McFadden)	0.218
R^2 (Cox and Snell)	0.260
R^2 (Nagelkerke)	0.307

and sets class breaks which best group similar values and maximize the differences between classes.

According to this classification, 31.3 % of the study region (nearly 6,600 ha), falls into high and very high susceptibility classes. Considering that about 25 % of this area could be misclassified (Table 4), there still remain about 4,900 ha correctly classified. Comparing this value to the actual extent of landslides (2,326 ha) indicates the occurrence of about 2,600 more hectares of the terrain showing high and very high susceptibility for landsliding.

However, comparing our results to those from a previous study that used the same approach for a cuesta front area in the Moldavian Plateau (Grozavu et al. 2010) indicates that the logistic regression is less appropriate for mountainous areas, mainly due to the non-linearity of the relations between landslide occurrence and quantitative and qualitative terrain characteristics.

Table 4 Numbers and percentages of correctly classified points

From\to	0[a]	1[a]	Total	% correct
0[b]	1,407	582	1,989	70.74 %
1[b]	504	1,506	2,010	74.93 %
Total	1,911	2,088	3,999	72.84 %

[a] points corresponding to predicted landslide-free (0) and landslide area (1)
[b] points corresponding to actual landslide-free (0) and landslide area (1)

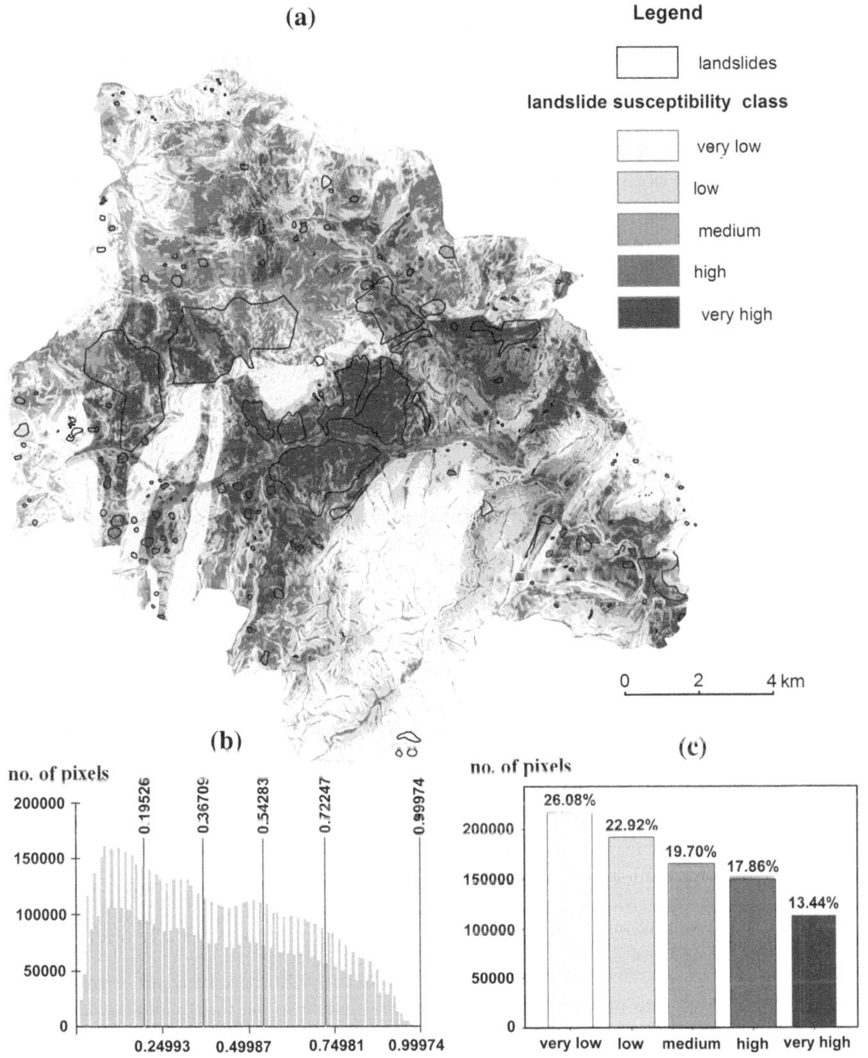

Fig. 6 Classes of the landslide susceptibility index for the study area: **a** spatial distribution, **b** classification method, **c** histogram of landslide susceptibility classes

5 Conclusions

This study has indicated that logistic regression approach is an adequate tool for the evaluation of landslide susceptibility and that the application of GIS techniques facilitates data processing and spatial visualization of the results.

Application of this model to a complex mountainous environment, characterized by a high structural diversity and lithologic heterogeneity and constituting a favorable context for slope processes, reveals that over 30 % of the 210 km^2 region (about three times more than the mapped landslide area) displays high and very high susceptibility for landsliding.

The model indicates that present and future landslides are mainly determined by slope inclination class, altitude, soil classes, distance to drainage network and geology.

Considering the spatial resolution (i.e. 5×5 m) of the classified landslide susceptibility map obtained from logistic regression, this model may be a tool for landslide susceptibility analysis at a large scale.

The study also emphasizes the importance of a correct characterization of the processes leading to landsliding for producing reliable susceptibility zonation map.

The resulting map allows delineating zones where precaution measures should be implemented, establishing standards and requirements for the use of land on and around slopes that are likely to fail and, also, assessing the risk that a proposed use of land will affect the slope stability of the studied area.

Acknowledgments This study was carried out with financial support from the project POSDRU/89/1.5/S/49944 coordinated by „Al. I. Cuza" University of Iaşi, Romania. The authors wish to thank the reviewers whose comments and corrections were extremely useful for the improvement of the chapter.

References

Aleotti P, Chowdhury R (1999) Landslide hazard assessment: summary review and new perspectives. Bull Eng Geol Environ 58:21–44

Anbalagan R (1992) Landslide hazard evaluation and zonation mapping in mountainous terrain. Eng Geol 32:269–277

Ayalew L, Yamagishi H, Ugawa N (2004) Landslide susceptibility mapping using GIS-based weighted linear combination, the case in Tsugawa area of Agano River, Niigata Prefecture, Japan. Landslides 1:73–81

Ayalew L, Yamagishi H, Marui H, Kanno T (2005) Landslides in Sado Island of Japan: Part II. GIS-based susceptibility mapping with comparisons of results from two methods and verifications. Eng Geol 81:432–445

Bai SB, Wang J, Lü GN, Zhou PG, Hou SS, Xu SN (2010) GIS-based logistic regression for landslide susceptibility mapping of the Zhongxian segment in the three Gorges area, China. Geomorphol 115:23–31

Barredo JI, Benavides A, Hervas J, van Westen CJ (2000) Comparing heuristic landslide hazard assessment techniques using GIS in the Tirajana basin, Gran Canaria Island, Spain. Int J Appl Earth Obs Geoinf 2:9–23

Bălteanu D (1979) Procese de modelare a versanţilor declanşate de cutremurul din 4 martie 1977 în Carpaţii şi Supcarpaţii Buzăului. Stud Cercet Geol Geofiz Geogr Ser Geogr 26
Bălteanu D, Chendeş V, Sima M, Enciu P (2010) A country-wide spatial assessment of landslide susceptibility in Romania. Geomorphol 124:102–112
Binaghi E, Luzi L, Madella P (1998) Slope instability zonation: a comparison between certainty factor and fuzzy dempster-shafer approaches. Nat Hazards 17:77–97
Borgogno Mondino E, Giardino M, Perotti L (2009) A neural network method for analysis of hyperspectral imagery with application to the Cassas landslide (Susa Valley, NW-Italy). Geomorphol 110:20–27
Carrara A, Cardinali M, Detti R, Guzzetti F, Pasqui V, Reichenbach P (1991) GIS techniques and statistical models in evaluating landslide hazard. Earth Surf Proc Landf 16:427–445
Castellanos Abella EA, Van Westen CJ (2008) Qualitative landslide susceptibility assessment by multicriteria analysis: A case study from San Antonio del Sur, Guantánamo, Cuba. Geomorphol 94:453–466
Corominas J, Moya J (2008) A review of assessing landslide frequency for hazard zoning purposes. Eng Geol 102:193–213
Dumitrescu I, Săndulescu M, Bandrabur T (1970) The geological map of Romania, Covasna sheet, scale 1:100,000. Com de Stat al Geologiei, Bucharest
Fell R, Corominas J, Bonnard C, Cascini L, Leroi E, Savage WZ (2008) Guidelines for landslide susceptibility, hazard and risk zoning for land use planning. Eng Geol 102:85–98
Grozavu A, Mărgărint MC, Patriche CV (2010) GIS applications for landslide susceptibility assessment: a case study in Iaşi County (Moldavian Plateau, Romania). In: Brebbia CA (ed.) Risk Analysis VII & Brownfields V, WIT Press, Southampton, pp 393–404
Guzzetti F, Carrara A, Cardinali M, Reichenbach P (1999) Landslide hazard evaluation: a review of current techniques and their application in a multi-scale study, central Italy. Geomorphol 31:181–216
Ichim I, Rădoane M, Rădoane N, Miclăuş C, Grasu C (1996) Sediment budget of the Putna drainage basin (Vrancea). Rev Roum Geogr 40:125–132
Ichim I, Rădoane M, Rădoane N, Grasu C, Miclăuş C (1998) Dinamica sedimentelor. Aplicaţie la râul Putna-Vrancea, Editura Tehnica, Bucharest
Komac M (2006) A landslide susceptibility model using the analytical hierarchy process method and multivariate statistics in perialpine Slovenia. Geomorphol 74:17–28
Lee S, Ryu JH, Won JS, Park HJ (2004) Determination and application of the weights for landslide susceptibility mapping using an artificial neural network. Eng Geol 71:289–302
Malczewski J (1999) GIS and multi-criteria decision analysis. Wiley, New York
Mantovani F, Soeters R, Van Westen CJ (1996) Remote sensing techniques for landslide studies and hazard zonation in Europe. Geomorphol 15:213–225
Micu M, Bălteanu D (2009) Landslide hazard assessment in the Bend Carpathians and Subcarpathians, Romania. Z Geomorphol 53(3):49–64
Nefeslioglu HA, Duman TY, Durmaz S (2008) Landslide susceptibility mapping for a part of tectonic Kelkit Valley (eastern Black Sea region of Turkey). Geomorphol 94:410–418
Soeters R, Van Westen CJ (1996) Slope instability recognition, analysis, and zonation. In: Turner KA, Schuster RL (eds) Landslides: Investigation and Mitigation. Transport Research Board Special Report 247, National Research Council, Washington, DC
Süzen ML, Doyuran V (2004a) Data driven bivariate landslide susceptibility assessment using geographical information systems: a method and application to Asarsuyu catchment, Turkey. Eng Geol 71:303–321
Süzen ML, Doyuran V (2004b) A comparison of the GIS based landslide susceptibility assessment methods: multivariate versus bivariate. Env Geol 45:665–679
Thiery Y, Malet JP, Sterlacchini S, Puissant A, Maquaire O (2007) Landslide susceptibility assessment by bivariate methods at large scales: application to a complex mountainous environment. Geomorphol 92:38–59
Tufescu V (1966) Subcarpaţii. Editura Ştiinţifică, Bucharest

Ursu A (2006) The use of Geographic Informational Systems in the study of natural hazards. Geogr Tech 2:113–120

Van Den Eeckhaut M, Marre A, Poesen J (2010) Comparison of two landslide susceptibility assessments in the Champagne–Ardenne region (France). Geomorphol 115:141–155

Van Westen CJ, Van Asch TWJ, Soeters R (2006) Landslide hazard and risk zonation: why is it still so difficult? Bull Eng Geol Environ 65:167–184

Yin KL, Yan TZ (1988) Statistical prediction model for slope instability of metamorphosed rocks. In: Bonnard C (ed) Proceedings, 5th International Symposium on Landslides. Balkema, Rotterdam

GIS Evaluation of Erosion-Sedimentation Risk Caused by Extreme Convective Rainstorms: Case Study of the Stonávka River Catchment, Czech Republic

Boris Šír, Peter Bobáľ and Jozef Richnavský

Abstract Timeliness of erosion-sedimentation processes is evident at both local and global scale. In Central Europe, problems of landscape management are often discussed in the relation to rainfall-runoff phenomena and the resultant processes of soil erosion and sediment deposition. Nowadays, these issues are well elaborated theoretically and with the use of information technologies (IT) and geographical information systems (GIS) potential, the evaluation of predisposition of a given area to erosion is quite easy. IT and GIS offer the effective use of a wide range of erosion models. In this chapter, a USPED (Unit Stream Power based Erosion/Deposition) model (Mitasova et al. 1996) is applied to the analysis of a rainstorm event in June 2009 within the Stonávka River catchment, Czech Republic. In comparison to the well-known USLE (Wishmeier and Smith 1965, 1978) model and its newer versions, the USPED model is more suitable for the use at a catchment scale and besides erosion, it is also able to calculate the rate of deposition of eroded material. Because of the convective character of the causal rainfall, the Onstad-Foster's equation (Onstad and Foster 1975) was used to derive the *R* factor describing an erosive effect of rainfall/surface runoff.

1 Introduction

Soil, the basic natural wealth and resource, is a contact zone among other units of the physico-geographical sphere and it is an active factor in the process of substance and energy exchange among them. Therefore, it is necessary to protect the pedosphere, as an integral component of the environment, and to increase its productive potential (Buzek 1983).

B. Šír (✉) · P. Bobáľ · J. Richnavský
Faculty of Geology and Mining, VŠB, Technical University of Ostrava,
17. listopadu 708 33 Ostrava, Czech Republic
e-mail: sir.boris@centrum.cz

J. Kozak et al. (eds.), *The Carpathians: Integrating Nature and Society Towards Sustainability*, Environmental Science and Engineering,
DOI: 10.1007/978-3-642-12725-0_5, © Springer-Verlag Berlin Heidelberg 2013

Soil erosion is one of the main processes endangering the pedosphere. Although it is a natural process, it is becoming a serious threat with an increasing intensity of land use at both local and global scale. Thus, soil erosion at its present extent is mainly a result of human activities and not a product of natural processes (Schmidt 2000).

Because man has faced soil erosion since the beginning of his agricultural activities, the process of erosion is well described theoretically. Good theoretical knowledge of erosion processes has led to the development of a great number of erosion models, that nowadays, with the availability of information technologies (IT) and geographical information systems (GIS), become an effective tool for the landscape management.

Soil erosion and the related processes depend on a range of factors. Because of erosion solution, this complex of factors is more or less reduced in the models to a selection of easily describable parameters: hydro-meteorological situations, qualities of soil and land cover and topography. Nowadays, we arc able to represent these parameters, including their spatial variability, in a digital format thanks to geographical information systems. GIS became a very suitable tool for the evaluation of erosion-sedimentation hazard.

Rainfall and the resultant surface runoff are the basic determinants and a triggering mechanism of soil erosion. Many authors tried to describe the relation between erosion and rainfall characteristics (see Morgan 2005). Some works show that most of the erosion processes are connected with medium-intense rainfall-runoff situations of a medium return period, while others associate greater importance with less frequent, high-intensity rainfalls. Undoubtedly, some regional aspect plays a significant role here. Moreover, impact of intense rainstorms on the total soil loss due to erosion is increasing together with climate change and growing extremity of hydro-synoptic situations. High-magnitude events themselves are becoming significant natural phenomena of the present time and it is necessary to consider an increasing frequency of their occurrence in the future.

This chapter presents an evaluation of the soil erosion-sedimentation risk caused by extreme convective rainstorm of June 2006 within the Stonávka River catchment. USPED model together with Onstad-Foster's formula for the determination of rainfall-runoff erosivity factor were used as main modeling tools. The erosivity factor was calculated using rainfall radar images of the evaluated meteorological situation. The modeling was done in the geographic information systems environment.

2 Study Area

The catchment of the Stonávka River in the Moravian-Silesian region, Czech Republic (Fig. 1), was selected as the study area. The Stonávka River is a left-side tributary of the Olše River and it is a stream of third order. The Stonávka rises on the northern slopes of the Moravian-Silesian Beskydy Mountains at an altitude of about 750 m and in its upper part, it has a character of a mountain stream. Its

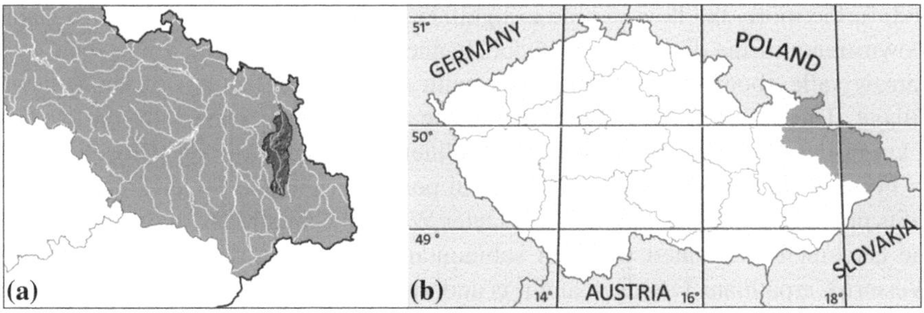

Fig. 1 Catchment localization within (**a**) the Moravian-Silesian region and (**b**) the Czech Republic

confluence with the Olše River is located at an altitude of 220 m within the town of Karviná, after 33 km of the river length. The watershed elevation range is about 700 m (Fig. 2c). The catchment is about 131 km^2 in area.

In its upper course, where the river runs through the front zone of the flysch Moravian-Silesian Nappe, the dynamics of the Stonávka are determined by a high channel gradient. In the lower course, the channel gradient is much more gentle

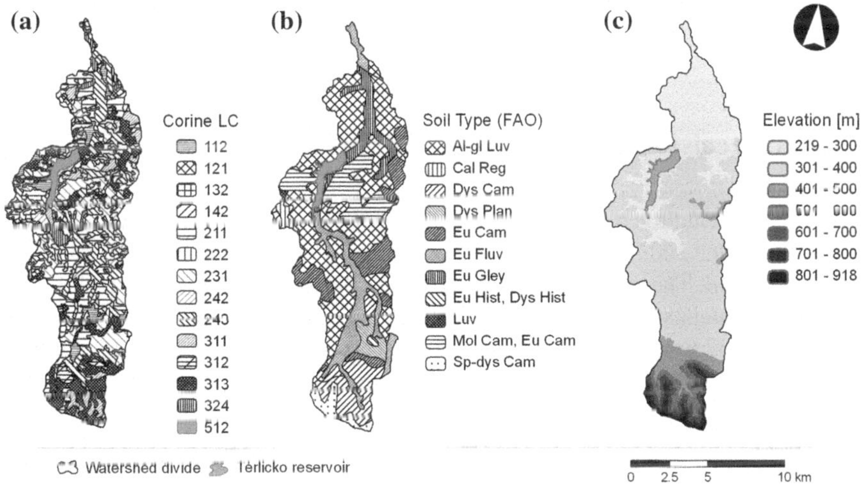

Fig. 2 Land cover (**a**), soils (**b**) and elevation (**c**) of the Stonávka river catchment. CORINE land cover: *112* discontinuous urban fabric; *121* industrial and commercial units; *132* dump sites; *142* sport and leisure facilities; *211* non irrigated arable land, *222* fruit trees and berry plantations; *231* pastures; *242* complex cultivation patterns; *243* land particularly occupied by agriculture, with significant areas of natural vegetation; *311* broad-leaved forest; *312* coniferous forest; *313* mixed forest; *324* transitional woodland-shrub; *512* water bodies. Soil type of food and agriculture organization: *Al-gl Luv* Albo-gleyic Luvisol; *Cal Reg* Calcaric Regosol; *Dys Cam* Dystric Cambisol; *Dys Plan* Dystric Planosol; *Eu Cam* Eutric Cambisol; *Eu Fluv* Eutric Fluvisol; *Eu Gley* Eutric Gleysol, *Eu Hist*, *Dys Hist* Eutric Histosol, Dystric Histosol; *Luv* Luvisol; *Mol Cam*, *Eu Cam* Mollic Cambisol, Eutric Cambisol; *Sp-dys Cam* Spodo-dystric Cambisol

and, furthermore, the flow regime is significantly influenced by the Těrlicko Dam. Downstream of the dam, the Stonávka channel meanders through a flat, relatively narrow valley bounded by high slopes, with scattered settlements of the Stonava village. The natural runoff conditions in the lower course, especially on the last 3 km of the river, are also significantly influenced by coal mining (Brosch 2005).

From geological and geomorphological point of view, the Stonávka catchment in its upper part belongs to the flysch Outer Western Carpathians. In its lower part, the catchment is located within a submontane depression developed within the Western Carpathians Foredeep and it is underlain mainly by Quaternary sediments. The soil cover of the watershed is quite varied, with cambisols and podzols dominating in the upper parts of the catchment, whereas in its lower parts there occur various subtypes of cambisols but also several types of luvisols, dystrict planosols, gleysols, calcaric regosols and fluvisols (Fig. 2a). The land cover is also quite varied with 14 categories of CORINE Land Cover classification (Fig. 2b). The higher areas of the Beskydy Mts. are covered with forest and mountain pastures, whereas the lower submontane areas are used more intensively.

According to Quitt's climatic classification (1971), the Stonávka catchment belongs to the areas CH6, CH7 (cold regions) and MT2, MT9, MT10 (regions of moderate climate). The catchment is characterized by medium water availability and runoff generation conditions, with the highest runoff in March and April. The specific runoff from the catchment is 6–10 l s^{-1} km^{-2}, the runoff from the catchment can be evaluated as heavily fluctuating and the runoff coefficient varies from 0.21 to 0.7.

3 Characteristics of Meteorological Situations

Erosion-sedimentation processes in the catchment were modeled for a rainfall-runoff event of 6th June 2009. Specifically, it was 2 h-long rainfall which was falling from 6:15 to 8:15 PM. The event was characterized by the total precipitation of 11.3 mm, the average intensity of 5.7 mm h^{-1} and the maximum recorded intensity in the catchment of 37.6 mm h^{-1}. The centre of the precipitation cell was moving through the lower part of the catchment, which was most affected by the rainfall. Figures 3 and 4 indicate spatial distribution of rainfall-runoff parameters of the event.

4 Methods

The first mathematical approach to describe soil erosion by water was the universal soil loss equation (USLE) developed by Wischmeier and Smith (1965). Nowadays, there is a wide range of soil erosion models available which can be divided into three groups: (i) models computing only long-term soil loss without any consideration of transport and deposition processes (e.g. USLE and its modifications (RUSLE—revised USLE)); (ii) empirically based erosion models additionally

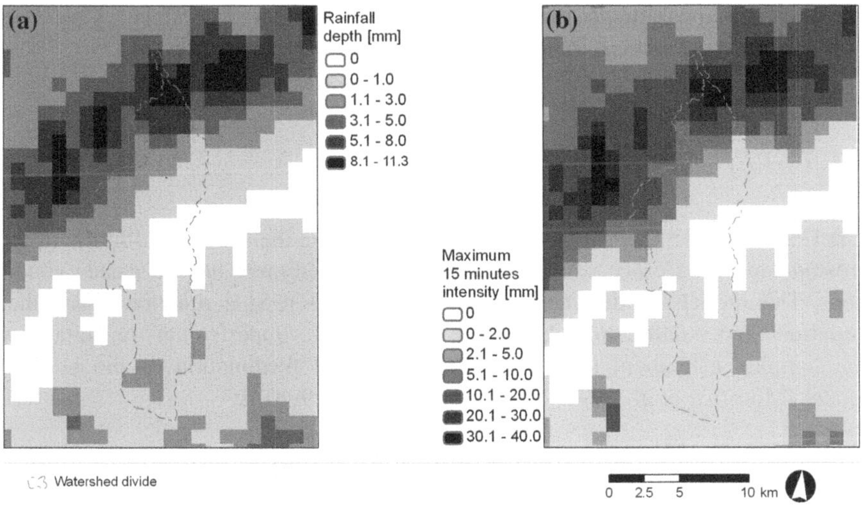

Fig. 3 Rainfall depth (**a**) and maximum 15-min rainfall intensity (**b**) of the 6th June 2009 event estimated from rainfall radar images (original spatial resolution, cell size = 1 km)

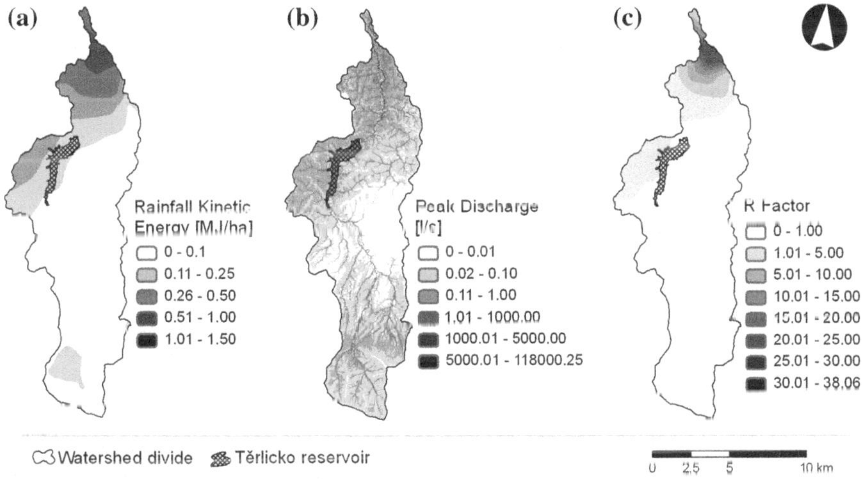

Fig. 4 Spatial distribution of rainfall kinetic energy (**a**), overland flow peak discharge (**b**) and *R* factor values (**c**) within the Stonávka River catchment for the rainfall event of 6th June 2009

simulating transport and deposition processes (e.g. USPED); and (iii) process-oriented erosion models simulating the effect of raindrop splash, sheet runoff and further processes (e.g. EROSION-3D, WEPP, LISEM) (Deinlein and Böhm 2000).

In the present study, a USPED model (Unit Stream Power based Erosion/Deposition) (Mitasova et al. 1996) in combination with RUSLE (Renard et al. 1991) and Onstad-Foster's formula (Onstad and Foster 1975) were used as a mathematical apparatus to simulate the erosion-sedimentation effects of selected rainstorms. As a working tool, the capabilities of GIS were used.

4.1 USPED model

The USPED model is a simple tool used to determine the space distribution of the erosion/deposition processes as the divergence of sediment flow in steady conditions. This model is a derivate of the well-known (R)USLE equation or its other variations and modifications but removes their basic imperfection—unsuitability for complex catchment terrain. The mathematical formulation of the USPED model (Mitasova et al. 1996; Mitas and Mitasova 1998) is:

$$ED = div q_s = K_t[grad h] \cdot S \sin(b) - h[k_p + k_t] \tag{1}$$

where ED is erosion/deposition rate; q_s is sediment flow rate; K_t is coefficient of transportability; h is water depth estimated from the upslope area; S is unit vector in the direction of the steepest slope; b is slope gradient; k_p is profile curvature of relief; and k_t is tangential curvature of relief.

The (R)USLE equation and its other modifications and variants consider only the erosion along the lines of surface runoff and they are not able to express the effect of flow convergence/divergence resulting from curvature of relief (Warren et al. 2000). Because of this, they often predict erosion in the areas where deposition of eroded material occurs in stated conditions.

The USPED model uses the basic form of RUSLE equation (Renard et al. 1991) which is:

$$E = R \cdot K \cdot L \cdot S \cdot C \cdot P \tag{2}$$

where E is average annual soil loss; R is factor of erosivity (an erosive effect of rainfall/surface runoff); K is factor of soil erodibility; L is factor of slope length; S is factor of slope gradient (product of L and S is so called topographic factor, LS); C is factor of protective effect of vegetation; and P is factor of erosion control.

A specific connection between the RUSLE equation and the USPED model is that the product of factors K, C and P in the former can be replaced in the USPED model with K_t parameter, i.e. a coefficient of transportability. The USPED model develops the final equation replacing factor LS by the product of a factor of contributing area and a factor of slope gradient in the form:

$$LS = R \cdot K_t \cdot A^m (\sin \beta)^n \tag{3}$$

where A is contributing area; β is slope gradient; and m and n are coefficients (if rill erosion prevails, $m = 1.6$, $n = 1.3$; if sheet erosion prevails, both m and $n = 1$).

By implementing this form of *LS* factor and combining factors *K*, *C* and *P* into one parameter K_t, we can call *E* term of the RUSLE equation as a sediment transport capacity of flow, which mathematical expression is as follows:

$$T = R \cdot K_t \cdot A^m \cdot (\sin \beta)^n \tag{4}$$

After that the USPED model calculates the rate of erosion/deposition as a divergence of sediment transporting flow for each cell of a grid. Negative divergence means erosion prevailing over sedimentation in a given cell of a raster and vice versa. The final mathematical notation of the USPED model is given by the equation:

$$ED = \frac{\partial(T \cdot \cos \alpha)}{\partial x} + \frac{\partial(T \cdot \sin \alpha)}{\partial y} \tag{5}$$

where α is aspect of relief.

5 *R* Factor Derivation

As a basic data for *R* factor derivation, the radar images of 15-min rainfall depth were used. They allowed to derive spatially distributed values of unknown parameters of the Eqs. 6, 7, 8, 9 and 12. It was done in the GIS environment using map algebra tools which enable a wide range of computational operations with the grid data format.

Nowadays a number of methods are developed to quantify the impact of erosive effect of rainfall or surface runoff. For example, the RUSLE model, which is not designated for individual rainfall-runoff events, works with a kinetic energy of rainfall and with its maximum 30-min intensity, called index EI30. One of the frequently used forms of this model suitable for the usage in individual rainfall-runoff events is a modified universal soil loss equation (MUSLE) (Williams 1975; Williams and Berndt 1977). This equation replaces the factor of erosive rainfall effect with the factor of surface runoff (Williams 1975 in Boardman and Favis-Mortlock 1998) in the form:

$$R = 11.8 \cdot \left(Q \cdot q_{pk}\right)^{0.56} \tag{6}$$

where Q is surface runoff depth; and q_{pk} is peak discharge of the event.

Characteristic features of the hydro-meteorological events of an intense rainstorm type are not only rapid and considerable surface runoff but also the high intensity and kinetic energy of rain. It is obvious that for an evaluation of erosion-sedimentation processes caused by rainstorm, it is appropriate to take into consideration also the parameters of causal rainfall. An equation developed by Onstad and Forster (1975) seems to be appropriate for this purpose. The Onstad-Foster's model is one of the methods of R factor determination for a single event.

It seems to be a good tool for predicting soil erosion caused by rainstorm, because it takes into consideration the combined effect of kinetic energy of rainfall and surface runoff of storm water. Its mathematical notation (Onstad and Foster 1975 in Boardman and Favis-Mortlock 1998) is as follows:

$$R = 0.64 \cdot E \cdot I + 0.45 \cdot \left(Q \cdot q_{pk}\right)^{0.33} \tag{7}$$

where E is kinetic energy of rainfall; and I is maximum intensity of rainfall per time unit.

It is possible to derive the kinetic energy of rainfall from a whole set of equations (see e.g. Morgan 2005). In the present study, the following equation was used (Brown and Foster 1987 in Morgan 2005):

$$E = 0.29\left(1 - 0.72e^{-I/20}\right) \tag{8}$$

Parameter I is often considered to be the maximum 30-min rainfall intensity, which is quite frequently questioned in literature. For instance, Stocking and Elwell (1973 in Boardman and Favis-Mortlock 1998) recommend it only for bare soil. For covered soil, it is recommended to use in the calculations the maximum 5–15-min rainfall intensity (Morgan 2005). In this study, a 15-min value of maximum intensity was used.

Concerning the parameters of surface runoff, the depth of runoff was calculated for every DEM cell according to the classic SCS CN (Soil Conservation Service Curve Number) method with the correction of the values of CN curves by slope gradient. The basic equation of the used method (Dingman 2008; Bedient et al. 2008) is as follows:

$$Q = \frac{(p - 0.2 \cdot S)^2}{(p + 0.8 \cdot S)} \tag{9}$$

where Q is surface runoff volume; p is rainfall depth; and S is storage parameter.

Parameter S can be calculated according to the following formula (Dingman 2008; Bedient et al. 2008):

$$S = 25.4 \cdot \left(\frac{1000}{CN} - 10\right) \tag{10}$$

The adjustment of the CN curve values alongside slope gradient was done according to the formula of Nietsch et al. (2000):

$$CN_{2S} = \frac{(CN_3 - CN_2)}{3} \cdot [1 - 2\exp(-13.86slp)] + CN_2 \tag{11}$$

where CN_{2S} is adjusted value of a CN curve for given slope gradient; CN_2 is value of a CN curve for average moisture conditions; and CN_3 is value of a CN curve for a catchment saturated to the value of field capacity (Nietsch et al. 2000).

The peak discharge of surface flow through a DEM cell was obtained using a well-known rational method (Maidment 1993), which mathematical notation is as follows:

$$q_{pk} = \frac{(C \cdot i \cdot A)}{3.6} \tag{12}$$

where C is runoff coefficient; i is rainfall intensity; and A is contributing area.

6 Derivation of *K*, *C* and *P* Factors

The classification of the catchment area according to the values of K factor, expressing the erodibility of soil (Table 1), was done by deriving the factor values for soil types and individual main soil units of estimated pedologic-ecological unit (BPEJ, in Czech "bonitované půdně-ekologické jednotky") according to Janeček (2002).

Table 1 Values of *K* factor for soil types distinguished by food and agriculture organization (FAO) classification

Soil type (FAO system)	*K* factor
Eutric fluvisol	0.31
Eutric gleysol	0.30
Luvisol	0.31
Eutric cambisol	0.29
Dystric cambisol	0.28
Mollic cambisol, eutric cambisol	0.23
Albo-gleyic luvisol	0.30
Eutric histosol, dystric histosol	0.30
Calcaric regosol	0.25
Spodo-dystric cambisol	0.20
Dystric planosol	0.34

Table 2 *C* factor values for CORINE land cover categories recorded in the Stonávka catchment

CORINE land cover category	*C* factor
112	0.30
121	0.30
132	0.30
142	0.01
211	0.20
222	0.35
231	0.05
242	0.25
243	0.10
311	0.09
312	0.04
313	0.07
324	0.01

The evaluation of the catchment area in respect to the values of *C* factor, i.e. protective effect of vegetation (Table 2), was done by deriving the factor values for the categories of CORINE Land Cover classification according to Mohaupt-Jahr (2004). The erosion control in the catchment was not considered, and thus the value of *P* factor was set at 1.

7 Results

Erosion-sedimentation processes were modeled for two settings of the USPED model, reflecting two settings of *LS* factor calculation (see Eq. 3), separately for sheet and rill erosion. The outputs for sheet erosion are presented in Fig. 5 and those for sheet deposition in Fig. 6.

With respect to time aspect, the USPED model is static, similarly as all its input data. The event of 6th June 2009, with its 2-h long rainfall, distinguishes by a high intensity of precipitation, culminating in 37.6 mm h^{-1}. The main centre of the rainfall was located over the northern, lower part of the catchment, whereas the montane part of the catchment was affected by less intensive rainfall. The spatial

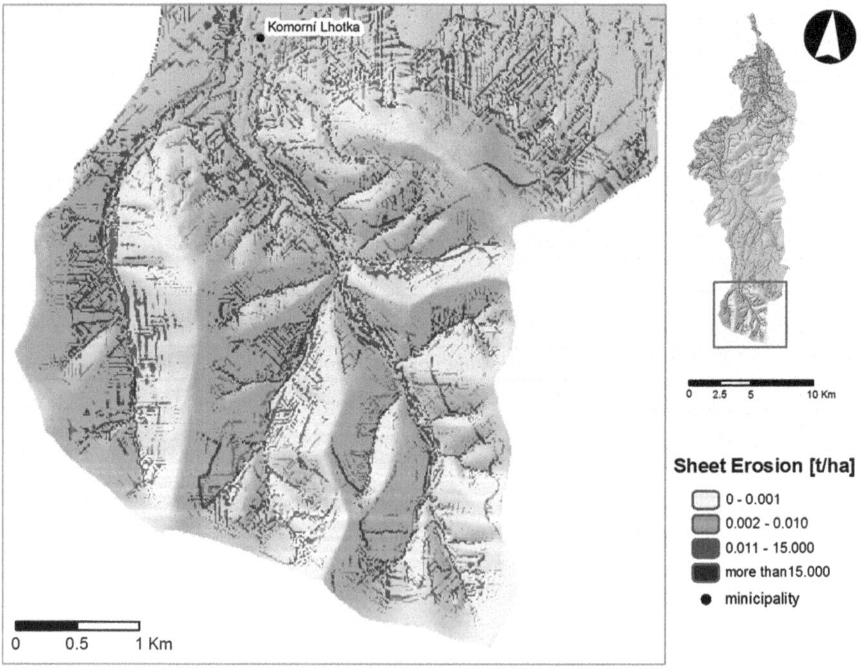

Fig. 5 The amount of sheet erosion within the southern part of the Stonávka catchment (Moravian-Silesian Beskydy Mts.) predicted by the USPED model for the rainfall event of 6th June 2009

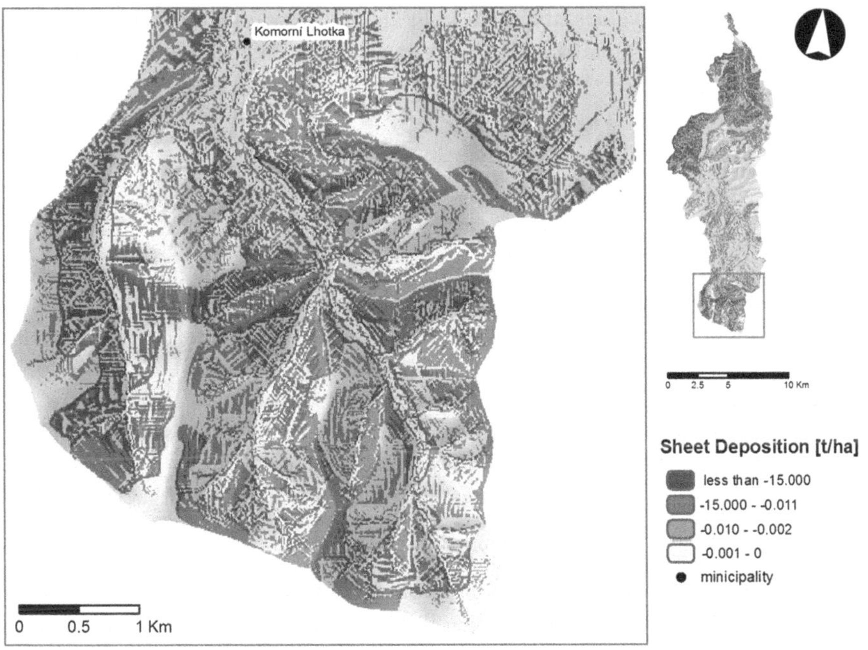

Fig. 6 The amount of sheet deposition within the southern part of the Stonávka catchment (Moravian-Silesian Beskydy Mts.) predicted by the USPED model for the rainfall event of 6th June 2009

distribution of erosion rate mimics this pattern. The maxima of erosion are situated in the area of the main precipitation centre, although it reaches high values also on the slopes of the Moravian-Silesian Beskydy Mts., where the impact of precipitation and runoff dynamics is usually high. Here, the dynamics of runoff is given by a significant relief gradient of the Beskydy Mts. Nappe. A considerable influence of topography can be observed also on the asymmetric ridges eastwards from the Těrlicko Dam. The northern, steeper slopes are more endangered by erosion than more gentle slopes with the southern aspect. Considering the influence of other factors on the results of erosion modeling, we can observe in the areas affected by storm water the occurrence of isolated areas with a lower rate of soil erosion in comparison with their surroundings. The lower rates of erosion mostly reflect high values of C factor, as they are associated with the occurrence of considerably paved surfaces in industrial and business areas. On the contrary, in the areas minimally affected by the rainfall, the occurrence of areas relatively endangered by erosion can be explained by the combination of soils predisposed to erosion and an intensive agricultural use of the land.

The analysis of sedimentation processes indicated that the zones with a high potential for eroded material deposition occur at the toe of slopes and on valley bottoms. The rate of sedimentation is the highest there. The results of modeling indicate that also the flatter parts of slopes exhibit material deposition. Across the whole range of their values, the areas with erosion prevailing over deposition are more fragmented into smaller cells of a raster, that are localized in the areas of the highest concentration of substance flow.

8 Discussion

The results obtained from the model are indicative of the spatial distribution of erosion-sedimentation phenomena rather than of the absolute amounts of erosion and deposition at particular locations. It is obvious from the character of the outputs of the (R)USLE equation and the USPED model which have to be understood as reflecting the potential for erosion/sedimentation risk to occur at a given place. The uncertainties arise from the essence of the models as well as from the input data and local features of the studied catchment.

In the case of input data, the greatest uncertainty is linked with meteorological data acquired by meteorological radars. These data represent estimates rather than actual values of precipitation depth/intensity. The radar data were not adjusted. The raster layer of radar precipitation totals has the spatial resolution of 1 km^2. During the pre-processing of input data, this grid was converted into a point layer of precipitation depths. This layer was then interpolated into a grid with a finer resolution. That process removed quite high discretization of the values of rainfall depth into the 1×1 km squares and thus more realistic spatial distribution of rainfall depth was reconstructed.

The calculation of R factor using the Onstad-Foster's equation also offers space for discussion. As shown in Eq. 7, this formula is composed of two parameters: the erosive effect of rainfall and the erosive effect of the surface runoff. The significance of both parameters for the value of their multiples in the equation can be considered. In this chapter a common form of the equation was used.

Another uncertainty can arise from the use of the rational method of peak discharge estimation. This method is usually recommended for small catchments. Brooks et al. (2003) claim that the method allows successful estimation of peak discharge of the runoff from small (less than 1,000 ha), relatively homogeneous catchments. Other authors (e.g. Bedient et al. 2008; Ward and Trimble 2004) recommend this method for the catchments with an area of a few square kilometers. However, Maidment (1993) mentions that in Australia consistent flood estimates have been obtained for drainage basins up to 250 km^2 in area. Of course, regional conditions must be considered. The Stonávka River catchment has an area of 131 km^2 but with its shape and type of drainage network, peak discharges of overland flow through each cell of slope can be estimated successfully. It is so because the catchment area of particular tributaries to the Stonávka River satisfies the condition formulated by Brooks et al. (2003). Also if we compare the values of peak discharges obtained by the rational method with the long-term mean discharge of the Stonávka River, which is about 1.5 $m^3\ s^{-1}$, then it is obvious that the results of the calculations are realistic. Although other, more sophisticated methods of the peak discharge estimation such as some distributed hydrological models are available, the results of calculations performed with the rational method are relatively satisfactory, considering the time and data required for distributed hydrological models such as MIKE SHE or TOPMODEL.

An interesting problem has been noticed during data pre-processing and their adjustment for the needs of the model. Although the Těrlicko Dam was considered in DEM by indicating a horizontal water surface in the area of the reservoir, the flow accumulation algorithm was unable to notice this fact in either ArcGIS or GRASS GIS software. Thus the Těrlicko Dam was not included in the interpretations and it was masked by a water dam polygon in the visualizations.

Considering the two settings of LS factor for rill or sheet erosion, it seems the most suitable to use the coefficients for rill erosion. In the case of rill erosion, the maximum values of erosion and deposition predicted by the model are of the order of tonnes per ha. In the case of sheet erosion, they are of the order of thousands of tonnes per ha, although such values occur in isolated areas of a size of several grid cells. In the real terrain of a catchment, with uneven relief and a variable mosaic of land cover, the sheet erosion will be probably limited to the slopes with a smooth surface and a very poor vegetation cover and rill erosion will be certainly more frequent.

9 Conclusion

All the discussed uncertainties are not related only to the USPED model but also to commonly used (R)USLE equations. On the contrary, the USPED model is able to take into consideration the complexity of catchment relief and to incorporate erosion and sedimentation processes together. Another advantage of the USPED model is that it can be applied to a single event as well as to the studies of annual soil loss. So, it can be considered an appropriate and better alternative of these older models.

Acknowledgments We would like to acknowledge the SGS project SV5480051/2101. If it is not said differently, all the used data were provided by the FLOREON + project performed by the team of researchers of the VSB-Technical University of Ostrava for the Moravian-Silesian region Council. The vector representation of estimated pedologic-ecological unit (BPEJ) was provided by the Research Institute for Soil and Water Conservation, Prague, Czech Republic. Radar rainfall images were provided by Czech Hydrometeorological Institute for the need of FLOREON + project.

References

Bedient PC, Huber WC, Vieux BC (2008) Hydrology and floodplain analysis. Prentice Hall, London

Boardman J, Favis-Mortlock D (eds) (1998) Modelling soil erosion by water. NATO ASI series, series I: global environmental change, vol 55. Springer, Berlin

Brooks KN, Ffolliott PF, Gregersen HM, DeBano LF (2003) Hydrology and the management of watersheds. Blackwell Publishing, Ames
Brosch O (2005) Povodi odry. Anagram, Ostrava
Brown LC, Foster GR (1987) Storm erosivity using idealized intensity distributions. In: Morgan RPC (2005) Soil erosion and conservation. Blackwell Publishing, Malden
Buzek L (1983) Eroze pudy. Pedagogical Faculty of Ostrava, Ostrava
Deinlein R, Böhm A (2000) Modeling overland flow and soil erosion for a military training area in southern Germany. In: Schmidt J (ed) Soil erosion. Application of physically based models. Springer, Berlin
Dingman SL (2008) Physical hydrology. Waveland Press, Long Grove
Janeček M (2002) Ochrana zemedelske pudy pred erozi. ISV Nakladatelstvi, Praha
Maidment DR (ed) (1993) Handbook of hydrology. McGraw-Hill Professional, London
Mitas L, Mitasova H (1998) Distributed soil erosion simulation for effective erosion prevention. Water Resour Res 34:505–516
Mitasova H, Hofierka J, Zlocha M, Iverson LR (1996) Modeling topographic potential for erosion and deposition using GIS. Int J GIS 10:629–641
Mohaupt-Jahr B (ed) (2004) Workshop CORINE land cover 2000 in Germany and Europe and it's use for environmental applications. Federal Environmental Agency (Umweltbundesamt), Berlin, 20–21 Jan 2004
Morgan RPC (2005) Soil erosion and conservation. Blackwell Publishing, Malden
Nietsch SL, Arnold JG, Kiniry JR, Williams JR, King KW (2000) Soil and water assessment tool theoretical documentation. version 2000. Texas Water Resources Institute & Blackland Research and Extension Center, Temple
Onstad CA, Foster GR (1975) Erosion modeling on a watershed. Trans Am Soc Agricult Eng 18:288–292
Quitt E (1971) Mapa klimatickych oblastí CSR, 1:500000. Geographical Institute of Czechoslovak Academy of Science, Brno
Renard KG, Foster GR, Weesies GA, Porter JP (1991) RUSLE—revised universal soil loss equation. J Soil Water Conserv 46:30–33
Schmidt J (ed) (2000) Soil erosion. Application of physically based models. Springer, Berlin
Ward AD, Trimble SW (2004) Environmental hydrology. Lewis Publishers, Boca Raton
Warren SD, Senseman GM, Block PR, Ruzycki TS, Wilcox DD (2000) Soil survey for Camp Guernsey, Wyoming using new-generation multi-dimensional soil erosion modelling. Colorado State University, Ft. Collins
Williams JR (1975) Sediment-yield prediction with universal equation using runoff energy factor. Present and prospective technology for predicting sediment yield and sources. In: Proceedings of the sediment yield workshop, USDA Sedimentation Lab. Oxford, 20–30 Nov 1972, pp 244–252
Williams JR, Berndt HD (1977) Sediment yield prediction and utilization of rangelands. Documentation and user guide. US Department of Agriculture, ARS 63, Washington
Wischmeier WH, Smith DD (1965) Predicting rainfall erosion losses. Agriculture handbook 282. USDA-ARS, Washington
Wischmeier WH, Smith DD (1978) Predicting rainfall erosion losses: a guide to conservation planning. Agriculture handbook 537. US Department of Agriculture, Washington

Assessment of the Impact of Agricultural Land Utilization Practices on Soil Losses

Pavol Kenderessy

Abstract This chapter presents results of a scenario-based simulation of soil loss on agricultural land. This method was used to simulate the soil loss under various stages of crop growth and management practices during the vegetation period. Scenarios were designed to capture the development of agricultural crops and related soil parameters throughout the year and the effect of soil erosion-control measures such as grassed waterways, buffer strips and size of plots. Soil loss was simulated using a physically based soil erosion simulation model Erosion 3D. This model was used to indicate the main areas of soil loss and to simulate the erosion rates before and after implementing soil conservation measures. Results showed that the applied practices can effectively reduce the rates of soil loss. They also confirm that simulation models such as Erosion 3D can provide the information needed for adequate localization and scaling of site-specific practices of soil conservation.

1 Introduction

Practical assessment of soil erosion is a difficult task. Soil erosion occurs at varying rates depending on natural conditions and the type of farming. In addition, many difficulties are associated with monitoring of this process. In many cases, direct measurements of soil losses are conducted on small experimental runoff plots where the relevant hydraulic conditions leading to erosion cannot be completely reproduced. For the same reason, runoff plot measurements cannot be directly transferred to natural slopes and catchments without considering the differing hydraulic conditions. Thus, the technology for estimating rates of soil erosion has emerged as a major tool to overcome these difficulties. The application of physically based models

P. Kenderessy (✉)
Institute of Landscape Ecology, Slovak Academy of Sciences, Štefánikova 3
814 99 Bratislava, Slovakia
e-mail: pavol.kenderessy@savba.sk

J. Kozak et al. (eds.), *The Carpathians: Integrating Nature and Society Towards Sustainability*, Environmental Science and Engineering,
DOI: 10.1007/978-3-642-12725-0_6,

represents the recent trend in soil erosion research together with the development of GIS and remote sensing techniques. Experimental activities using ANSWERS, AnnGNPS98 models were applied to model overland flow, erosion processes and river sedimentation (Hlavčová and Macura 1993). The most remarkable application of physical modeling, ERDED model, was presented in the work of Hofierka and Šúri (1996). Other studies presented by Slovak authors were carried out outside Slovakia. The SMODERP model has been verified on experimental plots in Japan (Janský 2001) and the USPED model has been successfully applied in modeling erosion hazard in army training sites in the USA (Michael et al. 2005).

Operational models for regional assessments should be based on simple data requirements, must consider spatial and temporal variability in hydrological and soil erosion processes and must be applicable to a variety of regions with a minimum of calibration. This study aims to assess the applicability of the Erosion 3D model for erosion risk assessments at the landscape scale and the evaluation of the impact of soil cover on soil loss.

2 Materials and Methods

2.1 Study Area

The study area represents a typical region of the loess plateau situated in the south-western part of Slovakia. According to the morphogenetic soil classification system of Slovakia, the prevailing soil types on the moderate slopes and the loess plateau can be characterized as chernozems (Šály 2000). Haplic luvisols and regosols prevail on steeper slopes. Gentle slopes dominate, only locally exceeding 4° (up to 8°–12°). Because high-quality soils prevail, agriculture is well developed. This fact significantly influenced the landscape structure: arable land dominates the landscape. The region is oriented towards an intensive agricultural production, with the absence of industrial facilities in the catchment. A small but typical catchment 2,260 ha in size has been selected as a case study (Fig. 1).

2.2 Erosion Modeling

A general scheme of methodology used in the study is shown in Fig. 2.

To predict soil loss in the study area, the physically-based soil erosion model Erosion 3D was used. The model calculates the amount and directions of overland flow based on the slope and aspect of the considered land surface, and the infiltration rate which is estimated by an infiltration with the approach of Green and Ampt (1911). The algorithm for calculating the spatial distribution of flow paths

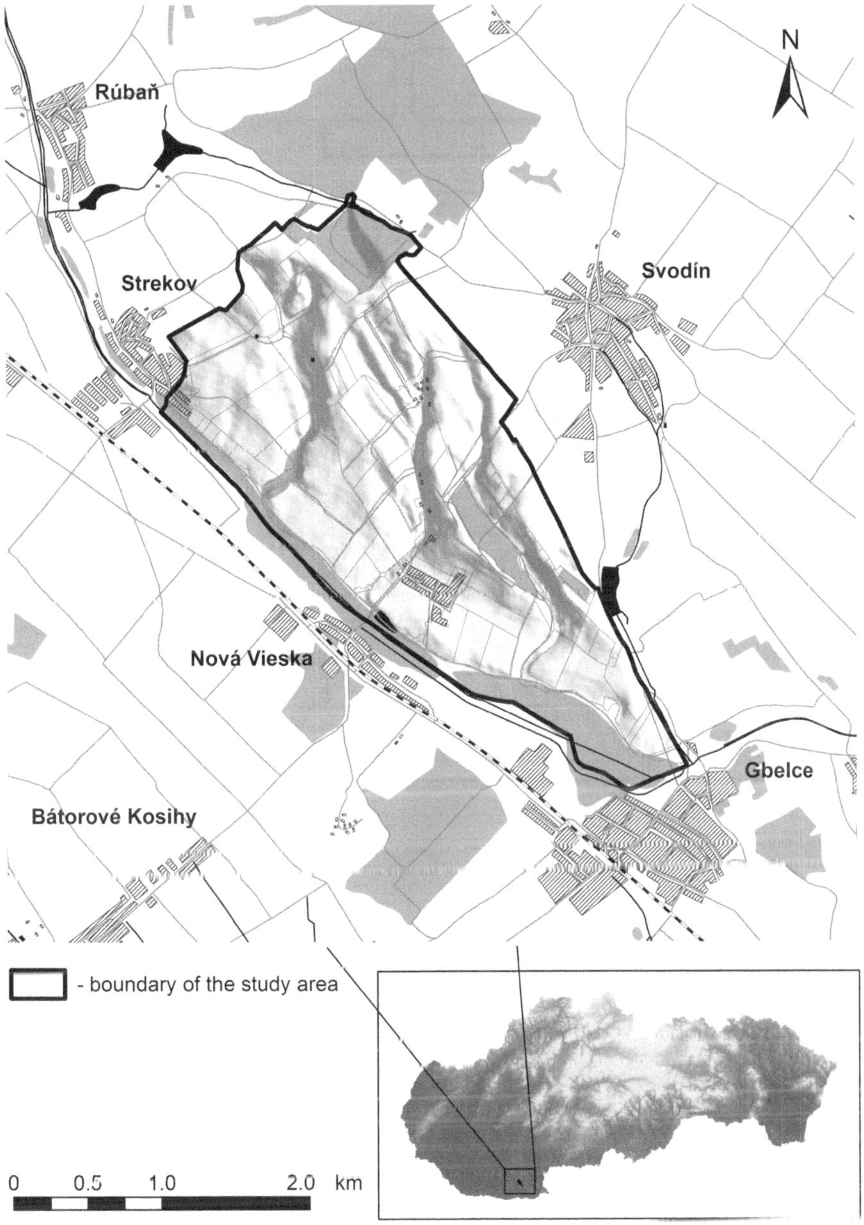

Fig. 1 The study area

uses a raster-based digital elevation model (Schmidt and Werner 2000). The application of Erosion 3D model requires information on site-specific relief and soil and rainfall conditions. This information is supplied to the model using the following parameters:

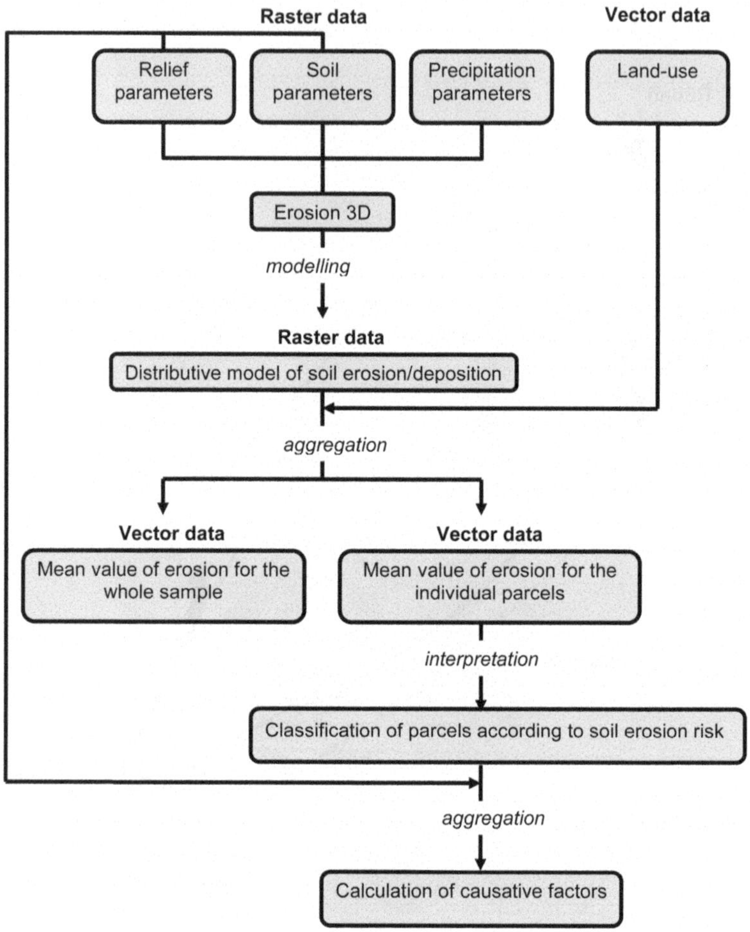

Fig. 2 The methodology of erosion risk assessment used in this study

- relief parameters:
  - Digital Elevation Model (regular squared tessellation/matrix),
- soil parameters:
  - bulk density (kg m^{-3}),
  - initial soil moisture content (%),
  - organic carbon content (%),
  - erodibility (N m^{-2}),
  - roughness coefficient (Manning n),
  - canopy cover grade (%),
  - grain size distribution (fine clay to coarse sand according to DIN classification) (%),

  - correcting factor for hydraulic conductivity,
  - related precipitation measuring points,

- precipitation parameters:

  - duration of rainfall (min),
  - rainfall intensity (mm min^{-1}).

The values of all input parameters are assumed to be spatially uniform below the scale of grid resolution which has been set to 1 m. The same applies also to DEM resolution.

3 Input Parameters

Topography

The required relief parameters are computed in a pre-processor module. The input parameter is represented by Digital Elevation Model (DEM) with 10 × 10 m resolution. A DEM file was processed in ArcGIS/Spatial Analyst tools using the spline interpolation method of point elevation data. The following morphographic and hydrographic parameters are computed from a digital terrain model by means of digital relief analysis (Wickencamp et al. 2000):

- DEM after removal of sinks,
- aspect,
- inflow table,
- flow accumulation,
- watershed,
- main flow direction.

Following spatially distributed soil parameters were adjusted according to cell size and extend of DEM.

Soil parameters

The number of soil parameters, which are required for the computation, is kept low for the sake of the easy-to-use model. Soil characteristics are regarded as homogeneous for each grid cell. The soil parameters consist of the 9 files mentioned above, with each file containing one single parameter for each grid cell. The soil data were obtained in three ways.

First, interpolation from a set of point data has been used for data with constant values, which means that they do not vary significantly according to crop rotation or management. This method was applied to data for soil texture, which was interpolated using the Universal Kriging interpolation method. The point data represented the soil particle content and were obtained from the digital database of Complex Soil Survey (CSS) taken between 1961 and 1970 and managed by the Soil Science and Conservation Research Institute. Because the sum of the content

Table 1 KA4 (Bodenkundliche Kartieranleitung) classification scheme of soils occurring in the study area

Soil symbol	Soil class	Clay (%)	Silt (%)	Sand (%)
Lu	Silty loam	17–30	50–65	5–33
Ls2	Weak sandy loam	17–25	40–50	25–43
Ls3	Medium sandy loam	17–25	30–40	35–53
Lt2	Weak clayey loam	25–35	30–50	15–45

Source von Werner (2003), modified

of three fractions must equal 100 %, after the interpolation and recalculation, we obtained three continuous grids representing the percentage of clay, silt, and sand. The accuracy of the interpolation was validated by field measurements. The mean deviation between interpolated and measured data varied from ± 1.7 to ± 2.41 %. The highest deviation may be related to the redistribution of soil material due to management activities, but it had no significant influence on the modeling results. Based on these results, the soils were classified according to KA 4 (Bodenkundliche Kartieranleitung) (Table 1; also used in the model input parameters catalogue). Other input soil parameters were then determined according to texture class.

Next, in many cases, interpolation methods do not reflect natural conditions of the agricultural areas considered. The model provides an opportunity to enter soil parameters via a patch-related procedure which reflects the variability in soil properties within field boundaries. This procedure was applied in the case of parameters such as soil bulk density, initial water content and organic carbon content. The soil samples for the estimation of bulk density and initial water content were taken from 16 sample points and analyzed in a laboratory. The sample points were localized so that they represented all combinations of textural soil classes (KA4) and crops within given crop rotation systems at the level of individual parcels. To reflect the seasonal variability of soil properties, the samples were collected repeatedly in May, June and October. The estimated values for bulk density varied from 1,356 kg m^{-3} for vineyards on Ls3 and Lt2 soils to 1,654 kg m^{-3} for summer barley on Lu soils in June. These data were then compared with literature values. According to Bedrna (2002), optimal values for loamy soils are within the range of 1,500–1,700 kg m^{-3}. From this point of view, the soils in the study area were in the optimal range, although the values obtained for the soils under crops such as corn and cereals (1,100–1,300 kg m^{-3}) and sunflower (1,200–1,300 kg m^{-3}) fell outside the optimal range. Concurrently, values for initial water content were estimated. These values could not reflect the high spatial and temporal variability of this parameter, therefore the mean values related to given soil textural classes were calculated. Similarly, the mean value for organic carbon content was calculated at 1.536 % Cox. This value was then compared with the mean value of 1.515 % Cox calculated from data in the digital database of Complex Soil Survey (CSS), which confirmed its accuracy.

Finally, resistance to erosion, hydraulic roughness and the percentage of soil cover were obtained from tabular data in the parameter catalogue for various soils, surface conditions and management options. This catalogue enables the user to compile the necessary input parameters and to apply "Erosion 3D" to a wide variety of modeling tasks. These data are derived from rainfall experiments performed for a large number of soils with high erosion susceptibility during the "Soil Erosion Research Program" that was carried out in the years 1992–1995 at the Technical University in Freiberg.

Precipitation parameters

As mentioned above, Erosion 3D is an event-related erosion model in which the amounts of erosion and deposition are calculated for a distinct period of a precipitation event. This temporal resolution can range between 1 and 15 min. Precipitation data described typical rainfall events with a return period varying between 1 and 100 years and duration between 5 and 180 min. These data were obtained from the database of Slovak Hydrometeorological Institute.

3.1 Simulation Scenarios

The erosion model was applied to simulate soil loss under 3 main soil cover scenarios. These scenarios were designed so as to describe the spatial and temporal variability of land cover, soil properties and related erosion processes throughout the entire vegetation period. The assessment of these scenarios was based on the actual crop rotation system and representative precipitation data. Relief parameters and textural classes were set as constants. Furthermore, all parameters related to impermeable surfaces such as settlements and roads and also forest vegetation were set as constants since they show little spatial and temporal variation. In total, 15 scenarios were assessed. The 3 main scenarios represented the soil cover in April (scenario A), June (scenario B) and October (scenario C), and 12 partial scenarios represented hypothetical situations when the whole area would be covered by one crop: summer barley (scenarios A1, B1, C1), winter wheat (scenarios A2, B2, C2), spaced growing crops (A3, B3, C3) and oil rape (A4, B4, C4). These partial scenarios could then be used to determine the most suitable site for individual crops. Furthermore, a version of each scenario was also constructed to include application of erosion mitigation measures (A1_n, A2_n, etc.).

Scenario A describes land cover in April when most agricultural fields are not well protected by low vegetation cover. Under these circumstances, the soil is susceptible to more frequent erosion processes, especially during heavy rainfalls or snowmelt events. In early spring, the soil is also more saturated with water which decreases its infiltration capacity. This effect could be partly prevented by conservation measures, which improve the physical properties of soils, and thus minimize the impact of rainfall.

Scenario B refers to June, when the vegetation cover of soil is rather high. However, the positive effect of spring tillage operations have decreased and other processes such as compaction and soil aggregate disintegration prevail. Despite these facts, the occurrence of erosion events is low due to the highly developed soil cover.

Scenario C refers to October when the percentage of soil cover decreases due to harvesting activities. If compared with the spring vegetation stage, the residues remaining on the surface or within the root zone play an important role in compensating the negative effect induced by heavy machinery on soil properties. After implementing conservation practices, the situation becomes similar to scenario A, but with the absence of snowmelt runoff and a lower intensity but higher frequency of rainfall.

Scenarios n represent main and partial scenarios coupled with selected soil erosion measures. These measures were selected according to the available literature (Janeček 1992; Demo 1998; Fulajtár and Janský 2001; Jambor 2002; Uhlířová 2005) and the existing legislation (Act n. 220/2004), namely.

4 Determining the Optimum Size of the Parcels

The best land use is based on rearrangement or adjustment of both the size and the shape of parcels and their layout according to topography. The main goal is to decrease the length of parcels along slope and to minimize erosion rates in order to preserve the minimal economical size and accessibility to the parcels. Optimal recommended parameters are as shown by technical standard (Slovak technical standard 2000) (Tables 2 and 3).

Table 2 A proposed parcel size in the study area according to the technical standard STN 75 4501 (2000)

Slope	Length (m)	Width (m)	Size (ha)	Erosion
0°–3°	750	400	30	No erosion
3°–7°	550	250	10–20	Medium
7°–12°	400	250	5–10	Strong
>12°	Destined to meadows and pastures		Arbitrary	Extreme

Table 3 Optimal geometrical parameters of parcels in the study area according to the technical standard STN 75 4501 (2000)

Parameter	Units
Parcel size on flatland	30–50 ha
Parcel size on slopes	5–10 ha
Minimal economical parcel size	2 ha
Optimal length	400–500 m
Minimal length	200 m
Minimal width	50 m

Source Slovak Technical Standart (2000), modified

5 Crop Management

The basic principle of crop management is to choose the appropriate site for each crop and to implement practices that aim at minimizing the impact of soil erosion. This method is based on the mitigating effects of different agricultural crops on erosion. One of the most effective practices consists in adopting close-growing crops. The alfalfa-grassland mixture mitigates erosion up to 1/100 and alfalfa itself up to 1/50 in comparison with bare soil. Cereals are effective in reducing erosion from 1/5 up to 1/20, depending on sowing and harvesting times. Spaced growing crops and root crops mitigate erosion only up to 1/2 and thus their cultivation is recommended on flat land with slopes <3°. For parcels with moderate slope gradients, between 3° and 7°, it is recommended to enhance the erosion-mitigating effect of spaced growing and root crops by applying an adequate crop rotation system or by cultivating different crops (cereals). On the slopes steeper than 10°–12°, the arable land should be substituted by meadows.

6 Conservation Practices

In general, the protection offered by close-growing crops needs to be supported by practices that will reduce the amount of water and soil losses. The most important of these conservation practices are contour tillage, contour stripcropping, buffer stripcropping, terracing and minimum (conservation) tillage.

The proposed measures created a base for the erosion mitigation scenario (Fig. 3). The main goal was to change continuous slope length and the effect of higher slope value by the orientation of parcels along contour lines. Reflecting the efficiency of the tillage operations, the goal was to preserve the actual parcel size. Where the above-mentioned measures could not be applied, some conservation measures in the form of 5–7 m wide vegetation and infiltration strips were proposed. Concurrently, the completion of vegetation borders and buffer strips was proposed. On some parcels active protection measures were suggested to delimit meadows or forests or to convert to crops with better erosion-mitigating effect such as close-growing crops and fodder crops. According to proposed measures, some of the input parameters (soil cover, surface roughness) were then modified.

This proposed structure represents a basis for the set of erosion mitigation simulations for the rainfall events with a return period of 100 years. Considering other rainfall events, no scenarios were proposed since output values could be predicted from the previous results.

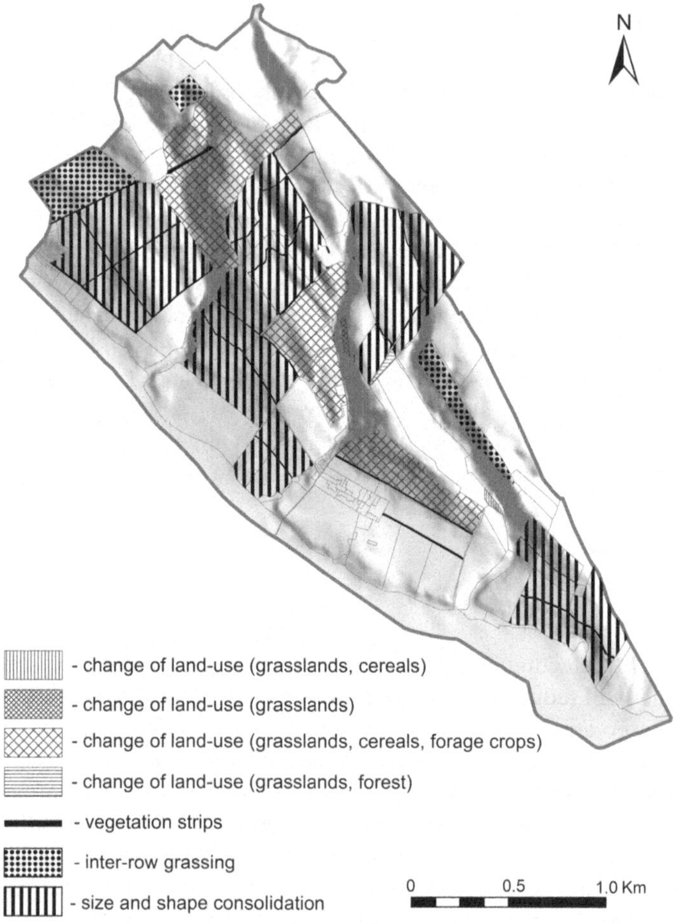

Fig. 3 Erosion mitigation scenario for the study area

6.1 Model Outputs

The model produces raster-based, quantitative estimates of soil loss, soil deposition and the sediment delivery into the surface water system. The following data are provided for each grid cell:

- parameters related to area:
  - erosion and deposition for a chosen grid cell (t ha^{-1}),
  - erosion, deposition and net erosion for the watershed draining into a chosen grid cell (t ha^{-1}),

- parameters related to cross-section of flow:
  - runoff ($m^3\ m^{-2}$),
  - sediment delivery (kg m^{-2}) and sediment concentration (kg m^{-3}).

A total number of 725 simulations were run based on the scenario criteria. Results of each simulation were exported in the form of a grid model representing the net erosion for each grid element. These results were then aggregated and the mean value of net erosion was calculated for the whole study area.

6.2 Validation

The process of model validation is generally based on the comparison between simulated and measured data. Since there were no direct measurements available for the study area, only qualitative assessment could be made. This included the comparison of soil profiles located on erosion-prone slopes identified during the erosion simulation. Each slope transect included 3 points located at the top, middle and the bottom of a given slope. In total, 5 transects were designed. Using a simple hand auger, the thickness of diagnostic soil horizons and the stratification of soil profile were examined. The presence or absence of diagnostic horizons, their thickness and transformation of the profile stratification should reflect the impact of erosion or deposition processes.

7 Results

The following results were obtained:

1. The highest mean values of erosion were recorded for rainfall events with a return period of 100 years, with a maximum value for the 15 min duration rainfall of 25.2 mm and an intensity of 1.68 mm min^{-1} (Fig. 4). The lowest values were recorded for the rainfall events with a return period of 1 year (Fig. 6).
2. The highest mean value of erosion of 2.51 t ha^{-1} was recorded for rainfall events with a return period of 100 years, with maximum value for the rainfall with duration of 15 min and partial scenario A1, in April with summer barley (Fig. 4).
3. The peak erosion of 43.24 ± 3.47 t ha^{-1} was recorded for rainfall events with a return period of 100 years, with a maximum value for the rainfall with duration of 15 min and also in partial scenario A1 during April with summer barley (Figs. 5, 6).

Using the basic zonal statistics tools to estimate the mean net erosion value for each parcel, parcels most susceptible to erosion were identified (mean value of net erosion >5 t ha^{-1}). The weighting coefficient of 1 was assigned to each parcel with

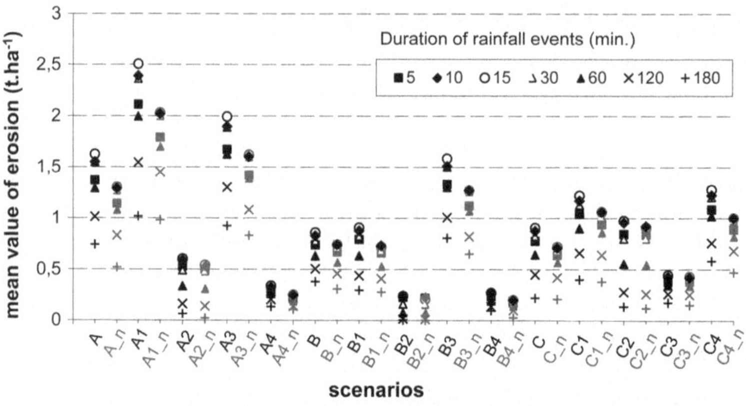

Fig. 4 Mean values of erosion in the study area calculated by erosion 3D model for rainfall events with a return period of 100 years. Codes of scenarios: *A*—April; *B*—June; *C*—October; *1*—summer barley; *2*—winter wheat; *3*—wide-sown crops; *4*—oil-rape; *_n*—scenarios with application of erosion mitigation measures

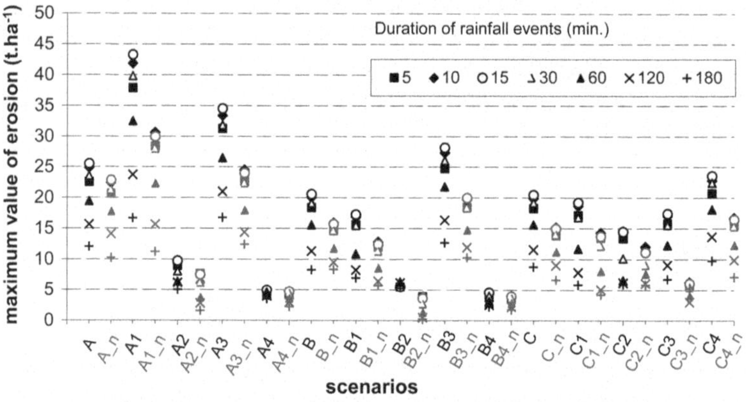

Fig. 5 Maximum values of erosion in the study area calculated by erosion 3D model for rainfall events with a return period of 100 years. Explanation of scenarios as in Fig. 4

the erosion value >5 t ha^{-1}. This calculation was applied to all the main and partial scenarios. After summation of the weighting coefficients (with results ranging from 1 for small to 10 for high susceptibility to soil erosion), all the parcels were classified according to erosion risk. The proportion of the total study area in particular classes of erosion risk is shown in Fig. 7.

The results show that approximately 58 % of the study area is at low erosion risk (classes 1–3), while 37 % can be classified as subjected to moderate risk of erosion (classes 4–7) and only 5 % is at high erosion risk (classes 8–10). For the parcels identified as highly threatened by erosion, the set of soil and geomorphological parameters was extracted by aggregating the mean and maximum

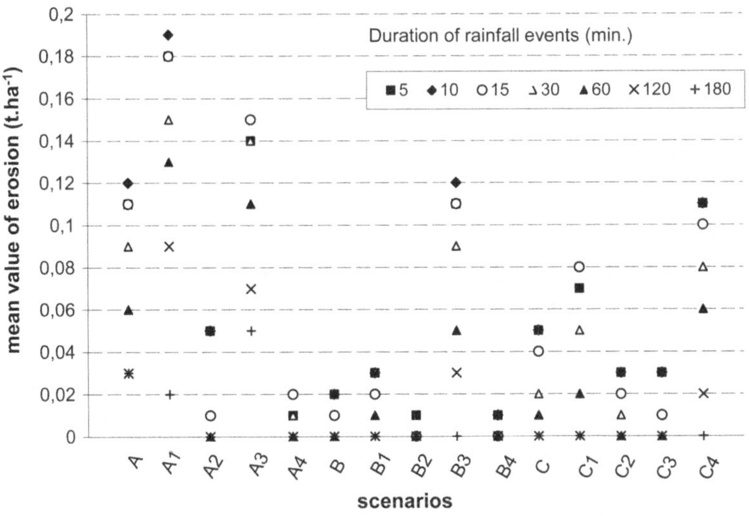

Fig. 6 Mean values of erosion in the study area calculated by erosion 3D model for rainfall events with a return period of 1 year. Explanation of scenarios as in Fig. 4

Fig. 7 Proportion of the total study area in particular classes of erosion risk

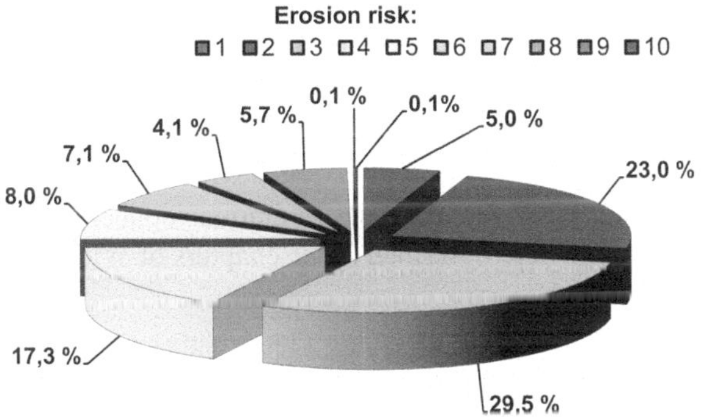

values of these parameters within each parcel. Based on these results, the most susceptible landforms are characterized by the following combination of soil and geomorphological parameters:

- maximum slope length: 1,000–1,500 m,
- slope: 7°–12°,
- vegetation cover of soil: between 0 and 10 %,
- surface roughness: 0.015–0.02 s $m^{-1/3}$,
- erosion susceptibility: 0.023–0.038 N m^{-2}.

The effectiveness and ability of the proposed measures to mitigate soil erosion was assessed by calculating the difference between the values for the main scenarios and the erosion mitigation scenarios. The maximum difference for the main erosion values reached 0.48 t ha^{-1} in scenario A1 for summer barley, with rainfall duration of 15 min and a return period of 100 years. This scenario produced a reduction in soil loss of 18.94 %. Regarding the maximum value, the difference for the same scenario was 13.30 t ha^{-1}, representing a reduction of 59.06 %.

8 Discussion

8.1 *Application of the Mathematic-Physical Modeling Approach (Erosion 3D) for Modeling the Soil Erosion in the Model Area*

Erosion 3D model represents an ideal compromise to combine the demands for fewer input parameters and a desired quality of simulation of hydrological and erosion processes. According to literature (Schmidt et al. 1999; Schmidt 2000; Michael et al. 2005; Weigert and Schmidt 2005; Schob et al. 2006), this model showed good results in identifying the erosion-prone areas as the sources of sediment entering surface waters. Its main advantages are:

- it is an adequate tool on which land-use planners can base figures needed in discussions on important protection efforts,
- input parameters can be obtained from literature or they can be easily estimated in comparison to other models,
- it takes into account the deposition of eroded sediments.

Some disadvantages and limitations concerning the application of the Erosion 3D model were described in the chapters of Wickencamp et al. (2000) and Schmidt (1996).

8.2 *Modeling Outputs and Their Interpretation*

This model produces raster-based, quantitative estimates of soil loss, soil deposition and sediment delivery into surface water systems. For the purpose of this work, the net erosion has been estimated for each grid cell. From this data, the mean value, the maximum value and standard deviation of the amount of erosion have been calculated. The highest values have been identified for the April and October scenarios, especially where crops are characterized by low canopy cover (A1—in April with summer barley, A3—wide-sown crops, C4—oil rape, etc.). In these periods of the year, the important determining factor for soil erosion is the high water saturation of the soil complex. The lowest values of erosion were recorded in simulations for the summer period (B2—in June with winter wheat, B4—June for oil rape, etc.). During this period, vegetation cover of soil reaches its highest values, whereas the soil is rather dry and thus most of the precipitation is infiltrated. Low values were also recorded for scenarios related to April and October, but with crops generating high soil cover (A4—April oil rape, A2—April winter crops and C3—October wide-sown crops).

Using the basic zonal statistics tools (calculation of mean net erosion value for each parcel), most erosion-prone parcels were identified and the map of the susceptibility of cultivated parcels to soil erosion was produced. For these parcels, the set of soil and geomorphological parameters was extracted. The following results were obtained:

- the parcel size did not have a significant impact on the occurrence of erosion processes. Most of the erosion-prone parcels had an optimal size (0–5 ha) for erosion mitigation. The most threatened parcels were characterized by steeper slopes (7°–12°) and the length between 500 and 1,500 m. Therefore, it seems that parcel shape and orientation are more significant factors than parcel size,
- as for vegetation cover, the most threatened parcels are those with low vegetation cover (0–10 %), but some of the parcels are in the category with 50–70 % of vegetation cover. On these parcels, factors other than the degree of vegetation cover have been considered,
- similarly concerning erosion susceptibility and surface roughness, most of the erosion-pronc parcels were characterized by low values of these parameters,
- considering bulk density of soil, threatened parcels were characterized by values varying between 1,340 and 1,644 kg m^{-3}. Variation of the values within this range did not reasonably affect the output results. This statement is also supported by the fact that the measured values of bulk density are within the optimal limit for clayey soil (1,400–1,700 kg m^{-3}) and thus did not significantly affect soil loss.

To minimize the impact of soil erosion on the most susceptible landforms, a set of best management practices consisting of a combination of passive measures, such as the orientation and size of parcels, infiltration vegetation strips, grassed waterways, and also active measures was proposed. This proposed structure represents a basis for the set of erosion mitigation simulations which were then compared with main and partial scenario simulations without the application of soil erosion-control measures. The results of such comparison support the fact that the proposed measures effectively reduced the impact of soil erosion without significant modification of the parcel's accessibility and cultivability. The most effective measures were: modification of the spatial structure of the agricultural landforms, and modification of parcel size, orientation and shape. These results also demonstrated that simulation models such as Erosion 3D can provide necessary information for the adequate localization and scaling of site-specific measures.

8.3 Validation

Evaluation of the transformation of the stratification of soil horizons allowed to verify the presence of erosion/deposition processes. The eroded forms located on steeper slopes were characterized by the consistent change of luvisols to regosols. This process resulted in the absence of surface A horizon with a high content of organic matter and the denudation of light colored subsurface B horizon. At footslopes, the soil profiles were characterized by the absence of pedogenetic material and other diagnostic horizons, except for the dark thick surface horizon created by the continuous accumulation of fine soil material. Both the eroded and accumulated forms were characterized by poor discrimination of the diagnostic horizons and homogenization of soil profile. The presented method could successfully validate the presence of both the eroded and accumulated forms. However, it cannot sufficiently prove whether these forms originated as a result of erosion processes or by the combined action of erosion and tillage activities.

9 Conclusion

This chapter presents a methodology of evaluation of erosion risk using a physically based modeling approach. To capture the spatial and temporal variability of erosion processes in the study area, simulations were based on modeling scenarios. Scenarios were designed so as to reflect the variability of vegetation cover, management practices and soil properties throughout the vegetation period. The results of erosion mitigation scenario showed significant decrease in erosion values, especially in comparison with scenarios with the highest soil loss. Based on the mutual comparison of individual scenarios, it was possible to demonstrate their impact on soil erosion in the study area. The simulation results were also used to classify the area according to erosion risk and to identify its main causative factors.

Results of such erosion modeling should be validated using experimental data. This validation is based on the comparison of experimental (measured) and simulated data. With the lack of available experimental data for the study area, it was necessary to use the so called indicative method to validate the results. Therefore, the simulation outputs could be used only for the assessment of erosion risk and not for the precise quantification of eroded soil. The indicative method was based on assessment of the transformation of the soil profile. However, this method could not sufficiently prove whether the erosion/accumulation forms were created by water erosion alone or also by tillage.

Despite these limitations, the Erosion 3D model provided adequate information necessary for the optimal localization and scaling of site-specific measures and also for the assessment of their impact on soil erosion.

Acknowledgments The publication of this chapter was supported by the project of Slovak Research and Development Agency APVV-0240-07 'Model of representative geosystems on local level'.

References

Act No 220/2004 College of the National Council of the Slovak Republic, on protection and exploitation of agricultural land

Bedrna Z (2002) Environmental pedology. Bratislava, VEDA

Demo M (1998) Usporiadanie a využívanie pôdy v poľnohospodárskej krajine. Slovak Agricultural University in Nitra and Soil Science and Conservation Research Institute (SSCRI), Bratislava

Fulajtár E, Janský L (2001) Vodná erózia pôdy a protierózna ochrana. Soil Science and Conservation Research Institute (SSCRI) and Faculty of Nature Sciences of Comenius University, Bratislava

Green WH, Ampt GA (1911) Studies in soil physics: 1. The flow of air and water through soils. J Agricult Soils 4:1–24

Hlavčová K, Macura V (1993) Modelling of overland flow and erosion processes. Slovak J Civil Eng 1:3–8

Hofierka J, Šúri M (1996) Modelling spatial and temporal changes in soil water erosion. Geogr J Inst Geogr SAS 48:255–269

Jambor P (2002) Water and wind erosion on soils in Slovakia. In: Proceedings of the Soil Science and Conservation Research Institute, vol 25. Bratislava

Janeček J (ed) (1992) Erosion control on agricultural land. Metodika ÚVTIZ, Praha

Janský L (2001) Overenie modelu SMODERP na vybraných lokalitách v Japonsku. In: Fulajtár E, Janský L (eds) Soil erosion and soil conservation. Soil Science and Conservation Research Institute and Faculty of Nature Sciences of Comenius University, Bratislava

Michael A, Schmidt J, Enke W, Deutschlander T, Malitz G (2005) Impact of expected increase in precipitation intensities on soil loss—results of comparative model simulations. Catena 61:155–164

Šály R (ed) (2000) Morfogenetický klasifikačný system pôd Slovenska. Soil Science and Conservation Research Institute and Societas Pedologica Slovaca, Bratislava

Schmidt J (1996) Entwicklung und Anwendung eines physikalisch begrundeten Simulationsmodells fur die Erosion geneigter landwirtschaftlicher Nutzflachen. Berliner Geographische Abhandlung

Schmidt J (2000) Soil erosion: application of physically based models. Springer, Berlin

Schmidt J, von Werner M (2000) Modeling the sediment and heavy metal yields of drinking water reservoirs in the Osterzgebirge region of Saxony (Germany). In: Schmidt J (ed) Soil Erosion: application of physically based models. Springer, Berlin

Schmidt J, von Werner M, Michael A (1999) Application of the Erosion 3D model to the CATSOP watershed, The Netherlands. Catena 37:449–456

Schob A, Schmidt J, Tenholtern R (2006) Derivation of site-related measures to minimise soil erosion on the watershed scale in the Saxonian loess belt using the model EROSION 3D. Catena 68:153–160

Slovak Technical Standart (STN) 75 4501 (2000) Conservation of agricultural soils—basic regulations. Slovak Standarts Institute

Uhlířová J (2005) Erosion control in agro-environmental programmes. In: Proceedings from the conference of soil scientists in Slovakia, Čingov. Fourth soil science days. Soil Science and Conservation Research Institute, Bratislava, 14–16 June 2005

Von Werner M (2003) EROSION 3D manual—A soil erosion and sediment yield model 1994. Arbeitsgruppe Boden der Geologischen Landesämter und der Bundesanstalt für Geowissenschaften und Rohstoffe der Bundesrepublik Deutschland, Bodenkundliche Kartieranleitung

Weigert A, Schmidt J (2005) Water transport under winter conditions. Catena 64:193–208

Wickencamp V, Duttmann R, Mosiman T (2000) A multiscale approach to predicting soil erosion on cropland using empirical and physically based soil erosion models in a geographic information system. In: Schmidt J (ed) Soil Erosion: application of physically based models. Springer, Berlin

The Effects of Climate Change on Element Content and Soil pH (Síkfőkút DIRT Project, Northern Hungary)

János Attila Tóth, Péter Tamás Nagy, Zsolt Krakomperger, Zsuzsa Veres, Zsolt Kotroczó, Sándorné Kincses, István Fekete, Mária Papp, Ilona Mészáros and Viktor Oláh

Abstract In the *Detritus Inputs and Removal Treatment* (DIRT) field experiments established at the Síkfőkút Site (northern Hungary) in October 2000, an experimental approach was applied to study long-term effects of litter quality and quantity on pH and nutrient mobility of soil in a *Quercetum petraeae-cerris* forest. Our previous results have suggested that decreases in organic matter content, total N, Ca^{2+} and Mg^{2+} concentrations in the soil are the consequence of the reduction in forest litter production induced by climate change and the resultant soil degradation over a longer period. An eight-year litter manipulation demonstrated a relation between soil pH and Mg^{2+} and Ca^{2+} concentrations in the soil. The reduction of litter production resulted in a decrease of the soil pH which occurred due to the decreasing Mg^{2+} and Ca^{2+} inputs into the soil, and consequent reduction of soil buffering capacity against the acidifying effects of the acidic intermediates of litter decomposition and humus compounds. If the litter production increases as a result of climate change, it should be accompanied by increasing C, total N, Ca^{2+} and Mg^{2+} contents in the soil and soil pH, with a positive effect on the soil organic matter and fertility. However, this scenario is rather unlikely as our results indicated decreasing litter production in the measurement period.

J. A. Tóth (✉) · Z. Krakomperger · Z. Veres · Z. Kotroczó · M. Papp · I. Mészáros · V. Oláh
Faculty of Science and Technology, Debrecen University, Egyetem tér 1, Debrecen H-4032, Hungary
e-mail: tja@tigris.unideb.hu

P. T. Nagy · S. Kincses
Department of Agrochemistry and Soil Science, Debrecen University, Böszörményi út 138, Debrecen H-4032, Hungary

I. Fekete
College of Nyíregyháza, Institute of Environmental Sciences, Sóstói út 31./b, Nyíregyháza H-4400, Hungary

J. Kozak et al. (eds.), *The Carpathians: Integrating Nature and Society Towards Sustainability*, Environmental Science and Engineering,
DOI: 10.1007/978-3-642-12725-0_7,

1 Introduction

The long-term ecological research Síkfőkút Project was started in 1972 as a part of IGBP (International Geosphere-Biosphere Program), and later it continued within the MAB (Man and Biosphere) program. The study site was designated to investigate dynamics of a temperate-zone *Quercetum petraeae-cerris* forest which is typical of Hungary (Jakucs 1973). The study area is situated 6 km northeast of the city of Eger close to the southern hill-country of the Bükk Mountains (47°55' N and 20°46' E) at an elevation of 320–340 m a.s.l. In 1976 the entire forest area, including the Síkfőkút site, became protected as a nature reserve. As a consequence of lacking silvicultural management in the past decades, the Síkfőkút forest is assumed to have recently become natural forest. Based on data from an intensive and long-term ecological research during the past 37 years, effects of climate change (warming, drought) on the species composition, structure and fitness of the forest are well recognized. Such information is particularly important in the case of the Síkfőkút Project as better understanding of the ecosystem helps to improve silvicultural practices undertaken elsewhere on local and regional scales. It is also important in the context of environmental policy and socio-economics. As a result of climate change, species composition and structure of the Síkfőkút *Quercetum petraeae-cerris* forest have changed considerably. This was shown by a strong decrease in the abundance of Sessile oak (*Quercus petraea*) in favor of Turkey oak (*Quercus cerris*) and *Acer* species (Kotroczó et al. 2007). This shift in tree species composition led to the qualitative and quantitative alteration of leaf litter production and it has clearly affected related soil properties (Tóth et al. 2008). Considering the phenomena described above, the present study focuses on how the alteration of litter production induced by climate change modifies physical, chemical and biological soil properties. To answer this question, the approach of DIRT (*Detritus Inputs and Removal Treatment*) (Neilson and Hole 1963) was applied with field experimental plots set up to determine the long-term effects of changing amounts and quality of litter on soil properties. By contributing to DIRT, the Síkfőkút Project is an associated member of the ILTER (International Long-Term Ecological Research) DIRT Project which consists of four American (Harvard Forest, Bousson Forest, Andrew Forest and Michigan Forest) and two European (University of Bayreuht and Síkfőkút) LTER sites creating an international and intercontinental research network. Our previous litter decomposition experiments have shown the effects of changing amounts of litter on the soil water dynamics, changes in bacterial and fungal flora enzyme activities and soil respiration (Tóth et al. 2007a, b, 2008). The aim of the present study is to demonstrate and analyze respective effects on soil organic matter, nitrogen (NO_3^--N, NH_4^+, organic N, and total N), phosphorous and mineral nutrients content and soil pH. This chapter evaluates litter decay experiments conducted within the DIRT framework with regard to nutrient dynamics and soil-pH after 8-year duration.

2 Materials and Methods

Inside the Síkfőkút forest stand, 7 × 7 m permanent experimental plots were set up in 2000 in accordance with the protocol used in the DIRT plots established in the USA (Table 1).

During the experiment six different treatments were applied: Control (C), No Litter (NL), No Root (NR), No Input (NI), Double Litter (DL) and Double Wood (DW) (Kotroczó et al. 2008a, b; Kotroczó et al. 2010). Each treatment was conducted in three replicates. Samples of 100 g of soil material were taken randomly on each plot from 0–5 and 5–15 cm depths using an Oakfield soil sampler (G model). Soil extracts were prepared using two different solvents. Ammonium-lactate/-acetic acid buffer solution (0.1 M; pH = 3.7) was used for the extraction of soluble and easily exchangeable nutrients (Egnér et al. 1960). Calcium-chloride (0.01 M $CaCl_2$) was used for the extraction of easily soluble nutrients (Houba et al. 1990). Aliquots of 5 g of air-dried, sieved and homogenized soil material were extracted with 100 cm^3 of the buffer solution or with 50 cm^3 of the calcium-chloride solution during 2 h long shaking and filtering.

Concentrations of Ca^{2+} and Mg^2 in the extracts were determined by means of atomic absorption spectrophotometer (AAS) (SpectrAA-20 Plus, Varian Australia Pty Ltd). Concentration of phosphorus was measured spectro-photometrically by the phospho-molibdovanadate method (VIS SP-850 Plus spectrophotometer, Metertech, Taiwan). Nitrogen species were determined spectro-photometrically by means of continuous flow analyzer (CFA) system (SA-2000 type Skalar photometer, Breda, The Netherlands). Organic matter content was determined after dry combustion (VARIO EL CNS elementary analyzer, Vario, Germany) according to the method of Nagy (2000). Soil pH was determined in 0.01 M $CaCl_2$

Table 1 Treatments applied in the open-field experiment at the Síkfőkút project, Hungary

Treatments	Description
Control (C)	Normal litter inputs. Average litter amount typical of the forest site
No Litter (NL)	Above-ground inputs are excluded from plots. Leaf litter was totally removed by rake. This process was repeated continuously in every year
Double Litter (DL)	Above ground leaf inputs are doubled by adding litter removed from NO LITTER plots
Double Wood (DW)	Above-ground wood debris inputs are doubled by adding wood to each plot. Annual input of wood litter was measured in boxes placed at the site and double amount of that value was applied in the case of every DW plot
No Roots (NR)	Roots are excluded by inserting impenetrable barriers in backfilled trenches to the top of C horizon of soil. Root resistant plastic foil was placed around the plot to the depth of 1 m, hindering the roots developing outside the plot to get into the NR plot. Trees and shrubs were eradicated when the plot was established, and plant roots decayed with time
No Inputs (NI)	Above-ground inputs are excluded from plots and the below-ground inputs are also eliminated as in NO ROOTS plots. This treatment is the combination of NR + NL plots

suspensions (EBRO PHT 3140 digital pH meter with a combined glass electrode). Results were statistically analyzed with the Sigma Stat software (v3.1.) using one-way analysis of variance.

3 Results and Discussion

3.1 Soil Carbon Content

The effect of removed litter on the soil carbon content was more pronounced than the effects of double litter input. In comparison with the Control plots, soil carbon content decreased after the leaf litter removal at both soil depths, while Double Litter treatments caused increased carbon content only at the depth of 0–5 cm. In the deeper soil layer (5–15 cm), Double Litter treatment did not cause accumulation of carbon during the first 8 years of the experiment (Fig. 1).

These results suggest two hypotheses:

1. a decrease in litter production induced by climate change would result in decreasing soil organic matter content over a longer period, which can impair the soil water, temperature and nutrient storage capacity;
2. if climate change enhances litter production, the resultant increase of soil organic matter content might gradually improve the soil properties (temperature insulation, humus and cation content).

Our previous results indicated decreased litter production as a consequence of ongoing climate change (Tóth et al. 2007b; Fekete et al. 2008). This supports the first hypothesis for the soil of the Síkfőkút site. However, it should be added that generalization is difficult because warmer climate might result in higher wood and

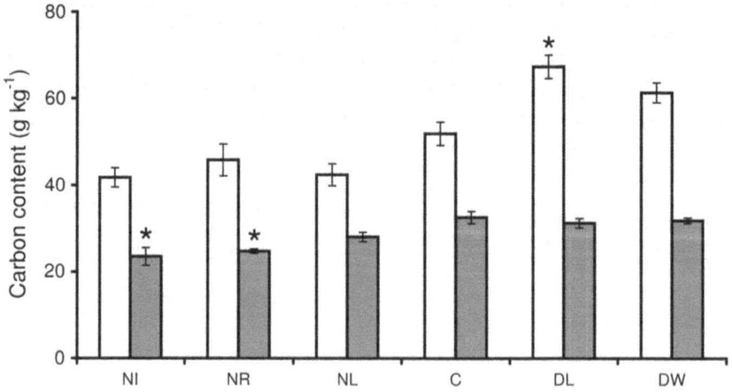

Fig. 1 The effects of litter input manipulation on soil carbon content. *NI*: no input, *NR*: no root, *NL*: no litter, *C*: control, *DL*: double litter, *DW*: double wood, *white* column: 0–5 cm soil depth, *grey* column: 5–15 cm soil depth, *: significantly different from control ($p < 0.05$)

leaf litter production as it would in the case of more humid ecosystems (Kotroczó et al. 2008a, b; Krakomperger et al. 2008; Varga et al. 2008).

3.2 Nitrogen Forms

At 0–5 cm depth a significantly increased NO_3^--N content was measured on DL plots, while a decrease in this parameter was recorded on NI plots (Fig. 2). Other types of litter manipulation treatments, however, had no significant effects on the NO_3^--N content of soil. A lack of clear tendency is not surprising with the highly soluble and mobile nitrate: beside seasonal fluctuations, its momentary concentration in soil is strongly influenced by several processes such as nitrification, denitrification, uptake by plants and leaching. NH_4^+-N content in the upper 0–5 cm of soil decreased on the plots where litter was excluded and increased on the plots with double litter amount (Fig. 3). The highest values were recorded on DL plots. In the 5–15 cm soil layer, only the DW treatment resulted in a significant change in relation to control values (Fig. 3). All types of litter exclusion treatments resulted in a marked decrease in organic-N content at 5–15 cm depth and a lack of significant decrease at the depth of 0–5 cm (Fig. 4). In the 0–5 cm layer, litter addition either increased (on DL plots) or slightly decreased the organic-N (on DW plots). Both treatments induced a decrease in organic-N at 5–15 cm depth (Fig. 4). Concerning total-N contents, effects of litter manipulation treatments were detectable only in the upper 0–5 cm of soil, while there was no difference at 5–15 cm depth (Fig. 5).

Recently formed litter pools assimilate more NO_3^- than NH_4^+ under ambient N deposition, but may lose the capacity to assimilate NO_3^- relative to NH_4^+ under potential future increases in N deposition (Micks et al. 2004). Global atmospheric composition and climate change effects on the C/N ratio in plants are thus likely to

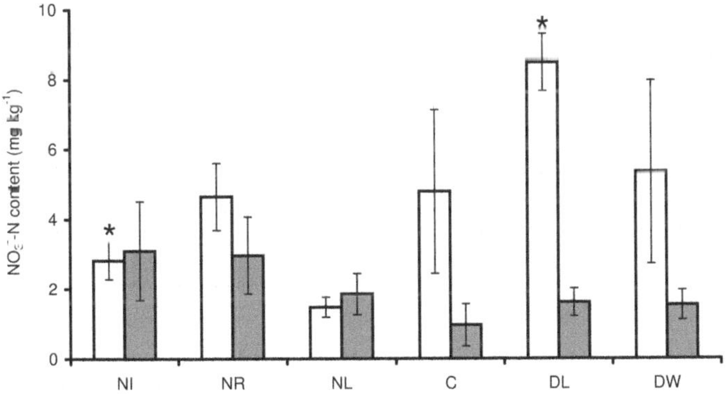

Fig. 2 The effects of litter input manipulation on soil NO_3-N content. Explanations as in Fig. 1

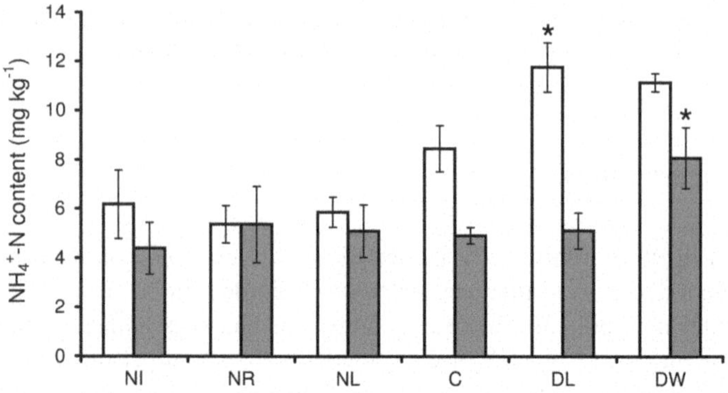

Fig. 3 The effects of litter input manipulation on soil NH_4^+-N content. Explanations as in Fig. 1

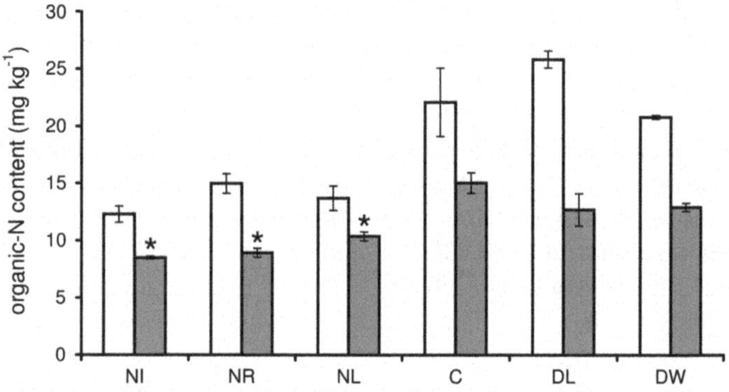

Fig. 4 The effects of litter input manipulation on soil organic-N content. Explanations as in Fig. 1

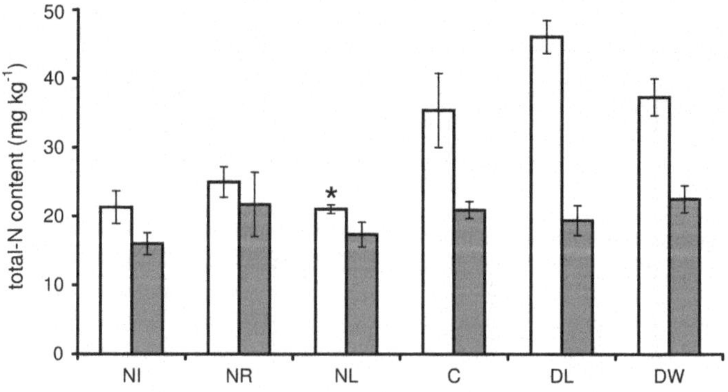

Fig. 5 The effects of litter input manipulation on soil total-N content. Explanations as in Fig. 1

be important when predicting possible second-order impacts of the enhanced greenhouse effect (Kunz et al. 1995).

3.3 Extractable Phosphorus Content

The upper 0–5 cm soil layer contained remarkably more phosphorus than the 5–15 cm layer on all plots, irrespective of the applied litter management (Fig. 6). The P content at both depths increased as a result of litter exclusion. No significant differences in the content of this element were recorded on plots with doubled litter input in comparison to control values (Fig. 6).

Several factors can explain the higher P contents on plots with litter exclusion. In NI and NR treatments, uptake of P by plants did not influence the P status of soil; for similar reasons the activity of phosphatase enzyme was also lower (Fekete et al. 2007; 2008), allowing P on these plots to remain organically bound. Higher soil acidity on the plots with litter exclusion (see below) might enhance the possibility of precipitation of P in the form of Fe^{-} and Al-phosphates.

3.4 Extractable Ca^{2+} Content

Calcium content of soil decreased at both depths as a result of litter exclusion. This decrease was significant on NI and NL plots (Fig. 7). Litter addition had an opposite effect because it increased the Ca^{2+} content in the upper 0–5 cm of the soil (significant on DL plots, Fig. 7). In the 5–15 cm layer, however, such an increase was undetectable (Fig. 7). It can be hypothesized that the potential decrease of litter production induced by climate change might lower the Ca^{2+} content of soil and, as a consequence, reduce buffering capacity and increase the

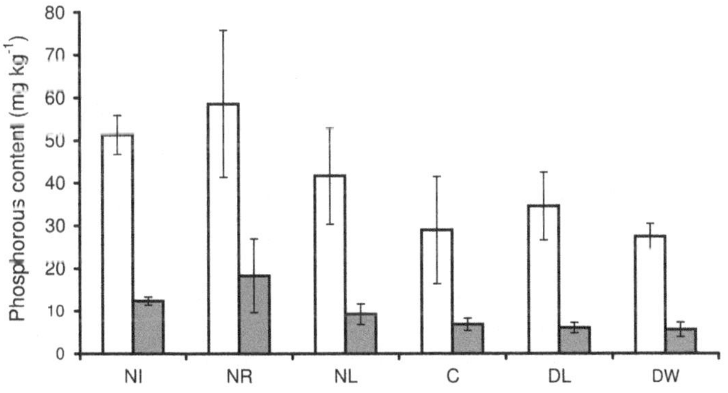

Fig. 6 The effects of litter input manipulation on soil phosphorus content. Explanations as in Fig. 1

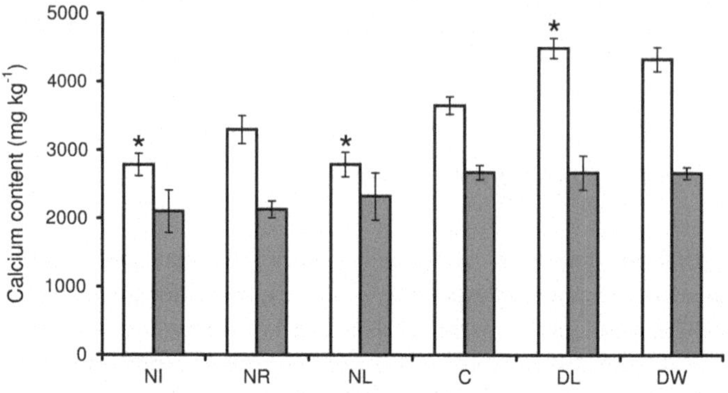

Fig. 7 The effects of litter input manipulation on soil Ca^{2+} content. Explanations as in Fig. 1

risk of acidification of the affected soil. In the case of enhanced litter production, this process might tend to enhance Ca^{2+} content and therefore improve the buffering capacity of the soil.

3.5 Mg^{2+} Content

Magnesium content of soil decreased at both 0–5 and 5–15 cm depths due to litter exclusion treatments and at the 0–5 cm layer differed significantly from control values for all three treatments (Fig. 8). Litter addition increased the Mg^{2+} content only in the 0–5 cm layer for DL treatments (Fig. 8).

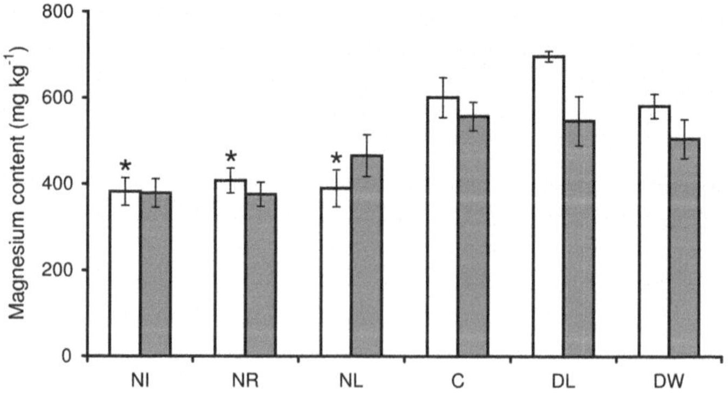

Fig. 8 The effects of litter input manipulation on soil Mg^{2+} content. Explanations as in Fig. 1

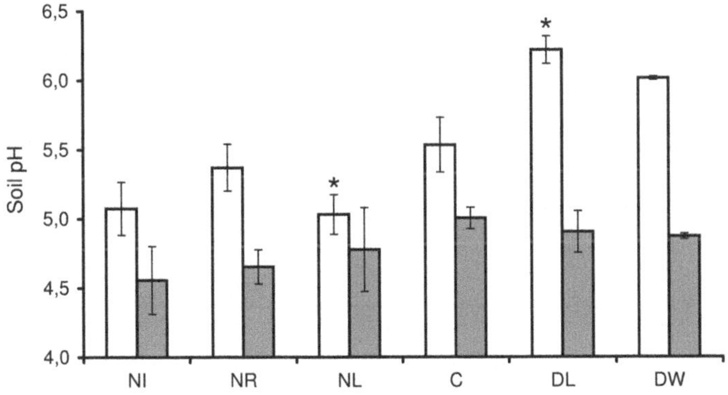

Fig. 9 The effects of litter input manipulation on soil pH. Explanations as in Fig. 1

3.6 Soil pH

Reduced litter input caused a decrease in soil pH (Fig. 9). This could be attributed to the depleted buffering capacity of the soil due to the restricted Mg^{2+} and Ca^{2+} inputs which could not compensate for the acidifying effect of acidic intermediates and humus compounds. The increase of litter input resulted in higher soil pH since it also enlarged Mg^{2+} and Ca^{2+} inputs and improved soil buffering capacity (Fig. 9). Changes in soil pH and the Mg^{2+} and Ca^{2+} concentrations in the soil showed positive relation (Figs. 10 and 11). Soil pH increased with increasing concentrations of Mg^{2+} and Ca^{2+}, although the relationship was stronger in the case of Ca^{2+} concentration. This confirmed similar results obtained by Finzi et al. (1998) who found a highly significant positive relation between soil pH and the concentration of extractable Ca^{2+}. At our site, the 5–15 cm soil layer is more acidic than the 0–5 cm layer. The Ca^{2+} has a stronger alkaline effect than Mg^{2+}.

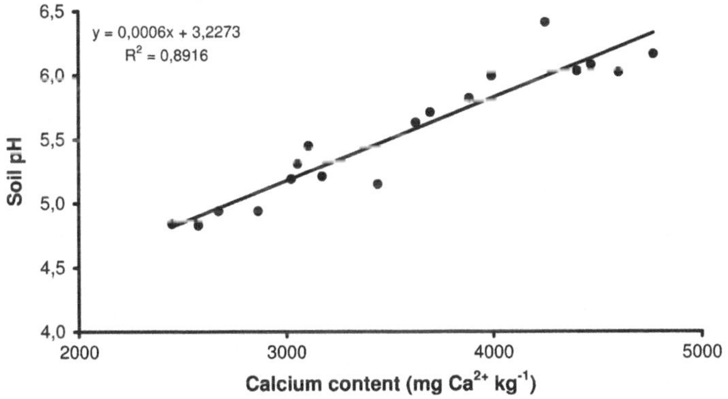

Fig. 10 Scatter plot and estimated regression relationship between soil pH and Ca^{2+} content under conditions of litter input manipulation

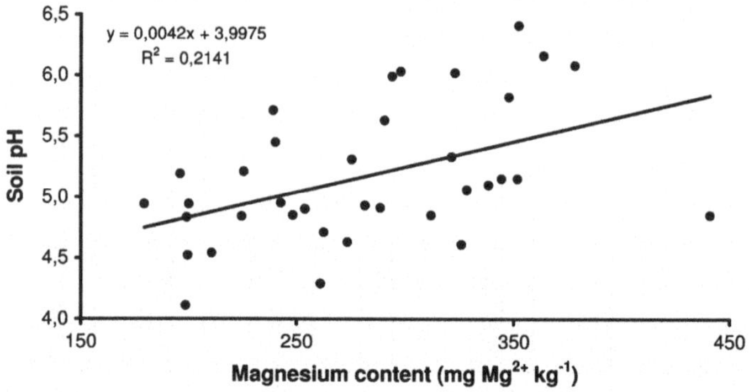

Fig. 11 Scatter plot and estimated regression relationship between soil pH and Mg^{2+} content under conditions of litter input manipulation

High Ca^{2+} concentrations in leaf litter coupled with a large quantity of leaf litterfall could increase the quantity of exchangeable Ca^{2+} in the surface layer (0–5 cm) of the soil (Finzi et al. 1998). These results suggest that the properties of litter in a forest could fundamentally influence the soil pH and, consequently, the nutrient mobility.

4 Conclusions

The effect of litter removal on the soil carbon content was larger than the effects of double litter input. In comparison with the control sample, soil carbon content decreased as a result of leaf litter removal at both examined soil depths, while Double Litter treatments increased carbon content only at the depth of 0–5 cm. In the deeper soil layers (5–15 cm), Double Litter treatment did not cause carbon accumulation during the first 8 years of the experiment. Only the carbon content of the upper soil layer increased by the surplus litter input.

Organic nitrogen content of the soil decreased at both soil depths under the influence of litter removal. In the case of the depth of 5–15 cm, all three litter withdrawal treatments (NI, NR, NL) caused a significant decrease in N content. Under the influence of doubled litter treatments, only Double Litter treatment increased the organic N content of the upper 0–5 cm soil layer, while decreasing N concentrations were recorded at the depth of 5–15 cm. The effects of the applied treatments on the total nitrogen content were recorded only in the upper 0–5 cm soil layer, and the content was significantly different from the control sample only for No Litter treatment.

Our measurements indicate that decreased litter production may decrease the Ca^{2+} content in the soil, with the resultant reduction of the buffering capacity of soil leading to its acidification.

Acknowledgments Kate Lajtha, Kristin Wanderbilt, Bruce Caldwell and Richard Bowden, American cooperative researchers of ILTER DIRT Project are thankfully acknowledged for their kind and wide-ranging help in the Síkfőkút DIRT Project.

References

Egnér H, Riehm H, Domingo WR (1960) Untersuchungen über die chemische Bodenanalyse als Grundlage für die Beurteilung des Nährstoffzustandes der Böden. II K Lantbr Högsk Ann 26:199–215

Fekete I, Varga Cs, Halász J, Krakomperger Zs, Krausz E (2008) Study of litter decomposition intensity in litter manipulative trials in Síkfőkút Cambisols. Cereal Res Comm 36 Suppl:1779–1782

Fekete I, Cs Varga, Kotroczó Zs, Krakomperger Zs, Tóth JA (2007) The effect of temperature and moisture on enzyme activity in Síkfőkút Site. Cereal Res Comm 35:381–385

Finzi AC, Canham CD, Van Breemen N (1998) Canopy tree-soil interactions within temperate forests: species effects on pH and cations. Ecol Appl 8:447–454

Houba VJG, Novozamsky L, Lexmond TM, van der Lee JJ (1990) Applicability of 0.01 M $CaCl_2$ as a single extraction solution for the assessment of the nutrient status of soils and other diagnostic purposes. Soil Sci Plant Anal 21:2281–2290

Jakucs P (1973) "Síkfőkút Project". Egy tölgyes ökoszisztéma környezetbiológiai kutatása a bioszféra-program keretén belül. MTA Biol Oszt Közl 16:11–25

Kotroczó Zs, Krakomperger Zs, Koncz G, Papp M, Bowden RD, Tóth JA (2007) A Síkfőkúti cseres-tölgyes fafaj összetételének és struktúrájának hosszú távú változása. Természetvédelmi Közl 13:93–100

Kotroczó Zs, Krakomperger Zs, Veres Zs, Vasenszki T, Halász L, Koncz JG, Papp M, Tóth JA (2010) Talajlégzés vizsgálatok tartamhatású avarmanipulációs modellkísérletben. Természetvédelmi Közl 15:328–337

Kotroczó Zs, L. Halász J, Krakomperger Zs, Fekete I, D. Tóth M, Vincze Gy, Varga Cs, Balázsy S, Tóth JA (2008a) Erdőtalaj szerves-anyag mennyiségének változása avarmanipulációs kísérletek hatására (Síkfőkút Project). Talajvédelem Suppl:431–440

Kotroczó Zs, Fekete I, Tóth JA, Tóthmérész B, Balázsy S (2008b) Effect of leaf- and root-litter manipulation for carbon-dioxide efflux in forest soil. Cereal Res Comm 36 Suppl:663–666

Krakompereger Zs, Tóth JA, Varga Cs, Tóthmérész B (2008) The effect of litter input on soil enzyme activity in an oak forest. Cereal Res Comm 36:323–326

Kunz RP, Schulze RE, Schole RJ (1995) An approach to modelling spatial changes of plant carbon: nitrogen ratios in southern Africa in relation to anticipated global climate change. J Biogeogr 22:401–408

Micks P, Downs MR, Magill AH, Nadelhoffer K, Aber JD (2004) Decomposing litter as a sink for ^{15}N-enriched additions to an oak forest and a red pine plantation. For Ecol Manag 196:71–87

Nagy PT (2000) Égetéses elven működő elemanalizátor alkalmazhatósága talaj- és növényvizsgálatokban. Agrokémia és Talajtan 49:521–534

Neilson GA, Hole FD (1963) A study of the natural processes of incorporation of organic matter into soil in the University of Wisconsin Arboretum. Wisc Acad Rev 52:213–227

Tóth JA, Krakomperger Zs, Kotroczó Zs, Koncz G,Veres Zs, Papp M (2008) A klímaváltozás hatása a Síkfőkúti cseres-tölgyes avarprodukciójára és talajdinamikai folyamataira. Talajvédelem Suppl.:543–554

Tóth JA, Lajtha K, Kotroczó Zs, Krakomperger Zs, Caldwel B, Bowden RD, Papp M (2007a) A klímaváltozás hatása az elhalt szerves anyag lebontási folyamataira. In: Mátyás Cs (ed) Erdő és Klíma V. NyME Sopron

Tóth JA, Lajtha K, Kotroczó Zs, Krakomperger Zs, Caldwel B, Bowden RD, Papp M (2007b) The effect of climate change on soil organic matter decomposition. Acta Silv Ligniaria Hung 3:75–85

Varga Cs, Fekete I, Kotroczó Zs, Krakomperger Zs, Vincze Gy (2008) The effect of litter on soil organic matter (SOM) turnover in Síkfőkút site. Cereal Res Comm 36 Suppl:547–550

Climate Changes in the Vertical Zones of the Polish Carpathians in the Last 50 Years

Anita Bokwa, Agnieszka Wypych and Zbigniew Ustrnul

Abstract The variability and extremes of air temperature and precipitation in the years 1951–2006 were analyzed for different vertical zones as well as for the W-E profile of the Polish Carpathians. Some results of the analysis were compared with those concerning the Eastern Alps. Data from 9 Polish and 18 Austrian stations were used. The variability of air temperature in the Polish Western Carpathians during the study period decreased with altitude whereas no relation between temperature variability and altitude was observed in the Alps. The rate of temperature increase per decade was much lower at Kasprowy Wierch than at the lower locations (0.14 and 0.21 K, respectively) and for the period 1956–2000 it was even statistically insignificant. Notably, the temperature at Kasprowy Wierch was much lower than the temperature at the similar altitudes in the Alps. No clear trends in precipitation can be observed in either mountain chains. The observed differences in the climate change patterns between both mountain chains are connected with their geographical features and the air temperature increase probably results from diversified effects of atmospheric circulation.

1 Introduction

In Europe, the research on climate change in mountains has been focused on the Alps as the Greater Alpine Region, i.e. the Alps and their surroundings, is representative for large parts of European mountain areas (Böhm et al. 2001). It shares three of the principal European climates: Atlantic-maritime in the west and

A. Bokwa (✉) · A. Wypych · Z. Ustrnul
Institute of Geography and Spatial Management, Jagiellonian University, Gronostajowa 7, 30-387 Kraków, Poland
e-mail: anita.bokwa@uj.edu.pl

Z. Ustrnul
Institute of Meteorology and Water Management, National Research Institute, Piotra Borowego 14 30-214 Kraków, Poland

J. Kozak et al. (eds.), *The Carpathians: Integrating Nature and Society Towards Sustainability*, Environmental Science and Engineering,
DOI: 10.1007/978-3-642-12725-0_8,

northwest, the more continental in the east and Mediterranean in the south of the region (Matulla et al. 2005). In that region, international climate research projects have been successfully conducted for several years (e.g. ALOCLIM, ALPCLIM, HISTALP). Recent studies have focused on the long-term variability of particular climate elements in the context of the global environmental changes (Auer et al. 2001; Matulla et al. 2005; Auer et al. 2007; Brunetti et al. 2009). The research results were included in the Fourth Assessment Report of the Intergovernmental Panel on Climate Change (Alcamo et al. 2007).

Unlike in the Alps, climate changes in the Carpathians, the largest mountain chain of Central and Eastern Europe, are relatively poorly recognized. The main limiting factor is a small number of meteorological stations, especially in higher parts of the Carpathians. There have been no research projects comprising the whole mountain chain, and the numerous local research activities do not yield comparable information. Additionally, there is no international climatological data base available as is the case in the Alps (Auer et al. 2007), and meteorological data may not be accessed freely in the Carpathian countries. Therefore, the present study is limited to the Polish Carpathians.

Obrębska-Starklowa et al. (1994a, b, c, 1995) studied climate change that occurred in the period 1950–1990 along the vertical profile of the Polish Western Carpathians, from Kraków to the peaks of the Tatra Mts., the highest range of the Carpathians. Mean summer temperature was found to be decreasing and the winter temperature to be increasing in the whole profile. Above 1,500 m a.s.l. the changes were of smaller magnitude and they were shifted in time in comparison with the lower parts of the mountains. The increase in air temperature in the 1980s was much larger than in the previous decades due to the intensification of oceanic influences. In the case of precipitation, the changes were much smaller and the trends were not statistically significant. Similar results were also obtained for the Tatra Mts. (Niedźwiedź 1992). The temperature increase in the Alpine belt was delayed by about a decade in comparison to low-lying stations. Before that a significant decrease in the mean annual temperatures had been noticed. These changes were attributed to the intensive air mass advection from the Atlantic in the period 1972–1990 (Niedźwiedź 1992). These findings do not completely accord with the 4th IPCC report which states that in the years 1977–2000, air temperature was increasing more rapidly in central and north-eastern Europe and in mountainous regions than in other parts of Europe, while slower changes were recorded in the Mediterranean region. Moreover, greater temperature increase occurred in winter than in summer (Alcamo et al. 2007). The differences between the findings of Obrębska-Starklowa et al. (1994a, b, c, 1995) and the IPCC report (Alcamo et al. 2007) may reflect the fact that the information about mountains in the IPCC report was mainly derived from the Alps where some factors controlling climate operate differently than in the Carpathians, even though both mountain chains are located in similar latitudes (Weber et al. 1997). The research in mountain regions of Central Europe demonstrated a west-east gradient in the temperature trends, with a stronger increase in the western part of the area.

The occurrence of extreme atmospheric phenomena is often included in the analysis of long-term climate variability, not only because of large economic losses and danger for human life (Moberg et al. 2006) but also due to the sensitivity of the phenomena to human activities. At the same time, many climatologists consider the occurrence of extreme phenomena as a significant indicator of possible climatc change. Worldwide analyses show significant changes in temperature extremes, especially daily minimum temperatures. According to Alexander et al. (2006), changes in daily maximum temperatures are less pronounced. European studies (Klein Tank and Können 2003) indicate that in the second half of the twentieth century changes in warm extremes contributed much more to the warming than those in the cold extremes.

The aim of this work is to establish the magnitude of climate change and variability along the vertical and W-E profiles of the Polish Carpathians in the period 1951–2006. The analyses concerning the vertical profile were performed only for the western part of the region. Two principal climate elements, i.e. air temperature and precipitation were considered in terms of the trends of mean values and the occurrence of extremes. For the vertical profile, the concept of vertical climatic zones in the Polish Western Carpathians and other mountains formulated by Hess (1965) was used. Data from Austrian stations located in the Eastern Alps were used for verification and comparison of the obtained results. This longitudinally elongated region is located in similar latitudes as the Western Carpathians and it supports many meteorological stations. Unfortunately, due to limited availability of the data from those stations, the comparison could be performed only partially.

2 Data and Methods

The data comprise daily air temperature and precipitation values for the years 1951–2006 from nine meteorological stations in the Polish Carpathians. All the data have been carefully checked and homogenized with the Standard Normal Homogeneity Test (Alexandersson 1986). Detected inhomogeneities (mainly connected with slight relocation of the stations) were verified and corrected where necessary. Most of the stations provided complete data series. For some stations, data from shorter periods were available (Table 1, Fig. 1).

The data allow to study climate change along the vertical profile but only in the western part of the Polish Carpathians (stations: Kraków-Balice, further referred as Kraków, Bielsko-Biała, Zakopane, Kasprowy Wierch), while the change in the W-E profile may be analyzed mainly for the lower part of the Carpathians (stations: Bielsko-Biała, Kraków, Nowy Sącz, Tarnów, Rzeszów, Lesko, Przemyśl). The detailed analysis for the W-E profile was conducted for Bielsko-Biała (representing the western part of the Carpathian Foothills) and Lesko (representing the eastern part). These stations have been chosen due to their location (similar altitude and

Table 1 Meteorological stations in the Polish Carpathians used in the study and the record period for which the data are available

No.	Station	Altitude (m)	N Latitude	E Longitude	Period
1	Rzeszów	195	50°07′	22°02′	1952–2006
2	Tarnów	209	50°01′	20°59′	1951–2006
3	Kraków-Balice	237	50°05′	19°48′	1951–2006
4	Przemyśl	279	49°48′	22°46′	1954–2000
5	Nowy Sącz	292	49°38′	20°41′	1951–2006
6	Bielsko-Biała	398	49°48′	19°00′	1951–2006
7	Lesko	420	49°28′	22°20′	1954–2006
8	Zakopane	857	49°18′	19°57′	1951–2006
9	Kasprowy Wierch	1991	49°14′	19°59′	1951–2006

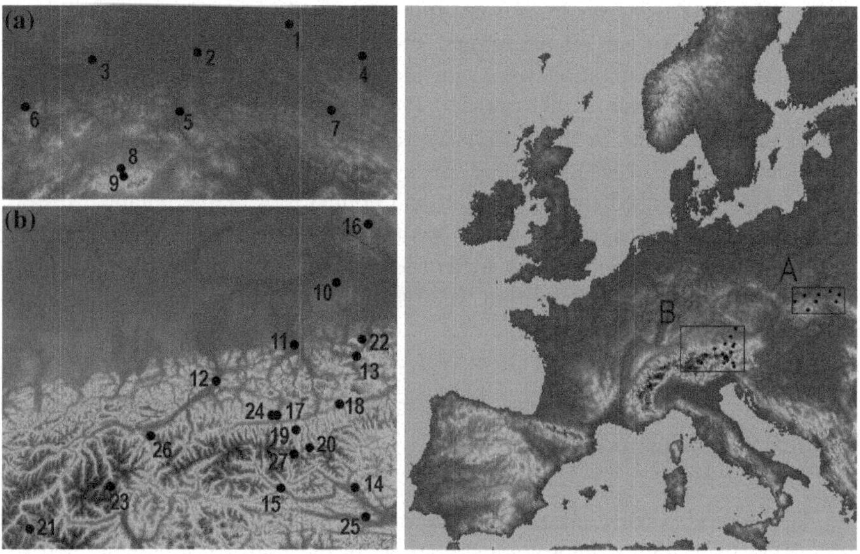

Fig. 1 Location of the examined meteorological stations from the Polish Carpathians (**a**) and the Eastern Alps (**b**); station numbers as in Table 1 (**a**) and Table 2 (**b**)

physiographic characteristics of the area) and data homogeneity (data series covering the whole period under investigation).

For mean annual and seasonal air temperature values as well as annual and seasonal sums of precipitation, decadal means and standard deviation values were calculated, linear trends were identified and the amounts of change in air temperature and precipitation per decade were determined. Additionally, deviations of particular values from the mean for a standard period 1971–2000 were calculated and tested for the presence of statistically significant linear trends. All indices were separately calculated for the periods 1956–2000 and 1951–2006. The first period was chosen to analyze the longest period possible for the whole data-set (some of the stations provide only shorter data series). However, because of the importance of the length of data series, e.g. for the identification of trends, we decided to use the longer period for stations where data were available.

The study of long-term changes in the occurrence of extreme phenomena was based on daily maximum and minimum air temperature values and daily precipitation totals. The daily data were analyzed regarding the long-term variability of extreme values: minimum winter (DJF) and maximum summer (JJA) daily air temperatures as well as daily precipitation totals and the frequency of their occurrence within the period under examination. The extreme cases were identified by the probability approach. To define the extremes, 10th and 90th percentiles of daily values were applied. All the values below the 10th percentile as well as above the 90th percentile were analyzed as extreme cases. Even though this method is not suitable for impact studies, it has a great advantage for detection of climate change as the changes can be compared across Europe between regions with different climates and topographical conditions (Klein Tank and Können 2003). The method was implemented to analyze essential climatological differences, particularly in thermal and precipitation conditions, over the whole Carpathians. Application of arbitrary threshold methods seems inappropriate in mountain areas.

In order to compare the climate change tendencies in the Western Polish Carpathians and in the Eastern Alps, analyses of mean annual air temperature values and annual sums of precipitation were performed for 18 Austrian meteorological stations (Table 2). The data, covering the northern part of the Eastern Alps, were derived from the HISTALP website (http://www.zamg.ac.at/histalp/index.html). This sub-region, defined in the HISTALP project as NE-E part of the Alps, exhibits the continental features of climate (e.g. Matulla et al. 2005; Auer et al. 2007). This part of the Alps is closest to the Polish Carpathians and geographical factors controlling climate in both areas may be considered similar. Additionally, the greater number of meteorological stations located at different altitudes provides more representative results for particular vertical climatic zones than in the Carpathians. For some stations both mean annual air temperature and annual sums of precipitation from the years 1951–2006 were available, and some provided only temperature data (Table 2). Unavailability of daily data from the Alpine stations precluded the comparison of extreme values of the climatic parameters between both regions.

The multi-annual changes of air temperature and precipitation in the Carpathians and in the Alps were referred to the variability of atmospheric circulation, defined for both regions with the same method, using the zonal and meridional indices (Ustrnul 1997). The annual values of the indices were calculated for the grid point 50°N, 20°E representing the Polish Western Carpathians and 47°30′N, 12°30′E representing the Eastern Alps.

A further analysis was performed for vertical zones distinguished according to the concept of Hess (1965) with modification of the present authors. The same mean annual air temperature may be found in the Alps about 400 m higher than in the Carpathians but at about 2,000 m a.s.l., the difference decreases to about 200 m. Hess (1965) defined the vertical climatic zones on the basis of mean annual

Table 2 Meteorological stations in the Alps used in the study and the data available

No.	Station	Altitude (m)	N Latitude	E Longitude	Data available
10	Ried	431	48°13′	13°29′	T, P
11	Salzburg-Flughafen	450	47°48′	13°00′	T, P
12	Kufstein	493	47°35′	12°10′	T, P
13	Bad Ischl	512	47°43′	13°38′	T, P
14	Lienz	659	46°50′	12°48′	T, P
15	Millstatt	719	46°48′	13°34′	T, P
16	Kollerschlag	725	48°36′	13°50′	T
17	Zell am See	766	47°20′	12°47′	T, P
18	Radstadt	858	47°23′	13°27′	T, P
19	Rauris	941	47°13′	13°00′	T, P
20	Badgastein-Böckstein	1100	47°06′	13°07′	T, P
21	Galtür	1587	46°35′	10°11′	T
22	Feuerkogel	1618	47°49′	13°43′	T
23	Obergurgel-Vent	1938	46°52′	11°02′	T
24	Schmittenhöhe	1973	47°20′	12°44′	T
25	Villacher Alpe	2160	46°36′	13°40′	T
26	Patscherkofel	2247	47°13′	11°28′	T
27	Sonnblick	3105	47°03′	12°57′	T

T mean annual air temperature, *P* annual sum of precipitation

air temperatures, while in this work the altitude criterion was used. We distinguished the following vertical zones:

1. the foreland zone: up to 400 m a.s.l. in the Carpathians and up to 800 m a.s.l. in the Alps,
2. the medium-elevation zone: 400–1,200 m a.s.l. in the Carpathians and 800–1,500 m a.s.l. in the Alps,
3. the high-elevation zone: above 1,200 m a.s.l. in the Carpathians and above 1,500 m a.s.l. in the Alps.

3 Long-Term Variability

3.1 Air Temperature

The research conducted hitherto in the Alps (Auer et al. 2007) indicated that particular sub-regions (NE, SE, SW, NW), including high-elevation mountain ranges, show small differences in long-term temperature variability. At the altitudes above 1,500 m, the same tendencies are observed as in rural areas located in the valleys and plains around the Alps. The whole Great Alpine Region has warmed twice as much since the late nineteenth century compared to the global average and the change has mainly reflected the temperature increase in recent decades (Auer et al. 2007).

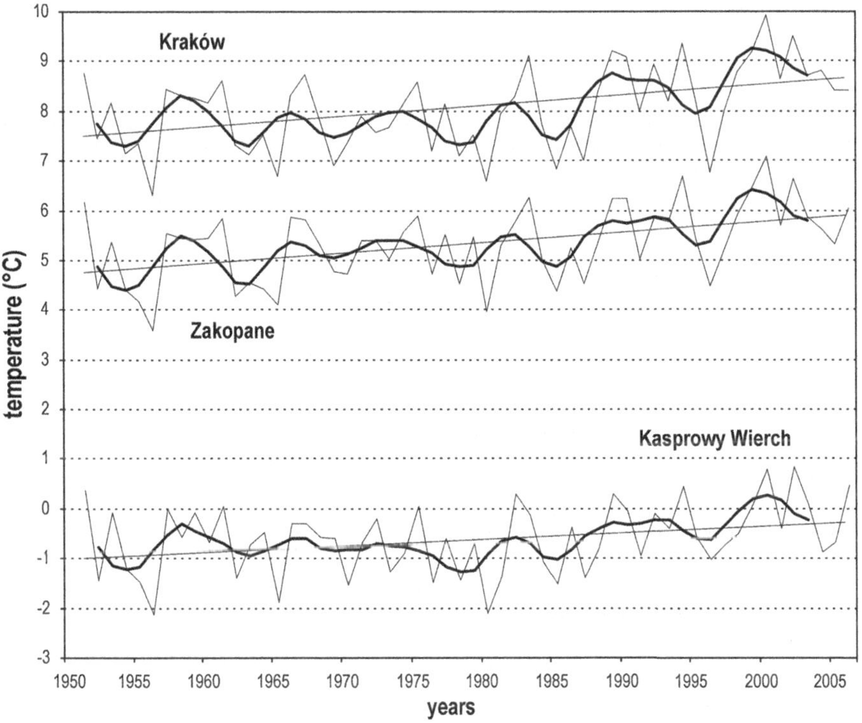

Fig. 2 Mean annual air temperature (°C) in Kraków, Zakopane and Kasprowy Wierch in the years 1951–2006, together with 5-year Gauss filter averages and linear trend lines

Figure 2 presents the multi-annual changes of mean annual air temperature in the vertical profile of the Western Polish Carpathians. Data from Bielsko-Biała are omitted as the values are very close to those from Kraków.

At all three stations, a statistically significant increase in air temperature occurred in the study period. However, in the highest zone the increase was the smallest. In Kraków, mean annual air temperature was rising at the rate of 0.21 K per decade, in Zakopane 0.20 K per decade and at Kasprowy Wierch only 0.14 K per decade (Table 3). The variability of air temperature, expressed with standard deviation, decreased with altitude. Additionally, at Kasprowy Wierch, where the average long-term (1951–2006) annual temperature amounts to –0.6 °C, mean annual air temperature exceeded 0 °C only ten times in the examined period, 6 of which took place in the last two decades.

The differences in the change of mean annual air temperature between the western and eastern parts of the Carpathians' lower zone may be exemplified by the stations Bielsko-Biała and Lesko (Fig. 3). The temperature increased at both stations but in the eastern part (Lesko), the increase at a low rate of 0.06 K per decade was not statistically significant, while in the western part (Bielsko-Biała), it was significant and equaled 0.19 K per decade (Table 3).

Table 3 Changes of air temperature (Δt, K per decade) and standard deviation values (σ) for annual air temperature series for the stations in the Eastern Alps and in the Polish Carpathians in the years 1951–2006 and 1956–2000. Statistically significant values are marked in bold ($p < 0.05$)

Stations in the Eastern Alps	1951–2006		1956–2000		Stations in the Polish Western Carpathians	1951–2006		1956–2000	
	Δt	σ	Δt	σ		Δt	σ	Δt	σ
Foreland zone									
Ried	**0.28**	0.80	**0.32**	0.77	Bielsko-Biała	**0.17**	0.81	**0.19**	0.83
Salzburg-Flughafen	**0.26**	0.79	**0.28**	0.75	Kraków-Balice	**0.21**	0.82	**0.21**	0.83
Kufstein	**0.31**	0.79	**0.35**	0.75	Nowy Sącz	**0.18**	0.76	**0.16**	0.77
Bad Ischl	**0.20**	0.72	**0.28**	0.71	Tarnów	**0.16**	0.78	**0.18**	0.80
Lienz	**0.23**	0.70	**0.20**	0.64	Rzeszów	**0.20**[a]	0.80	**0.19**	0.82
Millstatt	**0.21**	0.66	**0.23**	0.64	Przemyśl	–	–	0.04	0.81
Kollerschlag	**0.21**	0.74	**0.20**	0.69	Lesko	–	–	0.06	0.73
Zell am See	**0.25**	0.71	**0.26**	0.67	Mean W	0.19	0.82	0.20	0.83
Mean	0.24	0.74	0.27	0.70	Mean E	–	–	0.05	0.77
Medium-elevation zone									
Radstadt	**0.27**	0.75	**0.28**	0.69	Zakopane	**0.20**	0.75	**0.20**	0.74
Rauris	**0.24**	0.76	**0.27**	0.75					
Badgastein-Böckstein	**0.16**	0.64	**0.20**	0.63					
Mean	0.22	0.72	0.25	0.69					
High-elevation zone									
Galtür	**0.21**	0.67	**0.23**	0.65	Kasprowy Wierch	**0.14**	0.70	0.11	0.68
Feuerkogel	**0.23**	0.80	**0.27**	0.77					
Obergurgel-Vent	**0.23**	0.72	**0.29**	0.71					
Schmittenhöhe	**0.25**	0.80	**0.27**	0.76					
Villacher Alpe	**0.23**	0.75	**0.26**	0.73					
Patscherkofel	**0.26**	0.78	**0.26**	0.74					
Sonnblick	**0.22**	0.69	**0.25**	0.67					
Mean	0.23	0.74	0.26	0.72					

Mean W mean value for stations in the western part of the Polish Carpathians (Kraków, Bielsko-Biała); *Mean E* mean value for stations in the eastern part of the Polish Carpathians (Przemyśl, Lesko); [a] value for the period 1952–2006

The rates of temperature change and their significance differ between the periods 1951–2006 and 1956–2000 (Table 3). A comparison of the temperature values for particular years with the long-term average for the standard period 1971–2000 indicates that in the years 2001–2006 deviations were mostly positive at all stations, with 2002 and 2003 values ranked as highest in the study period (Fig. 4). These years of the examined period were essential to the observed trends.

For mean seasonal values of air temperature, deviations from the average for the period 1971–2000 as well as the linear trends were established. Winter and autumn trends were not statistically significant at any station, whereas spring and summer trends were not significant for Kasprowy Wierch, Lesko and Przemyśl.

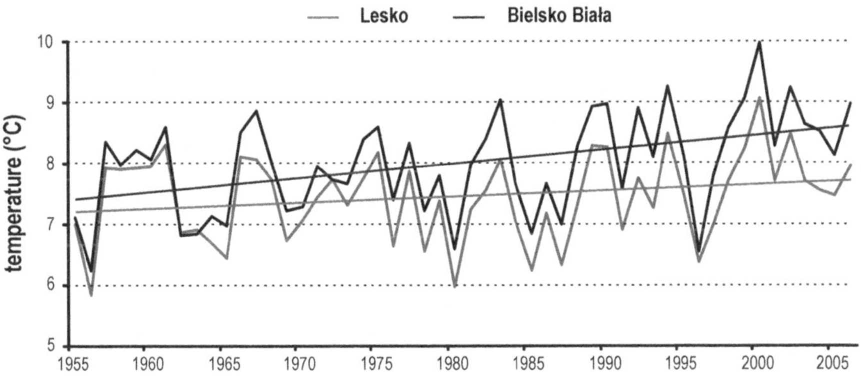

Fig. 3 Mean annual air temperature (°C) in Lesko and Bielsko-Biała in the years 1955–2006, together with linear trend lines

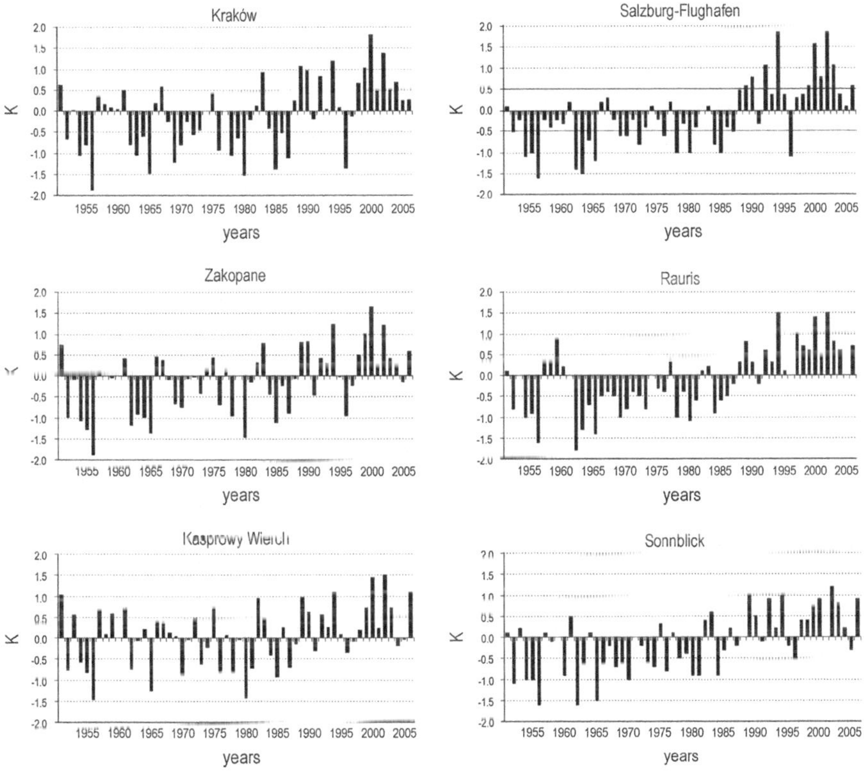

Fig. 4 Deviations of mean annual air temperature (K) in particular years of the period 1951–2006 from the average value for the standard period 1971–2000 at chosen stations in the Polish Western Carpathians and in the Eastern Alps, representing the vertical profiles

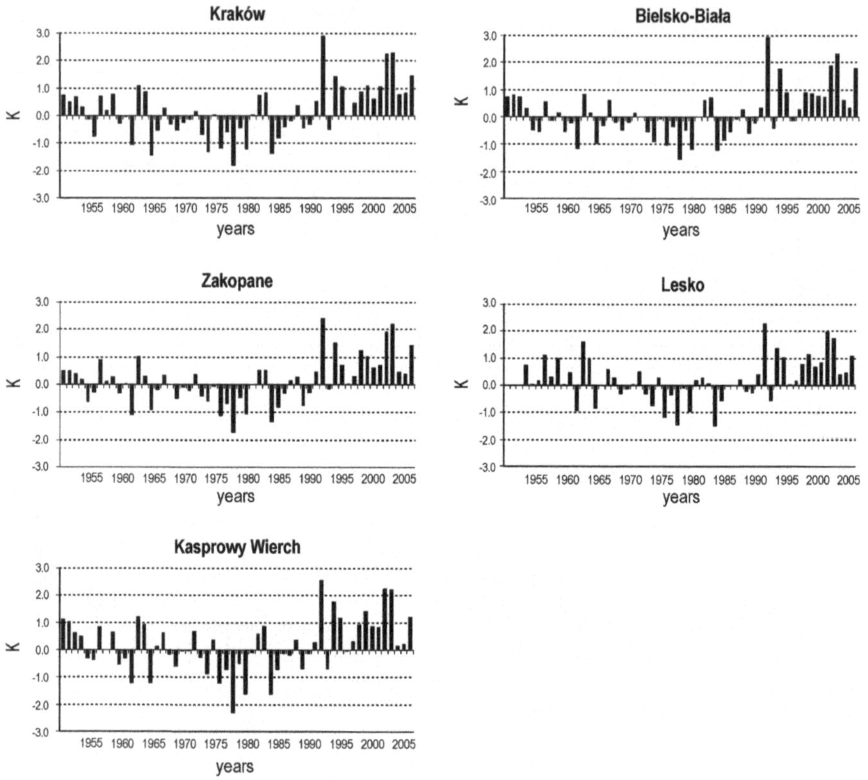

Fig. 5 Deviations of mean summer air temperature (K) in particular years of the period 1951–2006 from the average for the period 1971–2000 at the Kraków, Zakopane and Kasprowy Wierch stations representing the vertical profile of the Polish Carpathians and at the Bielsko-Biała and Lesko stations representing the W-E profile

Figure 5 shows the deviations of summer temperature at the stations representing the vertical and W-E profiles.

Contrary to the Carpathians, in the Alps the increase in temperature was significant at all examined stations and in both periods. There seems to be no clear relation between temperature variability, expressed with standard deviation, and altitude (Table 3).

In spite of the relatively large distance between both areas (about 600 km), the changes in mean annual air temperature were to a large extent synchronous, especially in the lowest vertical zone. However, in some years deviations of opposite sign occurred at the stations representing the same vertical zones in both mountain chains. These differences may be explained by the impact of atmospheric circulation conditions. At the end of the 1980s and the beginning of the 1990s, mean annual air temperature was generally higher in the Alps than in the

Carpathians, which seems to be related to the more southward (approx. 300 km) location of the Alps. The analysis of meridional and zonal indices for both areas showed significant differences in their values as well as the long-term variation. In the mentioned period the meridional circulation was more intensive in the Alps than in the Carpathians, while considerably higher values of the zonal index (westerly advection) were recorded in the Carpathians (Ustrnul 1997).

Additionally, mean decadal air temperatures were calculated for the Polish and Austrian stations (Table 4, 5). For the period 1956–2000, common for all stations, mean decadal air temperature values show similar thermal conditions in the foreland and medium-elevation zones of both mountain chains, but in the case of the high-elevation zone, a distinct difference can be observed. The temperature at Kasprowy Wierch was much lower than the temperature at similar heights in the Alps, including the height shift according to the concept of Hess (1965). The years with mean temperature considerably higher or lower than the average were not always synchronous in the Carpathians and the Alps.

The comparison of decadal air temperature values for Kasprowy Wierch, Villacher Alpe and Patscherkofel, i.e. the stations located at comparable altitudes according to the concept of Hess (1965), shows further differences. In the 1950s and 1960s the values at all three stations showed little change and for Kasprowy Wierch and Patscherkofel, they were almost identical. In the 1970s the temperature decreased at Kasprowy Wierch, while at the two other stations it increased which again proves that different large-scale processes control climate at both regions. In the 1980s and 1990s a temperature increase occurred at all stations, but in the years 2001–2006 its rate was higher at Kasprowy Wierch than at the two other stations. Another important environmental aspect is the increase of mean decadal air temperature above 0 °C. This threshold was crossed in the 1970s in Villacher Alpe and in the 1980s in Patscherkofel, while at Kasprowy Wierch the value was reached as late as in the years 2001–2006. In the Alps the increase in air temperature was of similar magnitude in the whole vertical profile, while in the Carpathians the increase at Kasprowy Wierch was much smaller than at the lower locations; for the period 1956–2000 it was even not statistically significant.

The differences in air temperature changes in the Western Carpathians and in the Eastern Alps over the period 1951–2006 seem to show that the decisive factors controlling climate are different for particular vertical mountain zones. In the foreland and the medium-elevation zones, local factors such as relief appear to be the most important. In both regions, the discussed zones are sheltered by convex landforms and the air masses brought by advection processes are therefore significantly transformed. This may be the reason for the similarity of thermal conditions in those zones. At higher altitudes, however, the modifications of air masses are much smaller and advection seems to be substantially more important than at lower altitudes.

Table 4 Mean decadal air temperature (°C) at the stations in the Polish Carpathians in the years 1951–2006

Station	1951–1960	1961–1970	1971–1980	1981–1990	1991–2000	2001–2006	1951–2006	1956–2000
Foreland zone								
Bielsko-Biała	7.8	7.6	7.8	8.1	8.4	8.6	8.0	7.9
Kraków-Balice	7.8	7.6	7.6	8.1	8.5	8.7	8.0	7.9
Nowy Sącz	–	7.9	8.0	8.3	8.5	8.7	8.2	8.2
Tarnów	8.3	8.2	8.2	8.7	8.8	9.0	8.5	8.5
Rzeszów	7.6[a]	7.5	7.6	8.1	8.3	8.6	7.9	7.9
Przemyśl	–	12.3	11.7	12.3	12.4	–	–	8.0
Lesko	–	12.1	12.0	11.7	12.1	12.4	–	7.4
Mean W	7.8	7.6	7.7	8.1	8.5	8.7	8.0	7.9
Mean E	–	12.2	11.9	12.0	12.3	–	–	7.7
Medium-elevation zone								
Zakopane	5.0	4.9	5.1	5.4	5.7	5.8	5.3	5.3
High-elevation zone								
Kasprowy Wierch	−0.8	−0.8	−1.0	−0.7	−0.3	−0.1	−0.6	−0.7

Mean W mean value for stations in the western part of the Polish Carpathians (Kraków, Bielsko-Biała); *Mean E* mean value for stations in the eastern part of the Polish Carpathians (Przemyśl, Lesko); [a] value for the period 1952–1960

Table 5 Mean decadal air temperature (°C) at the stations in the Eastern Alps in the years 1951–2006

Station	1951–1960	1961–1970	1971–1980	1981–1990	1991–2000	2001–2006	1951–2006	1956–2000
Foreland zone								
Ried	7.6	7.5	7.9	8.3	8.8	9.0	8.1	8.0
Salzburg-Flughafen	8.1	8.1	8.2	8.5	9.1	9.4	8.5	8.4
Kufstein	7.4	7.3	7.7	8.0	8.7	8.9	7.9	7.8
Bad Ischl	7.8	7.6	7.9	8.3	8.8	8.5	8.1	8.1
Lienz	6.5	6.5	6.7	6.9	7.4	7.9	6.9	6.8
Millstatt	7.1	7.0	7.2	7.4	8.0	8.1	7.4	7.4
Kollerschlag	6.4	6.3	6.5	6.7	7.2	7.5	6.7	6.6
Zell am See	6.0	5.9	6.2	6.3	7.0	7.3	6.4	6.3
Mean	7.1	7.0	7.3	7.6	8.1	8.3	7.5	7.4
Medium-elevation zone								
Radstadt	5.2	5.1	5.3	5.6	5.3	6.6	5.6	5.5
Rauris	4.9	4.3	4.6	5.0	5.7	5.8	5.0	4.9
Badgastein-Böckstein	5.1	4.9	5.4	5.5	5.9	5.8	5.4	5.4
Mean	5.1	4.8	5.1	5.4	6.0	6.1	5.3	5.3
High-elevation zone								
Galtür	2.3	2.2	2.2	2.7	3.2	3.3	2.6	2.5
Feuerkogel	3.0	2.8	3.0	3.5	3.9	4.0	3.3	3.3
Obergurgel-Vent	1.4	1.3	1.4	2.0	2.3	2.4	1.7	1.7
Schmittenhöhe	1.4	1.4	1.4	1.9	2.4	2.5	1.8	1.7
Villacher Alpe	0.0	−0.1	0.0	0.5	1.0	1.1	0.4	0.3
Patscherkofel	−0.6	−0.7	−0.5	0.0	0.4	0.6	−0.2	−0.2
Sonnblick	−5.9	−6.0	−5.8	−5.4	−5.0	−4.9	−5.5	−5.6
Mean	0.2	0.1	0.2	0.7	1.2	1.3	0.6	0.5

3.2 Precipitation

Contrary to the air temperature, precipitation shows more complicated spatial patterns in the Western Carpathians and in the Eastern Alps. There is no single measurement series representative for the whole Alpine region and considerable differences in the seasonal means as well as short- and long-term variability can be observed (Böhm et al. 2002). The same can be stated for the Carpathians. During the last three decades SW part of the Alps dried up to 10 %, while the NE region experienced 8 % increase in precipitation (Auer et al. 2007).

For precipitation, similar indices were calculated as for air temperature. For the Alpine stations, no precipitation data were available for the highest vertical zone. The data presented in Tables 6 and 7 show that the spatial variability of precipitation is much higher than in the case of air temperature. Moreover, the relation between precipitation and altitude is much less pronounced than in the case of air temperature. The decadal sums of precipitation show no clear differences between particular decades and no trends of changes over time. This is additionally proved with the data shown in Table 8. None of the estimated linear trends was statistically significant ($p < 0.05$). The annual precipitation sums show high inter-annual variability at all stations. The precipitation sums in the foothill zone of the Carpathians are much smaller than at the stations located at similar altitudes in the Alps which shows that the climate of the Carpathians is definitely more continental than the climate of the Eastern Alps. Additionally, in the Polish Carpathians alone, precipitation sums decrease from west to east as the climate of the region is transitional between the maritime and continental one.

3.3 Extreme Phenomena

The study focused on the number of days with the occurrence of particular extremes. Due to the spatial variability of the physiographic conditions in the Polish Carpathians, especially topography, threshold values for the extremes (distinguished with the probability method) vary (Table 9). In the period 1951–2006, the number of days with extreme maximum summer temperature (upper 10 % of the values) increased at a rate of about 1 day per decade. The tendency is observed over almost the whole area of the Polish Carpathians and in some parts (i.e. the summits of the Tatra Mountains, Eastern Carpathians foreland) it is somewhat larger (1.2 days per decade—Fig. 6). The inverse tendency is observed for minimum winter temperature. The number of days with extreme minimum temperature (lower 10 % of the values) decreased at a rate of approximately 1 day per decade (Table 10). The only station with a higher rate (1.5 days per decade) is Zakopane (Fig. 6); however, this may be related to the urban effect generated by the town of Zakopane rapidly developing over the last two decades. The number of days with extreme daily precipitation totals (upper 10 % of the

Table 6 Mean decadal precipitation sums (mm) at the stations in the Polish Carpathians in the years 1951–2006

Station	1951–1960	1961–1970	1971–1980	1981–1990	1991–2000	2001–2006	1951–2006	1956–2000
Foreland zone								
Bielsko-Biała	980	1,119	958	892	979	983	985	994
Kraków-Balice	614	735	690	628	669	663	667	675
Nowy Sącz	–	775	704	683	704	768	–	722
Tarnów	616	756	771	644	707	730	702	712
Rzeszów	554[a]	656	653	566	667	681	627	619
Przemyśl	–	723	666	614	671	–	–	663
Lesko	–	820	819	765	795	876	–	791
Mean W	797	927	824	760	824	823	826	834
Mean E	–	746	722	655	709	764	–	727
Medium-elevation zone								
Zakopane	1,084	1,147	1,160	1,069	1,092	1,155	1,115	1,118
High-elevation zone								
Kasprowy Wierch	1,559	1,787	1,906	1,711	1,784	1,792	1,754	1,761

Mean W mean value for stations in the western part of the Polish Carpathians (Kraków, Bielsko-Biała); *Mean E* mean value for stations in the eastern part of the Polish Carpathians (Przemyśl, Lesko). [a] value for the period 1952–1960

Table 7 Mean decadal precipitation sums (mm) at the stations in the Eastern Alps in the years 1951–2006

Station	1951–1960	1961–1970	1971–1980	1981–1990	1991–2000	2001–2006	1951–2006	1956–2000
Foreland zone								
Ried	1,000	940	890	959	988	1,047	965	952
Salzburg-Flughafen	1,166	1,153	1,120	1,203	1,227	1,169	1,173	1,173
Kufstein	1,303	1,319	1,268	1,275	1,356	1,372	1,311	1,305
Bad Ischl	1,657	1,651	1,623	1,727	1,741	1,801	1,693	1,684
Lienz	1,015	963	912	846	953	821	925	933
Millstatt	959	958	856	843	897	886	901	901
Zell am See	1,227	1,207	1,096	1,144	1,169	1,168	1,168	1,160
Mean	1,190	1,170	1,109	1,142	1,190	1,180	1,162	1,158
Medium-elevation zone								
Radstadt	1,254	1,276	1,147	1,207	1,189	1,097	1,202	1,208
Rauris	1,186	1,128	1,031	1,131	1,092	1,012	1,103	1,109
Badgastein-Böckstein	1,234	1,210	1,138	1,158	1,179	1,253	1,191	1,186
Mean	1,225	1,205	1,105	1,165	1,153	1,121	1,165	1,168

Table 8 Changes of precipitation sums (Δp, mm per decade) and standard deviation values (σ) for annual precipitation sums series for the stations in the Eastern Alps and in the Polish Carpathians in the years 1951–2006 and 1956–2000. All values are not significant statistically

Stations in the Eastern Alps	1951–2006		1956–2000		Stations in the Polish Western Carpathians	1951–2006		1956–2000	
	Δp	σ	Δp	σ		Δp	σ	Δp	σ
Foreland zone									
Ried	11.0	142.8	8.6	129.9	Bielsko-Biała	−10.1	166.0	-30.4	167.0
Salzburg-Flughafen	12.6	154.8	25.9	156.7	Kraków	2.3	118.9	−3.7	119.9
Kufstein	14.4	171.6	13.0	165.7	Nowy Sącz	−4.4	111.3	−15.0	112.1
Bad Ischl	31.4	248.0	31.2	234.3	Tarnów	11.2	134.2	−1.3	138.7
Lienz	−24.1	180.7	−16.8	171.4	Rzeszów[a]	16.7	126.6	22.0	133.8
Millstatt	−13.2	141.4	−19.7	151.6	Przemyśl	–	–	−1.7	133.8
Zell am See	−8.6	162.1	−6.7	153.7	Lesko	–	–	8.2	127.9
Mean	3.4	171.6	5.1	166.2	Mean W	−3.9	142.5	−17.1	143.5
					Mean E	–	–	3.3	130.9
Medium-elevation zone									
Radstadt	−21.8	158.6	−13.1	148.9	Zakopane	4.1	165.9	−8.6	161.3
Rauris	−16.9	149.3	−8.2	147.2					
Badgastein-Böckstein	1.2	145.8	−11.9	147.6					
Mean	−12.5	151.2	−11.1	147.9					
High-elevation zone									
–	–	–	–	–	Kasprowy Wierch	0.14	0.70	0.11	0.68

Mean W mean value for stations in the western part of the Polish Carpathians (Kraków, Bielsko-Biała); *Mean E* mean value for stations in the eastern part of the Polish Carpathians (Przemyśl, Lesko); [a] value for the period 1952–2006

Table 9 The 10th and 90th percentile temperature and precipitation thresholds for particular stations in the Polish Carpathians

Station	Thresholds of extremes		
	t_{min} (°C)	t_{max} (°C)	Daily precipitation totals (mm)
Rzeszów	−15.1	28.7	9.9
Tarnów	−14.9	29.0	10.4
Kraków	−13.9	28.7	9.8
Przemyśl	−13.6	27.6	10.6
Nowy Sącz	−14.5	29.2	11.0
Bielsko Biała	−12.8	27.3	13.7
Lesko	−13.6	27.3	11.3
Zakopane	−15.8	24.9	14.5
Kasprowy Wierch	−17.4	15.8	19.0

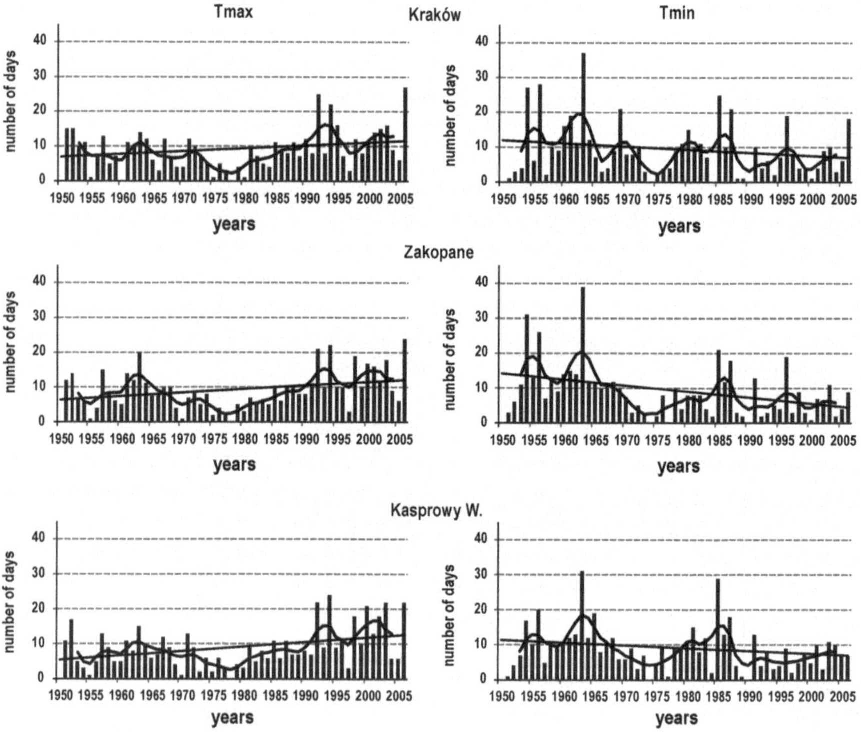

Fig. 6 Long-term variability of maximum summer (*left*) and minimum winter (*right*) air temperature extremes at particular stations in the Polish Carpathians (1951–2006)

values) does not show any significant tendency within the Polish Carpathians (Table 10). Depending on the station, either increase or decrease of the index is observed. Nevertheless, in each case the trend is not statistically significant.

Table 10 Observed changes in the number of temperature and precipitation extremes at particular stations in the Polish Carpathians

Stations	Δ (days/10 years)		
	tmax	tmin	Precipitation
Rzeszów	1.4	−1.0	0.5
Tarnów	0.8	−0.6	0.1
Kraków	0.7	−0.8	−0.1
Przemyśl	0.1	−2.0	−0.5
Nowy Sącz	1.2	−0.9	−0.1
Bielsko-B.	0.9	−0.6	0.0
Lesko	0.6	−0.9	0.6
Zakopane	0.9	−1.6	0.0
Kasprowy Wierch	1.2	−0.7	0.8
Areal mean	0.8	−0.8	0.2

The final results were compared with the Eastern Alps. Unfortunately, the analysis was constrained by the data accessibility and was eventually conducted only for two Alpine stations (Salzburg and Sonnblick) with the data from the European Climate Assessment and Dataset (Klein Tank et al. 2002).

Neither region shows any significant variability in the number of days with extreme precipitation totals, whereas some tendencies in the number of days with extreme temperatures can be seen. In the case of maximum summer temperature, the increasing tendency is slightly stronger in the Alps than in the Carpathians, mainly in the highest parts (Sonnblick: 2 days per decade). On the contrary, the number of days with minimum winter temperature shows similar decreasing tendencies in both regions (about 1 day per decade).

The analysis of particular extreme phenomena indicates that they occurred at various moments from 1951 to 2006. Many extremes were recorded in the last two decades (since 1990), however, a considerable number of extremes were also observed in the 1950s and 1960s. For both minimum winter and maximum summer temperatures, the magnitude of observed trends and variations can be treated as normal within the same climate type, especially when they refer to the whole study area. The obtained results do not show significant differences between the regions or vertical zones.

4 Conclusions

The increase in air temperature in the period 1951–2006 was higher and more spatially uniform in the Eastern Alps than in the Polish Western Carpathians. In the Carpathians the air temperature increase occurred only in the western part, while no significant trends were found in the eastern part. No significant changes or trends were found also in precipitation, either in the Alps, or in the Carpathians, regardless of the index used.

As similar tendencies can be observed all over the entire study area, it may be suggested that they are mainly caused by mesoscale factors, especially atmospheric circulation conditions. The studies completed for the Alps proved a similarity between the variability of air pressure and temperature at low elevations and even stronger correlation at high elevations. The authors attributed the conformability to the influence of atmospheric circulation on the regional climate (Böhm et al. 1998; Auer et al. 2001).

The considerations presented in the chapter are only an attempt to explain some aspects of the climatic differences observed between the Polish Western Carpathians and the Eastern Alps. Many factors can control these differences, e.g. the location of both mountain chains at various latitudes, climate continentality and oceanity, atmospheric circulation and the size of the mountain chains. The obtained results also confirm the local diversity of climate conditions and raise the problem of station representativeness. Data from a greater number of meteorological stations with fully homogenous dataset are necessary for more detailed analyses. Moreover, the stations should represent different vertical zones as well as different relief.

References

Alcamo J, Moreno JM, Nováky B, Bindi M, Corobov R, Devoy RJN, Giannakopoulos C, Martin E, Olesen JE, Shvidenko A (2007) Europe. In: Parry ML, Canziani OF, Palutikof JP, van der Linden PJ, Hanson CE (eds) Climate change 2007: impacts, adaptation and vulnerability. Contribution of working group ii to the fourth assessment report of the intergovernmental panel on climate change. Cambridge University Press, Cambridge

Alexander LV, Zhang X, Peterson TC, Caesar J, Gleason B, Klein Tank AMG, Haylock M, Collins D, Trewin B, Rahimzadeh F, Tagipour A, Rupa Kumar K, Revadekar J, Griffiths G, Vincent L, Stephenson DB, Burn J, Aguilar E, Brunet M, Taylor M, New M, Zhai P, Rusticucci M, Vazquez-Aguirre JL (2006) Global observed changes in daily climate extremes of temperature and precipitation. J Geophys Res 111:D05109

Alexandersson A (1986) A homogeneity test applied to precipitation data. Int J Climatol 6:661–675

Auer I, Böhm R, Schöner W (2001) Austrian long-term climate 1767–2000—multiple instrumental climate time series from Central Europe. Österr Beitr Meteorol Geophys 25:1–147

Auer I, Böhm R, Jurkovic A, Lipa W, Orlik A, Potzmann R, Schöner W, Ungersböck M, Matulla C, Briffa K, Jones PD, Efthymiadis D, Brunetti M, Nanni T, Maugeri M, Mercalli L, Mestre O, Moisselin J-M, Begert M, Müller-Westermeier G, Kveton V, Bochnicek O, Stastny P, Lapin M, Szalai S, Szentimrey T, Cegnar T, Dolinar M, Gajic-Capka M, Zaninovic K, Majstorovic Z, Nieplova E (2007) HISTALP—historical instrumental climatological surface time series of the Greater Alpine Region 1760–2003. Int J Climatol 27:17–46

Böhm R, Auer I, Schöner W, Hagen M (1998) Long alpine barometric time series in different altitudes as a measure for 19th/20th century warming. Preprints of the 8th conference on mountain meteorology, in Flagstaff, Arizona. AMS, Boston

Böhm R, Auer I, Brunetti M, Maugeri M, Nanni T, Schöner W (2001) Regional temperature variability in the European Alps 1760–1998 from homogenized instrumental time series. Int J Climatol 21:1779–1801

Böhm R, Auer I, Schöner W, Brunetti M, Maugeri M, Huhle C, Nanni T (2002) Regional precipitation variability in the European Alps 1803–1998 from homogenised instrumental time series. Proceedings of 9th AMS conference on mountain meteorology, Park City, UT, P2.12.40008, June 2002

Brunetti M, Lentini G, Maugeri M, Nanni T, Auer I, Böhm R, Schöner W (2009) Climate variability and change in the Greater Alpine Region over the last two centuries based on multi-variable analysis. Int J Climatol 29:2197–2225

Hess M (1965) Piętra klimatyczne w polskich Karpatach Zachodnich. Zesz Nauk UJ Prace Geogr 11

Klein Tank AMG, Können GP (2003) Trends in indices of daily temperature and precipitation extremes in Europe, 1946–99. J Climate 16:3665–3680

Klein Tank AMG, Wijngaard JB, Können GP, Böhm R, Demarée G, Gocheva A, Mileta M, Pashiardis S, Hejkrlik L, Kern-Hansen C, Heino R, Bessemoulin P, Müller-Westermeier G, Tzanakou M, Szalai S, Pálsdóttir T, Fitzgerald D, Rubin S, Capaldo M, Maugeri M, Leitass A, Bukantis A, Aberfeld R, van Engelen AFV, Førland E, Mietus M, Coelho F, Mares C, Razuvaev V, Nieplova E, Cegnar T, Antonio López J, Dahlström B, Moberg A, Kirchhofer W, Ceylan A, Pachaliuk O, Alexander LV, Petrovic P (2002) Daily dataset of 20th-century surface air temperature and precipitation series for the European climate assessment. Int J Climatol 22:1441–1453

Matulla C, Auer I, Böhm R, Ungersböck M, Schöner W, Wagner S, Zorita E (2005) Outstanding past decadal-scale climate events in the Greater Alpine Region analysed by 250 years data and model runs. GKSS-Report 2005/4

Moberg A, Jones PD, Lister D, Walther A, Brunet M, Jacobeit J, Alexander LV, Della-Marta PM, Luterbacher J, Yiou P, Chen D, Klein Tank AMG, Saladie O, Sigro J, Aguilar E, Alexandersson H, Almarza C, Auer I, Barriendos M, Begert M, Bergström H, Böhm R, Butler CJ, Caesar J, Drebs A, Founda D, Gerstengarbe FW, Micela G, Maugeri M, Österle H, Pandzic K, Petrakis M, Srnec L, Tolasz R, Tuomenvirta H, Werner PC, Linderholm H, Philipp A, Wanner H, Xoplaki E (2006) Indices for daily temperature and precipitation extremes in Europe analyzed for the period 1901–2000. J Geophys Res 111:D22106

Niedźwiedź T (1992) Climate of the Tatra Mountains. Mt Res Develop 12:131–146

Obrębska-Starkel B, Bednarz Z, Niedźwiedź T, Trepińska J (1995) On the trends of the climate changes in the higher parts of the Carpathian Mountains. Zesz Nauk UJ Prace Geogr 98:123–151

Obrębska-Starklowa B, Bednarz Z, Niedźwiedź T, Olecki Z, Trepińska J (1994a) Natural and anthropogenic fluctuations and trends of climate change in Southern Poland. Geogr Pol 62:7–22

Obrębska-Starklowa B, Bednarz Z, Niedźwiedź T, Trepińska J (1994b) Klimat Karpat w okresie globalnego ocieplenia i prognozowane zmiany gospodarcze. Probl Zagosp Ziem Górskich 37:13–38

Obrębska-Starklowa B, Bednarz Z, Niedźwiedź T, Trepińska J (1994c) Contemporary tendencies in the climatic changes of the Carpathian Mts and their foreland. Conf Papers 20:103–116

Ustrnul Z (1997) Variability of the atmospheric circulation on the Northern Hemisphere in the 20th century. Mat Badaw IMGW Ser Meteorol 27:1–208

Weber RO, Böhm R, Gajic-Čapka M, Zaninovic K, Brázdil R, Faško P (1997) 20th century changes of temperature in the mountain regions of Central Europe. Clim Change 36:327–344

Air Temperature Variability in the High Tatra Mountains

Anna Pribullová, Miroslav Chmelík and Jozef Pecho

Abstract Annual and monthly air temperatures at meteorological stations covering nearly the whole vertical range (696–2,636 m a.s.l.) of the High Tatra Mountains (Western Carpathians) were analyzed for the period 1961–2007. The long-term variability in the air temperature at different altitudes was evaluated using two independent data sources. Upper-air temperature measurements from an aerological station near to the examined ground stations were interpolated to correspond to the altitudes of the stations. In spite of systematic differences between ground and upper-air temperatures, a comparison between homogenized time series showed a high correlation and identical periods of increase or decrease in both data sets, with the same periodicities, identical coolest and hottest years, and similar shifts in annual and monthly temperature distributions. An increase in selected isotherm altitudes in the period 1991–2007 was found when compared to the climatic normal period 1961–1990. A statistically significant linear increase in annual temperature (0.21–0.30 °C/decade) was detected in both data sets. The most rapid increase in temperature was found for July, August, December and January. A less significant increase was detected for spring and no change or even negligible decrease in temperature was found for autumn. The altitudes of selected isotherms were calculated with two methods: (1) linear fitting and (2) cubic spline interpolation of the vertical temperature profile. Particular combinations of data sources (aerological and ground) and calculation methods (linear fit and cubic spline) gave slightly different altitudes for the selected isotherms. The method used to calculate the isotherm altitudes should be taken into consideration when

A. Pribullová (✉)
Geophysical Institute, Slovak Academy of Sciences, Dúbravská cesta 9
845 28 Bratislava, Slovakia
e mail: apribull@ta3.sk

A. Pribullová · M. Chmelík
Slovak Hydrometeorological Institute, Poprad-Gánovce 058 01 Poprad, Slovakia

J. Pecho
Slovak Hydrometeorological Institute, Jeséniova 17 833 15 Bratislava, Slovakia

J. Kozak et al. (eds.), *The Carpathians: Integrating Nature and Society Towards Sustainability*, Environmental Science and Engineering,
DOI: 10.1007/978-3-642-12725-0_9, © Springer-Verlag Berlin Heidelberg 2013

comparing results from previous studies. Linear increase in the altitude of selected isotherms of annual temperature, calculated for aerological and ground data, showed rates of 3.9–5.2 and 2.5–4.9 m/year, respectively. A significant linear increase in the position of selected monthly isotherms was found only for July and August. An increase in isotherm altitudes indicates a shift in the vertical climatic stratification.

1 Introduction

Analysis of spatial and temporal variability in climate characteristics is one of the main objectives of climatology. Air temperature and precipitation regimes are the basic phenomena determining the climate.

A global increase in air temperature was detected from ground measurements over the last century. The linear trends in air temperature revealed typical spatial and temporal variability. Linear temperature trends over the range 0.294–0.344 °C/decade at a significance level $\alpha < 0.01$ have been found over land in the northern hemisphere for the period 1979–2005 (IPCC 2007). Europe is a part of the region which displays a significant temperature increase in winter (especially Northern Europe), spring, and summer months; while a smaller increase in temperature has been detected for autumn (IPCC 2007).

Inconsistent temperature trends determined from ground, satellite and radiosonde measurements were some of the factors which did not support the global warming theory (IPCC 2001). The latest reanalysis of the satellite and aerological measurements shows the warming rates that are more consistent with each other and with the surface measurements in the lower and middle troposphere (IPCC 2007).

The shifts in climate zones related to modified isotherm positions and to a change in precipitation regime have been detected in the mountains. Changes in growing season, shifts of species to higher altitudes, and changes in tree line position have been the most important consequences of increasing air temperature in the mountains (IPCC 2007).

The last detailed analysis of basic climate characteristics (including selected isotherm position calculations) in the High Tatra Mountains (Western Carpathians) was done within the framework of Slovak–Polish scientific cooperation in the late 1960s with results published by Konček and Orlicz (1974). The climatological data from the period before 1961 were examined in that monograph. Hess (1965) also analyzed the isotherm altitudes in the High Tatras. An analysis of the air temperature regime in the European mountain areas (comparison between the Alps and the Carpathians) and its relation to the weather patterns above Central Europe was done by Niedźwiedź (1992). Weber et al. (1997) compared the differences in diurnal temperature variability between lowland and mountain areas of Central Europe.

The main aim of this study was to evaluate both the spatial and temporal variability of air temperature in the boundary layer of the High Tatra Mountains after 1961 (periodicities, changes in the temperature distribution during the period 1991–2007 in comparison to the climatic normal 1961–1990 and linear trends). Two independent data sources were used: (1) ground temperature measured at eight climatic stations representing the altitudes over the range of 696–2636 m, and (2) temperature measured by aerological radiosondes (launched at the Poprad-Gánovce aerological station located within the study area) and interpolated to the altitudes of the examined climatic stations.

The second aim was to estimate changes in the vertical distribution of the temperature in the investigated air layer and their significance by using different approaches to calculate the isotherm altitude.

2 Data and Methods

Mean annual and monthly temperatures were analyzed for the period 1961–2007 for which the data from ground measurements at all eight climatic stations (ground temperature) and from radiosonde measurements (upper-air temperature) were available.

The aerological station in Poprad-Gánovce is located at the altitude of 706 m (coordinates and altitudes of the examined ground stations are shown in Table 1 and the spatial model of the study area in Fig. 1). This station is located to the south-east of the High Tatras, about 15 km from the main ridge and about 6 km from the former aerological station at Poprad-airport, which operated prior to December 1977, and the climatic data were taken from this latter location up to that date. The difference between the altitudes of these two stations is 10 m.

Mercury thermometers located 2 m above the ground were used in air temperature measurements at the climatic stations in accordance with the WMO guides. The calibration laboratory of the Slovak Hydrometeorological Institute (SHMI) for the temperature and humidity measurement control was established in 2005 and was certified with the ISO 9001 norm. The declared temperature uncertainty is 0.1–0.2 °C.

Five series and about 20 different types of radiosondes were used during the study: Czechoslovak Metra and Russian A022 in the 1960s, Czechoslovak and German MARS and MARS/DFR in the 1970s and 1980s, Finish RS80 in the 1990s, Finish RS90 in the 2000s. The analysis of Luers and Eskridge (1998) showed that the monthly and annual data from these radiosondes can be used for climatological analysis of the temperature time series below 25 km altitude since the temperature error is less than ± 0.5 °C. One exception was the radiosonde A022, which had a temperature error of ± 0.8 °C.

Before a radiosonde was launched, radiosonde temperature was compared with the ground value. Differences of less than ± 1 °C between the ground and radiosonde temperatures are permitted and can be corrected, otherwise the

Table 1 Coordinates of climatic stations with their abbreviations and years with inhomogeneities detected in annual and monthly temperature time series

Station	Latitude (°N)	Longitude (°E)	Altitude (m)	Measurement type			
				Ground		Aerological	
				Corrected inhomogeneity	Uncorrected inhomogeneity	Corrected inhomogeneity	Uncorrected inhomogeneity
Lomnický štít (LST)	49.14	20.22	2,636	–	–	1992	–
Skalnaté Pleso (SKP)	49.18	20.23	1,778	–	–	1992	–
Štrbské Pleso (STP)	49.12	20.06	1,356	1992, 2005	1979	1992	–
Javorina (JAV)	49.31	20.14	1,009	–	1984	1992	–
Podbanské (PDB)	49.14	19.91	974	–	–	1992	–
Telgárt (TEL)	48.85	20.19	903	–	–	1992	–
Tatranská Lomnica (TAL)	49.16	20.29	829	1992	–	1992	–
Poprad airport (POP)	49.02	20.24	696	–	–	–	–
Poprad-Gánovce	49.03	20.32	706	–	–	–	–

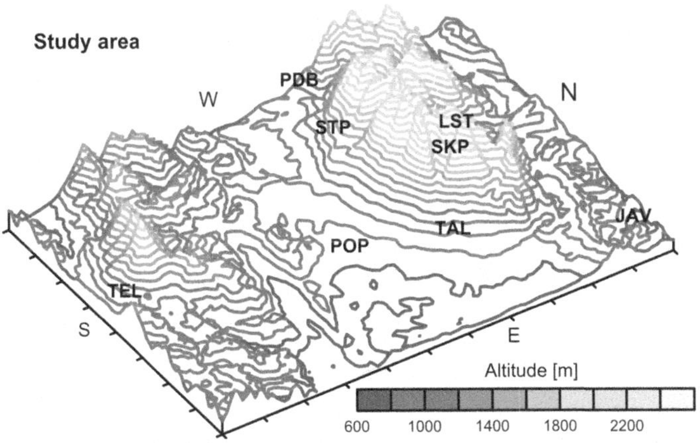

Fig. 1 Location of climatic stations in the High Tatra Mountains

radiosonde is considered as unreliable for launching. This, together with position characteristics of both stations, explains a very good agreement between the Poprad-Gánovce upper-air temperature and the Poprad-airport climatological daily temperature.

Annual and monthly means were calculated from daily temperatures. Daily ground temperatures were calculated as climatological daily means. The daily upper-air temperature was calculated as an average from measurements at 00 UTC, 12 UTC and 00 UTC for the next day.

Monthly means were included in the time series when at least 75 % of the daily data were available in a particular month. The annual temperature was calculated when less than one month of the data was missing.

Homogeneity of the monthly and annual time series was checked using the Standard-normal homogeneity test (SNHT) (Alexandersson and Moberg 1997; Moberg and Alexandersson 1997; Moberg and Bergström 1997). The SNHT test enables the identification of sudden changes in the temperature time series caused by e.g. a change in station position, instrumentation, or measurement methodology. The reference time series is constructed as a weighted average from some series at selected stations. Data from five ground stations which did not have any instrument replacement, a change in measurement method, or interruptions in measurements were used for the creation of the reference time series. The tested time series were compared to the reference time series. When some inhomogeneity in the tested time series was found, the calculated difference between the tested and the reference time series enabled us to correct the shift in the tested time series. The corrections were applied to data in the period before the year with a detected inhomogeneity.

The years with detected inhomogeneity are shown in Table 1. Corrections were made for inhomogeneities that corresponded to a change in station position or in the radiosounding system. Changes in the tested time series unrelated to the records in metadata were found at two stations (Javorina and Štrbské Pleso). These inhomogeneities are probably caused by a continuous alteration of the surroundings of the station and therefore could not be corrected easily since the differences between revised and reference time series varied with time. The inhomogeneity was more pronounced during the warm half-year at Javorina. It probably related to the growth of deciduous trees shading the station which caused a reduction in the expected increase in air temperature.

The autocorrelation function and Fourier spectral analysis were used to identify some periodic signals in the time series of the air temperature. The autocorrelation between the original time series and the same time series with different time lags was investigated. An increased correlation coefficient between the original and lagged time series indicated a periodic pattern. The Fourier transformation enabled us to express the time series as a discrete function of sinusoids and cosinusoids (Fourier series) of different frequencies. The most significant frequencies in the Fourier series can relate to periodicities in the investigated time series.

The altitude of a selected isotherm can be determined by different approaches. When temperature dependence on altitude is complicated and one or more temperature inversion layers appeared in the investigated air layer, several levels of the same temperature can be found in the temperature vertical profile and position of the selected isotherm can be determined only ambiguously.

Two methods for determining selected isotherm positions were used: linear fitting of the relation between the temperature and altitude, and cubic spline interpolation between stations (levels) with known temperature. The advantage of the first method is an unambiguous determination of the selected isotherm position, but in winter, when inversions often appear in the temperature vertical profile, the line does not represent the temperature profile well. Interpolation methods enabled us to determine the isotherm position also in the complicated vertical profiles. In the cases when several altitudes of selected temperature were detected in the vertical profile, the highest position was considered as the particular isotherm altitude.

The climatic normal period 1961–1990 was used in the study for comparison with the period 1991–2007.

3 Results and Discussion

3.1 Differences Between Ground and Aerological Data Sets

The basic characteristics (mean and variability expressed by standard deviation, STDEV) of the homogenized monthly and annual temperature time series obtained from ground measurements and from radiosondes were compared, and the correlation between both data sets was evaluated. A difference between the annual ground and upper-air temperatures was detected over the range from −1.54 °C

(SKP) to 0.58 °C (JAV) and a difference in the standard deviation was found to range from −0.08 °C (TEL) to 0.03 °C (SKP). The correlation between the ground and upper-air annual temperature time series was very high, with the correlation coefficient ranging from 0.792 (JAV) to 0.987 (POP).

The temperature differences, differences in variability expressed by STDEV and correlation coefficients between the monthly time series are presented in Fig. 2. During the warm half-year, air temperature registered at climatic stations was higher in comparison with the aerological data. During the cold half-year, the temperatures measured from the ground were lower than those registered at higher

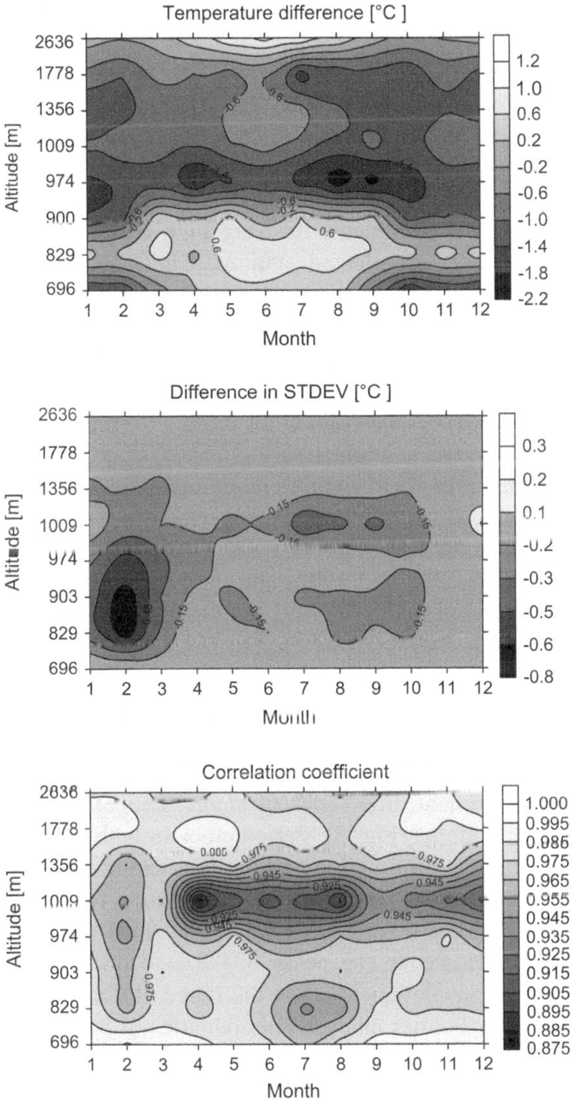

Fig. 2 Differences between ground and upper-air: (1) monthly temperature (*upper diagram*) (2) monthly temperature standard deviations (*middle diagram*) and (3) correlation coefficient between both data sets (*lower diagram*)

elevations. These differences result from radiative heating and cooling of the air close to the surface. However, the difference between aerological and ground temperature remains similar around a year at the highest altitudes.

A significant increase in the monthly temperature differences between the ground and aerological data can be seen at the altitude of station JAV. The concave terrain can contribute to the detected lower temperatures in comparison to other stations and to corresponding aerological data here (Hess 1965). The terrain effect should uniformly influence the difference between the tested and reference time series during the whole investigated period but this was not the case for the JAV station. The differences between JAV and reference time series increased with time which indicated inhomogeneity caused by the changing environment of the station.

A significantly higher STDEV for the aerological temperature in February was detected at altitudes around 800–1000 m. The correlation between both data series was very high, a decrease in correlation was detected at lower altitudes in July and also in winter. The inhomogeneity at station JAV is expressed also by a decrease in the correlation coefficient, especially in the warm half-year.

Figure 3 documents the very high correlation between the ground and aerological data at selected stations. The smoothed temperature time series show identical periods of decrease and increase for both data sets. Figure 3 shows the period 1961–1980 when the annual temperature did not exhibit a significant trend. This is in agreement with the temperature analysis of the High Tatras by Niedźwiedź (1992). The author described this period as more cloudy and rainy with more frequent advection of air masses from the Atlantic Ocean to Central Europe. The increase in temperature detected since the beginning of the 1990s and its possible relation to changes in regional circulation patterns should be investigated.

The analysis of ground and upper-air time series consistently indicated the same hottest years (1994, 2000, 2002, 2006, 2007, and at highly located stations LST, SKP also 1989 and 1982) and the same coolest years (1962, 1965, 1978, 1980, 1985, and at highly located stations LST, SKP also 1976).

3.2 Periodicities

The smoothed time series of annual temperature indicated the presence of periodical patterns (Fig. 3). Periodicities in both time series were identified using the autocorrelation function and the Fourier spectral analysis.

From the maxima of the autocorrelation function, several periodicities of 7 and 8 years were detected, as well as a third periodicity of 11 years for the LST station. However, the peaks of the autocorrelation function corresponded to low correlation coefficients over the range of 0.20–0.28 in annual temperature.

Using Fourier analysis, periodicities of 5.75 and 6.57 years were found in the annual temperature time series for all stations and a third periodicity of 3 years for the SKP and LST stations located at higher altitudes.

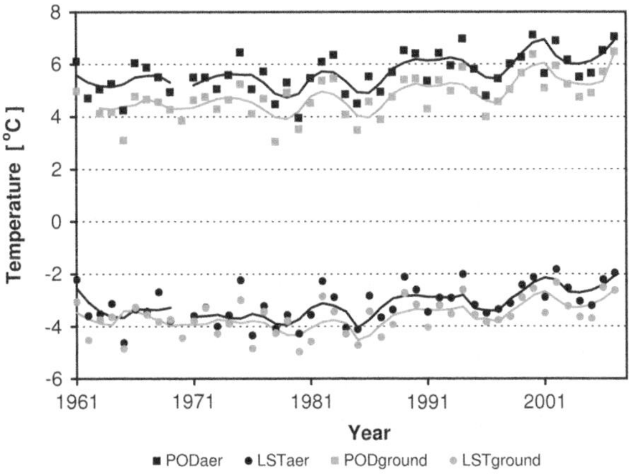

Fig. 3 Time series of ground (*gray*) and upper-air (*black*) annual temperature at the altitudes of stations PDB (*squares*) and LST (*dots*) smoothed by the robust locally weighted regression method (*continuous lines*) with quadratic smoothing function and smoothing half-window of 5 years

The periodicities detected in the monthly time series varied from one month to another and corresponded to low values of the autocorrelation function.

3.3 Differences Between 1961–1990 and 1991–2007 Periods

A shift from the climatic normal towards higher values was ascertained in the annual temperature distribution in ground and upper-air data for the period 1991–2007. Increases in mean temperature, the first and the third quartiles, the 10th and the 90th percentiles as well as the minima and the maxima of the distribution towards higher temperatures were found in both data sets (Table 2).

Differences between the means for the periods 1991–2007 and 1961–1990 ranged from 0.6 to 0.8 °C, but the error exceeded the values (Table 2). The inhomogeneity in the temperature time series at JAV station was manifested by lower differences between the periods 1991–2007 and 1961–1990. No significant change in the variability characterized by the STDEV was detected.

Higher mean temperatures in the summer (June, July, August) and winter months (December, January) were detected for the years 1991–2007 in comparison with the climatic normal period (Fig. 4). Differences were less significant in the spring and autumn.

Similarly, the largest shifts in the 10th percentile temperature towards higher values were detected during the summer (June–August) and winter (January, February) months (Fig. 5). A lower 10th percentile temperature was found in autumn, especially at the highest altitudes. A shift in the 90th percentile temperature towards higher values was detected from May to December. No change or lower 90th percentile temperature was detected for the period from January to March, especially at higher altitudes (Fig. 6).

Table 2 Differences (in °C) between mean annual temperatures (with standard errors), ST DEV, the 1st (Q1) and the 3rd (Q3) quartiles, the 10th (D10) and the 90th (D90) percentiles, the minima (MIN) and the maxima (MAX) during the period 1991–2007 and in the climatic normal period 1961–1990 at selected stations (ground) and at corresponding levels (aerological)

Parameter	Measurement	Station							
		LST	SKP	STP	JAV	PDB	TEL	TAL	POP
Mean annual temperature	Aerological	0.6 ± 0.7	0.6 ± 0.9	0.6 ± 0.9	0.6 ± 0.8	0.7 ± 0.9	0.7 ± 0.9	0.7 ± 0.9	0.6 ± 0.9
	Ground	0.6 ± 0.7	0.7 ± 0.8	0.6 ± 0.9	0.06 ± 0.9	0.8 ± 0.9	0.6 ± 0.8	0.7 ± 0.9	0.6 ± 0.9
ST DEV	Aerological	0.0	−0.1	0.0	0.0	0.0	−0.1	0.0	0.0
	Ground	−0.1	0.0	0.0	−0.1	0.0	0.0	0.1	0.0
Q1	Aerological	0.7	0.6	0.7	0.7	0.7	0.7	0.7	0.7
	Ground	0.8	0.9	0.6	0.2	0.8	0.7	0.6	0.7
Q3	Aerological	0.6	0.7	0.8	0.6	0.6	0.6	0.7	0.7
	Ground	0.8	0.7	0.6	0.0	0.9	0.3	0.6	0.6
D10	Aerological	0.7	0.3	0.6	0.7	0.6	0.6	0.7	0.8
	Ground	0.5	0.8	0.7	0.3	1.0	0.7	0.8	0.7
D90	Aerological	0.7	0.8	0.6	1.0	0.9	1.0	1.0	0.7
	Ground	1.0	0.6	0.3	0.4	0.8	0.7	0.5	0.7
MIN	Aerological	0.3	1.0	0.9	0.9	0.8	0.8	0.7	0.3
	Ground	0.9	1.0	0.6	0.1	0.9	0.7	0.7	0.4
MAX	Aerological	0.6	0.3	0.5	0.5	0.6	0.6	0.7	0.7
	Ground	0.4	0.6	0.6	0.2	1.0	0.7	0.8	0.8

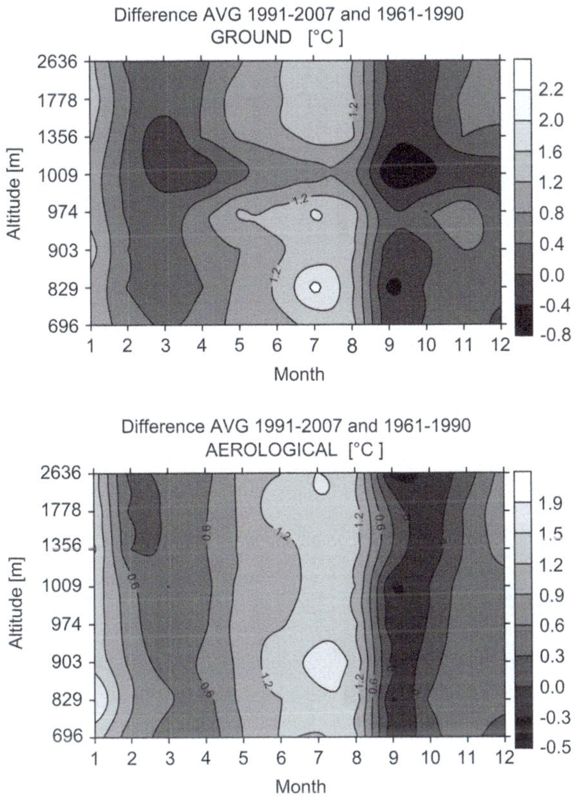

Fig. 4 Differences between mean (AVG) monthly temperature during the period 1991–2007 and the climatic normal period 1961–1990 for ground station data (*upper diagram*) and the radiosonde measurements interpolated to station altitudes (*lower diagram*)

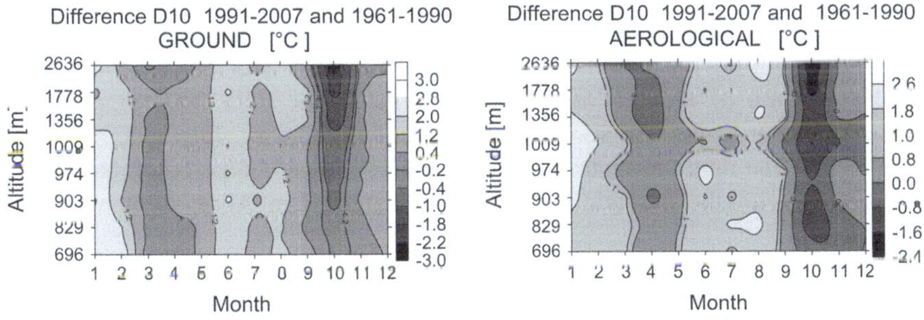

Fig. 5 Differences between the 10th percentiles of monthly temperatures during the periods 1991–2007 and 1961–1990 calculated from ground stations data (*left diagram*) and upper-air measurements (*right diagram*)

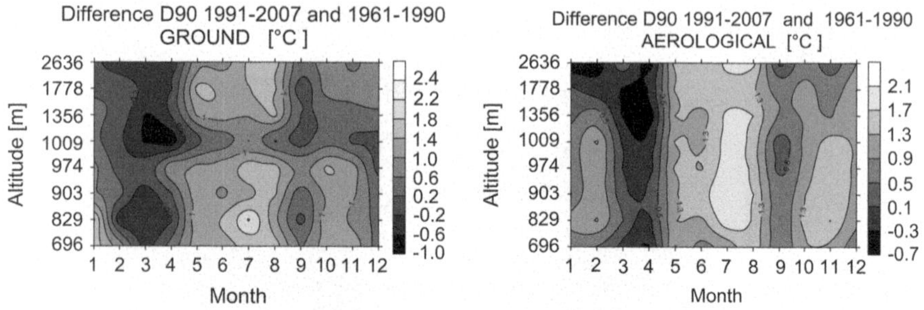

Fig. 6 Differences between the 90th percentiles of monthly temperatures during the periods 1991–2007 and 1961–1990 calculated from ground stations data (*left diagram*) and upper-air measurements (*right diagram*)

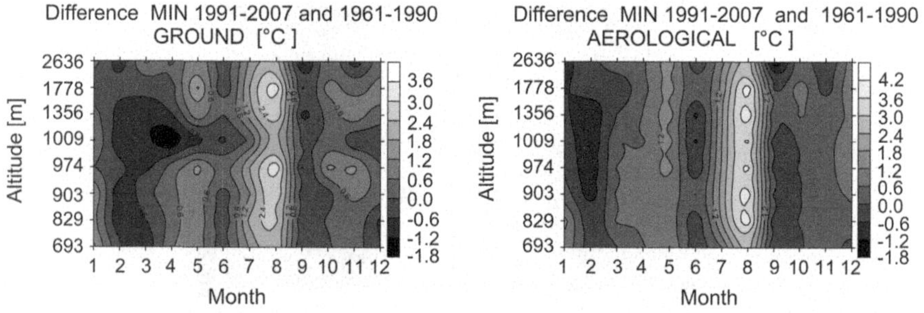

Fig. 7 Differences between the minimum monthly temperatures during the periods 1991–2007 and 1961–1990 calculated from ground stations data (*left diagram*) and upper-air measurements (*right diagram*)

An increase in the monthly temperature minima was detected in July and August, with the 1991–2007 minima exceeding the climatic normal by more than 3 °C at some altitudes (Fig. 7). No significant change in winter monthly minima was detected during the period 1991–2007 in comparison with the climatic normal.

The maxima of monthly temperature in the years 1991–2007 significantly exceeded values of the climatic normal in January, February, July and October in both ground and upper-air data series (Fig. 8). The lower temperature maxima, in comparison with the climatic normal, were detected in April, September and December.

Generally, both data sets showed a similar shift in the characteristics of the annual and monthly temperature distribution in the period 1991–2007 in comparison with the climatic normal.

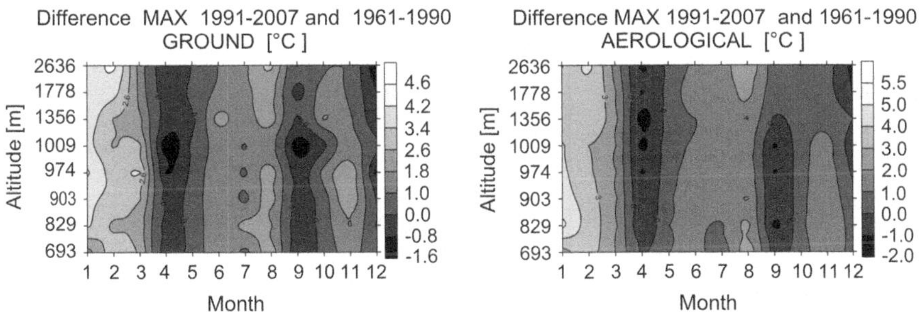

Fig. 8 Differences between the maximum monthly temperatures during the periods 1991–2007 and 1961–1990 calculated from ground stations data (*left diagram*) and upper-air measurements (*right diagram*)

3.4 Linear Trends

Data for nearly all stations and the corresponding upper-air levels indicate that during the period 1961–2007 mean annual temperature increased linearly at a rate of 0.21–0.30 °C/decade. Only at the ground station JAV, the trend was not significant (Table 3).

An analysis of the monthly time series showed that the linear trends varied during a year (Fig. 9). The trends of monthly temperature with significance levels $\alpha \leq 0.0025$ were found only for July and August. In winter, the increase in monthly temperatures was also rapid but the trend significance was low due to high variability in the data. The distribution in the monthly trends at different altitudes for every month followed a similar pattern for the ground and upper-air time series. In contrast to other stations, at station JAV no significant trends in ground temperature were found for the period April–September.

3.5 Changes in Isotherm Positions

Different phenomena affect the temperature close to the ground and in the free atmosphere.

Better agreement between the isotherm positions determined from the ground and aerological data was found in the warm half-year (mean standard error of the linear fit was 121 ± 56 m for aerological data and 126 ± 27 m for ground data in July). In the cold half-year, the altitudes of selected isotherms determined from the ground data were lower in comparison to the isotherm altitudes determined from aerological data (mean standard error of the linear fit was 250 ± 104 m for aerological data and 358 ± 190 m for ground data in December) (Fig. 10). Also the altitudes of annual temperature isotherms derived from the aerological data were systematically higher than those derived from the ground measurements (Table 4).

Table 3 Linear trends (in °C/decade) of the annual temperature measured at the ground stations and determined from aerological data interpolated to the altitudes of the ground stations, estimated for the homogenized series of measurements from the years 1961–2007

Measurement	Station							
	LST	SKP	STP	JAV	PDB	TEL	TAL	POP
Aerological	0.23 ± 0.07	0.25 ± 0.07	0.24 ± 0.07	0.25 ± 0.07	0.25 ± 0.07	0.25 ± 0.07	0.26 ± 0.07	0.25 ± 0.07
	(0.001)	(0.0005)	(0.0025)	(0.0005)	(0.0005)	(0.0005)	(0.0005)	(0.0005)
Ground	0.23 ± 0.07	0.23 ± 0.07	0.23 ± 0.07	0.23 ± 0.07	0.23 ± 0.07	0.23 ± 0.07	0.23 ± 0.07	0.23 ± 0.07
	(0.0025)	(0.0005)	(0.001)	(>0.4)	(0.0005)	(0.0025)	(0.0005)	(0.001)

Values in brackets indicate a significance level, α, of each trend

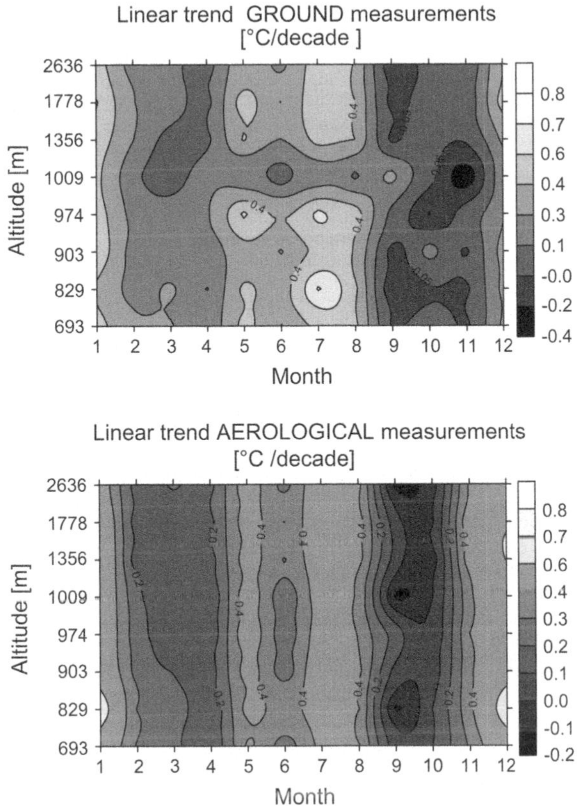

Fig. 9 Linear trends in monthly temperature calculated for the period 1961–2007 from the ground (*upper diagram*) and the aerological (*lower diagram*) measurements

Changes in the position of selected annual temperature isotherms are summarized in Table 4. The temperatures shown in the table are typical of the vertical profile of the examined stations; lower or higher temperatures cannot be found in the vertical profile in every year. A comparison of the altitudes of selected isotherms in the period 1991–2007 and during the climatic normal showed an upward shift of the isotherm position between 143 and 212 m according to the linear fitting method and between 75 and 137 m according to the spline interpolation method. The errors in determining the differences were large and comparable with particular values for the aerological and ground data due to the high variability in the isotherm altitudes.

Hess (1965) formulated vertical climate classification of the Tatra Mountains based on the altitudes of selected annual temperature isotherms and on the distribution of vegetation zones. The estimated altitudes of annual temperature isotherms on the southern slopes of the High Tatras presented by Hess (1965) are comparable to those shown in Table 4 (about 1,300 m for the isotherm 4 °C, about 1,700 m for the isotherm 2 °C, about 2,000 m for the isotherm 0 °C and about 2,300 m for the isotherm −2 °C) and fall within the range of uncertainty of the

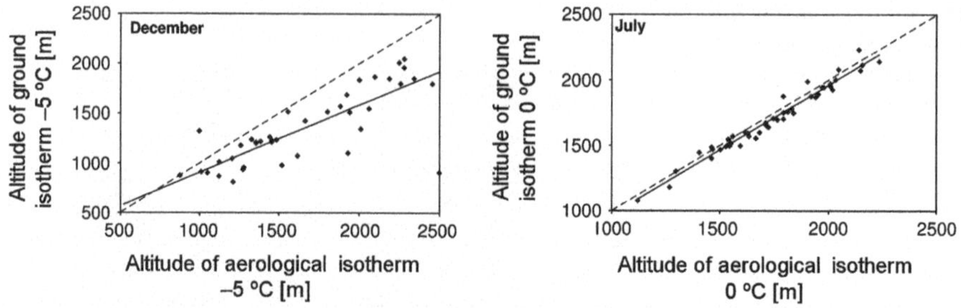

Fig. 10 Altitude of isotherms −5 °C in December (*left*) and 0 °C in July (*right*) determined from ground measurements as a function of the same isotherms altitude determined from aerological measurements. The dashed line represents the ideal agreement, and the solid line represents the determined linear relation between the ground and aerological data sets. Linear fitting of the vertical temperature profile was used to determine the isotherm altitudes

Table 4 Changes in the altitude of selected annual temperature isotherms (4, 2, 0, −2 °C) between the periods 1991–2007 and 1961–1990, rates of the changes as indicated by the 1961–2007 linear trends and their significance

Isotherm (°C)		−2	0	2	4
Linear fit					
Ground	Mean altitude (m)	2,371 ± 150	1,966 ± 134	1,561 ± 143	1,145 ± 137
	Difference between periods (m)	172 ± 174	146 ± 189	143 ± 181	199 ± 191
	Linear trend (m/year)	4.74	3.9	3.06	3.96
	Significance α	0.001	0.001	0.0025	0.001
Aerological	Mean altitude (m)	2,421 ± 154	2,036 ± 132	1,641 ± 134	1,247 ± 136
	Difference between period(m)	189 ± 173	212 ± 196	179 ± 164	169 ± 178
	Linear trend (m/year)	5.18	4.61	4.75	4.90
	Significance α	0.001	0.001	0.001	0.001
Spline interpolation					
Ground	Mean altitude (m)	2,383 ± 104	2,078 ± 108	1,739 ± 159	1,172 ± 232
	Difference between periods (m)	97 ± 131	101 ± 136	137 ± 192	75 ± 339
	Linear trend (m/year)	3.38	3.63	4.90	2.50
	Significance α	0.0025	0.0025	0.001	0.25
Aerological	Mean altitude (m)	2,412 ± 124	2,040 ± 132	1,662 ± 137	1,287 ± 134
	Difference between periods (m)	105 ± 155	120 ± 167	123 ± 173	122 ± 177
	Linear trend (m/year)	3.92	4.58	4.71	4.66
	Significance α	0.001	0.001	0.001	0.001

Changes in the isotherm altitudes calculated by linear fit and spline interpolation methods and results from the ground and the aerological data are presented separately isotherm altitude determined by the spline interpolation method. No significant change in isotherm positions was found in comparison with those indicated by the Hess (1965) analysis.

Konček and Orlicz (1974) presented mean altitudes of selected isotherms in the High Tatras calculated for the period 1931–1960. Values from stations located on concave and convex surfaces were presented separately (more stations were considered in their study). The mean altitude of 1,820 m for the 0 °C annual temperature isotherm was calculated for convex surfaces during this period. The altitude of 1,926 ± 126 m (for linear fit method) and 2,041 ± 100 m (for the spline interpolation method) for the 0 °C isotherm was found in the present study using ground measurements at all eight examined stations during the 1961–1990 climatic normal period. During the period 1991–2007 the altitude of this isotherm increased to 2,037 ± 119 m according to the linear fit method and to 2,142 ± 92 m according to the spline interpolation method. In turn, during the period 1931–1960 the altitude of the 10 °C isotherm in July was 1,640 m on convex surfaces. Our analysis using the spline interpolation method shows the position of this isotherm at 1,575 ± 92 m during the period 1961–1990 and at 1,819 ± 102 m during the period 1991–2007. Unfortunately, Konček and Orlicz (1974) did not indicate the calculation method (probably the interpolation method was used) or how the errors of the isotherm altitude were determined. Among the eight stations examined in this study only the surface at LST station can be considered as convex. The stations SKP and STP are located on slopes and the others lie at foothill locations with concave or flat surfaces. No comments on homogeneity control during the period 1931–1960 were provided by Konček and Orlicz (1974).

The incorporation of the JAV station must have affected the reliability of the isotherm positions calculated from ground measurements because of the inhomogeneous temperature time series from this station.

For the aerological data, the 1961–2007 trends in selected isotherm altitudes were significant ($\alpha = 0.001$), with the rate of change between 3.9 and 5.2 m/year. For the ground data, the trend significance was lower for some isotherms and the rate of change was between 2.5 and 4.9 m/year. Generally, the trends in the isotherm altitudes derived by a linear fit of the vertical temperature profile were steeper than those obtained using the spline interpolation method (Table 4).

Linear trends in selected monthly isotherm altitudes are summarized in Table 5. The variability in the altitude of monthly temperature isotherms was higher in comparison with annual temperature. Linear trends were the steepest (around 7 m/year) and statistically significant at $\alpha < 0.0025$ only for July and August. In winter, similar to summer monthly temperature trends, a steep increase in isotherm altitudes (5–6 m/year) was found, but the trends were not significant. The increase in isotherm altitude was small in the spring and an insignificant decrease was detected in the autumn. The correlation between isotherm altitudes obtained from ground and aerological data was the highest in the summer. The correlation decreased in spring, autumn and especially in winter due to an increase in the differences between the ground and aerological temperatures.

Table 5 Altitudes of selected isotherms of monthly (January, April, July, October) temperature, rates of the change in isotherm altitude and their significance as indicated by the 1961–2007 linear trends, and correlation coefficient between isotherm altitudes calculated from the ground and from the aerological data

Type of data	Isotherm parameters	Isotherm temperature (°C) in month			
		−6 January	−4 April	10 July	2 October
Ground	Mean altitude (m)	1,945 ± 440	1,745 ± 378	1,732 ± 344	1,961 ± 459
	Linear trend (m/year)	6.3	0.7	7.2	−0.9
	Significance α	>0.25	>0.25	0.001	>0.25
Aerological	Mean altitude(m)	1,893 ± 513	1,767 ± 257	1,651 ± 225	1,972 ± 400
	Linear trend (m/year)	5.4	3.6	7.4	−2.7
	Significance α	>0.25	>0.25	0.001	>0.25
Correlation coefficient		0.845	0.943	0.980	0.898

Isotherm altitudes were calculated by the cubic spline interpolation method

4 Conclusions

This chapter has presented changes in the annual and monthly temperatures that occurred in the High Tatra Mountains over the period 1961–2007. To estimate reliability of temperature changes, measurements at ground climatic stations located between 696 and 2,636 m a.s.l. were compared with aerological measurements at the appropriate altitudes. Homogeneity of both data series was checked by the SNH-test.

Although systematic differences between the ground and aerological temperatures were found at particular altitudes, the correlation between both datasets was very high and the same periodicities and the identical coolest and hottest years were found in the annual temperature time series.

A comparison between the mean annual temperature during the periods 1991–2007 and 1961–1990 revealed a consistent increase of 0.6–0.8 °C at all station altitudes for both datasets. However, the difference in the mean temperature between both periods may not be significant since its error was comparable to the mean values. Shifts towards higher values were also detected in selected quantiles of the temperature distribution. The changes detected in annual and monthly temperature distributions indicate a shift towards higher temperatures.

Periodicities in the annual temperature of the length of 6–8 years were detected in both datasets. The periodical signal could relate to some oscillation patterns affecting atmospheric circulation above Central Europe (Niedźwiedź 1992), but a more detailed analysis of this phenomenon is necessary.

Both data sets manifested similar statistically significant increases in annual temperature at the rate of 0.21–0.30 °C/decade, determined at a significance level $\alpha < 0.0025$.

An analysis of the monthly temperature time series showed that the highest increase in temperature and the steepest trend in monthly temperature typified the summer and winter months. A less significant increase in temperature was found in the spring, whereas no changes or even lower temperatures in comparison with the climatic normal were found for the autumn.

The increase in monthly temperatures was statistically significant at the significance level $\alpha < 0.0025$ only for July and August. Several days with extraordinary high temperatures occurred in July 2007, when the absolute maxima were exceeded by several degrees at the foothill stations. Similarly, December 2006 and January 2007 were the hottest months at several stations. A very high rate of temperature increase detected for July, August, December and January could have been partly influenced by the selection of the examined record period ending with extremely hot years 2006 and 2007.

The trends detected in annual temperature were comparable with the 1979–2005 trends calculated for land in the northern hemisphere (IPCC 2007). Less steep temperature trends in the autumn were revealed in the IPCC (2007) analysis of the temperature changes in Europe during the period 1979–2005.

An increase in the altitude of selected annual temperature isotherms was found for both datasets. The trends in the altitude of annual temperature isotherms were significant, with the rate of the change varying between 2.5 and 5.2 m/year. A significant altitude increase of monthly isotherms was detected only for July and August. The winter temperature isotherm altitudes increased rapidly, but the trends were not statistically significant. The definite determination of changes in the altitude of annual and monthly isotherms was difficult. An important result of the study is that largely different altitudes of annual temperature isotherms may be obtained by different methods. A direct comparison between absolute isotherm altitudes determined for different periods using different methods should be made carefully and the conclusions should be stated with respect to the variability of the data or the standard error of the measurement, which is not provided in some studies. On the other hand, significant consistent trends in isotherm altitudes determined from two independent data sets with different methods indicated a shift in vertical temperature stratification towards higher temperatures.

The analysis of both datasets showed a consistent increase in the temperature over the vertical range of the High Tatra Mountains, the most rapid in the summer and winter months.

The detailed temperature analysis presented in the paper focused on a small area (which spans one grid cell in global or regional analyses of interpolated air temperature—e.g. Haylock et al. 2008) but with a diverse topography. Comparisons between air temperature time series derived from local measurements and from large-scale spatial temperature analyses are planned.

Acknowledgments This work was supported by Slovak VEGA project Nr. 2-00-79-2011.

References

Alexandersson H, Moberg A (1997) Homogenization of Swedish temperature data. Part I. Int J Climatol 17:25–34

Haylock MR, Hofstra N, Klein Tank AMG, Klok EJ, Jones PD, New M (2008) A European daily high-resolution gridded data set of surface temperature and precipitation for 1950–2006. J Geophys Res 113:D20119

Hess M (1965) Piętra klimatyczne w polskich Karpatach Zachodnich. Zesz Nauk UJ Pr Geogr 11:1–267

IPCC (2001) Climate change 2001: the scientific basis. Cambridge University Press, Cambridge

IPCC (2007) Climate change 2007: the physical science basis. Contribution of Working Group I to the Fourth Assessment Report of the Intergovernmental Panel on Climate Change. Cambridge University Press, Cambridge

Končck M, Orlicz M (1974) Teplotné pomery. In: Konček M (ed) Klima Tatier. Veda, Bratislava

Luers JK, Eskridge RE (1998) Use of radiosonde temperature data in climate studies. J Clim 11:1002–1019

Moberg A, Alexandersson H (1997) Homogenization of Swedish temperature data. Part II: homogenized gridded air temperature compared with a subset of global gridded air temperature since 1861. Int J Climatol 17:35–54

Moberg A, Bergström H (1997) Homogenization of Swedish temperature data. Part III: the long temperature records from Uppsala and Stockholm. Int J Climatol 17:667–699

Niedźwiedź T (1992) Climate of the Tatra Mountains. Mt Res Dev 12:131–146

Weber RO, Böhm R, Gajic-Čapka M, Zaninovic K, Brázdil R, Faško P (1997) 20th century changes of temperature in the mountain regions of Central Europe. Clim Chang 36:327–344

Climate Trends in the Slovak Part of the Carpathians

Marián Melo, Milan Lapin, Ilana Kapolková, Jozef Pecho and Anna Kružicová

Abstract In this chapter, the temporal trends in air temperature and precipitation and territorial shifts in climate regions in Slovak part of the Carpathians during the twentieth century and at the beginning of the twenty first century are presented. The temperature series show an increasing trend at all meteorological stations in Slovakia. Climate has become warmer and more arid in the southern part of the Slovak Carpathians, particularly in the adjacent lowlands (e.g. the Danubian Lowland), while the northern part (the Orava Region) has become warmer and more humid (in terms of precipitation totals). Based on the Köppen and Konček climate classifications, climatic regions in Slovakia have been specified. According to these classifications, some shifts in climatic regions and sub-regions towards the higher altitudes and to the north were registered in Slovakia during this period. Selected results of the projection of climate change scenarios based on the outputs

M. Melo (✉) · M. Lapin
Faculty of Mathematics, Physics and Informatics, Comenius University in Bratislava, Mlynská dolina 842 48 Bratislava, Slovakia
e-mail: melo@fmph.uniba.sk

M. Lapin
e-mail: lapin@fmph.uniba.sk

H. Kapolková
Slovak Hydrometeorological Institute, Jeséniova 17 833 15 Bratislava, Slovakia
e-mail: hankakapolko@gmail.com

J. Pecho
Institute of Atmospheric Physics, Academy of Sciences of the Czech Republic, Boční II, 1401 141 31 Prague, Czech Republic
e-mail: pecho@ufa.cas.cz

A. Kružicová
Ministry of Environment of the Slovak Republic, Námestie Ľ. Štúra 1 812 35 Bratislava, Slovakia
e-mail: anna.kruzicova@enviro.gov.sk

J. Kozak et al. (eds.), *The Carpathians: Integrating Nature and Society Towards Sustainability*, Environmental Science and Engineering,
DOI: 10.1007/978-3-642-12725-0_10, © Springer-Verlag Berlin Heidelberg 2013

from General Circulation Models (GCMs) for the Slovak Carpathians up to 2100 are outlined. Scenarios based on three GCMs show further warming by 2–4 °C on this territory by the end of the twenty first century.

1 Introduction

The Western and Eastern Carpathians cover a major part of the Slovak territory. Lowlands are located only in the southwest (Záhorská and Danubian) and the southeast (Eastern Slovak Lowland) and are parts of Western Panonian and Eastern Panonian basins, respectively. The Carpathians form an extended, but relatively broad, upwarp in Slovakia with only a small proportion of the range typified by high-mountain relief (the Tatra Mts. and the Low Tatra Mts.). Highland and upland relief dominates a great part of the Slovak Carpathians. A characteristic form of the Carpathian relief are intramontane depressions which stand in sharp contrast to the adjacent mountain ranges.

A relatively small area of the Slovak Carpathians exhibits substantial differences in climate associated mainly with the complex orographic conditions and varied conditions of air circulation in the individual regions. The maritime influence on climate, both from the Atlantic and the Mediterranean Sea, decreases from the west towards the east. Cyclones fuelled by the warm Mediterranean air bring abundant precipitation, particularly to the southwestern and southern parts of the Slovak Carpathians, while cyclones of the Atlantic origin influence mostly their northwestern and northern parts. Due to greater continentality of the Eastern Carpathians, the winter air temperatures in this area are somewhat lower than in the western part of the Western Carpathians. In some mountain valleys and depressions (e.g. the Spišská Depression) with weak mean wind speed, the winter temperatures are lower by 2–4 °C as a result of the frequent occurrence of inversions (Konček 1964).

Global climate has become warmer in the last decades (IPCC 2007). Similar warming continues up to nowadays also in the Carpathian region. This chapter presents an analysis of temporal trends in air temperature and precipitation, as well as the territorial shifts in climate regions and sub-regions in the Slovak part of the Carpathians during the twentieth century and at the beginning of the twenty first century. In addition, selected results of climate-change scenarios projection, based on the outputs from General Circulation Models (GCMs) for the Slovak Carpathians region up to 2100, are outlined. Climate change may affect biodiversity, introduction of new species, spatial distribution and species composition within mountain areas. The presentation of such consequences of climate change for vegetation and fauna is also one of the aims of this chapter.

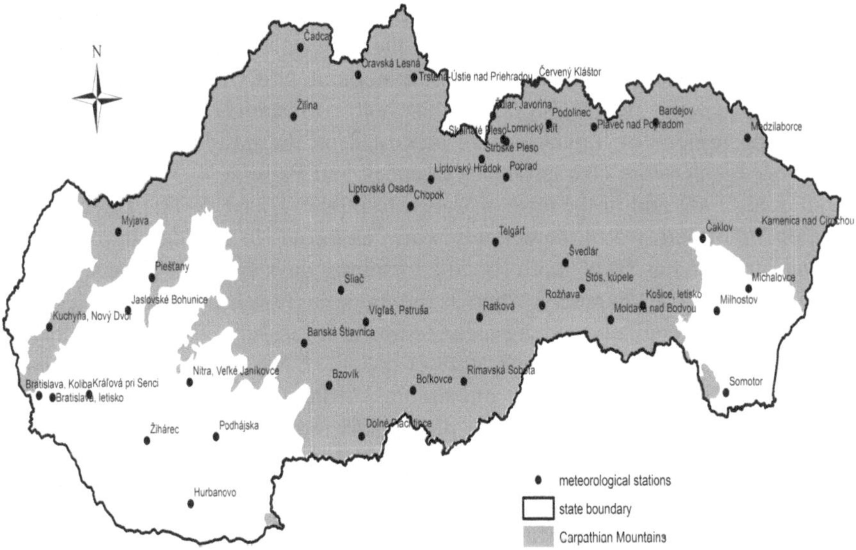

Fig. 1 Location of considered meteorological stations in the Slovak part of the Carpathians and in the neighboring lowlands in Slovakia

2 Data and Methods

The data from the Slovak Hydrometeorological Institute in Bratislava (air temperature, precipitation totals and wind speed) for 47 Slovak meteorological stations have been analyzed. These stations represent the area of the Slovak Carpathians and the neighboring lowlands in Slovakia, with the appropriate spatial and temporal homogeneity (Fig. 1). Several other stations have been included for a more detailed climate analysis of the Tatra Mts. region. Temperature and precipitation measurements in Košice and Liptovský Hrádok have high-quality and longest monitoring series in the Carpathian region, beginning in 1881.

In terms of Konček's and Köppen's climate classifications, climatic regions in Slovakia were specified as follows. The Konček's classification is based on the Konček's moisture index values together with some air temperature characteristics (Konček 1955, 1980; Lapin et al. 2002). The Konček's moisture index represents a simplified water balance characteristic. It is given by the formula:

$$Iz = 0.5R + r - 10T - \left(30 + v^2\right) \tag{1}$$

where R [mm] is the precipitation total in the growing period (April–September), frequently called the warm half-year; r [mm] is the precipitation total exceeding 105 mm on average for winter season (December-February), the negative deviations are excluded; T [°C] is the mean air temperature in the growing period; and v [m s^{-1}] is the mean wind speed measured at 14 h of MLT in the growing period.

The value of $Iz > 120$ indicates a very moist region, Iz ranging between 60 and 120 characterizes a moist region, Iz from the interval 0-60 is typical for a moderately moist region, Iz from −20 to 0 for a moderately dry region, Iz from −40 to −20 relates to a dry region and Iz less than −40 corresponds to a very dry region.

Climatic regions of Slovakia were identified on the basis of the Iz value and further air temperature characteristics (Končes and Petrovič 1957; Konček 1980; Lapin et al. 2002 and in the case of Czech Republic Tolasz 2007). Three regions were distinguished: warm, moderately warm and cold. In the long-term average, the warm region has 50 or more so-called summer days (with daily maximum air temperature ≥25 °C) annually, whereas the moderately warm region has less than 50 summer days annually and the mean temperature in July is 16 °C or higher. In the cold region, mean July temperature is below 16 °C. The warm region has 7 sub-regions (differing in mean air temperatures of January and Iz). The moderately warm region has 7 sub-regions (with differences in the mean air temperatures of January and July, and in Iz). The cold region has 3 sub-regions (all considered very humid and differing in the mean temperature of July, with the limits at 12 and 10 °C—see Lapin et al. 2002). The Konček's moisture index is a suitable indicator of the aridization trends in the land (Melo et al. 2007).

The second method of climate classification employed in this chapter is that based on the Köppen classification scheme. Monthly and annual normals of mean air temperature and precipitation totals represent the input variables in this scheme. Each region is categorized, in a symbolic form, by a sequence of two or three letters. In the Köppen classification of climate, the first letters "A–E" are the dominant (Blüthgen and Weischet 1980; Robinson and Henderson-Sellers 1999; Oliver 2008; Kottek et al. 2006).

Only "C", "D" and "E" climate types occur in the Slovak part of the Carpathians. The "C" climate (warm, temperate, rainy) is defined as one with the average temperature of the coldest month ranging between −3 and 18 °C, and with the average temperature of the warmest month exceeding 10 °C. The additional (second) letter "f" represents climate where no dry season occurs, the "w"—where winter is a dry season and the "s"—that with summer dry season. The "a" indicates that the warmest month has an average temperature over 22 °C, the "b"—the average temperature of the warmest month below 22 °C and at least 4 months warmer than 10 °C and the "c"—only 1–3 months with an average temperature above 10 °C. The "D" climate (cold boreal forest or snow) is similar to "C" but with the average temperature of the coolest month below −3 °C. The average temperature of the warmest month below 10 °C defines the "E" climate, that between 0 °C and 10 °C the "ET" (tundra) climate, and that below 0 °C the "EF" (perpetual frost or polar) climate.

For the analysis and delimitation of climatic zones according to the Köppen scheme, two thirty-year periods were selected: the standard period 1961–1990 and the years 1980–2009 as the latest possible period. Furthermore, the period 1991–2009 was also analyzed. This 19-year period was characterized by unusually warm climatic conditions as compared to any previous normal. Thus, this can indicate possible warmer climate in the Carpathians in the near future.

Climatic models are the most important source of information on the behavior of regional climate under the changed conditions in the global climate system. Therefore, they are generally considered as the best tool for assessing the response of the climate system to changes in the radiative forcing (Melo et al. 2009). The climate change scenarios presented herein are based on the outputs from three different global coupled atmosphere–ocean general circulation models (GCMs), that best characterize the observed climate corresponding to the period 1951–2000 in Slovakia:

- model data from the Goddard Institute for Space Studies in New York (GISS 1998 model under compounded 1 % CO_2 increase experiment with tropospheric sulfate aerosol changes) (Russell and Rind 1999),
- model data from the Canadian Centre for Climate Modeling and Analysis in Victoria, B.C. (CGCM2 model under two SRES emissions scenarios, A2 and B2, and one emissions scenario IS92a) (Boer et al. 1992; McFarlane et al. 1992; Flato and Boer 2001),
- model data from the Met Office Hadley Centre in Exeter, UK (HadCM3 model under emissions scenario IS92a) (Murphy and Mitchell 1995).

For the design of regional climate change scenarios, we have taken into account the model outputs from the nearest GCM gridpoints around Slovakia in the Central European region. The gridpoint resolution at these global GCMs varies between 250 and 300 km. Climatic parameters in individual gridpoints represent therefore some mean values for the area of about 100,000 km^2. The topography of central Europe in these GCMs is simplified, smooth and unrealistic. The Alps and the Carpathians are presented as one flat hill in central Europe without the Danubian Lowland in Western Panonian Basin. Many smaller mountain ranges, individual mountain ridges, small valleys and depressions are invisible in such a topography scheme. The resolution of global climatic models is not sufficient for the detailed regional climatic analysis. As a solution to this problem, the regional GCM output modifications (downscaling) have been performed. Statistical downscaling consists of the development of statistical relationships between locally observed climatic variables and outputs from global GCM experiments (both means and variability). Dynamical downscaling uses a detailed regional meteorological model nested into the global model system. Here the statistically based method recommended by the IPCC and designed for Slovakia by Lapin and Melo (2004) and Lapin et al. (2006) has been applied. This method is based on measured data from meteorological stations in Slovakia in the control period (1951–2000).

Time series of monthly data from the outputs from these models have been used to elaborate and design climate change scenarios for the conditions of strengthened greenhouse effect on the territory of northern Slovakia in the twenty first century. The GCMs provide data representing a sort of spatial averages around the grid points. These data exhibit lower temporal and spatial variability than those recorded at meteorological stations. As the first step, the interpolation of data from four GCMs gridpoints near to Slovakia to Liptovský Hrádok was applied, with weights corresponding to the gridpoint distance from this locality. Downscaling of the model outputs for the selected locality was achieved by the use of measured data from the meteorological station Liptovský Hrádok in the "control" or "reference" period 1951–2000. The means and variability (standard deviations in the case of air temperature, variation coefficients in the case of precipitation totals) of the modified model outputs are then in relatively good accordance with the observed data series. Assuming only insignificant variations in the transforming relation between the measured and modeled data (means and variability), we can also similarly modify the model outputs for the close future (up to 2100). Some modification of this method was also applied in former studies (Lapin and Melo 1999, 2004; Lapin et al. 2001, 2006; Melo 2005).

3 Air Temperature and Precipitation Trends

The global mean surface temperature increased by 0.74 ± 0.18 °C over the last 100 years (1906–2005) as estimated by a linear trend (IPCC 2007). The increase in air temperature in the Slovak part of the Carpathians over the years 1881–2009 is illustrated by the Košice and Liptovský Hrádok time series (Fig. 2). The Košice meteorological station (230 m a.s.l.) is representative for the region of the Košice Basin (southeastern Slovakia) and the Liptovský Hrádok station (640 m a.s.l.) for the Podtatranská Basin (northern Slovakia). Annual means of air temperature were increasing nearly continuously and in phase at both stations over this period. The linear increases of temperature in Liptovský Hrádok and Košice are approximately

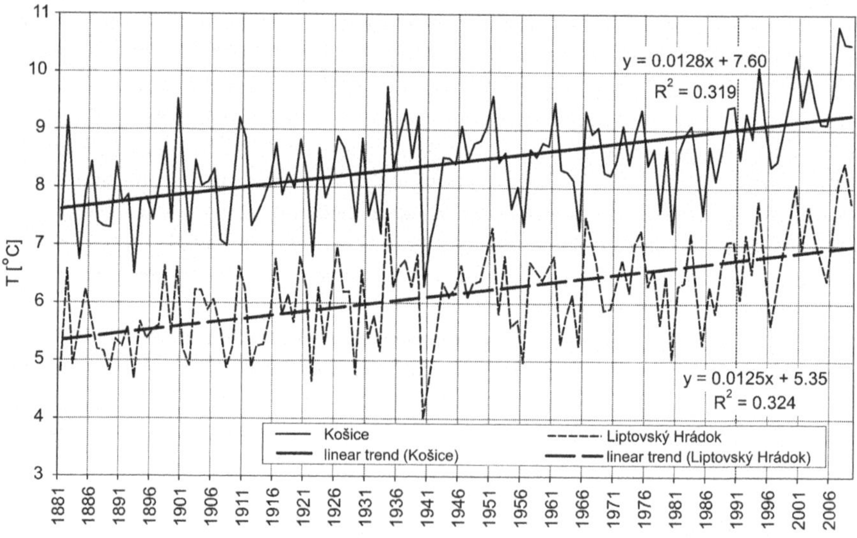

Fig. 2 Mean values of annual air temperature (T) in Košice and Liptovský Hrádok in the period 1881–2009

the same, amounting to about 1.6 °C in the period 1881–2009. The temperature series show an increasing trend in all the seasons of the period 1881–2009, the greatest in winter (2.1 °C for Liptovský Hrádok, 1.9 °C for Košice) and the smallest in autumn (1.1 °C for Liptovský Hrádok, 1.2 °C for Košice). The linear increase of temperature in spring is 1.5 °C for Liptovský Hrádok and 1.7 °C for Košice, whereas in summer it is about 1.7 °C at both stations. Similar increasing trends of mean annual air temperature during the twentieth and at the beginning of the twenty first centuries were found for all other stations, although with shorter record periods. The linear trends of 60-year series are significant at 95 % level if the coefficient of determination $R^2 > 0.0625$ and for 100-year series if $R^2 > 0.0380$ (Lapin and Melo 1999).

The trends of maximum and minimum daily temperatures were also studied. The average series have been analyzed by the linear regression analysis. Specifically, the means of daily air temperature maxima and minima at selected stations in the Slovak part of the Carpathians over the period 1951–2009 (the period 1961–2009 at some stations) were used, from which the corresponding characteristics were calculated for the whole year and each season. The daily range of temperature is defined as a difference between the daily maxima and daily minima. Brázdil et al. (1995) stated that according to the data averaged for 11 representative non-urban stations in the Czech Republic and 14 stations in the Slovak Republic in the period 1961–1992, the mean air temperatures, mean daily maxima and mean daily minima exhibited an increasing trend in winter, spring, summer and over the year, and a decreasing trend in autumn. Most of the stations exhibited a statistically not significant change in mean daily temperature between −0.16 and 0.15 °C per decade, while a significant increasing trend (0.2–0.3 °C per decade) was found for the Slovak mountain station at Lomnický štít (2,635 m a.s.l.). Brázdil et al. (1996) analyzed changes in maximum and minimum daily temperatures in central and southeastern Europe (Germany, central Switzerland, Poland, Czech Republic, Slovakia, Austria, Hungary, Slovenia, lowland of Croatia, Bulgaria). A rate of the linear increase in mean annual maximum daily temperatures in central Europe during the period 1951–1990 was slightly lower than that of mean minimum daily temperatures (0.52 and 0.60 °C per decade, respectively). This was reflected in a small decrease in the daily temperature range by −0.08 °C per decade.

In the Slovak Carpathians, the observed rise in mean temperatures corresponds with the rises in mean maxima and minima. In the period 1951–2009 all stations in the area exhibited an increasing trend in the annual averages of both maximum and minimum daily temperature (Figs. 3, 4). The upward trend has been found also for all seasons with only some exceptions (mainly in autumn at some stations). Whereas changes in the mean daily maxima and minima were more or less concomitant at all stations, greater differences were found in the daily range of air temperature. More pronounced increase in the mean maxima than the mean minima caused the daily range of mean air temperature to increase with time at the following stations: Boľkovce (214 m a.s.l.), Sliač (313 m a.s.l.), Telgárt (901 m a.s.l.), Poprad (695 m a.s.l.), Skalnaté Pleso (1,778 m a.s.l.), Lomnický štít

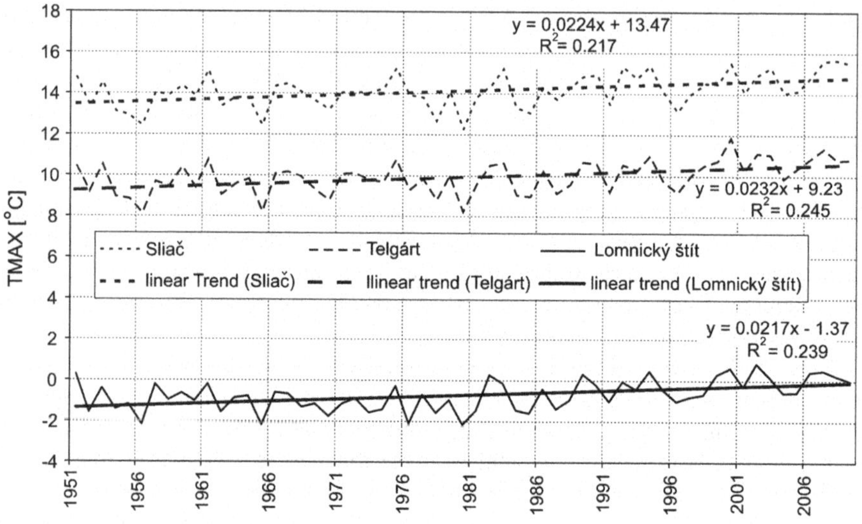

Fig. 3 Annual averages of daily temperature maxima (TMAX) at Sliač, Telgárt and Lomnický štít in the period 1951–2009

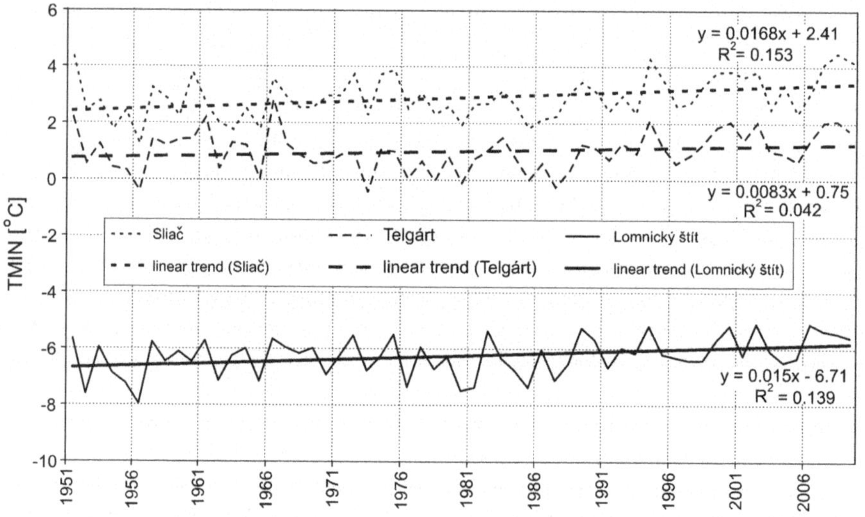

Fig. 4 Annual averages of daily temperature minima (TMIN) at Sliač, Telgárt (the trend is not statistically significant at 95 % level) and Lomnický štít in the period 1951–2009

(2,635 m a.s.l.) and Medzilaborce (308 m a.s.l.) (Fig. 5). Conversely, a decreasing trend of the temperature range, caused by a greater increase in the mean minima than the mean maxima, was recorded at Bratislava-Koliba (286 m a.s.l.), Trstená-Ústie nad Priehradou (598 m a.s.l.), Oravská Lesná (780 m a.s.l.), Liptovský

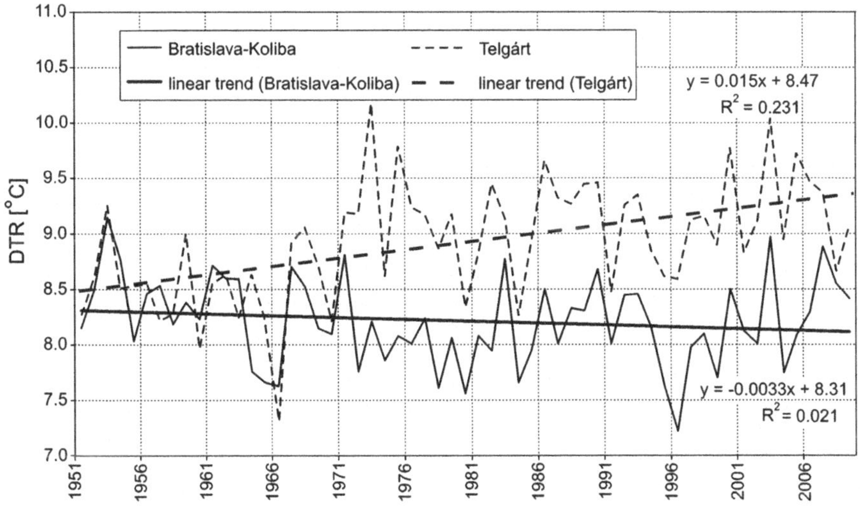

Fig. 5 Annual averages of the daily range of air temperature (DTR) in Bratislava-Koliba (the trend is not statistically significant at 95 % level) and Telgárt in the period 1951–2009

Hrádok (640 m a.s.l.) (Fig. 5). Similar changes in the daily range of air temperature were observed at individual stations also for each season, apart from a few exceptions. The analysis of mean maximum and minimum temperatures in the Slovak Carpathians for the period 1951–2009 does not confirm the decreasing daily range of air temperature in central Europe reported by Brázdil et al. (1996) for the period 1951–1990. The daily range of temperature can be negatively related to changes in cloudiness and air humidity (Lapin and Faško 1994). This can disguise its expected decrease at rising greenhouse effect. In fact, relative humidity and cloudiness decreased significantly over the period 1961–2009 at nearly all stations in Slovakia (up to 5 % in the south), which in turn influenced significantly the daily range of temperature.

The precipitation series (annual precipitation totals) features a slight linear decrease during the period 1881–2009: about 5 % in Košice (this trend is not statistically significant at 95 % level—Fig. 6) and about 10 % in Liptovský Hrádok (Fig. 7). While in Liptovský Hrádok the decreasing trend of precipitation was found for all seasons (from −6 % in winter to −13 % in spring), in Košice it was clear for winter (−9 %), spring (−17 %) and autumn (−17 %), whereas summer was typified by the increasing trend of precipitation totals (+14 %). Generally, the precipitation trends at stations in the Slovak Carpathians are rather different in direction in comparison with the trends of air temperature. For example, the northwestern part of the Slovak Carpathians (the region of Orava) was typified by an upward trend of precipitation during the twentieth century and at the beginning of the twenty first century.

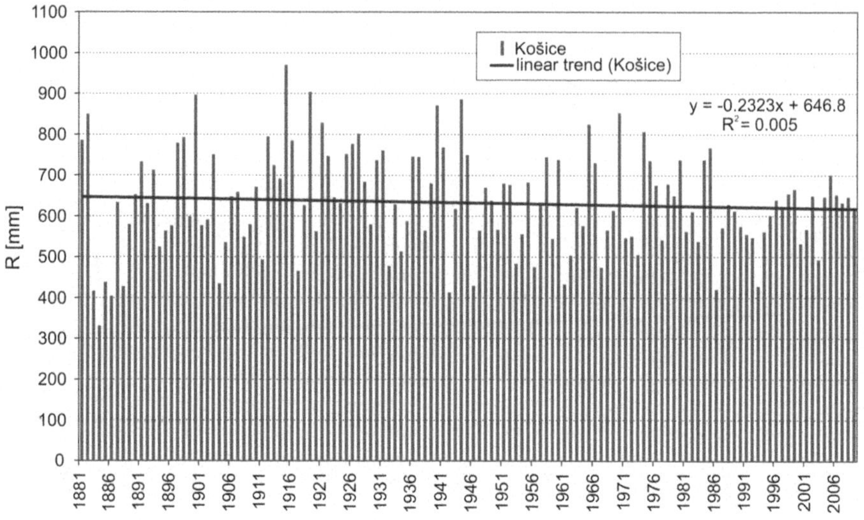

Fig. 6 Annual precipitation totals (R) in Košice (the trend is not statistically significant at 95 % level) in the period 1881–2009

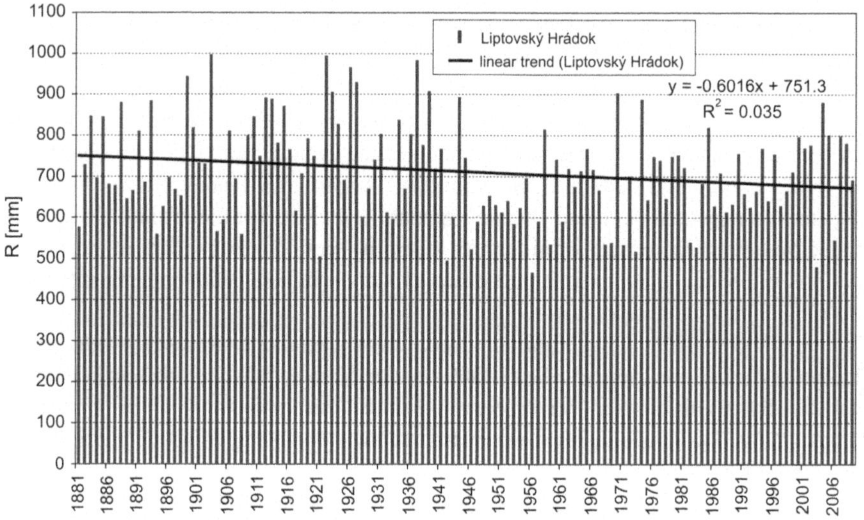

Fig. 7 Annual precipitation totals (R) in Liptovský Hrádok in the period 1881–2009

4 Spatial and Temporal Climate Trends Based on the Končekʼs Climate Classification

Climatic regions and sub-regions of Slovakia based on the Končekʼs climate classification for the periods 1901–1950 and 1961–1990 were presented by Konček (1980) and Lapin et al. (2002). These analyses showed some territorial shifts in climate regions and sub-regions in the Slovak part of the Carpathians during the twentieth century. During this period the aridization trends were detected mainly in the warmer, southern part of the Slovak Carpathians (Southern Slovak Basin, Košice Basin). Even new climatic sub-regions appeared here in the years 1961-1990: warm, moderately dry sub-region with cool winter in the Košice Basin (previously this area was a warm, moderately humid sub-region with cool winter), warm, dry sub-region with cool winter in the central and southeastern parts of the Southern Slovak Basin (previously it was a warm, moderately dry sub-region with cool winter), warm, dry sub-region with mild winter in the southwestern part of the Southern Slovak Basin (previously it was a warm, moderately dry sub-region with mild winter) and warm, moderately dry sub-region with mild winter in the northwestern part of the Southern Slovak Basin (previously it was a warm, moderately humid sub-region with cool winter). Climate has become warmer and more arid mainly in the adjacent lowlands (Danubian Lowland, Eastern Slovak Lowland). Generally, the Slovak Carpathians exhibit great variation in climatic sub-regions over a short distance due to altitudinal differences and complex topography conditions. More shifts of climatic regions and sub-regions towards higher altitudes were registered here during this period (1901–1950 and 1961–1990); for further details, see Konček (1980) and Lapin et al. (2002).

These shifts have continued at the beginning of the twenty first century. Figures 8 and 9 illustrate changes in climate types distinguished according to the Končekʼs classification in the southeastern part of the Tatra Mts. and the Poprad Basin between the periods 1951–1980 and 1980–2009. While in the first period only the cold region (with 3 sub-regions: C1—moderately cool, C2—cool mountainous and C3—cold mountainous) was recorded here, in the years 1980–2009 the extent of moderately warm region (sub-region M3—moderately humid) increased encompassing the Poprad Basin. The boundary between the moderately warm and the cold climatic regions shifted about 150 m higher. Moderately cool sub-region (C1) shifted about 100 m higher and cool mountainous sub-region (C2) about 155 m higher between the considered periods (Figs. 8, 9). Such a change was caused by the significant increase in the mean temperature of July between these two periods: from 15.5 to 16.4 °C in Poprad (695 m a.s.l.), from 9.5 to 10.4 °C at Skalnaté Pleso (1,778 m a.s.l.) and from 3.6 to 4.5 °C at Lomnický štít (2,635 m a.s.l.). As a result of higher temperature, we can expect the upward shift of tree line in the high-mountain part of the Carpathians in the near future.

From the values of Končekʼs moisture index (*Iz*) at some Slovak Carpathian stations for individual 21-year periods, from 1951–1971 to 1989–2009 (Fig. 10), it is

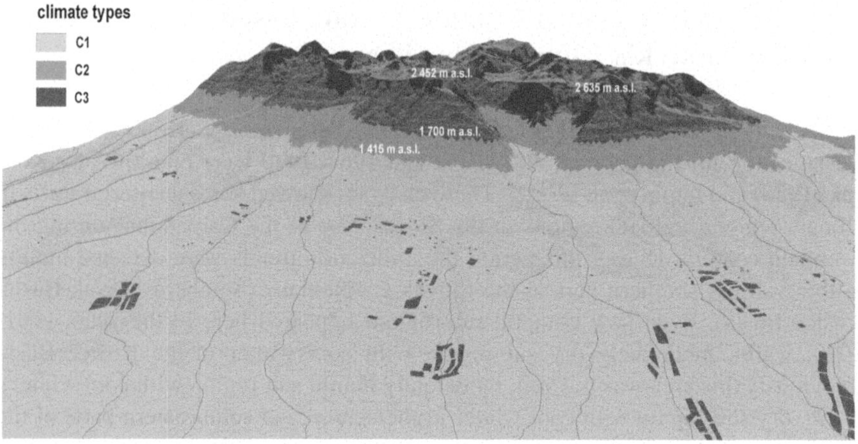

Fig. 8 Climate types according to the Konček's classification in the southeastern part of the Tatra Mts. and the Poprad Basin in the period 1951–1980

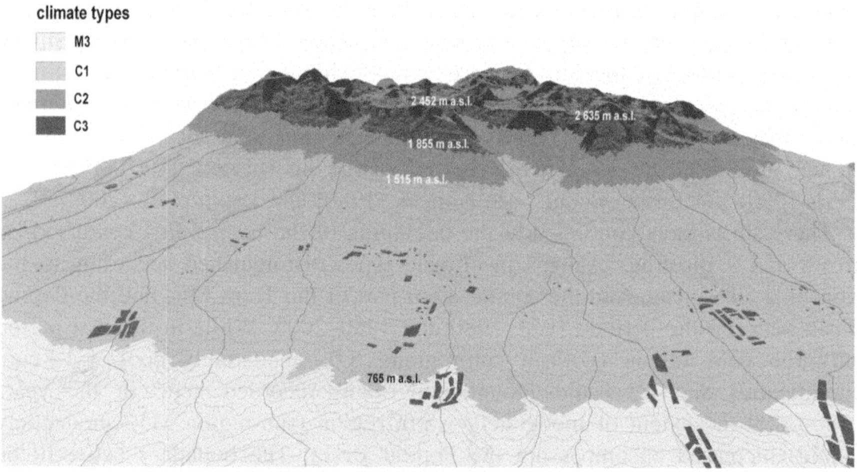

Fig. 9 Climate types according to the Konček's classification in the southeastern part of Tatra Mts. and the Poprad Basin in the period 1980–2009

evident that the stations at lower altitudes (in the southern part of the Slovak Carpathians) are remarkably more arid and the stations at higher altitude (in the northern part of the Slovak Carpathians) are slightly more humid in the same periods. While Boľkovce (214 m a.s.l.) is typified by dry and moderately dry climate and Bratislava-Koliba (286 m a.s.l.) by moderately moist and moderately dry climate, Oravská Lesná (780 m a.s.l.) exhibits very moist climate (cold region,

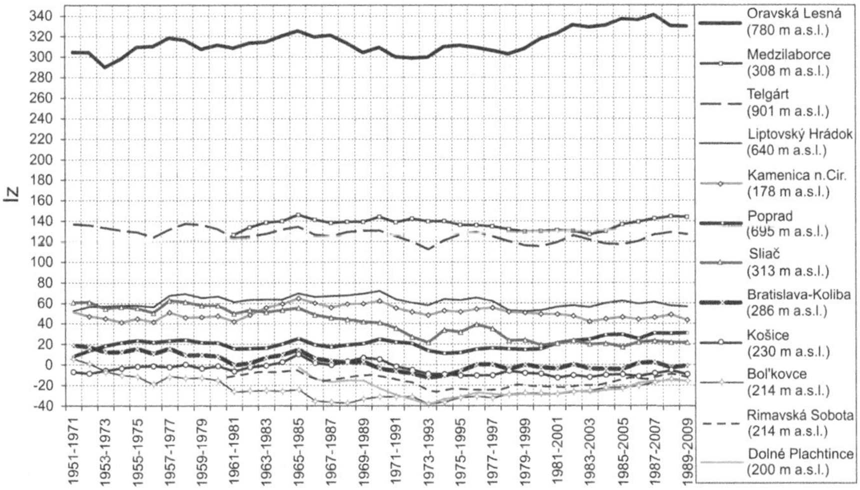

Fig. 10 Values of the Konček's moisture index (*Iz*) for some meteorological stations in the Slovak part of the Carpathians for individual 21-year moving periods from 1951–1971 to 1989–2009

moderately cool sub-region) in the same period (1951–2009). However, the Poprad station (695 m a.s.l.), located at relatively high altitude, is typified by a relatively low value of *Iz* (moderately moist climate) reflecting the lee effect of the Tatra Mts.

Temporal climate trends based on the Konček's moisture index in the years 1951–2009 (Fig. 10) show that the climate of the northern part of the Slovak Carpathians has become more humid during this period (mainly because of increased precipitation). In the central part of the Slovak Carpathians, warm, high-precipitation conditions lasted until the mid-1970s, while in the southern part such conditions ended yet in the mid-1940s. After the mid-1940s in the southern part and after the mid-1970s in the central part, warmer, low precipitation period started, culminating in a dry and very warm period between 1988 and 1994. At present warmer conditions with increasing (in particular convective) precipitation are typical of the Slovak Carpathians.

The value of *Iz* does not provide a complete picture of the actual changes of climate in Slovakia (especially in the southern part) in recent years (after 1994) because the precipitation regime has changed remarkably over time. Annual precipitation totals are slightly higher compared to the previous periods. However, there are fewer days with low precipitation on one hand and on the other precipitation intensity is higher. The most precipitation outflows during short rainy events and hence it does not significantly influence the soil moisture in this region. Therefore, it seems appropriate to modify the Konček's moisture index in the future to account for these new facts.

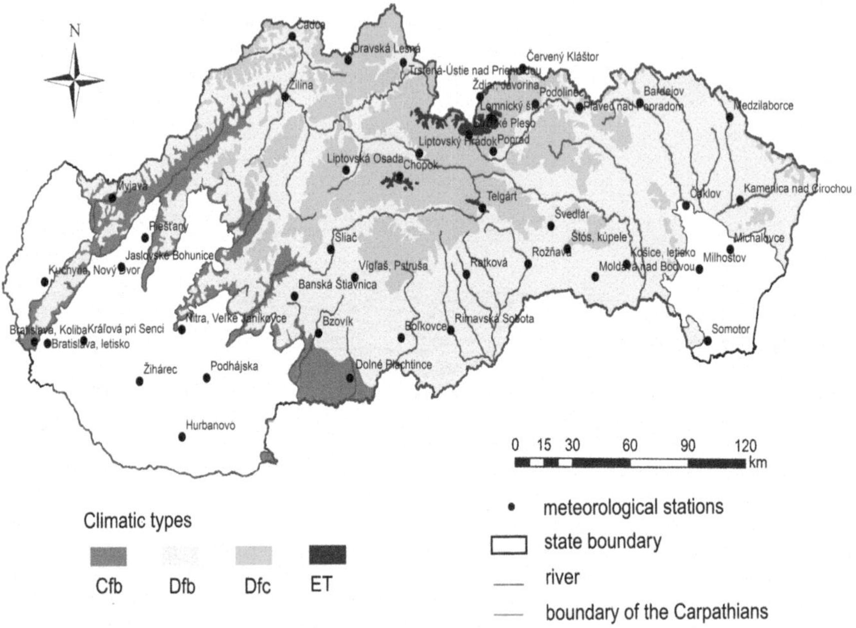

Fig. 11 Climatic regions according to the Köppen's classification in the Slovak part of the Carpathians in the period 1961–1990

5 Spatial and Temporal Climate Trends Based on the Köppen's Climate Classification

Thirty-one subtypes in the Köppen's climate classification have been recognized (Kottek et al. 2006), of which only four occur in the Slovak Carpathians: Cfb, Dfb, Dfc, ET. Figures 11, 12, 13 show the delimitation of climatic zones in this region according to the Köppen's classification in three different periods: 1961–1990, 1980–2009 and 1991–2009. In the years 1961–1990 most of the Slovak Carpathians featured the cold boreal forest climates Dfb and Dfc (Fig. 11). The warm temperate and humid zone Cfb occurred only in the southwestern part of the mountains with the lowest altitudes. On the contrary, the cold zone ET could be found only in the highest parts of the Carpathians (approximately above 1,700 m a.s.l. in the Tatra Mts. and the Low Tatra Mts.). In the recent 30-year period (1980–2009), the upward shifts of climatic zones are evident (Fig. 12). Furthermore, the warm temperate and humid zone Cfb expanded towards the eastern part of the Slovak Carpathians. In the years 1991–2009, extension and additional upward shifts of climatic zones were recorded (Fig. 13). In this period the mountain tundra climate ET remained only in the highest peaks of the Carpathians (the Tatra Mts. and the Low Tatra Mts.). For example, while the Skalnaté Pleso station (1,778 m a.s.l.) exhibited the ET climate during the period 1961–1990, in

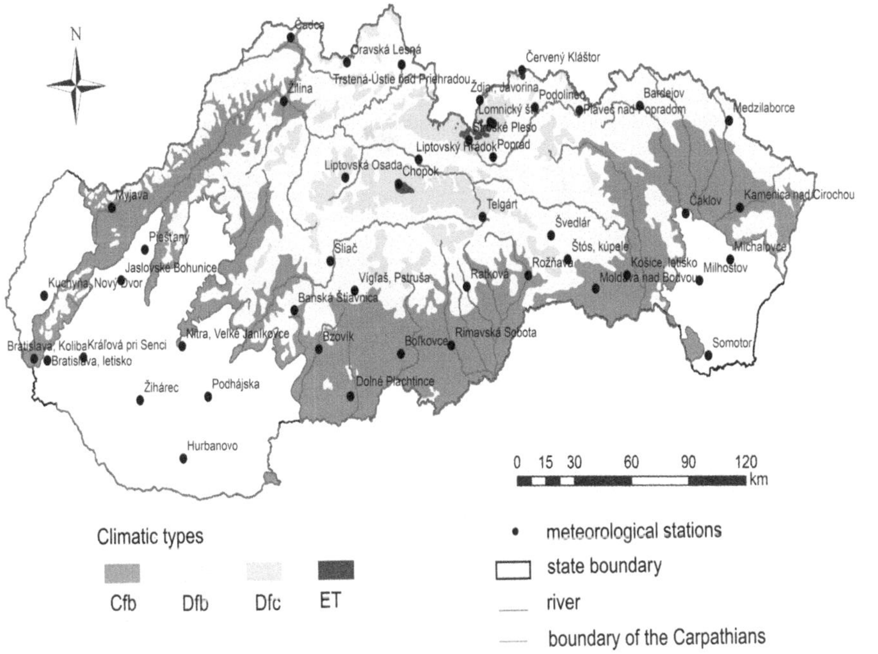

Fig. 12 Climatic regions according to the Köppen's classification in the Slovak part of the Carpathians in the period 1980–2009

both recent periods 1980–2009 and 1991–2009 the warmer Dfc climate was established. This reflected the increase of mean air temperature in July and August from 9.3 and 9.4 °C in the years 1961–1990 to 10.4 °C for both months in the years 1980–2009 and 10.9 and 10.7 °C, respectively, between 1991 and 2009. It is known that the value of T = 10 °C in the warmest month is an important boundary between the climatic regions in the Köppen's classification.

In general, the largest shifts recorded between the periods 1961–1990, 1980–2009 and 1991–2009 were those from the climatic region Dfc to Dfb and from Dfb to Cfb. Figures 11, 12, 13 indicate an increase in the proportion of Cfb climate and a decrease in that of Dfc climate within the territory of the Slovak Carpathians over the three considered periods.

6 Climate Change Scenarios

In this section, the climate change scenarios designed in the form of time series of air temperature and precipitation totals for Liptovský Hrádok over the twenty first century are presented. Based on all GCMs outputs, we note that the temperature in Liptovský Hrádok will continue to increase during the whole century. The

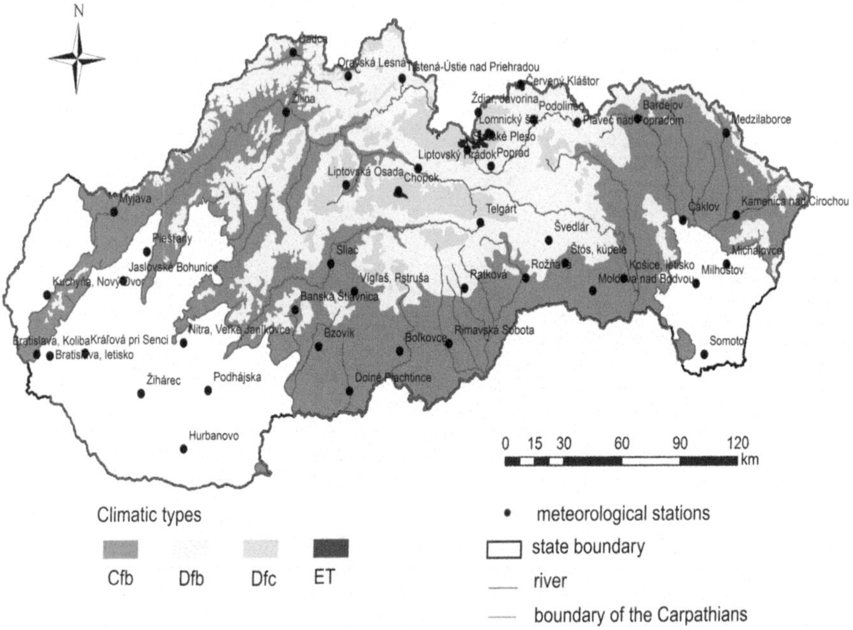

Fig. 13 Climatic regions according to the Köppen's classification in the Slovak part of the Carpathians in the period 1991–2009

temperature growth according to the U.S. GCM model GISS 1998 is the lowest. On the contrary, the most pronounced air temperature increase is anticipated by the UK HadCM3 model. A temperature increase is also predicted by an optimistic new IPCC B2-SRES emission scenario (with an expected decrease of GHGs emission to 5.2 Gt C in 2100), although it is lower than predicted by the older IPCC IS92a scenario for the same Canadian model (CGCM2). The forecast of pessimistic IPCC A2-SRES emission scenario (with an expected increase of GHGs emission to 28.9 Gt C in 2100) is similar to IS92a based on the same model CGCM2 (Fig. 14).

In the case of precipitation, the results obtained from these climate models are rather different (Fig. 15). In the twenty first century we can expect a decrease in annual precipitation totals in Liptovský Hrádok according to GISS 1998, HadCM3 and CGCM2 (A2-SRES) models, and some increase according to CGCM2 (B2-SRES) and CGCM2 (IS92a) models. Significantly large and opposite trends are expected in different seasons of the year. The amount of seasonal precipitation is very likely to increase in winter (especially in the northern part of the Slovak Carpathians), while it is expected to decrease in summer (particularly in the southern part of the region). These results are similar to the IPCC findings. In the twenty first century annual precipitation totals are very likely to increase in most of the northern half of Europe and decrease in most of the Mediterranean area. In central Europe, precipitation is likely to increase in winter but decrease in summer

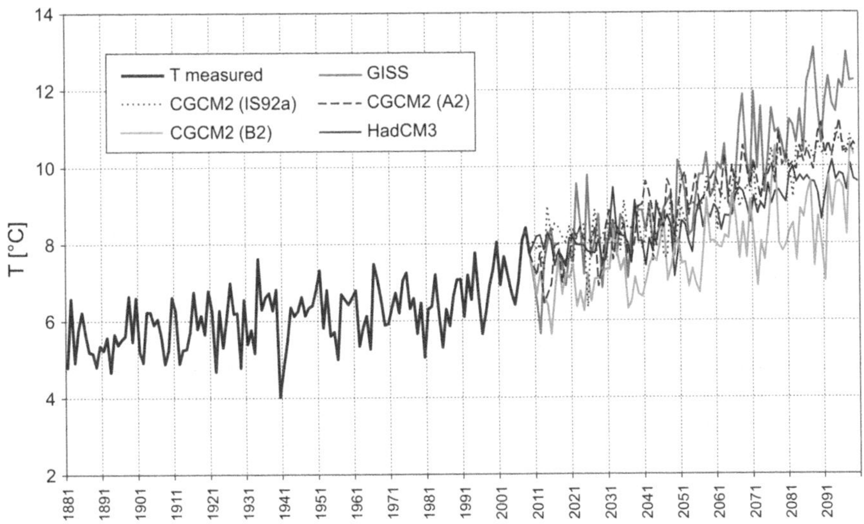

Fig. 14 Mean values of annual air temperature (T) in Liptovský Hrádok measured in the period 1881–2009 and projected for the period 2010–2100 according to the scenarios based on different GCMs (after modification)

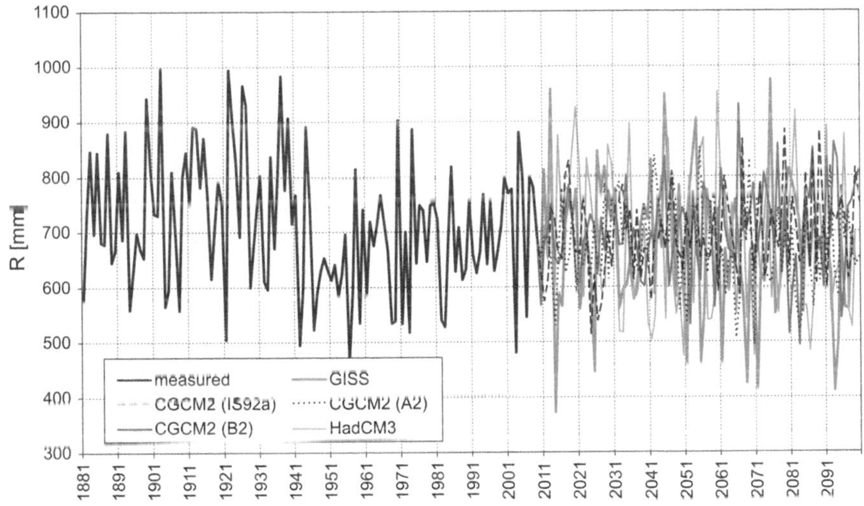

Fig. 15 Annual precipitation totals (R) in Liptovský Hrádok measured in the period 1881–2009 and projected for the period 2010–2100 according to the scenarios based on different GCMs (after modification)

(IPCC 2007). Climate scenarios for Slovakia based on the GCMs downscaling and statistical modification (using measured data in the control period, Lapin and Melo 2004) are also in good agreement with the current predictions of climate change in

Europe based on the regional modeling efforts under the project PRUDENCE (Bartholy et al. 2007). New elaboration of regional general circulation models (RCMs) for Slovakia undertaken in 2011 (KNMI and MPI RCMs based on German ECHAM5) confirmed the above-mentioned climate evolution.

7 Conclusions

Climate is changing very rapidly in the Slovak part of the Carpathians. Evidently, it has become warmer and more arid in the southern part of the Slovak Carpathians and mainly in the adjacent lowlands (e.g. the Danubian Lowland), while the northern part (the Orava Region) has become warmer and more humid (in terms of precipitation totals) during the twentieth century and at the beginning of the twenty first century. Other notable shifts in climatic regions and sub-regions towards the higher altitudes and to the north were registered in Slovakia during this period as well.

Mountains are an important part of the Earth's system. Climate change affects all landscape components. Responses are different in various timescales, with fauna responding first and followed by vegetation. However, species usually migrate rather slowly. For instance, surveys of plant species on mountain peaks in the Alps indicate an upward migration of alpine plants by 1–4 m per decade during the twentieth century (Barry and Chorley 2003). As a result of higher temperature, we can expect similar upward shifts of vegetation zones and tree lines also in the Carpathians. Scenarios based on the three adopted climate model outputs show further warming on this territory approximately from 2 to 4 °C for the period 2071–2100 compared to the 1951–1980 normal. The expected change of annual precipitation totals is different in this region depending on the model used. However, the amount of seasonal precipitation is very likely to increase in winter and to decrease in summer. We anticipate that in the next decades the Pannonian migration route of fauna and flora (xerophile and thermophile species) from the Mediterranean refuges to the Western Panonian Basin and also to the southern and southwestern parts of the Slovak Carpathians will increase in significance. The results of the present study are in good accordance with the IPCC (2007) findings and confirm the previous indications that the Carpathians region is one of the most vulnerable to the future climate change.

Acknowledgments This work was supported by the Grant Agency of the Slovak Republic under the Project VEGA No. 1/0063/10 and by the Slovak Research and Development Agency under the contract No č. APVV–0015-10. In this chapter, we used model data from the Canadian Centre for Climate Modeling and Analysis in Victoria, B.C., from the Goddard Institute for Space Studies in New York, from the Met Office Hadley Centre in Exeter, UK and measured data from the Slovak Hydrometeorological Institute in Bratislava.

References

Barry RG, Chorley RJ (2003) Atmosphere, weather and climate. Routledge London, New York
Bartholy J, Pongracz R, Gelybo GY (2007) Regional climate change expected in Hungary for 2071–2100. Appl Ecol Environ Res 5:1–17
Blüthgen J, Weischet W (1980) Allgemeine Klimageographie. Walter de Gruyter, New York
Boer GJ, McFarlane NA, Lazare M (1992) Greenhouse gas-induced climate change simulated with the CCC second-generation general circulation model. J Clim 5:1045–1077
Brázdil R, Budíková M, Faško P, Lapin M (1995) Fluctuation of maximum and minimum air temperatures in the Czech and the Slovak Republics. At Res 37:53–65
Brázdil R, Budíková M, Auer I, Böhm R, Cegnar T, Faško P, Lapin M, Gajic-Čapka M, Zaninovič K, Koleva E, Niedzwiedz T, Ustrnul Z, Szalai S, Weber RO (1996) Trends of maximum and minimum daily temperatures in Central and Southeastern Europe. Int J Climatol 16:765–782
Flato GM, Boer GJ (2001) Warming asymmetry in climate change simulations. Geophys Res Lett 28:195–198
IPCC (2007) Climate change 2007: the physical science basis. Contribution of Working Group I to the Fourth Assessment Report of the IPCC
Konček M (1955) Index zavlaženia. Meteorol Zpr 8:96–99
Konček M (1964) Sketch of climatic conditions in Slovakia. Geogr Čas 16:160–184
Konček M (1980) Climatic regions. In: Landscape atlas of the Slovak Republic. Ministry of Environment of the Slovak Republic, Slovak Environmental Agency, Bratislava-Banská Bystrica
Konček M, Petrovič Š (1957) Klimatické oblasti Československa. Meteorol Zpr 10:113–119
Kottek M, Grieser J, Beck C, Rudolf B, Rubel F (2006) World Map of the Köppen-Geiger climate classification updated. Meteorol Zeitschr 15:259–263
Lapin M, Faško P (1994) Daily air temperature range and relative air humidity correlation at Hurbanovo. In: Brázdil R, Kolář M (eds) Contemporary climatology. Masarykova Univerzita, Brno
Lapin M, Melo M (1999) Climatic changes and climate change scenarios in Slovakia. Meteorol Čas 2(4):5–15
Lapin M, Melo M (2004) Methods of climate change scenarios projection in Slovakia and selected results. J Hydrol Hydromech 52:224–238
Lapin M, Damborská I, Melo M (2001) Downscaling of GCM outputs for precipitation time series in Slovakia. Meteorol Čas 4(3):29–40
Lapin M, Faško P, Melo M, Šťastný P, Tomlain J (2002) Climatic regions. In: Landscape atlas of the Slovak Republic. Ministry of Environment of the Slovak Republic, Slovak Environmental Agency, Bratislava-Banská Bystrica
Lapin M, Damborská I, Drinka R, Gera M, Melo M (2006) Scenarios of climatic element daily values for Slovakia until 2100. Meteorol Čas 9.149–156
McFarlane NA, Boer GJ, Blanchet JP, Lazare M (1992) The Canadian climate centre second-generation general circulation model and its equilibrium climate. J Clim 5:1013–1044
Melo M (2005) Warmer periods in the Slovak mountains according to analogue method and coupled GCM. Croat Meteorol J 40:589–592
Melo M, Lapin M, Damborská I (2007) Končekov index zavlaženia ako možný indikátor aridizačných trendov v krajine. Acta Hydrol Slov 8:224–230
Melo M, Lapin M, Damborská I (2009) Methods for the design of climate change scenario in Slovakia for the 21st century. Bull Geogr Phys Geogr Ser 1:77–90
Murphy JM, Mitchell JFB (1995) Transient response of the Hadley Centre coupled ocean-atmosphere model to increasing carbon dioxide. Part II: spatial and temporal structure of response. J Clim 8:57–80
Oliver JE (ed) (2008) Encyclopedia of world climatology. Springer, Dordrecht

Robinson PJ, Henderson-Sellers A (1999) Contemporary climatology. Pearson Education Limited, Harlow

Russell GL, Rind D (1999) Response to CO_2 transient increase in the GISS coupled model: regional coolings in a warming climate. J Clim 12:531–539

Tolasz R (ed) (2007) Climate Atlas of Czechia. Univerzita Palackého, Olomouc, ČHMÚ Praha

Natural Factors Affecting the Chemical Composition of Water in the Catchment of Wołosatka Stream (High Bieszczady Mts.)

Janusz Siwek, Bartłomiej Rzonca, Barbara Jaśkowiec, Joanna Plenzler and Eliza Płaczkowska

Abstract The aim of this study was to identify the natural factors determining spatial differences in spring water chemistry in the flysch Carpathians using the Wołosatka catchment (High Bieszczady Mountains, SE Poland) as an example. A Principal Component Analysis (PCA) was used to find the regularity in the variance of common and biogenic ion concentrations among the surveyed springs. The PCA identified three factors that altogether explain 85 % of the variance. The first factor explained 51 % of the variance and was best correlated with concentration of Ca^{2+}, Mg^{2+} and HCO_3^-. Generally, the fundamental factor shaping the ions concentration is the lithology of aquifer. The second factor, correlated mainly with the concentration of SO_4^{2-} and Cl^-, seems to explain the variation in water chemistry resulting from varying elevation of springs. The elevation controls climate conditions and plant communities in the alimentation area. Generally, higher concentrations of sulfates are characteristic of the upper part of the catchment, while the highest concentrations of chlorides are characteristic of springs located on the valley floors. The third factor, reflecting mainly the concentration of nitrates, seems to be related to the role of shallow groundwater circulation in the alimentation of springs after rainfall events.

1 Introduction

The chemical composition of water in a catchment is shaped by an array of factors, the most important of which is usually the bedrock lithology. Yet, even in the catchments with a fairly uniform lithology, spatial differences in the chemistry of spring water can be observed. Such differences arise from different types of plant cover, relief, snow cover depths as well as water circulation patterns at different

J. Siwek (✉) · B. Rzonca · B. Jaśkowiec · J. Plenzler · E. Płaczkowska
Institute of Geography and Spatial Management, Jagiellonian University, Gronostajowa 7 30-387 Kraków, Poland
e-mail: j.siwek@uj.edu.pl

J. Kozak et al. (eds.), *The Carpathians: Integrating Nature and Society Towards Sustainability*, Environmental Science and Engineering,
DOI: 10.1007/978-3-642-12725-0_11, © Springer-Verlag Berlin Heidelberg 2013

elevations (Sickman et al. 2001; Mouser et al. 2005; Rzonca et al. 2008; Żelazny et al. 2011). The purpose of this chapter is to identify the principal natural factors determining spatial differences in spring water chemistry in the flysch catchment of Wołosatka Stream (High Bieszczady Mountains, Polish Carpathians).

2 Study Area

The High Bieszczady Mountains are the highest range within the Polish part of the Eastern Carpathians. Hydrochemical research was conducted in the upstream part of the Wołosatka Stream catchment (9.5 km^2) in the Bieszczady National Park (Fig. 1). Currently, the area is under little human pressure and, for that reason, it is a good place to study natural factors affecting the chemical composition of spring water in the flysch Carpathians.

The catchment is underlain by Oligocene flysch formations of the Silesian Nappe. The principal constituents of the bedrock are thick-bedded, medium- to coarse-grained Otryt Sandstones and so-called fine-rhythmic flysch, consisting of thin and thick sandstone beds and clay-marl shales. The flysch formations are deformed into NW-SE-running folds. Faults are an important structural feature in the area. The largest faults have a drop of about 1,000 m and run in NE-SW direction (Tokarski 1975; Haczewski et al. 2007).

Low overall permeability of the flysch results in the occurrence of numerous small springs and wetlands. A 2007 hydrological survey in the study area revealed 227 natural outflows of groundwater, mostly in the form of springs and bog-springs. The majority of the outflows were small (under 0.5 $dm^3\ s^{-1}$). However, several larger outflows with yields exceeding 1 $dm^3\ s^{-1}$ were also recorded (Rzonca et al. 2008; Siwek et al. 2009). The total groundwater runoff from the catchment was estimated by Plenzler et al. (2010) at 4.8 $dm^3\ s^{-1}\ km^{-2}$).

Despite the relatively uniform geology and small area, the catchment is rather diverse in terms of physical and chemical characteristics of spring water (Table 1). Total mineralization of water (both groundwater and surface water) ranges from 30 to 315 mg dm^{-3} and its pH varies between 6.8 and 8.2 (Siwek and Rzonca 2009). The most common water type is "three-ion water" with HCO_3^-, Ca^{2+}, and Mg^{2+}. Four-ion water is also found at a number of locations, where the fourth ion is sulfate (Rzonca and Siwek 2009).

Table 1 Spatial variation in the chemical composition of water among 79 springs and bog-springs in the catchment of upper Wołosatka Stream (after Rzonca and Siwek 2009)

	Ca^{2+}	Mg^{2+}	Na^+	K^+	HCO_3^-	SO_4^{2-}	Cl^-	NO_3^-	SC_{25}
Minimum	5.5	1.2	0.8	0.2	15.7	2.1	0.2	0.16	46
10th percentile	14.5	2.5	1.1	0.4	39.9	8.8	0.3	0.59	97
Median	26.4	5.5	1.5	0.7	89.1	14.7	0.4	1.82	170
90th percentile	37.5	9.2	2.7	0.9	142.7	19.8	0.5	3.48	234
Maximum	56.1	13.8	4.5	1.9	241.4	24.1	1.4	5.25	322

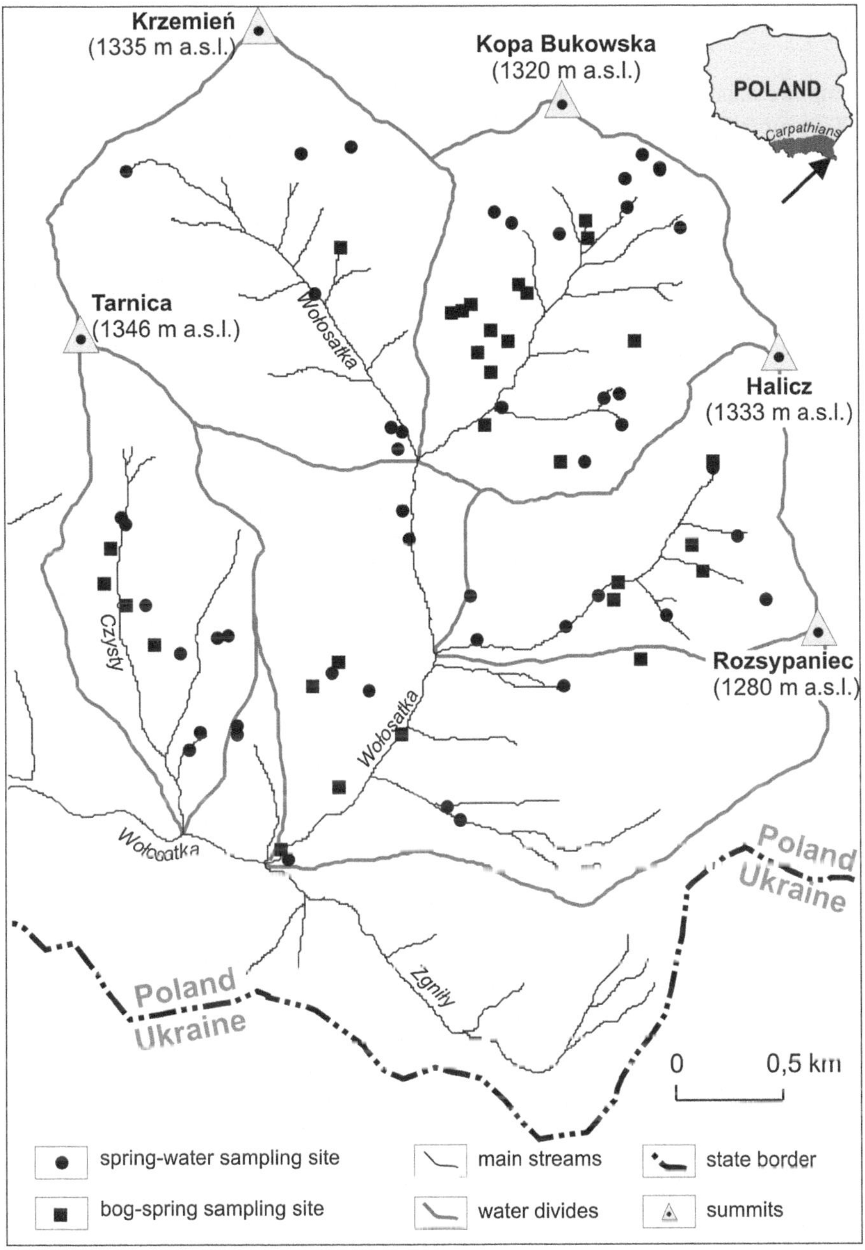

Fig. 1 The study area and the types of sampling sites

3 Study Methods

Sampling was conducted in three periods: Sept. 2–3, Oct. 1–3, and Oct. 16–17, 2007 with different weather conditions. The total precipitation in 10 preceding days ranged from 6.8 mm (Sept. 2–3) to 30.3 mm (Oct. 1–3). Investigations in the north-eastern part of the study area (Oct. 1–3) were conducted during a period of higher soil moisture and lower air temperatures relative to the conditions accompanying investigations in the remaining study area. Seventy-nine sites, representing springs and bog-springs, were analyzed. Spring discharge, temperature, specific conductivity (SC_{25}) and pH of water were measured in the field. Water samples were taken to a laboratory, where the content of several common ions (Ca^{2+}, Mg^{2+}, K^+, Na^+, HCO_3^-, SO_4^{2-}, Cl^-) and biogenic ions (NH_4^+, NO_2^-, NO_3^-, PO_4^{3-}) was analyzed with Dionex ICS-2000 ion chromatograph.

A Principal Component Analysis (PCA) was used to identify independent factors explaining the spatial variability in ion concentrations. The PCA was conducted on the base of normalized results of the hydrochemical analyses. The principal components with eigenvalues greater than one were only retained. The parameters of springs such as elevation, discharge, neighboring plant community and spring type were used as independent variables in the interpretation of factor scores. A similar approach to hydrochemical interpretations of spring waters on Reunion Island was presented by Join et al. (1997). Siwek and Chełmicki (2004) used PCA factor scores to identify the geology- and land use-related spatial patterns of spring water quality in the Małopolska Upland, Poland. In turn, Karimi et al. (2005) used a PCA based on hydrochemical and isotopic characteristics of springs to assess the spatial variability of karst aquifer in Iran.

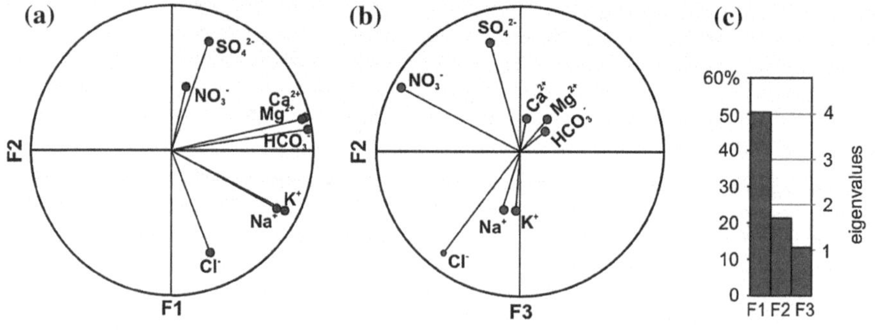

Fig. 2 Results of a principal component analysis performed on spatial variation in the chemical composition of spring and bog-spring water in the Wołosatka Stream catchment. (**a**, **b**) Correlation circles identifying the positions of concentrations of particular ions in relation to the first three PCA axes. (**c**) Histogram of eigenvalues

4 Results

The PCA identified three factors that together explain 85 % of the variance in the ion concentrations among the water samples analyzed. The main principal factor (F1) explained 51 % of the variance (Fig. 2c) and was positively correlated with the concentrations of Ca^{2+}, Mg^{2+}, HCO^{3-}, Na^{+} and K^{+} (Fig. 2a). Positive scores of this factor were characteristic of springs from the south-western part of the catchment, while the negative scores mainly of those from the north-eastern part (Fig. 3—F1).

The second factor (F2) explained 21 % of the variance (Fig. 2c) and was positively correlated with concentrations of SO_4^{2-} and negatively with those of Cl^- (Fig. 2b). In general, high positive scores of this factor were characteristic of

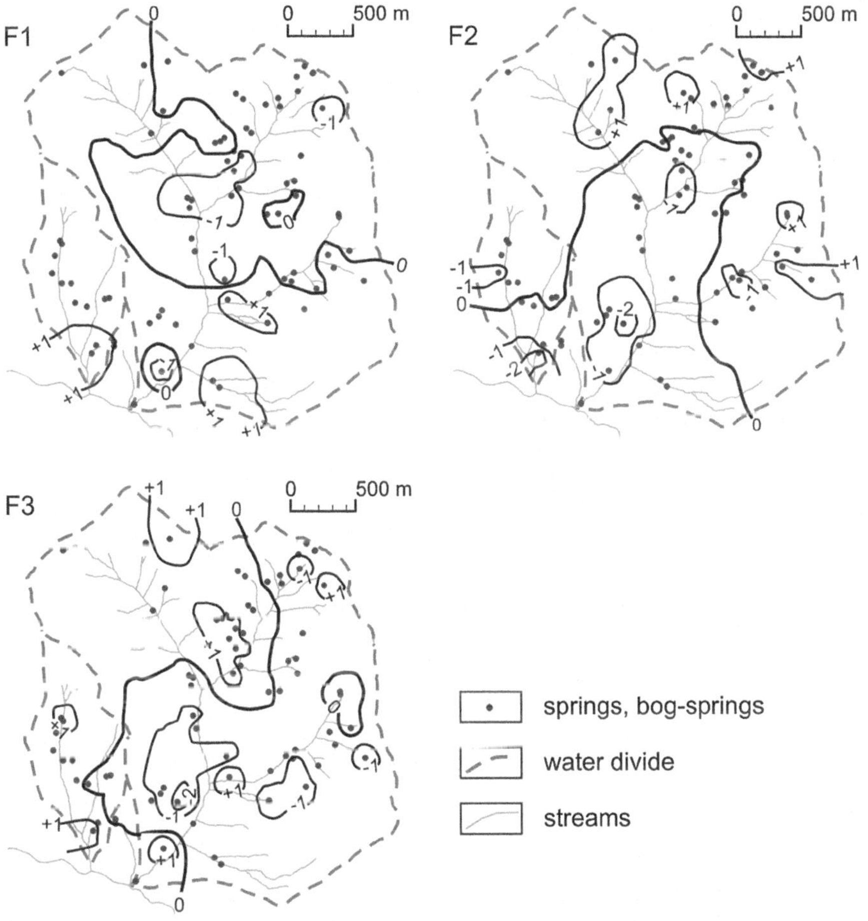

Fig. 3 Spatial patterns of the interpolated values of PCA factor scores within the Wołosatka Stream catchment

the sites located within the upstream part of the catchment, while most springs located on the valley floors usually exhibited negative scores (Fig. 3—F2).

The third factor (F3) explained almost exclusively the variance of one parameter—concentration of nitrates. To some extent it was also correlated with concentration of chlorides (Fig. 2b).

5 Discussion

The first factor (F1) of the PCA reflected the fundamental influence of the bedrock material on water chemistry in the catchment, which is strongly related to the duration of groundwater circulation. The highest values of the factor are characteristic of the springs supplied with water of the longest circulation time. Generally, the positive values of the first factor were characteristic of the western and southern parts of the catchment where water circulates in accordance to the direction of flysch strata. In these parts of the catchment, high mineralization of water is most probable to occur because the geological setup prolongs the contact time of water with the bedrock. In contrast, groundwater circulation on the eastern slopes is determined by the discordance of bedrock structure and relief. The drainage system in this region is highly dispersed, which is determined by the alternation of partially weathered, water-bearing sandstones and impermeable shales of fine-rhythmic flysch. Very low storage capacity of those systems leads to short residence time of groundwater and, thus, to lower mineralization of spring water than in the remaining parts of the catchment.

Interesting conclusions can be drawn from the interpretation of two other factors identified, even though the latter explain a relatively small proportion of the total variance. The positive scores of the second factor (F2) typified sites located in the upper part of the catchment, while sites located on the valley floor usually exhibited negative scores (Fig. 3—F2). As this factor is positively correlated with concentrations of SO_4^{2-} and negatively with those of Cl^-, it means that sulfate concentrations increase and chloride concentrations decrease with increasing elevation of springs. This phenomenon could be generally explained by a greater influence that fresh meteoric water exerts on the chemical composition of water in the springs situated in the upper parts of the catchment. In the Bieszczady Mts. annual precipitation totals increase with elevation at a rate of 83 mm per 100 m (Rzonca and Siwek 2011). Unfortunately, no gauging station records the rainwater quality in this area. Information about chemical composition of rainwater can be obtained from the studies performed in other parts of the Carpathians. It contains only trace amounts of chlorides (below 1 mg dm^{-3}), whereas the concentration of sulfates may vary between 1 and 44 mg dm^{-3} (Bytnerowicz et al. 1999, 2005; Żelazny et al. 2005).

The third factor (F3) primarily reflected the variation in nitrate concentrations. While the variation across the catchment is generally very low (between 0.16 and 5.25 mg dm^{-3}), there does exist a certain spatial trend. Generally, higher nitrate

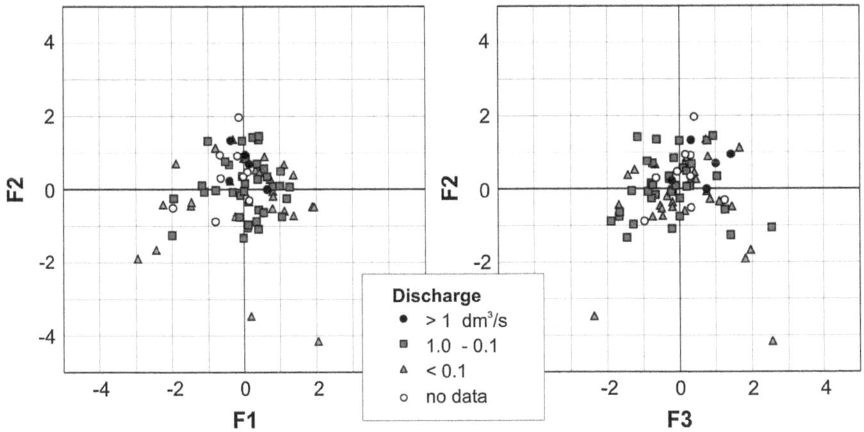

Fig. 4 Positions of the investigated sites classified according to spring discharge on the PCA factorial planes

concentrations were recorded in the eastern part of the study area. Probably the factor F3 reflects the role of throughflow in the alimentation of springs. This type of water contains larger quantities of organic matter converted into inorganic substances. The differences in the role of shallow water circulation in springs alimentation may by the result of varied weather conditions during the sampling, because the highest F3 factor scores were found in the area which was investigated in the period preceded by heavy rains.

It must also be noted that none of the three factors identified was directly correlated with spring type (springs or bog-springs), spring location in relation to relief forms, and the rate of spring discharge (Fig. 4). The only clear regularity in

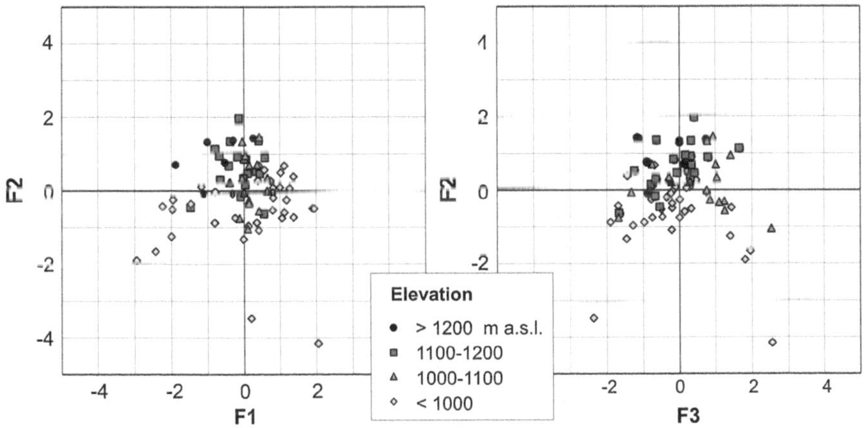

Fig. 5 Positions of the investigated sites classified according to spring elevation on the PCA factorial planes

the spatial pattern of spring-water chemistry was related to spring elevation; however, in the whole population of sampled springs, the correlations between concentrations of particular ions and elevation were not significant. Results of the PCA showed that most of the springs located on upper parts of slopes were usually characterized by high positive scores of factor F2, whereas most springs located on the valley floors had negative scores of this factor (Fig. 5).

6 Conclusions

Even small, unmanaged catchments in the flysch Carpathians, featuring uniform lithology, can exhibit spatially diverse chemistry of spring water. The fundamental factor (F1) shaping the proportions between different ions is lithology of the aquifer and duration of groundwater circulation in the catchment. Differences in the water chemistry can also be related to some extent to spring elevation (F2). Different elevations feature different climate conditions, different plant species and different conversion mechanisms of organic matter into inorganic compounds. Also precipitation totals are higher at higher elevations which results in different water circulation patterns there. However, it must be emphasized that only a certain proportion of the total variance in spring water chemistry is shaped by the elevation factor.

References

Bytnerowicz A, Godzik S, Poth M, Anderson I, Szdzuj J, Tobias C, Macko S, Kubiesa P, Staszewski T, Fenn ME (1999) Chemical composition of air, soil and vegetation in forests of the Silesian Beskid Mountains, Poland. Water Air Soil Poll 116:141–150

Bytnerowicz A, Badea O, Popescu F, Musselman R, Tanase T, Barbu I, Frączek W, Gembasu N, Surdu A, Danescu F, Postelnicu D, Cenusa R, Vasile C (2005) Air pollution, precipitation chemistry and forest health in the Retezat Mountains, Southern Carpathians, Romania. Environ Poll 137:546–567

Haczewski G, Kukulak J, Bąk K (2007) Budowa geologiczna i rzezba Bieszczadzkiego Parku Narodowego. Akademia Pedagogiczna w Krakowie, Kraków

Join JL, Coudray J, Longworth K (1997) Using principal components analysis and Na/Cl ratios to trace groundwater circulation in a volcanic island: the example of Reunion. J Hydrol 190:1–18

Karimi H, Raeisi E, Bakalowicz M (2005) Characterising the main karst aquifers of the Alvand basin, northwest of Zagros, Iran, by a hydrogeochemical approach. Hydrogeol J 13:787–799

Mouser PJ, Hession WC, Rizzo DM, Gotelli NJ (2005) Hydrology and geostatistics of a Vermont, USA Kettlehole Peatland. J Hydrol 301:250–266

Plenzler J, Bajorek J, Jaśkowiec B, Kołodziej A, Rzonca B, Siwek J, Wójcik S (2010) Podziemny odpływ jednostkowy w Bieszczadach Wysokich. Przegl Geol 58:1147–1151

Rzonca B, Siwek J (2009) Skład chemiczny wód zródeł i młak w zlewni górnej Wołosatki (Bieszczady Wysokie). In: Bogdanowicz R, Fac-Beneda J (eds) Zasoby i ochrona wód. Obieg wody i materii w zlewniach rzecznych. Fundacja Rozwoju Uniwersytetu Gdańskiego, Gdańsk

Rzonca B, Siwek J (eds) (2011) Hydrologia Bieszczadów. Zlewnie Sanu i Solinki powyżej Jeziora Solińskiego. Instytut Geografii i Gospodarki Przestrzennej Uniwersytetu Jagiellońskiego, Kraków

Rzonca B, Kołodziej A, Laszczak E, Macior E, Plenzler J, Płaczkowska E, Rozmus M, Siwek J, Ścisłowicz B, Wójcik S, Ziółkowski L (2008) Źródła w zlewni górnej Wołosatki w Bieszczadach Wysokich. Przegl Geol 56:772–779

Sickman JO, Leydecker A, Melack J (2001) Nitrogen mass balances and abiotic controls on N retention and yield in high-elevation catchments of the Sierra Nevada, California, United States. Water Resour Res 37:1445–1461

Siwek J, Chełmicki W (2004) Geology and land-use related pattern of spring water quality. Case study from the catchments of the Malopolska Upland (S. Poland). Geol Acta 2:167–174

Siwek J, Rzonca B (2009) Tło hydrochemiczne wód w zlewni Górnej Wołosatki w Bieszczadach Wysokich. Biul Państw Inst Geol 436:469–474

Siwek J, Kołodziej A, Laszczak E, Macior E, Plenzler J, Płaczkowska E, Rozmus M, Rzonca B, Ścisłowicz B, Wójcik S, Ziółkowski L (2009) Geologiczne i geomorfologiczne uwarunkowania wykształcenia sieci hydrograficznej w zlewni górnej Wołosatki (Bieszczady Wysokie). Kwart AGH Geologia 35:249–261

Tokarski AK (1975) Mapa geologiczna południowo-zachodniej części centralnej depresji oraz jej obrzeżenia w okolicy Ustrzyk Górnych, Geologia i geomorfologia okolic Ustrzyk Górnych. Polskie Karpaty Wschodnie, Stud Geol Pol 48

Żelazny M, Chełmicki W, Jaszczyńska B, Michno A, Pietrzak M, Raczak J, Święchowicz J (2005) Dynamika związków biogennych w wodach opadowych, powierzchniowych i podziemnych w zlewniach o różnym użytkowaniu na Pogórzu Wiśnickim. Instytut Geografii i Gospodarki Przestrzennej Uniwersytetu Jagiellońskiego, Kraków

Żelazny M, Astel A, Wolanin A, Małek S (2011) Spatiotemporal dynamics of spring and stream water chemistry in a high-mountain area. Environ Poll 159:1048–1057

Environmental and Land Use Determinants of Stream Water Chemistry During Flood Events in Small Carpathian Foothill Catchments in Poland

Joanna P. Siwek, Mirosław Żelazny and Wojciech Chełmicki

Abstract The goal of the research was to determine which factors control changes in the chemical composition of stream water during floods of different types: storm floods, frontal precipitation floods, and snowmelt floods. Three catchments in the Carpathian Foothills of woodland, agricultural and of mixed land use were examined and the data were explored with R-mode factor analysis. Three factors driving changes in the chemical composition of stream water during floods of particular types were identified for each catchment. The first factor (hydrological) is related to the magnitude of flow, the second factor (meteorological) is linked with air temperature and soil temperature, and the third factor (circulation) is related to mechanisms of water transfer to stream channels (surface runoff, shallow throughflow). The fourth factor (anthropogenic), related to the dilution of highly polluted pre-event waters with less polluted event waters, was identified only for the anthropogenically altered, agricultural and mixed-use catchments.

1 Introduction

20 years of economic transformation (1990–2010) in the countries of the former communist bloc have prompted significant changes in land use in the Carpathian Mountains region, i.e. an increase in forest cover and turning some proportion of

Wojciech Chełmicki – deceased in October 2011

J. P. Siwek (✉) · M. Żelazny · W. Chełmicki
Institute of Geography and Spatial Management, Jagiellonian University,
Gronostajowa 7 30-387 Kraków, Poland
e-mail: joanna.siwek@uj.edu.pl

J. P. Siwek
Faculty of Tourism and Recreation, University of Physical Education,
Jana Pawła II 78 31-571 Kraków, Poland

J. Kozak et al. (eds.), *The Carpathians: Integrating Nature and Society Towards Sustainability*, Environmental Science and Engineering,
DOI: 10.1007/978-3-642-12725-0_12,

arable land into pasture and meadows (Kozak et al. 2007). The transformation of land use patterns affects certain elements of the environment, including the hydrologic regime of rivers, river channel morphology (Wyżga 2008), as well as the chemical composition of water, and therefore water quality (Stålnacke et al. 2003; Fucik et al. 2008). This article presents the results of research on factors affecting the chemical composition of water of the streams draining catchments located in the Carpathian Foothills (Southern Poland) and characterized by diversified land use. In a paper published earlier (Siwek et al. 2011), a correlation analysis was used to verify whether there exist relationships between water discharge and flood type on one hand, and chemical composition of stream water on the other. In this paper, a factor analysis is used for identification of main processes controlling changes in stream water chemistry during flood events in the catchments studied earlier. The results of this research may be useful for the assessment of hydrological and hydrochemical effects of changes in land use in similar areas located in other parts of the Carpathian Mountains.

The chemical composition of river water during floods is influenced by many factors characterizing an entire catchment. Among environmental factors, two groups may be identified: (i) factors related to the characteristics of a catchment, as well as (ii) factors related to hydro-meteorological conditions present during particular flood events. Earlier research indicated that in the first group of factors, a very important role is played by geological structure (Walling and Webb 1980; Caissie et al. 1996; Holloway and Dahlgren 2001), elevation and geomorphology (Edwards 1973; Kendall et al. 1999; McGlynn et al. 1999; Stottlemyer 2001), soil and bedrock characteristics (Bazemore et al. 1994; Hinton et al. 1994; Sandén et al. 1997) and the size of a catchment (Brown et al. 1999).

Among the hydro-meteorological factors found in many research papers, the most attention is paid to conditions directly preceding a flood event (soil moisture content or freezing of the soil) (Walling and Foster 1975; Foster 1978; Caissie et al. 2006; Laudon et al. 2004), the intensity and form of precipitation (Cameron 1996; Suzuki 1995), and changes in stream discharge (Kendall et al. 1999; Abesser et al. 2006). The chemical composition of stream water during floods is influenced not only by environmental factors but by a host of other factors as well. A very important role is played by the anthropogenic factors—land use and land management in the catchment. All these factors so intertwined co-exist and together they determine the water circulation pattern in the catchment. Relatively little attention is paid to the role of land use in the catchment and how it impacts the chemical composition of water during floods. However, this issue is well recognized in the context of seasonal changes and changes over the years. One of the few papers on the subject resulted from the research conducted by Poor and McDonnell (2007) in three small catchments with different land use (woodland, agricultural, urban) in Oregon, USA.

The goal of the research presented herein was to identify the natural and anthropogenic factors that determine stream water chemistry during floods in three streams in the Carpathian Foothills. An attempt was made to answer the following questions:

1. Which factors play the most important role in determining stream water chemistry?
2. Are there differences in the sets of factors determining stream water chemistry between catchments of different land use?
3. Are there differences in stream water chemistry between floods of different type?

To answer these questions, three catchments of different land use (woodland, agricultural, and mixed land use) were selected and from every stream draining the catchments, water samples were collected during several flood events of different types: storm floods, frontal precipitation floods, and snowmelt floods. The obtained hydrochemical data were statistically analyzed with the use of R-mode factor analysis in order to identify factors explaining the observed patterns of stream water chemistry. This method is often used in hydrochemical research. According to Evans et al. (1996), the use of factor analysis allows the identification of dominant processes that cause changes in the chemical composition of water. This method is most often used to identify factors that determine the spatial variability of water chemistry (Love et al. 2004; Dragon 2006; Panda et al. 2006). It is also used to find factors responsible for temporal changes in stream water chemistry (Cameron 1996; Evans et al. 1996). This statistical procedure was used in the investigations of stream water chemistry during floods by Miller and Drever (1977), Bernal et al. (2002) and Abesser et al. (2006), as well as others. However, the data used in their papers were collected during floods caused by rainfall only. The data sets did not include floods caused by snow melting.

2 Study Area

Research on the dynamics of chemical composition of river water was conducted in the catchment of the Stara Rzeka (22 km^2), located at the northern edge of the Carpathian Foothills. Within this catchment, characterized by mixed land use, two small nested catchments were selected: the catchment of Leśny Potok (0.48 km^2), entirely covered by forest, and the catchment of Kubaleniec (1.03 km^2) which is almost entirely used for agriculture (Fig. 1).

The Stara Rzeka catchment is located at an elevation of 216.5–361.5 m a.s.l. Its higher, southern part is underlain by relatively resistant flysch rocks (sandstone and shale) of the Silesian tectonic unit (Fig. 1) and it is mostly wooded. The lower, northern part is mostly agricultural and cut in less resistant materials (sandstone, claystone, shale, clay, gypsum, marl clay and salt series) of the Sub-Silesian

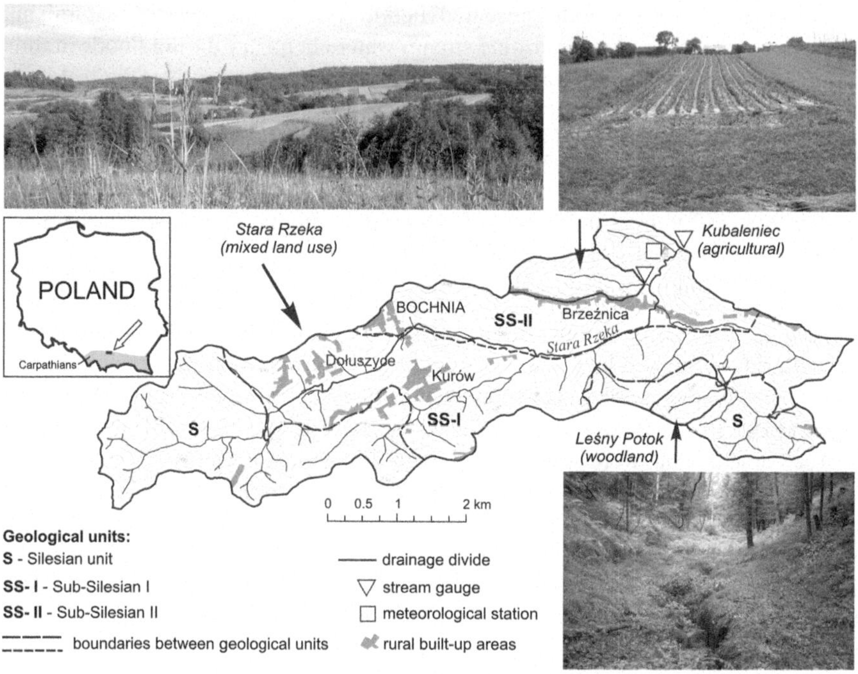

Fig. 1 Study area

tectonic units (Fig. 1). The Sub-Silesian units are covered by the argillaceous Miocene deposits. All the pre-Quaternary units are covered with loess-like deposits. The soils are of the following types: Haplic Luvisols, Stagnic Luvisols, Cambic Luvisols and Eutric Gleysols (Skiba et al. 1998). The width of the Stara Rzeka valley increases from a dozen or so meters in the river headwaters to 250 m in the downstream reach. The river channel is cut into the alluvial sediments of the valley as deep as 4 m. The mean annual discharge of the Stara Rzeka is 157 $dm^3\ s^{-1}$. Discharge varies significantly, from several $dm^3\ s^{-1}$ up to 20 $m^3\ s^{-1}$, due to negligible ground water retention (Chełmicki 2005).

The catchment of Leśny Potok is located at an elevation of 256.9–341.9 m a.s.l. Tree stand is composed of species typical for *Pino-Quercetum* forests with admixture of beech and fir. The valley is V-shaped. The width of the valley floor is between several and 20 m. Many small, side valleys are connected to it. Steep slopes of the valley (10–15 ^{o}) are unsuitable for agriculture. The mean discharge of the stream is slightly over 1 $dm^3\ s^{-1}$. Maximum discharge in the years 2003–2004 was 193 $dm^3\ s^{-1}$.

The catchment of Kubaleniec is located at an elevation of 223.4–296.0 m a.s.l. Arable land accounts for 69 %, meadows and pasture for 20 %, and forests just for 0.5 % of its area. Typically in this catchment, fields of arable land are narrow, and their width ranges from several to less than 100 m. Fields gently slope down from the water divide towards the stream. This is conducive to the formation of surface runoff and the erosion of soil. At the depth of 20–100 cm within the soil profile, there typically occurs a poorly permeable argillic horizon which outcrops in some places on the surface due to soil erosion (Skiba et al. 1998). The mean discharge of the stream is nearly 3.5 $dm^3\ s^{-1}$. Maximum discharge in the years 2003–2004 was almost 250 $dm^3\ s^{-1}$.

Up until the late 1980s, the main source of water for farm and household use was backyard wells that were dug to a depth of 5–20 m. After a water intake facility was constructed on the nearby Raba River, farms and households began to obtain water by a water main. This caused an increased use of water and consequently an increased amount of sewage to be disposed of. Streams and even roadside ditches were polluted with sewage wherever sewage was disposed of improperly due to a lack of a sewer system. Finally, the fertilization of fields and meadows occupying flat valley floors with natural fertilizer (liquid manure) is an important source of biogenic substances.

3 Study Methods

The research on the dynamics of stream water chemistry was conducted in the years 2002–2004. During that period, water samples were collected at terminal gauging stations in the catchments of Stara Rzeka, Leśny Potok and Kubaleniec (Fig. 1). Twenty-nine flood events were monitored during the entire research period. Seven of the 29 floods were caused by prolonged frontal rainfall. Other seven floods were caused by intense storm rainfall, and the remaining 15 floods were of snowmelt-type. The detailed characteristics of hydro-meteorological conditions causing the analyzed floods were described by Siwek et al. (2011). Water samples were collected manually at intervals ranging from several minutes up to several hours, depending on the flow dynamics and the duration of a given flood event. For that reason, the number of samples collected during every flood event ranged from 6 to 23. The collected water samples were analyzed for specific conductance (SC), concentrations of main ions (Ca^{2+}, Mg^{2+}, Na^+, K^+, HCO_3^-, SO_4^{2-}, Cl^-), and concentrations of biogenic substances (NH_4^+, NO_2^-, NO_3^-, PO_4^{3-}). The water samples were filtered through SARTORIUS (0.45 μm) filters. The following analytical methods were used to test for individual ion types: acidimetry (HCO_3^-), argentometry (Cl^-), spectrophotometry—using a Merck SQ 118 machine (Mg^{2+}, SO_4^{2-}, NH_4^+, NO_2^-, NO_3^-, PO_4^{3-}), and flame photometry—using a JENWAY PFP 7 machine (Ca^{2+}, Na^+, K^+). All the analyses were performed according to Polish Standards. Further details were described by Jaszczyńska (2005).

The rate of discharge was determined at gauge stations based on water stages measured continuously (float-type recorder, until May 2003) and in 10-min intervals (pressure-type water level sensors, after May 2003), with stage-discharge curves created for every site.

R-mode factor analysis (extraction method: principal components) was applied in order to determine the factors influencing the dynamics of variables, which describe the properties of water during flood events in the catchments of different land use. This method reduces the complexity of a large data set to a smaller set of factors, which are not correlated with each other (Davis 1973). The advantage of this method is that it allows each factor to be interpreted separately. This means that each factor can be associated with some specific source or process (Drever 1997). A matrix of factor scores, one of the most important parts of factor analysis output, was used as well. Factor scores provide a measure of the relationship between each observation (each sample) and the factors being analyzed (Shaw and Wheeler 1997).

The rate of discharge, air temperature, and temperature of the ground at a depth of 5 cm were measured during floods and taken into consideration. The values of two latter parameters were obtained from a meteorological station located in the Stara Rzeka catchment (see Fig. 1). The Kolmogorov–Smirnov test and Liliefors test were used to check for the normality of variable distribution. Variables not normally distributed were logarithmically transformed to achieve a normal distribution. All variables were then standardized. The Keiser criterion was used to separate out the factors and a significance of $p < 0.05$ was used across all calculations. The following categories were applied to factor loadings: high—over 0.75 and moderate—between 0.4 and 0.75. The higher the factor loading, the stronger the relationship between the given variable and the selected factor. The same classification was used in the papers on similar topics by Evans et al. (1996) and Bernal et al. (2002).

4 Results and Discussion

Basic characteristics of the chemical composition of water for each stream examined during baseflow and flood periods are shown in Table 1. They show very clearly different patterns of changes of certain ions (rise or fall of concentrations) depending on the catchment's land use.

A factor analysis identified four independent factors that determine changes in the chemical composition of water during floods in both the Kubaleniec (agricultural) and Stara Rzeka (mixed use) catchment. Three factors were identified for the catchment of Leśny Potok (woodland). According to Drever (1997), the smaller the number of factors that explain the bulk of the observed variability, the easier it is to identify the processes responsible for the variability. Every factor explained at least 5 % of the variability in the properties being analyzed. Considered together, the identified factors explained over 80 % of the variability in each of the streams of interest (Table 2).

The first factor explained the largest part of the variability in Leśny Potok (over 60 %), almost 50 % in Kubaleniec, and 43 % in Stara Rzeka. In the case of the first factor, called here "hydrological", there was a strong inverse relationship between discharge on the one hand and SC and the concentration of main ions, except K^+, on

Table 1 Main physico-chemical characteristics of stream water during baseflow (B) and flood periods (F); median values (specific conductance, SC, in μS cm^{-1}; ion concentrations in mg dm^{-3})

	Leśny Potok (woodland)		Kubaleniec (agricultural)		Stara Rzeka (mixed)	
	B	F	B	F	B	F
pH	7.85	7.61	7.66	7.46	7.71	7.65
SC	433.00	239.00	833.00	350	489.50	378.00
Ca^{2+}	40.91	22.24	91.45	38.59	60.70	39.90
Mg^{2+}	14.98	9.99	18.80	9.31	15.01	10.77
Na^{+}	10.70	5.57	42.67	14.35	17.96	13.98
K^{+}	1.61	2.11	3.93	5.51	4.59	5.08
HCO_3^{-}	199.83	100.60	258.69	92.72	213.50	118.95
SO_4^{2-}	58.70	38.35	94.30	57.90	46.85	49.15
Cl^{-}	7.24	5.11	73.27	20.45	19.45	17.82
NH_4^{+}	0.02	0.08	0.13	0.20	0.10	0.35
NO_2^{-}	0.02	0.05	0.08	0.09	0.12	0.15
NO_3^{-}	3.40	5.85	14.70	8.65	5.40	10.50
PO_4^{3-}	0.09	0.14	0.16	0.35	0.22	0.29

the other hand. This was true for all the streams in question. This type of dependence indicates that the first factor is related to the process of dilution of highly mineralized pre-event water (old water) with less mineralized event water (new water). Dilution is judged by many researchers to be the main process driving changes in water chemistry in rivers and streams during flood events (Edwards 1973; Walling and Foster 1975; Foster 1978; Bhangu and Whitfield 1997). Caissie et al. (1996) believe that what is being diluted are so-called geologically controlled ions. The ions had originated in the process of bedrock leaching. According to Caissie et al. (1996), geologically controlled ions are characterized by a strong relationship between their concentrations and discharge as well as by a strong mutual relationship among their concentrations. In the streams of interest, the strongest relationship was identified between discharge, SC, and the concentration of Ca^{2+}, Mg^{2+} and Na^{+} ($-0.90 >$ factor loadings > 0.90). In the agricultural catchment, the Cl^{-} ion also belonged to this group of variables. Its factor loading was very high (-0.99). This was due to the presence of salt deposits (halite) of the Bochnia salt series in the catchment's bedrock (Olewicz 1973).

In the woodland catchment, the first factor also influenced changes in concentrations of NO_2^{-}, NO_3^{-} and PO_4^{3-}. The relationship between discharge and biogenic compounds in the catchment of Leśny Potok was as follows: the greater the discharge, the higher the concentration of ions. This type of relationship indicates that event water carries biogenic substances into stream channels. In the woodland catchment, the largest quantity of water is delivered to stream channels during flood events by shallow throughflow. The role of surface runoff in this woodland catchment is marginal—it shows up only along rare footpaths and on the wet floor of a stream valley. It is worth noting that the same factor, which is responsible for the delivery of biogenic substances to the stream channel, also

Table 2 Results of the factor analysis showing factor loadings for three examined catchments of different land use in the Carpathian Foothills

	Log[a]	Leśny Potok (woodland)			Kubaleniec (agricultural)				Stara Rzeka (mixed)			
		F1	F2	F3	F1	F2	F3	F4	F1	F2	F3	F4
T_{air}	a; b; c		0.84			0.85				0.84		
T_{ground}	a; b; c		0.89			0.92				0.90		
Q	a; b; c	−0.94			0.91				0.75	(−0.49)		
pH		0.93			(−0.41)		(−0.67)			(0.45)		
SC	b	0.99			−0.99				−0.99			
Ca^{2+}	b	0.94			−0.98				−0.94			
Mg^{2+}	b	0.93			−0.98				−0.97			
Na^{+}	b	0.98			−0.98				−0.97			
K^{+}	b; c		(−0.58)	(−0.65)		(0.53)	(0.50)	(−0.42)		(0.47)	(−0.70)	
HCO_3^{-}	b	0.95			−0.89				(−0.71)	(0.54)		
SO_4^{2-}	b; c	0.84			−0.85				−0.77			(0.41)
Cl^{-}	b; c	0.87			−0.99				−0.89			
NH_4^{+}	a; b; c		(−0.67)				(0.44)	(0.66)		(−0.46)	(−0.43)	(−0.63)
NO_2^{-}	a; b; c	−0.79				(0.72)		(0.53)	(0.47)	(0.67)		
NO_3^{-}	a; b; c	(−0.61)	(−0.57)		(−0.52)		(0.44)	(0.41)			(−0.57)	(0.53)
PO_4^{3-}	a; b; c	(−0.72)		(−0.52)		0.75	(0.40)			(0.41)	(−0.64)	
Eigenvalue		9.64	3.08	1.32	7.96	3.13	1.50	1.39	6.85	3.59	1.77	1.17
Percent of explained variance		60.24	19.24	8.26	49.75	19.59	9.37	8.69	42.82	22.42	11.04	7.34
Percent of cumulative variance		60.24	79.48	87.74	49.75	69.34	78.71	87.40	42.82	65.24	76.28	83.62

[a] Logarithmic values: a—Leśny Potok; b—Kubaleniec; c—Stara Rzeka; loadings within the range of 0.40–0.75 are given in parentheses, loadings <0.40 are omitted

determines the concentrations of geologically controlled ions. In the case of biogenic substances, this factor works in a different manner. It does not cause the dilution of their concentrations, as it is in the case of geologically controlled ions, but it helps to deliver them to stream channels. This is a natural circulation mechanism for biogenic compounds in a woodland catchment.

No such relationship exists between biogenic substances and discharge in the catchment strongly affected by human pressure. In the agricultural catchment of Kubaleniec, the "hydrological" factor affects the concentration of NO_3^-. However, it operates in the opposite direction than in the woodland catchment—when the rate of discharge during a flood is high, the concentration of NO_3^- decreases. This means that pre-event water is being diluted during flood. Prior to a flood event, specifically during baseflow, the stream water in the agricultural catchment contains more NO_3^- than does flood water (on average: 18.5 and 10.9 mg dm^{-3}, respectively). This can be related to high concentrations of nitrogen compounds of anthropogenic origin present in groundwater in the agricultural catchment. It is a common practice among local farmers to pour natural liquid manure onto the flat valley floor in the Kubaleniec catchment. The high concentration of NO_3^- at baseflow is also related to the fact that sewage from farms and households located within the watershed (point sources) flows to the stream channel.

The first factor clearly differentiates water samples collected in each of the investigated streams into samples associated with the smallest surface runoff (specific discharge <5 dm^3 s^{-1} km^{-2}) and samples associated with the largest surface runoff (specific discharge >50 dm^3 s^{-1} km^{-2}) (Fig. 2). Absolute factor scores are largest just in such cases. Thus, the first factor allows to distinguish among flood events in which the share of event water is the smallest and those in which it is the largest. Therefore, this factor may be designated a "hydrological factor", which is related to the influx of water into stream channels. Abesser et al. (2006) obtained similar results for three catchments in the Southern Uplands in the United Kingdom—the most important factor causing changes in the concentration of Mn and Fe during several autumn rainfall-induced flood events emerged to be a factor related to changes in the rate of discharge. Bernal et al. (2002) analyzed the relationship between the concentration of NO_3^- and dissolved organic content (DOC), and hydro-meteorological factors controlling hydrological responses in a small catchment in Catalonia (Spain). They concluded that in the case of NO_3^-, the most important role was played by the magnitude of the storm event, while in the case of DOC, antecedent moisture conditions were the main determinant.

The second factor, called here "meteorological", explained approximately 22 % of the variability in the Stara Rzeka, 20 % in Kubaleniec, and 19 % in Leśny Potok. This factor was associated with air temperature, soil temperature, and the concentration of biogenic compounds and K^+. In the woodland catchment, this relationship was as follows: the higher the temperature of the air and soil, the smaller the concentration of biogenic compounds (NH_4^+, NO_3^-) and K^+. In the agricultural and mixed use catchment, the opposite was true: the higher the temperature of the air and soil, the higher the concentration of biogenic compounds (NO_2^-, PO_4^{3-}) and K^+.

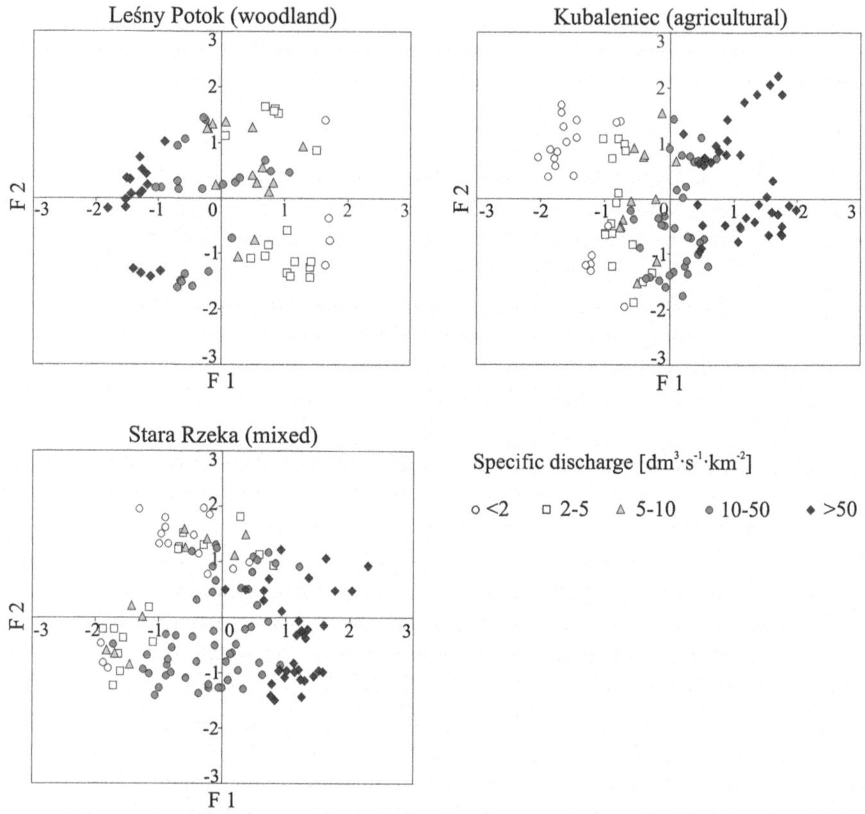

Fig. 2 Results of factor analysis for three investigated catchments of different land use: positions of water samples classified according to specific discharge value on the first factorial plane

Concentrations of biogenic compounds and K^+ in the woodland catchment of Leśny Potok are lower during rainfall-induced summer floods (higher air and soil temperatures) than during snowmelt-induced winter floods (lower temperatures). This most likely results from a larger intake of these compounds by plants in the summer. Moreover, during winter floods in the woodland catchment, snow cover is a significant source of nitrogen ions, primarily NH_4^+. Investigations performed by Żelazny (2005) in the catchment of Stara Rzeka indicated that the weighted average concentration of NH_4^+ in precipitation water is higher in the winter season than in the summer season (1.82 and 1.48 mg dm^{-3}, respectively). In agricultural catchments, the significance of the supply of nitrogen compounds stored in snow cover is smaller due to their large influx from sources related to agriculture and household activity. It should be noted that Bernal et al. (2002) analyzed the woodland catchment of Fuirosos (Spain) and did not find the influence of seasons on shifts in NO_3^- concentration during floods. It is likely that this is related to smaller intra-annual differences in temperature in Mediterranean catchments than in the catchments from the moderate climate zone.

In the agricultural catchment of Kubaleniec, the concentration of biogenic compounds and K^+ during rainfall-induced summer floods is higher than during snowmelt winter floods. In the case of PO_4^{3-} and K^+, this is a result of the presence of larger amounts of suspended matter in overland flow during summer floods in comparison with winter floods. The ions are absorbed by suspended matter (Johnson et al. 1976; Likens et al. 1994; Evans and Davies 1998). This has been confirmed by Święchowicz (2005) in the catchment of the Stara Rzeka. The weighted mean concentration of suspended matter in river water during summer floods was 1,500 mg dm^{-3}, while during winter floods, it was 851 mg dm^{-3}. In contrast to the woodland catchment, the surface runoff in the agricultural catchment plays a very important role in the delivery of water to the stream channel during floods (Siwek et al. 2009). Here, surface runoff forms very quickly along footpaths and dirt roads. In catchments located in the Carpathian Foothills, this process is intensified by plowing along slopes (see Fig. 1) and by the presence of a poorly permeable argillic horizon at small soil depths. Concentrations of biogenic compounds and K^+ during summer floods are higher than during winter floods, as a result of the infiltration of waters in the shallow throughflow zone, whereas during winter floods throughflow waters do not migrate because of frozen soil. This is also related to some agricultural practices undertaken in the summer season, such as the application of organic and artificial fertilizers. Primarily in the summer local farmers pour natural liquid manure onto the valley floor in the Kubaleniec catchment. In this way, farmers enrich shallow groundwater in biogenic compounds and K^+.

In the analysis of the second factor, it is important to note that the factor loading of the HCO_3^- ion in Stara Rzeka is high. The relationship between HCO_3^- and air and soil temperature indicates that the concentration of the HCO_3^- ion will be higher during summer floods (higher air and soil temperatures) than during winter floods (lower temperatures). This reflects the fact that during summer season relatively large amounts of HCO_3^- ion are delivered by shallow throughflow moving through the soil. Shallow layers of soil are one of the most important sources of bicarbonate (Hem 1985). In winter, water in the soil does not migrate simply because the soil is frozen.

The second factor differentiates between water samples collected during summer floods (storm-induced and prolonged rainfall-induced) and samples collected during snowmelt winter floods (Fig. 3). The largest difference has been identified between storm floods and snowmelt floods (high absolute factor scores). This is caused by the largest difference in the mechanism of water circulation during floods of these types, and consequently in the ways of delivering ions to stream channels. Storm floods are caused by intense rainfall, which usually lasts several hours. Conditions for water infiltration are relatively good in such situations. On the other hand, snowmelt floods take place less rapidly, as they are caused by melting of snow cover. In some cases, rainfall occurs at the same time. Snowmelt

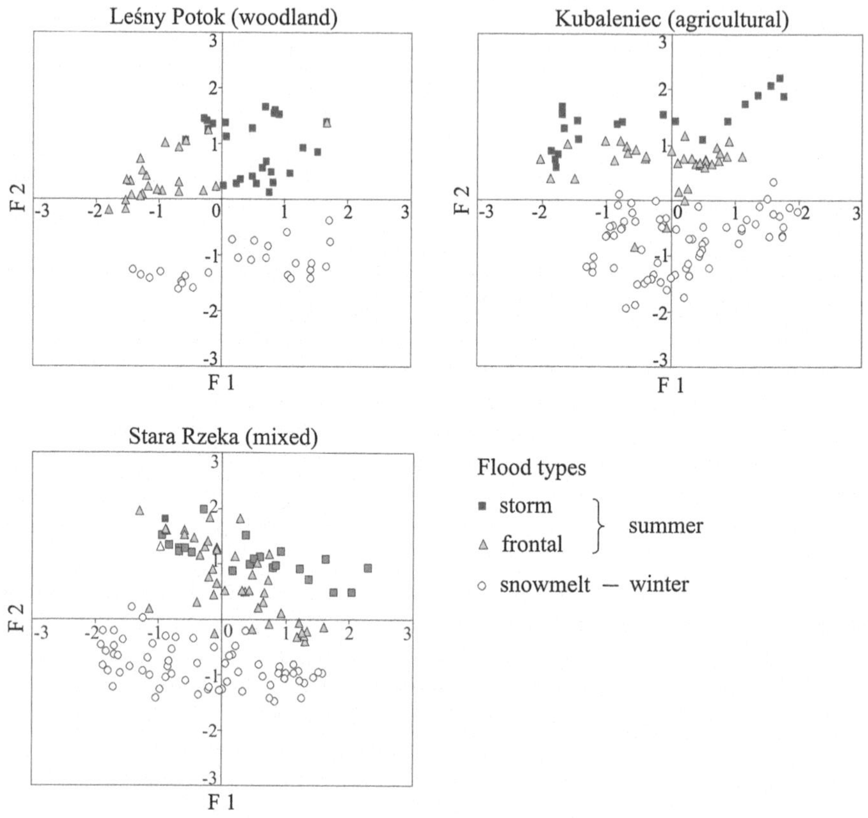

Fig. 3 Results of factor analysis for three investigated catchments of different land use: positions of water samples classified according to flood type on the first factorial plane

flood events lasted much longer than storm flood events, anywhere from several days up to 20 days. During such flood events, conditions for infiltration are not favorable, mostly because the soil is deeply frozen. Infiltration is also limited if the soil cover is not frozen but saturated with water. The second factor is strongly associated with the temperature of the air and the soil during floods. Thus, it may be designated a meteorological factor (seasonal factor).

The third factor, called here "water circulation", explained 11 % of the variability in the catchment of the Stara Rzeka, 9 % in Kubaleniec, and 8 % in Leśny Potok. This factor was responsible for changes in the concentration of K^+ and most biogenic compounds in all the studied streams. The relationship was as follows: the more K^+, the more biogenic compounds. In the Kubaleniec catchment, this factor was also responsible for changes in pH, where the relationship was inverse with respect to K^+ and biogenic compounds. Generally, the third factor determined the concentration of ions which were delivered primarily by shallow throughflow (K^+ and NO_3^-) and surface runoff (NH_4^+ and PO_4^{3-}). It should be noted that there exists an inverse relationship between the concentrations of K^+, NH_4^+, PO_4^{3-} and pH in the Kubaleniec catchment. Water pH changes significantly during floods in this catchment. During some flood events, water pH can change as much as one pH unit. This change is related to the fact that acidic rainwater mixes in with the water in the catchment. The pH of precipitation water in the Stara Rzeka catchment is, on average, 4.7 (Żelazny 2005). The decrease in pH of water in the river channel causes remobilization of K^+, NH_4^+ and PO_4^{3-} ions adsorbed onto river channel sediments. A low pH of atmospheric precipitation facilitates leaching of these ions from the soil cover. Hence, the third factor may be designated the factor of "short" and "shallow" water circulation. Similar factor, related to the leaching of Fe and Mn from the soil at low pH, was identified by Abesser et al. (2006) in three agricultural catchments in the United Kingdom.

The fourth factor, called here "anthropogenic", was identified in the agricultural catchment of Kubaleniec and the mixed use catchment of the Stara Rzeka. The absence of this factor in the woodland catchment suggests that it is related to processes involving anthropogenic activity. The analysis of relationships identified by the fourth factor is difficult and requires an analysis of changes in water chemistry during an individual flood event. In the catchment of Kubaleniec, the fourth factor is responsible for changes in nitrogen compounds, where the relationship is inverse with respect to K^+. This type of relationship exists in most cases during the falling limb of flood waves (Fig. 4). At that time Kubaleniec receives water primarily from shallow throughflow, which delivers K^+ ions to the stream channel. When a flood wave starts to recede, the proportion of water from shallow throughflow starts to increase. This causes a decrease in the concentration of nitrogen compounds, which are primarily delivered to stream channels by overland flow (e.g. NH_4^+) or come from household and agricultural activities (e.g. NO_3^-).

In the Stara Rzeka catchment, the fourth factor was responsible for changes in the concentration of NH_4^+, NO_3^- and SO_4^{2-}. The relationship was as follows: the

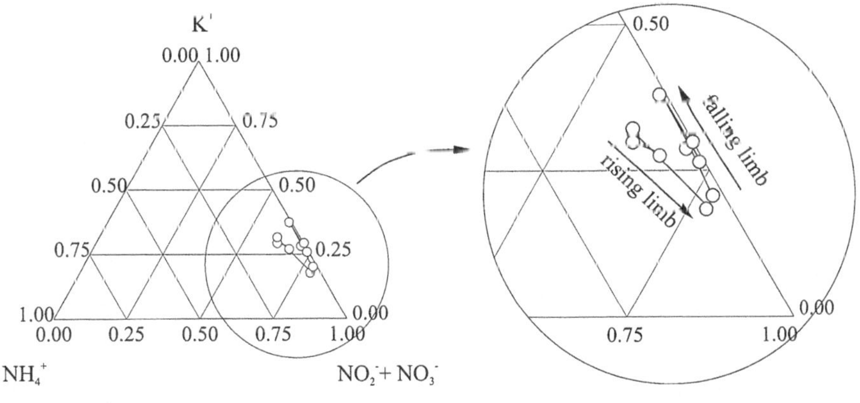

Fig. 4 Changes in mineral nitrogen and potassium content during the rising and falling limbs of the flood wave of 15–16 July 2002 in Kubaleniec stream

more SO_4^{2-} ions, the fewer NH_4^+ ions, and the more NO_3^- ions. This type of relationship results from the fact that rainwater reaching the catchment is characterized by a relatively high concentration of SO_4^{2-} (up to 40 mg dm^{-3}) (Żelazny 2005). The influx of this "new" water causes the dilution of the polluted waters of the Stara Rzeka River. A key pollutant in the Stara Rzeka is the NH_4^+ ion (Siwek et al. 2008, 2011). This, in turn, favors the process of nitrification, which is responsible for an increase in the concentration of NO_3^-. The fourth factor may be designated the "anthropogenic factor" associated with dilution of polluted stream water during floods.

Recent changes in land use in the Carpathians, including the transition from farmland to woodland, as well as climate change leading to more frequent and intense summer floods should bring positive results in terms of the amounts of biogenic compounds and potassium ions leaving the catchment during floods. Our research has shown that more biogenic compounds and potassium ions leave the agricultural catchment during summer floods versus winter floods. The opposite is true in the woodland catchment, as plants use up some of the biogenic compounds and potassium ions available in the catchment during the summer. The recent increase in forest cover in the Carpathians should reduce the loss of these ions by plant intake. This will also result in improved water quality in rivers and streams.

5 Conclusions

Four factors responsible for changes in stream water chemistry during floods were identified in each of the catchments affected by human pressure: an agricultural catchment and a mixed use catchment. In the woodland catchment, only three factors were identified. The first three factors can be classified as environmental factors, while the fourth factor can be classified as anthropogenic factor. Despite the fact that the first three natural factors were associated with the same processes in the three investigated catchments, they tended to affect water chemistry differently in the woodland catchment than in the agricultural and mixed-use catchments. This was especially true in the case of biogenic compounds and K^+. This reflected different concentrations of ions in particular components of runoff (baseflow, subsurface runoff, surface runoff) in the catchments of different land use. A summary of main findings is shown in Table 3.

The first and the most important factor identified in all the investigated catchments was related to the change in the rate of discharge (called the hydrological factor). This factor causes the dilution of ions originating from bedrock leaching in all the catchments (most of the main ions, except K^+). In the woodland catchment, the first factor also affected the concentration of biogenic compounds; their concentrations increased along with increasing discharge. Event waters, mainly subsurface runoff component, were the main source of biogenic compounds in the woodland catchment. In the agricultural catchment, however, an increase in water

Table 3 Identified factors controlling the chemistry of stream water in the investigated catchments of different land use in the Carpathian Foothills

Land use	F1 (hydrological factor)		F2 (meteorological factor)	F3 (circulation factor)	F4 (anthropogenic factor)
Woodland		Increasing NO_2^-, NO_3^- and PO_4^{3-} concentrations with increasing Q (*delivery of ions by event water*)	Decreasing NH_4^+, NO_3^- and K^+ concentrations with increasing air and soil temperatures (*biological uptake*)		–
Agricultural	Decreasing SC and concentration of main ions with increasing Q (*dilution process*)	Decreasing NO_3^- concentration with increasing Q (*dilution of contaminated pre-event water*)	Increasing NO_2^-, PO_4^{3-} and K^+ concentrations with increasing air and soil temperatures (*ion delivery via throughflow and surface runoff, limited due to winter freezing; use of organic and mineral fertilizers in the summer*)	Increasing concentrations of most biogenic compounds with increasing concentration of K^+ (*delivery of ions* via *shallow throughflow and surface runoff*)	Decreasing concentration of nitrogen-bearing ions (NH_4^+, NO_2^-, NO_3^-) with increasing K^+ concentration (*dilution of polluted Kubaleniec waters with throughflow waters arriving from increasingly flushed soils during the falling limb of a flood wave*)
Mixed-Use		Increasing NO_2^- concentration with increasing Q (*delivery of ions by event water*)			Decreasing NH_4^+ concentration and increasing NO_3^- concentration with increasing SO_4^{2-} concentration (*dilution of polluted pre-event waters with fresh floodwaters; nitrification process*)

discharge caused a decrease in the concentration of NO_3^- ion. This was the result of the dilution of strongly polluted baseflow water by event water.

The second factor was related to different temperatures of the air and the soil during different flood events (called the meteorological factor). This factor was responsible for the concentrations of biogenic compounds and K^+. In the woodland catchment, concentrations of these ions were higher during snowmelt winter floods than during rainfall summer floods (induced by storms and continuous rainfall). This was related to the increased intake of these ions by plants during the growing season. In the agricultural catchment and the mixed use catchment, the opposite was true. This predominantly results from intensified agricultural activities in the summer.

The third factor (called circulation factor) was responsible for the concentration of ions, which had originated primarily in shallow throughflow (K^+, NO_3^-) and surface runoff (NH_4^+, PO_4^{3-}). Water pH also possessed a relatively high factor loading and was inversely proportional to the concentration of the above mentioned ions.

The fourth factor was called the anthropogenic factor. It was identified in the agricultural and mixed use catchments only. This factor was related to the dilution of strongly polluted pre-event waters with less polluted event waters that originated from precipitation and shallow throughflow.

Acknowledgments The research project was funded by the Polish Committee for Scientific Research (Project 3 P04G 050 22).

References

Abesser C, Robinson R, Soulsby C (2006) Iron and manganese cycling in the storm runoff of a Scottish upland catchment. J Hydrol 326:59–78

Bazemore DE, Eshleman KN, Hollenback KJ (1994) The role of soil water in stormflow generation in a forested headwater catchment: synthesis of natural tracer and hydrometric evidence. J Hydrol 162:47–75

Bernal S, Butturini A, Sabater F (2002) Variability of DOC and nitrate responses to storms in a small Mediterranean forested catchment. Hydrol Earth Syst Sci 6:1031–1041

Bhangu I, Whitfield PH (1997) Seasonal and long-term variations in water quality of the Skeena River at Usk, British Columbia. Water Res 31:2187–2194

Brown VA, McDonnell JJ, Burns DA, Kendall C (1999) The role of event water, a rapid shallow flow component, and catchment size in summer stormflow. J Hydrol 217:171–190

Caissie D, Pollock TL, Cunjak RA (1996) Variation in stream water chemistry and hydrograph separation in a small drainage basin. J Hydrol 178:137–157

Cameron EM (1996) Hydrogeochemistry of the Fraser River, British Columbia: seasonal variation in major and minor components. J Hydrol 182:206–225

Chełmicki W (2005) Dynamika odpływu. In: Żelazny M (ed) Dynamika obiegu związków biogennych w wodach opadowych, powierzchniowych i podziemnych w zlewniach o różnym użytkowaniu na Pogórzu Wiśnickim. Instytut Geografii i Gospodarki Przestrzennej Uniwersytetu Jagiellońskiego, Kraków

Davis JC (1973) Statistics and data analysis in geology. Wiley, New York

Dragon K (2006) Application of factor analysis to study contamination of a semi-confined aquifer (Wielkopolska Buried Valley aquifer, Poland). J Hydrol 331:272–279
Drever JI (1997) The geochemistry of natural water. Surface and groundwater environments. Prentice Hall, Upper Saddle River
Edwards AMC (1973) The variation of dissolved constituents with discharge in some Norfolk Rivers. J Hydrol 18:219–242
Evans C, Davies TD (1998) Causes of concentration/discharge hysteresis and its potential as a tool for analysis of episode hydrochemistry. Water Resour Res 34:129–137
Evans CD, Davies TD, Wigington PJ Jr, Tranter M, Kretser WA (1996) Use of factor analysis to investigate processes controlling the chemical composition of four streams in the Adirondack Mountains, New York. J Hydrol 185:297–316
Foster IDL (1978) A multivariate model of storm-period solute behaviour. J Hydrol 39:339–353
Fucik P, Kvitek T, Lexa M, Novak P, Bilkova A (2008) Assessing the stream water quality in connection with land use in agricultural catchments of different scales. Soil Water Res 3:98–112
Hem JD (1985) Study and interpretation of the characteristics of natural water. US Geol Surv Water Supply Paper 2254:1–263
Hinton MJ, Schiff SL, English MC (1994) Examining the contributions of glacial till water to storm runoff using two- and three-component hydrograph separations. Water Resour Res 30:983–993
Holloway JM, Dahlgren RA (2001) Seasonal and event scale variations in solute chemistry for four Sierra Nevada catchments. J Hydrol 250:106–121
Jaszczyńska B (2005) Metodyka analiz chemicznych. In: Żelazny M (ed) Dynamika obiegu związków biogennych w wodach opadowych, powierzchniowych i podziemnych w zlewniach o różnym użytkowaniu na Pogórzu Wiśnickim. Instytut Geografii i Gospodarki Przestrzennej Uniwersytetu Jagiellońskiego, Kraków
Johnson AH, Bouldin DR, Goyette EA, Hedges AM (1976) Phosphorus loss by stream transport from a rural watershed: quantities, processes, and sources. J Environ Qual 5:148–157
Kendall KA, Shanley JB, McDonnell JJ (1999) A hydrometric and geochemical approach to test the transmissivity feedback hypothesis during snowmelt. J Hydrol 219:188–205
Kozak J, Estreguil C, Troll M (2007) Forest cover changes in the northern Carpathians in the 20th century: a slow transition. J Land Use Sci 2:127–146
Laudon H, Seibert J, Kohler S, Bishop K (2004) Hydrological flow paths during snowmelt: congruence between hydrometric measurements and oxygen 18 in meltwater, soil water, and runoff. Water Resour Res 40:W03102
Likens GE, Driscoll CT, Buso DC, Siccama DF, Johnson CE, Lovett GM, Ryan DF, Fahey T, Reiners WA (1994) The biogeochemistry of potassium at Hubbard Brook. Biogeochemistry 25:61–125
Love D, Hallbauer D, Amos A, Hranova R (2004) Factor analysis as a tool in groundwater quality management: two southern African case studies. Phys Chem Earth 29:1135–1143
McGlynn BL, McDonnell JJ, Shanley JB, Kendall C (1999) Riparian zone flowpath dynamics during snowmelt in a small headwater catchment. J Hydrol 222:75–92
Miller WR, Drever JI (1977) Water chemistry of a stream following a storm, Absaroka Mountains, Wyoming. Geol Soc Am Bull 80:286–290
Olewicz ZR (1973) Tektonika jednostki bocheńskiej i brzegu jednostki śląskiej między Rabą a Uszwicą. Acta Geol Pol 23:701–761
Panda UCh, Sundaray SK, Rath P, Nabak BB, Bhatta D (2006) Application of factor and cluster analysis for characterization of river and estuarine water systems—a case study: Mahanadi River (India). J Hydrol 331:434–445
Poor CJ, McDonnell JJ (2007) The effects of land use on stream nitrate dynamics. J Hydrol 332:54–68

Sandén P, Karlsson S, Düker A, Ledin A, Lundman L (1997) Variations in hydrochemistry, trace metal concentration and transport during a rain storm event in a small catchment. J Geochem Explor 58:145–155

Shaw G, Wheeler D (1997) Statistical techniques in geographical analysis. David Fulton Publishers, London

Siwek JP, Żelazny M, Chełmicki W (2008) Annual changes in the chemical composition of stream water in small catchments with different land-use (Carpathian Foothills, Poland). Soil Water Res 3:129–137

Siwek JP, Żelazny M, Chełmicki W (2009) The influence of water circulation on stream water electrical conductivity in catchments with different land use during flood periods (The Carpathian Foothills, Poland). IHP-UNESCO Tech Doc in Hydrol 84:75–80

Siwek JP, Żelazny M, Chełmicki W (2011) Influence of catchment characteristics and flood type on relationship between streamwater chemistry and streamflow: case study from Carpathian Foothills in Poland. Water Air Soil Pollut 214:547–563

Skiba S, Drewnik M, Klimek M, Szmuc R (1998) Soil cover in the marginal zone of the Carpathian Foothills between the Raba and Uszwica rivers. Pr Geogr 103:125–135

Stålnacke P, Grimvall A, Libiseller C, Laznik M, Kokorite I (2003) Trends in nutrient concentrations in Latvian rivers and the response to the dramatic change in agriculture. J Hydrol 283:184–205

Stottlemyer R (2001) Processes regulating watershed chemical export during snowmelt, Fraser Experimental Forest, Colorado. J Hydrol 245:177–195

Suzuki K (1995) Hydrochemical study of snow meltwater and snow cover, Biogeochemistry of seasonally snow-covered catchments. IAHS Publ 228:107–114

Święchowicz J (2005) Zróżnicowanie i zmiany koncentracji zawiesiny. In: Żelazny M (ed) Dynamika obiegu związków biogennych w wodach opadowych, powierzchniowych i podziemnych w zlewniach o różnym użytkowaniu na Pogórzu Wiśnickim. Instytut Geografii i Gospodarki Przestrzennej Uniwersytetu Jagiellońskiego, Kraków

Walling DE, Foster IDL (1975) Variations in the natural chemical concentration of river water during flood flows, and the lag effect: some further comments. J Hydrol 26:237–244

Walling DE, Webb BW (1980) The spatial dimension in the interpretation of stream solute behaviour. J Hydrol 47:129–149

Wyżga B (2008) A review on channel incision in the Polish Carpathian rivers during the 20th century. In: Habersack H, Piégay H, Rinaldi M (eds) Gravel-Bed Rivers VI—from process understanding to river restoration. Elsevier, Amsterdam

Żelazny M (ed) (2005) Cechy fizykochemiczne opadów atmosferycznych. Dynamika obiegu związków biogennych w wodach opadowych, powierzchniowych i podziemnych w zlewniach o różnym użytkowaniu na Pogórzu Wiśnickim. Instytut Geografii i Gospodarki Przestrzennej Uniwersytetu Jagiellońskiego, Kraków

Results of Monitoring Torrent-Control Hydrotechnical Structures in the Cârcinov River Catchment

Nicu-Constantin Tudose, Mihai-Daniel Niţă and Ion Clinciu

Abstract This study presents results of monitoring torrent-control hydrotechnical structures within the Cârcinov River catchment, which revealed the nature and frequency of events affecting the functioning of the structures. Among structure damages, the most common were: breakages (45), undermining of the body (16) and erosive damages (15). In the group of structure dysfunctionalities, the most common were: unsupervised installation of forest vegetation (48), clogging of the apron (19) and the deepening of the valley downstream the structure (16).

1 Introduction

Promoting multifunctional, sustainable and competitive silviculture within the context of preventing environmental damage comprises monitoring of the managed torrential watersheds in forested areas, maintenance of the structures constructed within the watersheds, rehabilitation of the structures affected by torrential flows and securing the systems they are part of (Conesa-García and García-Lorenzo 2009; Remaître et al. 2008; Clinciu et al. 2010).

In order to be systematic and permanent, the monitoring must have a scientific support on recognition of the nature, frequency and intensity of behavioral events during the functioning period of the structures. This is why this study considers the nature and frequency of behavioral events encountered on hydrotechnical structures in the Cârcinov River catchment after over 30 years of their functioning.

N.-C. Tudose (✉) · M.-D. Niţă · I. Clinciu
Transilvania University Brasov, Sirul Beethoven 1, 500123 Brasov, Romania
e-mail: cntudose@yahoo.com

N.-C. Tudose
Institute of Forest Research and Management, Closca 13, 500040 Brasov, Romania

J. Kozak et al. (eds.), *The Carpathians: Integrating Nature and Society Towards Sustainability*, Environmental Science and Engineering,
DOI: 10.1007/978-3-642-12725-0_13, © Springer-Verlag Berlin Heidelberg 2013

2 Study Area and Methodology

The Cârcinov River catchment is located in the southern part of the Meridional Carpathians (25°03′ E, 45°03′ N) in Romania. It is characterized by high torrentiality mostly due to low forest cover. Indeed, the name of the catchment, of Slavic origin, means "deforestation". The geology of the catchment supports its predisposition to torrentiality, with the area being underlain by cross-stratified marls and sandstones covered with a 3–5 m thick layer of loess deposits.

To collect data on the functioning and damages of hydrotechnical structures, a standard procedure was employed (Clinciu et al. 2010), comprising identification, registration and description of the behavioral events at component parts of the structures (body, wings, apron etc.).

3 Results

Two groups of behavioral events were investigated: (i) events affecting the operational safety and durability of the structures, and (ii) events that affect the functionality of the structures.

The events from the first group (damages) included: cracks, breakages, carrying away events, deformations, damages by water and sediment erosion, weathering, unfastening, infiltrations, undermining of the body or the apron of the structure and suffusion.

The events from the second group (dysfunctionalities) included: obstruction of the spillway, obstruction of the energy dissipating teeth, clogging of the apron, unsupervised installation of forest vegetation, clearing of the sediment trapped behind the structure, unattained sediment entrapment behind the structure and covering of the structures by alluvium.

According to the number of affected structures (NLA), the most common damages of the structures were: breakages (45), undermining of the body (16) and erosive damages (15). In turn, the most common structure dysfunctionalities were: unsupervised installation of forest vegetation (48), clogging of the apron (19) and channel deepening downstream the structure (16).

Regarding the number of affected parts of the structures (NPLA), the hierarchy of the damages was as follows: cracks (105), breakages (30), erosive damages (26) and infiltrations (23). Among the structure dysfunctionalities, the most common were: unsupervised installation of forest vegetation (140), clogging of the apron (19) and channel deepening/covering of the structures by alluvium (16).

The ratio between the number of affected parts of the structures and that of affected structures was encoded as NPLA/NLA. Among structure damages, the highest ratio (2.3) was recorded for cracks and carrying away events, followed by breakages and infiltrations (1.8). Among the structure dysfunctionalities, the highest NPLA/NLA ratio was recorded for unsupervised installation of forest vegetation (3.0) followed by covering of the structures by alluvium (1.3).

The research showed that the most affected parts of the structures were:

- overflowed body in 63 cases, apron in 38 cases and right wing in 37 cases, in the first group of events,
- upstream zone in 45 cases, downstream zone in 41 cases and apron in 33 cases, in the second group of events.

An overall analysis of all the events identified in this research, notwithstanding the group categorization, indicated the apron as the most commonly affected part of the structures (affected in 71 cases by 8 behavioral events) followed by overflowed body (affected in 65 cases by 10 behavioral events) and the right wing (affected in 52 cases by 9 events).

4 Conclusions and Practical Recommendations

The Cârcinov River catchment contains 58 torrent-control hydrotechnical structures that were constructed more than 30 years ago. Over the past 30 years, the most important torrential events occurring in the area resulted in the total loss of five structures and the partial loss of 15 structures.

Based on our findings, we recommend that annual maintenance works should be carried out regularly on each of the hydrotechnical structures present in the area. Moreover, we suggest that rehabilitation works should be carried out periodically, encompassing clearing the floating material and vegetation and leaving a free passageway for water to flow.

Acknowledgments This chapter was supported by the Sectoral Operational Program Human Resources Development, financed from the European Social Fund, and by the Romanian Government under the contract number POSDRU/6/1.5/S/6, and had a methodology support from ID_740/2008 research project financed by CNCSIS-UEFISCSU, in the frame of IDEI program. We thank Chris Reynolds and Alexandru Tudose for editorial suggestions to the paper.

References

Clinciu I, Petrițan I, Niță MD (2010) Monitoring of the hydrotechnical torrent control structures: a statistical approach. Environ Eng Manag J 9:1699–1707

Conesa-García C, García-Lorenzo R (2009) Effectiveness of check-dams in the control of general transitory bed scouring in semiarid catchment areas (South-East Spain). Water Environ J 23:1–14

Remaître A, van Asch ThWJ, Malet JP, Maquaire O (2008) Influence of check dams on debris-flow run-out intensity. Nat Hazards Earth Syst Sci 8:1403–1416

Part II
Forests

Forests in the Carpathian Mountains and Their Management

Andrzej Bytnerowicz

Forests are an integral component of the Carpathian Mountains in Central Europe and have many important functions for the entire Carpathian region. They are crucial for soil formation, slope stability, water retention and supply, preservation of biodiversity, carbon sequestration, conservation of endangered habitats, supply of corridors for migration of rare and endangered wildlife species, recreational use, and production of timber. Forest functioning and health depend on the effects of multiple interactive stressors, such as air pollution, climate change, outbreaks of pests and diseases, or improper management practices observed in the Carpathian Mountains. Consequences of these effects are changes in biogeochemical cycling, ecophysiological processes, regeneration and growth of trees, biodiversity, and water availability. In recent decades increases in temperature, changes in water regime, changes in land ownership, and new management practices have significantly affected the health, processes, and functioning of Carpathian forest ecosystems. In addition, recent extreme climatic events caused serious calamities, such as the 2003 drought that resulted in extensive dieback of sessile oak in the Hungarian part of the Carpathians, or the November 19, 2004, heavy windstorm resulting in a massive windthrow of Norway spruce stands in the Slovak part of the Tatra Mountains.

Nature-based forestry and returning to the natural composition of forest ecosystems (prior to the massive nineteenth-century establishment of nonindigenous Norway spruce monocultures), as well as multifunctional forest management, can greatly help improve the future condition of forest ecosystems in the Carpathian Mountains. This management approach is urgently needed given the massive dieback of Norway spruce monocultures and the rapidly declining condition of beech forests observed in vast areas of the Carpathians (Badea et al. 2004; Hlasny et al. 2010). These phenomena probably result from past soil acidification caused by long-term nitrogen and sulfur deposition affecting fine roots of trees and

A. Bytnerowicz (✉)
USDA Forest Service, Pacific Southwest Research Station, Riverside, CA 92507, USA
e-mail: abytnerowicz@fs.fed.us

J. Kozak et al. (eds.), *The Carpathians: Integrating Nature and Society Towards Sustainability*, Environmental Science and Engineering,
DOI: 10.1007/978-3-642-12725-0_14,

combination of other abiotic and biotic factors (Kurjak et al. 2010). By improving our understanding of forest functioning and conditions and applying this knowledge to sound management policy and practices, we can greatly help resolve recent problems facing the Carpathian forests.

This section of the book covers various aspects of forestry research, with an emphasis on understanding modern approaches to forest management in the Carpathians. Chapters included in this section deal with forests in different Carpathian countries: four in Ukraine, three in Slovakia, and one each in Poland, the Czech Republic, Romania and Hungary.

The forestry section begins with a chapter by Střelcová et al., who investigated physiology and growth of Norway spruce and European larch in the Tatra National Park in Slovakia. The authors found that the stem-circumference increment experienced accelerated growth through the end of June, at which point almost all of the annual growth in diameter had already occurred. Transpiration rates and stem circumference changes of both species were significantly correlated with micrometeorological factors. After leaves had been fully developed, transpiration in larch exceeded that in spruce and lasted as long as the needles were retained. Differences in transpiration rates between these two species support the formation of mixed multistory larch-spruce stands in an area that was devastated by a heavy windstorm on November 19, 2004.

In the second chapter Shandra et al. describe changes in forest cover and timberline position over the entire Carpathian mountain range from a perspective of climatic and land use changes. Between 1880 and 2000 forest cover for the entire region slightly decreased; however, there were significant differences between various regions. In the Western and Ukrainian Carpathians forest cover increased at all elevations, mainly due to the declining importance of agriculture and the increasing application of sustainable forestry practices. An opposite trend was seen in the Romanian Carpathians where the observed decline of forest cover at lower elevations could be attributed to wide-spread illegal logging in the past 20 years, while at higher, less accessible elevations, forest cover mostly increased. At timberline there was a substantial forest increase in the entire Carpathians similar to other European mountains.

Marinescu et al. analyzed changes in forests cover of the Parâng-Cindrel Mountains of the Southern Carpathians, Romania, over the last century by using available cartographic products, calculating various landscape metrics and analyzing morphological spatial patterns. The authors found three main periods related to major anthropogenic disturbances. In the first period in the beginning of the twentieth century, the most significant forest changes were mostly caused by expansion of the grazing and establishment of high-altitude settlements related to an intensive pastoral activity. The second stage of drastic reduction of forest cover was in the 1970s when large hydrotechnical constructions, increased timber harvesting rates and growth of tourist infrastructure occurred. The third stage occurred after 1989 due to a forest ownership change and lax institutional policies resulting in increase of forest harvesting for the purpose of large, short-term profit.

Ujházy et al. present the effects of various tree species on the understory of fir–beech forests of the Dobročský prales National Nature Reserve and its buffer zone. This area represents the most widespread case of tree-composition change in the Western Carpathians—replacement of mixed fir–beech forests by spruce plantations—and its ecological consequences. Frequencies of herb species distribution were affected both by beech and spruce. In general, number of herb species decreased with the increasing tree competition, especially by beech. The authors identified herb species typical for various types of forests. They found that spruce stands have a more diverse and species-rich herb layer; however, several species of the natural fir–beech forests are suppressed or excluded in such stands. Information about species composition shifts and the changed tree species of secondary stands is important for long-term ecological management.

Mázsa et al. describe a network of 16 forest reserves consisting of natural deciduous beech and oak forests in a hilly region of northern Hungary. The authors also present preliminary results of the current projects in the Vár-hegy Forest Reserve as a case study. The alteration of tree species composition was investigated based on the reconstruction of forest history in the previous 130 years (management period) and analyses of forest stand inventory. In addition, CO_2 sequestration changes of these forest stands were modeled, and carbon stored in the forest ecosystem was estimated since the clear-cutting in the 1880s. The authors point out that by studying the recently established forest reserves, information about the best strategies for current forest management can be obtained. Forest reserves can serve as reference areas for studying the natural succession of abandoned woodlands, analysis of the carbon cycle in semi natural forests, and in answering other various questions related to nature conservation, ecosystem functioning, and restoration ecology under changing climate.

Kulla et al. analyzed the spatial variability of geologic and climatic factors that pre-determine the natural potential of forest sites in Slovakia, as well as the relationships between these factors and the natural occurrence and productivity of the main tree species—oak, beech and spruce—in the Slovakian Carpathians. Several national classification systems that combine the existing and newly designed regional and site units were examined and compared in terms of variability of both the natural occurrence and productivity of these tree species. A site classification system composed of a maximum of 8 units on the regional level and 80 units on the site level appears to be sufficient for strategic forest management in natural conditions of Slovakia. The classification system, which considers both site productivity and natural tree species occurrence, is assumed to be a good basis for the development of a model that would integrate effects of climate change.

Two silvicultural chapters discuss recent attempts to improve management of the Carpathian forests. Pach et al. studied sycamore maple, little-leaf linden, and Scots pine as valuable admixtures in the beech and sycamore maple stands of the Bieszczady Mountains, linden forest in the Beskid Sądecki Mountains and scattered pine forests in the Polish Carpathians. The authors found that sycamore maple can be a highly productive admixture, a co-dominant, or even dominant species in beech forests and a protective species of the upper timberline. Little-leaf linden in stands where site conditions meet its requirements, especially in the lower part of the lower mountain zone, is a suitable admixture or even co-dominant species increasing forest productivity. In the Polish Carpathian forests three pine population groups were distinguished on the basis of morphological traits (lowland, foothill, and mountain groups), which varied also according to timber quality, stem, and crown traits, as well as Cu (copper), Fe (iron), and K (potassium) needle content. Pine is a very good nursing tree species, which enables other tree species, especially fir, to grow among pine trees or underneath their canopies.

Slodičák et al. describe results of long-term experiments aimed at improving the status of the declining Norway spruce monoculture stands in the Western Carpathians. Although broadleaves used to dominate in natural tree species composition of the Western Carpathians, Norway spruce stands currently cover more than one-half of the forested area. The higher proportion of spruce is a result of its massive planting in the mid-nineteenth century and a high demand for spruce wood. Current changes in precipitation amounts and patterns, as well as in temperature and solar radiation, indicate that the previous and current cultivation of Norway spruce cannot be sustained. Consequently, conversion of the current Norway spruce stands is imminent. Based on experiments in spruce stands in the Czech part of the Western Carpathians, the authors propose that carefully conducted thinning can support spruce–stand stability and is recommended as a major silvicultural practice for stabilization of its stands before their conversion. A new forestry concept based on an adequate thinning regime is discussed.

The next two chapters examine sustainable forest management (SFM) in the Ukrainian Carpathians. Elbakidze and Angelstam propose that to implement SFM policies, policy makers and land managers as well as different forest stakeholders need empirical information on how different SFM dimensions are understood and developed locally. Focusing on the trends of SFM implementation in the Ukrainian Carpathians, the authors analyzed the barriers and bridges at multiple levels, from the national to the local management unit. They identified three types of gaps—policy creation, implementation, and knowledge—that need to be bridged. The authors recommend accomplishing this by capacity building, collaboration, and applying a zoning approach at multiple scales, which will satisfy economic, ecological, and sociocultural dimensions of SFM.

Keeton et al. explore emerging models and innovative practices that offer guidance for implementing SFM in the western Ukraine. The authors propose that the contribution of SFM to biodiversity conservation depends on the establishment of fully representative and sufficiently extensive forest reserve systems. On managed forestlands, providing a better balance of stand ages and recently developed silvicultural practices should help to maintain ecosystem functions while ensuring a range of economic uses. Restoration of native species composition in areas dominated by spruce plantations is needed to enhance forest health and promote biodiversity conservation. Expanding forest sector participation in forest certification and carbon markets offers new opportunities and challenges: afforestation can potentially sequester large quantities of carbon and generate substantial economic benefits, while the relatively long rotations required under the Ukrainian forest code offer significant carbon storage benefits. Similarly beneficial is conservation of high biomass, old-growth Carpathian beech and spruce–fir forests. With climate change, a variety of stresses are predicted to increase, requiring adaptive responses. The challenge facing the Ukraine and other countries that manage areas of the Carpathians is to merge all these ideas into a holistic, landscape approach adapted to the context of transitional economies.

In the last chapter of this section, Zahvoyska and Bas provide information on how various stakeholders in the Ukrainian Carpathians perceive mountain forest ecosystem services (MFES). There are multiple stakeholders benefiting from MFES, and their interests often contradict each other. The paper provides insights into the implicit dynamics of stakeholders' perceptions and preferences concerning MFES through an analysis of forest values and a set of cognitive maps of preferences across various stakeholders. The dominant themes are: environmental, recreational, economic, educational, health care, emotional, and aesthetic. The set of cognitive maps of stakeholders' preferences regarding forest values was developed applying nonparametric statistical methods. The results indicate a consensus in perceptions of personal values across stakeholders with some contradictions in preferences of different stakeholder groups.

In addition to the chapters in this section, which provide only a portion of the ongoing forest research of the Carpathian Mountains, recommendations from two "Forestry" sessions at the first Forum Carpaticum (Kozak et al. 2011) are summarized below. These recommendations are based on approximately 20 talks and 40 posters presented during the conference and therefore provide a very comprehensive understanding of future forestry research needs for the Carpathian region.

Improved knowledge of long-term climatic and chemical changes in the Carpathian Mountains is essential to predict any future ecological, health, or growth changes in forests. Thus, a better understanding is needed about the interactive effects of multiple stressors on the ecophysiology of key forest tree species and on biodiversity, especially from a perspective of rare and endangered species and invasive plants. Better scientific knowledge is also needed on forest growth and health, as well as the capacity of forest ecosystems to provide goods and services under multiple interactive stressors. With regard to provisional ecosystem services, forests provide clean water for the entire Carpathian region, and therefore more knowledge is needed about the present and future effects of environmental pollution, climate change, and management practices on the Carpathian watersheds. For the provision of clean air, it is imperative to ensure that toxic gas emissions from the existing stationary pollution sources are strictly monitored and controlled. In addition, information about the effects of mobile air pollution sources, mainly automobiles, and proper planning and implementation of transportation routes is essential. Thus, a combination of various strategies, such as compliance with international air pollution standards, long-term goals of improved fuel efficiency, and reduced emissions of ozone precursors from cars is proposed. Forests also have a major role in carbon sequestration; therefore, there is a need for developing and testing new techniques for improved inventories of below—and aboveground biomass and carbon pools and the effects of various management practices on carbon balance.

In general, well-coordinated forest monitoring for the entire Carpathian region is needed to determine vulnerabilities and adaptation of forest ecosystems to the combined effects of climate change and the chemical environment. Therefore, priorities for policy and strategies leading to sustainability of the Carpathian forests should be established. To accomplish this goal, a comparison of forest management practices used in various countries and an exchange of knowledge is also needed. For a successful implementation of all those demands, it is essential to have well-informed decision makers and an engaged general public. Thus, understanding the public perception of the value of forests and ecosystem services that they provide in the Carpathian region is essential. These goals can be accomplished through properly designed and conducted surveys of various stakeholders and establishing strong and effective linkages between scientists, managers, policy makers, and the general public.

References

Badea O, Tanase M, Georgeta J, Anisoara L, Peiov A, Uhlirova H, Pajtik J, Wawrzoniak J, Shparyk Y (2004) Forest health status in the Carpahian Mountains over the period 1997–2001. Environ Pollut 130:93–98

Hlasny T, Grodzki W, Šramek V, Holusa J, Kulla L, Sitkova Z, Turcani M, Raczka G, Strzelinski P, Wegiel A (2010) Spruce forest decline in the Beskids. In: Hlasny T, Sitkova Z (eds) Spruce forest decline in the Beskids. National Forest Centre, Zvolen

Kozak J, Björnsen Gurung A, Ostapowicz K (2011) Research agenda for the Carpathians 2010–2015. Integrating nature and society towards sustainability. Institute of Geography and Spatial Management, Jagiellonian University, Kraków

Kurjak D, Konopka B, Priwitzer T, Kmet J, Ditmarova L (2010) Physiological aspects of spruce decline in the Beskids. In: Hlasny T, Sitkova Z (eds) Spruce forest decline in the Beskids. National Forest Centre, Zvolen

Growth and Water Balance Parameters of a Natural Spruce–Larch Forest in Tatra National Park, Slovakia

Katarína Střelcová, Dagmar Magová, Peter Fleischer and Erika Gömöryová

Abstract The transpiration and stem circumference changes of spruce (*Picea abies* [L.] Karst.) and larch (*Larix decidua* Mill.) trees were investigated during the vegetative period of 2009 in the Smrekovec research plot, which is located in Tatra National Park in Slovakia. The plot is situated in a primeval 125-year-old mixed forest (80 % spruce trees and 20 % larch trees) at an altitude of 1,250 m. Meteorological parameters and soil water potentials were also measured during the same period within the investigated plot. Whole-tree transpiration was continuously measured based on the stem-tissue heat balance method in five larch and five spruce tree samples. Stem circumference changes were continuously measured using automatic dendrometers. We also investigated the seasonal and diurnal changes in transpiration. Air temperatures from 6 to 10 °C appeared to control the initiation of cambial bole growth in both species at the beginning of June. The stem circumference increment experienced accelerated growth through the end of June, at which point 80 % of the annual growth in diameter had occurred. Diameter increment ceased to grow in early October. The transpiration rates and stem circumference changes of both species were significantly correlated with micrometeorological factors. On sunny days, we found a linear relationship between transpiration and stem circumference changes ($R^2 = 0.60$ for spruce trees and $R^2 = 0.56$ for larch trees). After the leaves were fully developed, transpiration in larch exceeded that in spruce and lasted as long as the needles were retained on larch trees. We suggest that the phenological state of the larch tree needles is important for regulating the timing of physiological processes, such as transpira-

K. Střelcová (✉) · D. Magová · E. Gömöryová
Department of Natural Environment, University in Zvolen, Technical T. G. Masaryka 24
960 53 Zvolen, Slovakia
e-mail: strelcov@vsld.tuzvo.sk

P. Fleischer
Research station of the Tatra National Park, Tatranská Lomnica, Slovakia

J. Kozak et al. (eds.), *The Carpathians: Integrating Nature and Society Towards Sustainability*, Environmental Science and Engineering,
DOI: 10.1007/978-3-642-12725-0_15, © Springer-Verlag Berlin Heidelberg 2013

tion and growth, and that the differences in transpiration rates between these species support the formation of mixed (larch–spruce) multistory stands in an area that was disturbed by a heavy windstorm in 2004.

1 Introduction

The water balance of forest ecosystems is an important factor that is affected by climate. The majority of water transport in such ecosystems occurs via transpiration. Tree transpiration is a key factor for water exchange because water fluxes are strongly linked to the passage of water through the stomata (Lagergren and Lindroth 2002). Forests, which cover large areas, make a major contribution to total energy and mass fluxes (Granier et al. 2000). Transpiration is an important component of water balance at high elevations. The magnitude and fluctuations of transpiration depend on forest stand characteristics and dynamics, plant physiology, and site climate conditions, which include the amount and the distribution of precipitation, soil moisture content, evaporative demands of the atmosphere that reflect the simultaneous influence of several meteorological factors (air temperature and humidity, soil water content, air circulation), and a host of other factors (Small and McConnell 2008; Střelcová et al. 2009a, b).

Transpiration is reduced under stress conditions (Čermák et al. 2007). Physiological processes, such as transpiration, are sensitive indicators of stress in plants, especially under extreme environmental conditions. It is difficult to identify all of the relevant factors that influence the water regime of forest stands (Centritto et al. 2011), although Schwalm et al. (2010) have recently shown that assimilation is more sensitive than respiration to drought at the ecosystem level. It is crucial to identify how transpiration in different species responds to various factors. Recent studies have confirmed interspecific variation in transpiration due to differences in the sensitivity of trees to drought (Hölscher et al. 2005). Actual rate of transpiration is limited by atmospheric evaporative demand and soil water content (Schume et al. 2005; Zweifel et al. 2005; Nadezhdina et al. 2007). Diurnal measurements of stem circumference changes can be a good indicator of tree water status and water use efficiency. The water status of the tree is reflected in the overall shrinkage or swelling of the stem. Wet conditions increase available water in the soil and result in an increase in diameter, while during drought, the diameter decreases (Ježík et al. 2007; Sevanto et al. 2008). The fluctuation in diameter follows the pattern of transpiration, and its daily amplitude reflects the balance between transpiration rate and water uptake from soil (Perämäki et al. 2001). Tree-specific patterns of water use in response to drought stress are influenced mainly by environmental conditions, root formation, and stomatal closure. Root formation significantly influences transpiration and the depletion of soil water (Gartner et al. 2009, 2011; Capuliak et al. 2010).

On November 19, 2004, natural larch–spruce forests in Tatra National Park (TANAP) were affected by a catastrophic windstorm. As a consequence of the altered microclimatic, soil moisture and microbial conditions caused by the windstorm, changes in transpiration, and other physiological processes of the dominant tree species were expected (Lichner et al. 2007; Střelcová et al. 2009a, b). During the 2004 windstorm, Norway spruce (*Picea abies* [L.] Karst.) was the most damaged tree species despite its flexibility in responding to extreme climate and soil conditions. The physiology of Norway spruce is influenced and limited by a range of abiotic and biotic factors and their interactions. These factors include drought, wind, snow, frost, insects, and management (Ditmarová et al. 2008). Norway spruce is adapted to a cool climate, it is not sensitive to low temperatures, and it is resistant to winter freeze desiccation and to physiological drought. However, it is susceptible to windthrow due to its shallow rooting system. European larch (*Larix decidua* Mill.) is also adapted to cool climates and to temperature fluctuations, but its resistance to adverse wind conditions is greater than that of spruce due to its winter deciduousness. The least injured species during the windstorm was larch due to its short crown and small surface area and due to the fact that its needles had fallen in the late autumn.

The rapid dieback of individual trees and of whole stands of trees in the Tatra Mountains region was also accelerated by extreme weather conditions connected with global climate change. Such changes also affect the physiological processes and the health of spruce stands (Hlásny and Sitková 2010; Centritto et al. 2011).

The estimation of annual values of transpiration in these forests is crucial for the calculations of stand and regional water balances in the TANAP area. Seasonal and diurnal dynamics of transpiration and sap flow rate are useful to explain the behavior of larch and spruce trees, which are species that are important for the stability and survival of these forests.

Because not enough information is available on the differences in the diurnal, seasonal, and annual dynamics of transpiration and sap flow rates in larch (broadleaf) and spruce (needled) trees or on their reactions to conditions in the boundary layer, our study focused on the assessment of some characteristics of these parameters measured directly at the whole-tree level. The purpose of this chapter is to compare the transpiration rates and stem circumference growth from bud burst to leaf shed of two coniferous tree species: spruce, a representative of evergreen conifers, and larch, a deciduous conifer.

2 Materials and Methods

The experimental area is located in Tatra National Park in northern Slovakia. The research plot was established by the Research Station of the TANAP (Fleischer 2008) as a reference, with intact forest after a windstorm on November 19, 2004. The plot is situated in a primeval spruce forest in Smrekovec (N 49°07′17.5″, E 20°06′16.4″). The mixed 125-year-old forest is situated at an elevation of

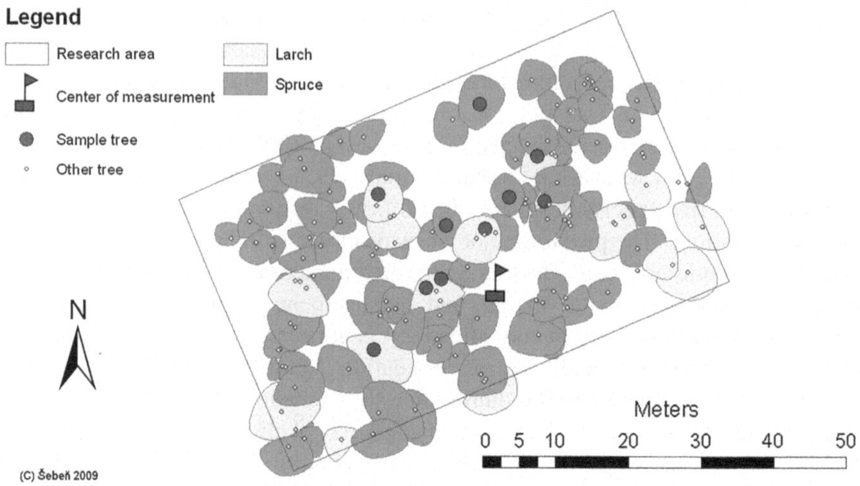

Fig. 1 Sample plot showing individual tree locations and the sample trees (*black circles*) for which sap flow and stem circumference changes were measured, Tatra National Park, Slovakia

1,250 m a.s.l. on a 10–20 % slope with a southeastern aspect. The studied forest stand belongs to the sixth forest altitudinal vegetation zone, which includes *Lariceto–Piceetum*-type forests. Spruce is the dominant tree species (80 %), and it is mixed with larch (20 %). During the study period, the experimental forest had a mean tree height (TH) 21.6 m and a diameter at breast height (DBH) of 31.7 cm. The total volume of the examined forest stand was estimated at 376 m^3/ha; the spruce volume was 319 m^3/ha of the forest stand, and the larch volume was 57 m^3/ha (Fig. 1). The soil type is Dystric Cambisol on moraine.

Transpiration rates and stem circumference (SC) changes were measured for five sample individuals of each of two coniferous species (*Picea abies* [L.] Karst. and *Larix decidua* Mill.) (Table 1). Transpiration was measured with a stem gage at the beginning (May 2009) and the end (October 2009) of the vegetative period. The stem circumference changes were recorded during the entire vegetative period from May to October 2009 at intervals of 20 min.

This locality is classified as a temperate cool region with the mean annual air temperature ranging from 3.5 to 4.0 °C with the mean annual precipitation varying from 900 to 1,100 mm. For a better comparison of the analyzed season with the long-term climate, we also used long-term climatic data from the Štrbské Pleso station of the Slovak Hydrometeorological Institute. The long-term average temperature from 1961 to 1990 was 3.7 °C; in the vegetative period (April–September), the average temperature was 9.2 °C. The long-term average annual precipitation total was 959 and 570 mm in the vegetative period (Fig. 2).

Air temperature and humidity were measured in the primeval spruce forest at 2 m above the ground. The soil temperature and water potential were measured at a depth of 15 cm. The global radiation and precipitation were measured in an open

Table 1 Mensuration variables for sample spruce and larch trees in 2009, Tatra National Park, Slovakia

Sample tree	SC (cm)	DBH (cm)	TH (m)
Spruce1	112	36.4	25.8
Spruce2	98.5	32.5	20.9
Spruce3	124	41	25.6
Spruce4	73.4	23.9	17.8
Spruce5	156.5	50	24.3
Larch1	138	44.9	29.2
Larch2	109	36.7	26.6
Larch3	99	31.8	23.2
Larch4	121.5	40	25.9
Larch5	94	31.5	24.3

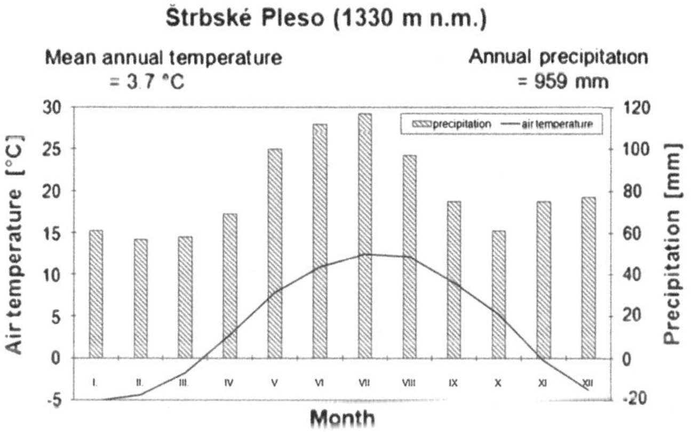

Fig. 2 Climate diagram for the meteorological station Štrbské Pleso (30-year averages, 1961–1990), Slovakia; "n.m." is an altitude

area, 1,500 m from the investigated forest stand. All data were measured at intervals of 10 min and were stored on a data logger (EMS Brno, CZ).

The transpiration of the sample trees was estimated by direct, nondestructive, and continuous measurements using the tree-trunk heat balance method (THB), with internal heating of xylem tissues and temperature sensing. The THB method involves heating a xylem segment using three electrodes inserted into the xylem. The arrangement of the measuring points followed the descriptions of Čermák et al. (1995, 2004), Kučera (2003) and Tatarinov et al. (2005) and the literature cited therein. This method is nearly independent of the radial sap velocity profile; the output value is calculated per cm of circumference using the effective width of the heating space. The final tree sap flow is calculated using the stem DBH (Kučera 2003). The sap flow rates were measured at intervals of 1 min, and 10-min averages were computed and stored on a data logger.

Table 2 Phenological phases of the vegetative organs of spruce and larch trees, Tatra National Park, Slovakia

Larch	
Spring phenophases	Autumn phenophases
1. Growing bud	4. The start of leaf yellowing
2. The first leaf	5. Leaf yellowing
3. Green (physiologically mature) leaf	6. The start of leaf fall
Spruce	
Spring phenophase: The first May sprouts	

The stem circumference changes were measured using an automatic dendrometer (DRL 26; EMS Brno, CZ). The dendrometer is designed for long-term registration of the tree trunk circumference via stainless steel tape that encircles the tree trunk. Its length variations are measured with a rotary position sensor and are stored in a data logger memory at regular intervals. The sensor is fixed to the trunk only by the strength of the tape; no sharp or invasive parts are used (Kučera 2007). The dendrometers were installed on 10 sample trees 2 m above the ground, and data were collected, computed, and stored in the data logger every 20 min.

Phenophases of vegetative organs of the larch and spruce trees were observed according the rules of the Slovak Hydrometeorological Institute (Table 2) (Střelcová et al. 2008).

To ensure proper data processing, we calculated hourly and daily averages and sums of the measured characteristics at varying time intervals.

To test the relationships between stem circumference changes or transpiration and selected climatic factors during the 2009 vegetative season, regression and correlation analyses were applied. Coefficients of correlation between the physiological characteristics of spruce and larch trees and the environmental conditions (daily mean air temperatures, daily sums of global radiation, and daily soil water potential) were tested for significance and reported at the 0.01 level.

The differences in stem circumference changes between spruce and larch were tested by adapting one-way analysis of variance in the case of the increment data for the entire vegetative period. The same approach was used to test the differences between spruce and larch transpiration at the beginning (total spring transpiration) and at the end of the vegetation season.

3 Results

Bioclimatic characteristics during the time of the experiment are presented in Table 3 and in Figs. 3 and 4. During the vegetative season, daily mean air temperatures in May, June, and October were within 2 standard deviations of the long-term averages (1961–1990). In April and September, the mean temperature was 2 °C higher than the long-term average. In July and August, the mean temperatures

Table 3 Bioclimatic and soil characteristics during the vegetative period in 2009, Tatra National Park, Slovakia

Char/Month	IV	V	VI	VII	VIII	IX	X	VP
AT (°C)	4.8 [a]	8.9	11.3	14.7 [b]	13.7 [b]	11.0 [a]	4.5	9.9
RAH (%)	75	78	84	83	86	88	94	84
P (mm)	18.4 [c]	74.2	168.2 [e]	92.8	160.8 [f]	38.2 [d]	102.6 [e]	655.2
GR ($kW{\cdot}m^{-2}$)	6.09	4.49	4.73	4.9	4.28	3.56	1.74	4.99
ST (°C)	–	6.1	8.0	11.3	10.9	8.8	4.1	8.5
SWP (kPa)	17.0	20.9	20.6	17.4	17.2	19.7	30.4	20.2

Char characteristics, *VP* vegetative period, *AT* air temperature, *RAH* relative air humidity, *P* precipitation, *GR* global radiation, *ST* soil temperature, *SWP* soil water potential

[a] Above-average month

[b] Substantially above-average month

[c] Far below average

[d] Below average

[e] Above average

[f] Substantially above average

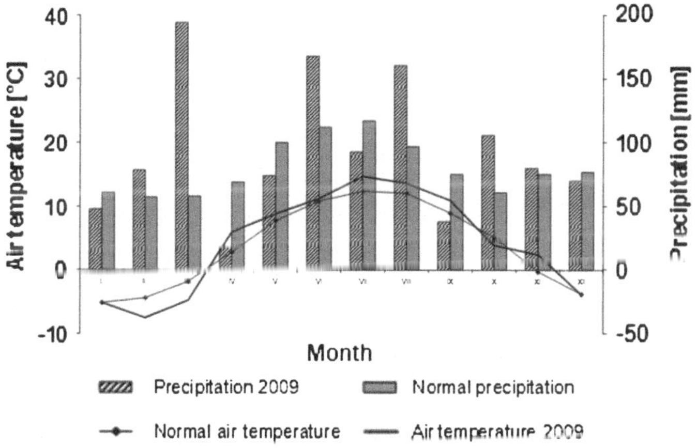

Fig. 3 Monthly precipitation and air temperature in 2009 compared to the long-term averages (30-year averages, 1961–1990), Tatra National Park, Slovakia

were higher than the long term averages by 2.3 and 1.6 °C, respectively. Compared to the long-term mean precipitation, April was extremely dry (26.6 % of normal precipitation), as was September (50.9 % of normal precipitation). In May and July, precipitation totals were within 2 standard deviations of the long-term mean totals. June and October were moist compared to the long-term mean precipitation (precipitation totals were greater than the means by 150.2 and 168 %, respectively). In August, the precipitation total was 165.7 % of the normal

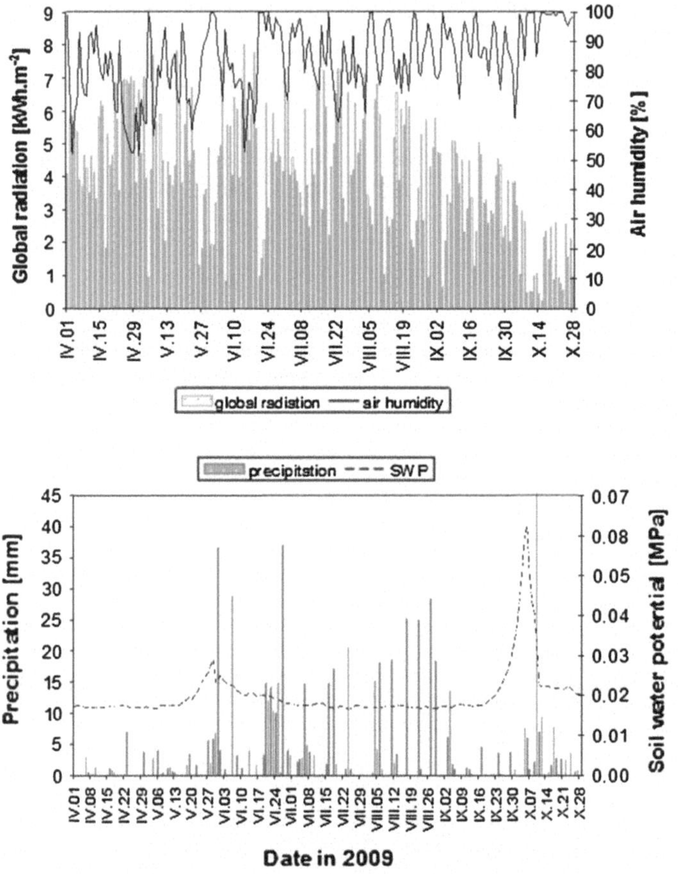

Fig. 4 Seasonal global radiation (*bar*) and relative air humidity (*solid line*) (*top*), soil water potential (*dashed line*) (absolute values) and precipitation (*bar*) (*bottom*) during the vegetative period in 2009, Tatra National Park, Slovakia

precipitation, indicating that this month was extremely moist, and high precipitation in August resulted in moist soil in September (Figs. 3, 4).

In general, from the climatological point of view, 2009 was warm with alternating periods of drought and heavy precipitation (Fig. 4). Below-average precipitation in April and September did not affect the soil water availability for trees (Fig. 4) because in April the soil contained water that had been stored from the March snow melt, and soil moisture in September was sufficient due to heavy August precipitation. Above-average April temperatures resulted in an early start of the spring phenological phase of "bud burst," which was observed on April 20, 15 days earlier than in previous years (Magová et al. 2010). This was followed by an earlier start of the second phenological phase, "first leaf," which was observed on May 7. Because the temperature in May was average at the beginning of the third phenological phase, "leaf green-up" was observed at the usual time on May

Table 4 Phenological phases of larch vegetative organs, Tatra National Park, Slovakia

Phenological phase	Start date
Bud burst	April 20
First leaf	May 7
Full leaf development	May 19
First yellowing	October 15
Complete leaf fall	November 16

19. Before this date, larch transpiration was restricted because the leaves of the trees were not fully developed. Their total surface area was therefore smaller than the surface area of spruce needles. The start date of each phenological phase is shown in Table 4.

On May 19, the third phenological phase, "full leaf development" was recorded for 100 % of the larch leaves. From the beginning of the sap flow and stem circumference measurements (May 6) until May 19, the calculated transpiration totals for spruce and larch were 526 and 344 l period^{-1}, respectively (Table 5). From May 20 until June 5, the date by which larch leaves were completely unfolded and stomatal transpiration was fully in progress, transpiration totals were 450 and 534 l period^{-1} for spruce and larch, respectively (Table 5). Daily sums of the individual tree transpiration values were proportional to the irradiance of the crowns and to the leaf area. Before the larch leaves were fully expanded, we observed lower daily total transpiration in larch than in spruce, although the differences were not significant (Fig. 5). We can therefore assume that the differences were caused by random effects and also partially by the fact that spruce, as an evergreen tree species, could transpire during the warm month of April, while the transpiration of larch was limited at that time because the leaves were still developing.

The phenological phase of "initial leaf yellowing" of larch occurred 10 days later than in previous years (October 15), probably because of the above-average temperatures in September. Similarly, the last phenological phase, "complete leaf fall," of larch was observed 6 days later than in previous years, on November 16. Until this date, average transpiration totals were 579 and 1,185 l period^{-1} for spruce and larch, respectively (Table 5). Significant differences between the transpiration totals of spruce and larch were detected through the first half of October (Fig. 5), when the leaves began to turn yellow.

Transpiration of forest stands depends on the interactive effects of a number of factors, including climatic conditions. Environmental factors affect transpiration mainly by influencing the gradient of water potential between the soil, the plant, and the surrounding air. Transpiration rate increased with decreasing air humidity, increasing air temperature, and increasing rate of air movement (Střelcová et al. 2009a, b). Global radiation, soil water potential, and water pressure deficit seem to be important climatic factors that influence transpiration. Daily transpiration of spruce and larch were positively and significantly correlated with global radiation (Table 6), which indicates sufficient soil moisture even in September, a month that

Table 5 The average transpiration of spruce and larch trees before full development of larch leaves, after full leaf development, and during the end of the vegetative period, Tatra National Park, Slovakia

Dates	Spruce (l period^{-1})	Larch (l period^{-1})
May 6–May 19	526	344
May 20–June 5	450	534
September 29–November 16	579	1,185

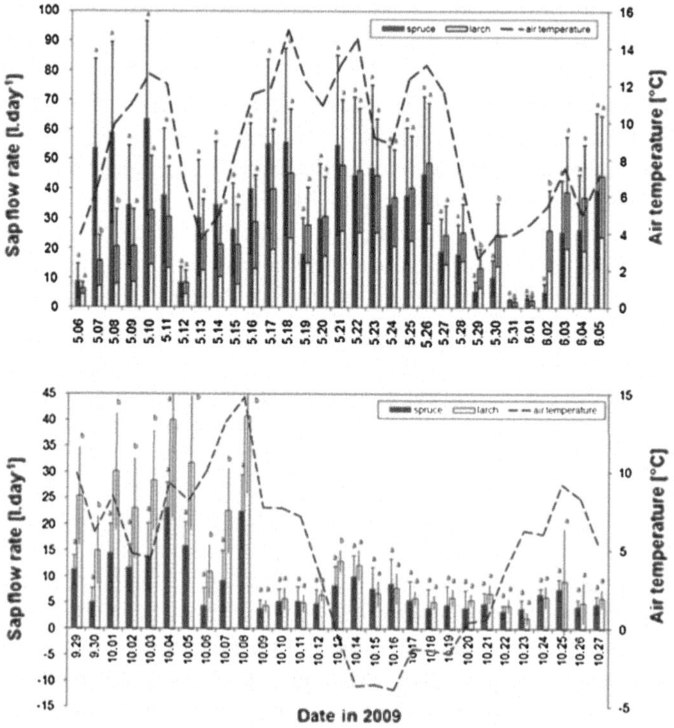

Fig. 5 Average daily total sap flow of the spruce and larch trees, with error bars representing one standard deviation, as well as daily air temperatures, at the beginning and at the end of the vegetative period in 2009, Tatra National Park, Slovakia

was characterized by low precipitation. The correlation coefficients of the relationships between daily transpiration and global radiation were 0.58 and 0.59 for larch and spruce, respectively (significance level $\alpha = 0.01$). These values indicate that when larch leaves are fully expanded, the correlation between sap flow and global radiation is high. Because temperature is closely related to global radiation, when decreased radiation was observed (on cloudy days), we also recorded decreased air temperature. In spring, the coefficient of correlation between larch transpiration and air temperature was 0.69, while at the end of the vegetative

Table 6 Regression and correlation analyses of spruce and larch sap flow values with respect to the meteorological parameters, Tatra National Park, Slovakia

DV	IV	NM	r
SFS	GR(1)	744	0.59[a]
SFL	GR(1)	744	0.58[a]
SFS	AT(1)	744	0.71[a]
SFL	AT(1)	744	0.69[a]
SFS	AT(2)	696	0.29[a]
SFL	AT(2)	696	0.39[a]

DV dependent variable, *IV* independent variable, *NM* number of measurement, *r* correlation coefficient, *SFS* sap flow of spruce, *SFL* sap flow of larch, *GR* global radiation, *AT* air temperature, (*1*) spring, (*2*) autumn

[a] Significance level $\alpha = 0.01$

season, the coefficient of correlation was 0.39 (both at a significance level $\alpha = 0.01$).

For spruce, the daily transpiration was closely correlated with air temperature in both spring ($r = 0.71$; $\alpha = 0.01$) and in fall ($r = 0.29$; $\alpha = 0.01$).

In some spruce and larch sample trees, we detected initial diameter growth at the beginning of June (Fig. 6). The dynamics of changes in stem circumference were similar for the two conifers over the course of the vegetative season. There was substantial rainfall in the last 10 days of June, to which larch responded more than spruce; this is shown in the species' average diameter increments. For both tree species, the greatest diameter increment was recorded in July, although in larch, distinct increment growth was first observed in June (Table 7). The increment growth of both spruce and larch was completed at the beginning of October. The stem circumference changes detected later were most probably caused by fluctuations in the water regime that were driven by water uptake from the roots and by transpiration, which caused stem shrinking and swelling.

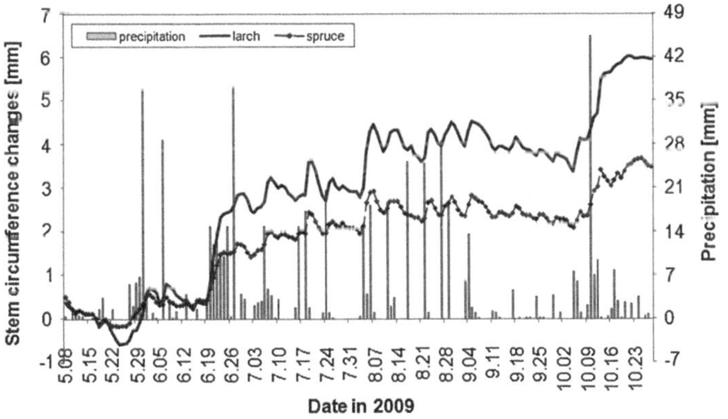

Fig. 6 Stem circumference changes of spruce and larch trees in relation to precipitation during the vegetative period in 2009, Tatra National Park, Slovakia

Table 7 Average monthly diameter increment of spruce and larch trees during the vegetative period in 2009, Tatra National Park, Slovakia

Month	Spruce		Larch	
	mm	%	mm	%
May	0.07	3.17	0.07	1.96
June	0.70	31.7	1.08	30.17
July	1.18	53.4	1.80	50.28
August	0.54	24.4	0.99	27.65
September	−0.05	−2.27	0.06	1.67
October	−0.23	−10.40	−0.42	−11.73
Total	2.21	100	3.58	100

The comparison of the stem circumference changes in spruce and larch revealed significant differences between the tree species only in the last 10 days of May (Fig. 6), which were characterized by no precipitation and a significant decrease in temperature. In the remaining monitoring period, we did not detect any significant differences in stem circumference changes between spruce and larch.

4 Discussion

Climate change may critically alter the biogeophysical and biogeochemical functions of forests. The three most important climate parameters that influence forests are precipitation, atmosphere and soil dryness, and temperature (Centritto et al. 2011). Studies by Škvarenina et al. (2009) showed a trend of increasing humidity during 1951–2007 at the northern Slovak stations (Štrbské Pleso).

Field measurements of water use at the whole-tree level are important for understanding both ecohydrology and plant physiological function. The analysis of the measurements of xylem diameter variations could therefore provide an alternative approach for estimating transpiration. Measurements of variation in xylem diameter could also be used to study xylem structure and function (Sevanto et al. 2008). Changes in temperature cause changes in the rates of physiological processes, including transpiration and tree growth. The amount of radiation captured by each individual leaf depends on the angle and the individual tree's position, its social position in the stand, and its crown shape, as well as foliage density (number of leaves per unit area).

Trees begin to transpire early in the morning, after sunrise. Transpiration reaches its maximum values at around noon (except on rainy and extremely cloudy days). After sunset, transpiration drops rapidly, and very low, almost zero, transpiration rates can be recorded at night. The time lag between the daily courses of transpiration and radiation ranges from 1 to 2 h. Based on this information, we can assume that the majority of the water stored in tissues is utilized for transpiration in the early morning hours (Čermák et al. 1995). Gartner et al. (2009) found that

Norway spruce reached its maximum transpiration rate at 11:30 a.m. during drought periods. After heavy rain falls, spruce trees reached their maximum transpiration rates in the afternoon (2–3 p.m.). In our study, no long drought periods occurred, so the soil was always sufficiently supplied with water. The transpiration of spruce reached a maximum between 2 and 4 p.m., while in larch trees, maximum transpiration was observed between 1 and 3 p.m.

The transpiration rate during the course of each day and within the growing season was highly dynamic for both species. Transpiration rates follow the seasonal radiation curve if the soil is moist (Střelcová et al. 2009a, b). If the soil water is limiting, the transpiration rate decreases with decreasing soil water availability (Schume et al. 2004). Calculated daily precipitation totals that were required to fulfill the storage capacity of the canopy were approximately 0.8–0.9 mm for a mountainous spruce forest (Holko et al. 2009). The rate of transpiration increased as the vapor pressure deficit (VPD) from plant to air increased. Střelcová et al. (2009a, b) observed that spruce trees were less sensitive to VPD than larch trees. The dependence of the transpiration rate on the VPD significantly differs between these tree species when VPD values are low. In the present study, no extreme drought periods occurred during the vegetative season, so we did not detect any significant reactions to soil water deficiency.

It is important to note that the trees at different dominance levels differed in their transpiration rates, which were influenced by their leaf areas, distribution patterns, and exposures to solar radiation. With respect to each tree's social status in the forest stand, the transpiration of subdominant individuals represents 10–30 % of the dominant trees' transpiration values in a mixed beech–spruce–fir stand in the Poľana Mountains (Střelcová and Minďáš 2002). As the leaves of subdominant individuals receive direct radiation only rarely, if at all, their water consumption for cooling, especially during clear sunny days, is several times lower than that of codominant and dominant trees (Střelcová et al. 2009a, b). In our study, the selected spruce and larch sample trees represented all forest stand layers. Average daily transpiration of the selected subdominant trees represented 20–40 % of the dominant tree transpiration measured in another research plot (in Smrekovec). In spring, before leaf development, the sum of transpiration in larch was only two-thirds of that in spruce. After larch leaves were fully developed, the transpiration of spruce was lower than that of larch, similar to the observations of Střelcová et al. (2009a, b). The transpiration rate of the spruce canopy was approximately two-thirds that of the larch canopy under the same conditions. At the end of the vegetative season, before leaf fall, spruce transpiration was only half that of larch.

Čermák et al. (2007) described a range of whole-tree transpiration rates from 150 to 300 l day^{-1} on clear days from late July to early October. The daily transpiration of individual trees was influenced by the sum of the daily radiation. In comparison, maximum daily stand transpiration ranged from 140 to approximately 200 l day^{-1} in a spruce monoculture at a mountainous site (Matyssek et al. 2009). In general, diameter increment growth depends on the combination of tree growth potential and external factors limiting growth (Mäkinen et al. 2003).

Temperature is the most important factor affecting growth at high altitudes. Rossi et al. (2007) found that cambial activity occurred from May through July or August, and Norway spruce was the last species to begin tracheid differentiation in treeline populations. Xylogenesis was initiated when the mean daily air temperatures were 5.6–8.5 °C and the mean stem temperatures were 7.2–9.0 °C. For this species, soil temperature was not the main factor limiting xylogenesis. Air temperature is considered to be a determining factor governing the onset and the growth intensity of cambial activity of 12 pine provenances (Savva et al. 2003). Moser et al. (2010) monitored European larch growth along two elevation transects and found that ring formation lasted from mid-May to the end of October, with the length of the growing season decreasing with increasing elevation from 137 to 101 days. The onset of the growing season changed by 3–4 days per 100 m elevation, but the end of the growing season appeared to be only minimally related to altitude. If associated with the monitored altitudinal lapse rate of −0.5 °C per 100 m, these results translate into a lengthening of the growing season by 7 days per degree Celsius.

In our study, we detected the onset of bole diameter growth at the beginning of June and its completion at the beginning of October. Over the whole growing season, there was a significant inverse ($r = -0.40$; $\alpha = 0.05$) relationship between stem circumference and air temperature. From July to August, the strength of this relationship increased ($r = -0.75$ and -0.72 for spruce and larch, respectively; $\alpha = 0.01$). With increasing global radiation, air temperature, and presumably evapotranspirational demand (Střelcová et al. 2009a, b), canopy transpiration also increased. The pull of the transpirational stream has been found to reduce bole diameter (Zweifel et al. 2005; Sevanto et al. 2008). Knott (2004) described a positive effect of precipitation and temperature on the bole diameter increment, while the dependence of the increment of particular trees on the climatic factors of global radiation, VPD, air temperature, and soil water potential, which are expressed by correlation coefficients, ranged from −0.4 to +0.75. We found a significant positive correlation between precipitation and bole diameter changes ($r = 0.30$; $\alpha = 0.01$). Sevanto et al. (2005) found that the amplitude of diurnal variations in xylem diameter was best correlated with photosynthetically active radiation and soil water content, but in this study, we did not find any correlation between these factors.

Lagergren and Lindroth (2004) found only a weak correlation between diameter and transpiration in birch (*Betula alba* L.) and Norway spruce; this was supported by Gartner et al. (2009). In the latter study, no significant relationship between transpiration and DBH was detected. In contrast, Sevanto et al. (2008) found that diurnal xylem diameter variations and transpiration were linearly correlated (daily average $R^2 = 0.61$–0.87) for the six deciduous species: red maple (*Acer rubrum* L.), black alder (*Alnus glutinosa* Miller), sweet birch (*Betula lenta* L.), European beech (*Fagus sylvatica* L.), red oak (*Quercus rubra* L.), and linden (*Tilia vulgaris* L.). Rossi et al. (2006) reported that dendrometers might provide estimates of the timing of the maximum increment growth rate within the growing season, as was also indicated in this study. In our study, the greatest diameter increment growth was observed on June 20, when the average daily air temperature fluctuated between 10–14 °C.

From spring to autumn, the timing of diurnal stem circumference changes varied with environmental conditions and transpiration. Zweifel and Häsler (2001) suggested that diameter changes along the stem correspond to a depletion of stored water by transpiration, while Perämäki et al. (2005) suggested that available soil water also plays a role in the reversible diurnal changes in bole diameter. The time lags between transpiration and stem shrinkage have been attributed to the time required for the release of stored water (Zweifel and Häsler 2001). Sevanto et al. (2002) found that less than one-third of whole-stem circumference changes resulted from changes in xylem, and two-thirds were caused by the changes in the living tissues outside of the xylem. The tree diameters reach their maximum before sunrise and in the early morning hours. Depleted water stores were recharged in the afternoon and at night until the early morning hours of the next day when transpiration resumed (Čermák et al. 2007; Čermák et al. 2008).

Perämäki et al. (2001) modeled transpiration based on the connection between water tension and xylem diameter shrinkage and showed a close relationship between transpiration and the changes in stem diameter in Scots pine.

5 Conclusions and Recommendations

The progressive massive dieback of woody plants, primarily Norway spruce, in central European regions is well known. Similar damage can be found in the nearby regions of Poland, the Czech Republic, and Germany. In our study, we compared the water regimes of spruce, an evergreen, and larch, a deciduous conifer.

In these species at the tree line, canopy transpiration depended on air temperature, global radiation, precipitation, and soil water potential. We investigated seasonal and diurnal changes in transpiration. At the end of May and at the beginning of June, we observed a decrease in transpiration due to higher precipitation and low global radiation, and there were few differences between the two species.

Air temperatures between 6 and 10 °C appeared to control the initiation of cambial bole growth in both species at the beginning of June. Diameter increment growth accelerated through the end of June, at which point 80 % of the annual diameter increment had occurred. Diameter increment formation ceased in early October. Subsequent changes in stem circumference were likely caused by stem swelling due to changes in stem water content. During periods of high transpiration, a reduction in stem circumference was observed. At night, when transpiration reached almost zero, stem circumference was at its maximum within each 24 h period.

After the leaves were fully developed, transpiration in larch exceeded that of spruce and lasted as long as leaves were retained on the trees. Vegetative dormancy began after leaf fall in larch and at approximately the same time in spruce. The phenological state of the leaves of larch trees is important for the timing of the physiological processes of transpiration and growth.

Climate changes directly influence forest ecosystems by means of a wide range of stressors. In the case of Tatra National Park, the effect of severe windstorms could be mitigated by favoring European larch. However, summertime watershed water outflow is likely to be lower in a larch forest than in a larch and spruce forest due to greater transpirational losses of larch. Differences in transpiration rates among stand layers and species support a plan to form mixed larch–spruce multistory stands in an area that was disturbed by a heavy windstorm in 2004. Even under longer expected periods of low air humidity and drought, a natural ecosystem with a mixture of the key regional coniferous species should adequately manage the water regime and serve as a model for human-assisted forest rehabilitation.

Acknowledgments This study was supported by the Slovak Research and Development Agency under contract Nos. APVV-0022-07 and APVV-0580-10 and APVV-0111-10 and APVV-0423-10.

References

Capuliak J, Pichler V, Flühler H, Pichlerová M, Homolák M (2010) Beech forest density control on the dominant water flow types in andic soils. Vadose Zone J 9:747–756

Centritto M, Togneti R, Leitgeb E, Střelcová K, Cohen S (2011) Above ground processes—anticipating climate change influences. In: Bredemeier M, Cohen S, Godbold D, Lode E, Pichler V, Schleppi P (eds) Forest management and water cycle. Ecol St vol 212, pp 31–64

Čermák J, Cienciala E, Kučera J, Lindroth A, Bednářová E (1995) Individual variation of sap-flow rate in large pine and spruce trees and stand transpiration: a pilot study at the central NOPEX site. J Hydrol 168:17–27

Čermák J, Kučera J, Nadezhdina N (2004) Sap flow measurements with thermodynamic methods, flow integration within trees and scaling up from sample trees to entire forest stands. Trees 18:529–546

Čermák J, Kučera J, Bauerle WL, Phillips N, Hinckley TM (2007) Tree water storage and its diurnal dynamics related to sap flow and changes in stem volume in old-growth Douglas-fir trees. Tree Physiol 27:181–198

Čermák J, Nadezhdina N, Meiresonne L, Ceulemans R (2008) Scots pine root distribution derived from radial sap flow patterns in stem of large leaning trees. Plant Soil 305:61–75

Ditmarová Ľ, Kmeť J, Leštianska A, Střelcová K (2008) Analysis of physiological parameters of spruce trees as indicators of spruce dieback in the Spiš region. Folia Oecol 35:1–5

Fleischer P (2008) Windfall research and monitoring in the High Tatra Mts., objectives, principles, methods, and current status. Contrib Geophys Geod 38(3):233–248

Gartner K, Nadezhdina N, Englisch M, Čermák J, Leitgeb E (2009) Sap flow of birch and Norway spruce during the European heat and drought in summer 2003. For Ecol Manag 258:590–599

Gartner K, English M, Leitgeb E (2011) Effects of climate on the vulnerability of Norway spruce stands—soil hydrological constraints for forest management in Austria's Lowlands. In: Bredemeier M, Cohen S, Godbold D, Lode E, Pichler V, Schleppi P (eds) Forest management and water cycle. Ecol St vol 212, pp 127–140

Granier A, Biron P, Lemoine D (2000) Water balance, transpiration and canopy conductance in two beech stands. Agric For Meteorol 100:291–308

Hlásny T, Sitková Z (eds) (2010) Spruce decline in Beskyds. National Forest Centre—Forest Research Institute Zvolen, Czech University of Life Scineces Prague, Forestry and Game Manegement Research Institute Jíloviště—Strnady. Zvolene

Holko L, Škvarenina J, Kostka Z, Frič M, Staroň J (2009) Impact of spruce forest on rainfall interception and seasonal snow cover evolution in the Western Tatra Mountains, Slovakia. Biologia 64:594–599

Hölscher D, Koch O, Korn S, Leuschner Ch (2005) Sap flux of five co-occurring tree species in a temperate broad-leaved forest during seasonal drought. Trees 19:628–637

Ježík M, Blaženec M, Střelcová K (2007) Intraseasonal stem circumference oscillations: their connection to weather course. Folia Oecol 34(2):105–115

Knott R (2004) Seasonal dynamics of the diameter increment of fir (Abies alba Mill.) and beech(Fagus sylvatica L.) in a mixed stand. J For Sci 50(4):149–160

Kučera J (2003) Sap flow meter—P4.2, instruction manual. Producer EMS Brno

Kučera J (2007) Dendrometer increment sensor—DRL 26, user's manual. Výrobca EMS Brno

Lagergren F, Lindroth A (2002) Transpiration response to soil moisture in pine and spruce trees in Sweden. Agric For Meteorol 112:67–85

Lagergren F, Lindroth A (2004) Variation in sapflow and stem growth in relation to tree size, competition and thinning in a mixed forest of pine and spruce in Sweden. For Ecol Manag 188:51–63

Lichner Ľ, Hallett PD, Feeney DS, Ďugová O, Šír M, Tesař M (2007) Field measurement of soil water repellency and its impact on water flow under different vegetation. Biol 62:537–541

Magová D, Střelcová K, Šudý M (2010) The changes of stem circumference and transpiration rate of spruce–larch forest by influcnce of environmental characteristics changes. Acta Fac For Tech Univ Zvolen 52(1):89–100

Mäkinen H, Nöjd P, Saranpää P (2003) Seasonal changes in stem radius and production of new tracheids in Norway spruce. Tree Physiol 23:959–968

Matyssek R, Wieser G, Patzner K, Blaschke H, Häberle KH (2009) Transpiration of forest trees and stands at different altitude: consistencies rather than contrasts? Eur J For Res 128:579–596

Moser L, Fonti P, Buntgen U, Esper J, Luterbacher J, Franzen J, Frank D (2010) Timing and duration of European Larch growing season along altitudinal gradients in the Swiss Alps. Tree Physiol 30(2):225–233

Nadezhdina N, Čermák J, Meiresonne L, Ceulemans R (2007) Transpiration of Scots pine in Flanders growing on soil with irregular substratum. For Ecol Manag 243:1–9

Perämäki M, Nikinmaa E, Sevanto S, Ilvesniemi H, Siivola E, Hari P, Vesala T (2001) Tree stem diameter variations and transpiration in Scots pine: an analysis using a dynamic sap flow model. Tree Physiol 21:889–897

Perämäki M, Vesala T, Nikinmaa E (2005) Modeling the dynamics of pressure propagation and diameter variation in tree sapwood. Tree Physiol 25:1091–1099

Rossi S, Deslauriers A, Anfodillo T, Morin H, Saracino A, Motta R, Borghetti M (2006) Conifers in cold environments synchronize maximum growth rate of tree-ring formation with day length. N Phytol 170:301–310

Rossi S, Deslauriers A, Anfodillo T, Carraro V (2007) Evidence of threshold temperatures for xylogenesis in conifers at high altitudes. Oecol 152:1–12

Savva YuV, Schweingruber FH, Vaganov EA, Milyutin LI (2003) Influence of climate changes on tree—ring characteristics of Scots pine provenances in southern Siberia (forest-steppe). IAWA J 24(4):371–383

Schume H, Jost G, Hager H (2004) Soil water depletion and recharge patterns in mixed and pure forest stands of European beech and Norway spruce. J Hydrol 289:258–274

Schume H, Hager H, Jost G (2005) Water and energy exchange above a mixed European Beech–Norway Spruce forest canopy: a comparison of eddy covariance against soil water depletion measurement. Theor Appl Climatol 81:87–100

Schwalm CR, Williams CA, Schaefer K, Arneth A, Bonal D, Buchmann N, Chen J, Law BE, Lindroth A, Luyssaert S, Reichstein M, Richardson AD (2010) Assimilation exceeds respiration sensitivity to drought: A FLUXNET synthesis. Glob Chang Biol 16:657–670

Sevanto S, Vesala T, Perämäki M, Nikinmaa E (2002) Time lags for xylem and stem diameter variations in a Scots pine tree. Plant Cell Environ 25:1071–1077

Sevanto S, Hölttä T, Markkanen T, Perämäki M, Nikinmaa E, Vesala T (2005) Relationships between diurnal diameter variations and environmental factors in Scots pine. Boreal Environ Res 10:447–458

Sevanto S, Nikinmaa E, Riikonen A, Daley M, Pettijohn JC, Mikkelsen TN, Phillips N, Holbrook NM (2008) Linking xylem diameter variations with sap flow measurements. Plant Soil 305:77–90

Small EE, McConnell JR (2008) Comparison of soil moisture and meteorological controls on pine and spruce transpiration. Ecohydrol 1:205–214

Střelcová K, Minďáš J (2002) Beech trees transpiration in relation to changing environmental conditions. Scientific Study 11/2000/A. Technical University in Zvolen

Střelcová K, Priwitzer T, Minďáš J (2008) Phenological phases and transpiration of European Beech in the mountain mixed forest. Meteorol J 11:21–29

Střelcová K, Kuřera J, Fleischer P, Giorgi S, Gömöryová E, Škvarenina J, Ditmarová Ľ (2009a) Canopy transpiration of mountain mixed forest as a function of environmental conditions in boundary layer. Biol 64:507–511

Střelcová K, Mátyás C, Kleidon A, Lapin M., Matejka F, Blaženec M, Škvarenina J, Holécy J (eds) (2009) Bioclimatology and natural hazards. Springer, Netherlands

Škvarenina J, Tomlain J, Hrvoľ J, Škvareninová J, Nejedlík P (2009) Progress in dryness and wetness parameters in altitudinal vegetation stages of West Carpathians: time-series analysis 1951–2007. Időjárás Q J Hung Meteorol Serv 113:47–54

Tatarinov F, Kuřera J, Cienciala J (2005) The analysis of physical background of tree sap flow measurement based on thermal methods. Meas Sci Technol 16:1157–1169

Zweifel R, Häsler R (2001) Dynamics of water storage in mature subalpine Picea abies: temporal and spatial patterns of change in stem radius. Tree Physiol 21:561–569

Zweifel R, Zimmermann L, Newbery DM (2005) Modeling tree water deficit from microclimate: an approach to quantifying drought stress. Tree Physiol 25:147–156

Influences of Climate and Land Use History on Forest and Timberline Dynamics in the Carpathian Mountains During the Twentieth Century

Oleksandra Shandra, Peter Weisberg, and Vazira Martazinova

Abstract The forest-tundra mountain ecotone, associated with the alpine timberline and treeline, is reported to have risen in elevation in many parts of the world as a result of global warming. The main goal of this study was to analyze the changes in timberline position over the Carpathian mountain range with respect to observed climate change and other global change issues, such as land use change. Global climate change has resulted in periods of significant warming in the Carpathian region, with the most significant of these occurring over 1975–2000. Forest cover change was analyzed for elevations above 1,000 m between 1880 and 2000 by comparing military maps of the Austro-Hungarian Empire and Landsat imagery. The whole region had a forest cover ratio of 72.8 % in 1880, decreasing to 71.2 % in 2000. There has been considerable forest increase at timberline (34 % of the total afforestation), indicating that substantial changes are happening at this ecotone, similarly to other European mountains; 12 % of total deforestation was observed also at timberline. Forest cover changes varied among regions, increasing by 8 % in the West Carpathians, by 1.4 % in the Ukrainian Carpathians, and decreasing in the North Romanian and South Carpathians by 6.1 and 0.7 %, respectively. Forest cover in the Ukrainian, North Romanian, and South Carpathians has declined at lower elevations, which can be attributed to widespread illegal logging in post-socialist times. At higher, less accessible elevations forest cover has mostly increased. In the West Carpathians forest cover has evenly risen

O. Shandra (✉) · V. Martazinova
Department of Long-Range Forecasting, Ukrainian Hydrometeorological Institute, Nauki Av. 37 030208 Kyiv, Ukraine
e-mail: alya.shandra@gmail.com

P. Weisberg
Department of Natural Resources and Environmental Science, University of Nevada, 1664N. Virginia St 89557 Reno, USA

Department of Physical Geography and Geoecology, National Taras Shevchenko University of Kyiv, Acad. Glushkova avenue, 2 GSP–680 Kyiv, Ukraine

J. Kozak et al. (eds.), *The Carpathians: Integrating Nature and Society Towards Sustainability*, Environmental Science and Engineering,
DOI: 10.1007/978-3-642-12725-0_16, © Springer-Verlag Berlin Heidelberg 2013

at all elevations—a result of the declining importance of agriculture and increasingly sustainable forestry practices. We consider these observed changes to be a result of favorable climate change coupled with ongoing land use change.

1 Introduction

Climate change influences many ecological processes world-wide and poses threats for high elevation ecosystems and rural economies. Ecological impacts from climate warming include altitudinal upper timberline shifts similar to forest advancement into arctic tundra. Rising timberlines are ecologically and economically important because (1) timberline rise is associated with declining alpine tundra habitat of conservation concern; (2) increased forest area has important implications for the ability of mountain forests to sequester carbon and assimilate carbon dioxide; and (3) timberline rise may provide a sensitive barometer for gauging the effects of global climate change on a particular region (Beniston 2003).

Recent timberline advance has been widespread over many mountainous regions in the Northern hemisphere, although there is disagreement as to the precise mechanisms that maintain timberline (Körner 1998; Grace et al. 2002), and timberline responses are not always consistent within and among regions with varying topography, climatic regimes and land use history (Holtmeier and Broll 2007). However, in the densely settled Carpathian Mountains impacts of climate change on timberlines are coupled with other global change influences, such as economic transition, changes in forest practice, and decline of traditional mountain farming, the latter having lowered the timberline by as much as 300–400 m (Kolischuk 1958). These influences may either mask or enhance the influence of climate change. Although many authors emphasize that agricultural abandonment is the main driving factor behind observed timberline change in European mountains (Gehrig-Fasel et al. 2007; Sitko and Troll 2008), climate change is regarded as a factor that will be more influential to timberlines in the future (Gehrig-Fasel et al. 2007; Holtmeier and Broll 2007), and trees would not establish in abandoned high-elevation pastures if climate was not suitable.

In many areas of the Carpathian Mountains, increases of timberline elevation and forest cover have been observed over different time periods (Kozak 2003; Wężyk and Guzik 2004; Dezso et al. 2005; Mihai et al. 2007; Sitko and Troll 2008; Kricsfalusy et al. 2008). At the same time, the Carpathian region has a history of intensive land use, which continues to the present day and is undergoing transformation, often resulting in deforestation (Plesnik 1978; Dezso et al. 2005; Mihai et al. 2007; Kuemmerle et al. 2009a, b). Our broad-scaled investigation addressed both aspects of timberline change, with attention paid to the extent of previous timberline depression and regional differences in timberline and forest cover change.

2 Study Area

Our study area covers all elevations above 1,000 m of the Carpathian mountain range, excluding territories that were not available or were not accurate on the 1880 Austro-Hungarian topographic maps, used to reconstruct historical timberline position. A detailed description of excluded areas is provided in the Materials and Methods section. Politically the study area is divided between Poland, Slovakia, Ukraine and Romania. However, in our study we divided the territory into four regions combining political, physiographic, and climatic considerations: West Carpathians (territory of Poland and Slovakia), Ukrainian Carpathians (Eastern Carpathians on the territory of Ukraine), North Romanian Carpathians (Bihor massif and Romanian part of Eastern Carpathians), and South Carpathians (Fig. 1). The territory studied has a total area of 23,260.8 km^2. Forest stands are comprised of beech, beech–fir, fir–spruce, and spruce (Grodzinska et al. 2004). Two main types of timberline are present: deciduous, formed mainly by beech, and coniferous, mainly formed by spruce. The deciduous timberline is predominantly of an artificial nature, having been formed after removal of superjacent coniferous forests. Only rarely are krummholz forms of beech at treeline observed as a transition to the alpine zone (Komendar 1966).

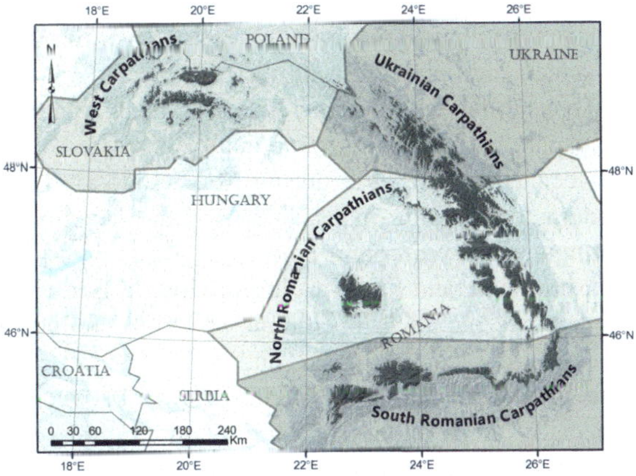

Fig. 1 Area and study regions of the investigation. Area inside which forest change was studied (with elevations over 1,000 m) is in dark gray

3 Materials and Methods

Changes in timberline position were determined by comparing topographic maps (1:75,000) from the Third Military Survey of the Austro-Hungarian Empire (issued circa 1880) with modern satellite imagery. Old maps have been used for similar research, but mostly at a regional scale (Kozak 2003; Kricsfalusy et al. 2008); satellite imagery has been widely utilized for vegetation change studies in the Carpathian region (e.g. Dezso et al. 2005; Mihai et al. 2007; Kuemmerle et al. 2009a).

Digital copies of maps were obtained from the V. Stefanyk Library in Lviv and downloaded from the server mapywig.org; altogether, 137 sheets were used. Georeferencing of the maps was divided into two stages: (1) preliminary rectification and mosaicking; (2) final georeferencing by ground control points (GCP's). The projection parameters of the survey vary from sheet to sheet. A substitute sinusoid projection for each map column has been used for georeferencing the whole set of maps from the Third Military Survey, which made it possible to follow the original map trapezoid shape with an accuracy of approximately 20 m (Molnar and Timar 2009). Instead of rectifying each column in a separate projection, we used a single Albers conic projection with a WGS84 datum for our study area, which gave us a comparable error of around 25 m. Mosaics of map sheets were generated for regions of the study area. Final georeferencing by the 3rd Order Polynomial function was carried out based on GCP's from 1:100,000 Soviet military topomaps and tourist maps of the Slovak republic. The average error for all the sheets was 74 m. During this stage territories outside the political border of the Austro-Hungarian Empire in Southern and South-Eastern Romania (territories of the Kingdom of Romania) were excluded from the study area due to their low geodetical accuracy. A vector layer of non-forest territory was created by on-screen digitizing from the maps.

To establish the modern position of the Carpathian timberline we mosaicked 13 atmospherically and topographically corrected Landsat ETM+ scenes (acquired between 2000 and 2002 during May–August). The mosaic was registered to the modern topographic maps. We used the Maximum Likelihood Classification to distinguish between eight land cover types based on training data from field GPS measurements and high resolution Google Earth imagery and photographs. Land cover types included: deciduous and coniferous forests; lakes; shrubs; grassland and pasture; recently logged; and rocks. Overall classification accuracy was 78 %. Due to a high rate of misclassification between coniferous forest and shrubs, especially those of the sub-alpine communities, the automatic classification was further improved by manually delineating the timberline by means of Google Earth high-resolution imagery and Tiles-on-Line instrument (Mitrich tools, http://mitrichtools.narod.ru/Eng/TilesOnLine_eng.htm). The corrected classification had a higher accuracy of 88 %, with a forest—non-forest classification accuracy of 96 %. A vector layer of treeless area was created, comparable to the vector layer from the old topographic maps. An intersection of the two layers gave us two types of forest cover change: afforestation (forest growing in 2000 on territories that were devoid of forest in 1880) and deforestation (areas without forest in 2000 that were forested in 1880). We studied forest cover changes over 50-meter elevational gradients for territory above 1,000 m at each study region. Within each elevational gradient, we calculated the ratio of area having undergone afforestation or deforestation to total gradient area. Nonforest areas that encircled a mountain top without large connections with valley nonforest territory were manually classified as areas above timberline; afforestation and deforestation polygons that were adjoining to this area were classified as afforestation and deforestation at timberline. The SRTM dataset from CGIAR-CSI was used as a DEM. Analysis was conducted in Envi 4.7 and ArcInfo 9.3 software.

Climatic grids from the CRU TS 2.1 dataset were used to map temperature trends over the Carpathian region for different periods of the twentieth century (Mitchell and Jones 2005). Annual temperature from this dataset is well correlated with weather station data for the territory of Ukraine (correlation coefficient 0.969) (Krakowska et al. 2008), which makes this dataset suitable for climate studies of this region. Following IPCC (2001), we calculated linear trend coefficients of average annual and summer (June, July, August) temperature over the time periods of 1901–2000, 1910–1945, 1946–1975, and 1976–2000 for the Carpathian region.

A particular interest of this study was to identify territories at the high and low extremes of anthropogenic lowering of the timberline, and to compare the position of anthropogenic timberline with climatic timberline. One of the first approximations used for establishing climatic timberline was the July 10 °C isotherm; later investigations showed it to be only a coarse approximation for the temperate latitudes of the Northern Hemisphere, more globally consistent indexes being growing season means (Körner 1998). For the Tatra Mountains, the July 10 °C isotherm well resembled the position of locations with timberline of a climatic nature (identified from private communication); therefore, we preferred this simpler approach to shorten lengthy calculation time necessary for establishing growing season means over our large study region. To establish the position of the 10 °C July isotherm, we used detrended ordinary kriging with a regionally established vertical temperature lapse rate to interpolate average July temperature (1961–1990) from 48 weather stations. Data were downloaded from the European Climate Assessment Dataset at the KNMI (Klein Tank et al. 2002). Cross-validation of the interpolated temperature returned an R^2 value of 0.97; the average mean difference was 0.16 °C. Upon comparing this isotherm with the 1880 and 2000 timberlines it became apparent that a sufficient part of territory, marked as forest in 1880, exceeded the elevation of the isotherm. The contours of many 1880 forest patches highly resemble those of shrubs on modern satellite imagery. It is probable that in these places the surveyors had mistaken *Pinus mugo* krummholz for forest. Therefore, areas where the 1880 timberline elevation exceeded both the 2000 timberline and the climatic threshold, and the contour of the 1880 timberline itself resembled shrub vegetation on the satellite imagery, were considered to be erroneous and were masked from analysis, summing up to 1.1 % of the total study area.

4 Results

4.1 Climate Change

Average annual temperature trends over the Carpathian region during the twentieth century (<0.2 °C/100 years) were smaller than the global average (0.6 °C/100 years, IPCC 2001) (Fig. 2a). Annual trend values were insignificant during 1910–1945, slightly negative during 1946–1975, and significant and positive during 1976–2000. Changes in summer temperature (Fig. 2b) show two periods of significant warming (1910–1945 and 1976–2000), spanning an intermediate period of significant cooling (1946–1975). Summer temperatures are especially important for trees growing at the timberline due to the short growing season at those elevations. Hence, global climate change has resulted in periods of significant warming in the Carpathian region, with the most significant of these occurring over 1975–2000.

4.2 Comparison of Timberline Elevation with the Climatic Timberline

From comparing the position of the 1961–1990 10 °C July isotherm with the modern timberline we conclude:

1. The current timberline runs at a much lower elevation than this climatic approximation. Only 4.7 % of the 2000 timberline length is higher than this boundary (mostly in the South Carpathians), and only 11 % of the total length is within 200 m.
2. For most of the territory, timberline elevations do not exceed the 10 °C isotherm. However, at certain locations, especially in the South Carpathians, both the 2000 and 1880 timberlines occur at elevations above the 10 °C isotherm, indicating that this climatic generalization is relative and is not equally representative of all the Carpathians.

4.3 Forest Cover Change Over 1880–2000

The elevational distribution of forest cover has changed over the 120-year study period differently among the different regions (Fig. 3). In the West Carpathians, forest cover has increased evenly until 1,600 m, at which elevation the forest covers in 1880 and 2000 roughly equal out. In the Ukrainian Carpathians, forest cover has increased between 1,200 and 1,650 m, but has decreased from 1,000 to 1,200 m. This situation is similar to the North Romanian and South Carpathians, the thresholds being at 1,425 and 1,800 m in the North Romanian Carpathians and

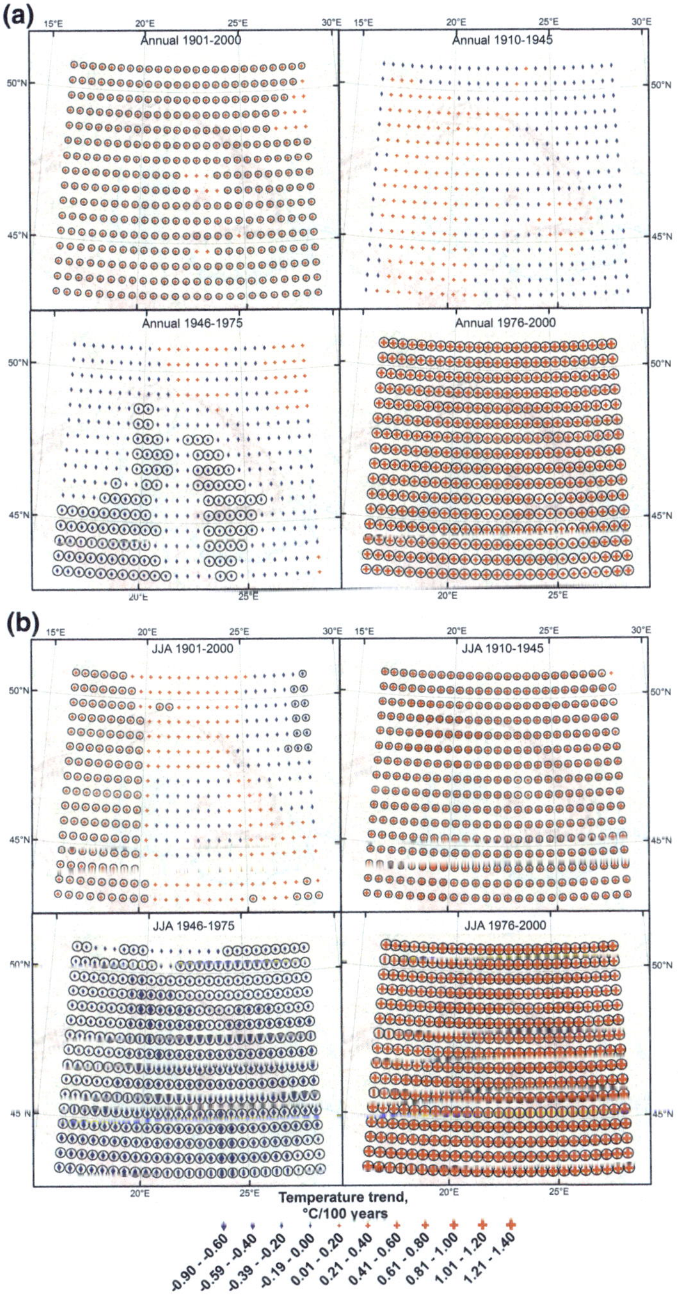

Fig. 2 Annual (**a**) and average June, July, August (**b**) temperature trends over the Carpathian region over time periods of 1901–2000, 1910–1945, 1945–1975, and 1975–2000. Trends significant at the 95 % level are circled (dataset CRU TS 2.1)

1,350 and 1,850 m in the South Carpathians. 100 % of the area is above the 10 °C July isotherm at about 1,700, 1,750, 1,850, and 1,950 m in the West, Ukrainian, North Romanian, and South Carpathians, respectively. Forest cover begins to decrease at about 400 m below this threshold in all four regions (Fig. 3).

There is much regional variance in the distance to climatic timberline. In the West Carpathians, the current timberline of some parts of the High Tatra Mountains is at a minimum distance from the potential one, suggesting minimal previous lowering. The largest distance of the timberline from the 10 °C isotherm is observed at locations with a deciduous timberline (700–800 m lower). In the Ukrainian Carpathians the timberline reaches the highest altitudes in the Chornohora and Gorgan Mountains, at some places intersecting the isotherm; similarly to the West Carpathians, the deciduous timberline is significantly below the isotherm (800–900 m lower). In the North Romanian and South Carpathians current timberlines come close to the climatic limit in the rugged slopes of the Piatra Crailui and Bucegi Mountains of the South Carpathians. The area of potential forest expansion at timberline, relative to the calculated 1961–1990 climatic timberline, totals 251.2 km^2 for the West, 542.8 km^2 for the Ukrainian, 358.8 km^2 for the North Romanian, and 539.9 km^2 for the South Carpathians.

A summary of forest cover change among the Carpathian regions is presented in Table 1.

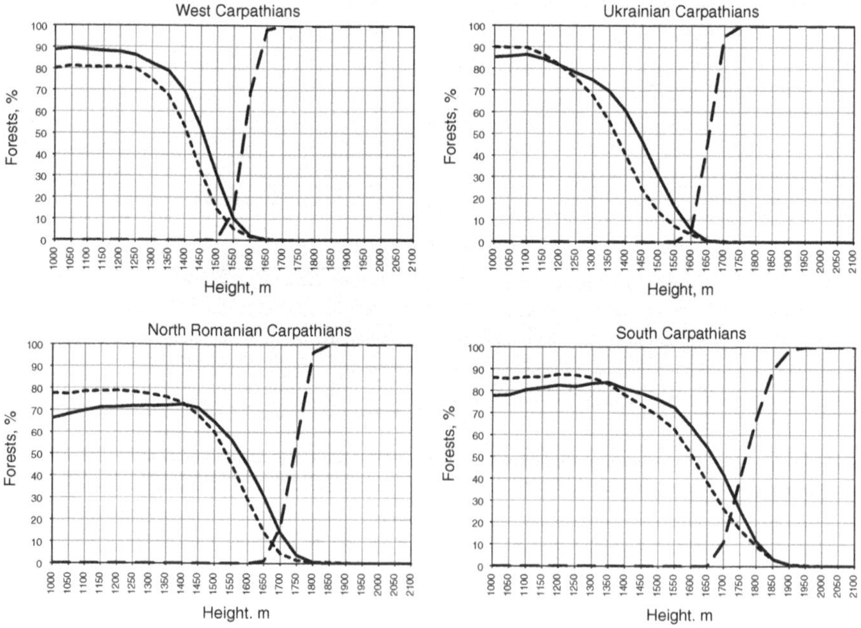

Fig. 3 Forest cover along elevational gradients in the four study regions of the Carpathians. Solid lines are values in 2000, shortly dashed lines—values in 1880: lines with long dashes represent areas above the approximated climatic timberline

Table 1 Forest cover change in the Carpathian mountains during 1880–2000

	All Carpathians	West Carpathians	Ukrainian Carpathians	North Romanian Carpathians	South Romanian Carpathians
Forest cover in 1880, S_1 (%)	72.8 (75.7)	67.1 (74.1)	75.5 (76.3)	74.0 (74.8)	71.6 (78.1)
Forest cover in 2000, S_2 (%)	71.2 (74.1)	75.1 (82.9)	76.9 (77.7)	67.9 (68.6)	70.9 (77.2)
Difference, S_2-S_1 (%)	-1.6	8.0	1.4	-6.1	-0.7
Afforestation (%)	9.6	12.9	9.3	9.5	8.3
Deforestation (%)	11.2	4.9	7.8	15.6	9.0

Note. Percentage values without parentheses are the ratio of area under variable to total study area above 1000 m. Values in parentheses are the ratio of area under variable to area between 1000 m and the climatic timberline, i.e. potential forest area

Forest cover here is presented in two ratios. The number without parentheses equals percentage of forest out of total study area above 1,000 m (the factual value), while the number in parentheses indicates percentage of forest out of total potential area suitable for forest, i.e. area between 1,000 m and the climatic isotherm (the relative value). It can be seen that in 2000 the relative forest cover is largest in the West Carpathians, while the factual forest cover is maximum in the Ukrainian Carpathians. While the relative value is more adequate for comparing forest cover between regions, the nature of the 10 °C isotherm is relative itself, and so we will further discuss only factual values. The forest cover in 2000 decreased by 1.6 % over the whole study area (elevations above 1,000 m) relative to 1880 (Table 1). There has been a major increase in forest cover over the North Carpathians (12.9 %), a minor increase in the Ukrainian Carpathians (1.4 %), and a decrease in the North and South Romanian Carpathians (−6.1 and −0.7 % respectively). Afforestation and deforestation values span from 8.3 to 12.9 % and from 4.9 to 15.6 % respectively, indicating large rates of land use transformation.

4.4 *Afforestation and Deforestation Over 1880–2000*

Afforestation and deforestation were split into afforestation and deforestation at timberline and other afforestation and deforestation, as described in Methods. A total of 34 % of all afforestation and 12 % of all deforestation occurred at timberline.

The elevational distribution of the forest change categories shows a net forest increase near the timberline (Fig. 4). The elevational peak of afforestation varies across the four study regions: 1,450 m in the West and Ukrainian Carpathians, 1,600 m in the North Romanian Carpathians, and 1,700 in the South Romanian Carpathians. These peaks occur at 100–250 m below the climatic timberline,

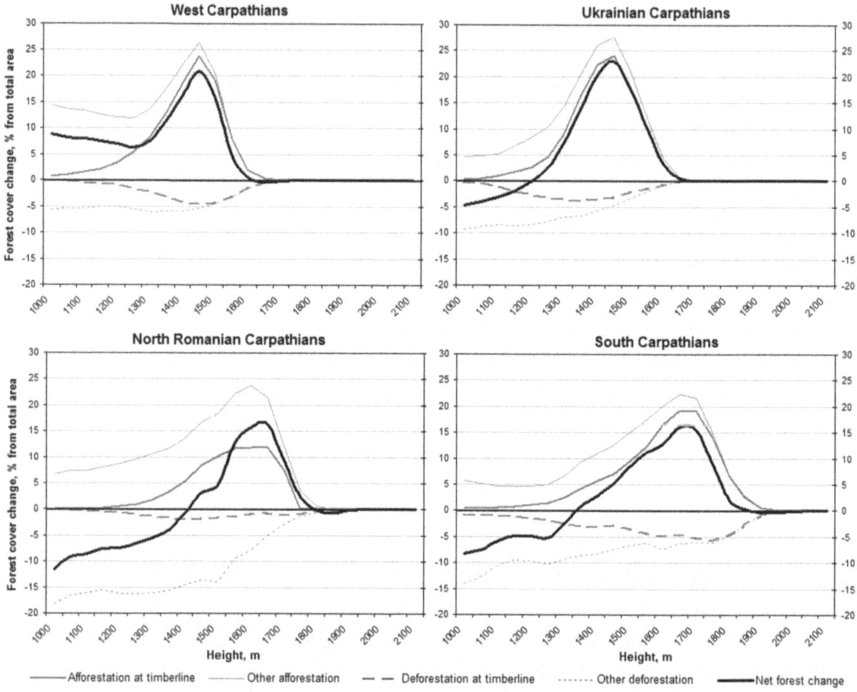

Fig. 4 Forest cover change (1880–2000) along elevational gradients in the four study regions of the Carpathians. Percentage values are shown as a ratio of area under variable to total area under elevational gradient

approximated by 10 °C. Deforestation exceeds afforestation at lower elevations in all regions except the West Carpathians. At the highest elevations, deforestation values roughly equal or are slightly larger than those of afforestation in all regions.

4.5 Land Cover and Slope Aspect Categories of Afforestation and Deforestation

According to the classification of land cover from the Landsat mosaic, coniferous forest boundaries have changed considerably over time whereas deciduous forest boundaries have remained relatively static. Afforested land has consisted of 25 % deciduous forest and 75 % coniferous forest. At timberline, the proportion of coniferous afforestation increases to 85 %, whereas only 15 % of afforestation is deciduous forest.

Deforested land has converted to the following land cover types:

grass: 68 %,
recently logged: 8 %,
shrubs: 24 %.

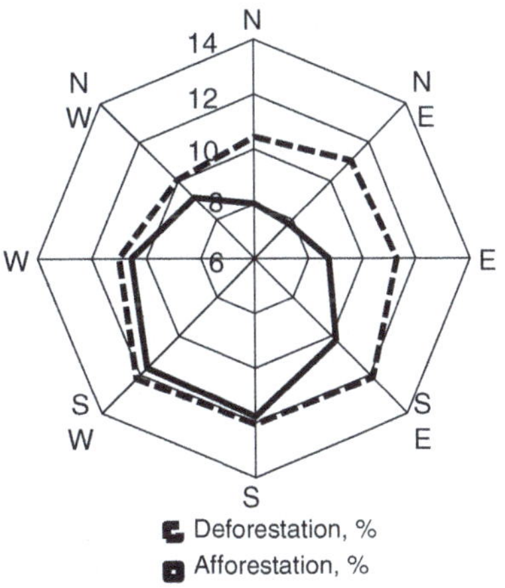

Fig. 5 Percent of area with afforestation/deforestation (1880–2000) according to slope aspect (*solid line*—afforestation; *dashed line*—deforestation)

Afforestation has occurred primarily on more southwesterly slopes, whereas deforestation has been distributed nearly evenly with respect to slope aspect (Fig. 5).

5 Discussion

Three main sources of uncertainty may influence our results. First, imprecision in georeferencing may lead to locational errors when comparing land cover class borders between time periods. Thus, some percent of the observed forest cover change may be attributed to spurious change. Nevertheless, such errors are expected to be random and unbiased such that, at timberline, spurious afforestation values would be canceled out by spurious deforestation values. Second, the choice of areas that are considered to be erroneous on the 1880 maps is subjective; yet this is a better alternative to including highly unlikely locations which will alter the data. Third, it is possible that Austro-Hungarian surveyors have made mistakes in delineating timberlines outside of the masked regions. For instance, avalanche tracks visible on satellite imagery are often not portrayed on the maps. This could lead to overestimated deforestation at the highest elevations.

Our approach of comparing timberline elevation to approximate elevation of climatic timberline using temperature isotherms has several potential applications. It can be used to map areas with greater influence of historical land-use on timberline position, as areas further removed from climatic timberline. The difference in elevation between actual and climatic timberlines along a particular slope facet can be used to construct an index of anthropogenic timberline lowering. One might hypothesize that this index would be proportional to potential climate change response of timberline in the absence of agricultural land use. Our approach can also be used to estimate the potential area for forest expansion given particular climate scenarios, for particular mountain ranges or sociopolitical regions, which should be of both ecological and economic importance.

Land use practices have lowered the timberline by different extents throughout the Carpathians region, with timberlines remaining close to the climatic limit only on steep and inaccessible locations. In average forest cover begins decreasing sharply at about 400 m below climatic timberline. There has been extensive forest cover change in all regions, with pronounced forest increase only in the West Carpathians. This suggests that the current timberline position in the Carpathians is strongly influenced by land use practices, and therefore that mountain forest responses to land use change may largely obscure potential responses to climate change.

As found by others (e.g. Garbarino et al. 2009), land use practices and land use abandonment near timberline in mountain forests are influenced by topographic variability. Greater proportions of afforestation areas on south and southwestern slope aspects) may reflect afforestation of previously cleared pastures, given that forest clearing for pastures is favored on warmer, sunnier slopes.

Changes in forest cover distribution over different elevations are probably related to socioeconomic changes (Fig. 3, Table 1). Commercial logging had become widespread only at the turn of the twentieth century (Kuemmerle et al. 2010) and it may be assumed that forest cover in 1880 was influenced greatly by local population density. The relative forest cover over the four regions in 1880 spanned from 74.1 to 78.1 %—a small difference of 4 %, while in 2000 the relative forest cover spanned from 68.6 to 82.9 %—a difference of 14.3 % (Table 1). Over time, the differences in forest cover among the four regions have increased, portraying their different paths of economic development and the resulting impacts on forest resources.

The increase in forest cover in the West Carpathians, especially over lower elevations, suggests that the declining importance of agriculture in mountainous areas, combined with increasingly sustainable forest management practices, have lead to an ongoing state of transition from open lands to forest in this territory (Kozak et al. 2007). In all, these results are consistent with previous case studies of forest cover change in this region (Kozak 2003). The other three regions, while demonstrating similarities in the profile of forest cover change with the Western Carpathians, differ from it by deforestation exceeding afforestation at the lower elevations. Many socioeconomic changes have occurred in this territory over the broad time span of the study period, and it is difficult to find a single explanation for the observed change. Changes are likely associated with rapid socioeconomic development of Romania and the resulting intensive deforestation after World War I, post-World War II renewal in Ukraine and Romania, and illegal logging in both Romania and Ukraine after the fall of the Soviet Union (Illegal logging in Romania 2005; Kuemmerle et al. 2009b). As found by Bouriaud (2005), occurrence of illegal logging in Central Europe is explained by poverty, reforms on land ownership and by weak law enforcement. According to the World Bank (2005–2010), Poland, Slovakia, Romania, and Ukraine rank 47, 40, 59, and 97th place respectively by GDP per capita (out of a total of 180). Thus, poverty can offer a partial explanation for the high deforestation of Romania and Ukraine over the twentieth century, while failing to explain the higher deforestation rates in Romania than in Ukraine. As stressed in Mihai et al. (2007), agricultural and pastoral activity is a significant part of the economical life of Romania. Yet, farmland abandonment, widespread after the collapse of the Soviet Union in Romania (Kuemmerle et al. 2009a) and Ukraine (Kuemmerle et al. 2010), has not yet resulted in widespread afforestation. Therefore, poverty and forestry policies, including widespread illegal logging, can be considered the main drivers of forest cover decline in the lower elevations of the Carpathian Mountains of Romania and Ukraine.

A key influence on timberline change in the Carpathians is expected to be the impact of seasonal grazing (Sitko and Troll 2008). Unfortunately, there exist no reliable statistics concerning changes of livestock numbers over the study period for Ukraine and Romania. Numbers mentioned in literature differ; there is also evidence that recent economic difficulties have driven the population to intensive usage of the mountainous pastures in Romania. Nevertheless, from the large afforestation rate at timberline we conclude that seasonal grazing has likely declined over all four regions.

One-third of all afforestation of elevations above 1,000 m occurred at timberline, indicating that substantial changes are happening at this ecotone, similarly to other European mountains (Grace et al. 2002; Gehrig-Fasel et al. 2007). Peaks of afforestation were observed at a specific elevation in all four regions; most of this afforestation is at timberline (Fig. 4). At the same time, 12 % of deforestation happened at timberline; also, at the highest elevations in three of the four regions deforestation exceeds afforestation slightly. This stands in contrast with forest cover change between 1985 and 1997 in the Swiss Alps, where afforestation exceeds deforestation at almost all elevations (Gehrig-Fasel et al. 2007). While our approach is not able to separate the effects of climate warming from land use change on forest increase near the timberline, we consider climate to be of influence. Trees growing at an artificial timberline experience harsh conditions similar to climatic timberlines (Holtmeier 2009). Timberline rise over much of the Carpathians depends on the combination of land-use changes leading to reduction in high-elevation pastures (i.e. agricultural abandonment) with periods of favorable climate for forest growth and regeneration. Overall, the profile of forest cover in 2000 is closer to the climatic timberline than in 1880 (Fig. 3).

Compared to afforestation and deforestation changes recorded for the Alps during 1985–1997 (Gehrig-Fasel et al. 2007), observed deforestation values in the Carpathians are higher at the treeline. Clearance of new forest for grazing land has been on-going in the Carpathians in recent decades, but is likely an outdated practice for the majority of the Alps where tourism takes an important place in the mountain economy (Motta et al. 2006).

The dominance of conifers in recently afforested areas (75 % of all afforestation, or 85 % in proximity to timberline) may indicate an advantage of the evergreen growth habit under the current timberline climatic regime, or may result from prior selective deforestation of coniferous forest. These results stand in contrast to recent research highlighting the transition from coniferous stands to mixed forest (Mihai et al. 2007); however, in a case study from the Chornohora

(Sitko and Troll 2008) the spruce timberline has risen substantially while the beech timberline has not. This could be the result of historical forest clearing practices that removed the higher-elevation bands of spruce forest, while leaving the beech forest intact, such that a novel, anthropogenic beech timberline was created. Thus the beech timberline may already be close to the climatic limit of its dominant species. Also relevant is the selective feeding preference of cattle for beech over spruce (Sitko and Troll 2008). We conclude that at timberline conditions are advantageous for the invasion of coniferous species throughout the whole Carpathian Mountains.

The timberline of the Carpathians represents a patchwork of forests that have variously expanded and declined in extent and elevational range, highlighting the complex nature of mountain forest responses to climate change in the context of rapidly evolving land use practices. Regional variation in forest cover change at timberline obscures climate change responses and can be attributed to sociopolitical differences in land use practices over time, including logging, sheep and cattle pastures, and subsequent abandonment of pastures.

Acknowledgments This work has been supported by an award from the CRDF foundation and Ministry of Education and Science of Ukraine (award #UKG2-2957-KV-08). We also thank two anonymous reviewers for their valuable suggestions and corrections, the server mapywig.org for hosting historical maps, the CRU and KNMI for access to climate data, and the United States Geological Survey for free access to Landsat images.

References

Beniston M (2003) Climatic change in mountain regions: a review of possible impacts. Clim Chang 59:5–31

Bouriaud L (2005) Causes of illegal logging in Central and Eastern Europe. Small-Scale For 4(3):269–291

Dezso Z, Bartholy J, Pongracz R, Barcza Z (2005) Analysis of land-use/land-cover change in the Carpathian region based on remote sensing techniques. Phys Chem Earth 30:109–115

Garbarino M, Weisberg PJ, Motta R (2009) Interacting effects of physical environment and anthropogenic disturbances on the structure of European larch (Larix decidua Mill.) forests. For Ecol Manag 257:1794–1802

Gehrig-Fasel J, Guisan A, Zimmermann NE (2007) Tree line shifts in the European Alps: climate change or land abandonment? J Veg Sci 18:571–582

Grace J, Berninger F, Nagy L (2002) Impacts of climate change on the tree line. Ann Bot 90:537–544

Grodzinska K, Godzik B, Fraczek W, Badea O, Oszlanyi J, Postelnicu D, Shparyk Y (2004) Vegetation of the selected forest stands and land use in the Carpathian Mountains. Environ Pollut 130:17–32

Holtmeier F (2009) Mountain timberlines. Ecology, patchiness, and dynamics. Springer Science, Dordrecht

Holtmeier F-K, Broll G (2007) Treeline advance—driving processes and adverse factors. Landsc online 1:1–22

Illegal logging in Romania (2005) In: Brandlmaier H, Hirschberger P (eds) WWF Danube Carpathian Programme. http://www.forestconsulting.net/Downloads/Publications/finalromaniaillegallogging.pdf. Accessed Nov 5 2012

IPCC (2001) Working group I: the scientific basis. Cambridge University Press, Cambridge
Klein Tank AMG, Wijngaard JB, Können GP, Böhm R, Demarée G, Gocheva A, Mileta M, Pashiardis S, Hejkrlik L, Kern-Hansen C, Heino R, Bessemoulin P, Müller-Westermeier G, Tzanakou M, Szalai S, Pálsdóttir T, Fitzgerald D, Rubin S, Capaldo M, Maugeri M, Leitass A, Bukantis A, Aberfeld R, van Engelen AFV, Førland E, Mietus M, Coelho F, Mares C, Razuvaev V, Nieplova E, Cegnar T, Antonio López J, Dahlström B, Moberg A, Kirchhofer W, Ceylan A, Pachaliuk O, Alexander LV, Petrovic P (2002) Daily dataset of 20th-century surface air temperature and precipitation series for the European climate assessment. Int J Climatol 22:1441–1453
Kolischuk V (1958) Suchasna verhnya meja lisu v Ukrainskyh Karpatah. Vydavnytstvo Akademii Nauk URSR, Kyiv
Komendar V (1966) Forposty gornyh lesov. Karpaty, Uzhgorod
Körner C (1998) A re-assessment of high elevation treeline positions and their explanation. Oecol 115:445–459
Kozak J (2003) Forest cover change in the Western Carpathians in the past 180 Years. Mt Res Dev 23:369–375
Kozak J, Estreguil C, Troll M (2007) Forest cover changes in the northern Carpathians in the 20th century: a slow transition. J Land Use Sci 2:127–146
Krakowska S, Palamarchuk L, Shedemenko I, Dukel G, Gnatuk N (2008) Verifikaciya danyh svitovogo klimatychnogo centru ta regional-noi modeli klimatu (REMO) schodo prognozu prizemnoi temperatury povitrya za kontrolniy period 1961-1990. Nauk Prac Ukr NDGMI 257
Kricsfalusy V, Mróz W, Popov S (2008) Historical changes of the upper tree line in the Carpathian Mountains (Ukraine). Mt Forum Bull 8(1) http://archive.mtnforum.org/rs/ol/counter_docdown.cfm?fID=1087.pdf. Accessed 17 July 2012
Kuemmerle T, Muller D, Griffiths P, Rusu M (2009a) Land use change in Southern Romania after the collapse of socialism. Reg Environ Chang 9:1–12
Kuemmerle T, Chaskovskyy O, Knorn J, Radeloff VC, Kruhlov I, Keeton W, Hostert P (2009b) Forest cover change and illegal logging in the Ukrainian Carpathians in the transition period from 1988 to 2007. Remote Sens Environ 113:1194–1207
Kuemmerle T, Olofsson P, Chaskovskyy O, Baumann M, Ostapowicz K, Woodcock C, Houghton R, Hostert P, Keeton W, Radeloff V (2010) Post-soviet farmland abandonment, forest recovery, and carbon sequestration in western Ukraine. Glob Chang Biol. doi:10.1111/j.1365-2486.2010.02333.x
Mihai B, Savulescu I, Sandric I (2007) Change detection analysis (1986–2002) of vegetation cover in Romania. Mt Res Dev 27:250–258
Mitchell TD, Jones PD (2005) An improved method of constructing a database of monthly climate observations and associated high-resolution grids. Int J Climatol 25:693–712
Molnar G, Timar G (2009) Mosaicking of the 1:75 000 sheets of the third military survey of the Habsburg Empire. Acta Geod Geoph Hung 44(1):115–120
Motta R, Morales M, Nola P (2006) Human land-use, forest dynamics and tree growth at the treeline in the Western Italian Alps. Ann For Sci 63:739–747
Plesnik P (1978) Man's influence on the timberline in the West Carpathian mountains, Czechoslovakia. Arct Alp Res 10:491–504
Sitko I, Troll M (2008) Timberline changes in relation to summer farming in the Western Chornohora (Ukrainian Carpathians). Mt Res Dev 28:263–271
Wężyk P, Guzik M (2004) The use of "Photogrammetry-GIS" (P-GIS) for the analysis of changes in the Tatra Mountains natural environment. In: Widacki W, Bytnerowicz A, Riebau A (eds) A message from the Tatra. Geographical Information Systems and Remote Sensing in Mountain Environmental Research. Institute of Geography and Spatial Management Jagiellonian University, USDA Forest Service, Kraków, Poland, Riverside, California
World Bank (2005–2010) List of countries by GDP per capita. http://en.wikipedia.org/wiki/List_of_countries_by_GDP_(PPP)_per_capita. Accessed 17 July 2012

Forest Cover Change in the Parâng-Cindrel Mountains of the Southern Carpathians, Romania

Emil Marinescu, Ioan Eustaţiu Marinescu, Alina Vlăduţ and Sidonia Marinescu

Abstract Forest cover change over the last century in the Parâng-Cindrel Mountains in the central part of the Southern Carpathians is analyzed. The Parâng-Cindrel Mountains stand out as one of the tallest Romanian Carpathian ranges characterized by massiveness and relatively wild forest areas preserved in spite of the intensive population and related human activities. Three main cartographic products addressing three major time stages were used: Austrian military maps from the end of the nineteenth century and the beginning of the twentieth century, topographic maps from the 1960s, and Corine Land Cover Data from 2006 that were the inputs to a unique cumulative map of the forest cover change over the last century. In order to quantify forest fragmentation, various landscape metrics were calculated and morphological spatial pattern analysis was performed. By analyzing and interpreting forest change, we found three main periods related to major anthropogenic disturbances. At the beginning of the twentieth century, the most significant forest changes appeared in the northeastern part of the study area (Northern Cindrel Mountains) where human influence was mostly caused by the need to extend the grazing area and by high-altitude settlements related to an intensive pastoral activity that reached its widest extension in the Southern Carpathians. The second stage of drastic reductions was in the 1970s when large hydrotechnical constructions appeared, especially the Lotru and Sebeş watersheds and adjacent logging, which increased timber harvesting rates and tourist infrastructure in the area. The third stage was is the post-socialist period due to a forest ownership change and lax institutional policies that caused an increase of forest harvesting for the purpose of large, short-term profit, which generated an over exploitation of the forests.

E. Marinescu (✉) · I. E. Marinescu · A. Vlăduţ · S. Marinescu
Department of Geography, University of Craiova, Al. I. Cuza Street, no. 13, 200585 Craiova, Romania
e-mail: emilmarinescu@central.ucv.ro

J. Kozak et al. (eds.), *The Carpathians: Integrating Nature and Society Towards Sustainability*, Environmental Science and Engineering,
DOI: 10.1007/978-3-642-12725-0_17,

1 Introduction

Forest cover change is the most expressive and significant form of landscape change over centuries. Romania, containing about half of the Carpathian Mountain range, has undergone striking variations of the forested areas over the last century, due to the fact that the Carpathians have always played a central role in Romanian life, both as a refuge and a source of vital necessities (food, timber, and water, as well as energy through coal mining deposits and hydrotechnical constructions).

Previous analyses of forest cover change of the area have been mostly linked to pragmatic forestry studies, which have focused on the fields of forest management, economics, genetics, ecology, dendrometry, and auxology. The literature addressing the forest cover changes in the Southern Carpathians were fragmentarily found in the general works regarding national forest history (Ivănescu 1972; Giurescu 1975) or partially in local studies on a smaller scale (Buza 2000; Stoiculescu 2003). This lack of information has been caused mainly by the fact that national geographical spatial data are not easily accessible, except to the employees of the research institutes involved in certain projects.

International forest cover change research encompasses numerous and comprehensive studies (Riitters et al. 2002; Kozak 2003; Kozak et al. 2007; Kopecká and Nováček 2009; Kümmerle et al. 2009) that focus on forest loss, causes and effects on habitat degradation, local spatial and temporal patterns of forest gains and losses that define forest fragmentation, change indicators, and various statistics of the forest landscape. It was observed that large forest patches have rapidly become more and more fragmented, resulting in permanent fragmentation with little remnant patches where the loss of habitat is not only quantitative but also qualitative.

The forest matrix can slowly or drastically change due to global warming or severe calamities (Kopecká and Nováček 2009; Giurgiu 2010). In addition, deforestation is not a newly observed practice, but its rate of increase is. Thus, correct management strategies with proper limitations that depend on local environmental factors are recommended. Often, political strategies are identified as possible solutions as well (Kümmerle et al. 2007, 2009).

Environmental patterns, especially high-density vegetation patterns as found in forests, are strong indicators of the degradation of critical ecological processes on entire ecosystems. Quantifying landscape patterns has been developed as an accurate indicator of degradation in any geographical study. Categorical maps are commonly analyzed through landscape metrics. Five metrics are considered to best describe the first five factors that account for 87 % of the variation in the landscape indicators: the Shannon-Wiener diversity index (SHDI), mean patch size (MPS), mean shape index (MSI), area weighted mean shape index (AWMSI), and nearest neighbor standard deviation (NNSD) (Louto 2000).

In order to conduct proper comparisons in time and space, forest pattern change was quantified as a particular case of landscape analysis and methods and metrics were borrowed from landscape ecology (Turner 1990; Jaeger 2000; Wu et al. 2002; Li and Wu 2004; Farina 2007). A particular approach that focused on forest pattern quantification and morphological spatial pattern analysis was developed recently (Vogt et al. 2007; Soille and Vogt 2009) and has given researchers a powerful tool of investigation.

This chapter discusses forest cover change analysis over the last century in the Parâng-Cindrel Mountains of the Southern Carpathians, focusing on complementary perspectives that analyze spatial–temporal relationships and evaluating fragmentation by using spatial morphology and patch quantification. Results are interpreted by means of historical analysis related to forest exploitation and property, a dichotomy applied differently by various forest management policies and mainly caused by the human influence on the forest. Results will be chiefly provided as relative figures in order to illustrate the temporal changes of the forest within the study area.

2 Study Area

The central part of the Southern Carpathians is represented by the Parâng-Cindrel Mountains, which stand out as one of the tallest and largest Romanian Carpathian ranges. With the highest altitude of 2,519 m, the range represents more than one-third of the Southern Carpathians (38 % or 5,530 km^2), which is considered the most important hydrographic and orographic range in the area (Fig. 1).

The most characteristic features of this range are its massiveness and relatively conserved forest areas, providing a vast alpine area and subalpine meadow despite the intensive population and related human activities. Within the study area, the main composition of the forest consists of *Fagus sylvatica* and *Picea abies*. These mountains have one of Romania's largest forest ecosystems, and thus they shelter important Carpathian hotspots of biodiversity.

Climate for the area shows the average annual temperature between 8 °C at the lower altitudes and −1.6 °C at altitudes around 2,000 m. Average annual precipitation ranges between 653 mm (Sibiu: 416 m) and 1,090 mm (Parângul Mare: 2,519 m). The difference between the average temperature of the southern and northern slopes is around 1–2 °C.

The fundamental method by which the forests were administrated in the study area is represented by the evolution of the property structure upon the forests throughout time (before 1863, 1864–1918, 1918–1947, 1948–1989, 1990–present) and by the application of the provisions of the forestry laws (1852, 1879 in Transylvania, and 1881, 1898, 1910, 1930 in the Kingdom of Romania, and 1962, 1996, 2008 in Romania). Before 1863, forests were in the property of the royal court and monasteries, but after the secularization law regarding the property of the monasteries (1863) was issued, the state became the owner of the forests.

Before 1918, in the northern part of these mountains, there were four categories of forests: alodial or nobiliary forests, forests of common use, "urbarial" forests, and forests in the property of a commune (Csucsuja 1998). An important fact that caused the massive reduction of the forests in the northeastern part of the study

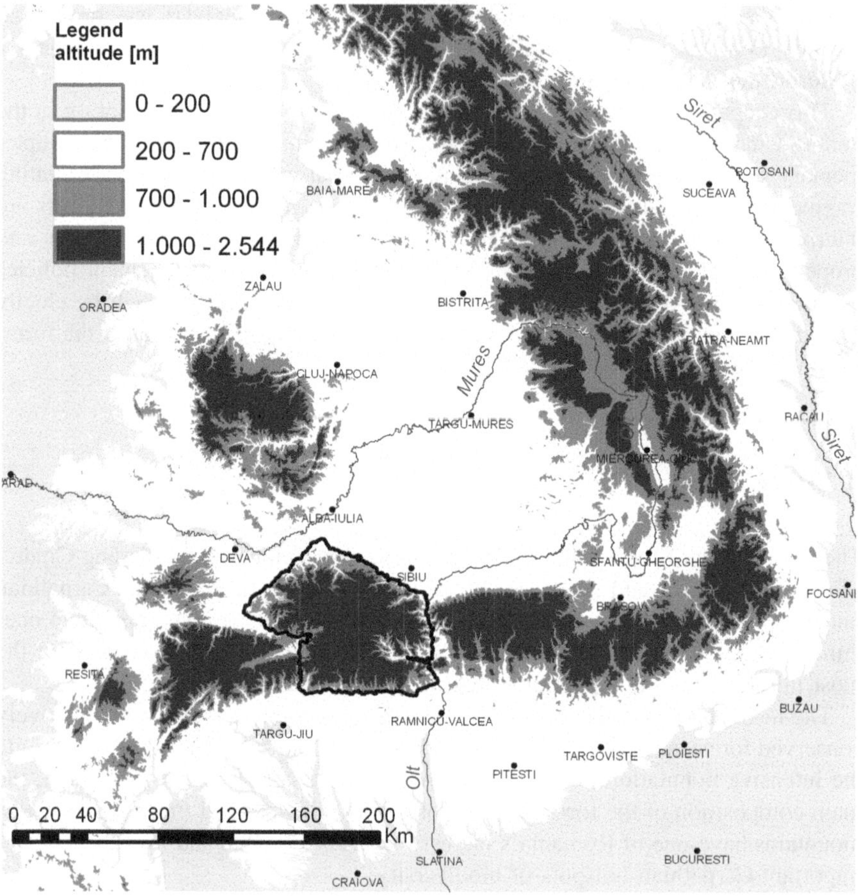

Fig. 1 Study area: the Parâng-Cindrel Mountains in the Romanian Carpathians

area was the construction of some wood processing plants, even before 1918 (Buza 2000). In 1870, workers had been brought from Baden (Germany) to work in the forests located near the Bistra watershed. During the same year, the Petreşti paper mill was built and it used wood coming from the forest exploitation area in the Sebeş valley. This valley was organized for log floating to build a dam at Oaşa Mare in 1880 (Giurescu 1975).

In 1890, the Voineasa settlement started to develop due to wood exploitation by companies from Austria that hired thousands of workers (Muică and Turnock 2003). Timber transportation using the floating system on the Lotru valley constantly improved, but at the beginning of the twentieth century this technique was almost abandoned in favor of railway and then road transport (Muică and Turnock 2003).

Until the middle of the twentieth century, transport was ensured by the forest railroad and by a system of large funiculars that were used especially between

1948–1954 and which transported wood from the Buila-Vânturariţa Mountains over the Căpăţânei Mountains to the Lotru watershed where it was loaded into forest trucks and transported to Brezoi on the Olt valley (Burdaşu 1971).

By the middle of the 20th century, in the central part of the Southern Carpathians, forest railroads had been used on the Lotru, Sadu, Cibin, Sebeş, Cugir, Sibişel, Orăştie, Strei, Gilort, and Bistriţa Vâlcii valleys, but after 1960 they were replaced by forest roads (Muică and Turnock 2003).

3 Materials and Methods

The time series of forest cover input data was established on the basis of the following cartographic materials:

- maps of the third military survey of the Austro-Hungarian empire from the end of the nineteenth and the beginning of the twentieth century at a scale of 1:200,000, published in 1918. On-screen digitizing was used (Crăciunescu 2006),
- Soviet topographic maps of Romania from the 1960s at a scale of 1:100,000, using on-screen digitizing (Crăciunescu 2010),
- Corine (European Commission Programme to Coordinate Information on the Environment) Land Cover 2006 (CLC2006) seamless data version 13 at a scale of 1:100,000 downloaded from the official site of the European Environment Agency. Classes that were considered include: broad-leaved forest (311); coniferous forest (312) and mixed forest (313) (Bossard et al. 2000).

In order to evaluate forest cover change we chose three approaches that compare sets of data addressing three points in time corresponding to the following cartographic materials used: morphological spatial pattern analysis (MSPA), landscape metrics, and raster overlay analysis of the on-screen digitized maps. The chosen approaches highlight the changes in a complementary and suggestive way. From the MSPA perspective (Figs. 2, 3) forest geometry and connectivity changes are especially highlighted. In addition, landscape metrics (Fig. 4) draw attention to patch number, size, and shape, providing a quantitative image that gives an overview on fragmentation and its dynamics. Finally, we performed a spatial representation of forest cover change (Fig. 5) for forest/nonforest surface area balance. The last method permitted us to interpret changes in relation to the terrain's physical reality.

Input maps provided the data required by the MSPA, which is a data mining approach capable of independently detecting a scale of seven segmentation classes that describe the geometry and connectivity of forest binary images. MSPA served as the theoretical basis GUIDOS 1.3 software (Vogt et al. 2007; Vogt 2010), and according to the mentioned reference these seven classes are defined as core: interior foreground excluding foreground perimeter; islet: disjointed foreground object and too small to contain core; loop: connected at more than one end to the

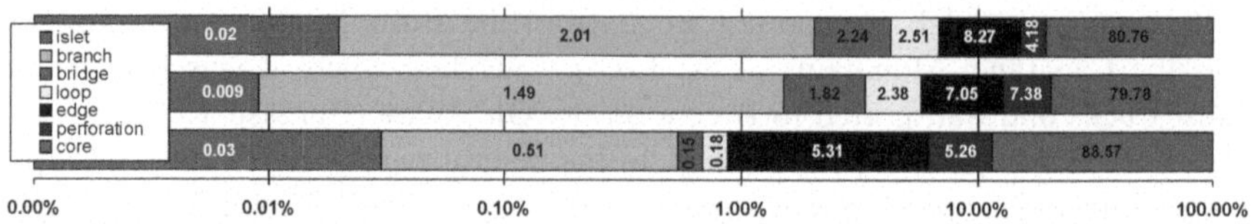

Fig. 2 The values of the MSPA classes for each input forest cover data in the Parâng-Cindrel Mountains, Romania. The seven classes are defined as: core (interior foreground excluding foreground perimeter); islet (disjointed foreground object and too small to contain core); loop (connected at more than one end to the same core area); bridge (connected at more than one end to different core areas); perforation (internal foreground object perimeter); edge (external foreground object perimeter); branch (connected at one end to edge, perforation, bridge, or loop)

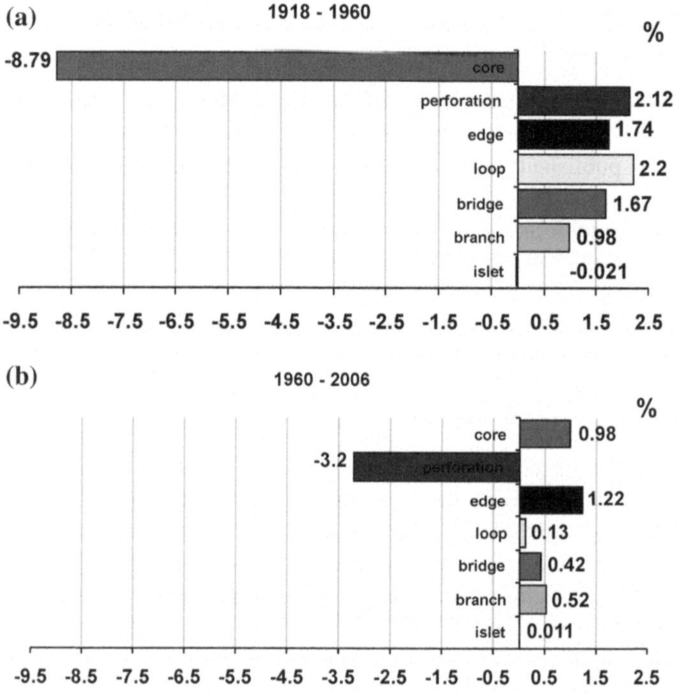

Fig. 3 Gains and losses of the MSPA classes between 1918–1960 (**a**) and between 1960–2006 (**b**), Parâng-Cindrel Mountains, Romania. See Fig. 2 for class definition

same core area; bridge: connected at more than one end to different core areas; perforation: internal foreground object perimeter; edge: external foreground object perimeter; and branch: connected at one end to edge, perforation, bridge, or loop.

Forest pattern quantification has been developed as necessary for any geographical forestry study, providing mathematical values in relation to the phases of forest fragmentation. Landscape metrics are readily applicable and useful to quantify forest cover changes. Landscape metrics are able to identify the distribution of resources and their dynamics over time within forest ecosystems.

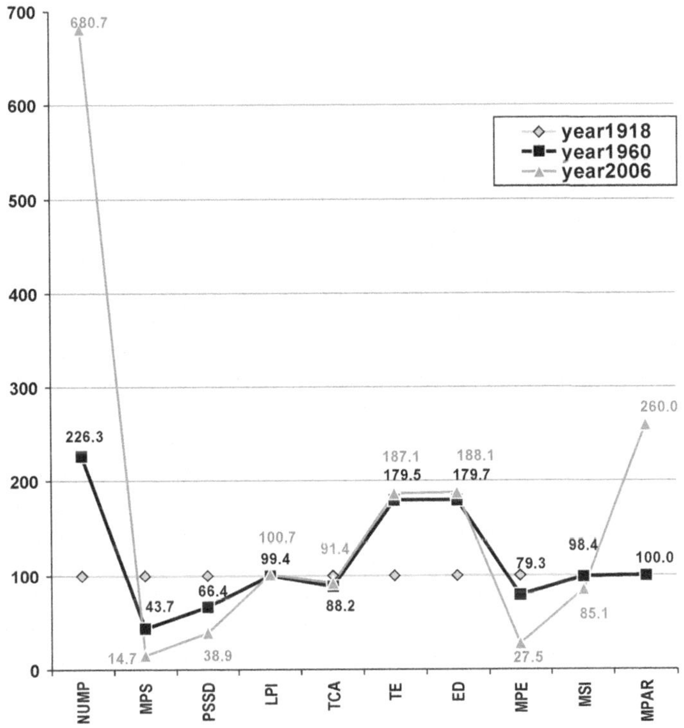

Fig. 4 Percent variation of forest metrics in 1960 and 2006 compared with the year 1918. The forest metrics are: *NUMP* number of patches; *MPS* mean patch size; *PSSD* patch size standard deviation; *LPI* largest patch index; *TCA* total core area; buffer 100 m; *TE* total edge; *ED* edge density; *MPE* mean patch edge; *MSI* mean shape index; *MPAR* mean perimeter-area ratio

The most common forest metrics were calculated using a powerful recent software, Vector-Based Landscape Analysis Tool Extension 1.1 (V-LATE) developed by the Spatial Indicators for European Nature Conservation (SPIN) Project (Lang 2005). We chose to analyze a set of 10 landscape metrics, that gave us basic information on patch number, patch size, patch shape, and core: number of patches (NUMP), mean patch size (MPS), patch size standard deviation (PSSD), largest patch index (LPI), total core area (TCA), total edge (TE), edge density (ED), mean patch edge (MPE), mean shape index (MSI), and mean perimeter-area ratio (MPAR). This intuitive landscape metrics set has various advantages: it can be used for comparison with other results, includes absolute and mean values, and comprises metrics that are often selected in fragmentation analysis.

To deal with the different scales of the input maps, we decided to perform the raster overlay analysis and the metrics comparison only at the smallest scale of the input data of 1:200,000. We adopted a simple and intuitive approach based on the results of some previous studies (O'Neill et al. 1996; Wu et al. 2002). The scale of the input maps determines the largest scale at which the data can be correctly analyzed and, subsequently, the raster cell size at which information is neither

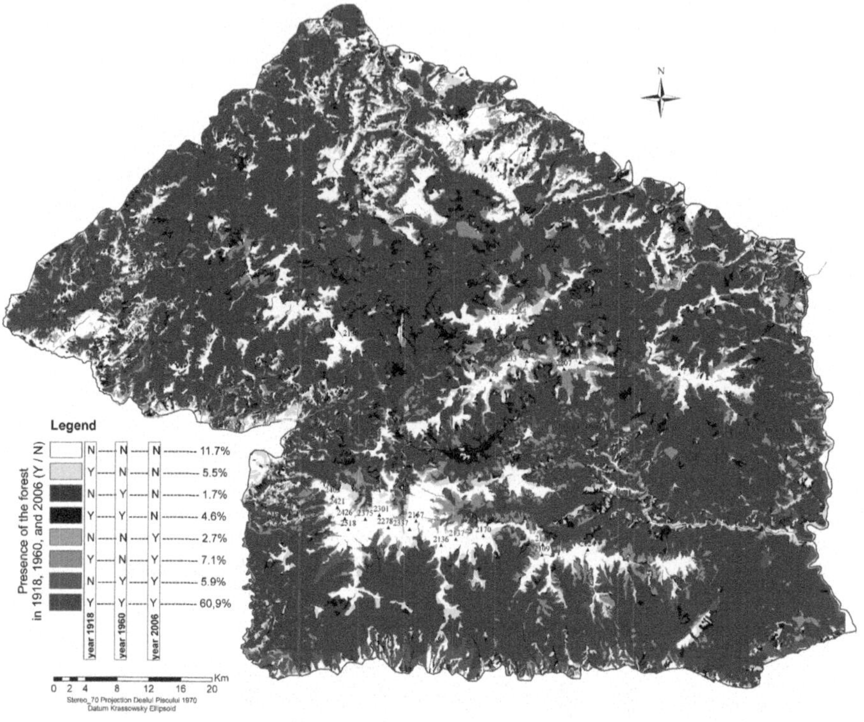

Fig. 5 Forest changes in the Parâng-Cindrel Mountains in the central part of the Southern Carpathians between 1918 and 2006

underestimated nor overestimated. In order to retain the smallest detail of the Austrian map, we used the maximum admissible cell size of 100 m. In addition, another imperative condition is that no detail of the more recent 1:100,000 scale maps, smaller than 1:200,000 scale accuracy, should be transferred by raster conversion. We depreciated all finer-scale information, which increased the cell size by means of the to-raster conversion tools. The majority resampling technique was then adopted to determine the output cell value. To conduct a proper comparison between the maps, the scale-depreciated maps were then again interpolated to 100 m while information remained as for the 1:200,000 scale. Scale depreciation was attained through Block Statistics type set to the Majority.

4 Results and Discussions

The MSPA statistics option permitted us to illustrate the forest variation (especially morphology and connectivity) through the quantification of the seven segmentation classes.

In order to attain a quantitative evaluation we provided the absolute values of the MSPA classes for each input forest cover (Fig. 2). A logarithmic chart type was selected due to its appropriateness to present data that cover a very large range. Figure 3 shows a complete MSPA data presentation in a bar chart with two comparisons that correspond to the periods 1918–1960 and 1960–2006, highlighting forest gains and losses from a spatial morphology perspective.

In 2006, core forest areas represented about 81 % of the forested areas, which was almost the same as in 1960 (80 %). That could be explained by sequential forest/nonforest use conversion in different zones of the study area. This finding is also confirmed by the classes of the cumulative map of forest changes over the last century that stand for 1960 forest/2006 nonforest and 1960 nonforest/2006 forest. Areas deforested in 1960 became forested in 2006 and vice versa. A more important variation was noticed in the period 1918–1960 when the core area decreased drastically as industrialization began to play an important role in Romania's economy. This significant core loss (8.79 %) is obvious in the chart that illustrates separate analyses of the first and second period (Fig. 3). The maximum core area appeared in 1918 when it was 88.5 % of the forest area.

Islet area was extremely small mainly in the adjacency and interior of the permanent settlement areas, denoting remains of larger forest trunks. In the period 1960–2006 the main modification identified through MSPA analysis was represented by perforation loss (3.2 %), which means a slight tendency of natural restoration of the forest cover in the last 50 years, naturally or artificially. This positive effect was generated by gap afforestations from the high-altitude rural settlement area where anthropogenic influence decreased in the 1970s and the 1980s as a result of a policy of industrialization. At that time, apart from traditional pastoral activity decline, there was a slight upward shift of the treeline, mainly on the southern slopes.

By comparing the gains and losses for the two periods (Fig. 3), we found some variations: the classes that indicate connectivity (bridge, loop, and branch) displayed a higher increase in the period 1918–1960, and bridge class was almost double.

The highest decrease rate was found in the loop class (from 2.2 to 0.13 %). These differences are explained by the different methods of forest exploitation: the traditional forest exploitation from the first period (1918–1960) when bridge, loop, and branch frequently appeared resulted in a striping aspect, versus the intensive forest exploitation from the second period (1960–2006) when, primarily in its three decades, clear-cut harvesting of large stands occurred. This policy of deforestation reduced the striping aspect characterized by high values of bridge, loop, or branch.

In order to attain a better evaluation of the forest areas and their spatial–temporal evolution, we opted for further investigation with landscape metrics (Fig. 4). For the sake of clearness, metrics are exclusively presented in percents, with the temporal reference as the year 1918. Chart analysis showed that serious fragmentation became visible only after 1960. But actually the process of fragmentation began in 1945, and by 1948, the year of the "great nationalization" occurred when the big forest owners responded to a new policy that the forests would be nationalized and quickly sold small areas of forest to the peasants. This fact along with the appropriation of the families of the war veterans, tripled the number of forest owners that owned surfaces smaller than 10 ha that soon were partially cut. In 1947, the private forests with areas under 10 ha represented 32 % from the forests and belonged to 97 % of the number of owners (Giurescu 1975), which resulted in a high fragmentation of the forest. MPS displayed a strong descending trend, reaching its minimum in 2006, which represents only 15 % from the value in 1918. The deviation from the mean in patch size showed that PSSD decreased to 66.4 % in 1960 and to 38.9 % in 2006, indicating less variability in patch size while small patch number increased.

The significant NUMP difference in the second interval of 1960–2006 was determined not only by its high increase in 2006 but also by its attenuation in 1960 due to different techniques of forest harvesting and timber removal in the remote areas. In the 1950s there was a maximum extension of the forest railways in the Romanian Carpathians. The use of special funiculars made the most inaccessible remote areas exploitable and thus the pressure was reduced in the lower areas that remained unfragmented.

The general trend was highly influenced by the increasing number of small patches (NUMP, MPS, PSSD) that affected all metrics related to it (in 2006 NUMP was three times larger than in 1918). Nevertheless, relative stability can be observed insofar as the largest patch maintained its large size. LPI displayed little variation over the century at ± 0.6 to 0.7 %. The percentage of the forest that comprises the largest patch remained constant. This finding denotes that the forest still displays a good spatial conservation status.

In the analysis of metrics we should take into account that the large patch shape showed a continuous lacing tendency but had a minor influence on the metrics because small patches were very numerous.

TCA constancy also showed that the total core area had preserved itself over the last century primarily due to patchiness that actually affected smaller patches and had only a minor effect on the largest patch.

The edge metrics showed that TE and ED increased to 180 % in 1960 but remained almost constant till 2006. This is most likely due to the increasing number of small patches and of patch lacing. MPE decreased to 79 % till 1960 and to only 27 % in 2006. Comparing the variation of MPS and MPE during the twentieth century, we can conclude that the trend in patch fragmentation displayed more patch size decreasing than patch edge one. In other words mean patch became more irregular. While MPS decreased to 14.7 %, MPE decreased to only 27.5 %. The same correspondence could be seen for MPAR, which remained constant in 1960 but presented a substantial change of 260 % in 2006.

Quantification of irregularity is given by MSI, average shape index of patches, whose minimum appeared in 2006 (85 %) while in 1960 was 98 % compared to the value of the reference year of 1918.

Forest spatial–temporal change is illustrated in a single cumulative eight-class map that comprises all possible combinations of changes in the period 1918–2006.

The represented classes are explained in the map legend and accompanied by the proper percent from the whole Parâng-Cindrel area (Fig. 5).

By analyzing the forest change map (Fig. 5), we found that the undisturbed forest represented more than 60 %. The minimum total forested area appeared in the 1960s (404,100 ha) and the maximum in 1918 (433,500 ha). In 2006, the forest area was determined to be around 423,200 ha, an amount comparable with the area in 1918.

Through the visual analysis of the output map in comparison with the 1960 map one can easily observe that the Corine Land Cover data introduced systematic errors. Due to quite similar responses of the coniferous forest and subalpine scrubs, errors of classification were observed in certain subalpine belts. This situation occurred mostly in the Parâng Mountains where some *Juniperus* populations were erroneously misclassified as forest stands, resulting in a false increase of the timberline in 2006. But, on the other hand, in the last five decades treeline uplift has been observed mainly on the southern slopes.

Forested areas in 2006 represented almost 77 % of the study area while 61 % had never been disturbed. In addition, 5.5 % of the study area represented forests that existed in 1918 but afterward suffered irreversible deforestation (Fig. 5). These areas experienced land use conversion at lower altitudes to permanent settlements and at higher altitudes to grazing areas. However, a relatively equal percent (5.9 %, Fig. 5) corresponded to areas that were forested after 1918 and remained also forested in 2006.

The important changes to the forest areas took place after the Land Law was issued in 1921, which permitted the creation of pasture by forest expropriation. That was the period when 5.5 % from the entire forest areas were lost. By analyzing the map, we found that the largest area that was deforested during this period was the spruce forest at the location where the timberline ecotone started. Thus, the upper limit of the treeline decreased around 200 m in altitude during the period 1925–1945, because of human influence, a fact that was somewhat compensated by the natural inversed phenomenon, which is the increase in altitude of the upper treeline (about 100 m) due to the warming of the climate. During 1921–1926 the highest period of exploitation was registered, with deforestation exceeding up to 40 % of the annual growing capacity of the forests (Ivănescu 1972).

Other major forest cover changes emerged after the nationalization that began in 1948, which was a period of intense forest exploitations with two intervals of maximum exploitation (1951–1955 and 1962–1975) (Săvulescu 2010). Part of the effects of these changes may be traced to the first interval of intensive exploitation in 1960 and the second interval of intensive exploitation in 2006. During the first interval (1951–1955) the forests were exploited by a Romanian-Soviet joint venture company to pay war reparations to the USSR.

A significant increase of the nonforest areas (4.6 %, Fig. 5) was the result of the deforestation that took place in the second period (1962–1975) and that corresponds to the period when the biggest hydrotechnical constructions were initiated within the Sebeş watershed (the Oaşa and Bistra Lakes), the Lotru watershed (the

Vidra, Petrimanu, Galbenu, Balindru, Jidoaia, and Malaia and Brădişor Lakes) and the Sadu watershed (the Negovanu Lake) at the same time when the big wood processing plants opened (Sebeş, Râmnicu Vâlcea, Tg. Jiu).

Forest cover changes that are visible in 2006 (Fig. 5) are partially the result of new management policies. After 1989, Romania oscillated between leaving the forests in the property of the state and *in integrum* restitution. In 1991, the right to property, regarding the appropriation of a maximum 1 ha of forest, was reinstituted, but in 2000–2005 new laws were issued that reinstituted the right to property (*in integrum* restitution).

By analyzing the map, we found a loss of about 6.3 % (1.7 % + 4.6 %, Fig. 5) from the forest areas within the central part of the Southern Carpathians during the period 1960–2006. The highest deforested areas from the study area were on the locations of the communal and compossesorate forests in the Lotru and Sebeş Mountains during the appropriations from 2000–2006. Forest restoration through plantation and natural regeneration during 1960–2006 reached 9.8 % (2.7 % + 7.1 %, Fig. 5), compensating for the previous loss.

The most compact core area is situated in the southeastern part of the Parâng-Cindrel Mountains, south of the Lotru valley, where human influence is low as a consequence of the lack of high-altitude settlements and of low road density. Additionally, because of the terrain inaccessibility, economically efficient forest harvesting cannot be done.

5 Conclusions

The analysis of the forest cover change over the last century in the central part of the Southern Carpathians revealed the negative effects of anthropogenic influence, which are a result of deforestation to provide raw material for the wood industry, to extend human settlements, and to facilitate hydrotechnical constructions and communication or exploitation routes.

By using the overlay analysis in which old Austrian military maps together with finer-scale topographic maps and newer digital data were used, we found that this method proved to be reliable and surprisingly accurate. For a clearer contribution of the policies of deforestation, an input map from 1990 is required.

In the last century, forest inventory showed a considerable loss of forest cover prior to 1960 and a slight increase after the 1960s; this finding contradicts the well-accepted hypothesis of continuous decreasing forest cover in Romania. The results of our study can be explained by the partial compensation of the natural regeneration of the Carpathians forest, the expansion of the upper forest limit as a result of global warming, and the steady continuity in reforestation due to forest management policy changes. It is likely that only after 2006 that important forest loss was found to be caused by the process of forest retrocession after the latest laws instituted in 2005.

The most important morphological feature of the Parâng-Cindrel forest is its well-preserved unitary trunk, although highly irregular, due to the existence of numerous dendritic middle-altitude drainage divides situated below the upper limit of the forest that ensures connectivity through the whole region.

However, fragmentation increases mainly in the northeastern and western areas and affects the smaller patches that increase in number rapidly, thus influencing all of the forest ecological phenomena.

References

Bossard M, Feranec J, Otahel J (2000) CORINE land cover technical guide—addendum 2000. European Environment Agency, Copenhagen

Burdaşu P (1971) Din istoricul exploatărilor forestiere de pe Valea Lotrului—studii vâlcene. Râmnicu Vâlcea

Buza M (2000) Munţii Cindrelului: studiu geoecologic. Editura Universităţii Lucian Blaga, Sibiu

Crăciunescu V (2006) Hărţile austriece (1910) reproiectate în Stereo 70, geo-spatial.org. http://earth.unibuc.ro/download/harile-austriece-1910-reproiectate-in-stereo70. Accessed 15 Aug 2010

Crăciunescu V (2010) Hărţile sovietice 1:100000 reproiectate în Stereo 70, geo-spatial.org. http://earth.unibuc.ro/download/hartile-sovietice-1-100-000-reproiectate-in-stereo70. Accessed 15 Aug 2010

Csucsuja I (1998) Istoria pădurilor din Transilvania 1848–1914. Presa Universitară Clujeană, Cluj-Napoca

Farina A (2007) Principles and methods in landscape ecology: toward a science of landscape. Springer, Heidelberg

Giurescu CC (1975) Istoria pădurii româneşti din cele mai vechi timpuri până astazi. Ceres, Bucureşti

Giurgiu V (2010) Pădurile şi schimbările climatice. Rev pădur 125(3):3–17

Ivănescu D (1972) Din istoria silviculturii româneşti. Ceres, Bucureşti

Jaeger JA (2000) Landscape division, splitting index, and effective mesh size: new measures of landscape fragmentation. Landsc Ecol 15:115–130

Kopecká M, Nováček J (2009) Forest fragmentation in the Tatra Region in the period 2000–2006. Landf Anal 10:58–63

Kozak J (2003) Forest cover change in the Western Carpathians in the past 180 Years: a case study in the Orawa region in Poland. Mt Res Dev 23(4):369–375

Kozak J, Estreguil C, Vogt P (2007) Forest cover and pattern changes in the Carpathians over the last decades. Eur J For Res 126:77–90. doi:10.1007/s10342-006-0160-4

Kümmerle T, Hostert P, Radeloff VC, Perzanowski K, Kruhlov I (2007) Post-socialist forest disturbance in the Carpathian border of Poland. Slovakia Ukraine Ecol Appl 17:1279–1295. doi:10.1890/06-1661.1

Kümmerle T, Chaskovskyy O, Knorn J, Radeloff VC, Kruhlov I, Keeton WS, Hostert P (2009) Forest cover change and illegal logging in the Ukrainian Carpathians in the transition period from 1988 to 2007. Remote Sens Environ 113:1194–1207. doi:10.1016/j.rse.2009.02.006

Lang DT (2005) V-LATE 1.1 Vector-based landscape analysis tool extension. http://arcscripts.esri.com/details.asp?dbid=13898. Accessed: 18 July 2012

Li H, Wu J (2004) Use and misuse of landscape indices. Landsc Ecol 19:389–399

Louto M (2000) Dimensions of the landscape structure, 3rd AGILE conference on geographic information science, Helsinki/Espoo, Finland May 25–27th http://plone.itc.nl/agile_old/Conference/2000-helsinki/61.pdf. Accessed 3 Aug 2010

Muică N, Turnock D (2003) The railway age in the Carpathian forests: a study of Romania. Geogr Pannonica 7:9–20

O'Neill RV, Hunsker CT, Timmins SP, Jackson BL, Jones KB, Riitters KH, Wickham JD (1996) Scale problems in reporting landscape pattern at the regional scale. Landsc Ecol 11:169–180

Riitters KH et al (2002) Fragmentation of Continental United States forests. Ecosyst 5:815–822. doi:10.1007/s10021002-0209-2

Săvulescu I (2010) Interrelaţiile dintre componentele fizico-geografice şi etajarea pădurii în Munţii Iezer: rezumatul tezei de doctorat. Universitatea din Bucureşti, Bucureşti

Soille P, Vogt P (2009) Morphological segmentation of binary patterns. Pattern Recognit Lett 30:456–459. doi:10.1016/j.patrec.2008.10.015

Stoiculescu CD (2003) Parcul National Potential Masivul Parâng—versantul nordic. Rev Geogr 9:102–113

Turner MG (1990) Spatial and temporal analysis of landscape patterns. Landsc Ecol 4:21–30

Vogt P (2010) GUIDOS 1.3. http://forest.jrc.ec.europa.eu/download/software/guidos. Accessed 22 July 2010

Vogt P, Riitters KH, Estreguil C, Kozak J, Wade TG, Wickham JD (2007) Mapping spatial patterns with morphological image processing. Landsc Ecol 22:171–177. doi:10.1007/s10980-006-9013-2

Wu J, Shen W, Sun W, Tueller PT (2002) Empirical patterns of the effects of changing scale on landscape metrics. Landsc Ecol 17:761–782

Tree Species and Management Effect on Herb Layer Species Composition in Mountain Fir-Beech Forests of the Western Carpathians

Karol Ujházy, Eva Križová, Peter Glončák, Blažena Benčaťová and Juraj Nič

Abstract The effect of tree species on herbs in forest understory was studied in the Dobročský prales National Nature Reserve in the Western Carpathians, as well as neighboring secondary stands at the same site. Herb species occurrence and dendrometric values were recorded on belt transects and analyzed by direct gradient analysis (RDA). The results of a permutation test showed that herb frequencies were significantly affected by both beech and spruce, regardless of plot position inside or outside of the Reserve. In general, frequencies of herb species occurrence reacted negatively to the increasing tree competition, particularly the density of trees (over 7 cm DBH) in their surroundings. Frequency of the occurrence of most herb species had negative correlations with increasing beech competition (especially with stem density). Several herb species were significantly associated with spruce (especially *Festuca altissima*, *Oxalis acetosella*, *Viola reichenbachiana*). A subset comprised of natural fir-beech and secondary spruce stands was also compared. *Dentaria enneaphyllos* was found as the best indicator of natural stands, whereas *Asarum europaeum*, *Luzula luzuloides*, *Moehringia trinervia*, and *Veronica officinalis* were characteristic for secondary spruce forests of the studied site. Spruce forests had a more diverse and species-rich herb layer; however, several species of original fir-beech forests were suppressed or excluded there. The consequences of such diversity change are also discussed in the chapter.

1 Introduction

Biodiversity has become one of the central topics of biology and ecology (Ehrlich and Wilson 1991) due to a dramatic global increase of species extinctions caused by human activities (Vačkář 2005). The majority of mountain areas in the Western

K. Ujházy (✉) · E. Križová · P. Glončák · B. Benčaťová · J. Nič
Department of Phytology, Faculty of Forestry, Technical University in Zvolen, T. G. Masaryka 24, 960 53 Zvolen, Slovakia
e-mail: ujhazy@vsld.tuzvo.sk

J. Kozak et al. (eds.), *The Carpathians: Integrating Nature and Society Towards Sustainability*, Environmental Science and Engineering,
DOI: 10.1007/978-3-642-12725-0_18,

Carpathians have been continuously forested. In the fourteenth century about 60 % of an area of Slovakia (including the Panonian lowlands) was forested (Michalko 1986). For example, the Poľana and Bukovské vrchy Mountains had the lowest percentage of forests by the end of nineteenth century (81 %, and 67 %, respectively) according to analysis of historical maps (Olah et al. 2006). Most of the present day Slovak forests still contain the original species diversity of the natural landscape; however, the majority of the forest stands are managed for timber production. Silviculture has, of course, a serious impact on forest ecosystems and thus it affects species diversity of extended areas. Manipulation of tree species composition is one of the main tools of forest management (Barbier et al. 2008). Thus, tree-species replacement is realized on a large scale. During the last centuries serious decline of originally common and typical plant species was documented in Central Europe after massive conversions of broad-leaved forests to coniferous monocultures (Emmer et al. 1998; Fanta 2007). Rapid diversity decline on the landscape level as a consequence of intensive forest management was observed also in the boreal zone of Europe (Pitkänen 1998). Higher species diversity within natural forests with natural structure of tree layer is expected in general (Bo Larsen 1995), as some species dependent on old-growth forest cannot survive in managed forests (Bengtsson et al. 2000). The majority of studies analyzing species diversity of forests consider only woody (or tree) species, which are directly affected by silviculture and harvesting. Less numerous studies are dedicated to changes in the entire range of plant diversity, and information about total diversity of all organism groups is missing.

Finding the balance between profitable management and natural diversity conservation is the goal of modern sustainable forest management (Mölder et al. 2008). To assess success or failure of management regimes designed to sustain species diversity, ecologists and forest managers look for measures of ecosystem change (Lindenmayer et al. 2000). While herb layer is not a direct subject of forest management, its composition is used to identify permanent site properties in the Western Carpathians (Zlatník 1959; Randuška et al. 1986), such as classification of forest ecosystems and mapping of biotopes for nature conservation (Stanová and Valachovič 2002). Indirect effects of changes in stand structure and especially by the change of tree composition (involving changes of soil properties and microclimate) on herb layer are frequently observed. It is known that each tree species has at least some specific influence on the forest environment. The influence of a particular tree species on the forest understory was studied by Kuuluvainen and Pukkala (1989) and Saetre (1999) and defined as the tree influence potential (IP). An overview of studies comparing understory vegetation richness and diversity as a consequence of upper-story species dominance can be found in Barbier et al. (2008). Both positive and negative effects of a particular tree species on understory diversity were reported.

Understanding the interactions between tree and herb species is necessary if herb species composition is to be used as site quality or ecosystem change indicators (for example, see analysis of indicator values of Ellenberg et al. 1992). An attempt to find the relationship between tree stand structure and herb species composition in both natural and secondary mountain forests typical to the Western Carpathians is presented in this chapter.

The objectives of this study were to provide evidence that tree species and stand structure affect species composition of the herb layer; to find herb species positively or negatively associated with a particular tree species; and to quantify changes of the herb layer between natural fir-beech and managed spruce forests and to find herb species indicators for them.

2 Study Site

Dobročský prales National Nature Reserve and its buffer zone represent the most widespread case of tree-composition change in the Western Carpathians: mixed fir-beech forests have been replaced by spruce plantations. The Reserve is situated on the granite bedrock of the Veporské vrchy Mountains in Central Slovakia (Western Carpathians) on gentle northern slopes about 900 m a.s.l. with deep Cambisols. Over 900 mm of annual precipitation enables the development of probably the most productive tree stands of Slovakia. Herb-rich communities of the Reserve belong to the *Dentario enneaphylli-Fagetum* Oberdorfer ex W. et. A. Matuszkiewicz 1960 *salvietosum glutinosae* Moravec 1974 subassociation, and the *Fagion* Luquet 1926 alliance. They are dominated by beech (*Fagus sylvatica*) and fir (*Abies alba*) with a regular admixture of sycamore (*Acer pseudoplatanus*), and less frequently with spruce (*Picea abies*), elm (*Ulmus glabra*), ash (*Fraxinus excelsior*), and maple (*Acer platanoides*). Small-scale stages of cyclic forest development have developed within the core area. Beech trees reach an age of about 230 years and the oldest firs exceed 400 years (Saniga 2002). The core area (about 50 ha) of the Reserve (about 100 ha) has not been managed since at least 1913 when it was established (Slávik et al. 2002).

Natural stands sharply contrast with even-aged managed stands about 80 (60–100) years old in their surroundings. The managed stands are quite homogenous without large clear-cuts or young stands, separated only by forest roads and logging lines and consist mainly of spruce plantations most likely of the second generation. Beech regeneration is frequent in the understory in small patches, forming a closed layer of about 5 m in height. About a quarter of the studied area is formed by secondary beech stands with the ash and sycamore admixture and less frequent mixtures of spruce and beech. Standard forest management typical for the Slovak mountain regions were applied there: periodical thinning eliminating suppressed or lower stem-quality trees was conducted everywhere, and selective "sanitary" cutting of individual spruce trees or small groups attacked by bark beetle or damaged by other harmful agents performed in the last decade. Recent management of the buffer zone aims to achieve a tree-species composition and stand structure similar to natural stands. However, the present structure of stands is still rather similar inside and outside of the buffer zone.

3 Methods

Two 20 m wide, 553 and 683 m long transects were led perpendicularly to the Reserve borders along contour lines, comprising approximately the same part in the Reserve and outside (Fig. 1). Transects were placed to represent both natural stands (comprising all developmental stages) in the Reserve and old secondary

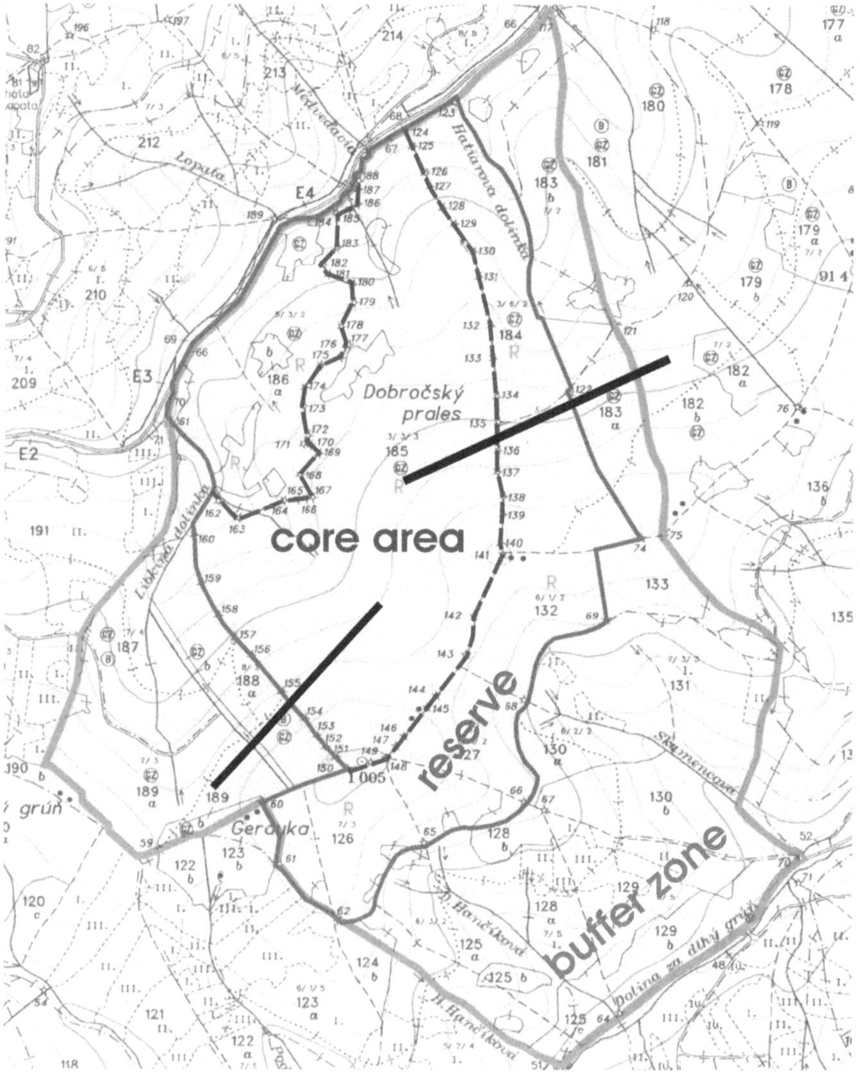

Fig. 1 Schematic map of the study area in the Dobročský prales national nature reserve, Western Carpathians, with transect positions (of 553 and 683 m length) crossing the opposite core area and reserve borders

forests with developed herb layer in its neighborhood. Total length of the transects was 1,236 m with an altitudinal amplitude of 80 m (about 925 m a.s.l.). On 94 square sample plots of 2.25 m^2 (with 13 m spacing), vascular plant species presence within nine 0.25 m^2 subplots and percentage of estimated cover in a whole plot were recorded.

Plot and tree positions were measured using the FieldMap technology (IFER 2008). Tree diameters in 1.3 m diameter at breast height (DBH) of thick trees (diameter over 7 cm) were measured in the whole transect area. Diameters and positions of thin trees ($\leq$ 7 cm) were recorded only in the central 10 m belt (Fig. 2). Coordinates of endpoints of both transects were measured by a global positioning system (GPS).

Spatial data were processed in the ArcGIS 9.2. A database of tree species, DBH, and coordinates were created. Thick trees belonging to the 10 m^2 plot neighborhood and thin trees within 5 m around the plot center were selected according to plot and tree coordinates. For each plot we calculated the tree IP index (Saetre 1999), defined as a circle with 10 m perimeter for thick and with 5 m perimeter for thin trees:

$$IP = \Sigma DBH \cdot e^{(-c \cdot r)} \tag{1}$$

where r is the distance from the center of the plot and the tree and c is constant, which was set to 0.5 according to the results of Glončák (2009). This index reflects both density and size of trees with respect to their distance from the center of a plot (see small square plot in the center, Fig. 2). It is based on the assumption that the influence of a tree to the chosen point of the forest understory increases with its size and decreases with its distance. Species frequencies were calculated according to species presence in the 0.25 m^2 subplots. Dendrometric variables were

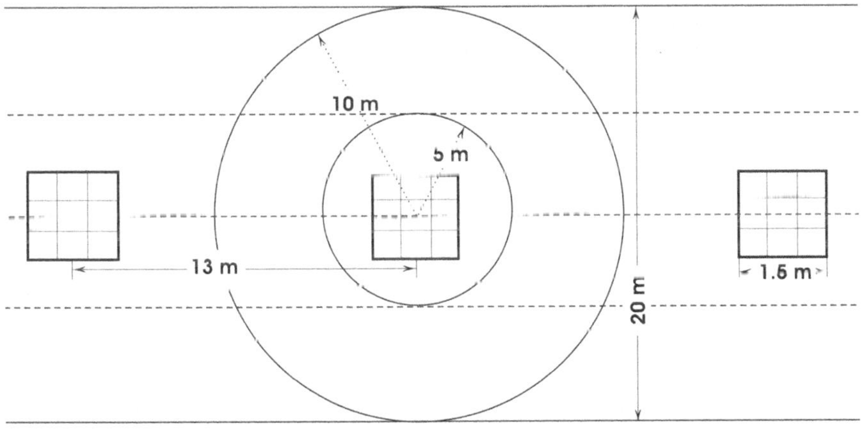

Fig. 2 Sampling design of the transects. Thick trees (over 7 cm DBH) were recorded within the whole transect width, thin trees only in the central 10 m belt. Trees within the circular area with the rectangular subplot in the center were considered for herb-tree relationship analyses

calculated for trees within circular plots: mean DBH, sum of DBH, and number of tree individuals for all trees, later separated for thin and thick trees of each species. Correlations between herb species frequencies and dendrometric values were calculated in the Statistica program using the Spearman rank coefficient.

Direct gradient analysis (RDA) was performed using the Canoco program (ter Braak and Šmilauer 2002) to find the relationship between dendrometric variables and herb-layer species. A Monte Carlo permutation test was performed to assess significance of variables.

Species preferences were analyzed using a phi coefficient, which ranges from −1 to 1. This coefficient is equal or close to zero when the species occurrence in the data set does not show any preference to any relevé group (cluster of vegetation samples). Higher values indicate that species occurrences are concentrated in the target group (Tichý and Chytrý 2006). After equalization of relevé groups in the JUICE program (Tichý 2002), phi values are entirely independent from the size of the group. Significance of species preferences was tested by Fisher's exact test within groups of a subset of 74 plots. Two contrasting stand types were defined (according to IP factor value of beech, fir, and spruce and according to position inside/outside of the Reserve): fir-beech natural forests inside the Reserve (41 plots) and secondary spruce-dominated stands outside of the Reserve (33 plots). Species constancy (C: percentage of species occurrence within a relevé group), species richness (SR: number of species in a relevé), and the Shannon-Wiener diversity index (H′) were calculated using the JUICE program (Tichý 2002). A t-test (in the MS Excel program) was used for testing the significance of differences between average values of the above mentioned values within two stand types.

Nomenclature of vascular plants followed Marhold and Hindák (1998). Species of *Dryopteris carthusiana* agg. were combined because of the problematic determination of juvenile plants (the most common of them was *Dryopteris dilatata*), similarly to that of *Senecio* sp., which comprises *S. ovatus* and *S. germanicus* (with *S. ovatus* prevailing). *Rubus hirtus* was identified as *R. hirtus* Waldst. et Kit. (Dostál and Červenka 1991). The term "noble broadleaves" is used for the group of nitrophilous tree species: *Acer pseudoplatanus, A. platanoides, Fraxinus excelsior*, and *Ulmus glabra* (*Acer pseudoplatanus* is most common).

4 Results

4.1 Species Preferences to Natural Fir-Beech and Secondary Spruce Stands

Several species occurred exclusively in spruce, but no one exclusively in fir-beech stands. However, *Dentaria enneaphyllos* were significantly concentrated in the Reserve, whereas *Asarum europaeum, Dryopteris carthusiana* agg., *Festuca altissima, Luzula luzuloides, Moehringia trinervia, Rubus idaeus, Senecio ovatus,*

Table 1 Species preferences analyzed within contrasting subsets of 74 plots in the Dobročský prales national nature reserve in the Western Carpathians. Only the species with the highest preference of particular stand types and the most frequent indifferent species are presented[a]

	Fir-beech primeval (41 plots)			Spruce secondary (33 plots)		
	MF %	C %	P	MF %	C %	P
Optimum in the fir-beech primeval forest:						
Dentaria enneaphyllos	25.7	61.0	62.4*	0.7	3.0	–
Mercurialis perennis	15.2	43.9	18.1	10.1	24.2	–
Salvia glutinosa	3.3	17.1	16.7	2.0	6.1	–
Dentaria bulbifera	16.5	53.7	16.6	17.5	39.4	–
Galeobdolon luteum	52.0	87.8	16.0	52.5	81.8	–
Geranium robertianum	23.0	46.3	15.6	16.2	33.3	–
Sanicula europaea	4.9	19.5	5.7	3.4	15.2	–
Paris quadrifolia	3.3	12.2	5.0	1.0	9.1	–
The most frequent indifferent species:						
Athyrium filix-femina	15.7	56.1	–	18.9	57.6	3.9
Impatiens noli-tangere	7.6	19.5	–	13.8	27.3	9.2
Mycelis muralis	1.4	7.3	–	2.0	15.2	8.2
Stellaria nemorum	7.3	22.0	–	7.4	27.3	6.2
Dryopteris filix-mas	3.3	22.0	–	5.1	27.3	6.2
Optimum in the secondary spruce forest:						
Viola reichenbachiana	14.6	43.9	–	45.8	97.0	50.5*
Festuca altissima	0.0	2.4	–	10.1	27.3	41.5*
Asarum europaeum	.	.	–	6.7	24.2	38.8*
Senecio nemorensis s.l.	1.4	7.3	–	8.4	30.3	30.0*
Moehringia trinervia	.	.	–	2.7	15.2	30.0*
Veronica officinalis	.	.	–	3.4	15.2	30.0*
Rubus idaeus	13.0	31.7		32.3	57.6	28.5*
Luzula luzuloides	.	.	–	3.4	12.1	26.6*
Dryopteris carthusiana agg.	1.9	9.8	–	3.4	27.3	26.1*
Prenanthes purpurea	0.8	2.4	–	6.4	15.2	23.1
Urtica dioica	1.9	14.6	–	10.4	30.3	22.4
Oxalis acetosella	61.8	85.4	–	92.3	100.0	19.7
Galium odoratum	48.0	70.7	–	60.3	87.9	16.7
Chrysosplenium alternifolium	.	.	–	2.4	6.1	18.6
Ajuga reptans	.		–	1.3	6.1	18.6
Rubus hirtus	.	.	–	1.3	6.1	18.6
Milium effusum	.	.	–	1.7	6.1	18.6
Hypericum maculatum	.	.	–	0.7	6.1	18.6
Circaea alpina	.	.	–	3.7	6.1	18.6

[a] Significant species preferences ($p \leq 0.05$) according to Fisher's exact test are indicated by an asterisk. MF: mean frequency in subplot; C: constancy; P: phi coefficient (values multiplied by 100 in the table)

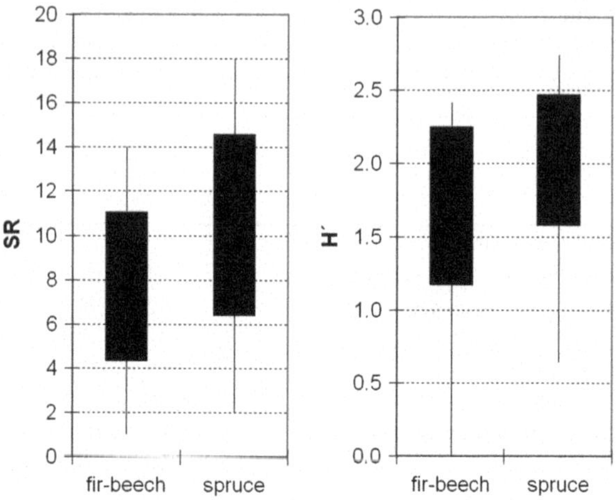

Fig. 3 Comparison of herb species richness (SR: number of herb species on a 2.25 m^2 plot) and diversity (H′: Shannon-Wiener index) of natural and secondary stands according to herb presence and frequency within two plot subsets (fir-beech: 41 plots; spruce: 33 plots) in the study area. Minimum, average ± standard deviation, and maximum values are shown

Veronica officinalis, and *Viola reichenbachiana* were found in secondary spruce stands. Among common species of both types, *Dentaria bulbifera, Geranium robertianum, Mercurialis perennis, Paris quadrifolia, Salvia glutinosa*, and *Sanicula europaea* had higher constancy and average frequency in fir-beech stands, while *Galium odoratum, Oxalis acetosella,* and *Prenanthes purpurea* were found in spruce stands (Table 1).

In total 53 herb species (including two *Rubus* species) were recorded on 74 plots of the subset. From that, 32 occurred in both stand types, 33 were recorded in the natural stands, and 53 species were in the secondary spruce stands. Also, average values of species richness were significantly higher in secondary spruce stands (10.5 on 2.25 m^2) than in natural stands (7.7 on 2.25 m^2). Similarly, averages of the Shannon-Wiener diversity index were also higher. Variance of herb-species richness was lower, whereas variance of diversity was higher in the fir-beech stands (Fig. 3).

4.2 Response of Herb Species to the Tree Species Influence

RDA of the whole data set (94 plots; excluding management type as position inside/outside of the Reserve, which was set as covariable in the analysis) shows that various herb species are differently associated with a particular tree species and with stand-structure characteristics. The Monte Carlo test results show that most of the dendrometric variables significantly affect herb layer species frequencies (Table 2). Frequencies of herb species are affected negatively by increasing density of thick trees in their surroundings (number of trees over 7 cm DBH in the circular plot; see Fig. 4). Increasing influence of beech is related to the significant decrease of frequencies of the majority of herb species (Fig. 5). Only *Veronica montana, Dentaria bulbifera, Asarum europaeum*, and *Isopyrum*

Table 2 Effect of dendrometric variables on herb layer variability analyzed by RDA analysis, Dobročský prales National Nature Reserve, Western Carpathians. Each variable was tested separately (marginal effect) by the Monte Carlo test. Occurrence in the Reserve was set as a nominal covariable

Variable	% Explained
N thick trees	5.5***
N beech	5.0***
IP beech	4.4***
N spruce	3.3**
Mean DBH	3.2**
Mean DBH thick trees	3.1***
N all trees	2.9**
IP all trees	2.1*
N fir	2.1*
IP spruce	2.0*
IP noble broadleaves	1.3 ns
IP fir	1.2 ns
N noble broadleaves	1.0 ns
N thin trees	0.6 ns

[1] Significant values are marked according to p-level (***0.001; **0.01; *0.05; ns: > 0.05). Other species are mostly *Corylus avellana,* with a few individuals of the *Sambucus nigra*, *S. racemosa*, *Lonicera nigra,* and *Sorbus aucuparia* species; noble broadleaves are *Acer pseudoplatanus*, *A. platanoides*, *Fraxinus excelsior*, and *Ulmus glabra*. IP: tree influence potential; N: number of individual trees on circular plot

thalictroides positively respond to the beech influence. *Dentaria enneaphyllos*, a typical species of natural stands, shows no relationship with beech influence, positive with the noble broadleaves, and negative with spruce (similarly with *Dentaria bulbifera, Veronica montana,* and *Isopyrum thalictroides*). *Dentaria enneaphyllos* was affected by tree density and less by tree influence (thus not shown in Fig. 4). Many species (*Galium odoratum* and *Mercurialis perennis* first of all) are positively associated with mean DBH (Fig. 4) as found for the optimum stage of fir-beech forest with a low number of big trees (lower IP factor value than in younger dense stands). Another large group of species is positively associated with spruce, especially *Festuca altissima, Oxalis acetosella, Rubus idaeus, Rubus hirtus,* and *Viola reichenbachiana*. The other tree species, fir and noble broadleaves, do not exhibit significant influence on the herb layer in our data set.

The RDA analysis shows that most of the dendrometric variables significantly affect species composition of herb layer. Despite the fact that species composition is affected significantly in general, particular herb species exhibit specific relationships with tree species. Thus, individual response of herb species to dendrometric variables was studied by correlation analysis. As shown in Table 3, the IP factor or number of trees on a plot (stand density) is less significant, regardless of tree species. The majority of the most frequent species are affected by both spruce and beech. Spruce show mostly positive correlations (primarily with *Festuca altissima*, *Oxalis acetosella*, and *Viola reichenbachiana* species), while beech show

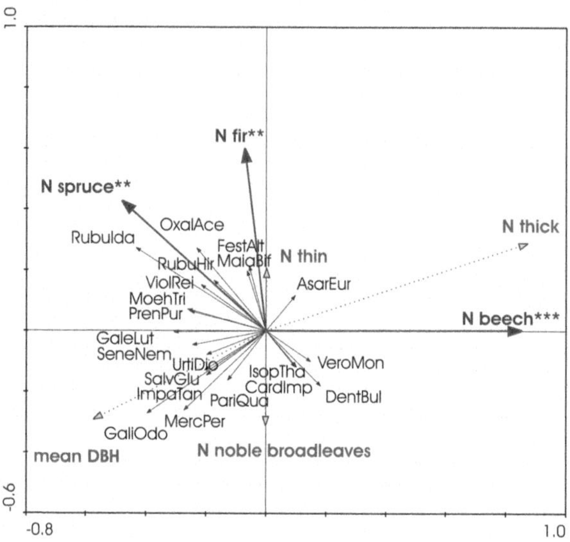

Fig. 4 Ordination diagram of RDA analysis calculated according to herb species frequencies, Dobročský prales national nature reserve, Western Carpathians. Significance level (see Table 2) of explanatory variables is shown (N: number of tree individuals). Supplementary variables are drawn by dotted line. The first four letters of genus and species names are used (see Table 1). First and second axes are shown (eigenvalues: 0.062, 0.031; species-environment correlations: 0.532, 0.607; percentage variance of species-environment relation: 60.2, 30.1)

Fig. 5 Species responses to tree-species influence according to RDA analysis (IP: tree influence potential; for details and abbreviations see the caption of Fig. 4) in the study area. First and second axes are shown (eigenvalues: 0.53, 0.018; species-environment correlations: 0.526, 0.474; percentage variance of species-environment relation: 63.4, 21.3)

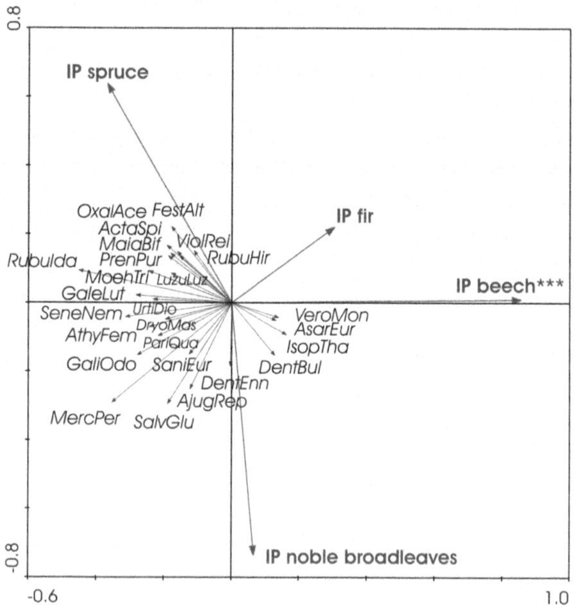

Table 3 Spearman correlations between herb-species frequencies and dendrometric values, Dobročský prales National Nature Reserve, Western Carpathians

	Galium odoratum	*Mercurialis perennis*	*Dentaria enneaphyllos*	*Dentaria bulbifera*	*Dryopteris carthusiana* agg.	*Festuca altissima*	*Asarum europaeum*	*Urtica dioica*	*Luzula luzuloides*	*Prenanthes purpurea*	*Moehringia trinervia*	*Rubus idaeus*	*Senecio* sp.	*Oxalis acetosella*	*Viola reichenbachiana*
IP trees	−0.16	**−0.44**	**−0.21**	−0.10	0.10	**0.26**	0.13	−0.06	−0.07	0.03	−0.06	−0.05	−0.07	0.04	0.04
IP beech	**−0.28**	**−0.37**	**0.22**	0.17	−0.13	**−0.28**	0.03	**−0.37**	**−0.34**	**−0.37**	**−0.39**	**−0.61**	**−0.46**	**−0.28**	**−0.35**
IP fir	**−0.26**	0.05	**0.38**	−0.02	−0.03	**−0.21**	**−0.21**	**−0.21**	−0.15	−0.16	−0.09	0.01	−0.13	−0.11	**−0.29**
IP nobles	0.19	0.15	0.20	**0.22**	0.01	−0.04	0.17	0.00	−0.10	0.00	−0.15	−0.17	−0.08	**−0.33**	**−0.25**
IP spruce	0.08	**−0.22**	**−0.47**	**−0.29**	**0.25**	**0.43**	0.17	**0.27**	0.17	0.18	**0.26**	**0.25**	**0.24**	**0.45**	**0.40**
N trees	**−0.35**	**−0.42**	−0.11	−0.11	0.05	0.16	0.14	−0.14	−0.07	−0.13	−0.08	−0.16	−0.19	−0.05	−0.12
N thick trees	**−0.36**	**−0.57**	−0.11	0.03	−0.04	0.06	**0.23**	−0.17	−0.06	−0.15	−0.11	**−0.31**	**−0.36**	−0.02	−0.05
N thin trees	−0.08	−0.01	−0.07	−0.19	**0.21**	**0.26**	−0.03	0.13	0.05	0.11	0.06	0.20	0.15	0.05	−0.01
N beech	**−0.36**	**−0.36**	0.06	0.15	−0.12	−0.14	0.03	**−0.25**	**−0.28**	**−0.33**	**−0.32**	**−0.55**	**−0.38**	**−0.28**	**−0.31**
N fir	**−0.26**	0.06	**0.37**	−0.03	−0.04	−0.20	−0.20	**−0.23**	−0.15	−0.20	−0.08	0.08	−0.10	−0.03	**−0.26**
N nobles	0.18	0.14	0.15	0.17	0.16	0.05	0.15	0.06	−0.04	−0.02	−0.10	−0.06	−0.06	**−0.28**	**−0.24**
N spruce	0.11	**−0.23**	**−0.47**	**−0.32**	**0.25**	**0.40**	**0.25**	**0.24**	0.20	0.20	**0.29**	**0.28**	**0.22**	**0.51**	**0.45**
sum DBH	−0.09	**−0.23**	−0.03	−0.01	−0.03	0.20	**0.21**	−0.14	0.02	−0.02	0.04	0.04	−0.10	0.16	**0.23**
mean DBH	**0.29**	**0.25**	0.12	0.11	−0.10	−0.02	−0.06	−0.02	−0.08	−0.05	0.05	0.16	0.05	0.15	**0.22**

Values significant at the 0.05 level are given in bold letters. IP: tree influence potential; N: number of individual trees on circular plot negative ones (especially with *Rubus idaeus* and *Senecio* sp.). Only frequencies of *Dentaria* species are positively correlated with increasing beech influence. There are also several significant correlations with fir and noble broadleaves. *Dentaria enneaphyllos* shows a relatively strong positive correlation with fir, which grows exclusively in the Reserve. However, frequencies of *Viola reichenbachiana* and *Galium odoratum* decrease under fir trees. Increasing density of noble broadleaves is unfavorable for *Oxalis acetosella* and *Viola reichenbachiana*.

In general, three groups of herb species can be distinguished:

- *Dentaria enneaphyllos* and *D. bulbifera* with negative relationships to spruce but positive to beech, fir, and noble broadleaves,
- *Galium odoratum* and *Mercurialis perennis* without positive reaction to any tree species but positively correlated with mean DBH,
- *Asarum europaeum, Dryopteris carthusiana* agg., *Festuca altissima, Luzula luzuloides, Moehringia trinervia, Oxalis acetosella, Prenanthes purpurea, Rubus idaeus, Senecio* sp., *Urtica dioica,* and *Viola reichenbachiana* with a positive relationship with spruce, negative to beech, and indifferent to mean DBH.

These findings mean that herb layer is best developed in old, sparse stands formed by a small number of thick trees. This situation is typical for the optimal stage of natural forests or for the oldest commercial stands without a dense regeneration layer.

5 Discussion and Conclusions

Our results confirm that tree species forming canopy of both natural fir-beech and secondary forests significantly affect herb layer. These findings are in agreement with those of van Oijen et al. (2005), Mölder et al. (2008), and Wulf and Naaf (2009) for various forest stand types. Spruce and beech have more effect on herb layer composition than fir and noble broadleaves. However, the lack of significant effect of noble broadleaves could be a consequence of their less frequent occurrence within the studied area. A review by Barbier et al. (2008) shows that the understory of coniferous forests provides less diversified vascular understories than broad-leaved forests in general. In our case, secondary spruce forests show higher species richness and diversity than original beech-dominated stands, which is consistent with other findings (Bürger 1991; Lücke and Schmidt 1997; Ewald 2000a). We cannot confirm strong inhibition of herb layer by spruce at the studied sites as it was reported by several other authors (Teuscher 1985; Simmons and Buckley 1992). However, the studied spruce stands were sampled in the optimal period for herb layer development (about 80 years of age and density reduced by thinning) with a partly opened canopy. Dense younger spruce plantations in the surroundings or spruce stands with closed beech undergrowth are often species-poor. More opened canopy in the old spruce stands in comparison with beech stands of the same age is a consequence of strictly geotropic growth of spruce crown. Spruce is not able to cover canopy openings as can beech, which causes phototropic growth (Otto 1994). We also suppose that selective thinning that reduces tree competition (density, canopy cover, litter amount) is favorable for herb layer development. There is already some evidence of this effect in the literature: increasing values of herb diversity and richness both in beech-dominated (Weckesser 2003; Schmidt 2005) and spruce-dominated forests (Abs et al. 2005; Heinrichs and Schmidt 2009) were found in managed forests (of the same site conditions) compared to unmanaged.

Interference between tree stand density and herb layer development is also reported from studies of unmanaged natural stands. A negative correlation between dendrometric variables (especially number of trees with DBH > 7 cm) and herb frequencies was found according to repeated sampling in the Dobročský prales (Ujházy et al. 2009) and the Badínsky prales (Ujházy et al. 2007) beech-dominated reserves in Slovakia. Beech was also found to be the strongest competitor for herbs in comparison with several other broad-leaved tree species in low-elevation mountain forests in Germany (Mölder et al. 2008).

The relationships found in this study between particular herb and tree species (especially of RDA ordination) are consistent with the recent findings of Máliš et al. (2010) from parallel plots (mostly in managed forests) in the broader area of the Veporské vrchy Mountains. *Festuca altissima, Prenanthes purpurea*, and *Luzula luzuloides* reached an optimum in the spruce-dominated stands there, whereas *Oxalis acetosella, Dryopteris carthusiana* agg., *Rubus idaeus*, and *Veronica officinalis* were found in the mixed forests with high spruce cover. *Oxalis acetosella* and *Dryopteris carthusiana* agg. are adapted to the spruce forest environment very well, as they are usually found among the most constant species of spruce forests of the Eastern Alps, Bohemian Massif, and Western Carpathians (Chytrý et al. 2002). Glonďák (2009) showed that the above mentioned species persist during the entire developmental cycle of natural spruce forests in the Nízke Tatry Mountains, since they can withstand strong competition of the dense spruce stands. It is interesting that these two species are the most frequently occurring herbs growing on the coarse woody debris (both of broadleaved and coniferous trees) in the Reserve. Increasing constancy and cover of *Oxalis acetosella* and *Maianthemum bifolium* in secondary spruce stands is reported also from the carbonate bedrock of the Slovenský raj (Križová et al. 2007) and the Pieniny Mountains (Benčaťová 2006) in northeastern Slovakia.

The affinity of *Dentaria* species to beech dominance is well known in syntaxonomy. *Dentaria bulbifera* is the characteristic species of the *Fagion* alliance (Wallnöfer et al. 1993; Jarolímek and Šibík 2008) and *Dentaria enneaphyllos* is the diagnostic species of the *Dentario enneaphylli-Fagetum* association (Moravec et al. 2000). The absence of *Dentaria enneaphyllos* in spruce forests and decrease of frequency of species, indicating a neutral soil reaction (*Mercurialis perennis, Sanicula europaea, Dentaria bulbifera, Salvia glutinosa,* and *Paris quadrifolia* with indicator values for reaction 8 and 7 according to Ellenberg et al. 1992), can be explained by acidification of topsoil by coniferous litter. Decrease of *Dentaria enneaphyllos* caused by anthropogenic acidification was reported from the Rychlebské hory Mountains (Hédl 2004) and from the Moravsko-Slezské Beskydy in northern Moravia (Šamonil and Vrška 2007). Anthropogenic acidification was not reported for the region of the Veporské vrchy Mountains. However, the effect of a spruce plantation on the topsoil reaction can be even stronger. Increase of the acidophilus, especially shallow-rooted species, in spruce plantations is widely recognized (e.g., Fajmonová 1974; Ewald 2000b; Chytrý et al. 2002; Šomšák and Balkovič 2002; Šomšák 2003). All species indicating acidic soils (indicator values 2–4) in the studied site were more frequent in spruce stands (*Oxalis acetosella, Festuca altissima, Polygonatum verticillatum, Gymnocarpium dryopteris*) or occurred only there (*Calamagrostis villosa, Hypericum maculatum, Veronica officinalis, Luzula luzuloides, Agrostis capillaris*).

In general, our findings from the model site of the Western Carpathian fir-beech forests show that herbs are negatively affected by increasing tree competition. Beech seems to be stronger competitor than other tree species for the herb layer species of the studied area. Thus, higher species richness was found in old secondary spruce stands. This can also be a consequence of an artificial reduction of tree density in commercial stands and of the great dispersal potential from the natural stands in the neighborhood. Herb-species response to tree competition is specific. Several indigenous fir-beech forest herb species were reduced in spruce forests (mainly *Dentaria enneaphyllos*) and higher richness was partly caused by the expansion of several oligotrophic or acidophilous species (*Luzula luzuloides, Veronica officinalis, Hypericum maculatum*), which do not occur in the original natural forests, or of the eutrophic species indicating soil nutrients and disturbances (*Moehringia trinervia*).

By comparing natural and secondary stands, we found a significant change of herb layer in the studied area. The study showed that tree species replacement was one of the main the reasons for this change. Shift of species composition and thus also of the forest site can be found through syntaxonomical or typological analysis, and these facts should be taken into account when secondary stands with changed tree composition are studied. Species richness is lower in the natural fir-beech stands due to the competition of undisturbed beech populations. Reduction of herb layer richness in the even-aged spruce stands aged 90 years old was not confirmed if compared with the mixed natural stands with complex structure (involving all developmental stages). However, the number of species alone cannot be the only or the best criterion to assess a change in biodiversity. Species that enriched spruce forest communities belong to the widely spread ones (with larger ecological amplitude), while the suppressed or outcompeted species are typical for the natural broad-leaved or mixed forests. Thus, their reduction on a broader scale can be considered a degradation of the original biodiversity. On the other hand, the occurrence of acidophilous species indicates a further direction in ecosystem development if spruce was repetitively planted at the site.

Acknowledgments We are grateful to Gabriela Chovancová, Patrícia Pekarovičová, and Michal Martinák for help in the field and to reviewers for their comments. The research was supported by the Slovak Grant Agency VEGA (project no. 1/0831/09).

References

Abs C, Fischer A, Heinrich J, Kugler F, Schöffmann C, Mayer P (2005) Pattern of plant diversity in natural and managed mountain forest. In: Commarmot B, Hamor FD (eds) Natural forests in the temperate zone of Europe—values and utilisation (proceedings, conference 13–17 October 2003 Mukachevo, Ukraine), Swiss Federal Research Institute WSL, Birmensdorf, Carpathian Biosphere Reserve, Rakhiv

Barbier S, Gosselin F, Balandier P (2008) Influence of tree species on understory vegetation diversity and mechanisms involved—a critical review for temperate and boreal forests. For Ecol Manag 254(1):1–15

Benčaťová B (2006) Geobotanická charakteristika lesných spoločenstiev a fytogenofond Pienin. Technical University in Zvolen, Dissertation

Bengtsson J, Nilsson SG, Franc A, Menozzi P (2000) Biodiversity, disturbances, ecosystem function and management of European forests. For Ecol Manag 132:39–50

Bo Larsen J (1995) Ecological stability of forests and sustainable silviculture. For Ecol Manag 3:85–96

Bürger R (1991) Immissionen und Kronenverlichtung als Ursachen für Veränderungen der Waldbodenvegetation im Schwarzwald. Tüexenia 11:407–424

Chytrý M, Exner A, Hrivnák R, Ujházy K, Valachovič M, Willner W (2002) Context-dependence of diagnostic species: a case study of the Central European Spruce Forests. Folia Geobot 37:403–417

Dostál J, Červenka M (1991) Veľký kľúč na určovanie vyšších rastlín I. SPN, Bratislava

Ehrlich PR, Wilson EO (1991) Biodiversity studies: science and policy. Science 253(5021):758–762

Ellenberg H, Weber HE, Dull R, Wirth V, Werner W (1992) Zeigerwerte von Pflanzen in Mitteleuropa. Scr Geobot 18:1–258

Emmer IM, Fanta J, Kobus AT, Kooijman A, Sevink J (1998) Reversing borealization as a means to restore biodiversity in Central-European mountain forests—an example from Krkonoše Mountains, Czech Republic. Biodivers Conserv 7:229–247

Ewald J (2000a) The influence of coniferous canopies on understorey vegetation and soils in mountain forests of the northern Calcareous Alps. Appl Veg Sci 3:123–134

Ewald J (2000b) The partial influence of Norway Spruce stands on understorey vegetation in montane forests of the Bavarian Alps. Mt Res Dev 20(4):364–371

Fajmonová E (1974) Niektoré výsledky štúdia druhotných a prirodzených ihličnatých porastov v pohorí Javorníky. Biol Bratisl 29:537–549

Fanta J (2007) Lesy a lesnictví ve střední Evropě. IV. Změny ve 20. století. Živa 2007(4):161–164

Glončák P (2009) Dynamika vegetácie prírodných horských smrečín. Technical University in Zvolen, Zvolen, Dissertation

Hédl R (2004) Retreat of Dentaria enneaphyllos in Rychlebské hory Mts (Czech republic), in perspective of habitat preferences. Biol Bratisl 59:417–423

Heinrichs S, Schmidt W (2009) Short-term effects of selection and clear cutting on the shrub and herb layer vegetation during the conversion of even-aged Norway spruce stands into mixed stands. For Ecol Manag 258:667–678

IFER (2008) FieldMap—tool designed for computer aided field data collection. IFER—Ústav pro výzkum lesních ekosystémů, s.r.o. http://www.fieldmap.cz. Accessed 17 July 2012

Jarolímek I, Šibík J (2008) Diagnostic, constant and dominant species of the higher vegetation units of Slovakia. Veda, Bratislava

Križová E, Ujházy K, Čaja R (2007) Zmeny fytocenóz dealpínskych bučín v dôsledku zmien drevinového zloženia. In: Neuhöferová P (ed) Problematika lesnické typologie IX. Česká zemědělská univerzita v Praze, Praha, Typologické hodnocení antropogenně ovlivněných lokalit

Kuuluvainen T, Pukkala T (1989) Effect of Scots pine seed trees on the density of ground vegetation and tree seedlings. Silva Fenn 23:159–167

Lindenmayer DB, Margules CR, Botkin DB (2000) Indicators of biodiversity for ecologically sustainable forest management. Conserv Biol 14(4):941–950

Lücke K, Schmidt W (1997) Vegetation und standortsverhältnisse in Buchen-Fichten-Mischbeständen des sollings. Forstarch 68:135–143

Mális F, Vladovič J, Čaboun V, Vodálová A (2010) The influence of Picea abies on herb vegetation in forest plant communities of Veporské vrchy Mts. J For Sci 56:58–67

Marhold K, Hindák F (eds) (1998) Checklist of non-vascular and vascular plants of Slovakia. Veda, Bratislava

Michalko J (ed) (1986) Geobotanická mapa ČSSR Slovenská socialistická republika. Veda, Bratislava

Mölder A, Bernhardt-Römermann M, Schmidt W (2008) Herb-layer diversity in deciduous forests: raised by tree richness or beaten by beech? For Ecol Manag 256:272–281

Moravec J, Husová M, Chytrý M, Neuhäuslová Z (2000) Přehled vegetace ČR Svazek 2 Hygrofilní, mezofilní a xerofilní opadavé lesy. Academia, Praha

Olah B, Boltižiar M, Petrovič F, Gallay I (2006) Vývoj využitia krajiny slovenských biosférických rezervácií UNESCO. Technical University in Zvolen, Zvolen

Otto HJ (1994) Waldökologie. Ulmer, Stuttgart

Pitkänen S (1998) The use of diversity indices to assess the diversity of vegetation in managed boreal forests. For Ecol Manag 112:121–137

Randuška D, Vorel J, Plíva K (1986) Fytocenológia a lesnícka typológia. Príroda, Bratislava

Saetre P (1999) Spatial patterns of ground vegetation, soil microbial biomass and activity in a mixed spruce-birch stand. Ecography 22:183–192

Saniga M (2002) Štruktúra pralesa. In: Slávik D, Burkovský J, Galvánková M, Glejdura S, Križová E, Kropil R, Pekarovič B, Rybár I, Saniga M, Šály R (eds) Dobročský prales, Národná prírodná rezervácia. ÚVVP LVH SR, Zvolen

Schmidt W (2005) Herb layer species as indicators of biodiversity of managed and unmanaged beech forests. For Snow Landsc Res 79(1/2):111–125

Simmons EA, Buckley GP (1992) Ground vegetation under planted mixtures of trees. In: Cannel MGR, Malcolm DC, Robertson PA (eds) The ecology of mixed-species stands of trees. Blackwell, Oxford

Slávik D, Burkovský J, Galvánková M, Glejdura S, Križová E, Kropil R, Pekarovič B, Rybár I, Saniga M, Šály R (2002) Dobročský prales. Národná prírodná rezervácia, ÚVVP LVH SR, Zvolen

Stanová V, Valachovič M (eds) (2002) Katalóg biotopov Slovenska. Daphne—inštitút aplikovanej ekológie, Bratislava

Šamonil P, Vrška T (2007) Trends and cyclical changes in natural fir-beech forests at the north-western edge of the Carpathians. Folia Geobot 42:337–361

Šomšák L (2003) Effect of secondary Spruce forests on Phytoenvironment in the Slovenské rudohorie mountains. Folia Oecol 30:41–59

Šomšák L, Balkovič J (2002) Cyclic succession and plant biodiversity within the secondary spruce forests in the Hnilec river watershed. Phytopedon Bratisl 1:45–51

ter Braak CJF, Šmilauer P (2002) CANOCO Reference manual and CanoDraw for windows users guide: software for canonical community ordination (version 4.5). Microcomputer Power, Ithaca

Teuscher F (1985) Fichtenforste im Mittelland. Schweiz Z Forstwes 136:755–761

Tichý L (2002) JUICE, software for vegetation classification. J Veg Sci 13:451–453

Tichý L, Chytrý M (2006) Statistical determination of diagnostic species for site groups of unequal size. J Veg Sci 17:809–818

Ujházy K, Križová E, Ondruš M, Vigoda M (2007) Short-term vegetation change in the fir-beech primeval forest (Badínsky prales, Central Slovakia). In: Abrudan VI (ed) Forest and sustainable development. Proceedings of the biennial international symposium, Editura Universităţii Transilvania, Braşov

Ujházy K, Križová E, Ondruš M, Murín M (2009) Short-term changes in the vegetation of fir-beech forests in the Dobročský prales virgin forest. Zpr Čes Bot Spol 44(24):59–77

van Oijen D, Feijen M, Hommel P, den Ouden J, de Waal R (2005) Effects of tree species composition on within-forest distribution of understorey species. Appl Veg Sci 8:155–166

Vačkář D (ed) (2005) Ukazatele změn biodiverzity. Academia, Praha

Wallnöfer S, Mucina L, Grass V (1993) Querco-Fagetea. In: Mucina L, Grabherr G, Wallnöfer S (eds) Die Pflanzengesellschaften Österreichs. Gustav Fischer Verlag, Jena, Stuttgart, New York

Weckesser M (2003) Die Bodenvegetation von Buchen-Fichten-Mischbeständen im solling—Struktur—Diversität und Stoffhaushalt. Cuvillier Verlag, Göttingen

Wulf M, Naaf T (2009) Herb layer response to broadleaf tree species with different leaf litter quality and canopy structure in temperate forests. J Veg Sci 20(3):517–526

Zlatník A (1959) Přehled slovenských lesů podle skupin lesních typů. Spisy Věd Lab Biocenol Typol Lesn 3:1–195

Strict Forest Reserve Research in the Margin of the Carpathians, the Vár-hegy Case-Study

Katalin Mázsa, Borbála Balázs, János Bölöni and Ferenc Horváth

Abstract Sixteen forest reserves are situated in the northern part of Hungary which belongs to the Carpathian region according to EURAC delimitation (Ruffini et al. 2006). These Hungarian forest reserves expand the natural forest remnant/forest reserve net of the Carpathians towards the lower hilly region, representing the deciduous beech and oak forest belts near their lower (xeric) distribution limits. This paper outlines the Hungarian forest reserves belonging to the Carpathian region and the preliminary results of current projects in the Vár-hegy Forest Reserve (Bükk Mts., Hungary) as a case study. The alteration of tree species composition was investigated here based on the reconstruction of forest history in the previous 130 years (management period) and analyses of forest stand inventory. In another project CO_2 sequestration changes of these forest stands were modeled since the clear-cutting in the 1880th and carbon stored in the forest ecosystem compartments was estimated. Our results show that the forest reserve stands are presently in a transition state from the managed forest towards a more natural mixed forest with several age-classes.

1 Introduction

Virgin and old-growth forest remnants play an outstand role in conserving the natural resources and the high level of biodiversity of the Carpathian region. Their first descriptions were published as early as in the nineteenth century (Fuchs 1861;

K. Mázsa (✉) · J. Bölöni · F. Horváth
Centre for Ecological Research Hungarian Academy of Sciences, Vácrátót, Alkotmány u. 2-4 2163, Hungary
e-mail: mazsa.katalin@okologia.mta.hu

B. Balázs
Department of Meteorology, Eötvös Loránd University, Budapest, Hungary

J. Kozak et al. (eds.), *The Carpathians: Integrating Nature and Society Towards Sustainability*, Environmental Science and Engineering,
DOI: 10.1007/978-3-642-12725-0_19,

Krauze 1898; Muzsnay 1899; Pulvermacher 1877). Conservation and research of old-growth forests have a long tradition in Central and Eastern European countries (Borlea 1999; Hort et al. 1999; Korpel 1995; Oszlanyi et al. 2004; Pruša 1985). As intensive forest management gained ground in the late twentieth century, the area of natural forests declined together with forest biodiversity. Consequently, natural forest remnants were shifted in the limelight once again and a European cooperation was initiated and set up in Europe, the COST Action E4: 'Forest Reserves Research Network' in 1995. The aims of the action were to promote co-ordination among countries and to focus research on 'natural' forests. The primary objectives were to create a European network of forest reserves, to collect ongoing research, to standardize research methodology and to create an accessible central data bank. The importance of the results from 'natural' forest research was highlighted for the application of ecologically oriented silviculture, for improved forest management and for the future planning of the forest protection network (Parviainen et al. 2000).

In Europe there are about 0.3 million ha of virgin forests (0.4 % of the total forest area) left in about 2500 strict forest reserves and other categories of protection in the temperate zone, mainly in the Balkan, Alpine and Carpathian geographic regions (Parviainen 2005). In the Carpathian region there are 217 virgin forest remnants or forest reserves situated in six countries: Czech Republic, Poland, Slovak Republic, Ukraine, Romania and Hungary (Barton 2010).

The designation of Hungarian Forest Reserve Network was launched in 1993. The aims of the Forest Reserve Program supervised by the State Secretary for Nature and Environment Protection are basically twofold: (a) to ensure the conservation of semi-natural forest stands, and (b) to get new sound knowledge, as natural development reference about the biodiversity, stand structure and ecological processes of our forest ecosystems (Horváth and Borhidi 2002). The total number of designated reserves in Hungary is presently 63. The northern part of Hungary belongs to the Carpathian region according to EURAC delimitation (Ruffini et al. 2006), see Fig. 1. Sixteen forest reserves are situated in the northern part of the country, see Table 1.

These reserves are the largest ones and represent the most characteristic (semi-natural) forest types of North Hungary. Forest reserve research has also been focused here, among others (Czájlik et al. 2003a, b, c; Kenderes et al. 2008; Ódor et al. 2006; Standovár et al. 2006). Here the main forest types are oak, oak–hornbeam or beech dominated mixed forests while in the upper belt of the Carpathians the majority of natural forest remnants are mixed fir-beech, or fir-spruce-beech stands. The Hungarian forest reserves expand the forest reserve net of the Carpathians towards the lower hilly region of the inner Carpathian Basin, both at the geographic and the vegetation level (Fig. 1.).

Ecologists agree that forests in the temperate zone will shift and alter due to the climate change. However the retreating low altitude/low latitude (xeric) limits of forests have been left largely unexplored until very recently (Mátyás 2010). Recent study indicates that climate change in Hungary may lead to drastic reduction in macroclimatically suitable sites for both beech and sessile oak forests in the next

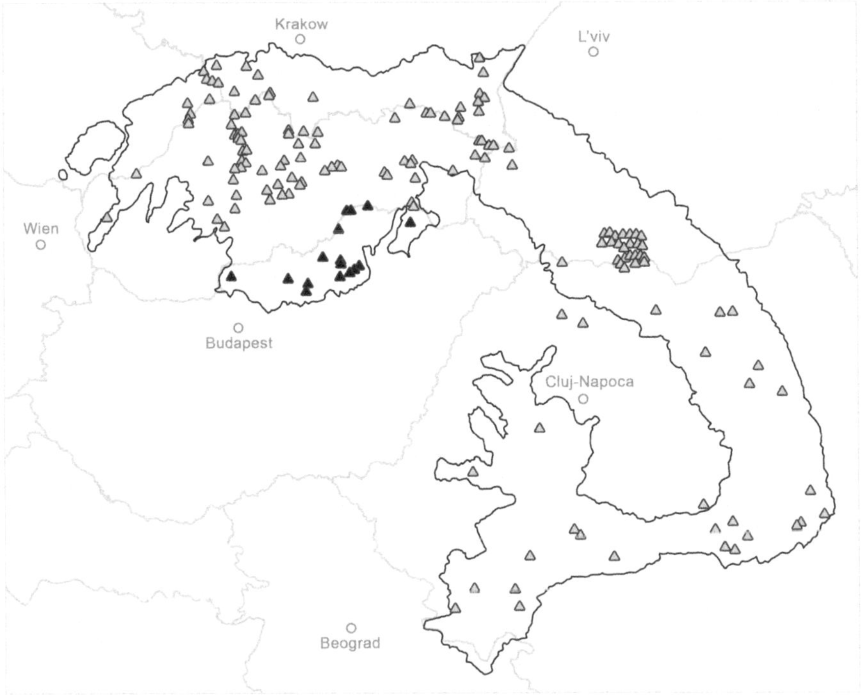

Fig. 1 Forest reserves or virgin forest remnants in the Carpathian region according to Barton (2010). *Black triangles* show the Hungarian ones. Delineation of the Carpathian region is based on Ruffini et al. (2006)

decades (Czúcz et al. 2011), however the actual pattern of forest cover strongly depends on the relief also. The border area of the deciduous forest and forest-steppe vegetation zone pass through Hungary (Borhidi 1961; Zólyomi 1967) see Fig. 2. Several forests reserves that are located in the southern part of Bükk Mts. (North Hungary) are close to the trailing edge of forest—forest steppe biome transition zone, hence suspected to strongly impact by climate change.

The total extent of the strictly protected core areas of the Hungarian forest reserves is cca. 3600 ha, each are surrounded by protected buffer zones (cca. 13.000 ha all together), comprising 0.8 % of the wooded land in Hungary. Most of the designated reserves have mature stands and they have been unmanaged since several decades (Mázsa et al. 2008). In the "Carpathian" part of the country 67 % of the area of strict forest reserves has developed spontaneously for at least 50 years or more (Fig. 3.). Although most Hungarian strict forest reserves were used and altered by human activities one exception is Kékes Forest Reserve. A recent forest historical study has proved that this site is a virgin forest remnant which had avoided exploitation, conversion and clearance as a manorial forest and hunting estate (Czájlik and Pászty 2009).

Table 1 Strictly protected forest reserves belonging to the Carpathian region in Hungary and their survey status

Name, location	Core area (ha)	Main forest type	Surveyed
Pogány-Rózsás, Börzsöny Mts.	91	Beech	
Csörgő-völgy, Mátra Mts.	51	Beech	
Kékes, Mátra Mts.	55	Beech	2005–2008
Hór-völgy, Bükk Mts.	61	Oak-hornbeam	
Kecskés-Galya, Bükk Mts.	87	Mixed sessile oak-turkey oak; downy oak	2010–2011
Vár-hegy, Bükk Mts.	94	Mixed sessile oak-turkey oak; downy oak; oak-hornbeam	2005–2008
Őserdő, Bükk Mts.	59	Beech	2008–2010
Leány-völgy, Bükk Mts.	57	Beech	
Paphárs-Kecskevár, Bükk Mts.	58	Mixed sessile oak-turkey oak; downy oak; oak-hornbeam	
Csókás-völgy, Bükk Mts.	144	Mixed sessile oak-turkey oak; downy oak;	
Nagy-sertéshegy, Eperjes-Tokaj Mts.	66	Beech	
Pataj, Heves hills	66	Beech	
Alsó-hegy, Aggtelek carst	113	Oak-hornbeam	
Haragistya-Lófej, Aggtelek carst	260	Beech, oak-hornbeam	2006–2010
Nagy-oldal, Aggtelek carst	234	Mixed sessile oak-turkey oak; downy oak; oak-hornbeam	
Kelemér-Serényfalu, Borsod hills	81	Oak-hornbeam	

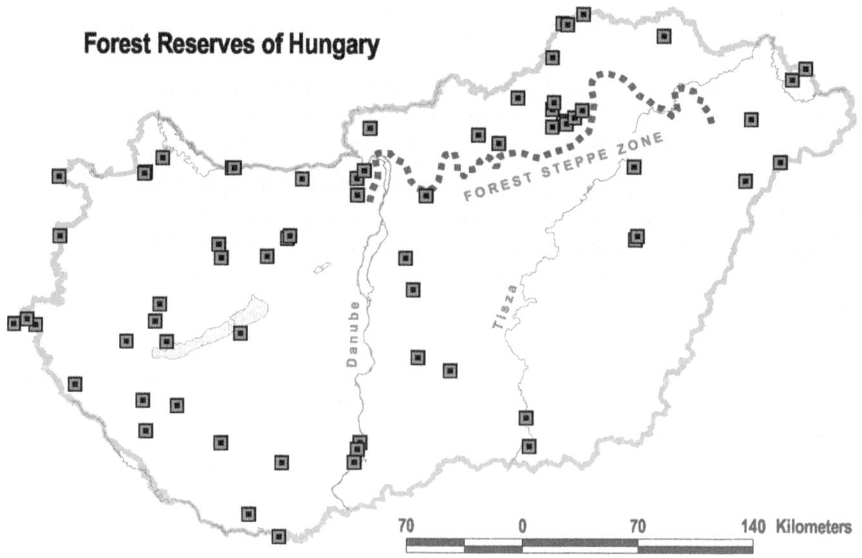

Fig. 2 Overview map of the Hungarian forest reserve network. *Dashed line* shows the border of the forest steppe zone

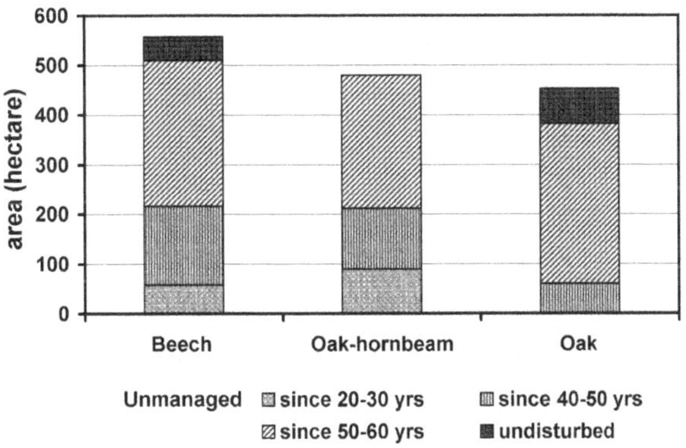

Fig. 3 Present management status of 16 strict forest reserves by main forest types belonging to the Carpathian Region based on a nationwide survey in 1998.
Oak = several types of mixed oak forest from mesic to dry, dominated by *Quercus petraea*, *Q. cerris* or *Q. pubescens*

The research team of the Centre for Ecological Research of the HAS coordinates the survey of the Hungarian forest reserves, and carries out more detailed research in the Vár-hegy Forest Reserve (Bükk Mts., Hungary). This paper shows the preliminary results of current projects in the Vár-hegy Forest Reserve as a case study. The following projects are run in the Vár-hegy Forest Reserve:

- reconstruction of forest history in the previous 130 years (management period) based on land-use history documents and analyses of forest stand structure;
- study of fine scale stand dynamic processes: alteration of canopy species composition after a large scale oak decline in the 1970/1980s;
- modeling the CO_2 sequestration changes/development of these forest stands since natural regeneration after clear-cutting in the 1880th and estimation of carbon stored in the forest ecosystem compartments.

2 Materials and Methods

2.1 *Case Study Site: The Vár-hegy Forest Reserve*

The Vár-hegy Forest Reserve is located in the south-western part of the Bükk Mts., approximately 10 km far from the town of Eger. Its geographical coordinates are: lat./long 47°54′N; 19°57′E. The 94 ha strictly protected core area comprises the upper one-third of the hill at elevations from 326 to 669 m a.s.l. The relief and the microclimate are highly diverse. The soil type composition of the area is varied, variations of rendzina and brown earths being the most characteristic (Bidló et al. 2004). The following Natura 2000 habitat types are the most important in the core area:

- 91H0 Pannonian woods with *Quercus pubescens* on the steep southern-south-eastern ridges with shallow soil,
- 91M0 Pannonian-Balkanic turkey oak—sessile oak forests on southern slopes,

- 91G0 Pannonic woods with *Quercus petraea* and *Carpinus betulus* on hilltops and fresh sites,
- 9130 *Asperulo-Fagetum* beech forests—at higher elevations with north-eastern exposition.

3 Methods of Reconstructing Forest History, Forest Inventory and Assessing Tree Age Classes

We collected forest use and forest management history documents of Vár-hegy area from the Heves county Archive and from the management plan archive of Heves county Forestry Board. Management plans and land-register maps dating from 1887, 1896, 1907, 1953, 1963, 1976, 1986, 1995 and 2006 were collected and other descriptions from the late nineteenth century (Gesztes 1887; Borovszky 1896–1914; Anonymus 1953–2005). Data from historical forest management plans and maps referring to the upper third of the Vár-hegy hill—the present area of the forest reserve—were considered in this study. Based on the documents we separated four historical period of regeneration of existing trees and regarded it as assessed four different age classes.

In the strict forest reserve area a fine scale stand survey was performed. The inventory concept was based on the COST E4 guidelines (Hochbichler et al. 2000) and earlier Hungarian experiences (Czájlik 2002). The methodology, infrastructure and service together are called "FOREST+n+e+t"—*monitoring network of forest stand dynamics and ecology* (Horváth et al. 2009). FOREST+n+e+t is a 50×50 m grid system of permanently marked field sampling points, where we perform forest stand structure, vegetation and soil inventory modules. We surveyed Vár-hegy Forest Reserve between 2005 and 2008 in cca. 400 sampling points. The stand survey methodology in the grid points combines fixed-area circular plot ($R = 8.92$ m), and horizontal point sampling according to (Bitterlich 1952) for standing trees, and line transect sub sampling methods for lying dead wood. For the surveyed trees (DBH > 5 cm) the following information was provided: position, species, DBH, height of selected trees, social status and health categories, decomposition stadium. This method samples the forest stand by a fine scale (50×50 m), while the traditional forestry inventory provides average data for management units which area size varies from 2 to 15 ha.

In this grid system about 8400 individual trees were sampled. Each single tree was ranked into cohorts according to their diameters at breast height (DBH) and social status categories. We regarded the different cohorts by size and history as assessed different age classes. We calculated the number of trees per hectare according to both tree species and assessed age groups by averaging the sampling points for the patch area.

4 Modeling of CO_2 Sequestration and Estimation of Carbon Stored in Different Pools

Based on the results of detailed site and stand survey, and historical forest management maps, we classified and subdivided the core area of the reserve into 28 different patches. Each of them could be considered homogenous according to the recent stand structure and forest use history. We have adapted a cohort-based carbon sequestration model, called CO2FIX 3.1 (Schelhaas et al. 2004) to reconstruct the development of each stands. We have developed, compared and evaluated several scenarios to simulate the carbon sequestration of the patches in the past 125 years, from 1880 (clearcutting and natural regeneration) to 2005 (stand survey) (Balázs et al. 2008). In this paper we calculated and used tree proportion of stands by assessed age categories (same as above) to evaluate forest changes, stand volume and tree species proportion by living and dead wood for the CO_2 sequestration model.

5 Selected Results

5.1 *Stand Structure Changes in the Past 130 Years from the Beginning of the Planned Management of the Forest Reserve*

5.1.1 Forest History

According to a title deed dating from 1261 the area had been an estate of the Episcopate in Eger from the thirteenth century until World War II. Later it was nationalized, and gained increasing nature conservation importance. The first management plan of the forest dates from 1887. Before that time, irregular cuttings were made to fulfill the needs of the manorial estate in terms of charcoal burning, lime burning, fire wood, and timber for the extensive agricultural estates. Most of the present reserve area was cut around 1880 (Gesztes 1887), however, some old seed dispersing trees have remained from the ancient forest stand. Although some of these old specimens were cut around the 1920s, those that were hard to approach survived. These old sessile and Turkey oak trees are mostly hollow, broken in the trunk or lying deadwood. The dominant 120–130 years old stand of the present core area of the forest reserve was regenerated around 1875–1885, comprising coppice oak to a greater and beech to a lesser extent (Borovszky 1896; Gesztes 1887). The aim of forest management according to the management plan was to ensure the dominance of sessile oak and to supplant Turkey oak and beech through tending cutting. Today, sessile oak trees, especially those of coppice origin, are declining forming the majority of deadwood in the reserve. The first management plan of the nationalized forest concerns the period between 1953 and 1963. During World War II uncontrolled wood trade and illegal cuttings must have prevailed, as urgent replacement plantings were ordered in the management plan. The present middle-aged, 55–70 year-old cohort of trees is evidence for the natural regenerative capacity of the late forest, as it originates from the replacement plantings and regeneration plantings of the 1950s and also from natural regeneration following illegal cuttings during the war period. The last management action in the forest reserve ordered by the management plan took place in the late 1960s. Clearcuttings were postponed according to the management plan in the 1960–1970s, being afraid of browsing damage of the over-populated game. As a result, the coppice oak stand of the reserve survived and matured. Later the most significant changes were the formation of the Bükk National Park Directorate in the 1980s and the designation of the area as a forest reserve in 1993. Beside the problems caused by the high amount of game, the disease causing oak decline throughout Europe appeared in the Vár-hegy stand as well in the 1970–1980s (Szepesi 1997). Consequently, sanitary cutting was performed in some stands in the 1980s. Since then, the forest has been left for free development.

5.1.2 Assessed Age Classes of Canopy Forming Trees

On the base of forest history and stand survey four regeneration period and assessed age classes were recognised for the main tree species—*Quercus petraea, Q. cerris, Q. pubescens, Carpinus betulus, Fagus sylvatica, Fraxinus excelsior* and *Acer campestre*. These are: more than 160 years old cohorts (AGE > 160): the existing remnants of old seed dispersing trees prior to the management system, AGE-130: the dominant 130 years old cohorts—mainly sessile oaks most of them with coppice originated around the first management plan (1880), AGE-60: the middle-aged cohorts originated from natural succession and replacement planting after the irregular cuttings around the World War II., and AGE-30: the young cohorts filling the gaps and openings after the oak decline of the 1970–1980s.

We found that the dominant canopy forming cohorts (AGE-130) and the age group of old seed dispersing trees (AGE > 160) are composed mainly of sessile oak, downy oak and Turkey oak. Old oak stands are apparently opening up and producing gaps (Mázsa et al. 2009). The ratio of oak species is less in the cohort originated in the World War II (AGE-60). This age class has a balanced mixture ratio of oak, hornbeam and other mixing species comprising about the same proportion, coupled by the appearance of field maple. The species composition shifting dominance towards hornbeam, common ash and field maple, especially in the young age group (AGE-30). The tree species composition of the four age classes shows a remarkable decrease in all oak species, but first of all in *Quercus petraea* in the middle-aged and young age classes, see Fig. 4. Parallel with the increasing role of natural succession (around World War II, and after the oak decline from the 1980s), the proportion of *Fraxinus excelsior*, *Acer campestre*, *Carpinus betulus* and other associate species has increased.

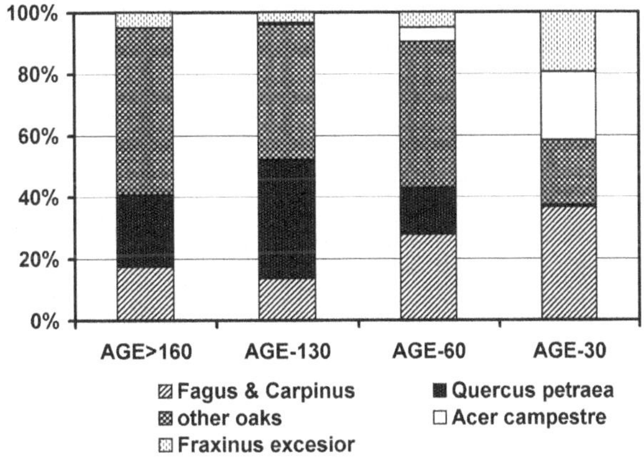

Fig. 4 Proportion of main tree species by age classes based on the average density for *Fagus sylvatica*, *Carpinus betulus*, *Quercus petraea*, other oaks as *Q. pubescens*, *Q. cerris*, *Acer campestre* and *Fraxinus excelsior* (Mazsa et al. 2009)

This change is conspicuous for *Quercus petraea*. Although the stand has been left for free development for about 30–50 years, the ratio of sessile oak is small in the regeneration layer and even smaller in the shrub layer. Gaps opening after the dieback of oak trees are colonized spontaneously mostly by common ash, field maple and hornbeam saplings. Similar shift in tree composition was found in the long-term study of Turkey oak—sessile oak woodlands in the frameworks of the closely located Síkfőkút Project (Kotroczó et al. 2007).

6 Carbon Sequestration of Vár-hegy Forest Reserve

According to our simulations, the overall amount of carbon accumulated in all living biomass (including living roots), total dead biomass (lying and standing dead trees, fine woody debris, litter and dead roots), carbon increments of soil (the starting carbon pool of soil was not considered), thinned wood products of the core area of Vár-hegy Forest Reserve is 18600 ± 720 tons per 94 ha at the end of the model simulation. The mean value is 200 ± 10 tons per hectare which is comparable to a countrywide average of managed forest in Hungary. Figure 5 shows the calculated carbon sequestration during the simulated 125 years (from 1880 to 2005). After clear cutting, the seed dispersing cohort provide a low initial stock, then carbon accumulation is increasing continuously. From the 1970s, a drop can be observed which turns again into an intensive increasing period from the 1990s. This is caused by the oak decline and after that the increment of the younger generations (AGE-60 and AGE-30). The enclosed pie chart presents the proportional distribution of living (L), dead (D) biomass, soil carbon increment (S) and carbon content of removed wood product (W) compartments of the system in 2005.

In case of partitioning these results by age groups of living biomass, we can find that the remnant trees (AGE > 160) store 6 %, dominant old trees (AGE-130)

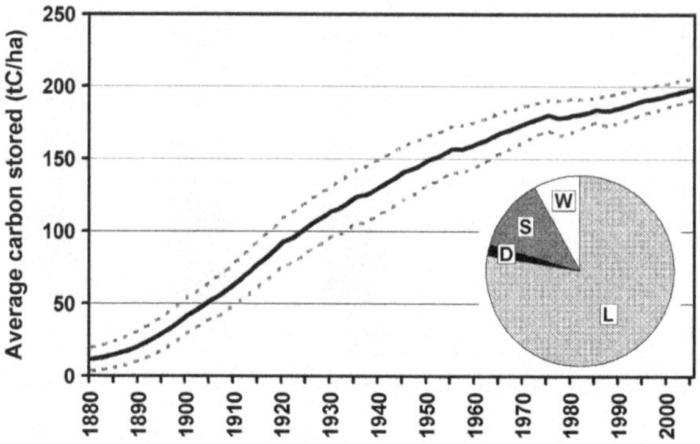

Fig. 5 Overall carbon sequestration of the Vár-hegy core area in the simulated period (1880–2005). *Thick black line* is the average of the results of different scenarios; *broken line* is the standard deviation. The pie chart inserted shows the proportion of carbon accumulated in different pools: living biomass (L), dead biomass (D), increments in the soil (S) and harvested wood products (W) summarized at the end of the simulated period

store 76 %, middle-aged trees (AGE-60) store 13 % and young trees (AGE-30) store 4 % of the total sequestered carbon. The amount of carbon stored of not-dominant cohorts is remarkable (23 % of total), which imply also to transition stage of the forest to a near-natural one.

7 Conclusion

The importance of forest reserve research, as was highlighted Europe-wide by the COST Action E4, lies in the hypothesis that by understanding the processes that take place in natural forests we get closer to the establishment and later switching to nature-oriented silviculture (Parviainen et al. 2000). This idea was later questioned by Brang (2005). He stated that research on virgin forests has had important implications on managed forests only in topics which objects could be studied only in virgin forests, such as nurse logs as a seedbed for tree seedlings or the maximum age and size of the trees. In several other cases like competitive interactions between tree species or forest dynamics after abandonment, managed forest, or a combination of managed and virgin forest should be studied (Brang 2005).

Our researches point out that study of recently established forest reserves which were designated only 20 years ago may answer questions beyond current forest management issues. Our results show that forest reserve can serve as reference area for the natural succession of abandoned woodlands under changing climate and forest management or reference area of carbon cycle in semi-natural forests. We believe that the natural forest remnants and forest reserves of various types and history form an outstanding network in the Carpathian region, the study of which could contribute to answering questions in the field of nature conservation, ecosystem functioning, climate change and restoration ecology.

Acknowledgments The Hungarian Forest Reserve Program is supported by the State Secretary for Nature and Environment Protection (Ministry of Rural Development).

References

Anonymus Felsőtárkány állami erdő üzemterve (1953–1963); (1963–1973); (1976–1985); (1986–2000); (1996-2005)—MGSZH Heves Megyei Erdészeti Igazgatósága, Eger, Vác

Balázs B, Horváth F, Mázsa K, Bölöni J (2008) Forest reserve as a model area for future climate forest restoration—a case study. Extended abstract, 6th European conference on ecological restoration, Ghent, Belgium, pp 1–4, 8–12/09/2008 (CD-ROM)

Barton ZS (2010) A Kárpátok őserdő-maradványai. In: Bartha D, Barton ZS, Oroszi S (eds) Őserdőleírások a Kárpát-medencéből. Erdtört Közl 81:223–267

Bidló A, Gucsik A, Heil B, Illés G, Juhász P, Kovács G, Varga ZS (2004) Termőhelyfeltárás a Vár-hegy erdőrezervátum területén. Research report, manuscript. Forest Reserve Archive 2004/D-006/1-2, KvVM—MTA ÖBKI, Vácrátót

Bitterlich W (1952) Die Winkelzählprobe. Ein optisches Meßverfahren zur raschen Aufnahme besonders gearteter Probeflächen für die Bestimmung der Kreisflächen pro Hektar an stehenden Waldbeständen. Forstwiss Cent 71(7–8):215–225

Borhidi A (1961) Klimadiagramme und Klimazonale Karte Ungarns. Ann Univ Sci Bp Sect Biol 4:21–50

Borlea GF (1999) Forest reserves and their research in Romania In: Diaci J (ed) COST action E4: forest reserves research network, Proceedings of the invited lecturers' reports 67–86

Borovszky S (ed) (1896–1914) Magyarország vármegyéi és városai. CD-ROM (2004) Arcanum; ISBN 963 9374 91 1

Brang P (2005) Virgin forests as knowledge source for Central European silviculture: reality or myth? For Snow Landsc Res 79(1/2):19–32

Czájlik P (2002) A módszerek részletezése vizsgálati szintenként. In: Horváth F, Borhidi A (eds) Aim, strategy and methods of the National Forest Reserve Research. Természetbúvár, Budapest

Czájlik P, Gálhidy L, Kenderes K, Mihók B, Ódor P, Standovár T, Tímár G, Kelemen K (2003a) Report on site-based permanent plot, second-phase and new mapping studies: Alsóhegy forest reserve. Nat-Man Proj Work Rep 53:17

Czájlik P, Kenderes K, Standovár T (2003b) Report on site-based permanent plot, second-phase and new mapping studies: Őserdö Forest Reserve. Nat-Man Proj Work Rep 52:11

Czájlik P, Kenderes K, Standovár T, Tímár G (2003c) Report on site-based permanent plot, second-phase and new mapping studies: Kékes Forest Reserve. Nat-Man Proj Work Rep 51:28

Czájlik P, Pászty G (2009) The Kékes Észak Forest Reserve and the surrounding region's history: survival of a virgin forest fragment. Results of the forest reserve research in Hungary. 3:5–114 http://www.botanika.hu/ER

Czúcz B, Gálhidy L, Mátyás CS (2011) Present and forecasted xeric climatic limits of beech and sessile oak distribution at low altitudes in Central Europe. Ann For Sci 68(1):99–108

Fuchs F (1861) Ungarns Urwälder und das Wesentlichste zur Kenntniss, Beurtheilung und Einführung eines rationellen Forstbetriebs in ungarischen Privatwaldungen für solche, die ohne Forstmänner von Fach zu sein, als Waldbesitzer oder deren Bevollmächtigte,

Rechtsfreunde. Oekonomiebeamte oder sonst wie in die Leitung der Forstverwaltung einen wesentlichen Einfluss nehmen. Georg Kilian's Universität-Buchhandlung, Pest

Gesztes K (1887) Az egri érseki uradalom felsőtárkányi erdőbirtokának (A és B üzemosztály) általános leírása (1887/88). Heves County Archive IV-415/34, Eger

Hochbichler E, O'Sullivan A, van Hees A, Vandekerkhove K (2000) Recommendations for data collection in forest reserves, with an emphasis on regeneration and stand structure. European community, directorate general for research 2000: Forest reserves research network COST action E4. EUR 19550

Hort L, Tesař V, Vrška V (1999) The Czech Republic Country Report In: Diaci J (ed) COST action E4: forest reserves research network, Proceedings of the invited lecturers' reports, 25–44

Horváth F, Borhidi A (eds) (2002) A hazai erdőrezervátum-kutatás célja, stratégiája és módszerei. A KvVM Természetvédelmi Hivatalának tanulmánykötetei 8. Természet-Búvár, Budapest

Horváth F, Balázs B, Bölöni J, Mázsa K (2009) "Forest-net"—monitoring strict forest reserves left for free development in Hungary. In: Kaennel Dobbertin M (ed) Long-term ecosystem research: understanding the present to shape the future. International conference Zurich, Switzerland, 7–10 Sept 2009. Abstracts. Birmensdorf. Swiss Federal Research Institute WSL

Kenderes K, Mihók B, Standovár T (2008) Thirty years of gap dynamics in a central European beech forest reserve. Forestry 81(1):111–123

Korpel Š (1995) Die Urwälder der Westkarpaten. Gustav Fischer Verlag, Stuttgart-Jena-New York

Kotroczó ZS, Krakomperger ZS, Koncz G, Papp M, Bowden RD, Tóth JA (2007) A síkfőkúti cseres-tölgyes fafaj-összetételének és struktúrájának hosszú távú változása. Termvéd Közl 13:93–100

Krauze K (1898) A luczfenyő tenyésztésének kérdése ősluczfenyvesekben szerzett tapasztalatok alapján. Erd Lapok 37:1085–1108

Mátyás CS (2010) Forecasts needed for retreating forests. Nature 464:1271

Mázsa K, Balázs B, Horváth F, Bölöni J, Aszalós R (2008) Transition of a managed forest towards a natural one-forest history and stand survey study in an oak forest reserve. 6th European conference on ecological restoration, Ghent, Belgium, pp 1–4, 8–12/09/2008 (CD-ROM)

Mázsa K, Horváth F, Balázs B, Bölöni J, Aszalós R (2009): A felsőtárkányi Vár-hegy erdőrezervátum faállományának korosztály viszonyai erdőtörténeti összefüggésben. Termvéd Közl 15:347–357 http://www.mbtktv.mtesz.hu/teljesszoveg.html

Muzsnay G (1899) A romániai őserdőkről. Erd Lapok 38:127–138

Ódor P, Heilmann-Clausen J, Christensen M, Aude E, van Dort KW, Piltaver A, Siller I, Veerkamp MT, Walleyn R, Standovár T, van Hees AFM, Kosec J, Matočec N, Kraigher H, Grebenc T (2006) Diversity of dead wood inhabiting fungi and bryophytes in semi-natural beech forests in Europe. Biol Conserv 131(1):58–71

Oszlanyi J, Grodzinska K, Badea O, Shaparyk Y (2004) Nature conservation in Central and Eastern Europe with a special emphasis on the Carpathian Mountains. Environ Pollut 130:127–134

Parviainen J (2005) Virgin and natural forests in the temperate zone of Europe. For Snow Landsc Res 79(1/2):9–18

Parviainen J, Bücking W, Vandekerkhove W, Schuck A, Päivinen R (2000) Strict forest reserves in Europe: efforts to enhance biodiversity and research on forests left for free development in Europe (EU-COST-Action E4). Forestry 73(2):107–118. doi:10.1093/forestry/73.2.107

Pruša E (1985) Die böhmischen und mährischen Urwälder—ihre Struktur und Ökologie. Verlag der Tschechoslowakischen Akademie der Wisenschaften, Praga

Pulvermacher G (1877) A karánsebesi m. k. erdőhivatal, illetőleg a föloszlatott román-bánáti határőrezred erdeinek leírása. Erd Lapok 16:649–651, 695–697

Ruffini F, Streifeneder T, Eiselt B (2006) Implementing an international mountain convention—An approach for the delimitation of the Carpathian convention area. European Academy, Bolzano/Bozen, p 119

Schelhaas MJ, van Esch PW, Groen TA, de Jong BHJ, Kanninen M, Liski J, Masera O, Mohren GMJ, Nabuurs GJ, Palosuo T, Pedroni L, Vallejo A, Vilén T (2004) CO2FIX V 3.1—A modelling framework for quantifying carbon sequestration in forest ecosystems, Wageningen, Alterra, Alterra-rapport 1068

Standovár T, Ódor P, Aszalós R, Gálhidy L (2006) Sensitivity of ground layer vegetation diversity descriptors in indicating forest naturalness. Community Ecol 7:199–209

Szepesi A (1997) Forest health status in Hungary. Environ Pollut 98:393–398

Zólyomi B (1967) Rekonstruált növénytakaró 1:1 500 000. In: Radó S (ed) Magyarország nemzeti atlasza. Kartográfiai Vállalat, Budapest 21–31

Regionalization of Forests in Slovakia Based on Their Natural Potential

Ladislav Kulla, Michal Bošel'a and Karol Burgan

Abstract The study presents the results of an analysis of the spatial variability of geologic and climatic factors that predetermine the natural potential of forest sites in Slovakia, as well as the relations between these factors and the natural presence and productivity of tree species. Several national classification systems that combine the existing and newly designed regional and site units were examined and compared in terms of explained variability of both natural presence and productivity of the main tree species in Slovakia: oak, beech, and spruce. Forest management data describing the presence and the productivity of these tree species were extracted from the sample file of the representative forest stands and utilized for the analysis. Maximization of explained variability and minimization of the number of units were predefined as desirable capabilities of a suitable classification system. The results show that including the regional level is a beneficial contribution to the classification system. Generally, combinations of site and regional units explain approximately 50–70 % of the total variability of the tree species' natural presence, but only about 15–25 % of the total variability of tree species production. A site classification system composed of a maximum of 8 units on the regional level and 80 units on the site level appears to be sufficient for strategic forest management in natural conditions in Slovakia.

L. Kulla (✉) · K. Burgan
National Forest Centre–Forest Research Institute Zvolen, T. G. Masaryka 22, SK-960 92 Zvolen, Slovakia
e-mail: kulla@nlcsk.org

M. Bošel'a
Faculty of Forestry and Wood Sciences, Czech University of Life Sciences in Prague, Kamýcká 129, Praha 6-Suchdol 165 21, Czech Republic

J. Kozak et al. (eds.), *The Carpathians: Integrating Nature and Society Towards Sustainability*, Environmental Science and Engineering,
DOI: 10.1007/978-3-642-12725-0_20, © Springer-Verlag Berlin Heidelberg 2013

1 Introduction

Currently, several schemes of forest type classification exist. In most European countries, the schemes have been developed for practical use in forestry and/or for scientific purposes (forestry, vegetation mapping, landscape analysis, etc.). Each system has been designed to serve a certain purpose and/or is based upon the tradition of a certain branch of science (Larsson 2001).

Generally, two principal approaches have been used for the classification of forest sites: direct and indirect. Direct classification is based on evaluation of climate, topography, and soil characteristics, where as indirect determination is based on typical composition and/or productivity of vegetation.

The ecological site classification (ESC), widely used especially in Scandinavia and North America (Kuusipalo 1985; Cleland et al. 1993), could serve as an example of direct classification. ESC classifies sites in terms of climatic factors and site quality, and determines the suitability of alternative tree species on the basis of the interconnection of key site factors and the ecological demands of tree species. Hills (1960) proposed physiographic site types, taking the terrain relief and parent rock primarily into account. He divided landscapes into several regions with the aim to achieve higher homogeneity. These more homogenous areas were further divided into smaller districts.

Most of the indirect classification systems in Europe are based on the Braun-Blanquet system (Braun-Blanquet 1964), which emphasizes floristic similarities of ground vegetation in order to aggregate hierarchical units. For the area of the Western Carpathians, an original concept of site classification was designed by Zlatník (1959, 1976). On the basis of his extensive research on primeval Carpathian forests, he suggested basic groups of forest site types distinguished along the gradients of altitude, soil quality, and soil water supply, which are indicated by ground vegetation. The ecological profiles of ground vegetation communities are assessed in compliance with principles formulated by Ellenberg (1974), and each unit is characterized by the typical main tree species composition. Zlatnik's system is reconstructive and is the basis for the geobotanic map of potential vegetation in Slovakia (Michalko 1986), as well as for further development of reconstructive forest typology in Slovakia and the Czech Republic.

Ecosystems can be considered within two fundamental dimensions: classification and regionalization. Classification refers to the process of ordering, synthesizing, or arranging of objects into groups or sets with regard to similarities or relationships between the variables or characters being classified. Regionalization refers to a separate mapping procedure needed for spatial delimitation of landscape segments that have a degree of internal homogeneity as well as features that contrast with those of adjacent areas (Bailey et al. 1978).

For example, Kilian et al. (1994) proposed a classification scheme for Austria consisting of nine main growth regions further divided into 22 ecoregions and seven vegetation zones. For Germany, Gauer and Aldinger (2005) proposed 82 growth regions divided into 610 climatic and geologically homogeneous growth districts.

In Slovakia, a detailed two-level system of site classification currently exists. The system is generally used for strategic forest management planning for proposed tree species composition and management alternatives. The country of Slovakia has been divided into 46 forest districts, which have been further divided into subdistricts and parts (Vladovič et al. 1994), resulting in a total of 81 regional units within the first regional classification level. The system of forest site types according to Hančinský (1972) represents the second site classification level. A forest site type is defined as a group of forest biocoenoses, original or changed, and their developmental stages, such as geobiocoenoses that are evolutionarily identical. For forest management purposes, management groups of forest site types (MGFT) were created by aggregating forest types with regard to altitudinal vegetation zones and similarity of soil properties. Currently, 187 MGFTs represent the second level of site classification for forest management purposes. The number of regional and site units is large, and hence too many possible combinations can be generated. Currently, the discussion on the adequacy of such a detailed system has become urgent.

The aims of this study were to examine the relations between site units and their natural potential as expressed by tree species natural presence and productivity; to document the advantages of two-level site classification comprising regionalization in comparison with a simple, one-level classification; and to test the opportunities for the simplification of a detailed site classification system in Slovakia without the loss of site-related information on natural tree species presence and productivity rates.

2 Materials

Forest management data about the presence and productivity of the main tree species in forest stands (compartments), taken from the central forestry information system covering the whole territory of Slovakia, were used for the analysis.

To test the natural presence of tree species, only forest stands with the highest level of conservation that belong to forest nature reserves, with forest stands older than 100 years, an area of over 1.0 ha, and stand density of more than 50 % of full crown cover, were selected for the sample file I. Given these criteria, more than 11,000 stands were chosen for analyses, representing all of the natural reserves in Slovakia.

To quantify tree species productivity, only forest stands with even-aged forests (coppice forests were also excluded) from 50 to 90 years were included in the sample file II, because of errors in determination of site index in young stands, as well as in stands affected by the final cutting. The next criteria considered were minimum stand area of above 1.0 ha and the proportion of tree species of at least 30 %. With respect to these criteria, the number of forest stands that were investigated for productivity was more than 50,000, representing managed commercial forests over the whole country of Slovakia.

In both cases, a minimum of 80 % site homogeneity was required to be included in the analysis data set: one site unit had to cover 80 % or more of the stand area.

A geological map of Slovakia produced by the Slovak Academy of the Environment (Biely et al. 1996) and raster climatic data from the period 1961–1990 with resolution of 90 m, which were adapted from the work of Fabrika (2008), were used for spatial identification of basic ecological factors.

Digital forest-site-type maps (created according to Hančinský 1972) and forest district maps of Slovakia (according to Vladovič et al. 1994) were utilized for the simulation of alternative classification designs by aggregating basic units.

3 Methods

In addition to one official forest-management classification system currently utilized, Forest districts (Vladovič et al. 1994), and one alternative system, Aggregated forest districts (Rizman and Feiková 2008), two newly designed regional classifications are proposed on the basis of this study (Table 1).

Geological-climatic regions have been suggested on the basis of the synthesis of geological factors, climatic factors, and some known anomalies in tree species spreading in the central high mountains of Slovakia (characterized by the lack of beech). As a first step, the forest districts were classified into four categories according to prevailing geological substrate:

1. crystalline rocks with Mesozoic overlaying,
2. flysch rocks,
3. volcanic rocks,
4. quaternary sediments.

Table 1 Basic characteristics of the examined regional units and their relation to the main selected factors and/or indicators

Code	Name of units	Number	Origin	Geology	Geomorphology	Regional climate	Tree species range
1	Forest districts	81	Vladovič et al. 1994	(+)[b]	+[a]	–[c]	+
2	Aggregated forest districts	16	Rizman and Feiková 2008	+	+	–	+
3	Geological-climatic regions	8	This study	+	–	+	+
4	Climatic regions	5	This study	–	–	+	+

[a] Factor principally taken into account; [b] factor partially taken into account; [c] disregarded factor

In the next step, mean values of air temperature, yearly precipitation, and water–climate balance were computed using raster data for the forest districts. In the third step, spatial differences in tree species productivity were analyzed. An index of relative productivity of oak, beech, and spruce was calculated for each forest district:

$$IRP_{ik} = \frac{\sum_{j=1}^{m} \left(\frac{SI_{ijk}}{SI_{ijs}} \cdot n_{ijk} \right)}{\sum_{j=1}^{m} n_{ijk}} \quad (1)$$

where IRP_{ik} is an index of relative production of tree species i in the regional unit k, SI_{ijk} is a mean site index of tree species i on a site unit j in the regional unit k, SI_{ijs} is a mean site index of tree species i on a site unit j in the country of Slovakia s, and N_{ijk} is a number of stands (cases) with presence of tree species i on a site unit j in the regional unit k.

A site index model developed by Halaj and Petráš (1998) was used to calculate the site index of tree species. In the model, the Korf growth function (Korf 1939) was used. To explain high growth of tree species with respect to different site conditions (site index), parameters of the Korf growth function were adjusted by site index. Finally, the site index is the mean height at the standard age (100 years). The final height-growth model has the following form:

$$h = B \exp\left[\frac{(a_3 + a_4)B \cdot 100^{a_1 + \frac{a_2}{B}}}{B(1 - a_1) - a_2} \left(t^{1-a_1-\frac{a_2}{B}} - 100^{1-a_1-\frac{a_2}{B}}\right)\right] \quad (2)$$

where h is a mean height of tree species in a forest stand, B is a site index, $a1$–$a4$ are regression parameters and t is an age of tree species in a forest stand.

Finally, the geological–climatic regions were designed by aggregation of forest districts by the synthesis of all considered attributes, using the visual assessment of similarities and spatial gradients of their mean values (Fig. 1).

Climatic regions have been suggested on the basis of the main climatic line postulated by Zlatník (1959) that divides the territory of Slovakia into two parts: the northern part influenced by a wet and cold Baltic oceanic climate with expansive spreading of spruce, and the southern part influenced by a warm and more continental Pannonian climate with expansive spreading of beech. The anomalies of tree species spreading in the central high mountains of Slovakia (discussed above) were also considered (Fig. 2).

The definitions and the principles of mapping of forest site types (FST) as well as the creation and the current role of MGFT in forest management have already been described above. Management groups (MG) represent hierarchically higher site units than MGFT and serve some forest management purposes, for example forest categorization.

A management group of sites (MGS) represents a newly designed site unit hierarchically between MGFT and MG, which partially differs from them due to

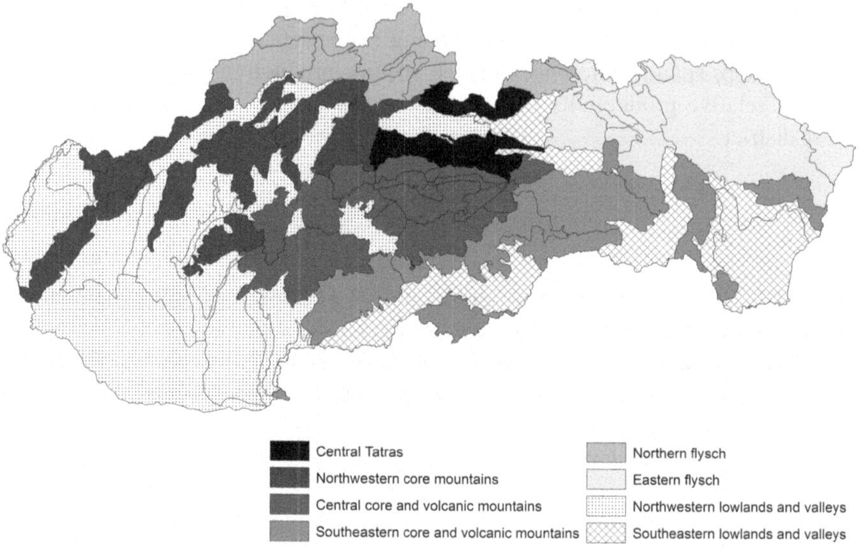

Fig. 1 Newly designed geological-climatic regions of Slovakia with marked borders of current forest districts

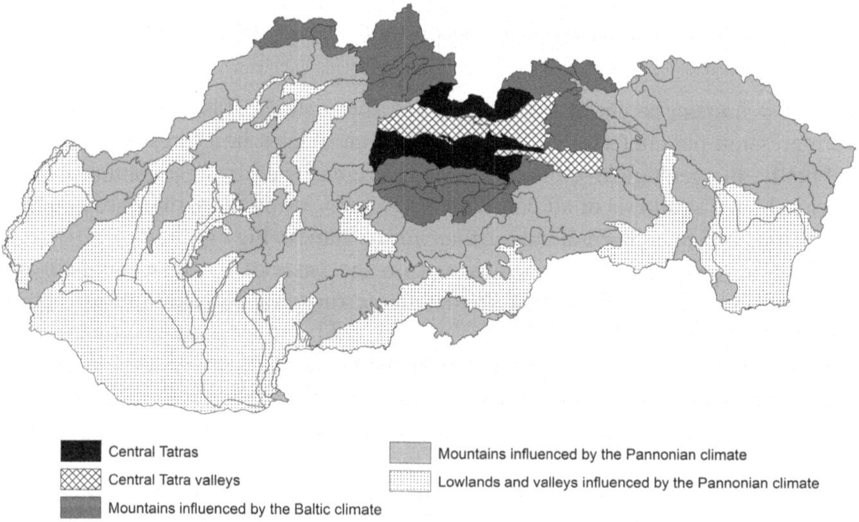

Fig. 2 Newly designed climatic regions of Slovakia with marked borders of current forest districts

the procedure of its creation. FSTs have been aggregated strictly according to the estimated gradients of the main ecological factors: altitudinal vegetation zone, trophic order, and hydrologic order, in the sense of Buček and Lacina (2000). Table 2 contains the list of characteristics of all of the examined site units.

Table 2 Basic characteristics of the examined site units and their relation to the main ecological factors and/or indicators

Code	Name of units	Number	Origin	Altitudinal zonality	Soil nutrients	Soil water	Tree species range	Phytoindication
a.	FST	365	Hančinský 1972	+[a]	–[c]	–	+	+
b.	MGFT	187	Aggregated a	+	+	+	+	–
c.	MGS	80	This study	+	+	+	–	–
d.	MG	32	Aggregated b	+	(+)[b]	(+)	–	–

[a] Factor principally taken into account; [b] factor partially taken into account; [c] disregarded factor

Different combinations of existing as well as newly proposed regional and site units were examined and compared in terms of natural presence of oak, beech, fir, and spruce in forest reserves (sample file I), and productivity of oak, beech, and spruce in mature even-aged stands before the start of final cutting (sample file II).

An increasing amount of explained variability of the analyzed attributes and a decreasing number of site units are regarded as favorable. In the case of natural tree species presence, a quantile frequency ranging from 25 to 75 % was used as a measure of variability. For each tree species, explained variability for tree species and variant of classification system EV_{ix} was calculated as follows:

$$EV_{ix} = \frac{Q_{is} - Q_{ix}}{Q_{is}} 100 = \frac{Q_{is} - \frac{\sum_{j=1}^{m}\sum_{k=1}^{o}(Q_{ijk}n_{ijk})}{\sum_{j=1}^{m}\sum_{k=1}^{o}n_{ijk}}}{Q_{is}} 100 \qquad (3)$$

where Q_{is} is a total quantile range of tree species i in Slovakia (%), Q_{ix} is a mean quantile range of tree species i achieved by classification system x (%), Q_{ijk} is a quantile range of tree species i inside a site unit j within a regional unit k (%) and N_{ijk} is a number of stands (cases) with presence of tree species i on a site unit j in regional unit k.

Standard deviation of the site index (mean height of the tree species population in the stands at the age of 100 years) was used for the analysis of potential productivity. Explained variability was calculated similarly according to Eq. (3), where the quantile (Q) was replaced by the standard deviation.

In both cases, the difference between the weighted mean of the attribute for a classification variant and the overall mean of the attribute for Slovakia expresses the portion of explained variability by the examined classification design. A higher proportion of explained variability along with a lower number of site units is desired and expected for accurate site classification and practical forest management. The efficiency of each variant was finally assessed by a ratio of explained variability to the square root of the number of analyzed site units, expressed relatively to the current classification system which served as a reference base:

$$EF_{ix} = \frac{q_{ix}}{q_{ir}} = \frac{\frac{EV_{ix}}{\sqrt{n_{ix}}}}{\frac{EV_{ir}}{\sqrt{n_{ir}}}} \tag{4}$$

where EF_{ix} is a relative efficiency of classification for individual tree species i and variant x, q_{ix} and q_{ir} are efficiency quotients for tree species and examined classification variants, EV_{ix} is an explained variability for tree species i and variant of classification variant x, EV_{ir} is an explained variability for tree species i and reference variant r (1b), and n_{ix}, n_{ir} are numbers of site units of classification variants.

4 Results

We found that regionalization significantly contributed to the amount of explained variability (Fig. 3) as an example of the current classification system (variant 1b). Hence, the results presented hereafter refer only to combinations of site and regional units. The explained variability of the analyzed attributes for natural tree species occurrence is much higher than for productivity specified by the site index, and the results show that this is a generally valid fact (Tables 3, 4). The results also

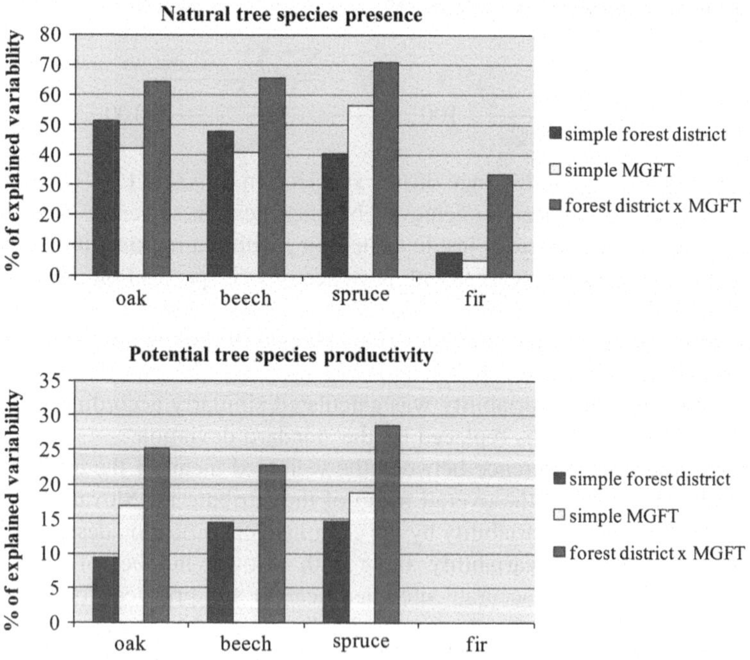

Fig. 3 Explained variability by simple and two-level alternatives of the current site classification system (variant 1b)

Table 3 Relative efficiency of different site classification designs in terms of potential tree species presence

	Sessile oak			European beech			Norway spruce			Silver fir			EF total
Variant	EV (%)[a]	N[b]	EF[c]	EV (%)	N	EF	EV (%)	N	EF	EV (%)	N	EF	
1b[d]	64.0	32	1.0	65.5	106	1.0	70.7	111	1.0	33.6	73	1.0	1.0
2a	52.3	25	0.9	66.4	88	1.1	65.2	90	1.0	36.6	68	1.1	1.1
2b	54.2	21	1.0	62.1	70	1.2	64.5	67	1.2	28.0	51	1.0	1.1
2c	59.1	18	1.2	60.1	59	1.2	62.2	54	1.3	18.1	41	0.7	1.2
2d	50.9	13	1.2	40.0	42	1.0	49.6	35	1.2	14.5	29	0.7	1.1
3a	51.0	23	0.9	67.5	87	1.1	65.9	96	1.0	36.6	69	1.1	1.1
3b	49.7	19	1.0	61.3	66	1.2	65.9	72	1.2	29.0	50	1.0	1.1
3c	51.1	15	1.2	58.7	55	1.2	63.6	60	1.2	18.5	40	0.7	1.2
3d	45.6	13	1.1	37.8	39	1.0	51.5	39	1.2	13.5	30	0.6	1.0
4a	41.7	17	0.9	55.3	63	1.1	61.4	85	1.0	35.9	62	1.2	1.0
4b	39.0	13	1.0	55.1	45	1.3	59.6	63	1.1	32.3	45	1.2	1.2
4c	35.7	10	1.0	54.0	36	1.4	58.1	51	1.2	21.3	33	0.9	1.2
4d	45.1	7	1.5	28.8	23	0.9	47.3	30	1.3	19.3	23	1.0	1.2

[a] Explained variability; [b] number of units; [c] relative efficiency; [d] current design used as a reference base

Table 4 Relative efficiency of different site classification designs in terms of potential tree species productivity

	Sessile oak			European beech			Norway spruce			EF total
Variant	EV (%)[a]	N[b]	EF[c]	EV (%)	N	EF	EV (%)	N	EF	
1b[d]	25.2	530	1.0	22.8	956	1.0	28.6	1,167	1.0	1.0
2a	21.5	452	0.9	19.0	843	0.9	26.7	950	1.0	0.9
2b	20.6	263	1.2	18.1	410	1.2	25.5	497	1.4	1.3
2c	19.4	199	1.3	17.6	314	1.4	25.1	350	1.6	1.4
2d	18.6	144	1.4	16.8	195	1.6	24.1	212	2.0	1.7
3a	20.2	373	1.0	18.9	653	1.0	25.4	831	1.1	1.0
3b	19.4	205	1.2	18.3	308	1.4	24.4	451	1.4	1.4
3c	18.5	148	1.4	18.0	226	1.6	24.1	297	1.7	1.6
3d	17.5	101	1.6	17.0	126	2.1	23.0	163	2.2	2.0
4a	17.8	181	1.2	14.2	339	1.0	21.3	563	1.1	1.1
4b	17.9	98	1.6	14.0	155	1.5	19.5	312	1.3	1.5
4c	16.4	69	1.8	13.8	109	1.8	19.2	187	1.7	1.8
4d	16.2	42	2.3	11.8	60	2.1	17.9	92	2.2	2.2

[a] Explained variability; [b] number of units; [c] relative efficiency; [d] current design used as a reference base

show that the efficiency of more aggregated classification variants, quantified by the proposed efficiency quotient *q*, is generally higher, particularly for the productivity of tree species. This means that with an increase of the units' number in the classification system, its real contribution to the quality of the obtained site-related information generally increases less than the square number of site units.

5 Discussion

In Slovakia, objective analyses related to FSTs and their actual differences in terms of tree species presence and productivity have been needed. Only some theoretical knowledge fundamentals about natural tree species composition in FSTs (Voločšuk 2001; Rizman 2007) and in the groups of forests site types (GFT) (Vladovič 2003) have been published. The results of our study from the forest reserves of Slovakia showed that along the ecological gradients, aggregated regional and site units (to aggregation level 3c) could describe natural variability of tree species presence comparable to the FSTs.

In the case of productivity, our results revealed huge variability of the site index of the analyzed tree species within the examined site units, resulting in a very low portion of its explained variability by site units. The height of forest stands and the site index derived from the height have been widely used for indirect site-quality quantification all over the world (Oliver and Larson 1996; Barnes et al. 1998; Socha 2008). Halaj et al. (1990) revealed high variability of production within the MGFT, when the variability of the site index within units was higher than the variability between them. A weak relationship between site units and tree species productivity was also found by other authors (Franz 1971; Šmelko et al. 1992).

At this time, the final digital versions of both FST and soil type maps of Slovakia are available. The primary focus during mapping was on ground vegetation. However, using ground vegetation as an indicator of site quality has several limitations, since vegetation composition is affected by many factors. In the case of forest ecosystems, these factors are primarily tree layer, especially its structure, and tree species composition (Križová and Ujházy 1998; Ewald 2000; Fischer et al. 2002; Bošel'a et al. 2007). To reduce the subjectivity of site unit classification multifactorial statistical methods can be applied that are able to combine climate, soil, and terrain characteristics (Host et al. 1996; McNab et al. 1999; Abella et al. 2003).

All classifications are artificial (Allen 1987). The inherent continuous nature of environmental gradients and complexity of vegetation responses ensures that obvious breaks along environmental gradients at which boundaries can easily be drawn are rare. Hence, the number of categories and boundaries between them should be as low as possible when there is clear evidence of different conditions that cause differences in vegetation responses (Bryan 2006).

Ecosystem boundaries have often been defined by consensus and thus are subjectively drawn with unclear choice and weighting of input data. On the other hand, regional ecosystem classification is important for understanding the potential distribution of species and the productivity of the landscape, and it provides a tool for interagency strategic planning (Host et al. 1996) as well. Austin and Smith (1989) suggest that it may be more useful to accept that there is no single optimal classification, but many good classifications at a variety of scales. Furthermore, there has also been the issue of climate change during the last decades, resulting in the need to incorporate climate conditions into the classifications, since climate change has a high impact on forest production as well (e.g., Fabrika 2008). Our study, however, is mainly focused on the spatial variability of site conditions with relation to tree species productivity and presence.

In the Carpathian region, national site classification systems differ significantly between the countries. In the Czech Republic, a two-level system is currently applied in forestry that is similar to Slovakia's and based on Zlatník's classification (Viewegh et al. 2003). The Polish system is much simpler: it consists of 21 site units, but only a few of them are relevant to the Carpathians. In Ukraine, the Russian approaches influenced site classification (Viewegh 2003). In Hungary, ten forest regions are distinguished, out of which two are relevant to the Carpathians, and site classification is currently based on the Braun-Blanquet system (Larsson 2001). An original site classification system has been developed in Romania (Biris and Veen 2005; Donita and Biris 2006).

From a holistic approach, the actual challenge for Carpathian research and land use should be to make the site classification systems more compatible, as well as to harmonize these systems with the European Union (EU) classification schemes (European Commission 1992; European Environment Agency 2006, 2008). This type of comprehensive system would be a good resource, given that the Carpathian forests grow on similar sites resulting in similar tree species composition with similar potential productivity. However, to develop a classification system universal for the entire forest conditions in the Carpathians, the large-scale regional differences of present climate change should be carefully considered and implemented. A classification system that considers both site productivity as well as natural tree species occurrence is a good basis for the development of a model that would integrate climate change.

Acknowledgments This work was supported by the Slovak Research and Development Agency under contract No. APVV-0632-07 and No. APVT-27-009304.

References

Abella SR, Shelburne VB, MacDonald NW (2003) Multifactorial classification of forest landscape ecosystems of Jocassee Gorges, southern Appalachian Mountains, South Carolina. Can J For Res 33:1933–1946

Allen BH (1987) Ecological type classification for California: the forest service approach. General Technical Report PSW-98: Pacific southwest forest and range experiment station. USDA Forest Service, Berkeley

Austin MP, Smith TM (1989) A new model for the continuum concept. Vegetatio 83:35–47

Biely A, Bezák V, Elečko M et al (1996) Atlas krajiny Slovenskej republiky. Geologická stavba. http://globus.sazp.sk/atlassr/. Accessed 26 Oct 2010

Bailey RG, Pfister RD, Henderson JA (1978) Nature of land and resource classification—a review. J For 76:650–655

Barnes VB, Zak RD, Denton RS, Spurr HS (1998) Forest ecology, 4th edn. Wiley, New York

Biris I-A, Veen P (eds) (2005) Virgin forests in Romania. ICAS and KNNV, Romania

Bošeľa M, Vladovič J, Merganič J (2007) Príspevok k poznaniu vzťahov medzi štruktúrálnou diverzitou stromovej a bylinnej etáže lesných ekosystémov. In: Križová E, Ujházy K (eds) Dynamika, stabilita a diverzita lesných ekosystémov. TU Zvolen, Zvolen

Braun-Blanquet J (1964) Pflanzensoziologie. Grundzüge der Vegetationskunde, 3rd edn. Springer, Vienna

Bryan BA (2006) Synergistic techniques for better understanding and classifying the environmental structure of landscapes. Environ Manag 37(1):126–140

Buček A, Lacina J (2000) Geobiocenologie II. MZLU Brno, Brno

Cleland DT, Hart JB, Host GE, Pregitzer KS, Ramm CW (1993) Ecological classification and inventory system of the Huron-Manistee National Forests. USDA Forest Service, Cadillac

Doniţă N, Biris I-A (2006) Creation of an ecological basis for sustainable forest management in Romania. An ICAS 46:81–91

European Commission (1992) Council Directive 92/43/EEC of 21 May 1992 on the conservation of natural habitats and of wild fauna and flora. OJ L 206, 22.7.1992

European Environment Agency (2006) European forest types. Categories and types for sustainable forest management, reporting and policy. EEA Technical Report 9/2006

European Environment Agency (2008) Indicative map of biogeographical regions. http://www.eea.europa.eu/data-and-maps/data/biogeographical-regions-europe-2008. Accessed 19 Nov 2012

Ellenberg H (1974) Zeigerwerte der Gefässpfanzen Mitteleuropas. Scr Geobot 9:1–97

Ewald J (2000) The influence of coniferous canopies on understorey vegetation and soil in mountain forests of the northern Calcareous Alps. Appl Veg Sci 3:123–134

Fabrika M (2008) Modelovanie produkcie lesa pod vplyvom klimatických zmien pomocou modelu SIBYLA. In: Lapin M, Nejedlík P (eds) Národný klimatický program Slovenskej republiky. Zväzok 12, MŽ SR, SHMU, Bratislava

Fischer A, Linder A, Abs C, Lash P (2002) Vegetation dynamics in Central European forest ecosystem (near-natural as well as managed) after storm events. Folia Geobot 37:17–32

Franz F (1971) Grundlagen und Verfahren standortbezogener Leistungsschautzung. Forstliche Forschunganstalt, München

Gauer HJ, Aldinger E (2005) Forest ecologically-based nature areas in Germany. Forest growth regions and growth districts with a 1:1 000 000 map. Verein für Forstliche Standortskunde und Forstpflanzenzuchtung e.V., Freiburg

Halaj J, Bortel J, Grék J, Mecko J, Midriak R, Petráš R, Sobocký E, Tutka J, Valtýni J (1990) Rubná zrelosť drevín. Príroda, Bratislava

Halaj J, Petráš R (1998) Rastové tabuľky hlavných drevín. Slovak Academic Press, Bratislava

Hančinský L (1972) Lesné typy Slovenska. Príroda, Bratislava

Hills GA (1960) Regional site research. For Chron 36:401–423

Host GE, Polzer PhL, Mladenoff DJ, White MA, Crow TR (1996) A quantitative approach to developing regional ecosystem classifications. Ecol Appl 6(2):608–618

Kilian W, Muller F, Starlinger F (1994) The forest growth regions of Austria: a land classification based on forest ecology. Forstliche Bundesversuchsanstalt Wien (Federal Forest Research Centre), Vienna

Korf V (1939) Příspěvek k matematické definici vzrůstového zákona hmot lesních porostů. Lesn Pr 18:339–379

Križová E, Ujházy K (1998) Dynamika horských smrečín Poľany (Slovenské stredohorie). Zesz Nauk Akad Rol w Krakowie 332(56):227–241

Kuusipalo J (1985) An ecological study of upland forest site classification in Southern Finland. Acta For Fenn 192:1–78

Larsson T-B (2001) Biodiversity evaluation tools for European forests. Ecological Bulletins 50. Wiley, New York

Michalko J (1986) Geobotanická mapa ČSSR. Veda, Bratislava

McNab WH, Browing SA, Simon SA, Fousts PE (1999) An unconventional approach to ecosystem unit classification in western North Carolina, USA. For Ecol Manag 114:405–420

Oliver CD, Larson BC (1996) Forest stand dynamics. Wiley, New York

Rizman I (2007) Poznatková báza o zastúpení drevín v lesných typoch Slovenska. Lesnícka typológia a zisťovanie stavu lesa vo väzbe na trvalo udržateľné obhospodarovanie lesov 2007. NLC Zvolen, Zvolen

Rizman I, Feiková Z (2008) Návrh oblastného členenia Slovenska pre potreby rámcového plánovania, hodnotenia stavu a vývoja lesov a tvorbu OPRL. Lesnícka typológia a zisťovanie stavu lesa 2008. NLC Zvolen, Zvolen

Socha J (2008) Effect of topography and geology on the site index of Picea abies in the West Carpathian, Poland. Scand J For Res 23:203–213

Šmelko Š, Wenk G, Antanaitis V (1992) Rast, štruktúra a produkcia lesa. Príroda, Bratislava

Viewegh J (2003) Klasifikace lesních rostlinných spoločenstev (se zaměřením na Typologický systém ÚHÚL). PEF ČZU, Praha

Viewegh J, Kusbach A, Mikeska M (2003) Czech forest ecosystem classification. J For Sci 49(2):85–93

Vladovič J, Burgan K, Dubeň Z, Dupkala J, Flachbart V, Greč V, Hatala N, Hudecová D, Kučera J, Lupták I et al (1994) Lesné oblasti Slovenska. Lesoprojekt Zvolen, Zvolen

Vladovič J (2003) Oblastné východiská a princípy hodnotenia drevinového zloženia a ekologickej stability lesov Slovenska. Lesnícke štúdie 53. Príroda, Bratislava

Vološčuk I (2001) Teoretické a praktické problémy ekologickej stability lesných ekosysté-mov. Vedecké štúdie. TU Zvolen, Zvolen

Zlatník A (1959) Přehled slovenských lesů podle skupin lesních typů. Spisy vědecké laboratoře biocenologie a typologie lesa LF VŠZ v Brně 3. Brno

Zlatník A (1976) Přehled skupin typů geobiocenů původně lesních a křovinných v ČSSR. Zprávy geografického ústavu ČSAV 13, Brno

Acer Pseudoplatanus L., *Tilia Cordata* Mill. and *Pinus Sylvestris* L. as Valuable Tree Species in the Carpathian Forests

Maciej Pach, Andrzej Jaworski and Jerzy Skrzyszewski

Abstract Sycamore maple, little-leaf linden, and Scots pine occurring in unique mixed stands, such as beech and sycamore maple in the Bieszczady Mountains, linden forest in the Obrożyska Reserve (the Beskid Sądecki Mountains), and scattered Carpathian pine forests, were the subject of this study. It was found that sycamore maple can be a highly productive admixture and a co-dominant or even dominant species in beech forests as a protective species of the upper timberline at altitudes between 930 and 1,160 m. Little-leaf linden in stands where site conditions meet its requirements, especially in the lower part of the lower mountain zone at altitudes up to 600–700 m, is a suitable admixture or even co-dominant species that increases productivity. In the Polish Carpathian forests three pine population groups were distinguished on the basis of morphological (cone) traits—lowland, foothill, and mountain group—which also varied according to timber quality, stem, and crown traits, as well as copper (Cu), iron (Fe), and potassium (K) content in needles. Pine is a very good nursing tree species that enables other tree species, especially fir, to occur among pine trees or underneath their canopies.

1 Introduction

Polish Carpathian forests are dominated mainly by silver fir (*Abies alba* Mill.), European beech (*Fagus sylvativa* L.), and Norway spruce (*Picea abies* L.) that together account for about 71 % of the state forests in this region (Rozwałka 2003). Each of these main tree species occupies suitable habitats in terms of

M. Pach (✉) · A. Jaworski · J. Skrzyszewski
Department of Silviculture, University of Agriculture in Kraków,
Al. 29-Listopada 46 31-425 Kraków, Poland
e-mail: rlpach@cyf-kr.edu.pl

J. Kozak et al. (eds.), *The Carpathians: Integrating Nature and Society Towards Sustainability*, Environmental Science and Engineering,
DOI: 10.1007/978-3-642-12725-0_21,

altitude, fertility, and climatic conditions. In addition to these, there are many other tree species that can serve as valuable admixtures or even co-dominant species in the Carpathian region.

The main admixture species in fir, beech, or spruce stands growing on mountain-rich sites is sycamore maple (*Acer pseudoplatanus* L.). It finds suitable conditions for growth in the Bieszczady Mountains (Głaz 1985), where together with beech forms many stands and constitutes rare forest associations like *Lunario-Aceretum* and *Phyllitido-Aceretum*, and two subassociations, *Dentario glandulosae-Fagetum lunarietosum* and *Dentario glandulosae-Fagetum allietosum ursini*, as well as two unique associations, *Sorbo-Aceretum carpaticum* and *Aceri-Fagetum* (Michalik 1993; Michalik and Szary 1993, 1997; Wilczek 1995; Matuszkiewicz 2001). The latter creates forests mostly near the upper timberline, and due to their inaccessibility (steep slopes) and the low merchantable value of beech and sycamore timber, they have not been harvested up to now and can be considered as primeval forests (Schuck et al. 1994).

On lower elevations (up to 600 m) little-leaf linden (*Tilia cordata* Mill.) can be found where together with beech, oak (*Quercus robur* L. and *Q. petrea* Liebl.), hornbeam (*Carpinus betulus* L.), spruce, fir, pine, and maple (*Acer pseudoplatanus* L., *A. platanoides* L. and *A. campestre* L.) constitutes the *Tilio-Carpinetum* forest association. These forests are in the "Las lipowy Obrożyska" Forest Reserve, now called the Obrożyska Reserve, which is situated near the Poprad River (Beskid Sądecki Mountains) in the vicinity of Muszyna town. The Reserve was created in 1919 and at present covers 101.74 ha (98.25 ha of forests), including 26.68 ha of primeval forests under strict protection (Plan Urządzania 1999). Little-leaf linden forest is considered a relict of linden stands, abundantly occurring in this part of the Carpathians in the postglacial climatic optimum (the Atlantic period 8000–5000 BC), and is unique in Poland and the Carpathians.

Scots pine (*Pinus sylvestris* L.) dominated stands, the majority of which are of nonnative origin, constitute about 17.5 % of the Carpathian forests, mostly in the uplands (up to 600–700 m). At higher elevations and on rich forest sites (such as mountain broad-leaved forest sites), pine stands occur as patches of admixture, often of native origin, in silver fir or beech stands. Among the pine stands some variation is observed, which can have some influence on the introduction of pine to forests. This diversity is the result of the mixing of Scots pine from northern and southern refugium, which took place at the beginning of the Holocene period in the Carpathian area (Tobolski and Hanover 1971; Ralska-Jasiewiczowa 2004).

The aim of this study was to determine the species composition and productivity of forest stands relevant to these three species, and to outline the possibilities of implementation of some conclusions drawn from the observation of these forests in deliberate forest management.

2 Materials and Methods

2.1 Field Measurements

In the primeval forests consisting of sycamore, maple, and beech (*Aceri-Fagetum* and *Dentario glandulosae-Fagetum* forest associations), occurring in the Moczarne Reserve of the Bieszczady Mountains, four permanent sample plots were established in 1993 (Table 1). On every plot diameter at breast height (DBH) and height measurements of trees ≥ 8 cm of DBH were measured using a standard calliper and Vertex III hypsometer. In 2003 measurements were carried out to determine changes in species composition, stand volume, and basal area.

In the strictly protected linden forests of the Obrożyska Reserve, three permanent sample plots were established in 1990 (Table 1). In 2000, stands growing on the plots were remeasured. Additionally, 26 circular sample plots (0.04 ha each) in a grid of 100 × 100 m were established in the strictly protected part in 1995. The control measurements on the small plots were carried out in 2005. Detailed descriptions of site conditions were presented in a paper by Jaworski et al. (2005). The scope of the measurements on large and small sample plots in linden forests was the same as in the case of sycamore stands. All established sample plots are permanently marked in the studied forests in order to perform the measurements every 10 years.

In 39 stands of *Pinus sylvestris* L., mostly autochthonous, the investigations of the occurrence and differentiation of the main tree species were conducted. In each pine stand the DBH and height measurement of a selected 20 mean sample trees (80–170 years of mean age), without visible injuries from the upper stand stratum, were carried out. Additionally, vitality, crown size, traits of branches, bark type and thickness, stem quality (including knots and knobs) of pine trees as well as number and DBH of species co-occurring with pine in stands were determined. A sample of 300 cones was collected from the bottom of each forest. Wood samples from the base of thicker branches and small branches with needles (1–3 years old) and those with 1-year-old cones were gathered from felled mean pine trees. Wood samples were taken from a height of 1.3 m up the trunk in order to determine radius increment.

2.2 Analyses

The measurements of stands carried out on the classical permanent sample plots in forests with linden (3 plots) and sycamore (4 plots) allowed the calculation of the number of trees, stand volume, and basal area using the "ZASOBY" computer program based on the Ukrainian and Moldavian volume tables for standing linden trees (Anonymous 1987) and Czuraj's tables (1991) for the remaining tree species.

Table 1 Location of sample plots and their site and stand characteristics (linden and sycamore forests)

	Tilia cordata Reserve			*Acer pseudoplatanus* in Bieszczady National Park			
Sample plot	Obrożyska 1	Obrożyska 2	Obrożyska 3	Moczarne 1	Moczarne 2	Rabia Skała 1	Rabia Skała 2
Mountain range	Beskid Sądecki			Bieszczady			
Area [ha]	0.25	0.50	0.50	0.25	0.33	0.25	0.20
Latitude	49°20′54″N	49°20′56″N	49°20′07″N	49°06′43″N	49°06′40″N	49°06′33″N	49°06′29″N
Longitude	20°52′20″E	20°52′24″E	20°52′22″E	22°28′70″E	22°28′87″E	22°27′23″E	22°27′35″E
Aspect	W	W	SW	NNE	E	NWW	NW
Slope [°]	16	11–30	14–19	18	25	18	12–18
Altitude [m a.s.l.]	510	520	515	1,010	930	1,120	1,160
Plant	*Dentario*			association *glandulosae-Fagetum*	*Tilio-*	*Carpinetum* *Dentario glandulosae-Fagetum allietosum ursini*	*Aceri-Fagetum*

The measurements of pine tree crown enabled the calculation of live crown ratio: ratio of crown width to tree height (Assmann 1968). The content of nitrogen (N), phosphorus (P), potassium (K), calcium (Ca), magnesium (Mg), iron (Fe) and copper (Cu) was determined in 1-year-old needles by the Kjeldahl method (N), vanadium method (P) and AAS method (K, Ca, Mg, Fe, Cu) (Ostrowska et al. 1991). Dry weight of needles was determined after drying in temperature 105 °C with accuracy 0.0001 g. From 300 cones gathered in every stand, 50 were randomly chosen and the following cone trait measurements were done: length, distance between top and largest diameter, largest diameter, width at middistance between top and largest diameter, width of top, scale length, width, and thickness. For each cone the number of scales and cone form, on the basis of scale morphology (*plana*, *gibba*, *hamata*, *reflexa*), were determined (Staszkiewicz 1993). On the basis of those measurements many cone shape indexes were computed and analyzed. Measurements were made of the setting angle of 1-year-old cones on branches from felled mean trees on each plot.

The usability of cone traits as morphological markers of pine trees was verified using an analysis of the principal components and discriminant analysis. The description of the provenance groups distinguished on the basis of those markers put special emphasis on the pine traits that are important for commercial and silvicultural/management purposes. The groups were compared with respect to the size, quality and growth of pine trees, the characteristics of the stem and crown, and content of nutrients. The analysis of variance ANOVA with covariates like age, stand, and site features (adjusted means are shown in tables) and Scheffe's post hoc test were used (StatSoft 1997).

3 Results and Discussion

3.1 Sycamore-Beech Stands in the Bieszczady Mountains

The volume of stands with sycamore maple growing on the sample plots in the Bieszczady Mountains depended on the altitude and site conditions. The highest volume (486.56 m^3 ha^{-1}) was recorded on the lowest established plot: Moczarne 2 (Table 2). The stand on the highest altitude analyzed (Rabia Skała 2 plot) was characterized by a much lower volume (128.62 m^3 ha^{-1}). *Acer pseudoplatanus* was a dominant tree species by basal area and volume on the Moczarne 1 plot and Rabia Skała 1 plot. Its share amounted to 61.8 and 57.0 % by basal area and 65.0 and 61.8 % by volume (Table 2) for each plot, respectively. On Moczarne 2 plot *Fagus sylvatica* predominated in species composition (68.0 % by basal area and 60.9 % by volume). The stand on the Rabia Skała 2 plot was composed of a similar percentage of sycamore maple and beech (Table 2). The highest sycamore maple trees reached 34 m in the stand growing on the Moczarne 2 plot, which was established at the lowest altitude (Table 1). The height of sycamore maple trees growing in the stand at an altitude of 1,160 m (Rabia Skała 2 plot) was 10 m at the most.

The values of volume and basal increment during the 10-year control period were higher for beech (volume from 16.17 on the Rabia Skała 2 plot to 49.14 m^3 ha^{-1} on the Moczarne 1 plot and basal area from 3.4 on the Moczarne 2 plot to 4.7 m^2 ha^{-1} on the Moczarne 1 plot) than for sycamore maple (volume from 2.45 on the Rabia Skała 2 plot to 15.90 m^3 ha^{-1} on the Moczarne 2 plot and basal area from 1.0 on the Moczarne 2 plot to 1.7 m^2 ha^{-1} on the Rabia Skała 1 plot) on all sample plots (Table 3). In each stand analyzed the total volume and basal area increased during the 10 years regardless of the stand development, site conditions, and altitude (Table 3). Bartkowicz et al. (2008) presented a detailed description of changes in the proportions of sycamore maple and common beech on the studied plots.

Sycamore maple can be a highly productive admixture, co-dominant, or even dominant species (standing volume from 65 at the highest elevation to 261 m^3 ha^{-1} at lower altitude) in mixed sycamore-beech stands growing at altitudes between 930 and 1,160 m (Table 2). The stands (*Aceri-Fagetum*) situated at the highest elevation (Rabia Skała 2) near upper timberline perform mainly protective functions (soil and water protection, biodiversity) due to the tree dwarf form (elfin wood), low timber quality, and mostly vegetative reproduction. These stands of natural or even virgin-type climax association characteristics are unique in the Polish Carpathians and should be under strict protection (Michalik and Szary 1997; Kucharzyk 2003).

Table 2 Species composition by volume and basal area (DBH ≥ 8 cm) of stands with sycamore in four plots in the Bieszczady National Park in 2003

Plot	Species	Number of trees ha^{-1} N	Volume [m^3 ha^{-1}] V	Basal area [m^2 ha^{-1}] G	Species composition [%] V	G
Moczarne 1	*F. sylvatica*	456	140.58	15.65	35.0	38.2
	A. pseudoplatanus	116	261.64	25.28	65.0	61.8
	Total	572	402.22	40.93	100.0	100.0
Moczarne 2	*F. sylvatica*	336	296.21	27.84	60.9	68.0
	A. pseudoplatanus	48	189.71	12.95	39.0	31.6
	A. platanoides	3	0.64	0.13	0.1	0.4
	Total	387	486.56	40.92	100.0	100.0
Rabia Skała 1	*F. sylvatica*	600	96.03	17.88	38.2	43.0
	A. pseudoplatanus	152	155.18	23.66	61.8	57.0
	Total	752	251.21	41.54	100.0	100.0
Rabia Skała 2	*F. sylvatica*	710	63.10	15.88	49.1	52.5
	A. pseudoplatanus	370	65.52	14.36	50.9	47.5
	Total	1,080	128.62	30.24	100.0	100.0

Table 3 Increment of volume and basal area of main tree species and total on studied plots in the Bieszczady Mountains in the 10-year control period

Plot	Species	Increment 1993–2003	
		Volume [$m^3 \cdot ha^{-1}$]	Basal area [$m^2 \cdot ha^{-1}$]
Moczarne 1	*F. sylvatica*	49.14	4.7
	A. pseudoplatanus	12.15	1.2
	Total	61.29	5.8
Moczarne 2	*F. sylvatica*	32.06	3.4
	A. pseudoplatanus	15.90	1.0
	Total	47.97	4.5
Rabia Skała 1	*F. sylvatica*	25.54	4.4
	A. pseudoplatanus	11.69	1.7
	Total	37.23	6.1
Rabia Skała 2	*F. sylvatica*	16.17	4.0
	A. pseudoplatanus	2.45	1.5
	Total	18.62	5.5

Similar natural stands consisting of beech and sycamore maple can be found in the Stužica Reserve in Slovakia (near the Polish and Ukrainian border), which have been studied since the 1930s (Korsuň 1938).

3.2 Linden Stands in Obrożyska Reserve

The volume and basal area of studied stands with linden (on three permanent plots) growing in the Obrożyska Reserve ranged between 761.21 and 54.77 (Obrożyska 3) to 860.54 m^3 ha^{-1} and 62.24 m^2 ha^{-1} (Obrożyska 2) (Table 4). *Tilia cordata* was definitely a dominant tree species in two stands (Obrożyska 1 and 2) when regarding its partition by volume and basal area (Table 3). On the Obrożyska 3 plot linden share was lower (73.3 % by basal area, 76.8 % by volume) because of the occurrence of a significant number of *Abies alba* trees (276 individuals per ha), mainly in the stand layer under the main linden canopy stratum (Table 4).

Considering data obtained from the 26 sample plots established in a 100 × 100 m grid (1 plot of 0.04 ha), we found that the average volume and basal area for the strict protected part of the linden forest were lower (568 and 43.86 m^3 ha^{-1}, respectively) than for the large, permanent plots (Tables 4 and 5). Except for the short-leaf linden, which was still dominating tree species, the percentages of other tree species in the species composition were significant in terms of volume and basal area (Table 5). Partitions of *Abies alba* and *Carpinus betulus* exceeded 10 % (15.3 % and 10.8 % by volume and 14.8 % and 13.2 % by basal area, respectively) (Table 5).

The highest volume and basal area increments were recorded on the Obrożyska 2 plot, where their values reached 95.78 m^3 ha^{-1} and 6.05 m^2 ha^{-1}, respectively, in the 10-year period (Table 6). In the remaining two stands the increments were

Table 4 Species composition by volume and basal area (DBH ≥ 8 cm) of stands with linden on three plots in the Obrożyska Reserve in 2000

Plot	Species	Number of trees·ha^{-1} N	Volume [m^3·ha^{-1}] V	Basal area [m^2·ha^{-1}] G	Species composition [%] V	G
Obrożyska 1	*T. cordata*	408	747.90	56.92	97.4	95.1
	A. alba	40	12.06	1.62	1.6	2.7
	C. betulus	100	6.20	1.01	0.8	1.7
	Others	20	1.61	0.30	0.2	0.5
	Total	568	767.77	59.85	100.0	100.0
Obrożyska 2	*T. cordata*	418	833.89	59.57	96.9	95.8
	A. alba	52	13.55	1.45	1.6	2.3
	C. betulus	42	10.12	0.90	1.2	1.4
	Others	12	2.98	0.32	0.3	0.5
	Total	524	860.54	62.24	100.0	100.0
Obrożyska 3	*T. cordata*	186	584.93	40.13	76.8	73.3
	A. alba	276	135.62	11.16	17.8	20.4
	C. betulus	58	32.44	2.92	4.3	5.3
	Others	8	8.22	0.56	1.1	1.0
	Total	528	761.21	54.77	100.0	100.0

Table 5 Species composition by volume and basal area (DBH ≥ 8 cm) on 26 plots (in 100 × 100 m grid) in the Obrożyska Reserve in 2005

Species	Number of trees·ha^{-1} N	Volume [m^3·ha^{-1}] V	Basal area [m^2·ha^{-1}] G	Species composition [%] V	G
T. cordata	219	335.83	25.55	59.1	58.3
A. alba	61	87.15	6.51	15.3	14.8
C. betulus	155	61.41	5.78	10.8	13.2
A. pseudoplatanus	50	33.32	2.52	5.9	5.7
F. sylvatica	21	20.96	1.31	3.7	3.0
P. abies	16	16.94	1.19	3.0	2.7
A. platanoides	10	5.50	0.43	1.0	1.0
Others (*Betula pendula, Larix europaea, Alnus glutinosa, Pinus sylvestris*)	5	6.88	0.57	1.2	1.3
Total	537	567.99	43.86	100.0	100.0

lower (Obrożyska 1: 79.98 m^3 ha^{-1}, 5.48 m^2 ha^{-1}; and Obrożyska 3: 81.75 m^3 ha^{-1}, 5.54 m^2 ha^{-1}) but still significantly high. Linden had the largest percentage of the total increments values. A relatively high partition of fir in the increments' values on the Obrożyska 3 was caused by its numerous occurrences in the second canopy stratum.

Table 6 Volume and basal area of the main tree species and total on studied plots in the Obrożyska Reserve in the 10-year control period

Sample plot	Species	Increment 1990–2000	
		Volume [$m^3 \cdot ha^{-1}$]	Basal area [$m^2 \cdot ha^{-1}$]
Obrożyska 1	*T. cordata*	77.17	5.2
	A. alba	1.73	0.2
	C. betulus	1.04	0.1
	Others	0.04	0.0
	Total	79.98	5.48
Obrożyska 2	*T. cordata*	91.81	5.7
	A. alba	2.69	0.2
	C. betulus	0.86	0.1
	Others	0.42	0.1
	Total	95.78	6.05
Obrożyska 3	*T. cordata*	53.77	3.4
	A. alba	24.46	1.9
	C. betulus	3.04	0.2
	Others	0.48	0.0
	Total	81.75	5.54
		Increment 1995–2005	
On the basis of 26 sample plots (0.04 ha)	*T. cordata*	25.81	2.20
	A. alba	10.22	1.28
	C. betulus	10.06	0.88
	A. pseudoplatanus	6.95	0.43
	F. sylvatica	6.42	0.24
	Others	3.57	0.26
	Total	63.03	5.29

The values of volume and basal area increment, determined on the basis of 26 sample plots of 0.04 ha each, were lower than found in the stands growing on the large sample plots (Table 6). The highest volume and 10-year basal area increment were recorded for little-leaf linden (25.81 m^3 ha^{-1} and 2.20 m^2 ha^{-1}, respectively). The percentages of the remaining tree species in the increment values were higher than on the large sample plots (Table 6). The changes in volume and basal area in the control period depending on stand growth stage were analyzed by Grałek et al. (2009). The highest growing stock, volume, and basal area increment were in the stand on the Obrożyska 2 plot, where the initial phase of the optimum stage were determined by Jaworski et al. (2005) according to Korpel (1995) criteria. On the remaining plots, the transition from the growing-up stage to the optimum stage (the Obrożyska 1) and the selection structure phase (the Obrożyska 3) were determined (Jaworski et al. 2005).

The growing stock per hectare of the studied linden forests in the Obrożyska Reserve is one of the greatest (760–860 m^3 ha^{-1} on large sample plots or 568 m^3 ha^{-1} on 26 small sample plots) among primeval and natural stands in the Polish part of the Carpathians. The differences between the volume based on 3 large and

26 small sample plots could be explained by the fact that the large plots were established in subjectively selected homogeneous fragments of forest representing specified stage and phase of development (Korpel 1995). The volumes determined on large plots were higher than the volumes per hectare of pure beech and mixed stands with fir, beech, and spruce (Dziewolski and Rutkowski 1991; Jaworski et al. 1994; Przybylska et al. 1995; Jaworski and Paluch 2001; Jaworski et al. 2001a, b; Jaworski and Kołodziej 2002). Therefore, the high productivity of the stands may be an effect of the advantageous terrain configuration (mostly west and south slope aspects, protected against the north and east cold winds), fertile soil (brown soils), and mild climate (the Reserve is situated in the Poprad River valley along which warm air inflow from the south frequently occurs) (Fabijanowski 1961).

3.3 Pine Stands

Most of the investigated pine stands (75 %) grew at an altitude of 300–600 m, on a warm aspect (south, southeast, and southwest) and on rich or medium-rich forest site types characterized as mountain broad-leaved forest (LG) and mountain mixed broad-leaved forest (LMG) (Siedliskowe Podstawy Hodowli Lasu 2004). Due to the more favorable light and warmth conditions among pine trees and underneath pine canopies, many other tree species accompanying pine were recorded; among those *Abies alba*, *Fagus sylvatica* and *Picea abies* occurred most frequently (Fig. 1). Their participation in the number of trees, volume, and basal area ranged from 5 to 23 %. The remaining percentage belonged to pine and 12 other mostly broad-leaved, tree species.

Pine stands were mostly even-aged, and their average age ranged from 80 to 170 years between stands. The investigated pine stands were in most cases characterized by normal vitality, long crowns, low quality of the trunk, and high volume (up to 600 m^3 ha^{-1}). In each of the analyzed stands a high stability of Scots pines was observed, which was determined based on the ratio of height to DBH of trees.

On the basis of pinecone traits as morphological markers, pine stands were classified into three population groups named a lowland group, a mountain group, and a foothill group. The groups did not have any specific shape of the cone (morphotype), and they differed only in the proportion of three-cone morphotypes called "lowland" (L), "mountain" (M), and "foothill" (F). The cones of morphotype F were the largest, the least full, and had not numerous but long, wide, and thick scales. The L-type cones were the fullest, almost as wide as F-type, with flat and numerous scales. The M-type cones were the smallest, but similar in proportions to F-type. The morphotypes also varied in the percentage of scale forms (*plana*, *gibba*, *hamata*, *reflexa*). The setting angle of 1-year-old cones was smaller in the foothill and mountain groups of pines than in the lowland group. The spatial distribution of pine stands classified into the three population groups is shown in Fig. 2.

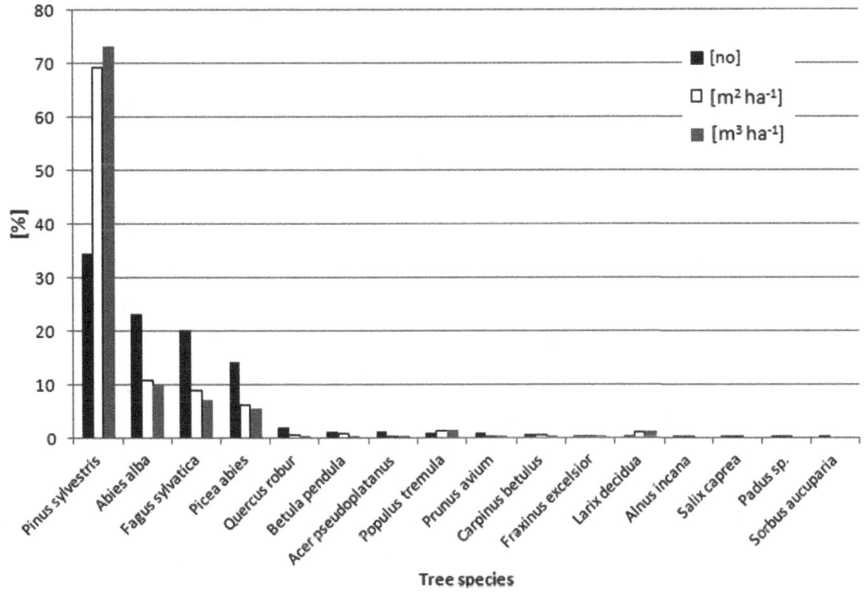

Fig. 1 Number, volume, and basal area percentage of tree species found among pine trees and underneath pine canopies

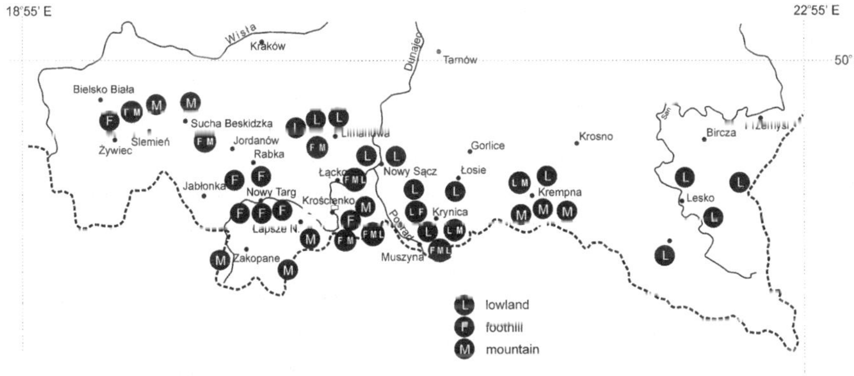

Fig. 2 Spatial distribution of pine stands classified into the three population groups (L, F, M) within the range of the Polish Carpathians

The lowland pine can create stands producing wood of high quality and commercial value if they grow on less fertile microsites and/or have a higher canopy density (Skrzyszewski 2004). When the sites are more fertile and/or the canopy density is lower, the lowland pine achieves high DBH increment and forms very stable stands, but the trees have thick branches and a limited ability for self-pruning (high branch age) (Table 7).

Table 7 Pine branch thickness, their age, and appearance of knots in population groups

Branch traits	Population group[a]	Adjusted mean	Homogeneous groups[b]		
Branch thickness [mm]	L	66	●		
	M	54		●	
	F	53		●	
Branch age [years]	L	53	●		
	M	45		●	
	F	40			●
Number of burl knots per tree	L	3.1	●		
	F	2.1		●	
	M	2.2		●	
Number of open knots per tree	M	2.5	●		
	L	2.3	●		
	F	1.0		●	

[a] L: lowland; M: mountain; F: foothill; [b] homogeneous groups are joined vertically by dots

Diameter increment of branches is very significantly correlated to the DBH increment (Skrzyszewski 2004). With respect to the future quality of the stem, the DBH of a tree should not exceed 7 cm at the age of 16 years and 30 cm at 80 years (diameter increment should be limited) (Skrzyszewski 2004).

The needles of lowland pines showed the highest Cu and Fe content (Table 8). This suggests that their photosynthesis process is more efficient (Starck 2002).

The age of the thickest branches (the lowest branches) showed that the lowland pines are the most shade tolerant (Table 7). The foothill pine has thin branches, a narrow crown, and smaller DBH increment, but the greatest height and quality of stem (number of burl and open knots) (Tables 7 and 9). Furthermore, these traits are maintained on fertile sites (mountain broad-leaved forest site) and at a low canopy density (Skrzyszewski 2004). The mountain pine forms stands of very good quality on part of the sites, but it exhibits the greatest needle loss (Table 9).

The mountain and foothill pines are more resistant to frost and drought damages (Starck 2002) due to the higher K content (Table 8).

Table 8 Nutrient contents of 1-year-old pine needles by population groups

Nutrient	Population group[a]	Adjusted mean	Homogeneous groups[b]	
K [%]	F	0.661	●	
	M	0.66	●	
	L	0.549		●
Cu [ppm]	L	15.91	●	
	M	14.66		●
	F	14.02		●
Fe [ppm]	L	151	●	
	F	113		●
	M	102		●

[a] L: lowland; M: mountain; F: foothill; [b] homogeneous groups are joined vertically by dots

Table 9 Stem and crown traits of sample pine trees in population groups

Stem and crown traits	Population group[a]	Adjusted mean	Homogeneous groups[b]		
DBH, D [cm]	L	44.5	●		
	F	41.8		●	
	M	41.1		●	
Mean annual DBH increment [mm]	L	3.6	●		
	F	3.4		●	
	M	3.3		●	
Height, H [m]	F	29.4	●		
	M	28.1		●	
	L	27.6			●
Coefficient of slenderness, H/D	F	72	●		
	M	70	●		
	L	62		●	
Crown spreading degree (narrow crown)	L	20	●		
Crown width/H [%]	M	19		●	
	F	18			●
Needle loss [%]	M	24	●		
	F	20		●	
	L	17			●

[a] L: lowland; M: mountain; F: foothill; [b] homogeneous groups are joined vertically by dots

Pine tree as an admixture species can play a stabilizing role (coefficient of slenderness height/diameter [H/D] is lower than 80) (Table 9) in stands prone to wind damage and can be a productive admixture or a co-dominant species in silver fir or beech stands. Additionally, the site conditions created by Scots pine are conducive to the abundant appearance of the main and admixture climax species, mostly *Abies alba* Mill (Fig. 2). This role of pine stands as a nurse crop enables the prolongation of the transformation period and the attainment of multistory, uneven-aged silver fir stands (Jaworski 1988).

4 Conclusions and Practical Implications

The results of this study showed that sycamore maple appeared to be a high productive species (high volume and basal area increment) and can play a valuable role as a productive admixture in the stands of the Bieszczady Mountains. The little-leaf linden stands in the Obrożyska Reserve had the highest volume per hectare among managed and primeval character stands in the Polish part of the Carpathians. Where site conditions meet its requirements, especially in the lower part of the lower mountain zone at altitudes up to 600–700 m, these stands are a suitable admixture or even codominant species, increasing the productivity and biological diversity of stands.

In the Polish Carpathian forests three pine population groups were distinguished on the basis of morphological (cone) traits—lowland, foothill and mountain groups—which also varied according to timber quality, stem and crown traits, as well as Cu, Fe and K content in needles. The pine populations group (lowland, foothills and mountain) should be considered during Scots pine introduction to forests stands as a codominant or admixture tree of different percentages (productive or nurse role) and as a nurse crop tree in open areas (before climax species).

Based on the results of our study, we suggest that increased introduction of little-leaf linden and sycamore maple as an admixture in forest stands should be followed by flexible methods of regeneration. The Swiss irregular shelterwood system is recommended as the best method for creating favorable conditions of light and shelter for regeneration in gaps while maintaining a suitable partition of both species within the stand.

The lower light demands of lowland Scots pine allows for its cultivation in small groups within fir stands with the result that in forest stands managed by the Swiss irregular shelterwood system Scots pine can be introduced into unregenerated gaps. In the case of existing Scots pine regeneration that constitutes a nursing admixture, it should be free from the influence of other species before the Scots pine trees are 10 years old (Skrzyszewski 2004).

The foothill Scots pine should be used to a greater extent in establishing nurse stands. In such stands, with the early reduced density caused by restructure felling, the Scots pine has the capacity to maintain high stem quality, especially on fertile sites such as mountain broad-leaved forest and mountain mixed broad-leaved forest. In addition, the higher frost resistance of mountain Scots pine allows its populations to be introduced, especially at high elevations including afforestation near the upper timberline. Because the mountain Scots pine grows mostly in small populations in only a few places in the Carpathians (Fig. 2), it is necessary to reestablish the higher elevation stands in the lower mountain belt of the Carpathians on the basis of the remaining stands.

References

Anonymous (1987) Normativno spravochnye materialy dlya taksatsii lesov Ukrainy i Moldavii. Urozhai, Kiev

Assmann E (1968) Nauka o produkcyjności lasu. PWRiL, Warszawa

Bartkowicz L, Jaworski A, Pach M (2008) Przypuszczalne mechanizmy zmian udziału jaworu i buka w bieszczadzkich drzewostanach o charakterze pierwotnym. Rocz Bieszcz 16:33–46

Czuraj M (1991) Tablice zasobności i przyrostu drzewostanów. PWRiL, Warszawa

Dziewolski J, Rutkowski B (1991) Tree mortality, recruitment and increment during the period 1969–1986 in a Reserve at Turbacz in the Gorce Mountains. Fol For Pol ser A 31:37–48

Fabijanowski J (1961) Roślinność rezerwatu lipowego ‘Obrożyska’ koło Muszyny. Ochr Przyr 27:109–159

Głaz J (1985) Drzewostany jaworowe w lasach państwowych. Sylwan 129(5):35–46
Grałek K, Jaworski A, Kołodziej Z, Majerczyk K (2009) Możliwości produkcyjne drzewostanu lipowego w rezerwacie 'Las lipowy Obrożyska'. Acta Agr Silv ser Silv 47:3–26
Jaworski A (1988) Ekologiczne podstawy projektowania składu gatunkowego odnowień. Zagadnienia wybrane. Akademia Rolnicza w Krakowie, Kraków
Jaworski A, Karczmarski J, Skrzyszewski J (1994) Dynamika, budowa i struktura drzewostanów w rezerwacie 'Łabowiec'. Acta Agr Silv ser Silv 32:3–26
Jaworski A, Kołodziej Z (2002) Natural loss of trees, recruitment and increment in stand of primeval character in selected areas of the Bieszczady Mountains National Park (South-Eastern Poland). J For Sci 48(4):141–149
Jaworski A, Kołodziej Z, Bartkowicz L (2005) Structure and dynamics of stands of primeval character composed of the little-leaf linden (*Tilia cordata* Mill.) in the 'Las lipowy Obrożyska' reserve (Southern Poland). J For Sci 7:283–304
Jaworski A, Kołodziej Z, Pach M (2001a) Skład gatunkowy, budowa i struktura drzewostanów w rezerwacie 'Śrubita'. Sylwan 145(6):21–48
Jaworski A, Kołodziej Z, Strząska T (2001b) Skład gatunkowy, budowa i struktura drzewostanów w rezerwacie 'Oszast'. Sylwan 145(4):5–32
Jaworski A, Paluch J (2001) Structure and dynamics of the lower mountain zone forests of primeval character in the Babia Góra Mt. National Park J For Sci 47(2):60–74
Korpel Š (1995) Die Urwälder der Westkarpaten. Gustav Fischer Verlag, Stuttgart
Korsuň F (1938) Taxační prozkum reservace Stužica, jawornik a Pop Ivan. Sborník VÚZ ÈSR 152(7):245–372, 438–448
Kucharzyk S (2003) Struktura i dynamika drzewostanów w strefie górnej granicy lasu w Bieszczadzkim Parku Narodowym. PhD Thesis, University of Agriculture in Kraków, Kraków
Matuszkiewicz JM (2001) Zespoły leśne Polski. PWN, Warszawa
Michalik S (1993) Zbiorowiska leśne Bieszczadzkiego Park Narodowego, ich waloryzacja i problemy ochrony. Rocz Bieszcz 2:51–62
Michalik S, Szary A (1993) Wschodniokarpacka jaworzyna ziołoroślowa *Aceri-Fagetum* J. et M. Bartsch 1940 w Bieszczadzkim Parku Narodowym. Rocz Bieszcz 2:185–192
Michalik S, Szary A (1997) Zbiorowiska leśne Bieszczadzkiego Parku Narodowego. Monografie Bieszczadzkie 1, Ośrodek Naukowo-Dydaktyczny Bieszczadzkiego Parku Narodowego, Ustrzyki Dolne
Ostrowska A, Gawliński S, Szczubiałka Z (1991) Metody analizy i oceny właściwości gleb i roślin. Katalog, IOŚ, Warszawa
Plan Urządzania Gospodarstwa Rezerwatowego 'Las lipowy Obrożyska' na okres gospodarczy 01.01.1999-31.12.2009
Przybylska K, Fujak J, Myćka P (1995) Dynamika zmian zasobów leśnych w rezerwacie 'Dolina Łopusznej' Gorczańskiego Parku Narodowego w okresie kontrolnym 1981–1992. Parki Narodowe i Rezerwaty Przyrody 4(3):23–31
Ralska-Jasiewiczowa M (2004) Late glacial and holocene history of vegetation in Poland based on isopollen maps, Szafer Institute of Botany. Polish Academy of Sciences, Kraków
Rozwałka Z (2003) Zasady hodowli lasu. DGLP, Bedoń
Schuck A, Parviainen J, Bücking W (1994) A review of approaches to forestry research on structure, succession and biodiversity of undisturbed and semi-natural forests and woodlands in Europe. EFI working paper 3, European Forest Institute, Joensuu, Finland
Siedliskowe Podstawy Hodowli Lasu (2004) ORW LP, Bedoń
Skrzyszewski J (2004) Charakterystyka morfologiczno-przyrostowa sosny zwyczajnej (*Pinus sylvestris* L.) w polskiej części Karpat i Sudetów. Zesz Nauk AR Kraków 300, Rozpr 411
Starck Z (2002) Gospodarka mineralna roślin. Rola składników mineralnych w roślinie. In: Kopcewicz J, Lewak S (eds) Fizjologia roślin. PWN, Warszawa

Staszkiewicz J (1993) Zmienność morfologiczna szpilek, szyszek i nasion. In: Białobok S, Boratyński A, Bugała W (eds) Biologia sosny zwyczajnej. Wydawnictwo Sorus, Poznań—Kórnik, pp 33–43

StatSoft Inc (1997) STATISTICA for windows. Computer program manual. Tulsa, OK

Tobolski JJ, Hanover JW (1971) Genetic variation in the monoterpenes of Scotch pine. For Sci 17:293–299

Wilczek Z (1995) Zespoły leśne Beskidu Sądeckiego i zachodniej części Beskidu Żywieckiego na tle zbiorowisk leśnych Karpat Zachodnich. Prac Nauk Uniw Śl w Katowicach 1490:5–130

Management of Norway Spruce Stands in the Western Carpathians

Marian Slodičák, Jiří Novák and David Dušek

Abstract An important part of the Western Carpathians is situated in the Czech Republic. This region is characterized by a high percentage of forest cover. Although broadleaves used to dominate in natural species composition, Norway spruce stands cover more than one-half of the present forested area. A higher portion of spruce is the result of a historical demand for wide utilization of spruce wood. However, current climatic studies have confirmed the decline of the spruce, due to changes in precipitation amounts and patterns as well as the change in temperature and solar radiation. It is apparent that cultivation of Norway spruce in this region in its previous and present extent cannot be sustained and therefore conversion of current Norway spruce stands is imminent. We propose that silvicultural measures (thinning) can support spruce stand stability and that conversion of these stands consequently can be less risky. The study presents new results from the long-term thinning experiments in Norway spruce stands in the Czech part of the Western Carpathians. A new forestry concept based on an adequate thinning regime is discussed.

1 Introduction

In the Czech Republic, the Moravskoslezské Beskydy Mountains, the Hostýnsko-Vsetínské vrchy Mountains and the Javorníky Mountains comprise an important part of the Western Carpathians (Fig. 1). The total forest area in this region is 196,894 ha with 76 % as coniferous species and 24 % broad-leaved. Although broadleaves dominated in the natural species composition (Jankovska 1995), Norway spruce (*Picea abies* Karst.) stands cover more than one-half of the present

M. Slodičák (✉) · J. Novák · D. Dušek
Research Station at Opočno, Forestry and Game Management Research Institute, Strnady, Na Olive 550, 517 73 Opočno, Czech Republic
e-mail: slodicak@vulhmop.cz

J. Kozak et al. (eds.), *The Carpathians: Integrating Nature and Society Towards Sustainability*, Environmental Science and Engineering,
DOI: 10.1007/978-3-642-12725-0_22,

Fig. 1 Location (*grey* area) of the Western Carpathians in the Czech Republic

forest area, which is slightly higher than the mean proportion of 48 % of this species in the country, according to the national inventory. This high proportion of spruce has resulted from a wide utilization of spruce wood in the past.

Norway spruce stands, which were established in the last century, grew relatively well in the previous decades, according to dendrochronological analyses (Šrámek et al. 2008). However, forest decline (drying, yellowing) has been observed here in recent years (Novotný et al. 2008; Šrámek et al. 2008; Mullerova et al. 2009). Symptoms of decline (defoliation and discoloration) have been found in all spruce stands cultivated in zones with an elevation of 400–600 m, mean temperature of 6.0–7.5 °C, and precipitation of 650–800 mm. These zones are areas where spruce is not the original main species, at least not in monocultures.

According to a local study (Šrámek et al. 2008), spruce stand disintegration is caused by a combination of three main influences:

1. serious nutrient disturbances consisting in nutrient depletion of upper soil horizons and minimal base saturation,
2. gradual change of climatic conditions,
3. massive occurrence of fungi (mainly genus *Armillaria*) and bark beetles (*Ips typographus, Ips duplicatus*).

These factors have caused trees to gradually decline and consequently die. The first sign of deterioration was observed in 2004 after an extremely hot and dry year in 2003. Climatic studies (Bagár 2007) supported the finding of the change in precipitation amounts and patterns as well as the change of temperature and solar radiation in the period 1961–2006, in comparison with long-term average values in 1901–1950 and with the calculated standard of 1961–1990.

The health condition of the forests in the Western Carpathians must be maintained by sustainable forestry management. The main principles of the new forestry concept in the region consist of conversion of current spruce monocultures to stands with natural species composition, and this process should be continued. Current spruce stands must be stabilized before beginning their conversion because canopy opening, as a start of conversion, is usually connected with potential stand disintegration in the unstable spruce monocultures.

The possibilities of preparing spruce stands for conversion can be evaluated using the long-term results of thinning experiments that were established in the Czech part of the Western Carpathians in 1960 in the framework of the study coordinated by the Forestry and Game Management Research Institute at Strnady that covered the entire territory of the Czech Republic.

The main focus of this study is the long-term effect of silviculture measures (high thinning and low thinning) on the stability of mature Norway spruce monocultures in the Western Carpathians. We tested the hypothesis that silvicultural measures (thinning) can support spruce stand stability and that the conversion of these stands is consequently less risky.

2 Materials and Methods

The Forestry and Game Management Research Institute at Strnady created a new experimental basis for thinning research in 1956–1957 (Pařez 1958). Our study on the establishment and evaluation of long-term thinning experiments is based on this methodology of standardized techniques used in forestry research. A total of 46 experimental plots were founded by the Institute in the Norway spruce and Scots pine stands of the Czech Republic, with 24 plots still being maintained. Our study focused on a group of 4 experimental plots established in young Norway spruce stands in 1960 in the Western Carpathians and maintained until today (Table 1).

The objective of the thinning experiment was to compare two basic methods of thinning: positive selection from above (H: high thinning, i.e., supporting the best quality trees by removing competitors and leaving lower tree classes left untouched) and negative selection from below (L: low thinning, i.e., removal of smaller trees from lower tree classes). Both treated variants are compared with control plots (C) without thinning.

It was expected that thinning applied to trees of 40–50 years of age will result in a deeper differentiation of diameter structure (e.g., better survival of smaller

Table 1 List of experimental plots in Norway spruce stands established in 1960 in the Western Carpathians

	Name	Age[a]	Comparative plots[b]	Coordinates[c]	Elevation (m)	Forest type group[d]
1	Frýdek-Místek	49	C, H	49°31′22″ 18°29′59″	730	*Piceeto-Fagetum acidophilum*
2	Ostravice	43	C, H, L	49°27′32″ 18°27′18″	570	*Piceeto-Abietum variohumidum acidophylum*
3	V. Karlovice I.	48	H, L1, L2	49°21′22″ 18°22′33″	794	*Piceeto-Fagetum eutrophicum*
4	V. Karlovice II.	48	C, H	49°21′22″ 18°22′29″	794	*Piceeto-Fagetum eutrophicum*

[a] Age in 1960
[b] C: control without thinning; H: high thinning (positive selection from above); L: (1, 2) low thinning (negative selection from below)
[c] Values in WGS-84 system
[d] According to the Czech forest ecosystem classification (Viewegh et al. 2003)

trees) and a higher diameter increment of all trees left after thinning and consequently in better static stability.

The basic area of partial comparative plots is a 0.25 ha^2. The experimental plots are surveyed every 5 years after completing their vegetative growth and include all trees measured by calipers over bark. The height of the stands is measured by the Blume-Leiss altimeter in representative groups of trees (30 individuals of all tree classes), and height curves are calculated to assess the mean and top height (i.e., the mean height of 100 thickest trees per hectare).

In this study, we used the plots established in 1960 for analyses of the diameter structure and diameter growth development of the evaluated spruce stands. Differences between variants were tested using the Kruskal–Wallis one-way ANOVA of the statistical software package UNISTAT (version 5.1). Unless otherwise indicated, test levels of $p < 0.05$ were used.

3 Results and Discussion

At the end of the observation period (2005), the diameter distributions differed for some of the comparative variants (Fig. 2). We found significantly lower amounts of thinner trees (diameter degrees of less than 20 cm and 21–30 cm) in plots with low thinning compared to the control plots. The number of thicker trees was not affected by thinning: differences between various treatments were insignificant.

Values of height/diameter (h/d) ratios continually decreased with diameter degrees in comparative plots (Fig. 2). Our study confirmed that the Norway spruce stands without thinning or thinned by positive selection from above (high thinning) were unstable, as large portions of individuals exceeded the h/d ratio of 100

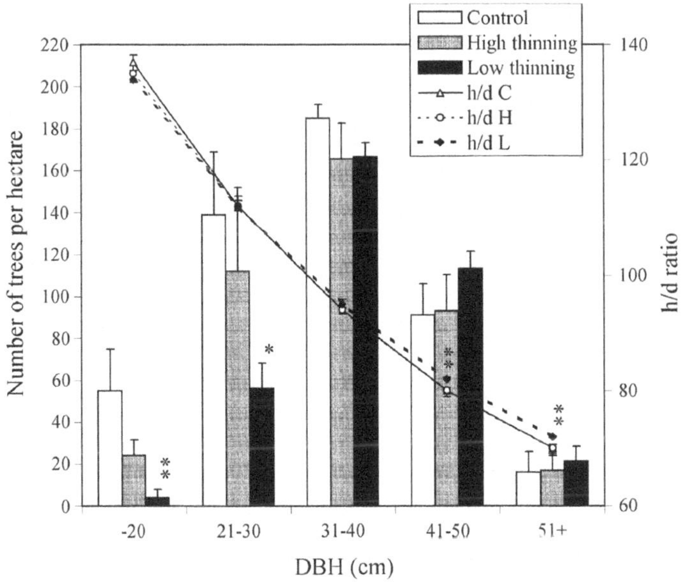

Fig. 2 Diameter distribution (*bars*) and height/diameter (h/d) ratio (*lines*) on comparative plots with different thinning regimes of plots founded in 1960 in the Western Carpathians [means with standard error (SE)]. *C* control, *H* high thinning, *L* low thinning. Statistically significant differences between variants: $*p < 0.10$, $**p < 0.05$

(diameter of up to 30 cm). These unfavorable parameters of static stability were probably caused by mutual competition for light, water, and nutrients, resulting in diameter growth depression of trees in the unthinned stands.

Generally, h/d ratio values higher than 100 imply low stability of stands (Lekes and Dandul 2000). Some studies (e.g., Mildner 1967) set the critical value to 90, especially in connection with snow damage in young stands. Lower values were recommended by Vicena et al. (1979) who mentioned an optimal h/d ratio of 79 or maximum (acceptable) of 83. A very detailed study on the basis of results from 85 research plots was done in Slovakia (Konopka 1999). This study determined the degrees of static stability of spruce stands on the basis of h/d ratio values as excellent ≤82, good 83–92, satisfactory 93–101 and dissatisfactory ≥102.

In our study we found significantly higher h/d ratio in thicker trees (diameter degrees over 40 cm) in plots with low thinning compared to the unthinned stands or stands with high thinning. But these differences are numerically small and the values of h/d ratio were 82 and lower in all variants. Thus, an excellent degree of static stability was found, according to the categories developed by Konopka (1999) mentioned above.

Observed effects of thinning on diameter structure were relatively small (significant only in lower diameter degrees in the case of low thinning; Fig. 2). Similar low effect of thinning in Norway spruce stands was reported by Eriksson (2006),

Makinen et al. (2006) and Pelletier and Pitt (2008). On the other hand higher positive effect of early thinning or wide spacing on stand stability is supported by many other studies (e.g., Pollanschütz 1974; MacCurrach 1991; Nielsen 1995; Brüchert et al. 2000; Slodicak and Novak 2006).

This study on the health condition of forest stands in the Czech part of the Western Carpathians (Šrámek et al. 2009) showed that younger stands up to the age of 40 years are less defoliated (mean defoliation 10–25 %) compared to older stands (mean defoliation 30–40 %). Therefore, the stabilization of these stands by proper treatment is urgent. The main objectives of thinning of young spruce stands in this region include:

- maintaining vitality and stability of dominant trees to increase the vitality of whole stands before conversion into stands with more natural species composition,
- reducing stand precipitation interception for better water management,
- creating a microclimate favorable to continual decomposition of litter (e.g., improvement of soil conditions, prevention of raw humus accumulation).

On the basis of our long-term results and from other published sources as discussed above, we recommend a new forestry concept focused on thinning of young Norway spruce stands in the Western Carpathians (Fig. 3). Thinning models for stands with prevailing Norway spruce are based on very heavy thinning at the thickest stage at a top height of 5 m (top height is the mean height of 100 thickest trees per hectare). First, heavily damaged trees with a defoliation of 60 % and higher and/or visible discoloration are to be removed by negative selection both from below and from above. Thereafter, trees with medium damage are removed from below and above, respectively. This first thinning is finished by traditional negative selection

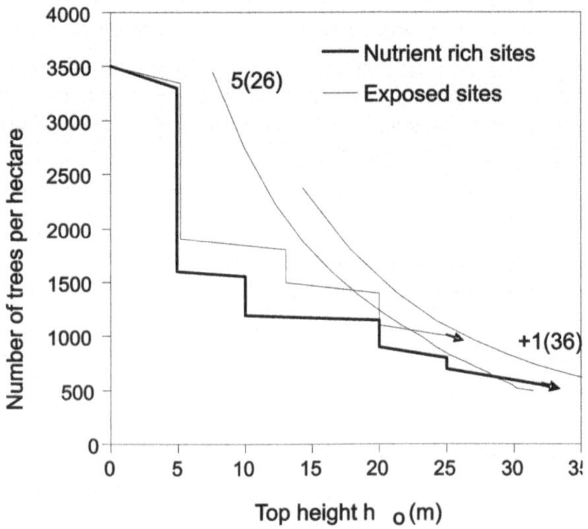

Fig. 3 Development of tree number according to top height (the mean height of 100 thickest trees per hectare) in proposed thinning programs for spruce stands on nutrient-rich and nutrient-poor sites in the Western Carpathians compared with number of trees modeled by growth tables (Černý et al. 1996) for site indexes +1 (36) and 5 (26)

from below, resulting in the recommended density. During the thinning, support of an admixture of shade-tolerant species (beech and fir) is also recommended.

After the second thinning, especially in relatively healthy stands, some aspects of the health status of stands can also be improved by a combination of low and high thinning. However, high thinning is recommended only for those spruce stands that have previously been thinned early and intensively.

Models are differentiated according to site type (nutrient-rich sites and exposed sites). Compared to the traditional management of density of spruce stands, our new models are more intensive at the young age. This is because young stands are relatively healthy, and intensive opening of the canopy is not as risky as in older stands. Additionally, higher thinning intensity is recommended on the nutrient-rich sites because of the higher growth potential of trees compared to exposed sites. Later (at the top height of 20 m or more), thinning intensity is decreased and corresponds with the density from growth tables (Černý et al. 1996).

4 Conclusion

On the basis of the presented analyses, we conclude that the h/d ratio was influenced only by low thinning in higher diameter degrees, but the observed differences were numerically small and excellent values (82 and lower) were shown for all variants. Diameter distribution of stands was influenced by treatment based on the low thinning only. We found significantly lower amounts of thinner, unstable trees in plots with low thinning compared to the control plots. Thus, in the Western Carpathian area we can recommend low thinning as a major silvicultural practice for stabilization of Norway spruce stands before their conversion.

In addition, on the basis of the literature results, an early start of thinning (compared to our experiments) in young spruce stands is suggested. The optimal period for the first thinning is the top height of 5 m on both nutrient-rich and nutrient poor sites. Thinning intensity is lower on nutrient-poor sites (only three treatments are proposed). According to the proposed programs, stand density should be decreased to about the top height of 20 m (on rich sites up to 25 m). Later, the density should follow the growth tables.

Acknowledgments This chapter was supported by the long-term project of the Czech Ministry of Agriculture, "Stabilisation of forest functions in anthropologically disturbed and changing environmental conditions" (MZE-0002070203). The authors are also grateful to Mr. Ian Hemingway (UK) for his helpful comments and language revision, and two anonymous reviewers for their constructive comments.

References

Bagár R (2007) Vývoj klimatických podmínek v oblasti Lesní správy Lesů České republiky Opava v letech 1961–2006 z dat ČHMÚ. Zpráva UHUL, Opava

Brüchert F, Becker G, Speck T (2000) The mechanics of Norway spruce [Picea abies (L.) Karst.]: mechanical properties of standing trees from different thinning regimes. For Ecol Manag 135:45–62

Černý M, Pařez J, Malík Z (1996) Růstové a taxační tabulky hlavních dřevin České republiky. (Smrk, borovice, buk, dub). IFER, Jílové u Prahy

Eriksson E (2006) Thinning operations and their impact on biomass production in stands of Norway spruce and Scots pine. Biomass Bioenergy 30:848–854

Jankovska V (1995) Forest composition changes in the Moravskoslezske Beskydy Mountains during the last millennium—a palaeoecological study. Folia Geobot Phytotaxon 30:375–387

Konopka J (1999) Ohrozenie lesnych porastov mechanicky posobiacimi abiotickymi cinitelmi. Lesn Cas 45:51–72

Lekes V, Dandul I (2000) Using airflow modelling and spatial analysis for defining wind damage risk classification (WINDARC). For Ecol Manag 135:331–344

MacCurrach RS (1991) Spacing: an option for reducing storm damage. Scott For 45:285–297

Makinen H, Isomaki A, Hongisto T (2006) Effect of half-systematic and systematic thinning on the increment of Scots pine and Norway spruce in Finland. Forestry 79:103–121

Mildner H (1967) Die Widerstandsfähigkeit von Fichtenjungbeständen gegenüber atmosphärischen Einwirkugen. Soz Forstwirtsch 17:57–59

Mullerova T, Cudlin P, Zid T, Cermak P, Rybnicek M, Kolar T, Konasova E (2009) Relationship between crown status and radial increment of Norway spruce forest stands under stress impacts in the Silesian Beskids Mts. In: Pribullova A, Bicarova S (eds) Sustainable development and bioclimate: reviewed conference proceedings. International scientific conference on sustainable development and bioclimate. Slovak Bioclimatological Society, Stara Lesna

Nielsen CCN (1995) Recommendations for stabilisation of Norway spruce stands based on ecological surveys. In: Coutts MP, Grace J (eds) Wind and trees. Cambridge University Press, Cambridge

Novotný R, Lachmanová Z, Šrámek V, Vortelová L (2008) Air pollution load and stand nutrition in the forest district Jablunkov, part Nýdek. J For Sci 54:49–54

Pařez J (1958) Návrh postupu při zakládání, sledování a vyhodnocování trvalých pokusných ploch se zvláštním zřetelem k pokusným plochám probírkovým a výnosovým. VÚLHM, Jíloviště-Strnady

Pelletier G, Pitt DG (2008) Silvicultural responses of two spruce plantations to midrotation commercial thinning in New Brunswick. Can J For Res 38:851–867

Pollanschütz J (1974) Ergebnisse eines 84 jährigen Fichten-Pflanzweiteversuches im Wienerwald. Allg Forstztg 29:816–818

Slodicak M, Novak J (2006) Silvicultural measures to increase the mechanical stability of pure secondary Norway spruce stands before conversion. For Ecol Manag 224:252–257

Šrámek V, Vejpustková M, Novotný R, Hellebrandová K (2008) Yellowing of Norway spruce stands in the Silesian Beskids—damage extent and dynamics. J For Sci 54:55–63

Šrámek V, Soukup F, Slodičák M (2009) Chřadnutí lesních porostů na LS Jablunkov. Les Pr 88:161–163

Vicena I, Parez J, Konopka J (1979) Ochrana lesa proti polomum. SZN Praha, Praha

Viewegh J, Kusbach A, Mikeska M (2003) Czech forest ecosystem classification. J For Sci 49:85–93

Sustainable Forest Management from Policy to Landscape, and Back Again: A Case Study in the Ukrainian Carpathian Mountains

Marine Elbakidze and Per Angelstam

Abstract To implement sustainable forest management (SFM) policies in actual landscapes, policy actors and managers exercising governance, and different forest stakeholders need to be provided with empirical information of how different SFM dimensions are understood and develop locally. Focusing on the state and trends of SFM implementation in the Ukrainian Carpathians we analyze the barriers and bridges at multiple levels from the national to the local management unit. First, we review the national Ukrainian policies relevant for forest and woodland landscapes, and describe how the involved stakeholders implement policies top–down. Using the Skole district in the Carpathian Mountains as a case study, we then describe the status of SFM dimensions, and evaluate the implementation process bottom–up. Interviews and analyses of official statistics show that three types of gaps need to be bridged: (1) a policy creation gap between the local level situation and ecological, economic and socio–cultural needs at the national and regional levels; (2) a policy implementation gap between the official definition of SFM, and how its different criteria and objectives are understood by forest stakeholders; (3) a knowledge gap between the need of a holistic transdisciplinary approach for SFM implementation, and the present sectoral approach to governance of forest landscapes and disciplinary research. Ways of bridging these gaps are capacity building, introducing arenas for collaboration, and applying a zoning approach at multiple scales to satisfy economic, ecological and socio–cultural dimension of SFM.

M. Elbakidze (✉) · P. Angelstam
School for Forest Management, Swedish University of Agricultural Sciences, P. O. Box 43, SE 739 21 Skinnskatteberg, Sweden
e-mail: Marine.Elbakidze@slu.se

J. Kozak et al. (eds.), *The Carpathians: Integrating Nature and Society Towards Sustainability*, Environmental Science and Engineering,
DOI: 10.1007/978-3-642-12725-0_23, © Springer-Verlag Berlin Heidelberg 2013

1 Introduction

Sustainable forest management (SFM) is a concept in transition from a focus mainly on sustained yield of wood to production of multiple goods, services and values (MCPFE 1993, 1995, 2001, 2007; Higman et al. 1999; Kennedy et al. 2001). Since the early 1990s the international forest policy discourse recommends that economic, ecological and socio–cultural values should be taken into account (Kimmins 1997; Salim and Ullsten 1999; Lindenmayer and Franklin 2002; Nilsson 2005). In Europe this is clearly reflected in the Pan-European forest policy process, which has set a suite of criteria and indicators to define these values in more detail at the Pan-European level (MCPFE 2001, 2007; Rametsteiner and Mayer 2004). Similar developments have occurred at the national level (Krott et al. 2000; Angelstam 2003; Balashenko et al. 2005). To implement SFM, Boyle et al. (2001) suggested a triad of activities including developing governance as a process of providing a vision and direction for sustainability, management as the operationalization of the vision, and monitoring of indicators representing agreed values to provide feedback by synthesizing observations to narratives of how the situation has emerged and unfolds. The tools required for each step are manifold (Higman et al. 1999; Nilsson 2005; Szaro et al. 2005), and need to be adapted to local and regional conditions.

While there is consensus at the international policy level about applying the idea of three pillars of sustainability—ecological, economic and socio–cultural—to forest management, the different sets of values shows large variation in the current focus and trajectories of development among countries and regions (e.g., Angelstam et al. 2005). Solberg and Rykowski (2000) stressed the need to acknowledge the range of conditions in different countries and regions when providing policy recommendations. As forest goods, ecosystem services and values have become increasingly globalised commodities, there is a need both to understand the local and regional footprint of international demands, and the suite of policy instruments and management approaches which are appropriate under different biophysical, economic and socio–cultural conditions, and systems for governance (Angelstam 2003). Understanding the effects of the international forest policy discourse and the globalization of the forest sector on the one hand, and traditional local, regional and national factors on the other, requires research based on multiple case studies (Elbakidze et al. 2010).

In Europe, the Carpathian Mountains are a unique laboratory for studying forest management units located along an European gradient between the West and East employing a diversity of top–down and bottom–up approaches. Implementation of SFM on the ground in the Carpathian Mountains region requires combination of different sets of tools to:

1. protect "a unique natural treasure of great beauty and ecological value, an important reservoir of biodiversity, the headwaters of major rivers, an essential

habitat and refuge for many endangered species of plants and animals and Europe's largest area of virgin forests" (Anon 2003b),
2. maintain traditional village systems (Elbakidze and Angelstam 2007),
3. develop multiple economic use of forest resources for local, regional and national development (Anon 2003b).

The Carpathian Mountains go across eight countries extending from north-eastern Austria, via the Czech Republic, Slovakia, northern Hungary, southern Poland and south-west Ukraine, into Romania and Serbia (Webster et al. 2001; Turnock 2002; Opelz 2004). Of the Carpathian countries all except Austria are in transition from socialist planned to market economy. These complex economic and political changes provide an excellent "pseudo-experimental" opportunity for multiple case studies using countries (Mikusinski and Angelstam 1998), regions and local landscapes (Angelstam et al. 2004a, b; Elbakidze and Angelstam 2007) as replicates.

Ukraine is located in the geographical centre of Europe with both zonal and azonal forest ecoregions (Mayer 1984), and represents a globally relevant range of gaps to be bridged to implement SFM in actual landscapes. During the socialist period until 1991 intensive development of the Ukrainian industry had a negative impact on the environment in several important forest regions due to air and soil pollution (Szaro et al. 2005) and unsustainable use of groundwater supply (Buksha 2004; Zibtsev et al. 2004). Clearing of forests for agricultural development during historic time and more intensive forest management during the socialism period led to forest fragmentation, and later a large proportion of planted (45.6 %) forests.

The Ukrainian Carpathians cover 3.5 % of Ukraine's area and 10.3 % of total area of the Carpathian Mountains. This region has forest resources of high economic value, and has retained both cultural and natural biodiversity, and many of Europe's last wilderness areas (Turnock 2002; Angelstam 2006). The ecoregion is also home to several ethnographic groups of Ukrainians—Lemko, Boiko and Hutzul—who have been shaping mountain landscapes for centuries and have created a rich cultural heritage (Anon 1983; Hajda 1998; Elbakidze and Angelstam 2007). Nowadays people in many parts of the Ukrainian Carpathians have experienced decreased standards of living due to disintegration of the planned economy developed during socialism, and ongoing transition to market economy under acute political and economic crisis in the country. The picture is, however, very complex, especially as most of the Carpathian ecoregion has been part of Austria–Hungary, Poland and the Soviet Union during the past centuries. Legal and illegal extraction of wood and non wood forest products has become a vital source of income for many people in the Carpathian Mountains. Additionally forests and woodlands provide subsistence for livelihood of a large village population, which lives in close proximity of forests (Elbakidze and Angelstam 2007).

To promote sustainability on national as well as regional and local levels Ukraine has joined the process of developing SFM ideas and principles oriented towards sustainable yield forestry, maintenance of forest biodiversity and socio–cultural values (MCPFE 2003; Anon 2006). The strategic objectives of the national forest policy are related to those enumerated in international agreements of sustainable development, sustainable use and protection of European forests. Ukraine has also signed the 17 resolutions of the Ministerial Conferences on Protection of Forests in Europe.

The aim of this study is to provide an example of a case study approach including the formulation of policy, management and feed-back based on how the situation in a concrete forest management unit, unfolds over time using quantitative and qualitative data. We evaluate the policy implementation process related to the sustainability of landscapes dominated by forest and woodland by using the Skole district in Ukraine's Carpathian Mountain region as a study area. To understand how SFM is defined, implemented and understood in the chosen case study, we first review the national policies relevant for the development of rural forest and woodland landscapes. We then describe the institutions and policy instruments translating policies "top–down" to the management unit we chose as case study. The resulting state of different local level dimension of SFM is then summarized. Finally, we discuss the implementation process "bottom–up", or how policy messages are fed back to the policy level again.

2 Study Area

Our study area is the 147,100 ha Skole district, which is situated on the north-eastern side of Eastern Carpathian Mountains in the upper part of the Dniester river basin and its tributaries Stryi and Opir in Lviv region. We used the Skole district as a "landscape laboratory" to understand how SFM has been implemented to satisfy ecological, economic and socio–cultural dimensions taking into account cultural heritage and natural landscape legacies of the region. Extending from 200 to 1400 m a.s.l. the Skole district ranges from cleared broad-leaved forest with agricultural land, villages and remnants of oak (*Quercus* spp.) forest to managed spruce (*Picea abies*) forests, remnants of natural beech (*Fagus sylvatica*) and beech–fir (*Abies alba*) forest to high altitude natural spruce forest (Holubets et al. 1983; Hensiruk et al. 1998). The Skole district has five state forestry enterprises (SFE), which are responsible for forest management and conduct commercial activities in 78 % of the forest area. The National natural park "Skolivsky Beskydy", which was created in 1999, covers the remaining 22 % of the total forested area. There are 55 villages and 1 town in this district. Recreational and tourism activities are connected with forests.

3 Methods

Working with a complex concept such as SFM on a landscape level requires special emphasis on finding platforms for integration among ecological, economic and socio–cultural values. This applies both to the relevant disciplines and landscape's actors and to the desired integration from policy to practice, and back again (Clark 2002). The landscape concept is a means of achieving this because it can be used both in the sense of territory and the sense of place (Head 2000). The latter emphasizes the interconnectedness of natural, social and cultural, as well as of temporal and spatial processes, in the evolution of a particular landscape. The landscape concept also reflects the need to expand the spatial scale of management, i.e., to move from smaller spatial units or objects to the scale of landscapes and regions, i.e., include micro, meso and macro levels (Lindenmayer and Franklin 2002; Elbakidze et al. 2010). Additionally, the corresponding social organizational scales must be considered (Manfredo et al. 2004).

To study the process of implementing SFM policies one must view natural and socio–cultural components in a temporally and spatially expanded context. Thus, we consider that a forest landscape forms a whole entity, where natural and cultural components are intermingled, and cannot be viewed as separate entities or processes (e.g. Elbakidze and Angelstam 2007).

We analyzed national forest legislation to understand the compliance with the international SFM discourse. To understand the official institutional arrangement, expert interviews were done with all local stakeholders responsible for management of forests as well as with the heads of local and regional communities in the study area. We made semi-structured open-ended interviews with directors and the chief foresters of all five state forest enterprises in the Skole district and with the director of the Skolivsky Beskydy national nature park (total 11 interviews). Additionally, interviews with representatives of three local communities were made to understand how policy implementation was perceived locally. In total 14 interviews were taken in spring–summer 2006. The standard interview manual contained several groups of questions including personal data of the respondents, data about forest composition and structure, ownership patterns, company's attitudes to forest management and conservation, biodiversity status, changes in forestry, and logistics of forest practices. In 2007 we organized a round-table discussion concerning the ecological, economic and social–cultural dimensions of the current situation in the study area with 15 representatives of forest enterprises, forest business, and administrations of regional and local communities. Finally, analyses of published socio–economic statistic data for the Skole district (Anon 2004) and the forest inventory data from 2006 as well as forest management plans of the state forest enterprises were used to quantify the state and trend of ecological, economic and socio–cultural dimensions of SFM.

4 Results

4.1 The National Policy Level and Institutional Framework

Ukraine has joined the process of developing forest management ideas and principles along the lines of the global SFM discourse. The country has thus supported and signed many resolutions of the Pan-European Ministerial conferences on SFM.

Additionally, bilateral agreements about cooperation in sustainable forestry with neighboring countries (Austria, Poland, Russia, Slovakia) have been signed, as well as the Carpathian Convention. Today the principles of SFM are adopted into the national legislation and forest programs (Synyakevych 2004, 2005; Anon 2006). The official forest policy is thus to provide a balance between the conservation of forest ecosystems and the continuous, multi-purpose use of forests. In Ukraine legislative frameworks of forests and forest resource management are formulated in the Forest Code of Ukraine (Anon 2006), Law on the Environmental Protection of Ukraine (1991), the governmental program "Forests of Ukraine during 2010–2015" (Anon 2009) and other legislative documents and governmental regulations that play a fundamental role in developing environmentally sound forest operations.

The Forest Code (Anon 2006), stipulates that forests have primarily soil protective, water-conservation, air-cleaning and health-giving functions, while their economic use is considered as having limited importance. According to the political and legislative documents the main goals of forestry in Ukraine are (Hensiruk 1992; Buksha 2004; Zibtsev et al. 2004; Anon 2006, 2009):

- to conserve biological diversity in forests,
- to extend forest covered territory to an optimal level in all natural zones,
- to protect forest function and to limit forest exploitation,
- to improve social protection of forestry workers,
- to increase the resistance of forest eco-systems to negative environmental conditions,
- to improve forest management legislation according to international principles of SFM,
- to encourage the development of forest research and education.

All forests are divided into the following four categories: protective forests (to fulfill mainly water-and soil protection functions); recreational forests (to fulfill mainly recreational, sanitarian and health care functions); forests for nature conservation, scientific, historical and cultural purposes; forests for commercial use (Anon 2006). According to the Forest Code (Anon 2006), forests may be owned by the state, privately owned or owned by communities. Forests may also be leased out temporarily or permanently for different kinds of utilization. Permanent forest lease is allowed by state forestry enterprises and other organizations which have special departments to conduct forest management and provide special use of forests and forest resources for hunting, recreation, research and education. Parts of the State Forest Fund may be leased out for periods of 3–25 years to enterprises, organizations and private citizens both of Ukraine and other countries for multiple use of forests. A total of 68 % of the forested areas are under permanent holding of state forestry enterprises subordinated to the State Agency of Forest Resources, 23 % are managed by Ministry of Agriculture, and 9 % are managed by Ministry of Defense, Ministry of Transport, Ministry of Ecology and Natural Resources and other organizations (Zibtsev et al. 2004).

There is a division of the forests resources into resources of state importance (wood from final harvest and resin collection) and resources of local importance (all other products) (Anon 2006). All citizens have the right to walk in the forests, pick berries and mushrooms. Any other utilization is connected with a fee.

4.2 The Forest Policy Implementation Process

There are three main levels of forest policy implementation in Ukraine: national, regional and local. The Supreme Council is the central legislative body and defines the principles of state policy in the sphere of forest relations; passes laws regulating relations in this sphere; approves state programs related to the forests' health, protection, use and reproduction; and decides on other issues in the field of forest relations in accordance with the Constitution of Ukraine (Anon 2006).

The state body especially responsible for forestry in Ukraine is the State Agency of Forest Resources and its departments on regional and local levels. Driven by the transition towards market economy from 1991, the governance system in Ukrainian forestry has been restructured. Compared with the Soviet time forestry and the wood products industry were split into two distinct bodies in 1996 when the State Committee of Forestry replaced the Ministry of Forestry. The wood-processing sector was to a large extent privatized (Popkov et al. 2001).

The State Agency of Forest Resources ensures regeneration and improvement of the forest stock, to provide protection and conservation of forest as well as providing an organization for forest resources and their use (Anon 2006). The Supreme Council, the State Agency of Forest Resources and the Ministry of Ecology and Natural Resources are the main policy makers and the dominant actors in the Ukrainian forest sector (Anon 2006).

The Ministry of Ecology and Natural Resources is the main responsible governmental body to perform, coordinate and control all issues related to forests' health, protection, restoration and reproduction. It also participates in the development of national, state and regional (local) programs on conservation, protection, use and reproduction of forests; and approves the defined norms of forest resources' use; organizes an environmental assessment on the impact of industrial activities on forests (Anon 2006).

Each main actor at the national level has own representatives at the regional and local levels. For example, the State Agency of Forest Resources is represented by 25 regional forest management units. They oversee 230 local state forest enterprises, 14 state hunting and 50 forest hunting enterprises, seven strict natural reserves, four national nature parks and 16 wood industry, road-building, special forest protection, forest inventory and other enterprises (http://dklg.kmu.gov.ua, retrieved 2011-05-29). The practical implementation of the forest policy is carried out by state forest enterprises. Their functions include forest regeneration and management of the wood supply. The range of activities of the state forest enterprises differs among regions. The extent of these activities depends on the local availability of resources and production capacities (Nijnik and Oskam 2004; Anon 2006). Finally, the local governments have responsibilities to allocate forests for permanent or temporary use.

4.3 Skole District: Ecological, Economic and Socio–Cultural Dimensions

4.3.1 Legacies of the Historical Development

The dynamic history of the Carpathian Mountain region with a range of different governance and management paradigms has influenced forest landscapes in many dimensions. Understanding these legacies of the past is an important starting point for the implementation of SFM in this diverse region. Considerable deforestation of the Eastern Carpathians began about 7000 years ago when the mountains came under the influence of primitive agricultural activity (Kalynovych and Sytnyk 2003). In the pre-agricultural period, the Skole area was populated predominantly by Slavic tribes, which were engaged in hunting, fishing and gathering since the mid-Neolithic period (Portenko 1958). In the second half of the fifteenth century, Boikos, a tribe or ethnographic group of Ukrainian highlanders who inhabit both slopes of the middle Carpathians, began to settle in the Skole area. They introduced slash and burn farming. The use of fire led to a considerable decrease in coniferous species, such as Norway spruce and fir.

Local people maintained the traditional land use almost until the nineteenth century. The agricultural and forestry practices of that time were to a certain extent a prototype of an environmental-friendly and locally sustainable use of natural resources. The Boikos depended completely on the availability of local natural resources, and on the maintenance of an ecologically balanced environment with minimal use of imported resources, goods and energy (Anon 1983).

Historically, Poland and the Austro–Hungarian Empire on the one hand, and Russia on the other, divided today's Ukraine geographically into an eastern and a western multi-ethnic sphere. Continuous external political, economic and social influences have resulted in the decline of traditional forest and land use systems. The character and intensity of the use of natural resources in Skole area began to change in the nineteenth century. As a result of high demand for wood in West European countries, the forest industry began to develop. Forests were cut, and mostly exported as timber, which was transported by means of river (Hensiruk 1964; Trokhimchuk 1968). As a rule the wood harvested was not used efficiently. The areas cleared of forests were not reforested again. Only a small quantity of wood was processed at the same place where it was cut down (Hensiruk 1964).

Starting in 1874 the wood export situation improved as railroads begun to be built across the Carpathian Mountains. Large areas of beech forests were burned in order to produce potash, which was also exported. The Carpathian Mountains had thus become a source for various kinds of wood and wood products at the international market place. The demand for spruce wood on the world market and the rapid decrease of the supply prompted the owners of the forests to replace the deciduous forests with spruce. In 1882 this tendency was legalized by the Austrian government, which passed a resolution to replace beech, fir and other forests with Norway spruce forests of Austrian geographic origin (Hensiruk 1964). Only at the end of the nineteenth century, the first attempts were made to reduce forest exploitation and to restore forests. In 1894, the Austrian government passed a law regulating the use of forests and instituting responsibility to care for young trees (Hensiruk et al. 1998).

A complete change of political, social and economic relations in the country that had a profound influence on the ways in which natural resources were used was initiated in 1939 when the Western regions of Ukraine became part of the Soviet Union. The Soviet regime (1939–1991) had an especially disastrous impact on the local people's way of life and use of land. Private land property was expropriated, people were forced to emigrate, arable lands increased at the expense of wooded grasslands, and forestry became more intensive (Trokhimchuk 1968). The structure of land and forest properties was changed. Forests were now owned by state, private plots of land were joined into collective farms. Collectivization and mechanization left no space for the traditional way of life (Trokhimchuk 1968). The use of natural resources in Skole area during this period was shifted towards industrial use of forests with spruce reforestation, which was caused by the growing importance of forestry in the Carpathians in general, and within the Skole district in particular. This was accompanied by an increase in both harvesting and reforestation.

1.3.2 Ecological Dimensions

Changes in forest environments caused by long or intensive wood harvesting include loss of species (a compositional aspect); reduced amounts of dead wood, large trees, old and structurally diverse stands and intact areas (structural aspects); and altered processes (functional aspects) (Peterken 1996). The forests of Skole district have more than 200 years of forest management history. As a result of this long and intensive forest exploitation several elements of forest biodiversity have been altered.

Data from 2006 show that of the total Skole district area (147,100 ha) forests occupied 71 %, agricultural land 25 %, urban areas 2 %. According to Holubets and Odynak (1983), natural forests of the Beskyd area with increasing altitude were made up by beech, beech–fir, spruce–beech–fir, fir–spruce. Pure spruce forests at lower altitude were not found. Intensive forest exploitation led to reduction of beech and beech–spruce–fir forests in Skole area. Oak forests in the valleys were reduced to a minimum already in the seventeenth century. At present time, monocultural spruce plantations prevailed. The dominant species were Norway spruce (59 %), beech (30 %), and fir (5 %). Middle-aged and young

Fig. 1 The bark-beetle is one of the main "destroyers" of spruce forests in the Ukrainian Carpathians. The tracks of bark-beetle in the bark of Norway spruce (photo credit: Marine Elbakidze)

stands covered more than 70 %, and premature and mature stand a total of 28 % of forested area, most of which were in protected areas. According to unpublished forest enterprise data and interviews with state forest enterprise directors, between 30 and 60 % of the forested area under their management suffered from insects and root rot, which cause death to Norway spruce stands (Fig. 1). To conclude, human forest activity has created ecologically unsustainable forests, the economic potential of which is not fully utilized.

The protective and protected forests, in which final felling was prohibited, covered 61,700 ha, or 61 % of the forested area. This included forests within the national nature park "Skolivsky Beskydy" and also forests along riverbanks, shelterbelts along railways and roads, green belts of settlements, and forests around sub-alpine meadows. Inappropriate logging and road building techniques continued to be one of the greatest obstacles to SFM in the Skole district. Cutting streamside buffer zones, skidding across rivers and up riverbeds, point-source pollution, and the reliance on obsolete or inappropriate timber transport technology characterized the logging practices. Decreased site productivity, soil compaction, sheet and gully erosion, mass movement, sedimentation, decrease in water quality and fish habitat were manifestations of these poor logging practices (Bihun 2005). Habitat loss and fragmentation was thus evident for both terrestrial and aquatic ecosystems.

Nevertheless, specialized and area-demanding species representing natural forest landscapes occurred in the study area. Of special interest were black stork (*Ciconia nigra*), lesser spotted eagle (*Aquila pomarina*), capercaille (*Tetrao urogallus*), brown bear (*Ursus arctos*), wolf (*Canis lupus*), lynx (*Lynx lynx*), wild cat (*Felis silvestris*), badger (*Meles meles*), pine marten (*Martes martes*), otter (*Lutra lutra*), and European bison (*Bison bonasus*). There were also seven breeding species of owls and nine breeding species of woodpeckers (Anon 2003a, b). This indicates the biodiversity status was not as deteriorated as in many West European countries (Angelstam et al. 2004a, b). An important factor contributing to improved ecological sustainability in the area was the creation of the national nature park "Skolivsky Beskydy" in 1999. The park's territory covers 24.3 % of the total area of Skole district, including 7.5 % of the total district's area of strict and regulated nature protection management. The national park was a refuge for 29 animal species from the Ukrainian Red List, including 22 animal species from the red list of the Bern convention and five species from the European Red List (Anon 2003a, b).

Natural reforestation of abandoned agricultural land in the valley bottoms was a widespread present phenomenon in Skole district. Marginal lands of former collective farms, which were not used any more for grazing and crop production are been covered by forests due to natural succession dynamic. The effects of natural reforestation on biodiversity were diverse and complex including both loss of important habitat types in the cultural landscape and initiation of secondary succession of use for other species (Mikusinski et al. 2003).

4.3.3 Economic Dimensions

The number of inhabitants in the Skole study area was 48,900, including a rural population of 35,800 (73.2 %), and with 26,500 people of working age. Since 1989 the local population had decreased by 2,600 people. The average population density was 33 person/km^2 (Anon 2004). The number of employed people in the district was 17,600 (66 % from total people in workable age and 49 % from total population), including 6,043 employees of state enterprises (22.8 % of the total people in workable age or 12.4 % of the total population) (Anon 2004). The main individual employers in the Skole area were educational foundations (6.5 % employed people from total number of workable age), forestry sector (3.6 %) and health service (3.2 %).

The forest in the Skole district belonged to the state. The State Agency of Forest Resources controlled 63.3 % of the study area's forests. The former collective farm forests (26.3 %) were under the jurisdiction of the Ministry of Agriculture. The Ministry of Defense and other ministries managed the remaining forest area. There were five state forest enterprises, one national natural park "Skolivsky Beskydy" (SBNNP) and five state hunting enterprises. The total area under the management of the state forest enterprises was 80,912 ha (or 80 % of forested area). The per capita non-state (private and commons) land distribution in the Skole district is 2.99 ha/person, of which 0.26 ha/person is arable land.

According to the 2006 forest inventory data, the total growing forest stock was 29.98 million m^3, including 20.41 million m^3 of conifers (68.1 % of total growing stock), 6.14 million m^3 of mature and over-mature stands (20.4 % of total growing stock), including 3.90 million m^3 of conifers. The average stock per hectare of forest land was 298.3 m^3, in mature and over mature stands 431.3 m^3. The annual wood increment in the area was 4.0–5.3 m^3 per hectare (Fig. 2). The

Fig. 2 The productive beech forests (*Fagus sylvatica*) remain in the Ukrainian Carpathians. The height of beech trees can exceed 45 m (photo credit: Marine Elbakidze)

area of forest available for final harvesting (from the second group) amounted to around 33,000 ha with additional 35,410 ha (from the first group) where final harvesting was allowed. On average 1.5 m^3/ha of timber was harvested from the final fellings and 1.7 m^3/ha from intermediate fellings, which corresponded in average 30 % of annual wood increment.

According to the interviews with directors of the state forest enterprises in Skole district, a total of 190,134 m^3 of timber was harvested in 2003, including 85,797 m^3 (45.1 %) from final harvest operations. The volumes of harvested wood have been increasing since 1998 (141,334 m^3) mainly due to increasing amount of timber from intermediate harvest operations; from 57,008 m^3 in 1998 to 104,337 m^3 in 2003. The total clearcut area in 2003 was 1,815 ha and forest regeneration was made on 587 ha. A low level of investments in the forestry sector was the main reason for a low level of regeneration activities after harvesting during the last years.

About 65–70 % of wood harvested in the district was exported as round wood, thus adding only limited value to local community economic development. According to interviews with directors of the state forest enterprises in the Skole district they paid stumpage fee as permanent forest users to the central budget.

The Skole district was actively used for recreational purposes. There were 12 resorts and tourist national level centers, three motels, and 75 small regional level recreational centers. The recreational and tourism activities were connected with forests, and depend ultimately on the stability of forest ecosystems and quality of forest resources. Forests used for recreational activities were mostly under the management of state forest enterprises, which had to invest money for development recreational facilities. The risk for conflicts between forest logging operations and different recreational uses of forests was obvious. The national natural park "Skolivsky Beskydy" could be a main tourist destination, but due to the lack of funds the park administration was unable to efficiently develop infrastructure for tourism.

The ongoing economic transition has caused a decrease in monetary incomes caused by inflation and inefficient economy, unsatisfactory social protection, and

Fig. 3 Forests for building construction and as fuel wood has been important for local people in the Ukrainian Carpathians for centuries (photo credit: Marine Elbakidze)

high level of unemployment. The forest sector in Skole district played therefore an important direct role in the livelihood of local people. The economical crisis during the transition period had made local people's physical survival directly depended on the local use of natural resources. The role of forests as a source of fuel wood and non-wood products for self-subsistence food production has been increased (Elbakidze and Angelstam 2007) (Fig. 3), as well as the illegal exploitation of forest resources (illegal cutting, poaching etc.).

4.3.4 Socio–Cultural Dimensions

Local people have kept their traditional land use practices, which play an important role for the maintenance of cultural landscape biodiversity and rural development. Non-wood forest products (NWFP) such as mushrooms, berries, honey, medicinal herbs, floral greenery, birch sap, resin and wild game are part of the social fabric and livelihood of Ukrainian culture (Bihun 2005; Elbakidze and Angelstam 2007), especially in forest-dependent communities, like the Skole district. The conflict between forest industry and vital interests of local people was due to increase of harvested timber and the conflicting sustainable production of NWFP.

The Skole district has a rich history. The restoration and protection of historic sites of regional and national value have been increasing since 1991, and the Skole district has been recognized as an integral part of Boyko's ethnographic area in the Carpathians (Pavliuk et al. 1996). The support of traditional Boyko's land use, which is closely connected to forests, should be a milestone in a regional program of SFM (Fig. 4).

The privatization of arable and forested land (with restrictions) by local people that began after the collapse of the socialist system, has increased the social and cultural value of forests, which are becoming family's heritage for generations.

Fig. 4 The traditional village system found in the Ukrainian Carpathians is characterized by a centre-periphery zoning from houses, gardens, fields, mowed and grazed grasslands to forests (i.e., the ancient system with domus, hortus, ager, saltus and silva). Village Volosyanka in the Skole district of Ukraine's west Carpathian Mountains illustrates this situation (photo credit: Marine Elbakidze)

This process was of exceptional significance for people in the Western part of Ukraine where the old generation still has feeling of ownership, and memories about political and social events.

5 Discussion

5.1 Evaluation of Obstacles and Gaps

Our analysis of the Skole district case study concurs with previous studies (e.g., Krott et al. 2000), indicating that there are many obstacles in the process of implementing SFM from policy to landscape. This has created different kinds of gaps between the aims of policies and results on the ground.

A first set of obstacles is associated with policy creation and related to the current transition from command and control to market economy approaches in forestry (Raiser 1997; Kallas 2000; Levintanous 2002; World Bank 2002), and other changing values (Mayers and Bass 1999; Kennedy et al. 2001). Altogether this has led to a wide range of challenges in countries in transition (Krott et al. 2000; Pugachevsky et al. 2005).

The primary condition for successful implementation of forest policy and legislation is a functional collaboration among different forest stakeholders representing different societal sectors at different levels (Elbakidze et al. 2010). However, no mechanism for stakeholders' influence on the forest policy have yet been developed. According to the Forest Code (Anon 2006), citizens, their organizations, committee of self-governance have a right to discuss and participate in decision processes concerning use, protection and restoration of forests. However, in the Skole district there were no non-governmental organizations or informal institutions that could realize this right.

These obstacles create a policy creation gap between ecological objectives of forest policy on national level on the one hand, and social–economic needs on regional and local levels on the other. In a forest-dependent district, such as Skole, forests have to satisfy regional, ecological, economic and socio–cultural considerations. Officially, only 3.6 % of total population in Skole area was employed by forest enterprises. However, a much larger proportion of the population was directly and indirectly dependent on the access to fuel wood, building material and non-wood products. Additionally, illegal cutting was a problem. There was no official data about the amount of illegally harvested wood, but it was unofficially estimated that it amounts to 30 % of the officially harvested amount of timber.

A second set of obstacles relate to the policy implementation process. The most important one is the top–down system of policy implementation process. There are overlapping and unclear legal and institutional arrangements between governmental institutions with respect to forest policies (Solberg and Rykowski 2000). As a rule, because authority is assigned from the top to many stakeholders, the functions and responsibilities among them are overlapping and contradictory, which gives a space for unprofessional decisions and corruption (Nijnik and Oskam 2004).

Interviews with local stakeholders in the Skole district, and analyses of statistics allow us to make the following two conclusions. First, the state forest enterprises focus their activity mainly on economic use of forest resources, because they do not get financial support from the government for ecological and socio–cultural activities. Secondly, while the forest legislation and forest programs promote principles of SFM, the local level forest managers do not understand the SFM concept, why it is needed and why it should be used.

The forest enterprises are subject to a range of controlling and demanding organizations, and have few rights and many responsibilities. This system of relations is typical for an administrative system of governance. The income generated by local forest enterprises partly flows to the state budget, which is separated from them by time, space and institutions. Regional industries of manufactured wood products, which give working places for local people and generate income for regional economy, are still in an embryonic state. During Soviet time, the Skole district was a manufacturing centre for wood products drawing wood not only from Ukraine and Russia, but even from Brazil and India. When the forest product industry and forest management were segregated during the reformation of forestry sector in 1995, the manufacturing industry was privatized and then went bankrupt.

To develop SFM the forest enterprises have to introduce new rules. Data from the Skole district shows that the volumes of harvested wood have been growing mainly due to an increase in the amount of intermediate felling. There are different reasons for this. One is poor health condition of forests with large volumes of trees dying from bark beetle (*Ips* spp.) infestation, which is followed by sanitary cuttings. Another reason is a desire to get more income from intermediate felling avoiding payments for commercial wood. At the same time "sanitation cutting" is a carte blanche for clear-cutting practices that implies unregulated cutting under the guise of forest protection or silvicultural "smoke screen" allowing free reign to forest managers to cut at will (Bihun 2005).

The most economically viable enterprises, for example, the Skole forest enterprise "Skolivskyy derzlishosp" has refused to use budget financing and decided to cover all forest management expenses (plantation, tending, protection against illegal cutting, pests) from wood sale incomes. This is an evidence that forest enterprises could be profitable at least in the most forested areas, in spite of a range of obstacles.

A third set of obstacles for SFM implementation is related to the dominating theoretical and disciplinary scientific approach in forestry. Traditionally, Ukrainian silvicultural science and education were highly advanced. The currently poor economic performance in forestry is largely the consequence of a lack of proper communication, cooperation and reciprocity (Nijnik and Oskam 2004). For example, to improve ecological functions of forest ecosystems in Skole district, which was deteriorated during the course of history, it is necessary to apply contemporary ecological knowledge. However, collaboration between foresters and scientists is ineffective. As a result there is a knowledge gap between needs of interdisciplinary knowledge and holistic approach for SFM implementation, and what is currently applied.

5.2 Bridging the Gaps

Three types of gaps need to be bridged:

1. A policy creation gap between the local level situation and ecological, economic and socio–cultural needs at the national and regional levels.
2. A policy implementation gap between the official definition of SFM, and how its different criteria and objectives are understood by forest actors.
3. A knowledge gap between the need of a holistic transdisciplinary approach for SFM implementation and the present sectoral approach to governance and management of forest landscapes and disciplinary research aimed at supporting implementation of SFM.

While forestry in the traditional sense has a clear positive impact on maintaining forest resources, the impact on ecological and socio–cultural aspects in a local landscape are highly dependent on the economic status and history, and the systems for government and governance. In an increasingly complex and changing world there is a need for initiating relevant innovative research and development to disseminate existing and develop new tools in a toolbox for implementation, and as an interface between practice and policy.

To implement SFM policy in the Carpathian Mountains national forest programs should follow a broad inter-sectoral approach, including the formulation of policies, strategies, and plans of actions as well as their implementation, monitoring, and evaluation. The programs should be implemented in the context of the socio–economic, cultural, political, and environmental situation and be integrated with wider programs for sustainable land-use and with the activities of other sectors (Nilsson 2002).

As a country in transition, it is important to evaluate the heritage in forestry in Ukraine from the previous political systems to understand what should be changed or remain under the new political and economic conditions. The debate concerning the "socialist heritage" in Ukrainian forest management shows that it should be critically analyzed based on empirical studies for the future development of forestry. For example, according to Polyakov and Sydor (2006), the Ukrainian forestry during the Soviet time (especially in the second half of the twentieth century) "could be judged as a sustainable". They concluded that "Ukrainian forest management under a socialist centrally planned economy did a good job in providing environmental benefits from the forests to the citizens, as well as in preserving and multiplying forest resources". Some features of the Ukrainian forestry like "longstanding sound plantation policies and sound methods" which were implemented under socialism rule, "constitute positive heritage and need to be maintained in order to succeed in the transition to a market economy". Nijnik and Van Kooten (2006) presented an opposite view. They argued that under the command-control economy "the forest resources were excessively exploited and that inadequate attention was paid to silvicultural investments, despite official rhetoric to the contrary". However, in none of these studies neither was the large regional variation in Ukraine considered (Synyakevych and Soloviy 2002), nor was data describing local level indicators for different dimensions of SFM presented for different regions. To resolve this debate we argue that empirical research be made at multiple levels tracking the policy cycle from policy-making to actual forest landscapes and back again (Nilsson 2005). This approach should then be applied in a suite of regions representing different phases in the development of SFM, as well as different economic histories and governance legacies.

Focusing on the local level in the Skole district, this study indicates that there are poor working connections between managers from different sectors: forest management units, a national park, recreational zones and local villages. However, there are no real contacts between representatives for developing a common vision on local and regional development. Thus, even if forest programs are sufficient in a narrow sectoral context, they are not in a broader landscape context.

There are several ways of bridging these gaps. First, introduction of arenas for good governance which could provides a forum for involvement of a variety of stakeholders ranging from the land managers, the general public, and policy makers. Second, dividing land into different zones could help fulfilling economic, ecological and socio–cultural dimension of SFM. In Ukraine a zoning approach for forest development on regional and local level is implemented by dividing forests into four categories. However, in Eastern Europe, including Ukraine, foresters and geographers generally developed zoning concepts, and socio–cultural issues were not considered. Because there are different kinds of forests with different dynamics (Angelstam and Kuuluvainen 2004) and socio–cultural values, different kinds of zoning (Innes and Nitschke 2005) and subsequent management approaches are needed to maintain all kind of forest values (Fries et al. 1997). Third, to implement concepts for integrated natural resource management concepts such as Model Forest could be employed (Besseau et al. 2002). This represents a way of establishing a societal arena for a partnership among individuals and organizations sharing the common vision of SFM.

Acknowledgments This paper was initially developed as an outcome of the COST action E25 (European network for long-term forest ecosystem and landscape research). We are grateful for the support provided through Folke Andersson to take part in this network. Financial support was provided to Per Angelstam from "Stiftelsen Marcus och Amalia Wallenbergs Minnesfond". Robert Axelsson, Marius Lazdinis, Maria Nijnik and Camilla Sandström provided valuable comments on the manuscript. Special thanks to Maksym Polyakov and Tim Sidor for their constructive comments and edits of the manuscript.

References

Angelstam P, Mikusinski G, Rönnbäck B-I, Östman A, Lazdinis M, Roberge J-M, Arnberg W, Olsson J (2003) Two-dimensional gap analysis: a tool for efficient conservation planning and biodiversity policy implementation. Ambio 33(8):527–534

Angelstam P, Dönz-Breuss M, Roberge J-M (eds) (2004a) Targets and tools for the maintenance of forest biodiversity. Ecol Bull 51:427–454

Angelstam P, Kuuluvainen T (2004) Boreal forest disturbance regimes, successional dynamics and landscape structures—a European perspective. Ecol Bull 51:117–136

Angelstam P, Roberge J-M, Lõhmus A, Bergmanis M, Brazaitis G, Dönz-Breuss M, Edenius L, Kosinski Z, Kurlavicius P, Lārmanis V, Lūkins M, Mikusinski G, Račinskis E, Strazds M, Tryjanowski P (2004b) Habitat modelling as a tool for landscape-scale conservation—a review of parameters for focal forest birds. Ecol Bull 51:427–453

Angelstam P, Kapylova E, Korn H, Lazdinis M, Sayer JA, Teplyakov V, Törnblom J (2005) Changing forest values in Europe. In: Sayer JA, Maginnis S (eds) Forests in landscapes. Ecosystem approaches to sustainability. Earthscan, London

Angelstam P (2006) Maintaining cultural and natural biodiversity in Europe's economic centre and periphery. In: Agnoletti M (ed) The conservation of cultural landscapes. CAB International, Wallingford

Anon (1983) Boikivshchyna, istoryko-etnografichne doslidzhennia. Naukova dumka, Kiev

Anon (2003a) Litopys pryrody natsyonalnogo pryrodnogo parku Skolivsky Beskydy. Unpublished manuscript

Anon (2003b) Carpathian convention: framework convention on the protection and sustainable use of the Carpathians. http://mail.menr.gov.ua/publ/bernconven/docum/index.html. Accessed 20 Oct 2010

Anon (2004) Statystichnyy dovidnyk po L'vivs'kiy oblasti. Naukova dumka, Lviv

Anon (2006) Lisovyi kodeks Ykrainy. http://zakon2.rada.gov.ua/laws/show/3852-12. Accessed 10 May 2010

Anon (2009) Programa Lisy Ykrainy 2010–2015. http://www.lesovod.org.ua/node/4663. Accessed 30 Sept 2012

Balashenko S, Laevskaya E, Makarova T, Lizgaro V, Shcherbina A, Grigoriev E, Tarasenko V (2005) Review of Dnipro basin biodiversity legislation ensuring public participation and support. Water Qual Res J Can 6:68–82

Besseau P, Dansou K, Johnson F (2002) The international model forest network (IMFN): elements of success. For Chron 78:648–654

Bihun Yu (2005) Principles of SFM in the framework of regional economic development. Vistnyk Lvivs'kogo Universytetu Ser Geogr 32:19–32
Boyle M, Kay J, Pond B (2001) Monitoring in support of policy: an adaptive ecosystem approach. In: Munn T (ed) Encyclopedia of global environmental change. Wiley, London
Buksha I (2004) Forestry sector of Ukraine in transition to market economy. Ukrainian Research Institute of Forestry and Forest Melioration. http://www.unece.org/trade/timber. Accessed 10 March 2011
Clark TW (2002) The policy process. A practical guide for natural resource professionals. Yale University Press, New Haven
Elbakidze M, Angelstam P (2007) Implementing sustainable forest management in Ukraine's Carpathian Mountains: the role of traditional village systems. For Ecol Manag 249:28–38
Elbakidze M, Angelstam P, Sandström C, Axelsson R (2010) Multi-stakeholder collaboration in Russian and Swedish Model Forest initiatives: adaptive governance towards sustainable forest management? Ecol Soc 15(2):14
Fries C, Johansson O, Pettersson B, Simonsson P (1997) Silvicultural models to maintain and restore natural stand structures in Swedish boreal forests. For Ecol Manag 94:89–103
Hajda Y (1998) Turkivshchyna: pryroda i liudy. Ranok, Uzhhorod
Head L (2000) Cultural landscapes and environmental change. Arnold, London/Oxford University Press, New York
Hensiruk S (1964) Lisy Ukrajinskykh Karpat ta jikh vykorystannia. Naukova dumka, Kyiv
Hensiruk S (1992) Lisy. Naukova dumka, Kyiv
Hensiruk S, Nijnik M, Kopiy L (1998) Lisy Zahidnogo regionu Ukrayny. Naukove tovarustvo Shevchenka, Lviv
Higman S, Bass S, Judd N, Mayers J, Nussbaum R (1999) The sustainable forestry handbook. Earthscan Publications Ltd, London
Holubetst M et al (1983) Biogeonotstycheskyy pokrov Beskyd i ego dimanicheskie tendentsii. Naukova dumka, Kiev
Innes JL, Nitschke CR (2005) The application of forest zoning as an alternative to multiple-use forestry. In: Innes JL, Hoen HF, Hickey G (eds) Forestry and environmental change: socioeconomic and political dimensions. Report No. 5 of the IUFRO task force on environmental change. CABI Publishing, Oxfordshire
Kallas A (2000) The Estonia forest sector in transition: institutions at work. Interim Report IR-00-073. International Institute for Applied Systems Analysis, Laxenburg
Kalynovych N, Sytnyk O (2003) Istorija zaselennia Ukrajinskykh Karpat. In: Ekolohichni problemy Karpatskoho regionu. Pr Nauk Tov Shevchenka 9:44–53
Kennedy JJ, Ward TJ, Glueck P (2001) Evolving forestry and rural development: beliefs at mid-point and close of the twentiethth century. For Policy Econ 3:81–95
Kimmins H (1997) Forest ecology. A foundation for sustainable management, 2nd edn. Prentice Hall, Upper Saddle River
Krott M, Tikkanen I, Petrov A, Tunytsya Y, Zheliba Y, Sasse V, Rykowina I, Tunytsya T (2000) Policies for sustainable forestry in Belarus, Russia and Ukraine. Koninlijke Brill NV, Leiden
Levintanous A (2002) Russia—forest policy development and related institutional changes in the transition period. In: Forests and forestry in Central and Eastern European countries. The transition process and challenges ahead. Ministerial Conference for the Protection of Forests in Europe (MCPFE), Vienna
Lindenmayer DB, Franklin JF (2002) Conserving forest biodiversity. A comprehensive multiscaled approach. Island Press, Washington
Manfredo MJ et al (eds) (2004) Society and natural resources. A summary of knowledge. Modern Litho, Jefferson
Mayer H (1984) Wälder Europas. Gustav Fischer Verlag, Stuttgart
Mayers J, Bass S (1999) Policy that works for forests and people. Series overview. International Institute for Environment and Development, London
MCPFE (2003) Documents adopted at the fourth ministerial conference on the protection of forests in Europe, Vienna, 28–30 April 2003

Mikusiński G, Angelstam P (1998) Economic geography, forest distribution and woodpecker diversity in central Europe. Conserv Biol 12:200–208

Mikusinski G, Angelstam P, Sporrong U (2003) Distribution of deciduous stands in villages located in coniferous forest landscapes in Sweden. Ambio 33(8):520–526

MCPFE (1993) Resolution H1General guidelines for the sustainable management of forests in Europe. In: Second ministerial conference on the protection of forests in Europe. Helsinki, Finland, 16–17 June 1993. http://www.mcpfe.org. Accessed 22 Feb 2010

MCPFE (1995) Pan-European criteria and indicators for sustainable forest management. Annex 1 of the Resolution 1, Lisbon. http://www.mcpfe.org. Accessed 22 Feb 2010

MCPFE (2001) Criteria and indicators for sustainable forest management of the MCPFE. International expert meeting on monitoring, assessment and reporting on the progress towards Sustainable Forest Management. Yokohama, Japan, 5–8 November 2001. http://www.mcpfe.org. Accessed 22 Feb 2010

MCPFE (2007) Fifth ministerial conference on the protection of forests in Europe. In: Conference proceedings. Warsaw, Poland, 5–7 November 2007. http://www.mcpfe.org. Accessed 22 Feb 2010

Nijnik M, Oskam A (2004) Governance in Ukrainian forestry: trends, impacts and remedies. Agric Res Gov Ecol 3:116–133

Nijnik M, van Kooten CG (2006) Forestry in the Ukraine: the road ahead? Reply. For Policy Econ 8(1):6–9

Nilsson S (2002) Scenarios and forest policy responses. In: IUFRO symposium "Forest research—challenges and concepts in a changing world", Occasional Paper, 16

Nilsson S (2005) Experiences of policy reforms of the forest sector in transition and other countries. For Policy Econ 7:831–847

Opelz M (2004) Towards a Carpathian Network of protected areas. Final report. Bundesministerium für Umwelt, Naturschutz und Reaktorsicherheit, Berlin

Pavliuk S (ed) (1996) Skolivshchina. Institute of Regional Studies, Lviv

Peterken G (1996) Natural woodland: ecology and conservation in northern temperate regions. Cambridge University Press, Cambridge

Polyakov M, Sydor T (2006) Forestry in the Ukraine: the road ahead? Comment. For Policy Econ 8(1):1–5

Popkov M, Polyakova L, Storozuk V, Kyryluk S (2001) Osnovnye resultaty i tendentsii raszitiya lesnogo sectora Ukrainy v 1991–2000. Ukrainian–Sweden Project, Kiev

Portenko K (1958) Pozvonochnye Karpat. Nauchnaya mysl, Moscow

Pugachevsky A, Tkach V, Smirnov S (2005) Analysis of forestry use and management practices in the context of landscape and biodiversity protection in the Dnipro Basin. Water Qual Res J Can 6:54–67

Raiser M (1997) Informal institutions, social capital and economic transition: reflections on a neglected dimension. Working Paper 25, European Bank for Reconstruction and Development, London

Rametsteiner E, Mayer P (2004) SFM and Pan-European forest policy. Ecol Bull 51:51–57

Salim E, Ullsten O (1999) Our forests, our future. Report of the world commission on forests and sustainable development. Cambridge University Press, Cambridge

Solberg B, Rykowski K (2000) Institutional and legal framework for forest policies in the ECA region and selected OECD countries—a comparative analysis. Forest policy review and strategy development: analytical studies/issues paper. The World Bank, Washington

Synyakevych I, Soloviy I (2002) Modern forest policy problems in changing society. In: Scientific papers of agricultural university of Poznan. Forestry 5:87–95

Synyakevych I (2004) Ecologichna i lisova polityka. ZUKTS, Lviv

Synyakevych I (2005) Lisova polityka. ZUKTS, Lviv

Szaro R, Angelstam P, Sheil D (2005) Information needs for ecosystem forestry. In: Sayer JA, Maginnis S (eds) Forests in landscapes. Ecosystem approaches to sustainability. Earthscan, London

Trokhimchuk S (1968) Zmina landshaftiv Stryjsko-Sanskoji Verkhovyny v Ukrajinskykh Karpatakh za istorychnyj chas. Rukopys dysertatsiji, Lviv

Turnock D (2002) Ecoregion-based conservation in the Carpathians and the land-use implications. Land Use Policy 19:47–63

Uhnovskyy I (2003) Lisove gospodarstvo v Ukrayni: strategia rozvytku. Mizsvidomcha rada z pytan' rozbytku prodyktyvnyh syl, Kiev

Webster R, Holt S, Avis C (eds) (2001) The status of the Carpathians. A report developed as a part of the Carpathian Ecoregion Initiative. WWF, Vienna

World Bank (2002) Transition. The first ten years. Analysis and lessons for Eastern Europe and the Former Soviet Union. The World Bank, Washington

Zibtsev S, Kaletnik M, Savuschik M (2004) Forests and forestry of Ukraine in the transition period. In: Proceedings of the FAO/Austria expert meeting on environmentally sound forest operations for country. http://dklg.kmu.gov.ua. Accessed 29 May 2011

Sustainable Forest Management Alternatives for the Carpathian Mountains with a Focus on Ukraine

William S. Keeton, Per K. Angelstam, Yurij Bihun, Mykola Chernyavskyy, Sarah M. Crow, Anatoliy Deyneka, Marine Elbakidze, Joshua Farley, Volodymyr Kovalyshyn, Ivan Kruhlov, Bohdan Mahura, Stepan Myklush, Jared S. Nunery, Ihor Soloviy and Lyudmyla Zahvoyska

W. S. Keeton (✉) · S. M. Crow · J. S. Nunery
Rubenstein School of Environment and Natural Resources, University of Vermont, Burlington, VT 05405, USA
e-mail: william.keeton@uvm.edu

P. K. Angelstam · M. Elbakidze
Swedish University of Agricultural Sciences, 43, SE 739 21 Skinnskatteberg, Sweden

Y. Bihun
Shelterwood Systems, Jericho, VT 05465, USA

M. Chernyavskyy
Department of Ecology, Ukrainian National Forestry University, 103 General Chuprynka St, Lviv 79057, Ukraine

A. Deyneka
Lviv Regional Forestry Administration, State Forestry Committee of Ukraine, 8b Yavornytsoho St, Lviv 79054, Ukraine

I. Kruhlov
Faculty of Geography, Ivan Franko National University, 41 Doroshenko St 79020 Lviv, Ukraine

J. Farley
Community Development and Applied Economics, University of Vermont, Burlington, VT 05405, USA

I. Soloviy · L. Zahvoyska
Institute of Ecological Economics, Ukrainian National Forestry University, 103 General Chuprynka St, Lviv 79057, Ukraine

V. Kovalyshyn
Department of Forest Economics, Ukrainian National Forestry University, 103 General Chuprynka St, Lviv 79057, Ukraine

B. Mahura
Department of Forest Engineering, Ukrainian National Forestry University, 103 General Chuprynka St, Lviv 79057, Ukraine

S. Myklush
Department of Forest Inventory and Management, Ukrainian National Forestry University, 103 General Chuprynka St, Lviv 79057, Ukraine

J. Kozak et al. (eds.), *The Carpathians: Integrating Nature and Society Towards Sustainability*, Environmental Science and Engineering, DOI: 10.1007/978-3-642-12725-0_24,

Abstract Sustainable forest management (SFM) has been challenging in the Carpathian Mountain region of Europe. We explore emerging models and innovative practices that offer guidance on implementing SFM, based on recommendations developed through a scientific atelier held in western Ukraine. Information was gathered through technical presentations, site visits, unstructured interviews with stakeholders, and literature review. The contribution of SFM to biodiversity conservation depends on the establishment of fully representative and sufficiently extensive reserve systems. On managed forestlands, providing a better balance of stand ages and recently developed silvicultural practices, such as "close to-nature" and disturbance-based forestry, will help maintain ecosystem functions while providing a range of economic uses. Restoration of native species composition in areas dominated by spruce plantations will both enhance forest health and promote biodiversity conservation. Broader use of contemporary watershed management approaches is recommended, including delineation of riparian buffers, riparian forest restoration, ecologically informed design of transportation infrastructure, and investment in modernized harvesting machinery. Expanding forest sector participation in forest certification and carbon markets offer new opportunities and challenges. Certification of forestlands is expanding but has been limited by nonconformities. Ukrainian afforestation goals have the potential to sequester large quantities of carbon and generate substantial economic benefits as international carbon markets develop. The relatively long rotations currently required under Ukrainian forest code offer significant carbon storage benefits, as would conservation of high biomass, old-growth Carpathian beech and spruce–fir forests. A variety of stresses are predicted to increase with climate change, requiring adaptive responses. The challenge facing Ukraine and other Carpathian nations is to merge these ideas into a holistic, landscape approach adapted to the context of transitional, post-socialist economies.

1 Introduction

In the Carpathian Mountain region of Europe translating principles of sustainable forest management (SFM) into meaningful change and management strategies has proven challenging in transitional, post-socialist economies. Yet this context also provides a window of opportunity for experimentation, innovation, and adaptation. These are needed to transform general principles and theory into practical guidance for multi-functional forestry that is administratively and operationally feasible.

This review explores emerging models and innovative practices in forestry and economics that offer guidance on implementing SFM criteria in the Carpathians, with a focus on western Ukraine. Western Ukraine is representative of many of the complex forest management issues at play throughout the broader region. We discuss three (of several, Table 1) international SFM principles: (1) conservation of biological diversity, (2) maintenance of water resources, and (3) contribution to

Table 1 SFM criteria under the Montreal Process (12 non-European temperate and boreal forested nations) and Helsinki Initiative (38 European temperate and boreal forested nations, including Russia). These have been standardized using terminology shared by the two systems. Note that while the Helsinki process does not explicitly address legal systems at the criterion level, the importance of legal frameworks for SFM is manifest in a number of specific indicators

Criteria	Montreal process (1994)	Helsinki process (1994)
Conservation of biological diversity	Yes	Yes
Maintenance of soil and water resources	Yes	Yes
Contribution to global carbon cycles	Yes	Yes
Maintenance of ecosystem health	Yes	Yes
Maintenance of ecosystem productivity (wood and non-wood)	Yes	Yes
Provision of multiple, long-term socio-economic benefits	Yes	Yes
Legislative, institutional, and economic frameworks	Yes	No[a]

[a] In the Helsinki process, there are indicators for this theme associated with the other six criteria

global carbon cycles. Our recommendations are based on a synthesis of information and literature collected in a scientific atelier (or problem solving workshop) on integrating ecological economics and SFM in the Carpathian ecoregion, held in Lviv, Ukraine in the fall of 2007. The atelier brought together over 100 participants from Carpathian nations, elsewhere in Europe, and the U S, including scientists, economists, resource managers, and stakeholders from academia, governmental agencies, funding institutions, local communities, and non-governmental organizations. An initial set of challenges and opportunities for SFM were identified through technical presentations, brainstorming sessions, and literature review, conducted in preparation for and at the outset of the atelier. These were explored further through field visits to forestry operations and protected areas, and on-site meetings with forest managers and local stakeholders in the Ukrainian Carpathians. Qualitative data were collected through unstructured interviews with key informants and additional data were collected through programmatic document review. We integrated the data using triangulation (Flick 2009), allowing identification of cross-cutting themes, such as opportunities for SFM expressed by multiple experts and stakeholders (Table 2). The atelier method and process are described in Farley et al. (2009).

About 10 % of the Carpathian range lie in western Ukraine and contain 16.7 % of the nation's total forest area. Two million ha or 56 % of the Ukrainian Carpathians are forested, representing a resource of high economic, biological, and cultural value. In Ukraine there is a long history of well-trained, professional forest management within the State Forestry Enterprises (Nordberg 2007). However, forest management remains highly regimented, with most policy emanating from centralized planning at the ministerial level (Soloviy and Cubbage 2007). This contrasts with Romania, for instance, where forest administration has decentralized but involved substantial post-socialist land restitution and privatization

Table 2 Recommendations for SFM in the Carpathian Mountain region developed by participants in the Atelier on ecological economics and sustainable forest management, held in Lviv, Ukraine in September 2007

Topic	Recommendation
Watershed management	Implement watershed restoration programs, including riparian forest restoration on main floodplains
	Relocate logging roads away from streams and rivers
Forest health	Expand restoration of genetically non-endemic *Picea abies* plantations to native forest composition
	Utilize Carpathian Convention to reduce deposition of airborne pollutants
Ecological forestry	Expand experimentation with and use of ecologically-based silvicultural systems, such as "close to nature" and "natural disturbance-based" silviculture
Conservation and protected areas	Conserve remaining old-growth and high conservation value forests
	Expand protected areas system to encompass adequate representation of biological diversity
Management planning	Coordinate forest management planning at landscape scales (i.e. "matrix management")
	Maintain connectivity and habitat representation on managed forestlands
Certification	Expand participation in SFM certification; address non-conformity issues such as illegal logging
	Maintain high forest management standards over long-term to avoid certification suspension
Illegal logging	Improve socio-economic conditions at the community level, including employment opportunities
	Introduce effective legal regulation and enforcement
Payments for ecosystem services	Incentivize SFM by participating in international carbon markets and other "payment for ecosystem services" opportunities
Non-timber forest uses	Incorporate non-timber forest uses, such as ecologically sound recreation and non-timber forest products, into forest management planning
Infrastructure: roads and equipment	Increase investment in the forest sector to modernize forest infrastructure and harvesting machinery
Community participation	Promote community-based SFM initiatives, including projects designed by NGOs and public participation in forest governance
	Maintain and promote community-based forest management, such as traditional village systems
Sustainable development	Address increasing development pressures through land-use planning and promotion of ecologically-sensitive tourism
	Evaluate forest privatization proposals based on sustainability criteria

(Sikor 2003). Centralized forest governance and overall declines in forest sector investment since the collapse of the Soviet Union have limited innovation and policy reform at local, district, and regional administrative scales (Nordberg 2007). The Ukrainian forest code was amended in 2006 to address some of these constraints (see below). However, illegal logging continues to hinder forest management efforts, remaining prevalent both within and outside of protected areas in Ukraine (Kuemmerle et al. 2007, 2009). By one estimate using remotely sensed data, the area affected by illegal logging during this period was roughly equal in size to the total area of government sanctioned logging (Kuemmerle et al. 2009), though this estimate is contested within Ukraine. Bureaucratic inefficiency, inadequate public involvement processes, and a poorly developed non-governmental sector have also limited broader adoption of SFM principles in forest planning and governance (Soloviy and Cubbage 2007). These factors inhibit the ability of forest managers to respond effectively to new demands and economic opportunities, such as certification and payments for ecosystem services including carbon markets.

On the positive side, accords established by the Carpathian Convention (2003), expanding enrolment in SFM certification schemes, and international non-governmental organization initiatives (Björnsen-Gurung et al. 2009) are bringing new attention to this region. Yet certification projects are frequently challenged by non-compliance, including illegal logging, insufficient transparency in forest planning, lack of public involvement, violations of worker safety standards, inadequate attention to High Conservation Value forests, weak monitoring of rare, threatened, and endangered species, and the poor condition of forest road systems (Kovalyshyn and Pecher 2009). There may be opportunities for application of new SFM approaches, for instance building on existing social capital in traditional village systems (Elbakidze and Angelstam 2007), but this will require access and openness to new information, collaborative planning among actors from different sectors, investment in the forest sector, and experimentation with alternative silvicultural systems, including restorative approaches.

2 Integrating Ecological and Socio-Economic Objectives

In difficult economic circumstances sustainable development initiatives that build the social capital necessary for a long-term commitment to environmental protection become as important as the technical and scientific basis for forest management decisions. SFM begins with an understanding of the capacity of an ecosystem to produce a full range of ecosystem goods and services. However, initiatives must also support economic opportunities and social capital in communities striving to meet the basic necessities of life (Elbakidze and Angelstam 2007), and requires an economic system that rewards the provision of both market and non-market goods and services (Farley 2008, 2009).

The Carpathian region is struggling economically (Palang 2006) and yet subject to development pressures including growing tourism interest (Turnock 1999). The latter includes expanding transport infrastructure and sprawl, especially near tourism developments like ski areas. Some remote areas of the Carpathians retain traditional, village-based forest management systems that promote community engagement (Angelstam 2006; Elbakidze and Angelstam 2007). These cultural traditions were mostly superseded by the government controlled State Forest Enterprises established during the socialist period; subsequent political upheavals further disconnected communities from a shared, cultural connection to the landscape in some regions (Palang et al. 2006). However, in recent years innovative projects have helped re-establish a limited degree of public participation in SFM decision making at the community and regional levels (Foellmi and Schwitter 2009).

Carpathian forests bear the legacy of a long history of utilization within traditional village systems (Elbakidze and Angelstam 2007), intensive production-driven management dating to the Austro-Hungarian period in the nineteenth century, and more recent forest management systems introduced during the Soviet period. While most of the lowland temperate forests were cleared during the Austro-Hungarian period, much of the native beech (*Fagus sylvatica*) and mixed species forests were converted to Norway spruce (*Picea abies*), a species native to the Carpathians but planted ubiquitously on non-endemic sites and using non-local genetic varieties (e.g. Austrian genomes). This, together with even-aged, plantation style forest management practices, resulted in homogenized and simplified forest structure and composition at both stand and landscape scales, especially in areas not protected under the Soviet system (Stoyko 1998). This situation is not unique globally, bearing a striking resemblance to landscape scale changes occurring in southern Sweden during the nineteenth century (Björse and Bradshaw 1998) and the U S Pacific Northwest during the twentieth century for instance (Swanson and Franklin 1992). Now with conversion of some areas back to uneven-aged stand structures and mixed species composition underway in the Carpathians (Chernyavskyy 2009), SFM principles stressing restoration of ecological complexity are particularly germane.

There are strong economic incentives reinforcing the status quo, such as the financial efficiency of even-aged management, support from central budgets, and increasing demand for raw timber exports. Thus adoption of SFM principles by state forestry enterprises will depend on clear evidence of economic feasibility. This is no small challenge for advocates of major transformation within the forest sector. Market-based approaches providing economic incentives for SFM, such as certification, payments for ecosystem services, non-timber forest products, and value-added production, become particularly important in this context.

3 Conserving Biodiversity: Reserves and “Matrix Management”

The Carpathian Mountain complex is the largest in Europe, stretching 1,500 km from Serbia and Romania in the southeast, arching through western Ukraine, Poland, Hungary, and Slovakia, reaching the Czech Republic and Austria in the northwest (Fig. 1). It encompasses over 10 million ha of forestland dominated by spruce (*Picea abies*) and beech (*Fagus sylvatica*) cover types, with smaller proportions of fir (*Abies*

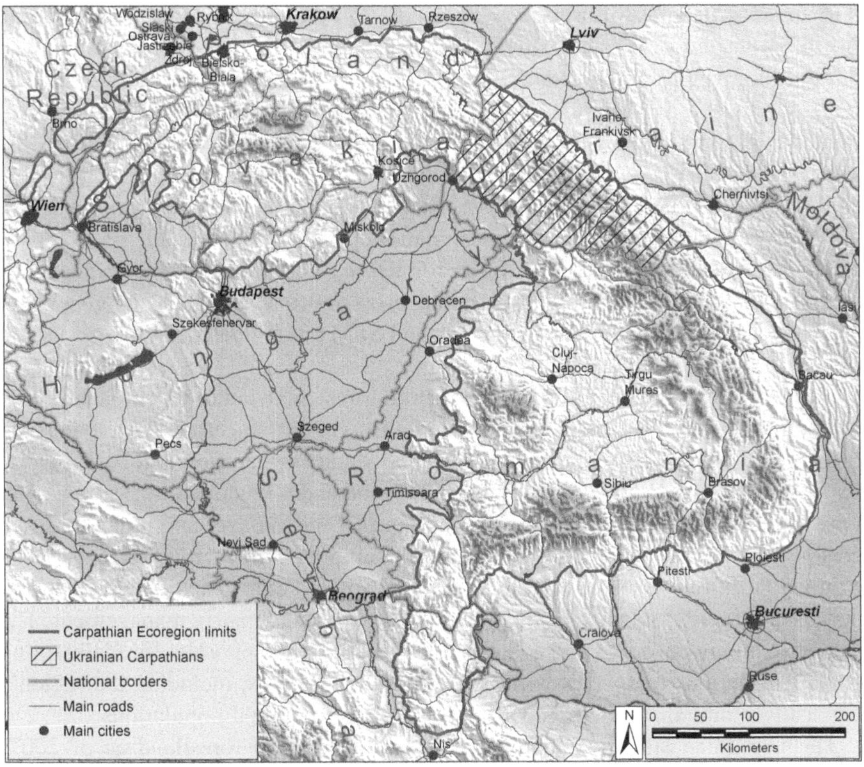

Fig. 1 Map of the Carpathian ecoregion, including political boundaries, cities, and major roads

alba), low elevation oak (*Quercus* sp.), and high elevation pines (*Pinus* sp.). The region is a focus of international attention due to its unique biological and cultural resources, now recognized as of global significance by UNESCO. The Carpathians harbour a full compliment of large European mammal species, including lynx (*Lynx lynx*), wildcat (*Felis silvestris*), river otter (*Lutra lutra*), gray wolf (*Canis lupus*), woodland bison (*Bison bonasus*), red deer (*Cervus elaphus*), and brown bear (*Ursus arctos*). Over 200 species of plants are endemic to this region, and stands of old-growth European beech larger than 10,000 ha are found only here, such as within the Uholka World Heritage Site in Ukraine.

Current conservation efforts in the Carpathians generally fit into three groups. The first focuses on establishment and better management of protected forest areas. The second includes those interested in SFM practices outside of core protected areas, and the third involves maintenance of traditional village systems providing high conservation value cultural woodlands. The tri-lateral East Carpathian Biosphere Reserve and bi-lateral Carpathian Biosphere Reserve, incorporate elements of all three approaches (Fall 1999), but suffer from lack of formalized transboundary cooperation mechanisms and insufficient institutional resources (Elbakidze and Angelstam 2009).

A topic of debate in Europe has been how to optimally allocate land among a mix of protected areas and actively managed areas to secure both natural and cultural biodiversity (Angelstam 2006). However, it is now generally recognized that reserves and SFM applied within a landscape approach are, in fact, complementary, and that both are required to achieve biodiversity conservation across large landscapes (Keeton 2007). Many conservation biologists argue that the first step is to design a functional reserve system containing the most complete and spatially redundant ecological representation possible (Noss and Scott 1997). Protected areas create greater flexibility for active forest management because they minimize risk associated with over reliance on any one approach (Lindenmayer and Franklin 2002). Thus protected areas and SFM are mutually advantageous and self-reinforcing at the landscape scale.

Sustainable management of semi-natural forest and cultural woodlands surrounding reserves is essential to meet conservation objectives. An emerging approach to landscape scale planning, called "matrix management", focuses on maintaining connectivity among reserves, watershed functionality, and habitat representation on actively managed timberlands (Lindenmayer and Franklin 2002; Keeton 2007). The Carpathian landscape, comprised of a mix of forestlands assigned to different levels of protection, intermingled human settlements, cultural areas and agricultural lands, is in many ways amenable to matrix management. The need to maintain corridors and ecological connectivity is important for maintaining viable populations of species with large area requirements, including the region's increasingly rare and geographically isolated large mammal populations.

The Ukrainian state-owned forest system ($\sim$97 % of forestland as of 2007) already provides the administrative foundation needed to design a comprehensive and complementary system of protected areas and SFM areas. Under recent (2006) forest code revisions state forests are allocated to four main groups. These are "protected forests" ($\sim$15 %) set aside for primarily for watershed protection; "recreation forests" ($\sim$10 %) managed for recreational uses and tourism; "protective forests" ($\sim$25 %) conserved for a variety of ecological, cultural, scientific, and aesthetic reasons; and "economic forests" ($\sim$50 %) managed primarily for commercial timber production (Anfodillo et al. 2008). These categories encompass a range of more specific designations, including strictly protected nature preserves (*zapovidniks*, IUCN category I), national nature parks and regional landscape parks (both similar to IUCN category V protected landscapes), and state forestry enterprises (comparable to the U S National Forest System). Since Ukrainian independence in 1991 the area of protected forest has more than doubled nationwide (Nordberg 2007). However, protected areas management remains constrained by lack of funding. Nevertheless, these land allocations provide the building blocks necessary for matrix management, particularly if illegal logging can be curbed.

In the Carpathians as in many regions of the world, there are important questions about how to design the most fully representative and sufficiently extensive system with the greatest likelihood of maintaining viable populations. Currently approximately 16 % of forests in the Carpathian bioregion (Anfodillo et al. 2008) and 17.6 % of the Ukrainian Carpathians specifically (Budyakova et al. 2005) are protected within core reserves (IUCN categories I–III). This is significantly higher than the 5 % average for European countries. An effort by the World Wide Fund for Nature is mapping the distribution of biodiversity and "High Conservation Value" forests throughout the Carpathian Range. This, together with on-going Gap Analyses, will allow more robust prioritization of areas for reserve status or special management consideration. There is also a need to develop mechanisms through which beneficiaries—at regional and larger scales—share the costs of provision (Farley 2009).

4 Silviculural Alternatives for the Carpathians: Providing a Broader Mix of Ecosystem Goods and Services

To complement a functional protected area network, matrix management requires planning and silvicultural practices in actively managed forests and woodlands that will sustain production of a broad range of ecosystem goods and services (Keeton 2007). An emerging approach in SFM internationally, termed "disturbance-based forestry", is particularly relevant to the Carpathians. The concepts are similar to European "natural dynamics forestry" (Angelstam and Kuuluvainen 2004) and "close to nature silviculture" (Chernyavskyy 2009). A paramount objective of disturbance-based forestry is to better conserve biological diversity by emulating the landscape dynamics and disturbance regimes to which organisms are adapted (North and Keeton 2008). The Carpathians have a natural disturbance regime dominated by wind events, ranging from frequent low intensity wind throw (e.g. gap creating disturbances) to moderate intensity and stand-replacing windstorms (Lavnyy and Lässig 2003; Nagel et al. 2006). Thus, innovative silvicultural systems emulating, at the landscape scale, a combination of gap dynamics and moderate intensity disturbances are directly applicable to this region. Examples include "the expanding gap" system first developed in Germany and now utilized in the northeastern U S (Seymour et al. 2005); the "structural complexity enhancement system" first tested in Vermont, U S (Keeton 2006); multi-cohort management systems that emulate the multi-aged stand structure associated with moderate intensity wind regimes (Hanson and Lorimer 2007); and variable retention harvest systems developed in the Pacific Northwest, U S (Franklin et al. 1997). Each of these helps achieve matrix management objectives by maintaining landscape heterogeneity and connectivity across managed forestlands, while also providing opportunities for timber revenue generation and a range of other economic uses.

However, more than three quarters of Ukraine's Carpathian forests are plantations (i.e. regenerated by planting) or regulated even-aged stands (Anfodillo 2008). About 72 % of the harvesting is clearcutting, 24 % is two and occasionally three-stage shelterwood cutting, and only 4 % of stands are managed using selection systems (Mahura et al. 2009). Variable retention systems or irregular shelterwoods (i.e. with retention of reserve trees over multiple rotations) are generally not employed in the Ukrainian Carpathian region, though they are actively used in the Polish Carpathians. Stand structure is less diverse in the resulting even-aged stands compared to multi-aged and natural forests. This holds true for the Norway spruce (*Picea abies*) plantations as well as younger beech and mixed species stands typical of the Carpathians. Compared to uneven-aged primary forests (Fig. 2), even-aged plantations typically have less vertical differentiation of the canopy (i.e. they are single layered), less horizontal complexity (e.g. gap structure), and lower densities of other important habitat characteristics, such as large trees and downed logs (Chernyavskyy 2005; Parpan et al. 2005). Thus, disturbance-based silviculture promoting redevelopment of complex stand structures provides a broader representation of habitat characteristics in managed stands (Franklin et al. 2002). If planned as an element of matrix management, together with reserves and careful scheduling and placement of harvest units, disturbance-based forestry accommodates both timber harvesting and production of non-timber forest products, as well as sustained ecosystem functioning. Disturbance-based forestry is not mutually exclusive of conventional harvesting methods; these of course would be part of the mix at landscape scales.

Mono-cultured plantations in the Carpathians have been particularly susceptible to mortality agents, such as root rots (e.g. *Armillaria* sp.; *Heterobasidion annosum*) and spruce bark beetle (*Ips typographus*), and have been stressed by airborne pollution, including acid deposition and heavy metals. Collectively these factors have contributed to extensive spruce dieback (Badea et al. 2004; Grodzki et al. 2004; Shparyk and Parpan 2004). Although dieback in Ukraine has been less severe than in other Carpathian countries, region wide about 40–50 % of both spruce and fir (*Abies alba*) has been damaged by heavy defoliation (Badea et al. 2004). Hence, there is a pressing need for broader adoption of restorative silvicultural practices, such as systems that convert stand structure from even to uneven-aged and reintroduce mixed species composition. These include sequential partial cutting approaches that promote development of multi-cohort, mixed species stands over time (e.g. Seidl et al. 2008). In Ukraine there are on-going trials of an approach termed "close to nature" silviculture that can be used for conversion purposes (Chernyavskyy 2009). This method employs group selection techniques, with small canopy openings placed around areas of desirable advanced regeneration. Ukrainian foresters have also developed a rapid restoration system involving clearcutting of dead and dying spruce (termed "sanitation cutting") followed by replanting with site-endemic species (Fig. 3). However, recent evidence suggests sanitation cutting often has been used for commercial purposes rather than strictly restorative objectives, exploiting an exemption from limitations on cutting unit size in the Ukrainian Forest Code (Kuemmerle et al. 2007, 2009). Moreover, it is important for restoration treatments to be planned strategically so as to restore high priority sites while minimizing fragmentation and watershed impacts.

Another concern pertains to the current age class distribution of Carpathian forests (Fig. 4), which is heavily skewed towards young to early mature (0–80 year old) plantations (Strochinskii et al. 2001; Anfodillo et al. 2008). This reflects a history of over-cutting during the socialist period, especially during the

Fig. 2 Examples of forest stand structure in relation to age. *Top left* young, managed beech forest in the Austrian Alps. *Top right* old growth beech forest in the Ukrainian Carpathians. *Lower left* mature Norway spruce plantation in the Ukrainian Carpathians, with some evidence of spruce decline. *Lower right* old-growth Norway spruce-silver fir forest in the Ukrainian Carpathians. Note the higher degree of structural complexity in the older forests, including vertically complex canopies, larger trees, and large woody debris both standing and downed. *Photo credits* W.S. Keeton

Fig. 3 Spruce decline and restoration in the Skole District of the Ukrainian Carpathians. *Left* declining Norway spruce stand with heavy mortality. *Top right* sanitation cut followed by replanting species endemic to the site. *Lower right* foreground shows a young, mixed species stand restored following sanitation cutting of declining spruce. This figure represents a forest restoration pathway, though there is evidence that sanitation cutting is being over-used as a way to circumvent regulations limiting clearcutting size. *Photo credits* W.S. Keeton

1950s (Nijnik 2004), resulting in conversion of landscapes from old to young forest dominated. Some have argued that the current forest age distribution will result in future timber supply limitations if harvest levels are not increased during the near term (Nijnik and van Kooten 2000). In our view, however, this argument discounts the current deficit of harvestable stands (100–120 year old) relative to sub-merchantable stands. Rather, the solution is to allow the merchantable growing stock to build up both passively and through stand improvement treatments, such as intermediate thinnings. Others have drawn similar conclusions, arguing that "harvest potential will increase in the future" and pointing out that the area of mature stands almost doubled from the late 1980s to late 1990s (Polyakova and Sydor 2006). Whereas in the 1990s more than 80 % of gross annual increment was harvested each year, currently legal harvest rates are 50–60 % of annual forest growth in the Ukrainian Carpathians (Anfodillo et al. 2008), suggesting that timber stocks are increasing, not decreasing. Far from limiting future timber availability, the current distribution provides an abundant source from which merchantable timber will recruit over the next several decades.

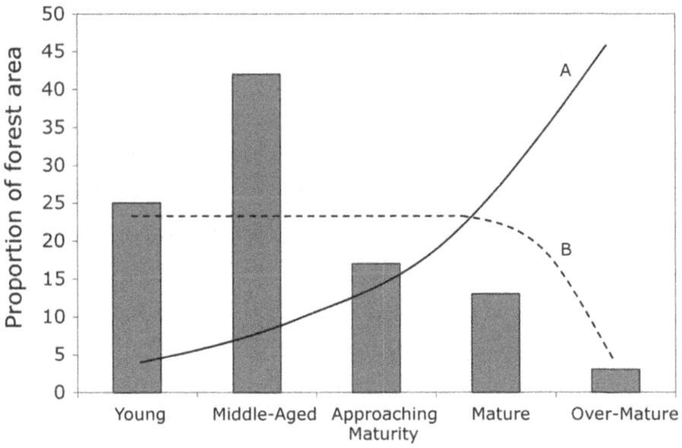

Fig. 4 The bars show the current age class distribution of commercial forests ("economic" group) in the Ukrainian Carpathians using data and classes from Anfodillo et al. (2008). The *solid line* (**A**) represents the probable age class distribution that would develop over time in the absence of human disturbances, and provides a benchmark for understanding the associated changes in forest biodiversity. The *dotted line* (**B**) represents a distribution that would favor sustained timber yield, but would not necessarily provide sufficient representation of late-successional forests. Age ranges for the dominant forest types are as follows: young, 0–40; middle-aged, 41–80; approaching maturity, 81–100; mature, 101–120; and over-mature, >120

Once sufficient merchantable stocking is achieved, careful stewardship to maintain a desired distribution (either balanced across age classes or tilted toward mature and late-successional forests) will sustain timber yields while maintaining a broader mix of ecosystem services (Fig. 4). Polyakova and Sydor (2006) argue that active management will be vital because so much of the forested landscape is in plantations or has developed from afforestation or forest regrowth on abandoned agricultural lands. Forest cover expanded at an annual rate of 0.1 % during the 1990s for the Carpathian region overall; forested area increased by 54,000 ha in the Ukrainian Carpathians from 1988–2004 (Anfodillo et al. 2008). Moreover, given the strong association between biodiversity and forest developmental stages (Stoyko 1998), shifting the landscape towards a better balance of stand ages would both provide a stable timber supply and ensure adequate representation of late-successional habitats. In addition, maintaining a margin of unharvested net growth provides numerous ecosystem services, such as development of stand structural complexity, riparian functionality, and carbon storage. There is unlikely to be a deficit of early successional forest habitat in the future due to forest regrowth on abandoned agricultural lands, which has increased markedly during the post-socialist period (Kuemmerle et al. 2006, 2008).

5 Conserving Freshwater Ecosystems and Watershed Functioning

The need to develop new approaches for integrated watershed management has been recognized internationally as a key element of SFM. The Carpathians have experienced several severe flooding events over the last decade, including particularly destructive floods in 1998, 2001 and 2008. While extreme precipitation is considered the primary cause, the magnitude of flooding has also been attributed to logging and land cover change in some cases (Shulyarenko 2002). For example, Dezso et al. (2005) concluded that recent floods in eastern catchments (Ukraine) of the Tisza river were not caused solely by heavy precipitation, but may also have been influenced by the 10–20 % forest cover loss that occurred from 1993–2001, although these linkages are debated in Ukraine. Thus, an important issue for SFM in the Carpathians is the application of integrated watershed management approaches designed to reduce flooding hazards, such as cumulative effects analysis, spatially-explicit planning, and collaborative governance including local community input (Naiman et al. 1997; Sabatier et al. 2005).

Emerging science describing forest ecosystem regulation of hydrologic and watershed functions is directly relevant to the Carpathian region. Scientists have documented important ecological interactions between entire catchments, riparian forests, and aquatic ecosystems (Naiman et al. 2005; Keeton et al. 2007). In many regions improved scientific understanding has led to regulations and changes in management practices designed to better protect freshwater ecosystems. For instance, delineation of riparian buffers and riparian forest restoration are now frequently employed as central elements of SFM (Gregory et al. 1997). Riparian buffers take different forms depending on context (Lazdinis and Angelstam 2005). Some are entirely off limits to logging or road construction, while others allow limited entry if deemed appropriate (Lee et al. 2004). Many regions employ zonations with varying intensity of permissible management, often including a strictly protected buffer zone immediately adjacent to the stream channel. Broader adoption and enforcement of similar approaches would help safeguard aquatic resources in the Carpathian Mountains. For example, lack of consistent riparian protections is frequently cited as a non-conformity impeding forest certification projects in Ukraine (Kovalyshyn and Pecher 2009). Reforestation and improved riparian protection is needed especially on cutover and degraded main floodplains, which are ubiquitous throughout the Carpathians.

Recently calls have been made throughout the Ukrainian forest sector for increased construction of logging roads (Mahura et al. 2009). This reflects the general disrepair and lack of investment in the existing forest infrastructure and limited forest access, particularly away from valley bottoms. Decommissioning deteriorating, poorly designed roads close to stream channels is an imperative, and the desire for greater access for forest management, fire control, and recreation are also understandable. A year 2000 law required the expansion of the hard-surface forest road network in the Ukrainian Carpathians to 10 km per 1,000 ha by 2010, but this target was not met despite some new construction. The road density today is ~3–5 km per 1000 ha. In our view, forest managers in the Carpathians should proceed cautiously with plans for new road construction. This is vital to minimize road system extent and landscape fragmentation, design well engineered roads with minimal erosive and hydrologic impacts, make use of temporary logging roads whenever possible, and avoid negative impacts to aquatic ecosystems, for instance by minimizing the number of stream crossings.

Much of the harvesting machinery in use in the Carpathians is obsolete by contemporary standards, such as decades-old tracked skidders that cause significant rutting, erosion, and damage to advanced regeneration. For example, by one estimate 60–85 % of harvest units in Ukraine experience significant soil damage due to current skidding practices (Mahura et al. 2009). Beginning in 2005, timber harvesting in mountain forests is permitted only with the use of low impact skidding systems and renovation of the Austro-Hungarian era narrow-gauge railway network. But implementing this policy will require substantial new investment in harvesting and transportation technology, such as mobile cable systems, wheeled skidders, and harvester/forwarder systems.

6 Forest Carbon Management: Opportunities and Incentives for SFM in the Carpathians

Rapidly developing international carbon markets have potential to incentivize aspects of SFM, though they are also fraught with challenges (Ruddell et. al. 2007). Under the Kyoto Protocol's Joint Implementation (J.I.) mechanism, developed countries (Annex I) can purchase credits from nations with transitional economies, including former socialist republics, to offset their greenhouse gas emissions above the cap set by the Protocol. Initially afforestation and reforestation were the primary forest sector opportunities for earning credits. Recent developments in carbon markets outside the Kyoto framework (primarily voluntary markets) have introduced credits for "avoided deforestation" (often termed REDD, or reduced emissions from deforestation and degradation) and "improved forest management" (IFM). The latter requires participants to substantiate that additional carbon will be stored with some measure of "permanence" over a baseline or "business as usual" scenario. There are the added difficulties of accounting for carbon stored in wood products, the "offsets" achieved by substituting wood for more energy intensive building materials, and "leakage" or the emissions from geographically displaced harvesting (Ray et al. 2009). Significant advances in voluntary markets have been made in the last several years to resolve these technical issues.

Ukraine and other Carpathian nations are positioned to immediately benefit from credits awarded for afforestation under existing carbon market mechanisms. In 2002 Ukraine launched a program called "Forests of Ukraine", with the objective of increasing forest cover from 15.6 % (10.9 million ha) to 16.1 % (11.3 million ha) of its total land area by 2015 (Soloviy and Cubbage 2007). According to Nijnik (2001) there are 2.3 million ha available for afforestation in Ukraine alone based on an analysis of current land use, terrain, and soils. This number declines to 1.7 million ha when areas requiring unacceptably high afforestation costs are excluded. Currently Ukrainian forests sequester about 180 million Mg of CO_2 annually through growth in existing forests (Soloviy and Yaremchuk 2001). Areas available for afforestation have the potential to store (i.e. in new forest biomass) an additional 200 million Mg of carbon after 40 years, and would be cost effective for earning carbon credits based either on in situ carbon storage or emissions offset by biomass fuel production (Nijnik 2001).

Ukraine has been signatory to the Kyoto Protocol since 2004, and emitted less greenhouse gas than permitted (by so-called "Assigned Amount Units"), a trend like to continue for some time under most scenarios of economic growth and energy efficiency (Victor et al. 2001). Consequently, forest sector participation in carbon markets would only supplement the already substantial market opportunity Ukraine is likely to enjoy under a successor to the Kyoto framework. Ukraine has been eligible to participate in J.I. sequestration projects, however, no J.I. projects have been conducted in the Ukrainian forest sector to date. Ukraine worked with the World Bank on a trial carbon management project involving afforestation on 15,000 ha of abandoned agricultural land near Chernobyl (World Bank 2006). However, the project was cancelled in 2007 due to contractual, land tenure, and ministerial disputes. Thus, Ukrainian participation in carbon markets remains fraught with challenges, many of them institutional, yet considerable potential remains, particularly under rapidly growing international voluntary market systems. Community support and benefit at the local level also will be important for the success of forest carbon projects in the Carpathians.

There may be opportunities for earning carbon credits through avoided deforestation in areas of the Carpathians experiencing land use pressures, such as recently privatized forests facing subdivision and development in Romania. These opportunities may increase if a $REDD^{++}$ program is adopted by a successor to the Kyoto Protocol and expanded outside of the tropics. Recent research suggests that structurally complex, old-growth Carpathian beech and spruce–fir forests store very large amounts of carbon (Szwagrzyk and Gazda 2007; Keeton et al. 2010a), 50 % more than mature stands and higher than the average reported for temperate old-growth forests globally (Keeton et al. 2010b). By one estimate there are over 322,000 ha of primary (or "virgin") forest left in the Carpathians (Anfodillo et al. 2008), though it is unknown how much of this remains outside of reserves. Thus, conservation of unprotected old-growth and primary forests would carry significant carbon benefits. A key challenge will be ensuring that a portion of the carbon revenue is returned directly to local administrative units for investment in projects that also benefit local communities.

While still evolving, IFM may add further potential for carbon market participation in the Carpathians. Innovative silvicultural options, including extended rotations and retention forestry, have been developed to enhance carbon storage in managed temperate forests (Nunery and Keeton 2010). These may provide economic incentives for SFM in the future. The relatively long rotations (e.g. 80–140 years) currently required under Ukrainian forest code already offer significant carbon storage benefits, but limit opportunities to create additionality (enhanced storage) over baseline levels. A concern with long rotations, however, is the increasing susceptibility to decline and dieback as spruce plantations age; indeed, some spruce stands have almost complete tree mortality by the time rotations periods are reached (Kovalyshyn and Pecher 2009). Thus, in the Carpathians there may be limitations to the carbon storage benefits typically associated with extended rotations, and restoration of endemic forest composition will be an important element of a comprehensive forest carbon management strategy.

Alternative silvicultural systems specifically intended to promote development of structurally complex, high biomass forests offer additional carbon sequestration opportunities, while also providing economically and ecologically sustainable timber harvests (Keeton 2006; Bauhus et al. 2009). These are most applicable to site-endemic beach, mixed hardwood-conifer, and native spruce-fir stands. Group selection (e.g. "close to nature silviculture") techniques designed to release advanced regeneration and restore native species composition in non-endemic spruce stands are an important element of this strategy (Chernyavskyy 2009).

7 Adapting to Climate Change

A major source of uncertainty for forest managers in the Carpathian region is how to anticipate and adapt to climate change. Regional-scale climate models for the Carpathians predict increases in mean annual temperature, concurrent with increased and decreased winter and summer precipitation respectively (Bartholy et al. 2007). Also predicted is greater irregularity of precipitation patterns, including increased frequency of extreme events (Bartholy and Pongracz 2007). A combination of summer drought and warmer winters may exacerbate insect pest risks and localized species extirpations, particularly for species already suffering geographic isolation and low genetic diversity (Levinsky et al. 2007). Interaction between climate impacts and other ecosystem stressors, such as airborne pollution, invasive species, and land-use change, has the potential to accelerate ecosystem shifts (Aber et al. 2001), and yet this topic appears poorly studied in the Carpathian region.

Adaptation science generally stresses management for ecosystem properties that increase resilience and reduce vulnerability, such as restoration and management for native species diversity and continuous forest cover. Another approach is to expand the representation of geophysical diversity within protected areas, thereby providing potential for difficult-to-predict species range shifts and formation of new assemblages, as occurred with past climate changes (Hunter et al. 1988). A universal challenge will be operationalizing and down-scaling the generalized adaptive strategies developed to date. We recommend this as a topic worthy of further investigation.

8 Conclusion

There is a long history of professional forest management in the Carpathians, but the region is facing new opportunities and challenges. International criteria provide a basic framework for SFM, but implementation approaches need to be adapted to the Carpathian context. Concepts such as matrix management and disturbance-based forestry, as well as multi-level governance incorporating local community involvement in planning and decision making processes, are readily adaptable to the Carpathian landscape. They would help provide a broader array of ecosystem goods and services, including biodiversity conservation and timber revenue, if employed in conjunction with regional level design and establishment of functional reserve networks. An adaptive approach to SFM will be essential due to the anticipated effects of climate change on Carpathian flora and fauna (Björnsen-Gurung et al. 2009). Recent research advances in the fields of watershed management, riparian forest conservation, and logging road system design and development of collaborative planning will help inform improved protections for aquatic ecosystems in the Carpathians. Market based approaches, such as forest certification, carbon markets, and payments for ecosystem services, may help incentivize sustainable forest management for a broad array of ecosystem goods and services. These represent significant opportunities for the Carpathian region. The challenge facing Ukraine and other Carpathian nations is to merge these ideas into a holistic, landscape approach that can be implemented within existing or reformed administrative frameworks.

Acknowledgments This paper resulted from scientific exchanges funded by the Trust for Mutual Understanding, the United States Fulbright Program, the Pennsylvania State University WIRA Program and the Marcus and Amalia Wallenberg Foundation. Two anonymous reviewers provided helpful comments. Special thanks to the Ukrainian National Forestry University, Znesinnya Regional Landscape Park, Skole State Forest Enterprise, Scolivsky Beskydy National Park, Khust State Forest Enterprise, Gorgany Nature Preserve, Khust Forestry College, the Swiss-Ukrainian Forest Development Project in Transcarpathia (FORZA), Ecosphere, and the Carpathian Biosphere Reserve for hosting events critical to the development of this publication.

References

Aber J, Neilson RP, McNulty S, Lenihan JM, Bachelet D, Drapek RJ (2001) Forest processes and global environmental change: predicting the effects of individual and multiple stressors. BioSci 51:735–751

Anfodillo T, Carrer M, Dalla Valle E, Giacoma E, Lamedica S, Pettenella D (2008) Programme Carpathian project: current state of forest resources in the Carpathians. Universita Delgi Studi Di Padova, Padova

Angelstam P (2006) Maintaining cultural and natural biodiversity in Europe's economic centre and periphery. In: Agnoletti M (ed) The conservation of cultural landscapes. CAB International, Wallingford

Angelstam P, Kuuluvainen T (2004) Boreal forest disturbance regimes, successional dynamics and landscape structures—a European perspective. Ecol Bull 51:117–136

Badea O, Tanase M, Georgeta J, Anisoara L, Peiov A, Uhlirova H, Pajtik J, Wawrzoniak J, Shparyk Y (2004) Forest health status in the Carpathian Mountains over the period 1997–2001. Environ Pollut 130:93–98

Bartholy J, Pongracz R (2007) Regional analysis of extreme temperature and precipitation indices for the Carpathian Basin from 1946 to 2001. Glob Planet Chang 57:83–95

Bartholy J, Pongracz R, Gelybo G (2007) Regional climate change expected in Hungary for 2071–2100. Appl Ecol Environ Res 5:1–17

Bauhus J, Puettmann K, Messier C (2009) Silviculture for old-growth attributes. For Ecol Manag 258:525–537

Björnsen-Gurung A, Bokwa A, Chełmicki W, Elbakidze M, Hirschmugl M, Hostert P, Ibisch P, Kozak J, Kuemmerle T, Matei E, Ostapowicz K, Pociask-Karteczka J, Schmidt L, van der Linden S, Zebisch M (2009) Global change research in the Carpathian Mountain region. Mt Res Dev 29:282–288

Björse G, Bradshaw R (1998) 2000 years of forest dynamics in southern Sweden: suggestions for forest management. For Ecol Manag 104:15–26

Budyakova T, Skrylnikov D, Pavelko A, Dovhanych Y (2005) National assessment of the national policy, legislative and institutional frameworks related to the Carpathian Convention: Ukraine. Regional Environment Center, EURAC, and Italian Ministry for the Environment, Land and Sea, Bolzano

Chernyavskyy MV (2005) The dynamics of virgin beech forests in the Ukrainian Carpathians. In: Commarmot B, Hamor FD (eds) Natural forests in the temperate zone of Europe: values and utilization. Proceedings of a conference, 13–17 Oct 2003, Mukachevo, Ukraine. Swiss Federal Research Institute, Birmensdorf

Chernyavskyy MV (2009) Forest stand dynamics and close to nature forestry. In: Soloviy I, Keeton WS (eds) Ecological economics and sustainable forest management: developing a trans-disciplinary approach for the Carpathian Mountains. Ukrainian National Forestry University Press, Lviv

Dezso Z, Bartholy J, Pongracz R, Barcza Z (2005) Analysis of land-use/land-cover change in the Carpathian region based on remote sensing techniques. Phys Chem Earth 30:109–115

Elbakidze M, Angelstam P (2007) Implementing sustainable forest management in Ukraine's Carpathian Mountains: the role of traditional village systems. For Ecol Manag 249:28–38

Elbakidze M, Angelstam P (2009) Cross-border cooperation along the eastern border of European Union: a review and approach to learning for sustainable landscapes. Cent Eur J Spat Landsc Plan 20(1):33–42

Fall JJ (1999) Transboundary biosphere reserves: a new framework for cooperation. Environ Conserv 26:252–255

Farley J (2008) The role of prices in conserving critical natural capital. Conserv Biol 22:1399–1408

Farley J (2009) Conservation through the economics lens. Environ Manag 45:26–38

Farley J, Zahvoyska L, Maksymiv L (2009) Transdisciplinary paths towards sustainability: new approaches for integrating research, education and policy. In: Soloviy I, Keeton WS (eds) Ecological economics and sustainable forest management: developing a trans-disciplinary approach for the Carpathian Mountains. Ukrainian National Forestry University Press, Lviv

Flick U (2009) An introduction to qualitative research. Sage Publications, London

Foellmi H, Schwitter R (2009) Forest resource planning for people with people: two-level planning for sustainable management. In: Soloviy I, Keeton WS (eds) Ecological economics and sustainable forest management developing a trans-disciplinary approach for the carpathian mountains. Ukrainian National Forestry University Press, Lviv

Franklin JF, Berg DR, Thornburgh DA, Tappeiner JC (1997) Alternative silvicultural approaches to timber harvesting: variable retention harvest system. In: Kohm KA, Franklin JF (eds) Creating a forestry for the twenty-first century: the science of ecosystem management. Island Press, Washington

Franklin JF, Spies TA, Van Pelt R, Carey A, Thornburgh D, Berg DR, Lindenmayer D, Harmon M, Keeton WS, Shaw DC, Bible K, Chen J (2002) Disturbances and the structural development of natural forest ecosystems with some implications for silviculture. For Ecol Manag 155:399–423

Gregory SV (1997) Riparian management in the twenty-first century. In: Kohm KA, Franklin JF (eds) Creating a forestry for the twenty-first century: the science of ecosystem management. Island Press, Washington

Grodzki W, McManus M, Knížek M, Meshkova V, Mihalciuc V, Novotny J, Turčani M, Slobodyan Y (2004) Occurrence of spruce bark beetles in forest stands at different levels of air pollution stress. Environ Pollut 130:73–83

Hanson J, Lorimer C (2007) Forest structure and light regimes following moderate wind storms: implications for multi-cohort management. Ecol Appl 17:1325–1340

Hunter ML, Jacobson GL, Webb T (1988) Paleoecology and the coarse-filter approach to maintaining biological diversity. Conserv Biol 2:375–385

Keeton WS (2006) Managing for late-successional/old-growth forest characteristics in northern hardwood-conifer forests. For Ecol Manag 235:129–142

Keeton WS (2007) Role of managed forestlands and models for sustainable forest management: perspectives from North America. George Wright Forum 24(3):38–53

Keeton WS, Kraft CE, Warren DR (2007) Mature and old-growth riparian forests: structure, dynamics, and effects on Adirondack stream habitats. Ecol Appl 17:852–868

Keeton WS, Chernyavskyy M, Gratzer G, Main-Korn M, Shpylchak M, Bihun Y (2010a) Structural characteristics and aboveground biomass of old-growth spruce-fir stands in the eastern Carpathian Mountains, Ukraine. Plant Biosyst 144:1–12

Keeton WS, Burrascano S, Pucko C, Blasi C, Bihun Y, Chen J, Chernyavskyy M, Commarmot B, Franklin JF, Gratzer G, Nunery JS, Shparyk Y, Spies TA, Swanson ME (2010b) A global analysis of temperate old-growth forests: commonality in carbon storage and co-varying ecosystem functions. Int For Rev 12(5):3

Kovalyshyn V, Pecher I (2009) Understanding the hurdles to sustainable forest management through FSC forest certification in the Transcarpathian region of Ukraine. In: Soloviy I, Keeton WS (eds) Ecological economics and sustainable forest management: developing a trans-disciplinary approach for the Carpathian Mountains. Ukrainian National Forestry University Press, Lviv

Kuemmerle T, Radeloff VC, Perzanowski K, Hostert P (2006) Cross-border comparison of land cover and landscape pattern in Eastern Europe using a hybrid classification technique. Remote Sens Environ 103:449–464

Kuemmerle T, Hostert P, Radeloff VC, Perzanowski K, Kruhlov I (2007) Post-socialist forest disturbance in the Carpathian border region of Poland, Slovakia, and Ukraine. Ecol Appl 17:1279–1295

Kuemmerle T, Hostert P, Radeloff VC, van der Linden S, Perzanowski K, Kruhlov I (2008) Cross-border comparison of post-socialist farmland abandonment in the Carpathians. Ecosyst 11:614–628

Kuemmerle T, Chaskovskyy O, Knorn J, Radeloff VC, Kruhlov I, Keeton WS, Hostert P (2009) Forest cover change and illegal logging in the Ukrainian Carpathians in the transition period from 1988 to 2007. Remote Sens Environ 113:1194–1207

Lavnyy V, Lässig R (2003) Extent of storms in the Ukrainian Carpathians. In: Proceedings of the international conference on wind effects on trees, Karlsruhe, 16–18 Sept 2003

Lazdinis M, Angelstam P (2005) Functionality of riparian forest ecotones in the context of former Soviet Union and Swedish forest management histories. For Policy Econ 7:321–332

Lee PC, Smyth C, Boutin S (2004) Quantitative review of riparian buffer width guidelines from Canada and the United States. J Environ Manag 70:165–180

Levinsky I, Skov F, Svenning JC, Rahbek C (2007) Potential impacts of climate change on the distributions and diversity patterns of European mammals. Biodivers Conserv 16:3803–3816

Lindenmayer DB, Franklin JF (2002) Conserving forest biodiversity: a comprehensive multiscaled approach. Island Press, Washington

Mahura B, Bihun Y, Deyneka A (2009) Opportunities and challenges in promoting sustainable timber harvesting in the Ukrainian Carpathians. In: Soloviy I, Keeton WS (eds) Ecological economics and sustainable forest management: developing a trans-disciplinary approach for the Carpathian Mountains. Ukrainian National Forestry University Press, Lviv

Nagel TA, Svoboda M, Diaci J (2006) Regeneration patterns after intermediate wind disturbance in an old-growth Fagus-Abies forest in southeastern Slovenia. For Ecol Manag 226:268–278

Naiman RJ, Bisson PA, Lee RG (1997) Approaches to management at the watershed scale. In: Kohm KA, Franklin JF (eds) Creating a forestry for the twenty-first century: the science of ecosystem management. Island Press, Washington

Naiman RJ, Decamps H, McClain ME (2005) Riparia: ecology, conservation, and management of streamside communities. Elsevier, San Diego

Nijnik M (2001) Mitigating climate change via afforestation in Ukraine. In: Essman HS, Pettenella D (eds) Forestry in Ukraine at the crossroads: analyses and ideas for a sustainable development. Afisha, Lviv

Nijnik M (2004) To an economist's perception on sustainability in forestry-in-transition. For Policy Econ 6:403–413

Nijnik M, van Kooten GC (2000) Forestry in the Ukraine: the road ahead? For Policy Econ 1:139–151

Nordberg M (2007) Ukraine reforms in forestry 1990–2000. For Policy Econ 9:713–729

North MP, Keeton WS (2008) Emulating natural disturbance regimes: an emerging approach for sustainable forest management. In: Lafortezza R, Chen J, Sanesi G, Crow TR (eds) Patterns and processes in forest landscapes—multiple use and sustainable management. Springer, The Netherlands

Noss RF, Scott JM (1997) Ecosystem protection and restoration: the core of ecosystem management. In: Boyce MS, Haney A (eds) Ecosystem management: applications for sustainable forest and wildlife resources. Yale University Press, New Haven

Nunery JS, Keeton WS (2010) Forest carbon storage in the northeastern United States: net effects of harvesting frequency, post-harvest retention, and wood products. For Ecol Manag 259:1363–1375

Palang H, Printsmann A, Konkoly Gyuro E, Urbanc M, Skowronek E, Woloszyn W (2006) The forgotten rural landscapes of Central and Eastern Europe. Landsc Ecol 21:347–357

Parpan V Shparyk Y, Parpan T (2005) Virgin and natural forests in Ukraine: state, diversity, and protection. In: Commarmot B, Hamor FD (eds) Natural forests in the temperate zone of Europe—values and utilization, conference proceedings Mukachevo 13–17 Oct

Polyakov M, Sydor T (2006) Forestry in Ukraine: the road ahead? comment. For Policy Econ 8:6–9

Ray DG, Seymour RS, Scott NS, Keeton WS (2009) Mitigating climate change with managed forests: balancing expectations, opportunity, and risk. J For 107(1):50–51

Ruddell S, Sampson R, Smith M, Giffen R, Cathcart J, Hagan J, Sosland D, Godbee J, Heissenbuttel J, Lovett S, Helms J, Price W, Simpson R (2007) The role for sustainably managed forests in climate change mitigation. J For 105(6):314–319

Sabatier PA, Focht W, Lubell M, Trachtenberg Z, Vedlitz A, Matlock M (eds) (2005) Swimming upstream- collaborative approaches to watershed management. MIT Press, London

Seidl R, Rammer W, Lasch P, Badeck FW, Lexer MJ (2008) Does conversion of even-aged, secondary coniferous forests affect carbon sequestration? a simulation study under changing environmental conditions. Silva Fenn 42:369–386

Seymour RS (2005) Integrating natural disturbance parameters into conventional silvicultural systems: experience from the Acadian forest of northeastern North America. In: Peterson CE, Maguire DA (eds) Balancing ecosystem values: innovative experiments for sustainable forestry, USDA Forest Service General Technical Report PNW-GTR-635

Shparyk YS, Parpan VI (2004) Heavy metal pollution and forest health in the Ukrainian Carpathians. Environ Pollut 130:55–63

Shulyarenko A, Yatsyuk M, Shulyarenko I (2002) Causes and peculiarities of recent floods on the Dniester River. In: Marsalek J, Watt WE, Zeman E, Siker F (eds) Flood issues in contemporary water management. NATO Science Series, Environmental Security, vol 71

Sikor T (2003) The commons in transition: agrarian and environmental change in central and Eastern Europe. Environ Manag 34:270–280

Soloviy IP, Cubbage FW (2007) Forest policy in aroused society: Ukrainian post-orange revolution challenges. For Policy Econ 10:60–69

Soloviy IP, Yaremchuk O (2001) Socio-economic and environmental aspects of afforestation in Ukraine. In: Essman HS, Pettenella D (eds) Forestry in Ukraine at the crossroads: analyses and ideas for a sustainable development. Afisha, Lviv

Stoyko SM (1998) Virgin ecosystems of the Carpathians and their significance for biological diversity conservation and maintenance of the sustainable development of forestry. In: Issues of sustainable development in the Carpathian region, vol 2, Rakhiv

Strochinskii AA, Pozyvailo YM, Jungst SE (2001) Forests and forestry in Ukraine: standing on the brink of a market ecology. J For 99:34–38

Swanson FJ, Franklin JF (1992) New forestry principles from ecosystem analysis of Pacific Northwest forests. Ecol Appl 2:262–274

Szwagrzyk J, Gazda A (2007) Above-ground standing biomass and tree species diversity in natural stands of Central Europe. J Veg Sci 18:563–570

Turnock D (1999) Sustainable rural tourism in the Romanian Carpathians. Geogr J 165:192–199

Victor DG, Nakicenovic N, Victor N (2001) The Kyoto Protocol emission allocations: windfall surpluses for Russia and Ukraine. Clim Chang 49:263–277

World Bank (2006) Ukraine forestry sector note: status and opportunities for development. The World Bank, Washington

Stakeholders' Perceptions of Mountain Forest Ecosystem Services: The Ukrainian Carpathians Case Study

Lyudmyla Zahvoyska and Tetyana Bas

Abstract The large part of the Ukrainian Carpathians is covered by forests. Mountain forests sustain human life through numerous goods and services. Some of them are complementary, others are mutually exclusive. There are a lot of stakeholders for the benefits of mountain forest ecosystem services (MFES) and their interests often contradict. The paper provides insight into implicit world of stakeholders' perceptions and preferences concerning MFES through a forest values universe and a set of cognitive maps of preferences across stakeholders, developed for the Ukrainian Carpathians. The forest values universe is designed using a Conceptual Content Cognitive Mapping method. This universe consists of nine dominant themes and more then 37 sub-themes. The dominant themes are: environmental, recreational, economic, educational, health care, emotional and aesthetical. The set of cognitive maps of stakeholders' preferences regarding forest values is developed applying non-parametric statistic methods. The results indicate a consensus in perceptions of personal values across stakeholders and some contradictions in preferences of different stakeholders groups.

1 Introduction

Forests sustain human life through numerous resources, functions and services of forest ecosystems, varying from such life-support functions as a food and a shelter provision and hazards mitigation to shaping cultural environment and emotional

L. Zahvoyska (✉) · T. Bas
Institute of Ecological Economics, Ukrainian National Forestry University, Generala Chuprynky Street 103, Lviv 79057, Ukraine
e-mail: zahvoyska@ukr.net

T. Bas
e-mail: tetyana_bas@ukr.net

J. Kozak et al. (eds.), *The Carpathians: Integrating Nature and Society Towards Sustainability*, Environmental Science and Engineering,
DOI: 10.1007/978-3-642-12725-0_25, © Springer-Verlag Berlin Heidelberg 2013

relaxation. More than four hundred million people throughout the world live in or around forests (Patosaary 2006). They enjoy and benefit from forest goods and ecosystem services. Therefore one of the forest decision-making priorities is meeting demands of all stakeholders by paying attention to the improvement of the livelihoods of forest-dependent people as a precondition to achieving Millennium Development Goals.

Around a half of the Carpathians is covered by forests. Most of the Carpathian forests are located in Romania (around 50 % of all Carpathian forests), Slovakia (20 %) and Ukraine (18 %). Like a green snake, the Carpathians lie on a background of deforested Europe on the map of world forest systems (Millennium Ecosystem Assessment 2005). All of the above mentioned benefits become more important when it comes to mountain forests.

Millennium Ecosystem Assessment (2005), often cited in modern ecosystem discourse, identifies four groups of ecosystem services: provisioning, regulating, cultural and supporting. Forest ecosystems provide all services generated by ecosystems, at least to some extent (Daly and Farley 2004). Being vital for human survive forest ecosystems are heavily impacted because of economic growth, growth of human population, climate change and environmental deterioration. Capacity of forest ecosystems and quality of their services strongly depends on forest management and governance.

The already mentioned diversity of mountain forest ecosystem services (MFES) determines a variability of users, consumers and stakeholders and interweaving of their sometimes conflicting and even mutually exclusive interests. Implicit world of these interwoven interests should be explored because they shape and drive forest practice and governance according to stakeholders' behavior caused by their values and insights.

MFES differ in their nature: some of them are complementary, and others are mutually exclusive. High ratio of non-marketed economic values like biodiversity conservation, carbon sequestration and watershed protection in total economic value of forest ecosystem services causes a strong and permanent bias of market transactions toward marketed ones and immediate financial income. Such peculiarities of their nature as non-excludability and non-rivalness (Daly and Farley 2004) make markets not relevant for forest ecosystem services provision and dangerous when "facts are uncertain, values in dispute, stakes high and decisions urgent" (Funtowicz and Ravetz 1991). Post-normal scientific paradigm which is expected to tackle arising challenges puts strong emphasis on values and preferences as instruments and indicators of decisions consistency and governance support. Policy and decision-makers' own value system and its cultural environment play an important role in forest goods and service consumption.

In contradiction to private goods and services, markets fail to measure value of ecosystem services, to signal their scarcity, or even to provide incentives to supply them (Costanza et al. 1997; Farley et al. 2009). Hence another mechanisms then market prices should drive environmentally-sound decision-making. Environmental economists make serious efforts to extend a scope of a market and to impose methodology of monetary valuation of nonmarket goods and services applying revealed (hedonic pricing, travel cost and averting expenditure methods) and stated (contingent valuation, choice experiment) preferences techniques (Arrow et al. 1993; Hanley and Spash 1998). The first group of the methods is applied for measuring a use value when the second one predominates for nonuse values. Despite the critical flaws over serious methodological limitations such as the "warm glow effect", rationally inconsistent responses, embedding effects, and payment for moral satisfaction (Hanemann 1991; Hausman 1993; Portney 1994; Bateman et al. 2002), contingent valuation is considered to be useful in case of non-marketed goods and services.

Nonexcludable and non rival forest ecosystem services in general and MFES particularly are closer to 'social states' than to the concept of 'monetary value', therefore environmentally-sound forest decision-making should be based on social preferences for different social states instead of prevailing cost-benefit analysis which misses all items without monetary values (Kant and Lee 2004). Hence social choice theory modified to address features of forest decision-making could provide relevant theoretical background for political decisions and economic instruments.

Despite scientists have identified ecosystem services (Costanza et al. 1997; Millennium Ecosystem Assessment 2005) and North American stakeholders' attitudes concerning MFES (Kearney et al. 1999; Kant and Lee 2004), preferences of Carpathian stakeholders remain undiscovered. Their investigation is important for understanding and predicting behavior trends concerning forest resource use and management in the region, as well as possible points of MFES use conflicts.

The paper aims to: (1) assist decision-makers in understanding stakeholders' perceptions concerning MFES by means of a forest values universe development; (2) provide decision-makers and stakeholders with relevant statistically significant cognitive maps for getting insight of stakeholders' perceptions and preferences, and finally (3) provide forest-related decisions and governance a needed consistency and support.

We assume that forest values universe to be constructed according to stakeholders' statements will cover nine tenths space of forest ecosystems services identified in the report "Millennium Ecosystem Assessment" (2005) but with different ranks and frequency of mention because of stakeholders' ignorance about a genuine complex nature and a real role of the systems under consideration. We assume that individual maps of preferences concerning MFES, described by respondents, will be more or less similar while groups' maps will significantly differ.

The paper is structured as follows: first, an overview of the methodology is presented together with the study area, subject of study, and methods used for data collection and analysis; then the forest values universe and the set of cognitive maps of preferences are presented and discussed; finally we conclude with some observations related to forest universe, cognitive maps and policy and decision making based on the research findings.

2 Methodology

Currently prevalent market-oriented stated and revealed preferences techniques for measurement environmental values rely on regression analysis, which estimates effects of independent variables (like income, age, gender, etc.) on the dependent one (willingness to pay) and depend a lot on the specification of an estimated model, its shape, pre-selected variables, and method of estimation. But modern theory of economic valuation of the environment does not provide strong recommendations resulting in high quality model with clear interpretation of the estimations because the real nature of the investigated complex systems remains a black box for an investigator (Meadows 2008).

Synergetic nature of ecological-economic systems gives rise to development and wider application of the social choice approach which involves both quantitative and qualitative techniques of public attitude examination. Qualitative research provides an in-depth understanding of human behavior and the reasons that govern it. The Q-method, multi-attribute utility analysis and the Conceptual Content Cognitive Mapping (3CM) method enable researchers to avoid market analogies and resulting from them an integration of non-meaningful monetary estimations into decision-making, which could lead us to wrong decisions from the inter- and intra-generation perspectives.

The Q-method enables scientists to identify relevant factors and their importance to respondents by means of the correlation analysis but in this case for qualitative (ordinal) variables like views and preferences of respondents. The Q-method is designed to study respondents' subjectivity reducing the many individuals' view-points of the subjects down to a few "factors", which represent shared ways of thinking. Hence, it combines the strengths of both qualitative and quantitative research methods and provides an insight of subjective viewpoints on the question/item under consideration (Stephenson 1963; Brown 2010). Application of the Q-methodology to woodlands, forestry and forest decision-making one can find in Nijnik and Mather (2008) and (Nijnik et al. 2009). They provide a comprehensive insight of stakeholders' perceptions, behavior and reasons that govern it.

An alternative method of aggregated multifaceted ecological indexes (Rozhko 2000) allows experts to quantify, compare and integrate recreational, aesthetical, scientific, cultural, historical and ecological values of inanimate components of a wooded landscape, such as a waterfall or a lake, a rock or a mountain, into the aggregative ecological index. Relevant scales facilitate the translation of qualitative estimations into quantitative, ordinal data. In the pilot research condacted for inanimate nature objects of National Nature Park "Skolivski Beskydy", located in the Ukrainian Carpathians, these indexes were compared with respondents' WTP for the objects conservation. Strong and statistically significant correlation between these variables shows that experts' assessment, aggregated in the multifaceted ecological indexes, and respondents' WTP coincide with a high accuracy. With a probability of $p = 0.94$ one can state that a unity increase in the aggregated index results in a 0.98 Ukrainian Hryvnya increment of WTP (Podolchak and Zahvoyska 2005). These findings let us state that aggregated multifaceted ecological indexes could be successfully used for elicitation, quantification and comparison of perceptions and estimations concerning environmental objects.

The third alternative approach, the 3CM method, is an open-ended technique for investigation of respondents' values and preferences concerning processes or phenomena in question (Kearney and Kaplan 1997; Kearney et al. 1999). Kant and Lee (2004) base the use of multi-group social choice method in order to determine stakeholders' preferences. This non-market oriented stated preference technique features such peculiarities of forest services as a joint production of goods and services, inter- and intra-generational distributional issues and erroneous substitution of values. Moreover, this approach deals much better with a continuum of use and non-use values and normative questions because it considers an individual as a rational and responsible citizen that does not follow only its own profit and pleasure maximization. It allows developing a values universe which summarizes values verbalized by respondents.

In terms of the 3CM method, respondents are supposed to identify all relevant aspects (values associated with MFES), to group (organize), label and rank them according to their own views and feelings. On the next stage an open-ended card sorting 3CM technique enables a respondent to make an assessment of somebody's or other group's knowledge structure or attitudes related to a particular object or process.

The 3CM procedure employed in the study consists of such three steps (Kearney et al. 1999):

- identification of values (aspects). The survey began with familiarizing a respondent with the method features and guarantee of privacy. All forest-related values (items) identified by the respondent were written on index cards (one item per card),
- organization of identified values. Respondents were asked to organize into groups (clusters) all values (s)he verbalized,
- Group labeling. Respondents were asked to explain the idea behind grouping and to provide a label to each group.

To conclude the 3CM procedure we asked our respondents to range all labeled groups from the most to the least important one.

Later on results of 3CM exercise (stakeholders' attitude) could be checked and analyzed in a more precise way using non-parametric statistic. Kant and Lee (2004) proposed a comprehensive approach for development of statistically significant set of cognitive maps of stakeholders' preferences:

- the Friedman test was used to check at a 5 % significance level a presence of significant differences in preferences regarding MFES for each stakeholder group. In other words can we state with 95 % probability that each stakeholders group has own set of preferences, i.e., some values are more important then others. For instance, does ranking of environmental values statistically differ from ranking local values?

- the real order of preferences was checked using the Sign test. This test let us elicit (at 5 % significance level) a relevant ranking of forest values, their related importance for each stakeholder group. Results of these tests enable us to develop stakeholders' cognitive maps of preferences which reflect individuals' or groups' attitudes towards MFES.
- stakeholders' perceptions about other groups preferences were treated in the same way using the Wilcoxon and the Kruskal–Wallis tests.

Procedure of forest values elicitation and statistical treatment is presented on Fig. 1.

The study area for this research was the city of Lviv, Ukraine, and mountains of Lviv region. Carrying out the 3CM task, respondents assigned ordinal values the following way: '1' means 'the first' place, the most important forest value, '2' means the second place, less important then the first one but more important then rest of themes and so on. Therefore data, collected and organized using the 3CM, were considered as ordinal data and were examined by non-parametric statistical methods.

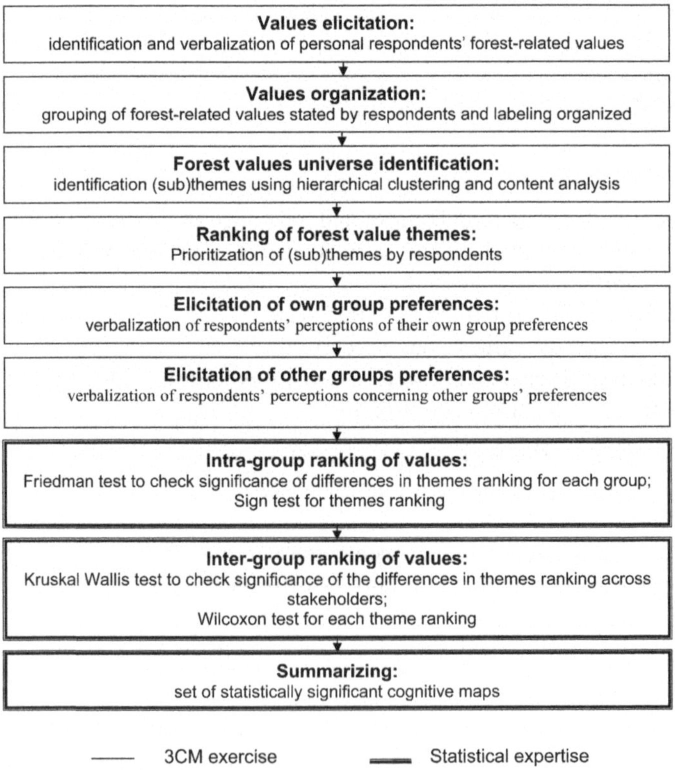

Fig. 1 Procedure of data collection and processing

In this study, the following criteria for identification of stake-holders group were taken into account: responsibility, impact, relationship, dependence, representation, and relevance (Hotulyeva et al. 2006). Hence our 3CM task was performed for these groups of stakeholders: local population, forest industry, environmental non-governmental organizations (ENGOs), and city population, beneficiaries of MFES.

A total of 105 respondents were approached. The response rate in the case study was 95 %. Twenty five representatives of each of four stakeholders groups completed the 3CM task. A modified snowball approach was applied: some respondents were selected from the initial contacts within each group. The interviews lasted from 40 min till 1.5 h.

3 Results and Discussion

In this chapter we present our findings concerning forest values identification and intra-group ranking of values using the Friedman test to check significance of differences in perceptions of forest themes for each stakeholder group and these themes ranking using the Sign test (Newbold et al. 2003).

3.1 Forest Values Universe

The forest values universe developed using 3CM consists of nine themes and 37 sub-themes (Zahvoyska and Bas 2009). The dominant themes are: environmental, recreational, economic, health care, tourist, aesthetical, local, educational, and cultural and emotional values. Frequency and spectral analysis of the forest values universe are presented at Table 1.

This table shows that the environmental values and little bit less recreational values were the most mentioned themes. Unlike them tourist values and health care received the smallest attention from the respondents. From the stakeholders perspective the forest values universe shows that local population verbalized the widest palette of MFES. Majority of empty cells (a kind of white spots) one can find in columns, which represent Environmental NGOs' and city inhabitants' perceptions.

More information about perceptions of MFES by stakeholders enables comparison of themes frequency in the forest values universe. For example city population underestimates all local values which are crucial to local population (Fig. 2).

Table 1 Forest values universe. Frequency analysis

Dominant themes	Sub-themes	Stakeholders			
		Local population	ENGOs	City inhabitants	Forest industry
1. Environmental	Air purification, Oxygen supply	84	96	100	84
	Climate regulation	24	32	12	24
	Biodiversity	36	52	36	32
	Water regulation	20	24	16	12
	Nutrient cycling	24	56	20	20
2. Recreational	Rest	48	25	96	80
	Hiking	36	40	36	32
	Picnics	40	40	52	32
	Pastime with friends	28	24	28	32
3. Economic	Income and benefits from forest industry spin off	80	88	48	68
	Timber and other marketed products	16	4	12	24
	Employment and relevant satisfaction	36	24	12	24
	Options for tourist and recreational business development	24	12	8	28
4. Local values	Non-wood forest products	100	48	36	48
	Wild animals meat and furs	24	–	4	–
	Firewood	80	8	4	16
	Fodder	44	–	4	–
5. Educational	Education and training	8	36	8	4
	Science and research	32	40	32	20
	Observations and monitoring	12	24	24	16
	Educational activities	24	36	16	20

(continued)

Table 1 (continued)

Dominant themes	Sub-themes	Stakeholders			
		Local population	ENGOs	City inhabitants	Forest industry
6. Health care and recovery	Health improving	32	56	68	40
	Medical herbs	24	40	44	40
	Vitamins	8	16	20	24
	Relaxation	20	12	20	28
7. Tourist	Hunting	4	4	8	8
	Rock-climbing	24	8	4	12
	Tourism	8	8	8	24
	Sports competitions	12	12	20	24
8. Aesthetic	Picturesque landscapes	28	28	48	20
	Wildlife observation	28	32	28	12
	Decorative items	24	8	32	12
	Scents and sounds	12	8	36	4
9. Cultural and emotional	Cultural and emotional	28	48	68	28
	Quietness, solitude, solitariness	28	28	60	44
	Inspiring, stimulation creative ability	24	58	52	12
	Attitude towards wildlife	40	28	36	24

The frequency of references to a particular value (row) by respondents of a particular group of stakeholders (column) is represented as a percentage

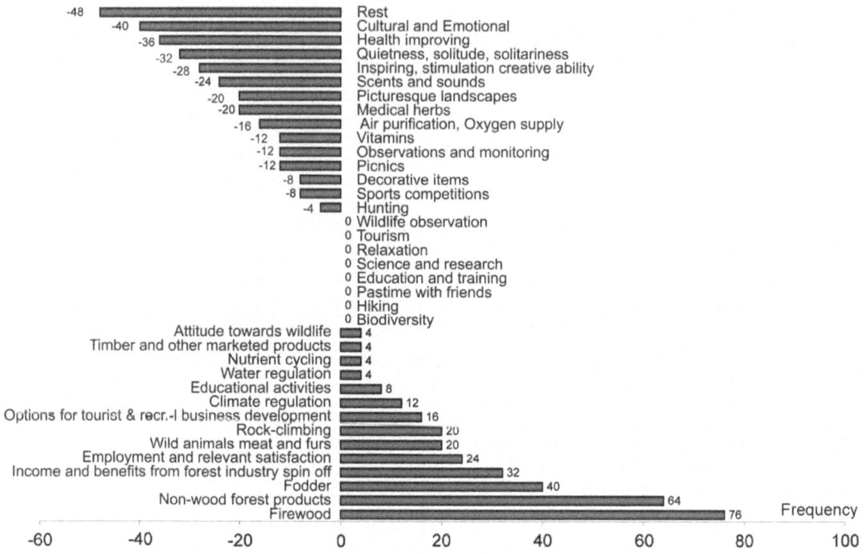

Fig. 2 Differences in the frequency of themes: local population against city population

3.2 Stakeholders' Preferences Regarding MFES

Investigation of the identified stakeholders' preferences regarding MFES with the non-parametric statistic methods (the Friedman test and the Sign test) allowed to generate statistically significant cognitive maps of preferences. Set of such maps which reflect perceptions of individuals and groups' values associated with MFES (both the own group and other groups) provided a comprehensive information for a comparison of values systems identified by respondents both from an individual and a group point of view.

Calculated Friedman statistics for checking statistical confidence of forest values differentiation range from 445.0 for ENGOs to 593.6 for the forest industry group whereas the critical value for a 5% significance level in case of nine themes equals to 15.51. They prove the fact of existing priorities regarding forest ecosystem services for each group of stakeholders.

A set of cognitive maps was developed using the Sign test (Table 2). Cognitive map of individuals' preferences is almost homogeneous. Most of the respondents appreciate the environmental and recreational values; and the cultural and emotional values take the third position. The city inhabitants and ENGOs are more sensitive to them. Economic services follow the previous items and hold the second (for local population and ENGOs) to the forth position (city inhabitants). The health care, educational and aesthetic values follow them. The last place belongs to the Tourist values. The interesting fact of the survey is that all stakeholders were unanimous in such ranking.

Table 2 Integrated cognitive map of stakeholders' preferences regarding mountain forest ecosystem services

Groups of stakeholders	Themes								
	Environmental	Recreational	Economic	Local	Educational	Health care	Tourism	Aesthetic	Cultural and Emotional
Local population	1/4	2/4	2/2	1/1	5/6	4/4	5/5	4/3	3/3
	3/4/3	5/4/3	2/2/2	1/1/1	6/6/7	6/5/6	6/5/6	4/3/4	4/3/4
Forest industry	1/3	2/1	3/1	5/6	5/5	3/3	5/4	5/6	4/2
	3/2/4	2/2/3	1/1/1	6/6/6	6/5/5	5/3/4	2/3/3	4/4/2	4/4/2
City inhabitants	1/1	1/2	4/4	6/6	4/5	2/1	5/3	3/3	1/2
	1/2/1	1/1/3	4/4/5	6/6/6	5/6/6	2/1/2	4/5/5	3/3/2	2/3/2
Environmental NGOs	1/1	1/2	2/4	5/6	3/4	4/3	6/5	4/1	2/2
	1/1/1	3/3/3	2/4/2	5/6/6	1/1/1	3/3/3	4/5/4	1/2/2	3/2/2

Carrying out the 3CM task respondents assigned ordinal values the following way: '1' means 'the first' place, the most important forest value, '2' means the second place, less important than the first one but more important than rest of themes and so on. Each set of cognitive maps is organised in two lines. In the first line an individuals'/groups' rankings of MFES are indicated. In the second line opinion of other stakeholders regarding appropriate group ranking

local population forest industry, city inhabitants, ENGOs

forest industry local population, city inhabitants, environmental NGOs

city inhabitants local population, forest industry, environmental NGOs

environmental NGOs local population, forest industry, city inhabitants

Participants' perceptions about the forest values from their own group's perspective are not so homogeneous: the first positions belong to the environmental and recreational values, but the local population considers them as the fourth item. All stakeholders rank the cultural and emotional values as a second item except the local population. The economic values are a little bit behind them: the forest industry and local population treat them as the most important, but two other stakeholders set them as the fourth ones. The health care and aesthetic values follow them, the tourist values seem to be of less importance, while the local and educational values are the last.

As for the stakeholders' perception of other groups' forest values, interest of the group the local population seems to be the most understandable for all other groups, but nevertheless they do not accept the crucial role of local values in their own group maps. The highest number of misunderstandings features the forest industry group. All other stakeholders believe that the tourism and aesthetic values should be more important for the forest industry, but both individual and group statements of this stakeholder do not meet these expectations. At the same time the forest industry's relatively high interest in the recreational, cultural and emotional themes is a bit unexpected in comparing with preferences of other groups. Also it is interesting to note that perceptions of the ENGOs are also obscure for other stakeholders. In particular, other stakeholders believe that educational values occupy the first place in the ENGOs' maps. However, these values are set on the third and fourth places in ENGOs' individuals and groups' cognitive maps accordingly. Also other stakeholders think that economic and tourist values should be more important for ENGOs. City inhabitants are more interested in the tourist values compared with other groups.

As one can see from Table 2 the most conflicts among stakeholders can be easily predicted for local values: the local population sets only these values on the first position in its maps whereas all other stakeholders are unanimous in their decision: the least interest and the last position. In our opinion a cause of this misunderstanding lies in asymmetric information and ignorance, and it should be corrected.

Another option for further development of the forest decision-making towards improvement of the well-being for the forest-dependent people is development of the tourist industry in the Ukrainian Carpathians and the local population expects that forest industry will induce it. Such development could serve as a step to mitigation of the above mentioned conflict and draw attention to the educational, aesthetic, cultural and emotional values which are of the least interest in stakeholders' maps.

In conditions of transition to a market economy, where there is no land market and where property rights are not well defined, conflicts among the stakeholders' interests are more common and a so called "political rent" and illegal agreements becomes a powerful instrument of their arrangement. Indeed, in this case short-term financial interests of individuals or groups do not drive society towards sustainable resource use. Therefore, exploration of society's attitude concerning vital MFES and resources provides an important message, helping to avoid the nontransparent decision-making and strong conflicts among stakeholders in conditions of weak institutions.

Statistically significant maps generated for the Ukrainian Carpathians case study highlight behavioral trends and value reasons for particular decisions lobbying in conditions of weak markets and ill-defined property rights, and can be used as a background for predicting possible and for understanding real-world conflicts and for their mitigation and resolution according to sustainability priorities. It could highlight strategies and priorities for development of the eco-efficient forest clusters in mountain regions for the common benefits of all stakeholders to achieve environmentally sound and just economic development.

4 Conclusions

The Millennium Development Goals point to a necessity of meeting demands of all stakeholders with a special attention to the improvement of the livelihoods of forest-dependent people. This provision is especially acute in a less developed economic environment where forest enterprises often are the only place of work for the local population. Mountain forest ecosystems, their health and quality strongly impact a lifestyle and welfare of the local population. Moreover often they provide even safety for it.

Recognition of social states nature of MFES forced us to apply a non-market stated preferences technique to elicit the respondents' knowledge and preferences concerning ecosystems role and values. Our findings obtained with the 3CM technique highlighted a quite broad universe of forest related values which cover all four groups of ecosystem services (Millennium Ecosystem Assessment 2005). Nevertheless the forest values universe confirms an assumption of Boyd and Banzaf (2006) that people are more convenient with ecosystems services which affect their wellbeing in a direct way (economic, local and recreational values) and are ignorant about supporting services like primary productivity, biogeochemistry etc., which are fundamental to their lives.

Our respondents have not mentioned traditional knowledge, which display 'ancient techniques and practices of a territory passed on through the generations and used for water harvesting, soil management, use and protection on natural resources, an extraordinary source of knowledge and cultural diversity from which appropriate innovations can be derived today and in the future' (Laureano 2006). This was a bit unexpected because in the case of urban forest we noted quite a strong interest to this item (Zahvoyska 2008). It seems that this invaluable cognitive resource is in danger. Therefore a global program launched by UNESCO on gathering/protecting historical knowledge and promoting/certifying eco-innovative practice based on re-proposal of traditional knowledge will be appreciated by society especially for the mountainous communities.

There is not strong difference between the forest values universe developed by the stakeholders from the Ukrainian Carpathians and those from the Northern

Ontario (Kant and Lee 2004). The forest values universes are organized in a slightly different way probably due to using different methods of data structuring. However, all of them reflect the breadth of environmental and socio-economic values regarding forest ecosystems resources and services.

The applied methods of the Conceptual Content Cognitive Mapping and the non-parametric statistics seem to be valuable tools for capturing and synthesizing views of various stakeholder groups regarding importance of various ecosystem services. Our findings obtained using these methods could be used by forest policy and decision making for identifying and mitigating conflicts in a forest resource use, for integrating stakeholders' preferences into multifunctional forest decision-making, and to obtain a public attitude and support for the effective forest governance.

In conditions of transition to a market economy, when land market and property rights are not well established, the stakeholders' interests conflicts are more common and so called "political rent" becomes a powerful instrument of their settlement. Indeed, in this case the short-term individuals' or groups' financial interests do not drive society towards sustainable development. Therefore such exploration of society's attitude concerning vital ecosystems resources and services provides society with an important message, which helps to avoid non-transparent decision-making and sharp conflicts among stakeholders.

References

Arrow KJ, Solow R, Portney PR, Leamer EE, Radner R, Schuman H (1993) Report on the national oceanic and atmospheric administration (NOAA) panel on contingent valuation. Fed Regist 58(10):4601–4614

Bateman I, Carson R, Hanemann M, Hanley N, Hett T, Jones-Lee M, Loomes G, Mourato S, Ozdemiroglu E, Pearce DW, Sugden R, Swanson J (2002) Economic valuation with stated preference techniques: a manual. Edward Elgar, Cheltenham

Boyd J, Banzaf S (2006) What are ecosystem services? The need for standardized environmental accounting units. RFF Discussing papers. Washington

Brown S (2010) The history and principles of Q methodology in psychology and the social sciences. http://facstaff.uww.edu/cottlec/QArchive/Bps.htm. Accessed 31 Aug 2011

Costanza R, D'Aarge R, de Groot R, Farber S, Grasso M, Hannon B, Limburg K, Naeem S, O'Neill R, Paruelo J, Raskin R, Sutton P, van den Belt M (1997) The value of the world's ecosystem services and natural capital. Nat 387:253–260

Daly H, Farley J (2004) Ecological economics. Principles and applications. Island Press, Washington

Farley J, Zahvoyska L, Maksymiv L (2009) Transdisciplinary paths towards sustainability: new approaches for integrating research, education and policy. In: Soloviy IP, Keeton WS (eds) Ecological economics and sustainable forest management: developing a trans-disciplinary approach for the Carpathian Mountains. UNFU Press, Liga-Press, Lviv

Funtowicz SO, Ravetz JR (1991) A new scientific methodology for global environmental issues. In: Costanza R (ed) Ecological economics: the science and management of sustainability. Columbia University Press, New York

Hanemann M (1991) Willingness to pay and willingness to accept: how much can they differ? Am Econ Rev 81(3):635–647

Hanley N, Spash C (1998) Cost-benefit analysis and the environment. Edward Elgar, Cheltenham
Hausman J (1993) Contingent valuation: a critical assessment. North Holland, Amsterdam
Hotulyeva M, Ye P, Vinnitchenko V, Cherp O, Yurkavichute A, Volostnov D, Dmitriyev A (2006) Stratyegucheskaya ekologicheskaya otsenka dlya razvitiya region-al'nogo i munitsipal'nogo planirovaniya. Ekoline, Moscow
Kant S, Lee S (2004) A social choice approach to sustainable forest management: an analysis of multiple forest values in Northwestern Ontario. For Policy Econ 6(3–4):215–227
Kearney A, Kaplan R (1997) Toward a methodology for the measurement of knowledge structures of ordinary people. The conceptual content cognitive map. Environ Behav 29(5):579–671
Kearney A, Bradley G, Kaplan R, Kaplan S (1999) Stakeholder perspectives on appropriate forest management in the Pacific Northwest. For Sci 45(1):62–73
Laureano P (2006) Traditional knowledge world bank for safeguarding ecosystems. In: Parotta J, Agnoletti M, Johann E (eds) Cultural heritage and sustainable forest management: the role of traditional knowledge. Proceedings of the conference held on 8–11 June 2006, Florence
Meadows D (2008) Thinking in systems: a primer. Chelsea Green Publishing, Vermont
Millennium Ecosystem Assessment (2005) Synthesis Report http://www.greenfacts.org/en/ecosystems/figtableboxes/map-forest.htm. Accessed 31 July 2011
Newbold P, Carlson WL, Thorne B (2003) Statistics for business and economics, 5th edn. Pearson Education, New Jersey
Nijnik M, Mather A (2008) Analysing public preferences for woodland development in rural landscapes in Scotland. Landsc Urban Plan 86(3–4):267–275. doi:10.1016/j.landurbplan.2008.03.007
Nijnik M, Zahvoyska L, Nijnik A, Ode A (2009) Public evaluation of landscape content and change. Land Use Policy 26(1):77–86. doi:10.1016/j.landusepol.2008.03.001
Patosaary P (2006) Foreword. In: Parotta J, Agnoletti M, Johann E (eds) Cultural heritage and sustainable forest management: the role of traditional knowledge. Proceedings of the conference held on 8–11 June 2006, Florence
Podolchak M, Zahvoyska L (2005) Ecological-economic valuation of landscapes. Sci Bull USUFWT 15(1):270–273
Portney PR (1994) The contingent valuation debate: why economists should care. J Econ Perspect 8:3–17
Rozhko I (2000) Recreational assessment of natural tourist complexes for tourists needs. Case of the Ukrainian Carpathians. Dissertation, Ivan Franko Lviv National University
Stephenson W (1963) Independency and operationism in Q-sorting. Psychol Rec 13:269–272
Zahvoyska L (2008) Analysis of Stakeholders' preferences regarding urban park. In: Cesaro L, Gatto P, Pettenella D (eds) The multifunctional role of forests. Policies, methods and case studies. EFI Proceedings 55, European Forest Institute, Joensuu
Zahvoyska L, Bas T (2009) Deeper insight of stakeholders' values and preferences regarding forest ecosystem services. In: Zahvoyska L, Jöbstl H, Kant S, Maksymiv L (eds) Building insights of managerial economics and accounting towards sustainable forest management: Proceedings of the IUFRO unit 4.05.00 international Symposium, UNFU Press, Lviv, 17–19 May 2007

Part III
Human Dimension

Carpathian Sustainability: Linking Local Actions and Regional Visions

Jacek Kozak

The Carpathian environment provides numerous opportunities for economic development, including vast resources of land for agriculture, forests, water, and various minerals that contribute to the development of the mining industry. Quite recently, the Carpathian Mountains have become an important tourist region, offering a variety of attractions in summer and winter. Because they are far less advanced in terms of human development than their partners from Western Europe, the Carpathian countries are frequently under political pressure for rapid economic growth. On the other hand, the European societies are becoming increasingly aware of the value of the Carpathian heritage that requires strict protection. This is why both national and local governments representing mountain populations must carefully plan their actions in this unique environment to avoid harmful effects on a multitude of services provided by the Carpathian ecoregion. Hence, there is a continuous discourse among conservationists, businesses, local governments, and mountain populations regarding optimal methods of development, requiring a compromise between various demands and constraints. Debarbieux and Price (2012) have reflected on this issue in the global context, noting the contradiction between considering the mountains and their diversity as an asset of great value at the global level and the rights of the mountain people to decide about the future of the land where they live.

This part of the book introduces various facets of human activities in the Carpathians that were the focus of the 1st Forum Carpaticum in 2010. They are presented here in 11 chapters a selection of numerous presentations and posters with a diversity of themes that required allocating them to five thematic sessions. The cultural landscape and biodiversity are the primary concerns of the first five chapters, while tourism and local industrial and urban development are discussed in the remaining six. Two chapters discuss the Slovak area of the Carpathians, five

J. Kozak (✉)
Institute of Geography and Spatial Management, Jagiellonian University, Gronostajowa 7
30-387 Kraków, Poland
e-mail: jacek.kozak@uj.edu.pl

J. Kozak et al. (eds.), *The Carpathians: Integrating Nature and Society Towards Sustainability*, Environmental Science and Engineering,
DOI: 10.1007/978-3-642-12725-0_26,

examine the Polish region, and two suggest a truly pan-Carpathian approach. Two studies were located in the Central European Mountains outside the Carpathians, but they are highly relevant to the Carpathian region.

Boltižiar and Olah present a thorough study of long-term land use and land cover changes in four Slovak biosphere reserves (Poľana, Tatry, Slovak Karst, and East Carpathians). The period covered in this study reaches as far back as the late eighteenth century: the time of the first full topographic survey of this part of Europe. The specific goal of this study, in addition to revealing the rates and patterns of land use and land cover change, was to verify the efficiency of the national nature protection system. The authors found that in the studied biosphere reserves the core zone is the most stable one, or slowly recovers to the natural state after the cessation of human activities; they note also, however, evidence of dynamic land use changes in legally protected areas, which is a combined effect of natural disturbances and human activities in the most attractive mountainous areas. This is especially visible in the Tatra Mountains, where areas of recent developments coincided with forests devastated by catastrophic winds in 2004.

Angelstam et al. discuss in detail various threads of biodiversity discourse and its significance to ensure the optimal forms of protection for the Carpathian environment. This chapter, coauthored by a team of scientists from all of the Carpathian countries, consists of three major parts. First, two major visions of biodiversity conservation are discussed: one emerges from the naturalness paradigm, and the other refers to the concept of a cultural landscape. To illustrate these two visions six case studies located in six Carpathian countries are presented: Börzsöny, Hungary; Tatra National Park, Slovakia; Bieszczady National Park, Poland; Skole region, Ukraine; Apuseni Mountains, Romania; and the Lower Morava region, Czech Republic. Finally, the authors discuss several consequences of biodiversity conservation for local and regional policies. In conclusion, Angelstam et al. emphasize the need to combine top-down planning with bottom-up implementation in conservation management. They also propose an exchange of knowledge and experience gathered over the long-term in different countries and regions (and, in fact they provide themselves a perfect example of such an exchange and prove its value). Finally, they recommend the development of transdisciplinary professions able to facilitate ecosystem management at the landscape scale. Similarly, the question of how to manage and maintain biodiversity is subsequently discussed by Ambroży and Grodzki in a short chapter that introduces a research project aiming at evaluating the risk of forest biodiversity and productivity loss in the context of climate change, attempting to link this knowledge with forest management practice.

The first three chapters highlight the importance of Carpathian cultural landscapes and biodiversity conservation, which is analyzed from various perspectives in the next two chapters. Policies that are relevant to the conservation of historical agricultural landscapes in Slovakia are thoroughly analyzed and discussed by Špulerová. In this chapter, the author defines historical agricultural landscapes as a special type of a cultural landscape, with specific forms of historical settlements and persisting traditional agricultural land use, claiming that these landscapes have irreplaceable ecological, cultural, and historical values. Although Špulerová presents many legal acts that may contribute to the protection of historical agricultural landscapes in Slovakia, she concludes that none of them specifically focuses on this type of cultural heritage. The consequence may be a loss of historical agricultural landscapes or a significant reduction of their values.

Likewise, the concept of cultural landscapes is elaborated by Bastian et al. in a study of historical landscape elements in the German Free State of Saxony. Though outside the Carpathians, the study area has similar legacies as other Central and Eastern European regions, and hence the results may be useful for researchers and stakeholders working in the Carpathian Mountains. The methodology to delineate and classify historical landscape elements proposed by the authors combines a detailed classification scheme with spatial operations and statistical analysis. In addition, as with Špulerová, the authors conclude that the conservation of historical landscape elements is difficult, because these elements are not focused on by the public and have no strong lobby. This is why several actions at local, national, and European levels are needed. The authors also suggest that besides voluntary (nongovernmental and nonprofit) activities, tourism can play an important role in the maintenance of historical landscape elements, as it stimulates a desire to preserve cultural landscapes for recreation.

A link between landscapes and tourism is further explored by Zawilińska, who analyzes various aspects of the functioning of landscape parks in the Polish Carpathians and their role in the development of sustainable tourism. Landscape parks cover almost a quarter of the Polish Carpathian area, yet the protection level is low. Several landscape parks are renowned tourist destinations, with a rich tourist infrastructure; however some parks are located in poorly developed regions, especially those that were depopulated after World War II. According to Zawilińska, tourism development guidelines for landscape parks should take into account these substantial differences. The author proposes a theoretical framework for sustainable development of tourism in landscape parks, pointing out several difficulties that may arise in its implementation: organizational and financial weakness of landscape park administrations due to the insufficiency of legal measures, lack of coordination among local governments, and a low level of cooperation with tourist organizations. These factors constrain the accomplishment of a coherent vision of the role of landscape parks in sustainable tourism development at the local level. Finally, Zawilińska notes specific shortcomings of the Polish spatial planning practice. Consequently, the landscape parks of the Polish Carpathians will continue to be an arena of a number of conflicts related to tourism development.

While Bastian et al. indicate only the role tourism may play in the maintenance of cultural landscapes, Zawilińska considers tourism a major driver that may either contribute to the protection of valuable Carpathian landscapes or accelerate their destruction. Factors, opportunities, and consequences of mountain tourism are presented in the subsequent three chapters. Navrátil and Pícha propose a methodology to assess factors influencing a willingness to recommend a visit to tourist attractions. They investigate various forms of water occurrence in the South

Bohemian Region. Similar to Bastian et al., the study area is located outside the Carpathians, yet the proposed methodology may be applied in other mountainous areas. Data for this study were collected via questionnaires completed at selected tourist attractions in the study area, and then analyzed using multiple linear regression. The authors conclude that a willingness to recommend a site for a visit is influenced by pull motivations (especially a pleasant natural environment), perceived quality, and on-site experience. They note that enhancing positive experiences of visitors would be the best way to attract new ones; however, the increased number of visitors should not exceed the capacity of the site.

Easy access to the attractive mountain sites is an important issue in a region stretching across several countries. It is discussed by Jaudas who presents the concept of *via carpatica*: a transnational hiking route connecting the most attractive regions of the Carpathians. As the author states, *via carpatica* is still a vision today, but in the future it may significantly help to develop sustainable tourism in the Carpathians and offer a number of opportunities for Carpathian societies. Jaudas lists several advantages of *via carpatica*: stimulation of the local economy and employment; rediscovery of natural, cultural, economic, and social resources; strengthening regional identity; and international networking. These advantages should be an incentive to turn this pan-Carpathian initiative into reality.

In the next chapter, Mika looks at yet another aspect of tourism, second-home development, and provides an overview of its spatial patterns in the Polish Carpathians. As in other countries of the region, there are several locations in the Polish Carpathians where second homes significantly modified the mountain landscape. Various factors influence the density and amount of second homes, in particular the attractiveness of the region and its accessibility from the main urban centers. Mika also points to the complex history of second-home development in Poland, especially in the communist period, when various legal restrictions and ambiguity led to a chaotic spread of second homes with several negative environmental impacts. In conclusion, Mika states that the urbanization of rural areas in the Polish Carpathians is one of the most visible consequences of second-home development and that in several mountain regions its intensity negatively impacts the quality and functioning of the natural environment, causing various environmental and functional conflicts.

The last two chapters discuss important questions of development opportunities and competitiveness, examining the role of industry in the Carpathian rural areas, and analyzing the importance of towns and cities. Dej and Micek attempt to determine the influence of dominant enterprises on rural communities in the Polish Carpathians, assessing first a general pattern for the whole study area, and then, on a basis of three case studies (the communes of Dydnia, Jasienica Rosielna and Czarny Dunajec), examining in detail local impacts of selected companies. They record significant variations among the communes in which dominant enterprises are present and conclude that the influence of a large company depends on the specific character of its activity. Moreover, they observe little evidence that the dominant enterprises in Carpathian rural communities act as the engines of local economies. Levels of local entrepreneurship and employment opportunities may significantly influence the relation between a dominant enterprise and the local community; for example, companies operating in attractive tourist destinations offer better employment conditions, as many opportunities exist at the local job market. However, the striking outcome of this study is that specific features of the mountain environment have in practice a very low impact on the significance of dominant enterprises for local rural communities and profiles of their activities.

In the last chapter, Więcław-Michniewska discusses competitiveness of towns and cities in the Polish Carpathians. A methodological framework is proposed that combines various environmental and socioeconomic factors into one synthetic indicator. Based on the results of this study, the author concludes that while the natural environment might have limited the potential for urban expansion in a number of cases, the skillful use of environmental assets may also improve urban competitiveness. This is evident in the case of well-known centers of mountain tourism and winter sports (e.g., Zakopane). Więcław-Michniewska notes that while the level of competitiveness in the western and eastern part of the Polish Carpathians does not currently differ significantly, this will likely change in the future due to a more peripheral position of the eastern part.

The discussions held at the 1st Forum Carpaticum in 2010 revealed several critically important aspects of the coupled human-environmental system of the Carpathians: conservation of biodiversity and cultural landscapes, land use and land cover change, land management, tourism, urban and rural development, and preservation of traditional knowledge (Kozak et al. 2011). All these issues and related questions have scale-dependent contexts: from local to global and from global to local. At the European level, the Carpathians, together with other mountain ranges, are increasingly considered "Europe's ecological backbone" (European Environment Agency 2010), an asset demanding a high priority given to the conservation of its cultural and natural values, as vividly discussed in the chapters by Boltižiar and Olah, Angelstam et al., Ambroży and Grodzki, Špulerová, Bastian et al., and Navrátil and Pícha. However, the local perspective, much more frequently, focuses on common concerns of living standards. At this level, mountains may provide opportunities for increasing income and welfare of societies from tourism as documented in the studies of Zawilińska, Jaudas, and Mika; reveal complex links between local communities and dominant enterprises as analyzed by Dej and Micek; or constrain socioeconomic development as suggested by Więcław-Michniewska. Thus, finding the right solutions to the sustainability of the Carpathian Mountains is not an easy task: it requires a continuous interaction of top-down and bottom-up approaches, thereby combining conservationist visions with the needs of local development.

References

Debarbieux B, Price M (2012) Mountain regions: a global common good? Mt Res Dev 32 Suppl:S7–S11

European Environment Agency (2010) Europe's ecological backbone: recognising the true value of our mountains. EEA Report 6/2010, Copenhagen

Kozak J, Björnsen Gurung A, Ostapowicz K (2011) Research agenda for the Carpathians 2010–2015. Integrating nature and society towards sustainability. Institute of Geography and Spatial Management, Jagiellonian University, Kraków

Land Use Changes of UNESCO Biosphere Reserves in the Slovak Carpathians Since the Late Eighteenth Century

Martin Boltižiar and Branislav Olah

Abstract UNESCO biosphere reserves in the Slovak Carpathians are part of a worldwide system of protected areas with three basic functions: nature conservation, research and sustainable development. The topic of the study was the identification of past land use and its changes since the late eighteenth century within Slovak biosphere reserves. Land use development was identified from historical maps and aerial photographs. The most stable areas (with unchanged land use forms) as well as the intensity of land use change were identified within the core zones of the biosphere reserves. In general, the results justified the delineation of Slovak biosphere reserves from the point of view of land use stability and indicated several similarities between the studied areas and other areas in the Carpathian Mts.

1 Introduction

Land use as a result of the mutual relation between human society needs and natural conditions has undergone significant changes during the last centuries. These changes were caused generally by technological development, legal territorial status or social preferences. Preserved historical data sources (historical maps)

M. Boltižiar (✉)
Department of Geography and Regional Development, Constantine the Philosopher University in Nitra, Trieda A. Hlinku 1 949 74 Nitra, Slovakia
e mail: mboltiziar@ukf.sk

M. Boltižiar
Institute of Landscape Ecology, Slovak Academy of Sciences, Bratislava, Branch Nitra, Akademická 2 949 01 Nitra, Slovakia

B. Olah
Department of Applied Ecology, Technical University in Zvolen, Masaryka 24 960 53 Zvolen, Slovakia

J. Kozak et al. (eds.), *The Carpathians: Integrating Nature and Society Towards Sustainability*, Environmental Science and Engineering,
DOI: 10.1007/978-3-642-12725-0_27,

combined with current modern methods and tools (GIS, remote sensing) enable to identify the land use history of a certain territory with a relatively high accuracy. This method has been widely applied in the recent landscape ecological and geographical research (e.g., Füleky and Major 1993; Skanes and Bunce 1997; Bürgi 1999; Cousin 2001; Petit and Lambin 2002; Bender et al. 2005). In Central Europe the research is based on the common scientific background and availability of historical maps (Kolejka 1987; Otahel' et al. 1993; Žigrai and Drgoňa 1995; Lipský et al. 1999; Kozak 2003; Olah 2003, 2009; Olah and Žigrai 2004; Kolejka and Marek 2006; Petrovič 2006; Šolcová 2009a, b).

UNESCO biosphere reserves (BR) compose a worldwide system of protected areas with the aim to synchronize biodiversity conservation, economic and social development and local cultural values. Achieving this goal is possible only through sustainable use of the landscape and its resources. Biosphere reserves thus offer a rare opportunity to study mutual relation between the society and nature, to describe long term land use development and to identify recent land use trends. The Sevilla strategy for biosphere reserves explicitly stresses the need to use BRs for research of sustainability indicators (UNESCO 1996).

Each BR is divided into 3 zones (core, buffer, transition zone) that correspond to local natural conditions and values (according to national nature conservation) and economic use. The core area should represent the most valuable parts of a BR dedicated to nature conservation and research activities. A buffer zone protects the core area and its exploitation is carried out in a sustainable way. The surrounding transition zone is an area where more intensive human activities (e.g., settlements, fields, roads) are located and it serves the everyday life of local inhabitants.

The Slovak national parks and protected landscape areas (as a national base for the UNESCO BR program) have only been protected since the 1950s and declared biosphere reserves since 1977 (Fig. 1). The selection of the new BR zones reflected knowledge on natural quality and conservation preference but the exact information on the previous land use of their parts was not available at that time. To fill this gap, we aimed to study long-term changes in the land use within the BRs' zones. The stability of the land use might be applied as a proxy indicator of ecosystems' ecological integrity. The results of a historical land use survey could therefore be applied in landscape and nature management (identification of possible unstable or sensitive areas), declaring new or adjusting the borders of existing conservation areas or BRs' zones. Moreover, as two of the Slovak BRs are bilateral (Tatry Mts. and Slovak Karst) and one is trilateral (East Carpathians), the identification of similar land use change trends within the Carpathian countries could be applied in closer and more effective cross-border cooperation in landscape management and biodiversity protection.

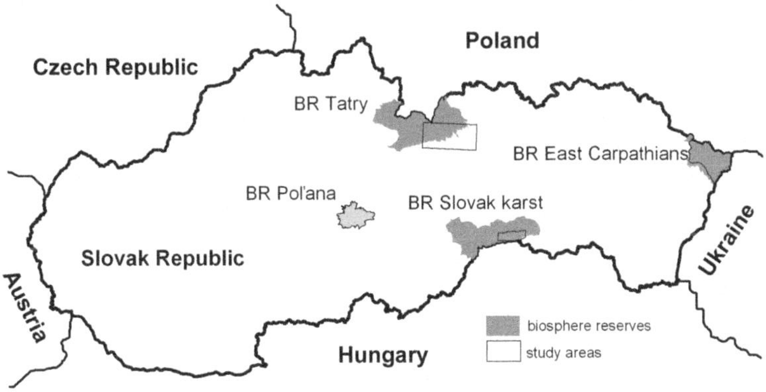

Fig. 1 The study areas of the UNESCO Slovak biosphere reserves

2 Study Areas

The Poľana BR is situated on the slopes of the Poľana stratovolcano and the Sihla plateau. Its altitude rises from 450 to 1,458 m a.s.l. and the total area is 20,760 ha. The territory (especially the southern part) is noted for its preserved dispersed settlements with unique traditional land use creating a rare landscape character. The BR is almost ideally zoned: the core area is circled by the buffer zone and the following transition zone.

The study area of the Tatry BR was chosen as a rectangle 32 km by 15 km, with the area 49,000 ha situated in the Popradská Basin and the High Tatra Mts., the elevation range is 600–2,656 m. Since it was not possible to analyse the whole Tatry BR, the study area was designed in a way that it includes all the BR zones and nearby towns and villages—therefore a representative part of the BR.

In the Slovak Karst BR, the Turnianska basin was chosen as a representative area for a land use development study. The study area covers the Horný vrch and the Dolný vrch karst plateaus, their slopes and the basin bottom with local villages. The total area is 8,285 ha and the altitude rises from 169 to 831 m a.s.l.

In the East Carpathians BR, the study area covers 83 % of the whole BR (the basins of the Starina, the Uličský and Zbojský streams). The total area is 34,220 ha and the elevation range rises from 232 to 1,180 m a.s.l.

3 Materials and Methods

Land use in the studied time periods was identified from historical military maps (Austrian Military Surveys 1772–1784, 1822–1854, 1900, and Czechoslovak Military Surveys 1956 and 1988) at scales of 1:25,000–1:28,800 (more in Olah 2009) and aerial orthophotographs (1949, 1987 and 2009) at scale of 1:5,000.

Historical maps were scanned and georeferenced using affine transformation (with RMS error ca. 200 m in the oldest and 5 m in the more recent maps). To be able to compare the identified land use forms from different sources that varied both in thematic and spatial resolution, the common broad land use categories (usually referring to the oldest map source) were applied: forests, woodland/shrubs (area covered with a mix of grasslands, trees and shrubs, solitary trees and/or bushes in fields), grasslands, vineyards, fields, built-up areas, water, wetlands, rocks and debris, alpine and subalpine vegetation. Land use changes were identified by overlaying historical land use layers. The overlaid land use vector maps created a spatial database of land use changes.

In order to quantify land use changes and to assess their multi-temporal trends each land use category was given a land use intensity coefficient:

1. Dominance of natural (autochthonous) species and ecosystem structure (forests or other natural ecosystems as natural grasslands),
2. Dominance of natural species with altered ecosystem structure or size (transitional woodland/shrub, non-forest woody vegetation),
3. Altered (originally not dominant or allochthonous) species but with natural ecosystem structure (secondary grasslands),
4. Prevailing introduced (allochthonous) species with managed ecosystem structure (arable land or permanent crops),
5. No vegetation or no natural ecosystems (built-up areas, open quarries).

The coefficients of land use intensity express the degree of anthropic impact (or vice versa naturalness) of land use categories regarding ecosystem dominant (plant) species and their structure (Kandrík and Olah 2010). The total land use change intensity was then calculated as a sum of partial intensity subtractions:

$$I_R = i_{2-1} + i_{3-2} + i_{n-m} \tag{1}$$

or

$$I_A = |i_{2-1}| + |i_{3-2}| + |i_{n-m}| \tag{2}$$

where I_R is a relative land use change intensity, I_A is an absolute land use change intensity, i_{2-1} is a partial land use intensity change between the 1st and the 2nd time horizon, m is the former time horizon, n is the later time horizon.

The land use change intensity was distinguished as relative or absolute. Relative intensity of land use refers to overall direction of land use changes. Positive values express land use intensification (transition to highly impacted ecosystems) and negative values refer to land use extensification (decreasing direct anthropic impact). Absolute intensity of land use change expresses the total amount of changes in land use regardless of their direction. This expression is useful to point out the least stable land use spots within the landscape.

4 Results

The land use development in the Poľana BR between 1782–2009 is described in Fig. 2. The most significant changes occurred in the area of forests and grasslands. The stable land use area (unchanged since 1782) covered 72 % of the total BR area (almost 95 % of it is forest, 5 % is grassland). The BR core area consists of stable forests (Fig. 3). The buffer zone is mostly covered with stable forests and grasslands while in the transition zone the less stable areas are situated. Generally, the spatial distribution of changes follows the criteria of BR zoning. The most stable and thus the most valuable ecosystems lay in the core area and less stable (changed) and more intensively used ecosystems lay in the transition zone.

Within the core area, the high intensity of change is very rare (Fig. 4). In the buffer zone, these areas coincide with the existing grasslands. The majority of the most changed areas are located within the southern and western part of the transition zone. These areas surround local villages and dispersed settlements; therefore their land use is the most intensive. The intensification of land use prevailed on locations with lower altitudes and slope inclinations, and S, SW and W aspects. On the other hand, the extensification of land use occurs mostly at higher altitudes, with steeper slopes and N, NE and E aspects.

The land use development of the Tatry BR is described in Fig. 2. The areas with unchanged land use cover 43 % of the studied territory (Fig. 5). 36.9 % of the stable polygons are fields and 23.6 % are forests. Another almost one-third of the

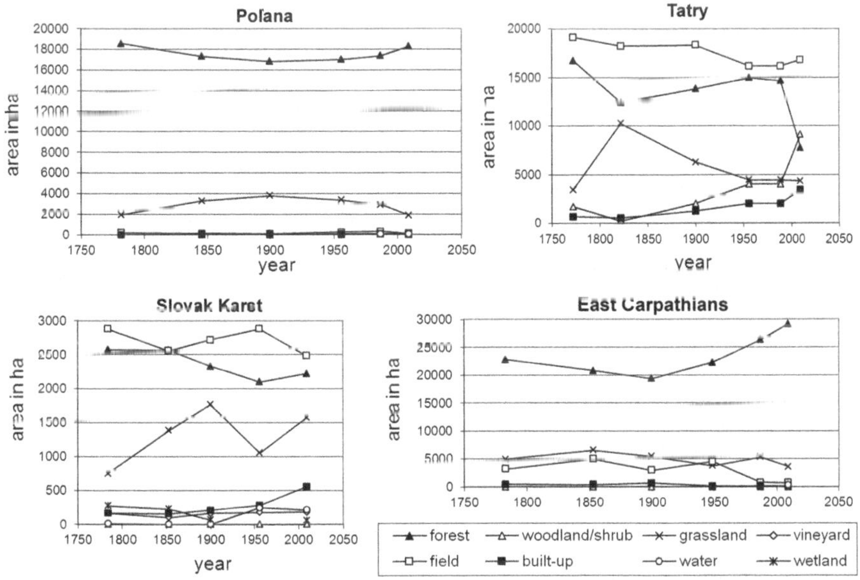

Fig. 2 Land use changes in the Poľana BR (1782–2009), Tatry BR (1772–2009), Slovak Karst BR (1784–2009), East Carpathians BR (1783–2009)

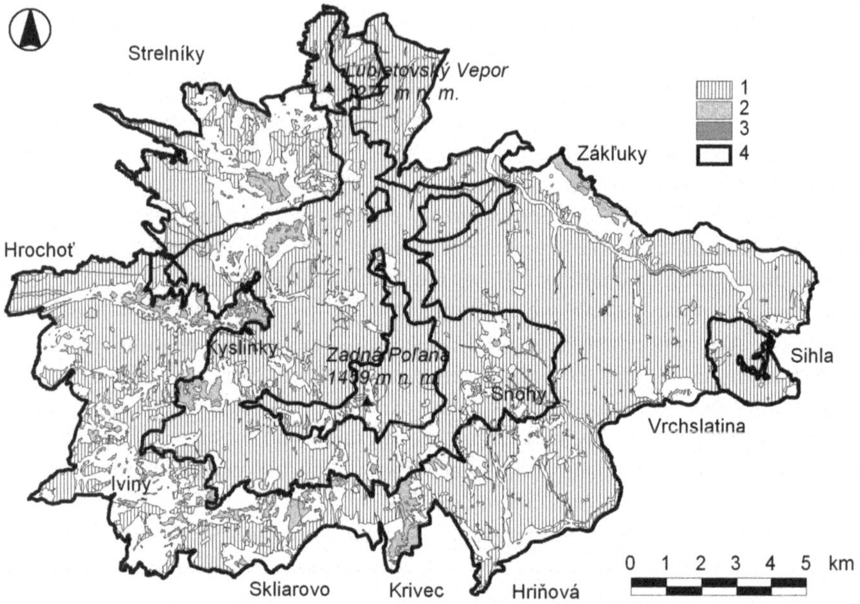

Fig. 3 Areas with stable land use in the Poľana BR between 1782–2009 (*1* forest, *2* grassland, *3* field)

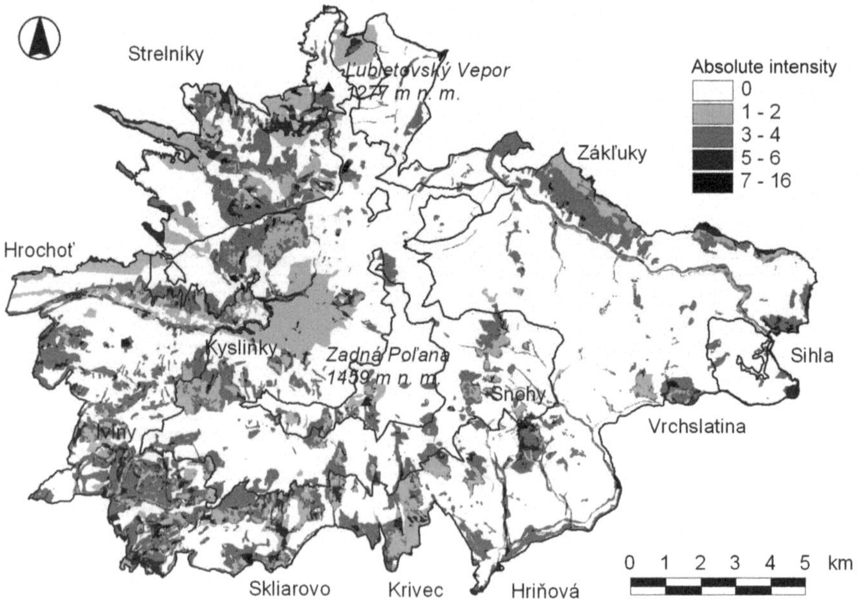

Fig. 4 Absolute intensity of land use change in the Poľana BR (1782–2009)

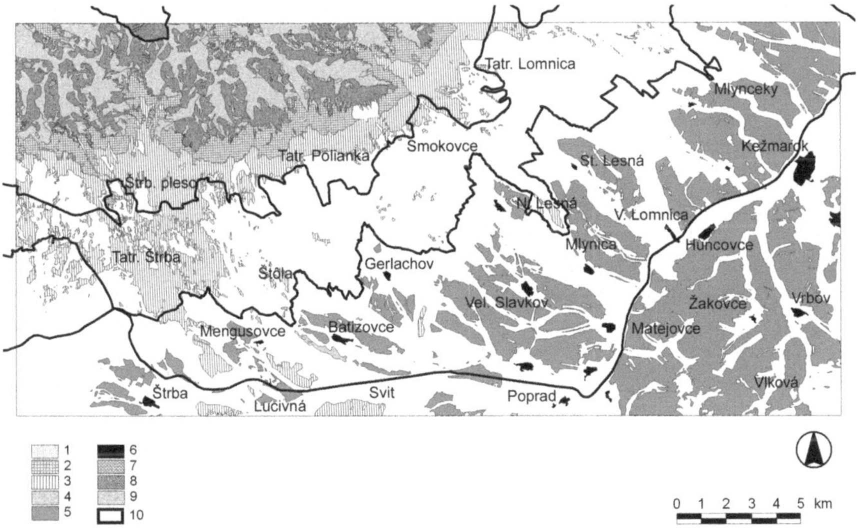

Fig. 5 The areas with stable land use in the Tatry BR in 1772–2009 (*1* alpine meadows, *2* subalpine vegetation, *3* forest, *4* grassland, *5* field, *6* settlements, *7* water, *8* rocks, *9* debris)

study area is the high mountain landscape covered by subalpine and alpine vegetation, rocks and debris. The stable built-up areas cover only 1.1 %.

Only minimal land use changes were recorded during the last 227 years in the core area where stable forests and high mountain areas prevail. Areas with at least one change surround settlements (Štrbské Pleso, Smokovce and Tatranská Lomnica) and documented nature catastrophes (the burnt areas under Slavkovský peak) and deforested areas in the eighteenth and nineteenth century (Fig. 6). Approximately one-third of the buffer zone is covered with stable forests, another part has been significantly changed. Most of the changes occur along the boundary between the buffer and transition zones. It is the area of frequent land use fluctuations among forests, shrubs and grasslands, and later forest disturbances due to strong winds (the most recent disaster covering 12,000 ha of forest occurred in 2004). More alarming are the changes toward more intensive (built-up) land use forms in and around the settlements. The basin bottom exhibits more intensive land use forms such as villages, towns and fields. The higher parts (up to 800 m a.s.l) are more extensively used (grasslands and forest), but areas between 800 and 1,000 m a.s.l. are again intensively used as tourism centers. Little or no changes were recorded above 1,300 m a.s.l.

Surprisingly, in the area not belonging to the BR changes were less evident (with the exception of the towns of Poprad and Svit) compared to the buffer and the transition zones. However, these zones did and probably will experience more land use changes due to the 2004 extreme windstorm in forests and consequent tourism development plans. The results showed a close association between the

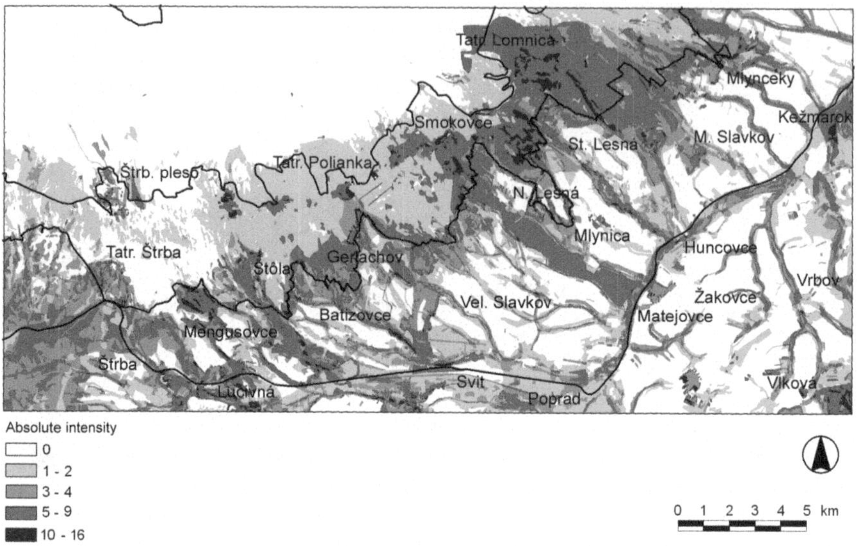

Fig. 6 Absolute intensity of land use change in the Tatry BR (1772–2009)

most labile areas and the occurrence of wind calamities and therefore could be used as an argument for the revitalization and management of the BR.

The overall view of the Turnianska basin (Slovak Karst BR) land use development is shown in Fig. 2. The most significant changes during the last 225 years occurred in the coverage of fields, forests, grasslands and shrubs. Other categories have remained almost at the same level since 1784, only built-up areas have recently increased. The areas without land use changes (Fig. 7) covered 34.6 % (2,865.7 ha) of the study area (out of this, forests were 49 % and fields 43 %).

The southern part of the core area as well as the surrounding buffer zone is almost entirely covered with stable land use areas (Fig. 8). The situation in the northern part is completely different. Only a part of the core area is unchanged, the rest has been changed at least once (forests-shrubs-grasslands). The situation within the buffer zone is very similar. The analysis of absolute land use change intensity (Fig. 8) shows that the most significant changes occurred in the geomorphologic transition zones between basin bottom and plateaus. Their occurrence very closely follows the borders of the BR zones, in fact justifying their correct demarcation. The core areas experienced least changes, the buffer zones experienced more, but these were mostly changes between forests and shrubs. The transition zone occupying mostly the lower part of the basin has been changing to either more or less intensive land use forms. The changed areas lie along the streams, around settlements and where former wetlands were turned into dams or drained to allow cultivation.

Generally, the intensification of land use prevails in the whole BR territory. The lower parts (up to 400 m a.s.l.) are the most dynamic areas. In this landscape type, very significant land use factors are slope inclination and aspect due to the karst

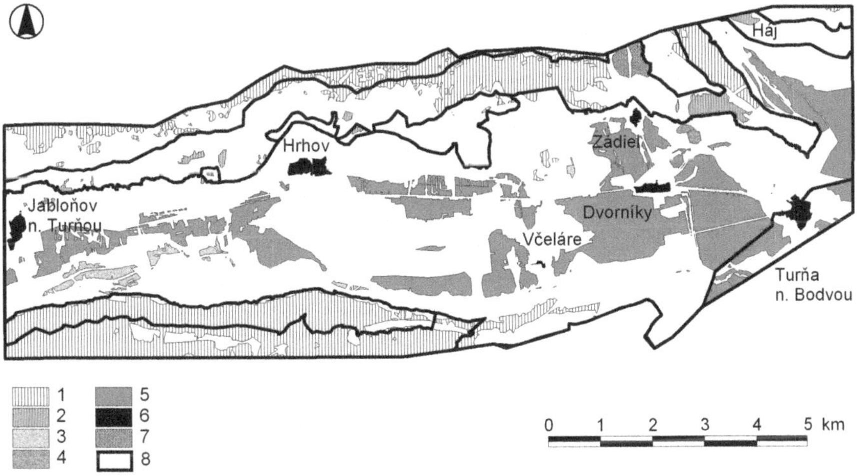

Fig. 7 The areas with stable land use in the Slovak Karst BR in 1784–2009 (*1* forest, *2* shrub, *3* grassland, *4* vineyard, *5* field, *6* settlements, *7* rocks, *8* BR zones)

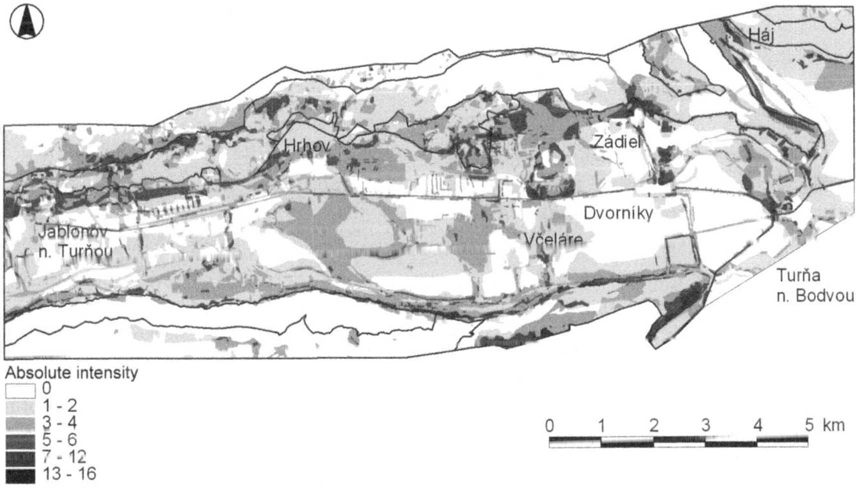

Fig. 8 Absolute intensity of land use change in the Slovak Karst BR (1784–2009)

relief and climatic conditions. The slopes with the northern aspect are covered with unchanged forest while the south facing slopes have turned to a mixture of xerotherm forests, shrubs and grasslands.

The overview of the land use changes in the East Carpathians BR is presented in Fig. 2. The most significant change was related to forests, and their area decreased until the beginning of the twentieth century and since then has increased continuously.

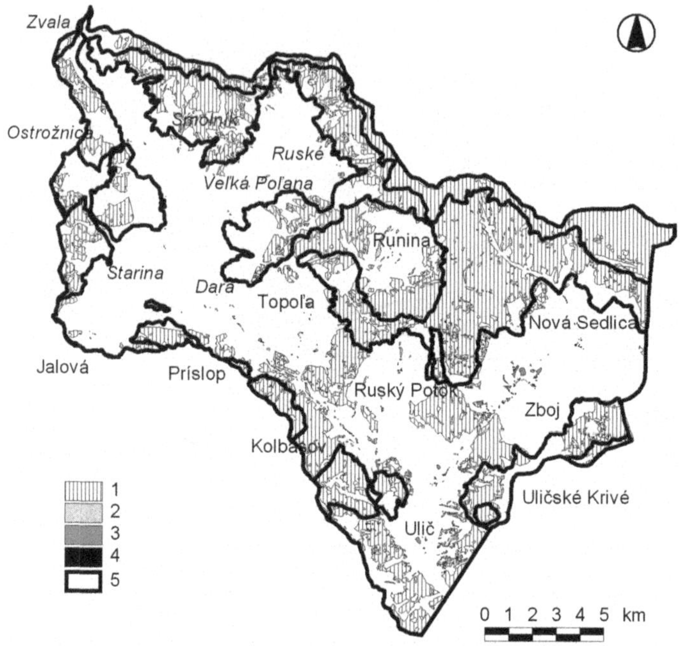

Fig. 9 Areas with stable land use in the East Carpathians BR in 1783–2009 (*1* forest, *2* grassland, *3* field, *4* settlements, *5* BR zones)

The unchanged land use areas (Fig. 9) cover 35 % of the BR area, 99 % of it is forest, the rest is divided into other categories. The core area along the Slovak-Polish border is almost unchanged. The other core areas consist of changed areas but are oriented towards less intense land use forms. In the buffer zone, stable forests prevail. Changes of land use occurred along the border between buffer and transition zones, however, several areas show also signs of extensification (Fig. 10). Most significantly changed areas lie within the transition zone. Its western part has been changed due to the forced emigration of local inhabitants and consequent land abandonment since the Starina water reservoir construction in 1987. Therefore, the changes follow the direction toward less intensive forms (fields-grassland-shrubs-forests). In the eastern part where settlements have been preserved, the land use changes occurred mostly around the villages. The closest areas became built-up or remain in cultivation; the remote parts have undergone extensification towards shrubs and forests (Olah et al. 2006).

In the East Carpathians, as in other studied BRs, the most important natural factors affecting land use changes were altitude and slope inclination. Aspect was less important. However, the majority of changes were directly caused by socio-economic factors—the construction of the water reservoir and marginal location of the whole area within the Slovak republic, coupled with an economic regression over the last decades. Thanks to these factors, the territory is one of the best preserved natural areas in Slovakia.

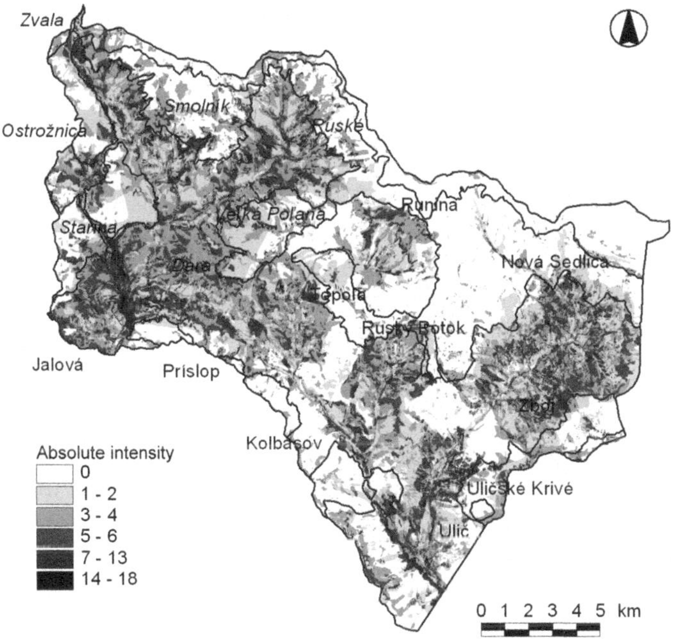

Fig. 10 Absolute intensity of land use change in the East Carpathians BR (1783–2009)

5 Discussion and Conclusion

Summarizing the long term land use study results we can conclude that the Poľana BR is a highly stable natural landscape. The more intensive land use forms lie outside its border. The area is covered by forests with grassland enclaves (with secondary succession due to the remoteness and low economic profitability of use in the last 50 years).

The study area of the Tatry BR is characterized by a highly stable basin bottom (with fields, settlements and rapidly developing towns) and an unchanged high mountain landscape. Most changes occurred in the transition zone between basins and mountains. In this zone, continuously growing tourism centers and periodical wind calamities can be considered as very significant land use change factors.

The Slovak Karst BR has been very stable for the last 200 years. The fertile basin bottom was ameliorated (wetlands turned to fields or water reservoirs), the plateaus and northern slopes remained covered with forests and southern xerotherm slopes were covered with secondary shrubs and grasslands.

Land use development of the East Carpathians BR has been significantly affected by two phenomena. First, the remote and marginal position of the whole territory led to creation of small villages and agricultural (cattle pasture) or forest usage of the landscape. Then, removal of 7 villages due to the construction of water reservoir in 1987 led to land abandonment and rapid secondary succession.

Despite their different natural and socioeconomic settings the land use trends in the Slovak BRs show certain similarities. First, the most stable land use was bound to low-altitude areas with low inclination on one hand (fields and settlements on basin and valleys bottoms) and to higher parts with steeper slopes and northerly aspects (high mountain vegetation, forests, shrubs, locally secondary grasslands) on the other. Most changes have occurred between those two locations. Next, intensification of land use prevailed at lower altitudes, extensification in higher and remote locations. The exceptions were only recorded where the land use was affected by new socio-economic phenomena such as tourism centre development in the Tatry BR and the water reservoir construction in the East Carpathians BR. Finally, the declared zones of all Slovak biosphere reserves fulfill their basic requirements from a land use stability point of view. The core areas represent the most stable territories, without any significant land use changes. The transition zones consist of relatively unstable land use areas (with more intensive use). The buffer zones are somewhere in the middle. This pattern is altered only in the Tatry BR and the Slovak Karst BR due to historical (deforestation and forest grazing) and current factors (windstorms and tourism development).

The studies of landscape changes from Slovakia or the other post-socialist Carpathian countries (the Czech Republic, Poland, Ukraine and Hungary) revealed similar long-term and recent trends: continuous land use intensification in lowlands (Cebecauerová 2007; Demek et al. 2008; Skokanová 2009), conversion of agricultural landscape since the 1950s (Kubeš 1994; Lipský et al. 1999) followed by land abandonment and forest expansion mainly in marginal areas after the 1990s (Bartoš et al. 1999; Chrastina and Boltižiar 2006; Kolejka and Marek 2006; Petrovič 2006; Björnsen Gurung et al. 2009; Chrastina 2009; Michaeli et al. 2009), and relatively stable mountain areas (Boltižiar 2006, 2007; Boltižiar et al. 2008) disturbed mainly by extreme windstorms (Falťan and Bánovský 2008).

The most recent transformation trends of the Slovak landscape correspond also with the pan-European land cover changes identified by the CORINE Land Cover methodology. Land use intensification (urban residential sprawl, construction of new industrial sites and transport infrastructure, and intense agriculture) and land abandonment with following afforestation occurring mainly on former permanent grasslands (EEA 2006, 2010).

The identified continuous increase of forest cover within the Slovak BRs supports findings of Kozak (2003) and Kozak et al. (2007) that reversal of trends from previous long period of deforestation to continuous forest expansion (forest transition) has been one of the dominant land cover change processes in the Carpathians since the beginning of the twentieth century. The forest expansion impacts overall carbon sequestration and forests are expected to be a significant carbon sink in the next 100 years (Kuemmerle et al. 2011). The latest results of CORINE Land Cover 2000–2006 change analysis show that the European forests, while still slightly expanding (0.1 % per year), are along with urban areas the most dynamic European land cover class with the annual turnover equal to 5.0 % of the total forest coverage (EEA 2010). The continuing land use polarization causes a loss of highly valuable landscape types (e.g., low intensity farmlands supporting biodiversity) and increases a pressure on the environment, human health and well-being.

The study results and their comparison with neighboring post-socialist countries and pan-European trends support the hypothesis that though local natural conditions pre-determine land use of the territory the land use and its changes are driven by the overall political and socioeconomic situation.

Acknowledgments The paper presents the results of two scientific projects, KEGA GP No. 023UKF-4/2011 "Terrain geoecological research as a base for creating of education equipment" and KEGA GP No. 025PU-4/2012 "Georelief and landscape structure".

References

Bartoš M, Těšitel J, Kušová D (1999) Marginal areas—historical development, people and land use. In: Kovář P (ed) Nature and culture in landscape ecology. Karolinum, Prague

Bender O, Boehmenr HJ, Jens D, Schumacher K P (2005) Analysis of land-use change in a sector of Upper Franconia (Bavaria, Germany) since 1850 using land register records. Landsc Ecol 20:149–163

Björnsen Gurung A, Bokwa A, Chełmicki W, Elbakidze M, Hirschmugl M, Hostert P, Ibisch P, Kozak J, Kuemmerle T, Matei E, Ostapowicz K, Pociask-Karteczka J, Schmidt L, van der Linden S, Zebich M (2009) Global change research in the Carpathian Mountain region. Mt Res Dev 29(3):282–288

Boltižiar M (2006) Changes of high mountain landscape structure in the selected area of Predné Meďodoly valley (Belianske Tatry Mts.) in 1949–1998. Ekol (Bratislava) 25(1):16–25. doi:10.4149_ekol_2006_01_16

Boltižiar M (2007) Štruktúra vysokohorskej krajiny Tatier (Veľkomierkové mapovanie, analýzy a hodnotenie zmien aplikáciou údajov diaľkového prieskumu Zeme. UKF, ÚKE SAV a SNK MAB, Nitra

Boltižiar M, Brůna V, Křováková K (2008) Potential of antique maps and aerial photographs for landscape changes assessment—an example of High Tatras. Ekol (Bratislava) 27(1):65–81. doi:10.4149_ekol_2008_01_65

Bürgi M (1999) A case study of forest change in the Swiss lowlands. Landsc Ecol 14:567–575

Cebecauerová M (2007) Analýza a hodnotenie zmien štruktúry krajiny (na príklade časti Borskej nížiny a Malých Karpát). Geogr Slovaca 24

Chrastina P (2009) Vývoj využitia krajiny Trenčianskej kotliny a jej horskej obruby. FF UKF, Nitra, Nitra

Chrastina P, Boltižiar M (2006) Kultúrna krajina SV okraja Bakoňského lesa v Maďarsku (súčasnosť v kontexte minulosti). Hist Geogr 34(1):175–188

Cousin SAO (2001) Analysis of land-cover transitions based on 17th and 18th century cadastral maps and aerial photographs. Landsc Ecol 16:41–54

Demek J, Havlíček M, Chrudina Z, Mackovčin P (2008) Changes in land-use and the river network of the graben Dyjsko-Svratecký úval (Czech Republic) in the last 242 years. J Landsc Ecol 1(2):22–51

EEA (2006) Land accounts for Europe 1990–2000. Towards integrated land and ecosystem accounting. EEA report 11/2006, European Environment Agency

EEA (2010) The European environment—state and outlook 2010: land use. European Environment Agency

Falťan V, Bánovský M (2008) Changes in land cover in the area of Vyšné Hágy—Starý Smokovec, impacted by the wind calamity in November 2004 (Slovakia). Morav Geogr Rep 3/2008(16):16–26
Füleky G, Major I (1993) Changes in the landscape during the last 200 years in the region of the Zala Valley and effect on the economic activity of the area. Landsc Urban Plan 27:265–267
Kandrík R, Olah B (2010) Land use development in the central part of the Spiš region (Slovakia) since the eighteenth century. Morav Geogr Rep 4/2010(18):28–38
Kolejka J (1987) Landscape-historical synthesis materials, methods and results. Ekol (ČSSR) 6(1):51–62. doi:10.4149_ekol_1987_01_51
Kolejka J, Marek D (2006) Sustainable land use convergence in border area in Central Europe. In: Vogtmann H, Dobretsov N (eds) Environmental security and sustainable land use—with special reference to Central Asia. Springer, Dordrecht
Kozak J (2003) Forest cover change in the Western Carpathians in the past 180 years. Mt Res Dev 23(4):369–375. doi:10.1659/mrd.0976
Kozak J, Estreguil C, Troll M (2007) Forest cover changes in the northern Carpathians in the twentieth century: a slow transition. J Land Use Sci 2(2):127–146
Kubeš J (1994) Bohemian agricultural landscape and villages, 1950 and 1990 land use, land cover and other characteristics. Ekol (Bratislava) 13(2):187–198. doi:10.4149_ekol_1987_01_51
Kuemmerle T, Olofsson P, Chaskovsky O, Baumann M, Ostapowicz K, Woodcock CE, Houghton RA, Hostert P, Keeton WS, Radeloff VC (2011) Post-Soviet farmland abandonment, forest recovery, and carbon sequestration in western Ukraine. Glob Change Biol 17:1335–1349. doi:10.1111/j.1365-2486.2010.02333.x
Lipský Z, Kopecký M, Kvapil D (1999) Present land use changes in the Czech cultural landscape. Ekol (Bratislava) 18(1):31–38. doi:10.4149_ekol_1999_01_31
Michaeli E, Ivanová M, Juhaščiková J (2009) Landscape structure changes in model village of Hlinné, Vyšný Žipov and Zlatník. Acta Fac Stud Humanit Nat Univ Prešoviensis. Prír vedy. Folia Geogr (Prešov) 14:98–109
Olah B (2003) Potential for the sustainable land use of the cultural landscape based on its historical use (a model study of the transition zone of the Poľana Biosphere Reserve). Ekol (Bratislava) 22(2):79–91. doi:10.4149_ekol_2003_02_79
Olah B (2009) Historical maps and their application in landscape ecological research. Ekol (Bratislava) 28(2):143–151. doi:10.4149_ekol_2003_02_79
Olah B, Žigrai F (2004) The meaning of the time-spatial transformation of the landscape for its sustainable use (a case study of the transition zone of the Poľana Biosphere Reserve). Ekol (Bratislava) 23(1):231–243. doi:10.4149_ekol_2004_01_231
Olah B, Boltižiar M, Petrovič F (2006) Land use changes' relation to georelief and distance in the East Carpathians Biosphere Reserve. Ekol (Bratislava) 25(1):68–81. doi:10.4149_ekol_2006_01_68
Oťaheľ J, Žigrai F, Drgoňa V (1993) Landscape use as a basis for environmental planning (case studies of Bratislava and Nitra hinterlands). In: Drgoňa V (ed) Geographical studies 2. University of Education, Nitra
Petit CC, Lambin EF (2002) Impact on data integration technique on historical land-use/land-cover change: Comparing historical maps with remote sensing data in the Belgian Ardennes. Landsc Ecol 17:117–132
Petrovič F (2006) The changes of the landscape with dispersed settlement. Ekol (Bratislava) 25(1):65–89. doi:10.4149_ekol_2006_01_65
Skanes HM, Bunce RGH (1997) Directions of landscape change (1741–1993) in Virestad, Sweden – characterised by multivariate analysis. Landsc Urban Plan 38:61–75
Skokanová H (2009) Application of methodological principles for assessment of land use changes trajectories and processes in the Czech Republic for the period 1836–2006. Acta Pruhoniciana 91:15–21
Šolcová L (2009a) Comparison of the Landscape Changes in Radobica, Veľké Pole and Píla During 1780–2007. Geogr Stud 13(2):87–92

Šolcová L (2009b) Present landscape structure of dispersed settlements (on example Veľké Pole). Geogr Inf 13(1):217–221

UNESCO (1996) Biosphere reserves: the Seville strategy and the statutory framework of the world network. UNESCO, Paris

Žigrai F, Drgoňa V (1995) Landscape-ecological analysis of the land use development for environmental planning (case study Nitra). Ekol (Bratislava) 14(1):97–112. doi:10.4149_e-kol_1995_01_97

Maintaining Cultural and Natural Biodiversity in the Carpathian Mountain Ecoregion: Need for an Integrated Landscape Approach

Per Angelstam, Marine Elbakidze, Robert Axelsson, Peter Čupa, Ľuboš Halada, Zsolt Molnar, Ileana Pătru-Stupariu, Kajetan Perzanowski, Laurentiu Rozulowicz, Tibor Standovar, Miroslav Svoboda and Johan Törnblom

Abstract Landscapes located in the periphery of economic development, such as in parts of the Carpathian ecoregion, host remnants of both near-natural ecosystems and traditional agricultural land use systems. Such landscapes are important both for in situ conservation of natural and cultural biodiversity, and as references for biodiversity restoration elsewhere in Europe. This paper first reviews the

P. Angelstam (✉) · M. Elbakidze · R. Axelsson · J. Törnblom
School for Forest Management, Swedish University of Agricultural Sciences,
SE-739 21 Skinnskatteberg, Sweden
e-mail: per.angelstam@slu.se

P. Čupa
Lower Morava Biosphere Reserve, Zamecke namesti 69 691 44 Lednice, Czech Republic

Ľ. Halada
Institute of Landscape Ecology SAS, Branch Nitra, Akademicka 2,
P.O.Box 23B 949 01 Nitra, Slovakia

Z. Molnar
Institute of Ecology and Botany of the Hungarian Academy of Sciences,
Vácrátót 2163, Hungary

I. Pătru-Stupariu · L. Rozulowicz
Faculty of Geography, University of Bucharest, 1, N. Balcescu Bd 010041 Bucharest,
Romania

K. Perzanowski
Carpathian Wildlife Research Station, Museum and Institute of Zoology, PAS, Ogrodowa 10, 38-700 Ustrzyki Dolne, Poland; and Catholic University of Lublin, Konstantynów 1H 20-708 Lublin, Poland

T. Standovar
Department of Plant Taxonomy and Ecology, Eötvös Loránd University,
Pázmány P. Sétány 1/c, Budapest H-1117, Hungary

M. Svoboda
Faculty of Forestry and Wood Sciences, Czech University of Life Sciences, Kamycka 129,
Praha 6 Suchdol 16521 Prague, Czech Republic

J. Kozak et al. (eds.), *The Carpathians: Integrating Nature and Society Towards Sustainability*, Environmental Science and Engineering,
DOI: 10.1007/978-3-642-12725-0_28,

contemporary understanding of benchmarks for biodiversity conservation in terms of ecosystems with natural disturbance regimes and pre-industrial cultural landscapes. Second, after providing a historical background, we review the challenges to natural and cultural biodiversity conservation and discuss current development trajectories. Third, we provide concrete examples from six Carpathian areas with different proportions of natural and cultural biodiversity. Fourth, we discuss the need for a diversity of management systems toward protection, management and restoration, spatial planning, and multi-sector governance for conservation of natural and cultural landscapes' biodiversity. Finally, we stress the need to encourage integration of management, planning and governance of social and ecological systems to maintain natural and cultural biodiversity. The natural vegetation of the Carpathian Mountains is mostly forests and woodlands. Natural disturbances as wind, snow, frost, fire and flooding as well as insects and fungi resulted in forests characterized by old and large trees, diverse horizontal and vertical structures, and large amounts of dead wood in various stages of decay. While some near-natural forests remain, in most of the Carpathian ecoregion pre-industrial cultural landscapes evolved. Human use created traditional village system with infield houses, gardens, fields, meadows and outfield meadows and pastures, and woodlands which not only provide ecosystem services but also represent cultural heritage. The maintenance of natural and cultural biodiversity may require active management of species, habitats and processes. However, designing management systems that emulate natural and cultural landscape's disturbance regimes is a major challenge requiring collaboration of private, public and civic sector stakeholders, and integration of social and ecological systems. Maintaining and restoring the traditional village system's social capital as well as functional networks of protected areas and implementing sustainable forest management in managed forests are thus crucial. The Carpathian ecoregion forms a quasi-experiment with new country borders that have created stark contrasts among regions regarding natural and cultural biodiversity. This ecoregion can therefore be seen as a landscape-scale laboratory for systematic studies of interactions between ecological and social systems to support the development of an integrated landscape approach to biodiversity conservation and cultural heritage.

1 Introduction

Since the emergence of the sustainable development discourse during the late 1980s, a range of international and national policies related to ecologically, economically, socially and culturally sustainable use of renewable natural resources have been formulated (e.g., Water Framework Directive 2000; Carpathian Convention 2003; Ioras 2003; Mayers and Bass 2004; Innes and Nitschke 2005). Stakeholders involved with management and governance of forests and cultural landscapes in rural regions are thus subject to the challenges of implementing ecological, economic and social objectives of sustainability policies on the ground (Norton 2005; Vucetich and Nelson 2010), and encouraging sustainable development as a societal steering process (Baker 2006). Traditional sustained yield forestry and agriculture are therefore required to supply a broad range of goods, ecosystem functions and landscape values rather than only wood, fibres, energy and food (e.g., Angelstam et al. 2005; Merlo and Croitoru 2005). This transition is closely linked to the conservation of biodiversity, i.e., the composition, structure and function of ecosystems (Noss 1990). While Central and Eastern Europe has a reputation of being dominated by polluted environments due to heavy industrial development during the period of socialism (Baker and Jehlicka 1998; Szaro et al. 2002), the biodiversity status is frequently better than in Western Europe (e.g., Puumalainen et al. 2003; Edman et al. 2011).

The implementation of sustainability policies on the ground needs to conceive landscape as an integrated social-ecological system with components, structures and processes at various spatial and temporal scales and different levels of societal organisation (e.g., Sauer 1925; Berkes et al. 2003; Dyakonov et al. 2007). The European Landscape Convention captures this at the Pan-European policy level (Anon 2000). At the same time, there is a strong request to satisfy specific market demands in terms of raw materials and bioenergy. In addition, uncertainties related to climate change, political and economical crises and economic globalisation need careful consideration.

In Europe, the Carpathian Mountains are "a unique natural treasure of great beauty and ecological value, an important reservoir of biodiversity, the headwaters of major rivers, an essential habitat and refuge for many endangered species of plants and animals and Europe's largest area of virgin forests" (Anon 2007; Borsa et al. 2009). This can be explained by the comparatively short history of modern development based on use of natural resources compared to most of Western Europe (Gunst 1989). The dominating potential natural vegetation of ecosystems in the Carpathian ecoregion is forest and woodland (e.g., Mayer 1984; Bohn and Neuhäusl 2000/2003). However, due to a long history of traditional land use in Central and Eastern Europe and limited modernisation, remote landscapes can still be viewed as a total phenomenon where man and the biophysical landscape have been integrated based on the use of landscape goods, ecosystem functions and values for product development (Angelstam 1997; Antrop 1997, 2005; Vos and Meekes 1999; Jongman 2002).

The maintenance of biodiversity therefore encompasses two main visions. The first vision involves biodiversity in dynamic forest and woodland ecosystems with reference to the concept of naturalness (Peterken 1996; Egan and Howell 2001) including natural disturbance regimes at the scales of stands and landscapes (e.g., Angelstam and Kuuluvainen 2004). Naturalness implies that compositional, structural and functional forest biodiversity indicators should represent naturally dynamic forest conditions (Noss 1990). This vision is widespread regarding mountain forest biodiversity, and to some extent implicit in near-to-nature silviculture and plantation forestry (Grabherr et al. 1998; Peterken 1999; Mason 2003).

The second vision is that of the pre-industrial agricultural landscape, which is an important aspect of Europe's cultural heritage (Agnoletti 2000; Jongman 2002;

Sauberer et al. 2004; Antrop 2005; Bezák and Halada 2010). These cultural landscapes include arable lands, wetlands, more or less wooded grasslands, and patches of woodland and forest as the results of traditional agroforestry and agrosilvopasturalism systems, which integrate small-scale agriculture, animal husbandry, and tree management by pollarding, lopping, coppicing and tree felling. Although influenced by human land use for a very long time, the pre-industrial cultural landscape included structural elements such as dead wood and large old trees that are typically found in naturally dynamic forests and woodlands (e.g., Jonsson and Kruys 2001). Consequently, remnants of the pre-industrial cultural landscape provide refuge for species adapted to a pristine or near-natural forest environment (Angelstam 2006). At the same time, they host species dependent on and favoured by cultural landscapes' semi-natural grasslands (Zechmeister et al. 2003; Bezák and Halada 2010).

Rapid changes in traditional land use patterns due to political and socio-economic changes (e.g., Angelstam et al. 2003b; Mikusiński et al. 2003; Bender et al. 2005) mean that the maintenance of cultural biodiversity are no longer automatically provided as a product of traditional land use (von Haaren 2002; Young et al. 2007). Imreh (1993) demonstrated for Transylvania, that villages had a very detailed and strict system of rules preventing too intensive land use from the Middle Ages until the eighteenth century. Long-term thinking, dominance of community interests upon individual interests and ecological process understanding were the most important characteristics of these rules.

During several hundred years of gradually intensified land use, the human footprint has resulted in gradients in landscape alteration from the centres of economic development into more remote regions (Mikusiński and Angelstam 1998, 2004; Konvicka et al. 2006). Already von Thünen (1875) noted that the type and intensity of land use was related to the distance from the market. The demand for timber, grain and other primary products was satisfied by imports from the periphery of the spreading industrial revolution (Gunst 1989), and reached Hungary, Romania and Ukraine for grain in the eighteenth to nineteenth century (Powelson 1994; Turner II et al. 1995), and then into forests (Fröhlich 1954). The exploitation of these resources depended on the facilities for transportation of bulky products such as railways and roads (Turnock 2001). As an example, the Hungarian export was initially restricted to live cattle herded to the destination countries until the mid-nineteenth century when the railways reached Hungary and grain replaced cattle for export (Gunst 1989).

From a biodiversity conservation perspective, these driving factors have led to gradual landscape changes that negatively affected specialised species (e.g., Tucker and Heath 1994; Törnblom et al. 2011a, b, c), habitat structure (Angelstam and Dönz-Breuss 2004), and processes in landscapes (Breitenmoser 1998; Szaro et al. 2002). This means that areas having the same stand scale forest structures could be affected differently depending on the landscape's location in relation to the centre and periphery of human economic development (Mikusiński and Angelstam 2004).

Whyte (1998) concluded that areas of retardation and tradition are still concentrated to northern Europe, the Atlantic periphery and mountain areas in Central Europe and the Mediterranean. Economic remoteness in Europe has thus both a West–East dimension, and lowland–mountain dimension. In the Carpathian Mountains the co-occurrence of the two dimensions explain why the region is still a hotspot for natural and cultural biodiversity (Miya 2000; Turnock 2002; Angelstam et al. 2003b; Opelz 2004; Oszlányi et al. 2004; Schmitt and Rákosy 2007; Reif et al. 2008). Understanding these legacies of the past is an important starting point for maintenance of biodiversity in the region.

We first describe the benchmarks of natural and cultural biodiversity visions. Second, we summarise the landscape history, review current trajectories of landscape development, and give concrete examples from six different landscapes in the Carpathian ecoregion. The discussion focuses on how management systems need to match natural and cultural disturbance regimes, spatial planning and that sectors governing landscape management need to be integrated. Finally, we advocate the need for establishing landscape governance and learning processes to maintain natural and cultural biodiversity in the Carpathian ecoregion. This involves the need for development of an integrated landscape approach for biodiversity conservation based on an improved understanding of both social and ecological mechanisms behind the different trajectories of landscape development.

2 Benchmarks for Biodiversity Conservation

2.1 Natural Disturbance Regimes

Conservation of biodiversity requires a range of disturbance regimes (Table 1) that result in ecosystems and environments to which species have adapted. As advocated within the natural disturbance regime paradigm for near-to-nature forest management (Hunter 1999), the management regimes chosen for different forest environments must tally with the ecological past of different forest types (Angelstam 2003).

Three main forest disturbance regimes are characteristic (e.g., Angelstam and Kuuluvainen 2004): (1) succession after stand-replacing disturbance from young forest to old-growth with shade intolerant species in the beginning and shade tolerant species later on, (2) cohort dynamics on dry sites, and (3) gap dynamics in moist and wet forest. Regarding the evolutionary background of the temperate deciduous forest, and thus the woodland conditions in cultural landscapes, the ideas revolve around both abiotic disturbances such as wind and the interaction between large herbivores and vegetation (Vera 2000; Bengtsson et al. 2003). In the Carpathian Mountains forest dynamic is dominated by gap dynamics in shade-tolerant beech and other broad-leaved forest (Keeton et al. 2010), succession after wind fall (Fig. 1), ice storms (Kenderes et al. 2007), and riparian cohort dynamic in flood-plain forests (Gurnell et al. 2009).

Table 1 List of abiotic, biotic and anthropogenic disturbances affecting the maintenance of natural and cultural biodiversity

Disturbance	Natural biodiversity vision	Cultural biodiversity vision
Wind	Uprooting creates dead wood, bare soil and special microhabitats	Dead wood is often removed and used as fuel
Flooding	Natural stream dynamics creates important aquatic and riparian habitat	Irrigation and draining often occur, as well as active flooding to benefit productivity of meadows and pastures
Fire	Larger patches, lower frequency	Smaller patches, higher frequency
Large herbivores	Domination of browsers	Domination of grazers
Insects and fungi	Important natural disturbances	Not important
Anthropogenic	Not important, unless restoration measures are needed	Vital, includes mowing, pasturing, pollarding, coppicing, shredding etc.

Fig. 1 A near-natural forest landscape in the Hungarian Börzsöny Mountains after windfall. Photo: Per Angelstam

The Carpathian Mountains host Europe's most extensive tracts of mountain forest, the largest remaining natural mountain beech and beech–fir forests ecosystems, and areas of old-growth forest remnants (Schnitzler and Borlea 1998; Opelz 2004; Oszlányi et al. 2004). As a consequence, the region hosts populations of large carnivores and herbivores that have become locally extinct or very rare elsewhere in Europe (Perzanowski et al. 2004; Rozylowicz et al. 2011), specialised vertebrates (Mikusiński and Angelstam 2004; Edman et al. 2011). Additionally, there are many endemic species (Webster et al. 2001; Oszlányi et al. 2004).

Furthermore the Carpathians contain some of the most intact, wild river systems in Europe. Many of the last flooded forests are found in the valleys of the Carpathians. The mountains form watershed areas for the Danube, Vistula, Oder and Dniester rivers. Moreover, the Carpathian Mountains form a 'bridge' between Europe's northern forests and those in the south and west. As such, they are a vital corridor for the dispersal of plants and animals.

2.2 Pre-Industrial Agricultural Landscape

Traditional pre-industrial management of grasslands, woodlands and forests by grazing, mowing and tree management with different intensities produced a structurally diverse landscape (Fig. 2).

Due to the occurrence of elements of naturally dynamic forests such as large old trees, dead wood, slow-growing trees in the cultural landscapes forest species may thus be present outside areas normally characterised as forest, e.g., in semi-natural wooded grassland with trees managed to provide leaf fodder, fruits and material for tools. Semi-natural habitats such as mountain pastures with diverse and rich flora, hay meadows, small arable fields with hedgerows and other structural elements are the result of centuries of traditional management of the land (Baudry et al. 2000). Species-rich and structurally diverse biotopes along fences and stone walls also provide habitats for forest species. Extensive grassland management favours light-demanding vascular plants and associated animal species; and traditional management of arable lands create favourable conditions for species depending on open space and field-forest edges (Baur et al. 2006; Bezák and Halada 2010).

To maintain cultural biodiversity the methods employed in the pre-industrial cultural landscape need to be considered. Without a deep understanding of local knowledge (e.g., lexical knowledge, perception of a landscape) on ecological patterns and processes, it will be difficult to combine local and scientific knowledge in landscape management for biodiversity conservation (Babai and Molnár 2009; Molnár and Babai 2009).

Fig. 2 A traditional pre-industrial cultural landscape in the Carpathian Mountains, Volosyanka in Ukraine. Usually centred on the village street with farm houses, traditional villages have a characteristic zonation from the centre to the periphery (Angelstam et al. 2003b; Mikusinski et al. 2003; Bender et al. 2005). These zones include: (1) built-up area with farm houses, a church or a building of a local administration, (2) vegetable and fruit gardens, (3) fields, (4) meadows for hay, (5) pastures and (6) forests, all of which satisfy different needs of land users (Elbakidze and Angelstam 2007). Photo: Per Angelstam

3 Challenges to Natural and Cultural Biodiversity Conservation

3.1 The Landscape History Background

Understanding landscape history is critical for natural and cultural biodiversity conservation (Marcucci 2000). During the times of the Hungarian Kingdom (1000–1918/1920) and the Habsburg Empire (1526–1918) most of the inner and northern slopes of the Carpathians were one geo-political unit (Kann 1974; Magosci 2002). Because of mining activities, large deforestation and intensive use of timber was typical in some regions already in the thirteenth–fourteenth centuries (e.g., central Slovakia). By the seventeenth century, the main river valleys were mostly deforested. Intensive forest exploitation began only in the eighteenth century. Focusing on sustained yield wood production, monocultures of Norway spruce were created in the different parts of the Carpathians. Additionally, wood was intensively used for potash production, iron and glass manufactory. The most intensive logging took place in the second half of nineteenth until the beginning of twentieth century. Thus, already in the nineteenth century, there was a clear economic development gradient from the centre to the periphery of the former Habsburg Empire, and remote regions were characterised as a traditional cultural landscape based on animal husbandry (Good 1994).

During the twentieth century, land use was regionally transformed several times for geopolitical reasons. After World War II, in the countries remaining under Soviet influence, forests became nationalised, more or less effectively managed according to long-term plans (Augustyn and Kozak 1997; Augustyn 2004, 2006). Forestry led to the further reduction of beech, fir and mixed forests (Hensiruk 1992).

Agriculture in the Carpathian Mountains saw a period of intensification with the breakdown of traditional farming and the replacement of small- and medium-sized private farms with larger-scale state or collective farms in most of the region apart from Romania (Rey et al. 2007) and Poland. As Carpathian countries were industrialized, large-scale rural–urban migrations occurred (Turnock 2007). The percentage of forest cover differed, however, considerably even in neighbouring countries (Kuemmerle et al. 2007, 2008).

The breakdown of communism in 1989 reversed some of these trends. Agricultural sectors collapsed as prices for agricultural products and inputs (e.g., fertilizer) were liberalized. Guaranteed markets within the former bloc of socialist countries were replaced by external competition (Palang et al. 2006). The result was widespread land use change, particularly the abandonment of vast areas of cropland and grasslands (Kuemmerle et al. 2008, 2009b).

The joining of the European Union by some Carpathian countries and the application of the Common Agricultural Policy (CAP) started to modify land use of these countries (Bezák and Halada 2010). During the transition from planned to market economy local people have frequently returned to their traditional land use practices. Non-wood forest products are part of the social fabric and livelihood, especially in forest-dependent communities.

3.2 Trajectories of Natural and Cultural Landscape Development

The natural forest and cultural landscapes in the Carpathian Mountains are presently developing in different directions. Concerning the development of natural forest biodiversity there are a diversity of trajectories, including:

1. protection of the remaining near-natural forests in reserves and national parks (Feurdean and Willis 2008);
2. intensification of forest harvesting (Kuemmerle et al. 2007, 2009a);
3. emerging ideas of close to nature silviculture (Fanta 1997; Brang 2005) in response to management and pollution legacies from communist times (Main-Knorn et al. 2009);
4. recreational and touristic use of forests (Abrudan and Turnock 1999).

Angelstam et al. (2003a) and Kuemmerle et al. (2007, 2008, 2009b) found that cultural landscapes developed along three different trajectories:

1. remained traditional (Vos and Meekes 1999; Jongman 2002; Antrop 2005; Elbakidze and Angelstam 2007);
2. changed due to intensified agriculture (Fearne 1997; Bezák and Halada 2010);
3. were abandoned with encroaching forest as a consequence of depopulation of rural areas (Kuemmerle et al. 2008).

Additionally, traditional village system may become disintegrated due to in-migration of non-native people (Baranyi et al. 2003; G-Fekete 2007).

Habitat loss and fragmentation are a unifying theme of the history of the European forests (Darby 1956) and cultural landscapes (Whyte 1998), and explain the local and regional extinctions of species, loss of habitat and alteration of landscape processes. Compared with Western European countries, the conservation status of many species considered endangered or threatened is remarkably better in Eastern and Central Europe than in most West European countries (van Swaay and Warren 1999; Angelstam et al. 2004b; European Environmental Agency 2010). By and large Carpathian forests maintain a relatively natural character. Stands with changed tree species composition or stands of non-native species are less abundant than in other mountain regions in Europe, and large intact forest massifs do occur (e.g., Soloviy and Keeton 2009). There is, however, a gradient from more altered forests in the west (e.g., the Czech Republic) to more intact in the east (e.g., Romania). The area of forests have increased in the Carpathians during the last decades, but air pollution and other environmental pressures have made forests sensitive to disturbances and large forest damages have started to appear in the western part (Szaro et al. 2002). For Romania, Schmitt and Rákosy (2007) found that increasing modern agriculture and abandonment of less productive sites affect butterfly diversity negatively. The same pattern applies to forest species depending on natural forest structures (Brang 2005), and area-demanding large mammals (Angelstam et al. 2004b).

The expansion of the EU brings both advantages and disadvantages to the natural and cultural landscapes in the Carpathian Mountains (Hodge 2001; Buza and Turnock 2004). EU's Common Market, policies and funding intensify threats to the natural and cultural biodiversity and long-term ecological sustainability of the Carpathian ecoregion as a whole. These include development of mass tourism facilities, transportation infrastructure, agricultural intensification as well as abandonment of traditionally farmed areas. At the same time, increasing EU integration is also driving the adoption and implementation of a number of progressive EU laws and policies (Blicharska et al. 2011). The countries have been aligning its national laws and policies to important pieces of the EU legislation. These harmonisation processes represent potentially powerful tools for biodiversity conservation and sustainable development.

However, the efforts of the EU to maintain natural and cultural landscapes are contradictory, as the EU distributes more and more of its budget to improve biodiversity and rural sustainability, and at the same time, provides increasing financial support for the modernisation in economically remote regions. The extensive plan to develop the transport infrastructure within the new EU member states is a good example. Unless effective mitigation measures are implemented (e.g., Deodatus and Protsenko 2010), this will subsequently result in a decrease in the functionality of existing habitat networks, and threaten the last remaining reference landscapes for both natural and cultural biodiversity. While it is certainly possible to satisfy some cultural heritage values and elements of biodiversity in the long term, the maintenance of sustainable rural landscapes and ecosystem integrity is a major challenge (Anon 2004). Remedy measures are needed to halt the ongoing decrease in landscape diversity. This applies both to ecological and social systems. The lack of landscape-scale incentives in EU agri-environmental schemes hampers planning and management for functional connectivity of habitat patches (Larsson 2004). Additionally, social systems' ability to develop collaboration among sectors at multiple levels of governance need to be strengthened.

3.3 Case Studies of Natural and Cultural Biodiversity Challenges

To illustrate the challenges that conservation of natural and cultural biodiversity in the Carpathian Mountain ecoregions are facing, we review the situation in a suite of particular landscapes representing a gradient from natural to cultural biodiversity challenges (Fig. 3, see also Table 1).

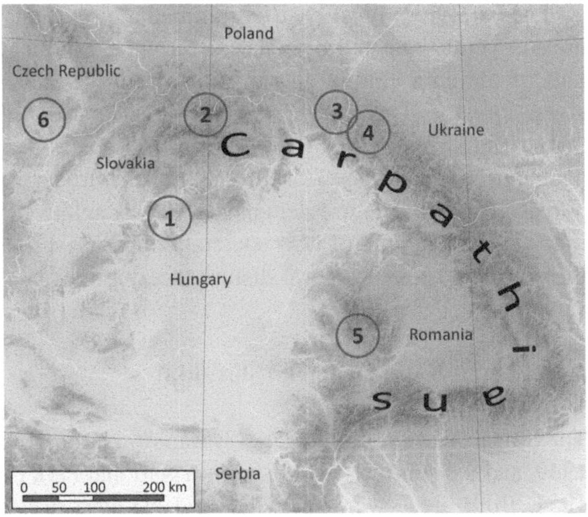

Fig. 3 Location of landscape case studies in the Carpathian Mountain ecoregions that represent a gradient from natural forest to cultural landscape visions for biodiversity conservation. *1*: Börzsöny Mountains in Hungary, *2*: Tatra National Park in Slovakia, *3*: Bieszczady Mountains in Poland, *4*: Skole and Turka in Ukraine, *5*: Apuseni Mountains in Romania, *6*: Lower Morava Biosphere reserve in the Czech Republic

3.3.1 The Börzsöny Mountains (Hungary)

The Börzsöny is a middle-range mountain region in Hungary (48°N, 19°E) forming the southern border of the western Carpathian Mountains. Located 60 km north of Budapest, the forested area has virtually no permanent settlements. The geomorphologic dichotomy between higher altitude (to 950 m a.s.l.) and valleys is manifested by the forest types. Forests are dominated by sessile oak (*Quercus petraea*), turkey oak (*Quercus cerris*) and hornbeam (*Carpinus betulus*). Natural disturbance includes ice break, windfall (Kenderes et al. 2007) and intensive deer browsing.

The Börzsöny Mountains have an unusual legal and governance status linked to its history of ownership and use. This predominantly state-owned forested landscape is part of the Danube–Ipoly National Park, but also managed by a state-owned forestry enterprise (Ipoly Erdő Zrt.) for timber production. This means that based on the duties defined by the conservation act the national park is in charge of the protection of biodiversity, whereas the forestry enterprise is responsible for timber production. This dual regulation is a constant source of conflicts between nature conservation and forestry. On the top of this, people from nearby Budapest are extremely sensitive to what happens to these forests in terms of cutting practices.

All these aspects (inherited age-class distribution of forests, conservation status, large-scale natural disturbances, and public awareness) led the forestry enterprise to change the predominant silvicultural practice characterised as uniform shelterwood system. This resulted in the current large coarse-grained mosaic with patches of a few hectares (from 0.5 to tens) with more or less even-aged stands, and a lack of biologically old stands, as standard rotation time has been 100–120 years.

The Királyrét Forest District initiated a large-scale experiment by leaving large areas severely affected by wind disturbance untouched, and managing more than half of the area aiming at transition from age-class forestry towards continuous cover forestry. The area (5,090 ha) has a management plan valid from 2007 to 2016. The strong belief is that with the successful implementation of this management approach based on the natural disturbance paradigm, a better reconciliation of the multiple use demands could be achieved. The Börzsöny Mountains is thus a good example of how societal choice drives forest management to emulate the consequences of natural disturbance regimes.

3.3.2 Tatra National Park (Slovakia)

The Slovak Tatra National Park (Tatranský Národný Park; TANAP) was founded in 1949, and the contiguous Polish Tatra National Park (Tatrzański Park Narodowy) in 1954. Both areas were included in 1993 into the UNESCO Biosphere Reserve (49°N, 20°E). Combined effects of air pollution, extreme weather and biotic agents have affected the forest condition in the Tatra National Park since early 1990s (Fleischer et al. 2005). On the one hand, natural disturbances such as strong winds are inherent part of the forest dynamics (Svoboda and Pouska 2008; Svoboda et al. 2010). They maintain or increase biodiversity and often change forest development towards more natural conditions by increasing the amount of dead wood and structural diversity (Jonasova and Prach 2004, 2008; Muller et al. 2008; Heurich 2009). On the other hand, they can cause serious social and political problems.

A windstorm in November 2004 felled 12,000 ha, and caused dramatic direct and indirect changes of land cover in the Tatra National Park. Subsequently, this facilitated the development of travel and tourism (new hotels, ski parks, etc.) (Kopecká and Nováček 2009). Despite the strict national and international conservation regulations, 93 % of the windbreak areas were commercially harvested, with only the most precious reserves saved for natural forest regeneration. Two years later salvage logging commenced also there. As the state environment authorities, influenced by the timber industry, were perceived to fail to protect even the most precious nature reserves from salvage logging, more than 1,000 people declared in 2007 the Ticha and Koprova valleys "Areas Protected by Citizens". Slovak Environment Inspectors declared that the logging had caused no harm to ecosystems.

This event can be linked to the history of forest management (e.g., Gąsienica-Byrcyn 1992). Commercial forestry with highly productive monocultures of fast-growing tree species is the base of classic sustained yield forestry as developed in Germany from the late eighteenth century, and was introduced to the western Carpathian Mountains in the nineteenth century. Forests dominated by beech and fir were thus replaced by Norway spruce plantations. Despite early pleas to grow mixed forests (e.g., Gayer 1886) Central European forestry developed further as a commercial activity oriented at high and sustainable timber yields and profits (Fanta 1997). During last three decades significant areas of mountain spruce forests in Central Europe suffered from forest dieback (Kubikova 1991) due to severe bark beetle outbreaks and windstorms (Muller et al. 2008; Hais et al. 2009). This has triggered discussions to restore the natural mixed tree species composition and structure (Fanta 1997; Kenk and Guehne 2001). The Tatra National Park is an example where restoration of near-natural forests composition, structure and function need to be considered. In addition there is a legacy of cultural landscape vision (Byrcyn 1992).

3.3.3 Bieszczady Mountains (Poland)

Cessation of the traditional management of cultural landscapes, and the disappearance of its biodiversity, may result in the rehabilitation and restoration of natural forest biodiversity. A particularly interesting case is the Bieszczady Mountains area (49°N, 23°E) in south-eastern Poland (Angelstam et al. 2003a) where, due to the resettlement of local population imposed in 1947, the "Vistula operation", the average population density decreased from about 65 people/km^2 in 1939 to less than 10 inhabitants/km^2 50 years later.

After World War II, the meadows and agricultural land went through a period of secondary succession within abandoned villages, then collective farms were established with vast monocultures of rape seed and oats. By the beginning of the 1990s, those farms went bankrupt and until Polish accession to the EU, those fields were mostly abandoned and woody vegetation encroached. In the last 5 years, most of this area went under private ownership, and with the beginning of EU subsidies, a majority of former fields are cultivated again as permanent hay meadows or pastures. The absolute majority of forests remain as government property. Therefore, the forest proportion that was about 40 % in pre-World War II period grew up to about 85 % by the end of twentieth century (Augustyn 2006). Currently, the ecological conditions resemble near-natural forest conditions.

Linked to this, the natural biodiversity is probably the highest in several centuries. Historical cultural biodiversity, however, is preserved mainly in the spatial arrangement of land cover in valleys, where still signs of former villages, fields and pastures are visible. The recent establishment of EU's Natura 2000 network of protected areas, which lacked proper consultations at community level, is not very well received by local inhabitants. There are numerous disagreements with restrictions regarding extension of residential areas and tourist infrastructure. In addition attempts to extend the area of Bieszczadzki National Park are opposed by both the forest administration and local community. Nevertheless, there is also a growing awareness of benefits and opportunities of nature and culture values, and a number of people involved in agrotouristic business and guiding of tourists is gradually increasing. Since the extraction of timber is not economically profitable anymore, the conversion of this area into a protected wilderness zone becomes an option. Logging would be performed only to cover local needs for firewood, and the forest could be managed towards possibly the highest biodiversity. To conclude, this case illustrates that, given sufficient time, restoration of natural biodiversity is indeed ecologically feasible.

3.3.4 Skole and Turka Raions (Ukraine)

The Skole and Turka local administrative units (raions) (49°N, 24°E) are situated in the westernmost part of Ukraine's Carpathian Mountains in the upper part of the Dniester river basin. In the fifteenth century, people began to settle and introduced land cultivation traditions, which created today's cultural landscape.

During the last several centuries, Skole and Turka were first a part of Austria-Hungary, then Poland and the Soviet Union. The forests became a source for wood and wood products in the international market from the middle of the nineteenth century as Austrian forestry was introduced. The demand for spruce timber prompted the owners of the forests to replace the natural deciduous beech forests with spruce forests. During the Soviet regime (1939–1991), private land property was expropriated. Forests were state owned and private land was joined into collective farms.

Today this part of the Carpathian Mountains hosts intact remnants of both near natural forests and traditional villages (Elbakidze and Angelstam 2007, 2013). The National nature park "Skolivsky Beskydy", created in 1999 in the Skole raion, covers almost 22 % of the total forested area. People have kept much of their material culture, architecture, costume and customs, and use this to attract visitors. Recreational and tourism activities are thus connected to both natural forests and cultural landscapes. The main industry in the area is forestry. The predominant state employers are educational foundations, forestry sector and health service.

Since 1991, when Ukraine became an independent state, the economic crisis has made local people's livelihoods directly dependent on the local use of natural resources. This has involved a return to their traditional agricultural land use practices. At the same time, forestry is being modernised and road building has commenced to make forests accessible for management (Elbakidze and Angelstam 2007, 2013).

Interviews with local politicians, managers and stakeholders involved with forest landscape issues and governance of natural resources in Turka illustrate the opportunities and obstacles for development based on natural and cultural biodiversity (Angelstam et al. 2009). Key development issues included:

1. harvest rates of forests, effects of logging on erosion and flooding events, access to fuel wood, effects of hauling wood on streams, and whether locals profit economically or not;
2. abandonment of the traditional village system associated to encroaching forest on abandoned fields, and thus reduced landscape attractiveness;
3. tourism as the main future new business sector, but limited by poor road access, lacking advertisement and investment opportunities;
4. degradation of villages as a socio-cultural units;
5. apprehension towards protected areas, as people do not want to be restricted in the use of the landscape.

To conclude, the Skole and Turka raions illustrate the need to maintain and strengthen natural and cultural biodiversity as infrastructures for local development, and to empower local stakeholders' ability to exercise governance.

3.3.5 Apuseni Mountains (Romania)

The Apuseni Mountains (47°N, 23°E) is an interesting part of the Carpathian mountains in terms of landscape, biodiversity and culture (Abrudan and Turnock 1999; Brinkman and Reif 2006). Studies of archaeology and vegetation history indicate that the human colonisation began more than 7000 year ago (Bodnariuc et al. 2002). However, the most extensive forest loss took place during the past 100 years.

The Apuseni Nature Park is located in the centre of the Apuseni Mountains, comprising a part of the Bihor and Vlădeasa massifs up to 1,880 m a.s.l., where three administrative units meet (Cluj, Bihor, and Alba counties). Feurdean and Willis (2008) showed that the landscape was continuously forested over the last 5700 years BP, but the forest composition and structure have been dynamic. While beech was the major tree species between 5200 and 200 years BP, Norway spruce forests appeared 400 years ago. During the last two centuries Norway spruce dominates as a result of selective forest clearance, intensive grazing and, more recently, plantations. This led to a large reduction in forest diversity and local extinction of many tree species. However, in most of the regions with lower altitudes up to about 1,300 m cultural elements are an essential component of the landscape. This involves multiple strategies grounded in local agriculture and based on a settlement network in which small hamlets predominate (Surd and Turnock 2000). Villagers have formed cultural landscapes rich in structures and vegetation types. Forests provide timber for construction and boards, firewood, wood pasture, berries and mushrooms. Unfertilized grassland occupies the steeper and less fertile soils, mainly providing pasture, while meadows are found on deeper soils fertilized with manure and harvested manually. Hence, consideration of both the natural and cultural legacies should be included in the management and conservation of landscapes (Feurdean 2010).

However, severe pollution problems associated with mining areas are a threat and forest and pasture zones are under pressure from villagers seeking to improve their incomes (Buza et al. 2001). At the same time, tourist pressure is growing.

The Apuseni Mountains illustrate the need for reclamation to cope with pollution from mining activity and mineral processing, management of building development, which is going on without consideration to natural and cultural landscape, and biodiversity conservation of both natural forest and cultural land scape legacies (Reif et al. 2008). This must the combined with sustainable solutions to problems of local community development (Buza et al. 2001). Dogaru et al. (2009) showed that high level of education increases peoples' awareness of environmental problems.

3.3.6 Lower Morava Biosphere Reserve (Czech Republic)

The Lower Morava Biosphere Reserve (LMBR) (49°N, 17°E) covers the unique combination of limestone cliffs of the Palava Hills—the westernmost outskirts of

Carpathians in the Czech Republic—the rare Central European lowland floodplains along the lower reaches of the Kyjovka, Dyje and Morava rivers. The LMBR is covered by managed alluvial forests, some 8,000 ha of continental floodplain meadows, and the largest European man designed landscape: Lednice-Valtice Cultural Landscape, a World Heritage landscape.

Land cover includes karst dry grassland, oak forests and Scots pine plantations, fishponds with fish farms and other standing water habitats, saline meadows and marshland, vineyards and other mostly intensively farmed agricultural land. Human activity formed the whole region for millennia. Most inhabitants in the LMBR engage in agriculture and small-scale industry with tourism as an alternative source of income.

The managing authority is the LMBR Public Benefit Corporation. It is the very first time in the Czech Republic that a BR is administered by a non-governmental organization. This concept of an independent and direct participation management is unique, as the rest of the Czech BRs are linked to official government protected areas and share responsibility for the management. In case of Lower Morava the founders of the Public Benefit Corporation came from a wide spectrum of society: representatives of local businesses, agriculture, industry, the Ministry of Environment and the largest nature conservation nongovernmental organization in the country. Local communities play a vital part in management via representatives in the BR's managing board.

Presence of various stakeholders with diverse interests within the reserve boundaries (the BR includes 20 sites designated by the Natura 2000 network, over 25 national categories of nature protection areas, the World Heritage Site and Czech historic zone, two Ramsar sites, and two Nature Parks) open a chance to co-operate on local, national and international levels in the fields of conservation and land management.

While co-ordinating projects the LMBR also serves as a platform, where different parties can seek compromise. This would never be possible without direct involvement of local people. To engage the locals in BR management and decision-making, the BR's managing board includes three regional community associations. To conclude, the LMBR has the advantage of receiving first hand feedback when proposing, for instance, new projects.

4 Discussion

4.1 Need for Diversity of Management Systems and Spatial Planning

The review of natural and cultural biodiversity visions, landscape history and trajectories of change, and the experiences from the case studies, clearly show that management for biodiversity conservation needs to consider both natural and

Table 2 Overview of six landscape case studies chosen with respect to visions for both natural forest and cultural landscapes

		Natural disturbance vision	Cultural landscape vision
Hungary	Börzsöny Mountains	++++	+
Slovakia	Tatra National Park	++++	+
Poland	Bieszczady Mountains	++++	++
Ukraine	Skole and Turka	++	+++
Romania	Apuseni Mountains	+	++++
Czechia	Lower Morava	+	++++

cultural dimensions of Carpathian landscapes (e.g., Oszlányi et al. 2004; Feurdean 2010) (Table 2).

The wide range of different even-aged, multi-aged and uneven-aged silvicultural systems (Matthews 1989; Puettmann et al. 2009) provides a high potential for emulating natural disturbance regimes by combining protection and management for both maintenance and restoration of forest biodiversity (Table 3). Similarly, if the traditional village system can be maintained, it will provide an important prerequisite and opportunity for ecological, economic and socio-cultural sustainable development (Parrotta et al. 2006) (Table 3).

Additionally, the spatial configuration of operational management needs to be considered. The European bison is a good example of population viability (Perzanowski et al. 2004) being dependent on active management for connectivity (Taylor et al. 1993) at the scale of landscapes and regions. For example in Poland, plans for the national network of ecological corridors were elaborated in 1998 (Liro 1998) and in 2005 (Jędrzejewski et al. 2005), but never implemented. In the Polish Carpathians only corridors linking summer and winter refuges of this species formally exist and are included into management plans of State Forests (Perzanowski et al. 2008). A large scale project leading to formal establishment of ecological corridors has been completed recently in Ukrainian Carpathians between Romanian and Polish borders (Deodatus and Protsenko 2010).

4.2 Governance by Multiple Sectors at Multiple Levels

Several land use sectors affect the composition, structure and function of individual landscapes, and thus biodiversity (Table 4). The mixture of natural and cultural biodiversity necessitates the formulation of strategies for the integration of conservation tools for habitat protection, management and restoration across management sectors and spatial scales within a geographical area. The term "governance" captures this issue. Governance can be understood as a "collective" or a shared set of responsibilities of public, private and civil society actors. It includes multiple actors at multiple levels and is thus often referred to and described as multi-level governance (Bache and Flinders 2004). When it comes to

Table 3 Diversity of different forest management systems and traditional village systems to satisfy ecological, economic and socio-cultural criteria of sustainable landscapes

Criteria	Objective	Forest management system			Traditional village system
		Cohort	Even-aged	Uneven-aged	
Ecological	Dry site biodiversity	Light, large trees, dead wood			Open habitat, often grazed
	Mesic site biodiversity		Successional stages from young to old		More or less wooded grasslands
	Wet site biodiversity			Gap phase dynamic	Wet grasslands and woods
Economic	Wood yield, food		Effective economic production		Livelihood
Socio-cultural	Recreation and health	Open forest with large trees	Not compatible	Continuous dense forest cover	Attractive landscape for tourism
	Cultural landscape	Grazed forests, wooded grasslands	Not compatible		Maintenance of social capital
	Urban green space	Open forest, lawns	Not compatible		Wooded grassland

Table 4 Overview of managing sectors affecting natural and cultural biodiversity in the Carpathian Mountains, and estimation of their opportunities for spatial planning

	Hierarchical planning approach involving strategic, tactical and operational steps?	Control of entire landscapes and regions?
Protected areas	Yes	No
Water management	No	No
Forest management	Yes	No
Traditional village systems	No	No
Tourism and recreation	No	No
Transport infrastructure	Yes	No

implementation of biodiversity conservation policies, actors at local to global governance levels affect policies and outcomes on the ground. Within a given sector or policy area there are several levels (Primdahl and Brandt 1997).

First, at the international policy level, the Convention on Biological Diversity's "Ecosystem approach" can be used as one starting point. The ecosystem approach is a strategy for the integrated management of land, water and living resources that promotes conservation and sustainable use in an equitable way. Application of the ecosystem approach will help to reach a balance of the ecological, economic and socio-cultural objectives of the Convention. The approach should be based on the application of appropriate scientific methodologies focused on levels of biological organisation, which encompass the essential processes, functions and interactions among organisms and their environment. It recognizes that humans, with their cultural diversity, are an integral component of ecosystems (e.g., Pirot et al. 2000). For forests, Sustainable Forest Management can be interpreted as an example of the ecosystem approach (Angelstam et al. 2004c). Thus, Natura 2000, being a system common for the entire EU and adjacent countries who have adapted it, becomes an important tool and part of an integrated approach towards sustainable management of natural resources since it is based upon assumed coexistence of nature and people (e.g., Stancioiu et al. 2010). The same integrated approach is required by the EU Water Framework Directive (2000). Both EU policies assume the coexistence of nature and people and focus upon maintaining both fundamental ecological processes and the appropriate state of natural resources on the one hand, and stakeholder participation on the other. However, in most of the cases (e.g., Romania) Natura 2000 network did not significantly improve the conservation status of species and habitats of European concern. A regional approach to conserving biodiversity and social acceptance is needed for the Natura 2000 network to comply with EU targets (Patroescu et al. 2006; Ioja et al. 2010).

Second, at the national level, policy instruments are then gradually developed, and may include legislation, information, subsidies, monitoring, education and vocational training. However, natural and cultural biodiversity is usually not maintained by formal institutions, rules and organisations, but rather informally by local people acting within different governance systems. Consequently, several policy areas with their respective planning traditions coincide: forestry, agriculture, tourism, transport infrastructure and the energy sector, as well as regional and urban planning. The newly introduced EU regulations may have the opposite effect, as they could destroy local land use traditions. For example, within 3 years, the quality rules for milk destroyed local animal husbandry in Ghimes, Romania (Molnár and Babai unpubl.).

Third, because different landscapes have different governance systems, it is important to understand how the actors' knowledge, attitudes and willingness to act correspond to the policy (Clark 2002; Angelstam et al. 2003c). The suite of policy instruments should ideally be adapted to the composition and structure of the actors in the actual landscape. The effects of policies on actual landscapes are thus indirect, and therefore subject to several potential barriers (Clark 2002; Rauschmayer et al. 2009).

Fourth, the effectiveness of the policy implementation process can be evaluated by the development of different indicators, which are monitored to measure change in local landscapes (Busch and Trexler 2003). However, results from monitoring

Table 5 Pros and cons of different approaches to formulate evidence-based performance targets about how much habitat is enough

	Comparisons	Natural experiments	Historical ecology
Landscape data	Good	Good	Variable
Species data	Good	Good	Limited
Sample size	Large	Very limited	Limited
Other aspects	Different species-habitat relationship in different regions may preclude valid comparisons	Same ecoregion	Hard to find relevant data about the occurrence of different species

should be compared with quantitative performance targets or other norms (Lammerts van Buren and Blom 1997; Angelstam et al. 2004d; Rauschmayer et al. 2009). This requires, for instance, systematic studies about how the amount and configuration of habitat affect the occurrence and viability of populations and species (Table 5, and e.g., Angelstam 2004; Angelstam et al. 2011; Bütler et al. 2004; Müller and Bütler 2010; Törnblom et al. 2011b).

Roberge and Angelstam (2009) presented a six-step procedure for identifying thresholds to be used in the determination of forest biodiversity conservation targets. These were:

1. stratify the forests into broad cover types as a function of their natural disturbance regimes;
2. describe the historical spread of different anthropogenic impacts in the forest that moved the system away from naturalness;
3. identify appropriate response variables (e.g., focal species, functional groups or ecosystem processes) that are affected by habitat loss and fragmentation (e.g., Roberge and Angelstam 2004);
4. for each forest type identified in step 1, combine steps 2 and 3 to look for the presence of non-linear responses and to identify zones of risk and uncertainty;
5. identify the "currencies" (i.e., species, habitats, and processes) which are both relevant and possible to communicate to stakeholders;
6. combine information from a suite of different indicators selected.

Using species as an example, this means that planners and managers need to understand that different species have different habitat affinities and quantitative requirements. To maintain ecological integrity and resilience usually requires landscape and regional perspective (Fig. 4). The maintenance of large intact forest areas is necessary for wide-range species such as large carnivores and herbivores (Breitenmoser 1998; Mikusiński and Angelstam 2004; Maanen et al. 2005). For bison, Perzanowski et al. (2004) concluded that this species have no chances for natural exchange of genes due to fragmentation and loss of large areas of natural habitats. The population viability thus depends on active conservation management (Kuemmerle et al. 2011).

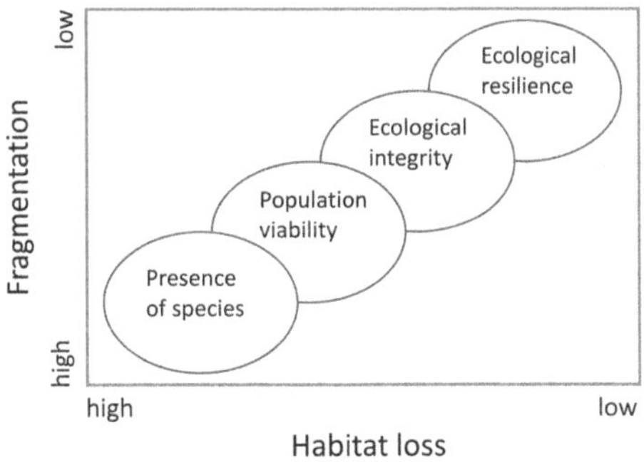

Fig. 4 Illustration of the increasing challenge in terms of mitigation the effects of habitat loss and fragmentation by conservation to achieve different levels of ambition for biodiversity conservation (see Angelstam et al. 2004a, c; Svancara et al. 2005; Tear et al. 2005)

The successful maintenance of all representative land cover types in a landscape to conserve natural and cultural biodiversity can be viewed as a series of partly overlapping and complementary "green infrastructures", each of which have different properties to which species are adapted. The required quality and extent of such habitat networks depends on the requirements of the species. For example, a species specialised on old-growth forest and with large area requirements will need more habitat area then a generalist with small area requirements. The umbrella species concept (e.g., Roberge and Angelstam 2004), whereby conservation management for specialised species confers protection to less specialised species in a particular habitat, is a useful concept as this means that knowledge about species' qualitative and quantitative habitat requirements can be used to formulate conservation targets. Additionally, large carnivores could act as flagship species and help to manage the landscapes and regions (Rozylowicz et al. 2011).

Thus, to steer towards agreed policy goals, there is a need for hierarchical planning with increasing resolution from broader ecoregional (i.e., international) to finer spatial and temporal scales (Angelstam and Andersson 2001; Angelstam et al. 2011). This often requires international co-operation across borders between adjacent regions in different countries (Opelz 2004; Elbakidze and Angelstam 2009). In addition there is a need for bottom-up approaches to engage the range of land owners, managers and actors in local landscapes (Angelstam et al. 2003c; Sayer and Campbell 2004).

4.3 Integrated Landscape Approach to Biodiversity Conservation

To conserve natural and cultural biodiversity a participatory and holistic approach is thus needed (Angelstam 1997; Reif et al. 2008; Pauleit et al. 2010), which requires interaction among different actors in a landscape. Apart from operational management of natural and cultural biodiversity and multi-level governance, we stress two additional important prerequisites to be satisfied. First, sustainability assessments should provide a strategic orientation to policy-makers, governors and managers (Weaver and Jordan 2008) such as landscape scale performance targets for the amount and configuration of habitats needed to maintain biodiversity (e.g., Angelstam et al. 2004a; Villard and Jonsson 2009). Second, platforms for multi-level governance are needed where owners, managers and stakeholders could develop solutions, resolve conflicts and together improve the level of sustainability within a landscape (Lickers and Story 1997; Baker 2006; Gilbert 2007; Elbakidze et al. 2010).

The term integrated landscape approach (e.g., World Forestry Congress 2009) captures the need to consider a larger functional geographical area when addressing sustainability, and to include both social and ecological systems and their interactions (Borrini-Feyerabend et al. 2004; Dudley et al. 2006; Singer 2007). De-constructing the landscape approach Axelsson et al. (2011) identified five core features:

1. focus on a large area of tens of thousands up to millions of hectares depending on the sustainability issues in focus;
2. collaboration among multi-level partners representing all societal sectors and fields of interest;
3. a commitment to sustainable development and an analytic approach to address sustainability;
4. production of new knowledge and knowledge management to identify useful traditional knowledge for socially robust solutions (Gibbons 1999; Daniels and Walker 2001);
5. sharing of knowledge and experience.

However, there are a number of barriers when attempting to apply a landscape approach for the conservation of natural and cultural biodiversity by a wise combination of management and non-intervention (Holling 1995; Soran et al. 2000; Gutzwiller 2002; Sandström et al. 2006; Lawrence 2009). The remedy has to combine multi-level solutions, and satisfy the economic and societal needs of the people in the long term. Successful approaches should be based on, and strive for transdisciplinarity, which implies participatory action research involving the stakeholders (local population, experts, administration, and politicians), and joint practical implementation of research findings (Reif et al. 2008; Axelsson 2010).

There are some important prerequisites to make this work. First, the land-use managers and planners need to acquire an attitude of "learning organisations" i.e., organisations must be flexible and allow personnel to work and learn at the same time (Lee 1993; Sayer and Campbell 2004). Next, researchers have to show true interest in contributing to practical and socially robust solutions outside the academic world. Finally, socially robust solutions to management and spatial planning as a collaborative learning process with local people need to be developed. Maps are often a useful way to communicate sustainability and planning issues and to get feedback from different stakeholders.

There are several approaches to establish such a dialogue among actors. The international model forest network, which forms a partnership between individuals and organisations sharing the common goal of sustainable forest management is one example (Besseau et al. 2002; Axelsson and Angelstam 2006; Axelsson et al. 2008). The UNESCO's biosphere reserve concept is another (UNESCO 2002). Both concepts imply that a management unit consisting of an actual landscape with its characteristic ecosystems, actors and economic activities is used as a site for syntheses, innovation, development and education. Ideally, what Boutin et al. (2002) termed adaptive management teams should be formed whereby researchers, land managers and policy-makers share decisions and responsibilities toward the success or failure of the strategy they jointly adopted.

5 Conclusions

Biodiversity conservation in Europe's landscapes is based on both natural and cultural visions. The introduction of sustained yield forest management and intensive agriculture generally lead to a reduction of the amount of dead wood, functional connectivity and intact areas of natural woodland and cultural landscapes. A major challenge is to identify and use as guidelines for management evidence-based performance targets for biodiversity conservation for agreed levels of ambition rather than negotiated targets such as forest certification, or the state of the environment in already managed and altered landscapes. It is also critically important that land management becomes spatially explicit at several spatial scales ranging from trees and stands to landscapes and regions. Additionally, participatory conservation management planning with societal arenas for combining top-down planning with bottom-up implementation is crucial. We argue in favour of a novel win–win oriented approach to research and development, which is based on exchanging knowledge and experience gathered over long time in different countries and regions. This will be of mutual benefit for both science and practice, and thus for continued sustainable use and conservation of natural resources providing a basis for human well-being and quality of life. Ultimately, acknowledging and adopting this perspective requires the gradual development of a new transdisciplinary profession able to facilitate ecosystem management at the landscape scale. This necessitates improved mutual feedback between the science, engineering and art of integrated natural resource management and governance.

References

Abrudan I, Turnock D (1999) A rural development strategy for the Apuseni Mountains, Romania. Geo J 46:319–336

Agnoletti M (2000) Introduction: the development of forest history research. In: Agnoletti M, Anderson S (eds) Methods and approaches in forest history. CABI Publishing, Wallingford

Angelstam P (1997) Landscape analysis as a tool for the scientific management of biodiversity. Ecol Bull 46:140–170

Angelstam P (2003) Reconciling the linkages of land management with natural disturbance regimes to maintain forest biodiversity in Europe. In: Bissonette JA, Storch I (eds) Landscape ecology and resource management: linking theory with practice. Island Press, Covelo and Washington

Angelstam P (2004) Habitat thresholds and effects of forest landscape change on the distribution and abundance of black grouse and capercaillie. Ecol Bull 5:173–187

Angelstam P (2006) Maintaining cultural and natural biodiversity in Europe's economic centre and periphery. In: Agnoletti M (ed) The conservation of cultural landscapes. CAB International, Wallingford

Angelstam P, Andersson L (2001) Estimates of the needs for forest reserves in Sweden. Scand J For Res Suppl 3:38–51

Angelstam P, Andersson K, Axelsson K, Elbakidze M, Jonsson BG, Roberge JM (2011) Protecting forest areas for biodiversity in Sweden 1991–2010: the policy implementation process and outcomes on the ground. Silva Fenn 45(5):1111–1133

Angelstam P, Boresjö-Bronge L, Mikusinski G, Sporrong U, Wästfelt A (2003a) Assessing village authenticity with satellite images—a method to identify intact cultural landscapes in Europe. Ambio 33(8):594–604

Angelstam P, Bütler R, Lazdinis M, Mikusinski G, Roberge JM (2003b) Habitat thresholds for focal species at multiple scales and forest biodiversity conservation—dead wood as an example. Ann Zool Fenn 40:473–482

Angelstam P, Dönz-Breuss M (2004) Measuring forest biodiversity at the stand scale—an evaluation of indicators in European forest history gradients. Ecol Bull 51:305–332

Angelstam P, Dönz-Breuss M, Roberge J-M (2004a) Targets and tools for the maintenance of forest biodiversity. Ecol Bull 51:1–510

Angelstam P, Edman T, Dönz-Breuss M, Wallis deVries M (2004b) Land management data and terrestrial vertebrates as indicators of forest biodiversity at the landscape scale. Ecol Bull 51:333–349

Angelstam P, Elbakidze M, Axelsson R, Törnblom J (2009) Sustainable forest management from policy to practice, and back again: landscapes as laboratories for knowledge production and learning in the Carpathian Mountains. In: Soloviy IP, Keeton WS (eds) Ecological economics and sustainable forest management: developing a trans-disciplinary approach for the Carpathian Mountains. Ukrainian National Forestry University Press, Lviv

Angelstam P, Kuuluvainen T (2004) Boreal forest disturbance regimes, successional dynamics and landscape structures—a European perspective. Ecol Bull 51:117–136

Angelstam P, Kopylova E, Korn H, Lazdinis M, Sayer JA, Teplyakov V, Törnblom J (2005) Changing forest values in Europe. In: Sayer JA, Maginnis S (eds) Forests in landscapes. Ecosystem approaches to sustainability. Earthscan, London

Angelstam P, Mikusiński G, Rönnbäck B et al (2003c) Two-dimensional gap analysis: a tool for efficient conservation planning and biodiversity policy implementation. Ambio 33:527–534

Angelstam P, Persson R, Schlaepfer R (2004c) The sustainable forest management vision and biodiversity—barriers and bridges for implementation in actual landscapes. Ecol Bull 51:29–49

Angelstam P, Roberge JM, Lõhmus A et al (2004d) Habitat modelling as a tool for landscape-scale conservation—a review of parameters for focal forest birds. Ecol Bull 51:427–453

Anon (2000) European landscape convention. European treaty series No. 176, Council of Europe

Anon (2004) New perspectives for EU rural development. The European Commission

Anon (2007) Handbook on the Carpathian convention. The Regional Environmental Center for Central and Eastern Europe, Szentendre

Antrop M (1997) The concept of traditional landscapes as a base for landscape evaluation and planning. The example of the Flanders region. Landsc Urban Plan 38:105–117

Antrop M (2005) Why landscapes of the past are important for the future. Landsc Urban Plan 70:21–34

Augustyn M (2004) Anthropogenic changes in the environmental parameters of Bieszczady Mountains. Biosph Conserv 6:43–53

Augustyn M (2006) Monografia rozwoju przemysłu drzewnego, jako podstawowego czynnika przekształceń środowiska leśnego Bieszczadów Zachodnich w XIX i pierwszej połowie XX wieku. PAN MiIZ, Ustrzyki Dolne

Augustyn M, Kozak I (1997) The trends of anthropogenic pressure in Polish and Ukrainian Carpathians. In: Perzanowski K, Augustyn M (eds) Proceedings of 2nd annual meeting of ICE PAS, Bieszczady

Axelsson R (2010) Integrative research and transdisciplinary knowledge production: a review of barriers and bridges. J Landsc Ecol 4(2):14–40

Axelsson R, Angelstam P (2006) Biosphere reserve and model forest: a study of two concepts for integrated natural resource management. In: Proceedings from the 1st VHU conference on Science for sustainable development: starting points and critical reflections, Uppsala, VHU

Axelsson R, Angelstam P, Elbakidze M (2008) Landscape approaches to sustainability. In: Frostell B et al (eds) Science for sustainable development—the social challenge with emphasis on the conditions for change. Uppsala, VHU

Axelsson R, Angelstam P, Elbakidze M, Stryamets N, Johansson K-E (2011) Landscape approach for sustainable development and sustainability: a practical interpretation. J Landsc Ecol 4(3):5–30

Babai D, Zs Molnár (2009) Népi növényzetismeret Gyimesben II.: termőhely- és élőhelyismeret. Bot Közl 96:145–173

Bache I, Flinders M (eds) (2004) Multi-level governance. Oxford University Press, Oxford

Baker S (2006) Sustainable development. Routledge, London and New York

Baker S, Jehlicka P (eds) (1998) Dilemmas of transition. The environment, democracy and economic reform in east central Europe. Frank Cass, London and Portland

Baranyi B, G-Fekete E, Koncz G (2003) A roma-szegregáció kutatásának területi szempontjai a halmozottan hátrányos helyzetű encsi és a sellyei–siklósi kistérségekben. Kisebbs Kut 2:344–362

Baudry J, Bunce RGH, Burel F (2000) Hedgerows: an international perspective on their origin, function and management. J Environ Manag 60:7–22

Baur B, Cremene C, Groza G, Rakosy L, Schileyko A, Bau A, Stoll P, Erhardt A (2006) Effects of abandonment of subalpine hay meadows on plant and invertebrate diversity in Transylvania, Romania. Biol Conserv 132:261–273

Bender O, Boehmer HJ, Jens D, Schumacher KP (2005) Using GIS to analyse long-term cultural landscape change in southern Germany. Landsc Urban Plan 70:111–125

Bengtsson J, Nilsson SG, Franc A, Menozzi P (2003) Biodiversity disturbances, ecosystem function and management of European forests. For Ecol Manag 132:39–50

Berkes F, Colding J, Folke C (eds) (2003) Navigating social-ecological systems: building resilience for complexity and change. Cambridge University Press, Cambridge

Besseau P, Dansou K, Johnson F (2002) The international model forest network (IMFN): elements of success. The For Chronic 78:648–654

Bezák P, Halada L (2010) Sustainable management recommendations to reduce the loss of agricultural biodiversity in the Mountain Regions of NE Slovakia. Mt Res Dev 30:192–204

Blicharska M, Angelstam P, Antonson H, Elbakidze M, Axelsson R (2011) Road, forestry and regional planners' work for biodiversity conservation and public participation: a case study in Poland's hotspots regions. J Environ Plan Manag 54(10):1373–1395

Bodnariuc A, Bouchette A, Dedoubat JJ, Otto T, Fontugue M, Jalut G (2002) Holocene vegetational history of the Apuseni Mountains, central Romania. Quat Sci Rev 21:1465–1488

Bohn U, Neuhäusl R (2000/2003) Karte der naturlichen Vegetation Europas. Maßstab 1:2,500,000. Teil 1: Erlauterungstext mit CD-ROM; Teil 2: Legende; Teil 3: Karten. Landwirtschaftsverlag, Munster

Borrini-Feyerabend G, Pimbert M, Farvar MT, Kothari A, Renard Y (2004) Sharing power. Learning by doing in co-management of natural resources throughout the World. IIED and IUCN/CEESP/CMWG, Cenesta, Tehran

Borsa M, Chifelea C, Egerer H et al (2009) Visions and strategies. In: The Carpathian Area (VASICA). Protection and sustainable spatial development of the Carpathians in a transnational framework. The Carpathian project, Druckwerker, Vienna

Boutin S et al. (2002) The active adaptive management experimental team: a collaborative approach to sustainable forest management. In: Veeman TS et al. (eds) Advances in forest management: from knowledge to practice. Proceedings from the 2002 sustainable forest management network conference, University of Alberta, Edmonton

Brang P (2005) Virgin forests as a knowledge source for central European silviculture: reality or myth? For Snow Landsc Res 79(1/2):19–32

Breitenmoser U (1998) Large predators in the Alps: the fall and rise of Man's competitors. Biol Conserv 83:279–289

Brinkmann K, Reif A (2006) Vegetation, landuse and landscape in the Apuseni Mountains, Romania. Bull Univ Agric Sci Vet Med 62:1–13

Busch DE, Trexler JC (2003) The importance of monitoring in regional ecosystem initiatives. In: Busch DE, Trexler JC (eds) Monitoring ecosystems. Island Press, Washington

Buza M, Dimen L, Pop G, Turnock D (2001) Environmental protection in the Apuseni Mountains: the role of environmental non-governmental organisations (ENGOs). Geo J 54:631–653

Buza M, Turnock D (2004) A research note: planning for the Carpathians. Geo J 60:135–142

Bütler R, Angelstam P, Ekelund P, Schlaepfer R (2004) Dead wood threshold values for the three-toed woodpecker in boreal and sub-alpine forest. Biol Conserv 119:305–318

Carpathian Convention (2003) Framework convention on the protection and sustainable development of the Carpathians. Kyiv

Clark TW (2002) The policy process. A practical guide for natural resource professionals. Yale University Press, New Haven and London

Daniels SE, Walker GB (2001) Working through environmental conflict- the collaborative learning approach. Praeger, Westport and London

Darby HC (1956) The clearing of woodlands in Europe. In: Thomas WL (ed) Man's role in changing the face of the Earth. University of Chicago Press, Chicago

Deodatus FD, Protsenko L (eds) (2010) Creation of ecological corridors in Ukraine. State agency for protected areas of the Ministry of environmental protection of Ukraine, Altenburg & Wymenga Ecological Consultants, Interecocentre, Kyiv

Dogaru D, Zobrist J, Balteanu D, Popescu C, Sima M, Amini M, Yang H (2009) Community perception of water quality in a mining-affected area: a case study for the Certej catchment in the Apuseni Mountains in Romania. Environ Manag 43:1131–1145

Dudley N, Schlaepfer R, Jackson W, Jeanrenaud JP, Stolton S (2006) Forest quality. Assessing forests at a landscape scale. Earthscan, London

Dyakonov KN, Kasimov NS, Khoroshev AV, Kushlin AV (2007) Landscape analysis for sustainable development: theory and application of landscape science in Russia. Alex Publishers, Moscow

Edman T, Angelstam P, Mikusiński G, Roberge J-M, Sikora A (2011) Spatial planning for biodiversity conservation: assessment of forest landscapes' conservation value using umbrella species requirements in Poland. Landsc Urban Plan 102:16–23

Egan D, Howell EA (2001) The historical ecology handbook. Island Press, Covelo

Elbakidze M, Angelstam P (2007) Implementing sustainable forest management in Ukraine's Carpathian Mountains: the role of traditional village systems. For Ecol Manag 249:28–38

Elbakidze M, Angelstam P (2009) Cross-border cooperation along the eastern border of European union: a review and approach to learning for sustainable landscapes. Cent Eur J Spat Landsc Plan 20(1):33–42

Elbakidze M, Angelstam P, Sandström C, Axelsson R (2010) Multi-stakeholder collaboration in Russian and Swedish model forest initiatives: adaptive governance towards sustainable forest management? Ecol Soc 15(2):14

Elbakidze M, Angelstam P (2013) Sustainable forest management from policy to landscape, and back again: a case study in the Ukrainian Carpathian Mountains. In: Kozak J, Ostapowicz K,

Bytnerowicz A, Wyżga B (eds) The Carpathians: Integrating Nature and Society Towards Sustainability, Springer-Verlag, Berlin, Heidelberg
European Environmental Agency (2010) Europe's ecological backbone: recognising the true value of our mountains. EEA report 6
Fanta J (1997) Rehabilitating degraded forests in Central Europe into self–sustaining forest ecosystems. Ecol Eng 8:289–297
Fearne A (1997) The history and development of the CAP 1945–1990. In: Ritson C, Harvey DR (eds) The common agricultural policy. CAB International, London
Feurdean A (2010) Forest conservation in a changing world: natural or cultural? Example from the Western Carpathians forests, Romania. Stud Univ Babeş-Bolyai Geol 55(1):45–48
Feurdean A, Willis KJ (2008) The usefulness of a longterm perspective in assessing current forest conservation management in the Apuseni Natural Park, (Romania). For Ecol Manag 256:421–430
Fleischer P, Godzik B, Bicarova S, Bytnerowicz A (2005) Effects of air pollution and climate change on forests of the Tatra Mountains, Central Europe. In: Omasa K, Nouchi I, De Kok L (eds) Plant responses to air pollution and global change. Springer, Tokyo
Fröhlich J (1954) Urwaldpraxis. Neumann Verlag, Radebeul and Berlin
Gayer K (1886) Der gemischte Wald, seine Begründung und Pflege, insbesondere durch Horst und Gruppenwirtschaft. Verlag von Paul Parey, Berlin
Gąsienica-Byrcyn W (1992) The history and present role of the Tatra National Park. Mt Res Dev 12(9):205–210
G-Fekete E (2007) Aprófalvaink átalakulóban. In: Zs Kocsis, T Csapó (eds) A kistelepülések helyzete és településföldrajza Magyarországon. Savaria University Press, Szombathely
Gibbons M (1999) Science's new social contract with society. Nature 402:11–17
Gilbert B (2007) Collaborative synergy in resource and environmental management. Doctoral dissertation, Dalhousie University, Nova Scotia
Good DF (1994) The economic lag of Central and Eastern Europe: income estimates for the Habsburg successor states, 1870–1910. J Econ Hist 54:869–891
Grabherr G, Koch G, Kirchmeir H, Reiter K (1998) Hemerobie österreichischer Waldökosysteme. Universitätsverlag Wagner, Innsbruck
Gunst P (1989) Agrarian systems of Central and Eastern Europe. In: Chirot D (ed) The origins of backwardness in Eastern Europe: economics and politics from the middle ages until the early twentieth century. University of California Press, London
Gurnell A, Surian N, Zanoni L (2009) Multi-thread river channels: a perspective on changing European alpine river systems. Aquat Sci 71:253–265
Gutzwiller KJ (2002) Applying landscape ecology in biological conservation: principles, constraints and prospects. In: Gutzwiller KJ (ed) Applying landscape ecology in biological conservation. Springer, New York
von Haaren C (2002) Landscape planning facing the challenge of the development of cultural landscapes. Landsc Urban Plan 60:73–80
Hais M, Jonasova M, Langhammer J, Kučera T (2009) Comparison of two types of forest disturbance using multitemporal Landsat TM/ETM plus imagery and field vegetation data. Remote Sens Environ 113:835–845
Hensiruk S (1992) Lisy. Naukova dumka, Kyiv
Heurich M (2009) Progress of forest regeneration after a large-scale Ips typographus outbreak in the subalpine Picea abies forests of the Bavarian Forest National Park. Silva Gabreta 15:49–66
Hodge I (2001) Beyond agri-environmental policy: towards an alternative model for rural environmental governance. Land Policy 18:99–111
Holling CS (1995) What barriers? What bridges? In: Gunderson LH, Holling CS, Light SS (eds) Barriers and bridges to the renewal of ecoystems and institutions. Columbia University Press, New York
Hunter ML (ed) (1999) Maintaining biodiversity in forest ecosystems. Cambridge University Press, Cambridge

Imreh I (1993) A természeti környezet oltalmazása a székely rendtartásokban. In: Várkonyi R, Kósa L (eds) Európa híres kertje, Történeti ökológiai tanulmányok Magyarországról. Orpheusz, Budapest
Innes JL, Nitschke CR (2005) The application of forest zoning as an alternative to multiple-use forestry. In: Innes JL, Hoen HF, Hickey G (eds) Forestry and environmental change: Socioeconomic and political dimensions. Report No. 5 of the IUFRO task force on environmental change. CABI Publishing, Oxfordshire
Ioja IC, Pătroescu M, Rozylowicz L, Popescu DV, Vergheleţ M, Zotta MI, Felciuc M (2010) The efficacy of Romania's protected areas network in conserving biodiversity. Biol Conserv 143(11):2468–2476
Ioras F (2003) Trends in Romanian biodiversity conservation policy. Biodivers Conserv 12:9–23
Jędrzejewski W, Nowak S, Stachura K et al (2005) Project of ecological corridors linking Natura 2000 sites in Poland. Mammals Research Institute Polish Academy of Science, Białowieża
Jonasova M, Prach K (2004) Central-European Mountain spruce (Picea abies (L.) Karst.) forests: regeneration of tree species after a bark beetle outbreak. Ecol Eng 23:15–27
Jonasova M, Prach K (2008) The influence of bark beetles outbreak vs. salvage logging on ground layer vegetation in Central European mountain spruce forests. Biol Conserv 141:1525–1535
Jongman RHG (2002) Homogenisation and fragmentation of the European landscape: ecological consequences and solutions. Landsc Urban Plan 58:211–221
Jonsson BG, Kruys N (eds) (2001) Ecology of woody debris in boreal forests. Ecol Bull 49
Kann RA (1974) A history of the Habsburg Empire 1526–1918. University of California Press, Berkeley
Keeton WS, Chernyavskyy M, Gratzer G, Main-Knorn M, Shpylchak M, Bihun Y (2010) Structural characteristics and aboveground biomass of old-growth spruce–fir stands in the eastern Carpathian Mountains, Ukraine. Plant Biosyst 144(1):148–159
Kenderes K, Aszalós R, Ruff J, Barton Z, Standovár T (2007) Effects of topography and tree stand characteristics on susceptibility of forests to natural disturbances (ice and wind) in the Börzsöny Mountains (Hungary). Community Ecol 8(2):209–220
Kenk G, Guehne S (2001) Management of transformation in central Europe. For Ecol Manag 151:107–119
Konvicka M, Fric Z, Benes J (2006) Butterfly extinctions in European states: do socioeconomic conditions matter more than physical geography? Glob Ecol Biogeogr 15:82–92
Kopecká M, Nováček J (2009) Forest fragmentation in the Tatra region in the period 2000–2006. Landf Anal 10:58–63
Kubikova J (1991) Forest dieback in Czechoslovakia. Vegetatio 93:101–108
Kuemmerle T, Hostert P, Radeloff VC, Perzanowski K, Kruhlov I (2007) Post-socialist forest disturbance in the Carpathian border region of Poland, Slovakia and Ukraine. Ecol Appl 17(5):1279–1295
Kuemmerle T, Hostert P, Radeloff VC, van der Linden S, Perzanowski K, Kruhlov I (2008) Cross-border comparison of post-socialist farmland abandonment in the Carpathians. Ecosystem 11:614–628. doi:1007/s10021-008-9146-z
Kuemmerle T, Chaskovskyy O, Knorn J, Radeloff VC, Kruhlov I, Keeton WS, Hostert P (2009a) Forest cover change and illegal logging in the Ukrainian Carpathians in the transition period from 1988 to 2007. Remote Sens Environ 113:1194–1207
Kuemmerle T, Müller D, Griffiths P, Rusu M (2009b) Land use change in Southern Romania after the collapse of socialism. Reg Environ Change 9:1–12
Kuemmerle T, Perzanowski K, Akcakaya HR, Beaudry F, Van Deelen TR, Parnikoza I, Khoyetskyy P, Waller DM, Radeloff VC (2011) Cost-effectiveness of strategies to establish a European bison metapopulation in the Carpathians. J Appl Ecol 48(2):317–329
Lammerts van Buren EM, Blom EM (1997) Hierarchical framework for the formulation of sustainable forest management standards. Principles, criteria, indicators. Tropenbos Foundation, Backhuys Publishers, AH Leiden

Larsson A (2004) Landskapsplanering inom jordbrukspolitik. Agraria 442, Swedish University of Agricultural Sciences

Lawrence A (2009) Forestry in transition: imperial legacy and negotiated expertise in Romania and Poland. Policy Econ 11(5–6):429–436

Lee KN (1993) Compass and gyroscope. Island Press, Washington

Lickers HF, Story PA (1997) Partnership building for sustainable development: a first Nation's perspective from Ontario. In: Bouman OT, Brand DG (eds) Sustainable forests: global challenges and local solutions. Food Products Press, Binghampton

Liro A (1998) Strategia wdrażania Krajowej Sieci Ekologicznej—ECONET. Fundacja IUCN, Warszawa

Maanen E van, Predoiu G, Klaver R, Soulé M, Popa M. Ionescu O, Jurj R, Neguş ş, Ionescu G, Altenburg W (2005) Safeguarding the Romanian Carpathian ecological network. A vision for large carnivores and biodiversity in Eastern Europe. A&W ecological consultants, Veenwouden, The Netherlands. ICAS Wildlife Unit, Brasov

Magosci PR (2002) Historical atlas of Central Europe. University of Washington Press, Seattle Revised and expanded edition

Main-Knorn M, Hostert P, Kozak J, Kuemmerle T (2009) How pollution legacies and land use histories shape post-communist forest cover trends in the Western Carpathians. For Ecol Manag 258(2):60–70

Marcucci DJ (2000) Landscape history as planning tool. Landsc Urban Plan 49:67–81

Mason WL (2003) Continuous cover forestry: developing close to nature forest management in conifer plantations in upland Britain. Scott For 57(3):141–149

Matthews JD (1989) Silvicultural systems. Oxford University Press, Oxford

Mayer H (1984) Die Wälder Europas. Gustav Fischer Verlag, Stuttgart

Mayers J, Bass S (2004) Policy that works for forests and people. Real prospects for governance and livelihoods. Earthscan, London

Merlo M, Croitoru L (2005) Concepts and methodology: a first attempt towards quantification. In: Merlo M, Croitoru L (eds) Valuing Mediterranean forests. Towards total economic value. CABI Publishing, Wallingford

Mikusiński G, Angelstam P (1998) Economic geography, forest distribution and woodpecker diversity in central Europe. Conserv Biol 12:200–208

Mikusiński G, Angelstam P (2004) Occurrence of mammals and birds with different ecological characteristics in relation to forest cover in Europe—do macroecological data make sense? Ecol Bull 51:265–275

Mikusiński G, Angelstam P, Sporrong U (2003) Distribution of deciduous stands in villages located in coniferous forest landscapes in Sweden. Ambio 33:519–525

Miya K (2000) Maramuresh. Humanitas, Bucarest

Zs Molnár, Babai D (2009) Népi növényzetismeret Gyimesben I.:növénynevek, népi taxonómia, az egyéni és közösségi növényismeret. Bot Közl 96:117–143

Muller J, Bussler H, Gossner M, Rettelbach T, Duelli P (2008) The European spruce bark beetle Ips typographus in a national park: from pest to keystone species. Biodivers Conserv 17:2979–3001

Müller J, Bütler R (2010) A review of habitat thresholds for dead wood: a baseline for management recommendations in European forests. Eur J Res 129:981–992

Norton BG (2005) Sustainability a philosophy of adaptive ecosystem management. The University of Chicago press, Chicago and London

Noss RF (1990) Indicators for monitoring biodiversity: a hierarchical approach. Conserv Biol 4:355–364

Opelz M (2004) Towards a Carpathian network of protected areas. Final report. Bundesministerium für Umwelt, Naturschutz und Reaktorsicherheit, Berlin

Oszlányi J, Grodzińska K, Badea O, Shparyk Y (2004) Nature conservation in Central and Eastern Europe with a special emphasis on the Carpathian Mountains. Environ Pollut 130:27–134

Palang H, Printsmann A, Gyuro EK, Urbanc M, Skowronek E, Woloszyn W (2006) The forgotten rural landscapes of Central and Eastern Europe. Landsc Ecol 21:347–357

Parrotta J, Agnoletti M, Johan E (eds) (2006) Cultural heritage and sustainable forest management: the role of traditional knowledge. MCPFE, Warsaw

Patroescu M, Ioja IC, Patroescu-Klotz I, Necsuliu, R (2006) Umweltqualitat in Rumänien. In: Kahl T, Metzeltin M, Ungureanu R (eds) Rumänien. Raum und Bevölkerung. Geschichte und Geschichtsbilder. Kultur. Gesellschaft und Politik heute. Wirtschaft. Recht und Verfassung. Historische Regionen. LIT Verlag, Münster-Hamburg-Berlin-Wien-London-Zürich

Pauleit S, Breuste J, Qureshi S, Sauerwein M (2010) Transformation of rural-urban cultural landscapes in Europe: Integrating approaches from ecological, socio-economic and planning perspectives. Landsc Online 20:1–10

Perzanowski K, Olech W, Kozak I (2004) Constraints for re-establishing a meta-population of the European bison in Ukraine. Biol Conserv 120:345–353

Perzanowski K, Wołoszyn-Gałęza A, Januszczak M (2008) Założenia do wyznaczenia ostoi żubra w Bieszczadach. Eur Bison Conserv Newsl 1:79–86

Peterken G (1996) Natural woodland: ecology and conservation in northern temperate regions. Cambridge University Press, Cambridge

Peterken G (1999) Applying natural forestry concepts in an intensively managed landscape. Glob Ecol Biogeogr 8(5):321–328

Pirot J-Y, Meynell PJ, Elder D (2000) Ecosystem management: lessons from around the world. A guide for development and conservation practitioners. IUCN, Gland and Cambridge

Powelson JP (1994) Centuries of economic endeavor, 4th edn. University of Michigan Press, Michigan

Primdahl J, Brandt J (1997) CAP, nature conservation and physical planning. In: Laurent C, Bowler I (eds) CAP and the regions: building a multidisciplinary framework for the analysis of the EU agricultural space. Institute National de la Recherche agronomique, Paris

Puettmann KJ, Coates KD, Messier C (2009) A critique of silviculture. Island Press, Washington (Managing for complexity)

Puumalainen J, Kennedy P, Folving S (2003) Monitoring of forest biodiversity: a European perspective with reference to temperate and boreal forest zone. J Environ Manag 67:5–14

Rauschmayer F, Berghöfer A, Omann I, Zikos D (2009) Examining processe or/and outcomes? Evaluation concepts in European governance of natural resources. Environ Policy Gov 19:159–173

Reif A, Rusdea E, Pacurar F, Rotar I, Brinkmann K, Auch E, Goia A, Bühler J (2008) A traditional cultural landscape in transformation the quest for sustainable development in the Apuseni Mountains, Romania. Mt Res Dev 28:18–22

Rey V, Groza O, Ianos I, Patroescu M (2007) Atlas de la Roumanie. Reclus, Montpelier and Paris

Roberge JM, Angelstam P (2004) Usefulness of the umbrella species concept as a conservation tool. Conserv Biol 18(1):76–85

Roberge J-M, Angelstam P (2009) Selecting species to be used as tools in the development of forest conservation targets. In: Villard M-A, Jonsson BG (eds) Setting conservation targets for managed forest landscapes. Cambridge University Press, Cambridge

Rozylowicz L, Popescu VD, Patroescu M, Chisamera G (2011) The potential of large carnivores as conservation surrogates in the Romanian Carpathians. Biodivers Conserv 20(3):561–579

Sandström UG, Angelstam P, Khakee A (2006) Urban comprehensive planning—identifying barriers for the maintenance of functional habitat networks. Landsc Urban Plan 75:43–57

Sauberer N, Zulka KP, Abensperg-Traun M et al (2004) Surrogate taxa for biodiversity in agricultural landscapes of eastern Austria. Biol Conserv 117(2):181–190

Sauer CO (1925) The morphology of landscape. Univ Calif Publ Geogr 2(2):19–53

Sayer JA, Campbell BM (2004) The science of sustainable development: local livelihoods and the global environment. Cambridge University Press, Cambridge

Schmitt T, Rákosy L (2007) Changes of traditional agrarian landscapes and their conservation implications: a case study of butterflies in Romania. Divers Distrib 13:855–862

Schnitzler A, Borlea F (1998) Lessons from natural forests as keys for sustainable management and improvement of naturalness in managed broadleaved forests. For Ecol Manag 109:293–303

Singer B (2007) How useful is the landscape approach? In: Patry M, Ripley S (eds) World Heritage forests. Leveraging conservation and the landscape level. Proceedings of the 2nd World Heritage forests meeting, March 9-11, 2005, Nancy, France. UNESCO World Heritage Centre, Paris

Soloviy IP, Keeton WS (eds) (2009) Ecological Economics and Sustainable Forest Management: Developing a trans-disciplinary approach for the Carpathian Mountains. Ukrainian National Forestry University Press, Lviv

Soran V, Biro J, Moldovan O, Ardelean A (2000) Conservation of biodiversity in Romania. Biodivers Conserv 9:1187–1198

Stancioiu PT, Abrudan IV, Dutca I (2010) The Natura 2000 ecological network and forests in Romania: implications on management and administration. Int Rev 12(1):106–113

Surd V, Turnock D (2000) Romania's Apuseni Mountains: Safeguarding a cultural heritage. GeoJ 50:285–304

Svancara LK, Brannon R, Scott JM, Groves CR, Noss RF, Pressey RL (2005) Policy-driven versus evidence-based conservation: a review of political targets and biological needs. Biosci 55(11):989–995

Svoboda M, Pouska V (2008) Structure of a Central-European mountain spruce old-growth forest with respect to historical development. For Ecol Manag 255:2177–2188

Svoboda M, Fraver S, Janda P, Bače R, Zenáhliková J (2010) Natural development and regeneration of a Central European montane spruce forest. For Ecol Manag 260(5):707–714

van Swaay CAM, Warren M (1999) Red data book of European butterflies (Rhopalocera). Council of Europe Publishing, Strasbourg (Nature and Environment 99)

Szaro RC, Bytnerowicz A, Oszlányi J (2002) Effects of air pollution on forest health and biodiversity in forests of the Carpathian Mountains. NATO Sci Ser 1: Life Behav Sci 345

Taylor P, Fahrig L, Henein K, Merriam G (1993) Connectivity is a vital element of landscape structure. Oikos 68(3):571–572

Tear TH, Kareiva P, Angermeier PL et al (2005) How much is enough? The recurrent problem of setting measurable objectives in conservation. Biosci 55(10):836–849

von Thünen JH (1875) Der isolierte Staat in Beziehung auf Landwirtschaft und Nationalökonomie. Wissenschaftliche Buchgesellschaft, Darmstadt

Tucker GM, Heath MF (1994) Birds in Europe: their conservation status. BirdLife international, BirdLife conservation series 3, Cambridge

Turner BL, Clark WC, Kates RW, Richards JF, Mathews JF, Meyer WB (1995) The earth as transformed by human action. Global and regional changes in the biosphere over the past 300 years. Cambridge University Press, Cambridge

Turnock D (2001) Railways and economic development in Romania before 1918. J Trans Geogr 9:137–150

Turnock D (2002) Ecoregion-based conservation in the Carpathians and the land-use implications. Land Use Policy 19:47–63

Turnock D (2007) Aspects of independent Romania's economic history with particular reference to transition for EU accession. Ashgate, Aldershot

Törnblom J, Angelstam P, Degerman E, Henrikson L, Edman T, Temnerud J (2011a) Catchment land cover as a proxy for macroinvertebrate assemblage structure in Carpathian Mountain streams. Hydrobiol Online. doi:10.1007/s10750-011-0769-2

Törnblom J, Degerman E, Angelstam P (2011b) Forest proportion as indicator of ecological integrity in streams using Plecoptera as a proxy. Ecol Indic 11:1366–1374

Törnblom J, Roberge J-M, Angelstam P (2011c) Rapid assessment of headwater stream macroinvertebrate diversity: an evaluation of surrogates across a land-use gradient. Fundam Appl Limnol 178:287–300

UNESCO (2002) Biosphere reserves. Special places for people and nature. UNESCO workshops, Paris

Villard MA, Jonsson BG (2009) Putting conservation target science to work. In: Villard MA, Jonsson BG (eds) Setting conservation targets for managed forest landscapes. Cambridge University Press, Cambridge

Vera FWM (2000) Grazing ecology and forest history. CABI Publishing, Wallingford

Vos W, Meekes H (1999) Trends in European cultural landscape development: perspectives for a sustainable future. Landsc Urban Plan 46:3–14

Vucetich JA, Nelson MP (2010) Sustainability: virtuous or vulgar? Biosci 60(7):539–544

Water Framework Directive (2000) 2000/60/EC of the European parliament and of the council of 23rd October 2000. Establishing a framework for community action in the field of water policy, Brussels

Weaver PM, Jordan A (2008) What roles are there for sustainability assessment in the policy process? Int J Innov Sustain Dev 3(1/2):9–32

Webster R, Holt S, Avis C (eds) (2001) The status of the Carpathians. WWF-The Carpathian ecoregion initiative report. WWF International, Vienna

Whyte ID (1998) Rural Europe since 1500: areas of retardation and tradition. In: Butlin RA, Dodgshon RA (eds) An historical geography of Europe. Oxford University Press, Oxford

World Forestry Congress (2009) XIII World forestry congress, Buenos Aires, Argentina forest development: a vital balance, findings and strategic actions. http://foris.fao.org/meetings/download/_2009/xiii_th_world_forestry_congress/misc_documents/wfc_declaration.pdf. Accessed 2 Sep 2012

Young J, Richards C, Fischer A et al (2007) Conflicts between biodiversity conservation and human activities in the Central and Eastern European countries. Ambio 36(7):545–550

Zechmeister HG, Tribsch A, Moser D, Peterseil J, Wrbka T (2003) Biodiversity ‘hot spots’ for bryophytes in landscapes dominated by agriculture in Austria. Agric Ecosyst Environ 94(2):159–167

Biodiversity and Climate Change, A Risk Analysis (BACCARA): Carpathian Case—Goals and Assumptions

Sławomir Ambroży and Wojciech Grodzki

Abstract BACCARA is a research project supported by the European Community's Seventh Framework Programme, planned for four years (2009–2012). The main goal of BACCARA is to build the tools that will enable forest managers and policy makers to evaluate the risk of European forest biodiversity and productivity loss under climate change. The paper is aimed to present the goals and approach of this project, with special regard to the Carpathian Case. The main outreach from BACCARA project will be the Guidelines "What-to-Grow" and "What-to-Combat", addressed to the forest managers and policy makers.

1 Introduction

BACCARA is a research project supported by the European Community's Seventh Framework Programme (contract no 226299), planned for four years (2009–2012) and coordinated by INRA Bordeaux (France).[1] Sixteen partners from 9 European countries (France, Germany, Italy, Netherlands, Poland, Spain, Switzerland, Sweden, United Kingdom) and China are involved; the Forest Research Institute is the only partner from Poland. Here the goals and approach of this project, with special regard to the Carpathian case study, are presented.

The main goal of BACCARA is to build the tools that will enable forest managers and policy makers to evaluate the risk of European forest biodiversity and productivity loss under climate change. BACCARA will construct a 3-dimensional

[1] for more details see www.baccara-project.eu

S. Ambroży (✉) · W. Grodzki
Department of Forest Management in Mountain Regions, Forest Research Institute, Fredry 39, Kraków 30-605, Poland
e-mail: S.Ambrozy@ibles.waw.pl

J. Kozak et al. (eds.), *The Carpathians: Integrating Nature and Society Towards Sustainability*, Environmental Science and Engineering,
DOI: 10.1007/978-3-642-12725-0_29, © Springer-Verlag Berlin Heidelberg 2013

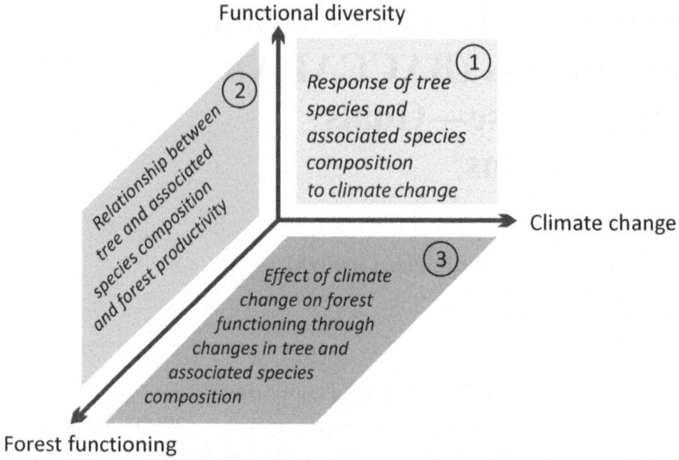

Fig. 1 Three-dimensional risk assessment model

risk assessment model (Fig. 1) linking climate change, functional diversity and forest productivity. The approach will be applied to the main European forest categories.

2 Carpathian Case Study

One of the study areas is located in the Polish part of the Carpathians. Investigations are conducted in mixed Norway spruce *Picea abies,* European beech *Fagus sylvatica* and silver fir *Abies alba* stands. Main study site "Krościenko" is located in the Radziejowa Massif (Beskid Sądecki) with altitudinal transect of four measurement blocks, each one every 200 m a.s.l. This transect is expected to cover local climatic variability and spectrum of local forest types and species. For comparison, some investigations are carried out on similar transect "Jeleśnia", located in pure Norway spruce stands covering the Pilsko Massif (Beskid Żywiecki).

The effect of climate change on forest biodiversity will be evaluated through a better understanding of the ecological processes that shape species composition and that are particularly sensitive to climate conditions. The field investigations will cover three trophic levels: plants, plant-associated organisms (insects, fungi), and natural enemies of insect herbivores. The comparisons between pure and mixed stands are the main research goal.

Level of plants. The objective is to analyse tree community patterns and trends along elevation gradients including studies of diversity, demography-regeneration patterns and optimum growth shift. Study of diversity will let us understand its changes along size classes and to determine which diversity model explains the altitudinal gradient. Study of demography-regeneration patterns refer to the role of the biotic and abiotic factors that limit the recruitment in different environmental conditions. By studying the demography we will assess the performance of the seedlings in terms of competition and climate. It will be estimated directly with demographic parameters (growth) and indirectly with the distribution of the seedlings, saplings and adults. The aim of the study on the optimum growth shift is to obtain the growth curves along the altitudinal gradient at two different time periods. The comparison of these curves will allow detecting the evolution of growth and the shift of the optimal growth altitude.

Level of plant-associated organisms (insects, fungi). The objective is to predict the impact of climate change on tree-associated species diversity and performance. We focus on species with a role in the functioning of forest in terms of net biomass production: antagonistic species (herbivorous animals and pathogens), and mutualistic species (mycorrhiza).

Forest health monitoring and inventory of forest pests and diseases are used to produce a list of pests and pathogens life history traits responding to climate. These data are used to determine how past forest pest and disease damage on key European tree species were dependent on weather conditions.

Sampling at different altitudes is used as spatial analogues of climate change, exploiting sharp gradients to identify the effects of climatic variables on the composition and abundance of pests and diseases communities.

The field investigations, done on a set of experimental plots, are focused on the determination of the species composition of herbivory insects and their natural enemies, mutualistic fungi (mycorrhiza) and fungal root pathogens associated to Norway spruce as main studied species.

A quantitative exploration of insect and fungi functional traits will allow the selection and grouping of species according to potential responses to climate change.

Level of natural enemies of insect herbivores. The general objective is to decipher the relationships between forest biodiversity and functioning through better understanding of the respective role of tree species richness and composition. We are going to provide new insights in the response of the natural enemies (insect parasitoids and predators, antagonistic fungi) to forest biodiversity and climate change. We compare the relative roles played by the richness (how many species?), the functional diversity (how dissimilar are they?), and the composition (which are they?) of tree species in the functioning of mixed forests.

The work is located mainly in the "Krościenko" transect on a set of 15 experimental plots at 4 various altitudes. There are 4 plots per altitude, with a gradient of species richness (spruce + 0–3 admixtures), except the altitude of 1,100 m a.s.l. where there are only 3 plots due to the limited species diversity. Sentinel logs are mainly used for insect collection.

Effect of species richness on the communities of natural enemies of herbivory insects will be evaluated using cambio- and xylophages as example.

3 Expected Outreach

The relationships between forest biodiversity and forest productivity will be investigated through a better understanding of the respective roles of tree species richness and composition in the studied trophic levels. The information will eventually be aggregated to predict the effect of climate change on forest productivity through changes in tree species composition. The main outreach from project BACCARA will be the Guidelines "What-to-Grow" and "What-to-Combat", addressed to the forest managers and policy makers.

Policy Instruments and Methods for the Protection and Maintenance of Historical Agricultural Landscapes in Slovakia

Jana Špulerová

Abstract Despite that there are no specific policies intended to preserve or manage the cultural value of agricultural landscapes in Slovakia, several national policy and legal documents are at least partially focused on historical agricultural landscapes, their conservation, management, and landscape planning. Some criteria for the evaluation of historical agricultural landscapes that are defined by legislative documentation include their cultural heritage value and significant landscape elements. Other criteria, such as characteristic landscape, architectural-technical elements, and agricultural land use elements are defined from a scientific point of view. However, many of the terms are unclear and undefined, and they can be treated differently by different stakeholders or groups of users. Therefore, other evaluation criteria and characteristics of historical agricultural landscapes, such as land value and cultural and natural value, should be clearly defined and incorporated in legislative documents. The first step to evaluation is knowledge about the current distribution of historical agricultural landscapes in Slovakia. A methodology for mapping historical structures of agricultural landscapes was developed in this study. The primary characteristics of historical agricultural landscapes were also studied, including the composition of land use elements, the definition of degree of land use, and the forms of anthropogenic relief and their features.

1 Introduction

Geologic, geomorphologic, and climatic diversity, as well as variability of culture and political and economic systems, are all factors that have affected the Carpathian Mountains in Slovakia and contributed to the diversity of this agricultural

J. Špulerová (✉)
Institute of Landscape Ecology, Slovak Academy of Sciences, Štefánikova 3, P.O.Box 254 814 99 Bratislava, Slovak Republic
e-mail: jana.spulerova@savba.sk

J. Kozak et al. (eds.), *The Carpathians: Integrating Nature and Society Towards Sustainability*, Environmental Science and Engineering,
DOI: 10.1007/978-3-642-12725-0_30,

landscape. The combined effects of nature and humans on the landscape can be termed "cultural landscapes" and are illustrative of the evolution of human society and settlement over time (UNESCO 2008). Many former cultural landscapes contain high-quality environmental resources, but the management regime they were developed under is no longer feasible economically. Modern society increasingly utilizes the landscape in a great variety of ways and for many purposes. This poses complex pressure on cultural landscapes, threatening landscape quality (Vos and Meekes 1999).

"Historical agricultural landscapes" create a special type of cultural landscape, and they are found in areas where past human activities marked the landscape with particular characteristics at the same time that traditional agricultural land use persisted. Historical agricultural landscapes can be described by characteristic features of the landscape, including the presence of specific forms of historical settlements and their scientific, cultural, ecological, aesthetic, and landscape significance. Historical agricultural landscapes are outlined in the Landscape Atlas of the Slovak Republic (Podolák et al. 2002). In Slovakia, they are represented by a mosaic of small-scale arable fields and permanent agricultural cultivations (Dobrovodská and Štefunková 1996). As the main sources of local biodiversity, they are linked to soil or stone features and margins of cross fields tracks, original meadows and pastures, grass-covered former arable fields abandoned after collectivization, and small wetlands or other low production or unfavourable areas (Ružičková et al. 1999). Historical agricultural landscapes that did not lose the shape of the cultural-historical countryside have been preserved mostly in less accessible, remote, and marginal areas with extreme natural conditions. These areas represent regions with specific combinations of natural and cultural diversity, including a high aesthetic value of the landscape (Štefunková and Cebecauer 2006).

Historical agricultural landscapes have irreplaceable ecological, cultural, and historical values. However, due to complex development trends in society, these landscapes have been marginalized, and they are currently no longer of interest from a production perspective (Štefunková and Dobrovodská 2009). These areas are now becoming rare and thus highly valuable in Europe as well. In most of the countries, they do not receive special protection; there is a misunderstanding and lack of communication between strict nature conservation and protection of this cultural heritage.

This paper analyzes existing policy instruments and methods that support the maintenance and protection of historical agricultural landscapes in Slovakia. Despite that these areas are not under special protection, they are part of the national, natural, cultural, and historical heritage that is reflected in some policy and legislative documents. Thus, there is a need to apply resources and tools to protect them and maintain them for future generations.

The study methods included analysis of existing policy documents aimed at management, conservation, and preservation of agricultural landscapes and the financial mechanisms for management and development of historical agricultural landscapes. The study also examined the criteria and methods for valuation of historical agricultural landscapes that are implemented in policy documents.

2 Results

2.1 Policy Instruments for the Management of Agricultural Landscapes

The Slovak Republic and its Ministries, including the Ministry of Agriculture (MA), the Ministry of Environment (ME), and the Ministry of Culture (MC), are national bodies, responsible for policies concerning the management, conservation, and preservation of landscapes.

The landscape in general is an object of several national policy documents. Although there are no specific policies intended to preserve or to manage cultural value in agricultural landscapes, some legal documents are at least partially or more focused on historical agricultural landscapes, their conservation, management, and landscape planning (Table 1).

The aim of the Act of NC SR No. 543/2002 on Nature and Landscape Protection is to support preservation of diverse living conditions and life forms; to create conditions for sustainability, restoration, and rational use of natural resources; to preserve the natural heritage and characteristic landscape features; and to reach and maintain ecological stability. The object of the Act is nature and landscape protection in the territory of the Slovak Republic (SR). The territorial protection specifies five levels of protection that are distinguished by the extent of

Table 1 Policy documents for management of agricultural landscape

Policy documents	Relevance to legislation to historical agricultural landscapes	
National law (acts)	Act of NC SR no. 543/2002 on nature and landscape protection	Low
	Decree of ME SR no. 492/2006 implementing the act on nature and landscape protection	Low
	Act of NC SR no. 24/2005 on impact assessment	Medium
	Act of NC SR no. 49/2002 on the protection of monuments and historic sites	Low
	Decree of the MC SR no. 16/2003 implementing the Act on the protection of monuments and historic sites	Low
	The Act of NC SR no. 220/2004 on conservation and use of agricultural land	Low
Conceptual framework programs	National Environmental Action Plan (NEAP II, 1999)	Medium
	Conceptual Framework of Nature and Landscape Protection (2006)	Low
	Conceptual Framework of Agricultural Development for 2007–2013	Low
	Rural Development Program of the SR for 2007–2013	Low
Documents implementing international strategies	Proposal of implementation of European Landscape Convention in the SR (2005)	High
	National Biodiversity Strategy of the SR (1997)	Low

restrictions. Currently, there are 9 national parks, 14 landscape protected areas, and 1,073 small, protected territories like protected sites and natural reserves in Slovakia. Territorial protection is mostly focused on habitats and threatened species (Izakovičová and Miklós 2007).

Some parts of the agricultural landscape with characteristic landscape features can be preserved as protected areas by definition, according to the Act on Nature and Landscape Protection:

- Protected Landscape Area: areas with fragmented ecosystems that are significant for conservation of biological diversity and ecological stability, with characteristic landscape features or with specific forms of historical settlements,
- point, linear, or other smaller ecosystems, their components or elements, generally not exceeding 50 ha in area, and of scientific, cultural, ecological, aesthetic or landscape significance may be designated by the regional office under a generally binding order as natural monuments,
- significant landscape elements are part of the territory that create the characteristic feature of the landscape or contribute to its ecological stability, including vegetation, alley, or balk.

The contents of a management plan of specially protected parts of nature and landscape are defined by Decree of ME SR No. 492/2006.

The Act of NC SR No. 49/2002 on the protection of monuments and historic sites governs conditions for the protection of cultural heritage monuments and historic sites in accordance with scientific knowledge and on the basis of international conventions in the field of European and world cultural heritage to which the Slovak Republic has acceded. The object of protection is Slovakia's cultural heritage, which means a set of movable and immovable property, including imported cultural works and thoughts, which have been included in Slovakia's cultural heritage. Works of visual, household, and folk art, monuments of folk architecture, relics of manufacturing, science, and technique, historical gardens, parks, and cultural landscapes are considered as part of the movable cultural heritage of humans. Historical agricultural landscapes could also be considered as part of the cultural heritage, but there is no specification for that.

The Decree of the MC SR No. 16/2003 that implements the Act on the protection of monuments and historic sites specifies details and content of urban-historical research documentation. Specifically, it contains requirements for:

- adequate functional land use respecting land value and monuments within the area,
- maintenance and regeneration of the historical ground plan in relation to monuments and land value,
- preservation of characteristic views and vistas,
- preservation, protection, and restoration of historical hedgerow and other cultural and natural values of the area.

The Act of NC SR No. 220/2004 on conservation and use of agricultural land is focused on conservation of attributes and services of agricultural land in general, including biological diversity, productivity, restoration ability, and the ability to fulfil services. Part II defines the principles of sustainable use of agricultural lands and their protection. The Act defines nine qualitative categories of agricultural soil. Preservation of characteristic landscape features of agricultural land is not emphasized in the document.

More attention is focused on agricultural landscapes in conceptual framework plans and strategies. The National Environmental Action Plan (NEAP II 1999) defines nature and landscape management and territorial development as the improvement of the quality of the urban and rural landscape environment, realization of cultural-social and environmental landscaping, and paying more attention to saving dilapidated immovable cultural monuments. NEAP was also included in the management plan of World Heritage localities and in the nomination project for new localities of the World Heritage.

Inclusion in the World Heritage List was facilitated by the Act of NC SR No. 49/2002 on the protection of monuments and historic sites. The Ministry of Culture (on its own initiative, on a proposal from the Monuments Board, other legal entity, or natural person) can make a proposal for the inclusion of a cultural heritage monument or historic site in the World Heritage List (Article 21).

The Conceptual Framework of Agricultural Development for 2007–2013 formulates the main strategic objectives, goals, and priorities of the agricultural department for the medium-term period by 2013. Besides the priority of agriculture and forest management competitiveness, the main priority is given to improve the environment and landscape quality.

The Conceptual Framework of Nature and Landscape Protection (2006) is aimed at nature and landscape protection in accordance with Act No. 543/2002. It is mainly focused on protected areas and habitats included in Council Directive 92/43/EEC on the conservation of natural habitats and of wild fauna and flora. In the chapter dealing with landscape protection, one of the objectives is to restore the ecological stability and character of agricultural landscapes by protection of landscape elements and by realization of land arrangement projects, which are slowly implemented.

Specific policy for the management of agricultural landscapes is provided by the Rural Development Program of the SR for 2007–2013. The list of tasks related to the protection and management of these landscapes are as follows:

- preventing abandonment of agricultural landscapes and moving out of the less favorable areas, by a continuing process of supporting agriculture in these areas,
- maintaining natural landscape potential and biodiversity protection by realization of specified agricultural practices,
- stimulating diversification of activities and developing conditions for stabilization of rural settlements and preservation of the cultural heritage.

New policies of landscape protection, management, and planning were developed by the European Landscape Convention (ELC) and implemented by the SR and the Ministry of the Environment in 2005. "Landscape protection" means actions to conserve and maintain the significant or characteristic features of a landscape, justified by its heritage value derived from its natural configuration and/or from human activity (Article 1d). One of the tasks from the general measures in Article 5 suggests the following:

- recognizing landscapes in law as an essential component of people's surroundings, an expression of the diversity of their shared cultural and natural heritage, and a foundation of their identity,
- establishing and implementing landscape policies aimed at landscape protection, management, and planning through the adoption of the specific measures set out in Article 6.

2.2 *Funding Opportunities*

Maintenance and preservation of historical agricultural landscapes and sustainable development of these areas is dependent on external funding. There are several funding schemes that can be relevant to these areas, such as rural development programs, European Union (EU) structural funds, programs of international cooperation, and other national grants and programs.

Rural Development Programs are linked to the Common Rural Development Policy of the European Union, which is aimed at improving:

- competitiveness of the agricultural and forestry sector,
- environment and the countryside,
- quality of life in rural areas and encouraging diversification of the rural economy.

The subsidies related to these programs are oriented toward ecological farming, support to management of agricultural landscapes in the areas of European significance, and the aid to farmers in Less Favoured Areas.

EU structural funds should promote an overall harmonious development and strengthen economic and social cohesion by reducing development disparities between the regions. For the 2007–2013 period, the instruments to pursue these objectives have their legal basis in a package of regulations adopted by the EU Council and the European Parliament in July 2006. The regulations of the European Regional Development Fund (ERDF) define its role and fields of interventions as the promotion of public and private investments helping to reduce regional disparities across the Union. ERDF supports also transboundary cooperation between the Slovak Republic and Czech Republic, Poland, Hungary, Austria, ENPI CBC SR–Ukraine–Hungary–Romania, Central Europe, Innovation and Environment regions of Europe sharing solutions (INTERREG IVC), and INTERACT II. The European Social Fund (ESF) is implemented via the European Employment Strategy, and it is focused on four key areas: increasing adaptability of workers and enterprises, enhancing access to employment and participation in the labor market, reinforcing social inclusion by combating discrimination and facilitating access to the labor market for disadvantaged people, and promoting partnership for reform in the fields of employment and inclusion. The Cohesion Fund contributes to interventions in the field of the environment and trans-European transport networks. Finally, the aim of the European Grouping of Territorial Cooperation (EGTC) is to facilitate cross-border, transnational, and/or interregional cooperation between regional and local authorities. The latter would be invested with the legal authority for the implementation of territorial cooperation programs based on an agreement between the participating national, regional, local, or other public authorities.

An important program of international cooperation supporting cross-border cooperation and regional development is the International Visegrad Fund, which promotes development of closer cooperation among the Visegrad Group (V4) countries: the Czech Republic, Hungary, Poland, and Slovakia. The Fund provides funding for common cultural, scientific, research, and educational projects, youth exchanges, promotion of tourism, and cross-border cooperation. Other national grants and programs are listed below:

- programs of the Slovak Environmental Agency: Rural Environmental Protection, Village of the Year, National Rural Network, and Traditional Folk Culture,
- Ekopolis Foundation and its activities in the area of Conservation of Natural and Cultural Heritage (e.g. Safe Energy, Block Grant TSD-ENGO, Nature Conservation and Tourism, Tatra Fund, Tree of the Year),
- Slovak Gas Holding foundation and its program "Heritage of Regions" focusing on projects promoting culture, local costumes, handicrafts, and traditions,
- regional programs and agencies, like the Carpathian Foundation, A-project.

Funding schemes and subsidies should contribute to biodiversity conservation, optimization of land use and improving the management of agricultural landscapes, and ultimately to rural and environmental sustainable development.

2.3 Evaluation Criteria and Methods

Analyses of policy instruments have showed that conservation and management of historical agricultural landscapes is an important part of landscape planning and management. Analysis of criteria and methods implemented in policy instruments, which can be applied for evaluation and characterization of historical agricultural landscapes, was the next phase of this study.

As a principal criterion for evaluation, the "cultural heritage value" is defined by the Act of NC SR No. 49/2002 on the protection of monuments and historic sites. The term "cultural heritage value" shall mean the aggregate value of important historic, social, rural, urban, architectonic, scientific, technical, visual art, artistic, and craft values for which the property or objects are subject to individual or territorial protection. The scientific terms are defined as:

- characteristic landscape features with specific forms of historical settlements and vegetation and landscape elements: alley, balk, meadows, monuments and historic sites, fruit-trees, etc.,
- architectural-technical elements: small architect elements, technical equipment, etc.,
- agricultural land use elements: field structure, terraces, size, shape, etc.,
- green elements: alleys, trees, hedges, orchards, etc.

The Decree of the MC of the SR No. 16/2003 Act implementing the protection of monuments and historic sites specifies the following criteria for research documentation:

- land value,
- preservation of characteristic views and vistas,
- cultural and natural value of the area.

The document implementing the European Landscape Convention in the SR (2005) describes the "landscape quality objective" for a specific landscape, which means the formulation by the public authorities of the aspirations of the public of the landscape features of their surroundings.

Following the Act of NC SR No. 24/2005 on impact assessment, the impact on the landscape can be evaluated on the basis of following criteria: landscape structure, land use, and scenic view.

Many of the terms are unclear and undefined, and they can be treated differently by different stakeholders or group of users. The need for definition of clear criteria and development of methods are reflected also in other tasks for landscape protection and implementation of the ELC in Slovakia:

- defining culture-historical potential and standardization of the Slovak landscape on the basis of characteristic environmental features, land use, culture-historical value, and applying the methodology of typological and individual regionalization of selected landscape segments,
- elaborating a methodology for landscape-historical potential (e.g. valuable characteristics of the landscape, historical landscape structure),
- involving stakeholders in the process of identification of landscape values and quality,
- realizing mapping of important landscape elements and the historical landscape,
- exchanging experience and methodological approaches applied to landscape identification and assessment with the other partner countries,
- evaluating preservation decree and changes of landscape, definition of the most threatened landscape types, their threats and carrying capacity of the area.

Even though historical structures have been the subject of several studies (Huba 1988; Jančura 2004; Olah et al. 2006; Petrovič 2007), there is still a lack of consistent scientific information, which can serve as a background for creation of relevant legislation measures. Compiling a complex scientific background about historical agricultural landscapes can provide valuable documentation for territorial planning on the regional and national level, resulting in a methodology for mapping historical structures of agricultural landscapes (Špulerová et al. 2009) and the realization of rural mapping of historical agricultural landscapes of Slovakia, which has been undertaken since 1999. The handbook for mapping historical structures of agricultural landscapes defines mapping key for understanding the term "historical structures of agricultural landscapes" and their main features recorded in the field, such as the degree of land use or forms of anthropogenic relief (Dobrovodská et al. 2010). This handbook introduces a methodology proposed for a complex inventory of historical structures of agricultural landscapes over the entire country of Slovakia in a record form. It consists of three parts: basic data, characteristics, and additional information. The basic data comprise the names of observers who completed the record sheet, the field mapping date, and location. The degree of land use, composition of land use elements, shape of plots of land, relief of plot curve, and forms of anthropogenic relief and their characteristic features, as well as maintenance of historical structures of agricultural landscapes, were identified as the main characteristics of historical structures in the agricultural landscape. Additional information, as the third part of the recording form, increases our knowledge of the historical, cultural, and natural value of historical structures in the agricultural landscape by acquiring more information about elements such as small architecture and significant species and habitats (Dobrovodská et al. 2010).

The outcomes of field mapping form a database of historical structures in the agricultural landscape. The statistical analysis method will be used to evaluate the database and to classify the historical structures in the agricultural landscape with regard to their character, parameters, and their natural and cultural conditions. These will be published in a catalog and the types of historical agricultural landscape will be the subject of our next evaluation process of the preservation decree and the landscape changes.

3 Discussion and Conclusions

The concept of an agricultural landscape as consisting of its historical or cultural heritage is not addressed specifically in any convention and charter, or in international or national laws (Pungetti and Kruse 2010). However, it is present in various policies that influence and sometimes conflict with each other. Likewise in Slovakia, protection of historical agricultural landscapes is not included in special natural protection activity according to the Act No. 543/2002 or any other specifically targeted legislation documents. Thus, there is a real risk that in the near future, due to forest succession and urban pressure, that decline or even irreversible loss of biodiversity linked to specific biotopes will occur (Špulerová 2008; Válkovcová 2008).

Some parts of agricultural landscapes with characteristic landscape features can be preserved as protected areas or significant landscape elements, as a result of their definition according to the Act of NC SR No. 543/2002 on Nature and Landscape Protection. However, the priorities of the State Nature Conservancy are to the preservation of species and habitats included in Habitat Directive. The objectives of the conceptual framework of nature and landscape protection (2006) are focused on reevaluation of the existing network of protected areas and reduction of their size. In addition, because the definition of characteristic landscape features or specific forms of historical settlements are not clear, some protected landscape areas, such as the Kysuce or Biele Karpaty, may be reduced since conservation is insufficient there. These areas represent historical agricultural landscapes with specific forms of historical settlements and a well-preserved mosaic of grassland and arable fields with characteristic soil-stone features. Agricultural landscapes such as this that include the presence or dominance of seminatural habitats with extensive farming should be part of the High Nature Value Farmland (Paracchini et al. 2008).

The aim for maintaining the historical agricultural landscape is not strict conservation, but preservation, protection, and restoration of cultural and natural value of the area, and preservation of characteristic views and vistas in accordance with sustainable development. Integration of sustainable development is dependent on policy tools and funding schemes in the various evaluation procedures (e.g. EU structural funds, programs of international cooperation, rural development programs, other national grants and programs) that should contribute to biodiversity conservation, optimization of land use and improving the management of agricultural landscapes, and ultimately to rural and environmental sustainable development.

Further studies are required in completing and refining the methodology and characteristics for valuing historical agricultural landscapes. The process has been initiated by creating a methodology and rural mapping of historical agricultural landscapes; the next steps of evaluation must provide comprehensive and comparable coverage, since the current results of our study only allow approximating the distribution of historical agricultural landscapes in Slovakia.

Acknowledgments This study was prepared within the grant project of the Ministry of Education of the Slovak Republic and the Slovak Academy of Sciences, No. 2/0114/10, "Identification of Purposive Landscape Features as the Basis of Landscape Ecological Research."

References

Dobrovodská M, Štefunková D (1996) Historické poľnohospodárske formy angropogénneho reliéfu v oračinovo-lúčno-pasienkarskej a vinohradníckej krajine. Acta Environ Univ Comen 7:85–91

Dobrovodská M, Špulerová J, Štefunková D, Halabuk A (2010) Research and maintenance of biodiversity in historical structures in the agricultural landscape of Slovakia. In: Landscape ecology—methods, applications and interdisciplinary approach. Institute of Landscape Ecology Slovak Academy of Sciences, Bratislava

Huba M (1988) Historické krajinné štruktúry. Ochranca prírody, odborná príloha spravodajca. Ochranca prírody, odborná príloha spravodajca MV SZOPK Bratislava

Izakovičová Z, Miklós L (2007) New concept of nature protection in the Slovak Republic. In: Chmielewski TJ (ed) Nature conservation management: from idea to practical results. PWZN Print, Lublin

Jančura P (ed) (2004) Historické krajinné štruktúry vo vzťahu k vývoju poľnohospodárskeho využívania zeme. Partner, Banská Bystrica

Olah B, Boltižiar M, Petrovič F, Gallay I (2006) Vývoj využitia krajiny slovenských biosférických rezervácií UNESCO. Vedecké štúdie 2/2006/B. TU a SNK MaB, Zvolen

Paracchini ML, Petersen JE, Hoogeveen Y, Bamps C, Burfield I, van Swaay Ch (2008) High nature value farmland in Europe. An estimate of the distribution patterns on the basis of land cover and biodiversity data. European Commission, Joint Research Centre, Institute for Environment and Sustainability, Ispra

Petrovič F (2007) Originalita rozptýleného osídlenia na Slovensku. Enviromagazín 3:24–25

Podolák J, Klinda J, Hrnčiarová T, Izakovičová Z, Lichard P, Maráky P, Mešša M, Zerola J (2002) Traditional land use and traditional culture. In: Hrnčiarová T et al (eds) Atlas of the landscape of the Slovak Republic. Ministerstvo životného prostredia SR, Bratislava

Pungetti G, Kruse A (eds) (2010) European culture expressed in agricultural landscapes. Perspectives from the Eucaland Project, Palombi Editori, Roma

Ružičková H, Dobrovodská M, Valachovič M (1999) Landscape-ecological evaluation of vegetation in relation to the forms of anthropogenic relief in the cadastre of Liptovská Teplička village, the Nízke Tatry Mts. Ekol (Bratislava) 18(4):381–400

Špulerová J (2008) Land use changes in the Veselovianka river catchment in the Horná Orava region. Ekol (Bratislava) 27(3):326–337

Špulerová J, Dobrovodská M, Štefunková D, Halabuk A (2009) Metodika mapovania historických štruktúr poľnohospodárskej krajiny. In: Kontrišová O et al (eds) Monitorovanie a hodnotenie stavu životného prostredia VIII. Zborník referátov, TU Zvolen, Zvolen

Štefunková D, Dobrovodská M (2009) Preserved European cultural heritage in agrarian landscape of Slovakia. Tájökológiai Lapok 7(2):283–290

Štefunková D, Cebecauer T (2006) Visibility analysis as a part of landscape visual quality assessment. Ekol (Bratislava) 25(Suppl 1):229–239

UNESCO (2008) Operational guidelines for the implementation of the world heritage convention Intergovernmental Committee for the protection of the world cultural and natural heritage, Paris

Válkovcová Z (2008) Evaluation of changes of non-forest woody vegetation in agriculturally utilised landscape within 50 years time horizon. Phytopedon 7(1):85–93

Vos W, Meekes H (1999) Trends in European cultural landscape development: perspectives for a sustainable future. Landsc Urban Plan 46:3–14

Historical Landscape Elements: Part of our Cultural Heritage—A Methodological Study from Saxony

Olaf Bastian, Ulrich Walz and Annette Decker

Abstract There is a wide spectrum of definitions for the terms 'landscape', 'cultural landscape' and 'traditional landscape'. Nevertheless, there is a widely accepted consensus that (historical) cultural landscapes and their typical elements are very important for modern human societies, because they supply many functions and services. Due to their importance and their rapid decline, efforts to register and to protect such traditional landscapes and historical landscape elements are increasing. A new multistage methodology for analyzing and evaluating historical landscape elements and whole remains of traditional landscapes developed in the framework of elaborating the Saxon landscape program is presented as well as documentation sheets for historical landscape elements exemplified on the type 'semi-natural grassland'.

1 Introduction

Due to the rapid economic and demographic changes in almost all parts of the world, traditional cultural landscapes and their characteristic elements are vanishing more and more. However, the awareness that such landscapes and landscape elements are worthy of protection is rising. In many countries, goal-oriented conservation and management efforts have been strengthened during the last years (e.g., Burggraaf and Kleefeld 1998; Pedroli 2000; Palang et al. 2006).

O. Bastian (✉) · U. Walz
Leibniz Institute of Ecological Urban and Regional Development,
Dresden, Weberplatz 1 01217, Germany
e-mail: o.bastian@ioer.de

A. Decker
Saxon State Office for Environment, Agriculture and Geology,
Halsbrücker Str. 31a, Freiberg 09599, Germany

J. Kozak et al. (eds.), *The Carpathians: Integrating Nature and Society Towards Sustainability*, Environmental Science and Engineering,
DOI: 10.1007/978-3-642-12725-0_31, © Springer-Verlag Berlin Heidelberg 2013

International and European agreements (e.g., the European Landscape Convention), laws and programs of many countries aim at maintaining valuable traditional cultural landscapes. For example, the German Nature Conservation Act postulates the protection of historical cultural landscapes. Historical landscape elements (HLE) play an important role also in the landscape program presently elaborated by the German Free State of Saxony.

After a survey of terminological and methodological aspects, approaches and results of the inventory of HLE are presented as it was applied for the elaboration of the Saxon landscape program.

2 The Terms Landscape, Cultural Landscape and Historical Landscape Element

The term landscape covers a broad spectrum of meanings and contexts. The political and colloquial dimensions, historical application of the term, its significance in different scientific disciplines but also the use of the term as metaphor and as a symbol contribute to the manifoldness of connotations of the term 'landscape'.

The decades-long discussions of the topic of 'landscape' have often been fairly controversial, and even today, there is no consensus as to the contents and meaning of the concept of landscape; rather, a number of quite different interpretation patterns exist. The understanding of landscape differs between languages, interest groups and academic backgrounds (Palang et al. 2006). According to Wascher (2005), landscape as a concept is rooted in a wide range of both social and natural disciplines—including geographic, ecological and aesthetic/cultural approaches—dating back to the early 1800s. Neef (1967) defined landscape as "a segment of the earth's surface characterized by a uniform structure, and the same structure of effects (process structure), of which the full integration of all geo-factors (geological subsoil, relief, soil, climate, water balance, flora, fauna, humankind and its works) of a site or a space consists." Analogously, the landscape concept of Haase and Richter (1980) describe "the contents and essence of a space pre-created by a natural process, and influenced and shaped by society, as a segment of the earth's shell. Landscape is a time–space structure determined by the metabolism between humankind and nature."

The understanding of landscape as an intermediate phenomenon between empirical facts and intellectual construct approaches highlight definitions such as that of the Council of Europe European Landscape Convention, Article 1, Council of Europe (2000): "an area, as perceived by people, whose character is the result of the action and interaction of natural and/or human factors"; or that of Fry (2000): "physical and mental reflection on the interactions between societies and cultures with their natural environment." Accordingly, landscape can be seen as a part (at different scale) of the Earth's surface, which is coined by natural conditions and formed by human influences to a different extent. It is perceived and felt by humans as characteristic, and it can be differentiated and classified according to defined rules (Bastian 2008a). Naveh (2010) recognizes landscapes "as tangible, spatially and temporally well defined ecological systems of closely interwoven natural and cultural entities of our living space. Ranging from the smallest discernible landscape cell or ecotope to the global ecosphere landscape, they serve as the spatial matrix and as the living space for all organisms, including humans, their populations and their ecosystems."

According to Leibenath and Gailing (2012) landscape can be defined in a 4-fold manner:

1. landscape as a physical space or as an ecosystem complex;
2. (cultural) landscape in the context of the relationship between man and environment;
3. (cultural) landscape as a metaphoric expression;
4. (cultural) landscape as a social construction or as an issue of communication.

Similarly to the landscape term, also to the term cultural landscape various meanings, theories and concepts are attributed (e.g., Schenk 2002; Solon 2008). Generally, cultural landscape is the result of the long-term interactions between natural conditions and human influence. Man shaped nature according to his existential, economic, social, cultural and mental needs. Cultural landscapes are changing more or less permanently, directly correlated with the development of human societies and the changes in humans' needs.

These processes of change can take place differently in terms of space and time (speed). Regarding a time axis, phases of continuity (stability) replace phases of changes (Burggraaf 1996; Schenk 2002). Cultural landscapes are economic goods, background for leisure activities, they proceed natural processes, and they contain specific land use patterns, forms and structures forming their identity (Konold 2006).

The specific traits of cultural landscapes are a high intensity and a long history of human influence (Naveh 1995; Haber 2001). Cultural landscapes exist in all regions of the world. One could ask, whether a landscape without human influence exists at all. Perhaps in some high mountain areas, in some deserts, in Antarctica, one will find landscapes, where humans did not change decisive qualities of the landscapes until now. Though, if one includes human activities into the concept, the term 'cultural landscape' becomes useless (Beierkuhnlein 2002). Also Schreiber (1999) and Haber (2008) refer to the tautological character of the term 'cultural landscape'. The sense of keeping this term anyway, is to emphasize:

1. the human influence, or
2. the cultural value of a landscape.

Traditional landscapes (historical landscapes) represent a specification of cultural landscapes. They are parts of the contemporary cultural landscape, where historical structures have not been removed or overshadowed by contemporary, modern methods of land utilization, and where many relics of the past outlived until today (Gunzelmann 1987; Antrop 2000; Moss and Okey 2004). The specific feature of such landscapes in comparison with the technologically transformed contemporary landscape is that it is still visible that they were originally transformed by people without the use of modern, powerful technology (Huba 2000; Fig. 1 and 2). Traditional landscapes represent a reflection of the complicated historical development of the landscape, a reflection of the co-existence and/or struggle between man and nature as well as reflection of cultural, social, political and economic transformations of the society.

Traditional landscapes have a distinct and recognizable structure, which reflect clear relations between composing elements, and they have significant natural, cultural or aesthetic values (Antrop 2000; Moss and Okey 2004). The coexistence (the simultaneity of non-simultaneous things) of features of several periods (temporal layers) is an essential characteristic of historical landscape (Becker 1998).

Fig. 1 Traditional rural landscape Zazrivá-Grune (Orava region, Slovakia) (Photo: O. Bastian)

Fig. 2 Extensive systems of hedgerows are characteristic for large areas of the Ore Mountains (Saxony, Germany) (Photo: O. Bastian)

Traditional landscapes are rich in historical landscape elements (HLE). To a certain but lower degree, such elements can be found also in 'normal', modern cultural landscapes. HLE represent a specific, historically determined and recently diminishing sub-category of overall landscape structures. They are documents of the cultural and economic life of former human generations in the landscape. Landscape elements can be qualified as historical, if they are not originating any more under the present conditions. In contemporary landscapes they have been preserved thanks to special regulations or due to the slowness of landscape change. As relics they remind, for example, of memorable events (e.g., court lime trees). They mark old trade routes (by avenues, narrow passes, milestones), but also borders (boundary stones, hedges, stone walls, castle walls). They also document economic systems (e.g., settlement and field forms, edges of fields, meadow orchards, vineyards, pollarded willows, fish ponds) (e.g., Konold 2006). Also semi-natural ecosystems can be categorized as HLE if their origin is not natural but anthropogenic, and if their management needs human management.

3 The Role of Historical Landscape Elements

Historical landscape elements are very important for our modern human society, because they meet many functions (landscape services). They:

- reflect the historical development of the landscape, the co-existence and/or struggle between man and nature as well as the cultural, social, political and economic transformations of the society;
- are a mirror of the technological development at that time. They enable insights into the needs and possibilities of our ancestors, and they give vivid examples of their history and culture;
- are important and expressive landscape-historical sources of knowledge;
- contribute to the individuality and the character of a landscape (together with the natural conditions and structures);
- tell history and stories, and they are connected with symbols (Fig. 3);
- carry collective memories and help that landscapes are highly regarded by people;
- essentially contribute to the regional identity (native region);
- are habitats of plants and animals and sometimes the last sanctuary of rare and threatened species (Figs. 4 and 5);
- raise the attractiveness of an area for recreation. They are also useful for education (e.g., extracurricular learning place), and they inspire arts.

The geographical location of Central Europe on the crossroads of trade routes, and its rich and eventful history, has been a precondition for richness, diversity and culturally significant historical manifestations. Already the postwar period after 1945, which was characterized by intensive industrialization, urbanization, collectivization and general automation of our life, caused an extremely rapid

Fig. 3 Crucifixes in the countryside are typical for the catholic Sorbian region of Upper Lusatia (Saxony, Germany) (Photo: O. Bastian)

Fig. 4 As they have been mostly established for centuries, fish-ponds (here in the Upper Lusatia region, Saxony, Germany) are historical landscape elements but also valuable biotopes (Photo: O. Bastian)

extinction of historical structures. Also after 1989 deep (and still maintaining) transition processes of the society in Eastern Central Europe lead to complex landscape transformations.

Fig. 5 Traditional land use systems like meadows with scattered fruit trees are historical landscape elements (Saxony, Germany) (Photo: U. Walz)

3.1 Inventory, Valuation and Conservation of Historical Landscape Elements

Due to the importance of traditional landscapes and HLE on one side, and their rapid decline on the other, it is an evident task to identify the most characteristic and valuable historical structures and to secure their 'survival'. Conservation, as well as effective management and utilization of at least a representative part of existing HLE is an important contribution to maintain our cultural heritage. For it, detailed information on the state of the cultural landscape and the stock of HLE is needed.

For several years, efforts to record HLE have been increased, in Germany from the 1980s. Various methodological approaches were developed, and in several German federal states, but also abroad, activities and programs were started, which are involving also voluntary work. The analysis of HLE can also contribute to delimit whole historical landscape areas (e.g., Burggraaf and Kleefeld 1998; Peters and Klinkhammer 2000).

It is important to organize the inventory, documentation and valuation of HLE according to a standardized methodology and to consider the specific tasks and questions (e.g., their role for nature conservation or for recreation and environmental education). Approaches and examples for the inventory of HLE are described e.g., by Wöbse (1999), Röhrer and Büttner (2008) and by Bund für Heimat und Umwelt (2008). Burggraaf and Kleefeld (1998) elaborated a general, Germany-wide manual. There are also specific manuals for some German states (Länder), e.g., for Bavaria (Bayerisches Staatsministerium für Landwirtschaft und Forsten 2001) and for Thuringia (Schmidt et al. 2006).

A crucial question is, if a landscape element is traditional or not? To answer this question, distinct criteria for HLE identification are necessary. Fundamental characteristics of cultural landscapes are (Nowak and Schulz 2004): a favorable state of conservation (of the historical landscape character), only little changes in the landscape structure and the scenery since 1950, the frequent occurrence of landscape features formed in the past, landscape elements evolved only recently are integrated into the historical landscape structure. The essential change of the landscape character and the scenery of the whole area, the technical transformation (land consolidation, vast growth of settlements), decay or loss of HLE are criteria of exclusion. Criteria for the assessment of HLE can be (Schmidt et al. 2006): rarity, the state of preservation, age, regional importance, experience value.

Due to the large spectrum of HLE, a classification similar to the biotope mapping is necessary. Peters and Klinkhammer (2000) mention requirements on the systematization of HLE. It should:

- elucidate historical and landscape contexts;
- be a useful completion of the catalogue of biotope types;
- suitable for the mapping (it should be possible to identify a HLE in the field without a thoroughly knowledge of its history);
- additional information should be easily to interpret;
- be open for new elements unknown until now;
- be suitable for a Geographical Information System (GIS).

Generally, the Procedure of the Inventory and the Assessment of HLE Consist of Several Steps (Peters and Klinkhammer 2000):

1. preliminary investigation: identification of HLE by the analysis of plans and maps (e.g., historical topographical maps), analysis of written sources and existing inventories (e.g., biotope mappings), enquiry of local experts, historical analysis of special HLE by studies of literature and in archives;
2. on-site-mapping in selected areas (selective mapping) or in whole areas (total mapping);
3. creation of a valuation framework (e.g., aesthetical value, historical role, functional aspects, need of protection) basing on the existing legislation and on regional targets (so-called leitbilder or visions, e.g., from regional plans);
4. description and evaluation of HLE with the help of a documentation sheet (examples for such sheets see e.g., Wöbse 1994; Peters and Klinkhammer 2000), photographical documentation.

The character of a landscape can be formed by the occurrence of HLE. From region to region, the type, the combination and the frequency of HLE is different, which gives the landscape a typical appearance. Thus, Burggraaf and Kleefeld (1998) identified cultural landscape areas in Germany. In the landscape program for the German state of Brandenburg, regional landscape areas are described on the basis of historical natural landscape elements (Peters and Klinkhammer 2000). Regarding the strength of the historical character, the regional plan for Central Hesse (Nowak and Schulz 2004) distinguishes between three categories of historical cultural landscapes. Their relative importance is differentiated concerning importance on national, supra-regional, regional and local scales.

4 HLE in the Landscape Program of Saxony: Methods

Also in Saxony, the various natural conditions and long traditions in land use have been the preconditions for richness, diversity and culturally significant historical manifestations. In pursuit of regional planning goals concerning the preservation and development of the cultural landscape, the Free State of Saxony pays attention to HLE in its landscape program (= landscape plan for a whole state in Germany) presently under development. By order of the Saxon State Office for Environment, Agriculture and Geology a new multistage methodology was developed by Walz et al. (2010, 2012) for analyzing and evaluating HLE and whole landscapes. This approach combines methods of the spatial overlay of geodata, the descriptive statistics and the spatial and hierarchical clustering. The aim was to deduce cultural landscape units and to allocate typical HLE for the different units. The analysis of HLE and the deduction of historical landscape areas were realized basing on local sub-districts as smallest reference units, according to the scale 1:50,000. It was used a statistical approach to assure the comparability and repeatability of the method and to reduce the subjective factor.

To identify and to evaluate (historical) cultural landscape areas, in a first step the spatial distribution of cultural landscape elements (40 types in 11 categories; see Table 1) was processed with the aid of a GIS. The classified frequency of the single element types per local sub-district is the result of this procedure. Sets of maps (incl. a synthesis map) show the distribution and the main areas of occurrence of the 40 HLE types.

In a second step, the single HLE were grouped by the method of hot spot analysis. This results in complex units characterized by the distribution of specific HLE types. In a third step, cultural landscape areas were established that differ in their diversity and character shaped by different HLE. Therefore, the areas of individual HLE types were combined by a cluster analysis, with the aim to aggregate similarly structured sub-districts (Fig. 6). In a fourth step, the local sub-districts were assessed regarding the degree they are coined by individual HLE types and priority areas were deduced.

As an additional project, for 16 selected types of HLE more precise information was collected and documented in detailed descriptions (Thiem and Bastian 2009): field-terraces, alleys, old dikes, extensively used arable fields, semi-natural grassland, pits, hedges, heaths, historical forestry systems, narrow passes, relic peat-cutting sites, stone walls, meadow orchards (Fig. 5), ponds, forest hide farmland, vineyards. The main focus was on such types that are also habitats for animals and plants. Therefore, they are important for landscape management on two counts. The descriptions are structured as follows:

- definition/features (description of the main characteristics);
- typology: description of varieties, sub-types and their characteristics;
- peculiarity/scenery: meaning for the landscape as a shaping element, landscape perception;

Table 1 Cultural landscape elements: categories and types (from Walz et al. 2010)

Agriculture	*Mining*
vineyard	relics of old mining on ore
meadow with scattered fruit trees	relics of old mining on black coal
forest hide farmland	relics of old mining on brown coal
stone ridge	relics of peat cut
field-terrace	former stone quarry or excavation of chalk
hedge	*Processing of food and materials*
mountain meadow	wind mill
wet meadow	water mill
damp meadow	other mill
heathland	*Military, safety, administration and representation*
old extensive agriculture	battle field
Forestry	rampart
old farm wood	landwehr, linear earthwork
pastural woodland	fortress, castle
medium forest	manor house
coppice	park
selection forest	*Building types*
Settlement types	local architectural style
one-street village	farm house
square village	*Fishery*
dispersed settlement	pond
Hunting	*Religion*
manorial hunting installation	monastery
Traffic	
narrow pass	
alley	
old roads before 1900	
railway before 1900	
narrow gauge railway	
old dike	

- origin and history of the HLE (types): short draft of the cultural and land use history, also changes in functions;
- occurrence and distribution in Saxony (and in Germany);
- classification as HLE;
- resilience/risks: which land use forms and other influences cause threats and losses?
- state of protection: is the element under preservation order (e.g., as biotope or historical monument) or not?
- significance for species, biocoenoses or habitats;
- conservation and management: appropriate land use models and measures;
- sources of information, references.

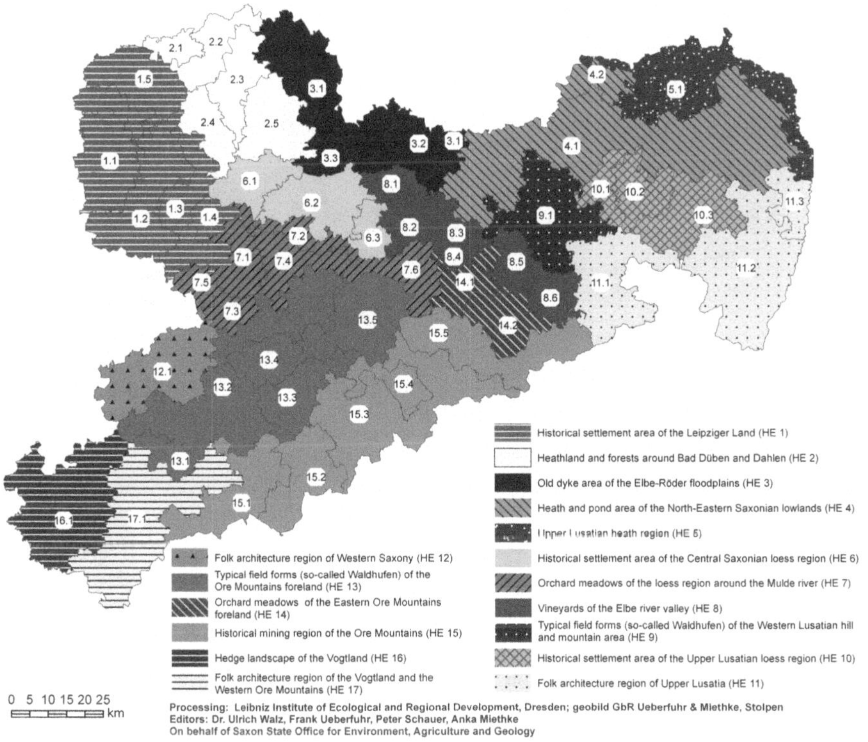

Fig. 6 Delineation and naming of cultural landscape units in Saxony (*Source* Walz et al. 2012)

5 Results

In this way, 17 cultural landscape areas and 53 sub-units could be defined, which can be distinguished by their more or less rich and diverse stock of HLE (Fig. 6). These areas were named for better identification. They can be used for landscape planning as reference base for general principle for spatial planning. For example, objectives of regional policy in conservation and management of historical landscape and elements can be addressed to such areas.

Also more detailed priority areas were deduced, which are shaped by HLE to a high extent. The criteria are the diversity (number of different HLE types per municipality) and the dominance of one (or several) type(s). For this, on the one hand the dominance of historical element types was evaluated. Element types are dominant, if they are very common (high number or coverage of specific element types) in single units. On the other hand, we investigated the imprint of the units by all HLE. Therefore all class values from step 1 were summarized. The result (see Fig. 7) shows landscape units, which are more imprinted by HLE than others. The interpretation should consider that a low number of element types with high

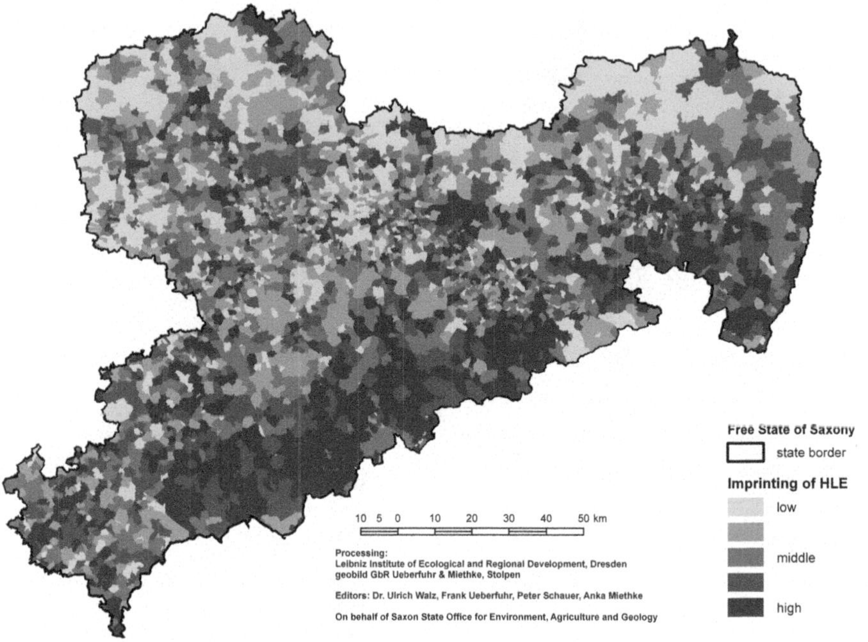

Fig. 7 Imprinting of landscape units in Saxony by HLE in Saxony (*Source* Walz et al. 2012)

dominance can imprint a landscape as well as a high number of different but for itself less dominant element types.

We could gather a lot of information for the selected HLE, and we have documented it in detailed descriptions (with a length of each 10–15 pages). As an example, we present the condensed version for the HLE 'semi-natural grassland' (Table 2).

6 Discussion

Changes in cultural landscapes are a matter of fact and directly correlated with the development of human societies. According to Palang et al. (2006) landscape loses its life-supporting and emotional role in people's thinking. Due to economic and technological changes, migration (daily, seasonal, permanent) today people are less connected to their land as they used to be in the past. In former times, land was the main or only source of income for the majority of people, so they were strongly associated with their property (land or landscape). Cultural landscapes are changing in Europe and in the whole world, in various speeds and to different extents, though. In the more industrialized Western Europe HLE have mostly been wiped out by urban sprawl (Antrop 2004), the loss of open spaces, and industrial

Table 2 Documentation sheet for HLE types, exemplified by the 'semi-natural grassland' (from Thiem and Bastian 2009, modified)

Criteria	Description
Definition	Grassland = a natural or an agricultural area used as meadow or pasture. Semi-natural grassland is not used intensively. Such a use is characterized by: - low input of capital and means of production (fertilizers, biocides, techniques, number of cattle) - no fundamental changes in site conditions (drainage, irrigation, relief leveling) - diversified pattern of vegetation types and management measures, small utilization units with irregular shapes and borders, long ecotones
Typology	There are quite different types of this HLE, depending on site conditions (water balance, trophy), management, floristic composition and vegetation (e.g., mesophilic meadows and pastures, mountain meadows, moist and wet grassland).
Peculiarity/scenery	Until the middle of the twentieth century, extensively used, rich-flowering meadows and pastures dominated the picture of the grasslands. Flowering meadows are very attractive still today, e.g., orchid meadows as well as mountain meadows with the colorful and smelling herbs. The different distribution of grassland types can contribute to the peculiarity and identity of a landscape essentially.
Origin and history	Artificial meadows could arise only after the introduction of the scythe. Probably our classical meadow-scythe stems from Tyrol (thirteenth to Fourteenth century) and came from the Alps to Central Europe. Also thanks to the wide-spread semi-natural grassland Central Europe had the ever highest biodiversity during the second half of the ninetieth century. Deep land use changes (intensification) caused the decline of such grassland in the second half of the twentieth century.
Occurrence and distribution in Saxony	Remains of semi-natural grassland can be found in all regions, esp. in the Ore Mountains above 500 m above sea level (mountain meadows) and in glacial landscapes of the lowlands (moist and wet grassland).
Classification as HLE	There was no natural grassland in Central Europe, except of salt grassland along the coasts and alpine meadows in the high mountains. All other grassland types have anthropogenic origin. Semi-natural grassland rich in species is a product of past forms of land use, especially of the small-scaled agriculture, which had to do with only few fertilizers and with the given natural site conditions. These land use forms and the related plant societies vanished during only few decades. Under the present prevailing conditions, such grassland types are not generated any more. The still existing remainders are threatened and can be maintained only by adapted management measures supported by financial subsidies. As HLE especially valuable are plots, which: - have been used as grassland for a long time (decades or centuries); - show a characteristic, highly-diverse combination of species adequate to the site; - are characterized by relatively natural site conditions; - contain structures typical for the former manual management, e.g., the once wide-spread small ditches managed by hand

(continued)

Table 2 (continued)

Criteria	Description
Resilience/risks	Still 40–50 years ago, semi-natural grassland was widely spread. Land use intensification has lead to changes in the species composition and finally, to grassland poor in species. Therefore, grassland valuable in terms of biodiversity became rare.
State of protection	In Saxony (but also in the whole Germany and in the EU) several habitat types are protected, generally, among them some types of semi-natural grassland, e.g., wet meadows rich in sedges and rushes, mesophilic grassland and mountain meadows rich in species. Some types of such grassland vegetation are protected as NATURA 2000 habitat types (Annex I). Also several plant and animal species of the annexes II, IV and V need semi-natural grassland as habitats, e.g., the Arnica *(Arnica montana)*, and the Dusky Large Blue *(Maculinea nausithous*; see Bastian 2008b).
Significance for species, biocoenoses or habitats	More than one third of all indigenous vascular plant species prefer grassland as habitat. Altogether more than 1,000 plant species can be found in the various grassland communities. Including orchards and alpine meadows the number increases by c. 1,000 additional species. Also the fauna is very rich.
Conservation and management	For the maintenance and restoration of such historical land use forms special measures are necessary. A fundamental precondition is the maintenance of the traditional use or its imitation by adequate management practices.

development. In less developed regions such as in the Carpathians traditional rural landscapes still survive.

The fast decline and the homogenization of our traditional cultural landscapes can be mitigated by the protection of HLE and whole parts of such landscapes. In the meantime, there are laws, programs and suitable methodical approaches for the inventory and evaluation of HLE and whole cultural landscapes. The methodological approaches developed in Saxony for analyzing and evaluating HLE and whole historical landscapes are a useful planning and management tool. It can be applied not only in Saxony or in Germany but also in other countries or regions, among them in the Carpathians.

As the investigations from Saxony show, the conservation of HLE, which are not in the focus of the lay public and having no strong lobby, is especially difficult, e.g., narrow passes, avenues, technical or industrial monuments. One of the main problems is the lack of use. The preservation of a specific state of the cultural landscape and HLE is difficult to imagine, though. One cannot restore the old links, as the contexts, processes, and functions in the landscape that have changed with time.

As the analysis of HLE in Saxony also show, it is very difficult, expensive and time consuming to establish a reasonably complete data basis for the HLE. A lot of information about single HLE types is collected by a multitude of local and regional institutions and by private persons. It is difficult to obtain such regional allotted information. However, for landscape planning purposes data for the whole country or at least for planning regions are necessary. For this, the establishment of a special GIS-supported cadastre for cultural landscapes, which is regularly managed and completed, could improve the situation on the long run. Such a database could integrate individual datasets from different sources and can be completed successively over time.

The definition and delimitation of HLE types proves an additional problem. It is not always easy to answer the question: is it a HLE or not? Sometimes it was not possible to prove if a special HLE-object has historical roots (e.g., some ponds). On the other hand, it is not easy to detect the traces of several HLE in the present landscape.

While dealing with HLE, the main focus is mostly on the inventory but aspects of planning are under-developed. That is why the aim to consider HLE in the Saxon landscape program must be highly appreciated.

In order to maintain the existing landscapes, effective actions taken by local, national and European authorities are needed (Palang et al. 2006). They include the appropriate assessment of natural and cultural values and their incorporation into feasible land management plans. These plans should also take into account the enhancement of the social and economic conditions of local populations. The preservation of culturally and environmentally-friendly landscapes is connected with the notion of sustainable development, and it depends upon the availability of financial resources and local participation. In this context, the European Union should play a significant role through its agri-environmental policy. Also the European Landscape Convention is very important (see Marschall and Werk 2007) (but unfortunately several countries—among them also Germany—have not signed this document, yet). Regional authorities should take care of regional cultural heritage using proper and coherent land use planning tools and techniques. The crucial question is (Palang et al. 2006): "How to maintain and preserve local landscape heritage as a part of a wider European heritage and keep a close link between humans and landscapes?"

A promising way is the revitalization of rural settlements, landscapes and all cultural-historical values by new activities, programs, projects of the renaissance of old ones to combine residential, economic and recreational functions (Fig. 8). As Huba (2000) stresses, this is not possible only via professional bodies, state institutions, etc. Success relies on the incorporation of volunteers, as well as the general public.

Voluntary (non-governmental and non-profit) citizens' activities in this field represent a challenge, not only for conservation, revitalization and sensitive utilization of HLE, but also for 'revitalization' of the mentality and positive patriotism of the people. It is the challenge for self-protection against pressures exerted by indolence, passiveness, unification, consumerism and general resignation on higher values.

Tourism can play an important role in the maintenance and restoration of HLE, because urbanization has as a consequence a strong desire to recover the landscape as a site for recreation. Tourism can also be a way to show people the diverse functions (life support, etc.) of landscapes.

HLE can fulfill many functions (services) for local people and tourists, for landscape interpretation (e.g., thematic trails, information brochures, internet presentations), teaching materials for schools, green classroom out of school, objects for arts, revival of historical management systems that can presented at feasts.

Fig. 8 Typical folk architecture in traditional weaver villages of Upper Lusatia (Saxony, Germany). Special efforts are necessary to maintain such nice old houses (Photo: O. Bastian)

7 Conclusions

The evident task of contemporary generations is to identify the most characteristic and valuable historical structures and to secure their survival. The conservation, maintenance, as well as effective utilization of at least part of the cultural heritage, in the form of existing historical landscape structures, is a great challenge. In a world of increasing globalization it is important for the identification of regions and the people living there. Also the preservation of cultural landscapes is a contribution, if not one precondition, for stopping the loss of biodiversity.

Suitable legislation and planning instruments, methodological approaches, financial resources but also people taking an interest are necessary not only to maintain HLE but also to explain the diverse functions of HLE, to revive them and to attribute new functions.

The future of traditional landscapes and HLE will depend on the question, whether the public can be made sensitive to the maintenance of this important part of our cultural heritage, and whether we will find ways to combine the conservation and management of these areas and elements with modern economic systems and lifestyles.

Acknowledgments We thank the Saxon State Office for Environment, Agriculture and Geology for supporting the projects presented here. We also thank Prof. Wolfgang Wende for his helpful remarks.

References

Antrop M (2000) Changing patterns in the urbanized countryside of Western Europe. Landsc Ecol 15:257–270

Antrop M (2004) Landscape change and the urbanization process in Europe. Landsc Urban Plan 67:9–26

Bastian O (2008a) Landscape classification—between fact and fiction. In: Lechnio J, Kulczyk S, Malinowska E, Szumacher I (eds) Landscape classification. Theory and practice, Wydział Geografii i Studiów Regionalnych Uniwersytetu Warszawskiego, Warsaw

Bastian O (2008b) Adopting the precautionary principle in designing and managing NATURA 2000 areas (exemplified by the conservation of the butterfly Maculinea nausithous in a rural landscape north of Dresden (Saxony)). J Landsc Ecol 1:14–22

Bayerisches Staatsministerium für Landwirtschaft und Forsten (2001) Historische Kulturlandschaft. Mater Ländlichen Entwickl 39, München

Becker W (1998) Die Eigenart der Kulturlandschaft. Bedeutung und Strategien für die Landschaftsplanung. PhD Thesis, Berlin

Beierkuhnlein C (2002) 7.8 Historical landscapes and landscape elements. In: Bastian O, Steinhardt U (eds) Development and perspectives of landscape ecology. Kluwer, Dordrecht, Boston, London

Bund für Heimat und Umwelt (2008) Kulturlandschaftliche Informationssysteme in Deutschland. Erfassen—Erhalten—Vermitteln. Bonn

Burggraaf P (1996) Der Begriff „Kulturlandschaft"und die Aufgaben der „Kulturlandschaftspflege"aus der Sicht der Angewandten Historischen Geographie. Nat Landschaftskd 32:10–12

Burggraaf P, Kleefeld K-D (1998) Historische Kulturlandschaft und Kulturlandschaftselemente. Angew Landschaftsökol 20

Council of Europe (2000) European landscape convention. European Treaty Ser 176

Fry G (2000) The landscape character of Norway—Landscape values today and tomorrow. In: Pedroli B (ed) Landscape Our Home—Essays on the culture of the European landscape as a task. Zeist, Stuttgart

Gunzelmann T (1987) Die Erhaltung der historischen Kulturlandschaft. Angewandte historische Geographie des ländlichen Raumes mit Beispielen aus Franken, Bamberger Wirtschaftsgeogr Arb 4

Haase G, Richter H (1980) Geographische Landschaftsforschung als Beitrag zur Lösung von Landeskultur- und Umweltproblemen. Sitzungsber Akad Wiss DDR, math., nat. tech., 5N:23–51

Haber W (2001) Kulturlandschaft zwischen Bild und Wirklichkeit. Forschungs- u. Sitzungsber Akad Raumforsch Landesplan (Hannover) 215:6–29

Haber W (2008) Naturschutz in der Kulturlandschaft—ein Widerspruch in sich? Laufener Seminarbeitr 1(08):15–25

Huba M (2000) The challenge of historical landscapes in Slovakia. In: Pedroli B (ed) Landscape Our Home—Essays on the culture of the European landscape as a task. Zeist, Stuttgart

Konold W (2006) Den Mythos des Ursprünglichen entlarven—Historische Kulturlandschaften. Politische Ökol 24:25–28

Leibenath M, Gailing L (2012) Semantische Annäherung an “Landschaft” und “Kulturlandschaft”. In: Schenk W, Kühn M, Leibenath M. Tzschaschel S (eds) Suburbane Räume als Kulturlandschaften. Forschungs- und Sitzungsberichte der ARL 236, Hannover

Marschall I, Werk K (2007) Die Europäische Landschaftskonvention. Ziele, Inhalt sowie ihre derzeitige landschaftspolitische Bedeutung in Deutschland. Nat R 29:719–722. doi:10.1007/s10357-007-1359-y

Moss M, Okey B (2004) A landscape-ecological appraisal of the diversity and relevance of cultural landscapes. In: Jongman RHG (ed) The new dimensions of the European landscape. Wageningen UR Frontis Series 4, Springer, Dordrecht

Naveh Z (1995) Interactions of landscapes and cultures. Landsc Urban Plan 32:43–54

Naveh Z (2010) Ecosystem and landscapes—a critical comparative appraisal. J Landsc Ecol 3:64–81

Neef E (1967) Die theoretischen Grundlagen der Landschaftslehre. Haack, Gotha

Nowak B, Schulz B (2004) Landschaftsräume der Planungsregion Mittelhessen. Landschaftskundliche Grundlagen für die Landschaftsplanung. Regierungspräsidium Gießen, Obere Naturschutzbehörde. http://www.hessen.de/irj/RPGIE_Internet?cid=c2b63d6b89dfc501f22faae314b86bd4. Accessed 20 October 2010

Palang H, Printsmann A, Konkoly Gyuro E, Urbanc M, Skowronek E, Woloszyn W (2006) The forgotten rural landscapes of Central and Eastern Europe. Landsc Ecol 21:347–357

Pedroli B (ed) (2000) Landscape Our Home—Essays on the culture of the European landscape as a task. Zeist, Stuttgart

Peters J, Klinkhammer B (2000) Kulturhistorische Landschaftselemente. Naturschutz Landschaftsplan 32:147–152

Röhrer A, Büttner T (2008) Dokumentation historischer Kulturlandschaftselemente im Biosphärenreservat Rhön: Notwendigkeit und Nutzen—Verfahren und Beispiele. Beitr Reg Nachhalt 5:58–76

Schenk W (2002) Landschaft und Kulturlandschaft—getönte Leitbegriffe für aktuelle Konzepte geographischer Forschung und räumlicher Planung. Petermanns Geogr Mitt 146:6–13

Schmidt C, Meyer H–H, Schottke M, Zeigerer A (2006) Kulturlandschaft Thüringen—Arbeitshilfe für die Planungspraxis. University of Applied Sciences, Erfurt

Schreiber K-F (1999) 2.1 Ökosystem, Naturraum, Landschaft, Landschaftsökologie—eine Begriffsbestimmung. In: Bastian O, Schreiber K-F (eds) Analyse und ökologische Bewertung der Landschaft, Spektrum Akadem. Verlag, Berlin

Solon J (2008) Typy krajobrazu kulturowego Polski. In: Lechnio J, Kulczyk S, Malinowska E, Szumacher I (eds) Klasyfikacja krajobrazu. Teorie i praktyka. Problemy Ekologii Krajobrazu, Wydział Geografii i Studiów Regionalnych Uniwersytetu Warszawskiego, Warsaw

Thiem K, Bastian O (2009) Steckbriefe für ausgewählte landschaftsprägende historische Kulturlandschaftselementtypen im Freistaat Sachsen. Mscr., unpubl., Saxon State Office for Environment, Agriculture and Geology, Freiberg

Walz U, Ueberfuhr F, Schauer P, Halke E (2010) Ableitung und Bewertung von Kulturlandschaftsgebieten für das Landschaftsprogramm Sachsen. Nat Landsch 85:17–23

Walz U, Ueberfuhr F, Schauer P, Halke E (2012) Historische Kulturlandschaft Sachsens. Schriftenreihe Landesamt für Umwelt, Landwirtschaft und Geologie 33, Dresden https://publikationen.sachsen.de/bdb/artikel/15690. Accessed 3 March 2013

Wascher D (ed) (2005) European landscape character areas—typologies, cartography and indicators for the assessment of sustainable landscapes. Final Project Report. Alterra Report No. 1254

Wöbse H–H (1994) Die Erhaltung historischer Kulturlandschaften und ihrer Elemente. Deutscher Heimatbund. Umwelt und Kulturlandschaft, Bonn, In

Wöbse H–H (1999) "Kulturlandschaft"und "historische Kulturlandschaft". Inf Raumentwickl 5+6:269–278

Landscape Parks and the Development of Tourism in the Protected Areas of the Polish Carpathians

Bernadetta Zawilińska

Abstract Protected mountain areas are usually highly attractive from the perspective of developing tourism. Tourism poses a chance for economic development of these areas, but also damages the natural environment. It is therefore of paramount importance that principles governing the management of tourism in protected areas be defined so as to reconcile the requirements of environmental protection with the needs of local communities and tourists' expectations. This chapter presents a theoretical model for the application of the concept of sustainable tourism development in protected areas and discusses difficulties that arise in harmonious development of Carpathian landscape parks. One of the reasons is the lack of stakeholders' cooperation as well as the lack of comprehensive long-term planning. In the landscape parks, efforts should be undertaken to introduce the practice of joint management, based on broad cooperation between the administrative bodies of landscape parks, local governments, representatives of local businesses, non-governmental organizations and other parties oriented towards the development of tourism.

1 Introduction

The Carpathians are a very attractive and one of the most frequently visited tourist regions in Poland. In terms of environmental value, the region is one of the most precious in the country, and thus to a considerable extent legally protected, with various types of areas under protection. By providing income to the residents, tourism stimulates the economy. The conviction that for tourism to flourish the natural environment must remain in a good condition actually supports environmental protection. At the same time, the increasing number of tourists and the development of various, frequently harmful, forms of tourism degrade

B. Zawilińska (✉)
Department of Regional Economy, University of Economics, Rakowicka 27, 31-510 Kraków, Poland
e-mail: zawilinb@uek.krakow.pl

J. Kozak et al. (eds.), *The Carpathians: Integrating Nature and Society Towards Sustainability*, Environmental Science and Engineering,
DOI: 10.1007/978-3-642-12725-0_32,

environmental assets and deteriorate the comfort of recreation, as well as cause unfavorable social phenomena. Consequently, tourism poses both an opportunity and a threat to the protected Carpathian areas.

Landscape parks appear to be in an exceptionally complex situation. According to guidelines (Baranowska-Janota and Korzeniak 1991; Kistowski 2004), these parks ought to be an area of harmonious socio-economic growth, remaining in concert with the protection of animate and inanimate natural values and cultural heritage. However, there are numerous barriers to their appropriate development, associated mainly with the multi-functionality of these areas and the range of often conflicting interests. Undoubtedly, it is possible to reconcile the requirements of environmental protection with tourists' needs and residents' interests; nevertheless, to achieve this, implementation of sustainable development principles is required.

The aim of this work is to present a theoretical model of sustainable tourism development in landscape parks of the Polish Carpathians and to highlight the difficulties in implementing harmonious development in practice. The empirical section is based on inventory research of tourist values and facilities, observation of the forms and spatial patterns of tourism, interviews with landscape park authorities, local governments, the owners of accommodation facilities and other stakeholders, and questionnaire surveys in the offices of the administrative units carried out since 2005 (Zawilińska 2007, 2010a, b).

First, the position of landscape parks in the structure of all protected areas in the Polish Carpathians is explained, focusing on differences in the development of their tourist function within the respective areas. Then, a theoretical model of sustainable tourism development in protected areas is presented. Finally, the problems of implementing sustainable tourism within the Carpathian landscape parks are discussed.

2 Problems Associated with the Development of Tourism in Mountainous and Protected Areas

The problems of tourism and environmental protection are closely linked. While tourism frequently stimulates the formation of protected areas, it does at the same time threaten environmental assets of these areas. The threat tends to grow with the increase in demand for recreational stays in the areas of high environmental value (Ceballos-Lascurain 1996; Jafari 2000; Kurek 2004). The relationship between tourism and nature conservation are particularly strong in mountain areas, surely as they are among the regions most frequently visited by tourists worldwide. At the same time, these areas, being valuable in terms of the environment, are largely protected by the law.

The problems accompanying the development of tourism in mountainous and protected areas rank high among the issues discussed in the scientific literature.

Numerous studies have looked at the impact of tourism on both the natural environment and socio-economic aspects of mountain regions (e.g., Bätzing 1988; Wachter 1993; Hunter and Green 1995; Khan 1996). Tourism is perceived as both a factor contributing to the economic development of usually less developed areas and a factor that brings about unfavorable social and cultural changes.

Due to the shortage of mountainous areas in Poland (areas elevated above 500 m a.s.l. occupy only 3.1 % of Polish territory; Concise Statistical Yearbook of Poland 2010), tourism exerts a strong impact on both the Sudetes and the Carpathians; in these areas, tourism is largely concentrated within the boundaries of protected areas. Tourism in the Carpathian region has been studied mostly by researchers from Kraków who focused on evaluating the attractiveness of tourist assets (Warszyńska 1989), land use and tourist flow issues. A typology of tourist resorts was produced (Jackowski 1981), together with the evaluation of the importance of tourism in the region and its socio-economic influence (Kurek 1990).

As the transformation of the political system unfolded, more attention was given to socio-economic aspects of tourism and its role in stimulating the economy of the Carpathian region. New issues came to be discussed, e.g. those associated with local government activity for the benefit of developing tourism (Mika and Pawlusiński 2002), the transition in the model of functioning of entities in a tourism economy (e.g. in the health resort sector—Groch 1997; Pitrus 2006), as well as new trends in tourism in the Polish Carpathians (e.g., the development of agrotourism, spa & wellness services—Mika and Pawlusiński 2006; Kurek 2008; Faracik et al. 2009; Szpara 2010). An increase in the interest directed at the problems of transboundary tourism has also been noted (Ptaszycka-Jackowska and Baranowska-Janota 2003; Mika 2008). A separate trend in research is represented by studies on the environmental effects of tourism development in the Carpathians. This research was mainly conducted in areas under various forms of legal protection (Wpływ narciarstwa i turystyki pieszej na przyrodę Masywu Pilska 1995; Jodłowski 2004; Mika 2004; Jodłowski and Wójcik 2007).

The principles of tourism development in protected areas in the Polish Carpathians have been proposed, among others, by Jagusiewicz (1979), Baranowska-Janota (1989), Baranowska-Janota and Korzeniak (1991), and Ptaszycka-Jackowska and Baranowska-Janota (1996). These authors also examine the difficulties of planning and managing tourism in areas protected by law. Management of tourism development in protected areas of various categories was also discussed by Eagles et al. (2002) and Phillips (2002), who emphasized the need for sustainable tourism development in such areas.

Sustainable tourism is usually understood as a phenomenon in which the activities undertaken by tourists do not cause damage or changes in the natural environment that are difficult to reverse, but provide benefits for tourists, local communities living in tourist areas and territories, as well as people and the organisations providing tourist services (Durydiwka et al. 2010). An outline of the programme of sustainable tourism development was laid out in the publication Agenda 21 for the Travel and Tourism Industry (1997). For the theoretical guidelines to be implemented, a more detailed definition is required, along with the formulation of principles and instructions for a particular region, as no universal model of development exists. Hunter (1997) believes there is not one but numerous paths to sustainable tourism development, each depending on a particular situation and formed by, among other things, demand for and supply of tourist services, the needs and expectations of local communities and factors associated with natural resources.

Hunter (1997) noticed considerable differences in local determinants of tourism development and relations between natural, social and economic environments in various regions of the world, as well as a wide spectrum of existing approaches to sustainable tourism. Using interpretations of the general concept of sustainable tourism, he defined four concepts of sustainable tourism development. These concepts differ in terms of the level of tourist development and the prominence of sustainable tourism in a particular region, and refer to the areas varying from the perspective of environmental and socio-economic conditions. Hunter's concepts can also be adapted for protected areas and zones with various protection status and land use intensity within their borders (Zawilińska 2010b). With respect to protected areas, the only concept to be rejected is the first one—"tourism imperative", which should be applicable solely in the areas of degraded natural environment and very poor world regions (Table 1).

"Product-led tourism" zones ought to be located exclusively within the borders of the areas characterised by a low or intermediate protection regime (landscape parks, for example) and should only cover urban areas of large tourist resorts. In such a situation, sustainable development proceeds in a tourism-dominated economy, with the prevalence of commercial, and usually mass-practised, forms of tourism. In these zones, it is essential to maintain the economic significance of tourism and, at the same time, remove its negative social and environmental effects.

"Environment-led tourism" zones cover areas of towns and villages as well as all areas outside built-up areas, with the exception of those with the highest environmental value. These zones are the most diversified in terms of desirable operations and directions of tourism development that depend on previous forms of land use. In environmentally valuable areas, extensively utilized for the purposes of tourism, the environmental model of tourism ought to be preserved

Table 1 Concepts of sustainable tourism development in national and landscape parks

Concept of tourism[a]	Sustainability position[a]	Protected areas		Zone in a protected area
"Tourism imperative"	Very weak	–	–	–
"Product-led tourism"	Weak	–	Landscape parks	Urbanised areas, major tourist resorts
"Environment-led tourism"	Strong	National parks		All areas except built-up areas, towns and villages
"Neotenous tourism"	Very strong			Nature reserves

[a] The concepts of tourism and sustainability positions adopted from Hunter (1997)

(governed by the consideration of the natural environment and mainly based on individual forms of tourist activity). In built-up areas of towns and villages, as well as those undeveloped areas where tourist flow concentrates, tourism development ought to be governed by consideration of the natural environment and tourism as a product; it also ought to be a factor in economic development, while preserving environmental and cultural assets.

"Neotenous tourism" zones cover the most environmentally valuable park areas (mainly nature reserves and strict protection zones in national parks). In these areas, tourism should be minimized. It ought to be limited to short-term stays by primarily individual tourists or those moving in small groups, exclusively within the strict borders of designated areas, and who have come mainly for cognitive purposes.

Aside from defining the principles of tourism development in particular protected areas and separating zones varying in the manner and intensity of land use, the questions of how the process of sustainable tourism is to be implemented in a given area and who is to control the process should be addressed. In the scientific literature, an opinion that has gained attraction is that sustainable development in protected areas and the management of these areas ought to be conducted with a broad cooperation between the administrative bodies of protected areas, local governments and local communities (Borrini-Feyerabend 1999; Phillips 2003; Borrini-Feyerabend et al. 2004).

In Poland, the formation of local partnerships and the implementation of joint management is a particularly important issue in areas protected as landscape parks (IUCN V category) since they are inhabited areas that are relatively intensively used. Elsewhere, this management model has been introduced in many protected areas whose protection status is similar to that of Polish landscape parks and has proven to be fairly beneficial (Ptaszycka-Jackowska 2000; Jones and Burgess 2005).

3 Landscape Parks in the Network of Protected Areas in the Polish Carpathians

Due to their inaccessibility and negligible economic usefulness, mountainous regions have been exposed to degradation to a lesser extent than lowlands (Kurek 2004), therefore they usually constitute some of the most environmentally valuable areas in the highly developed countries of Europe. A similar situation is evident in the Carpathians. The great environmental value of these areas has caused a considerable proportion of their surface area to be included under legal protection. Protected areas of various categories cover approximately 70 % of the Polish Carpathians (14,000 km^2). The most environmentally valuable parts of the mountains have been granted protection as national parks, nature reserves and landscape parks (Fig. 1, Table 2). Protected landscape areas have the highest share in the protected area (37.5 %); however, the protection level is relatively low. The most recent form of environmental protection in the Carpathians is represented by NATURA 2000 sites, primarily demarcated in regions which have previously been granted legal protection, predominantly as national parks, nature reserves and landscape parks.

Other forms of environmental protection, such as ecological sites, documentary sites and nature and landscape-protected complexes do not occupy significant area and are not as important to the development of tourism as are national parks, landscape parks and nature reserves. They are important mainly in terms of environmental tourism and environmental education.

Several protected areas in the Carpathians have international status, being part of the UNESCO biosphere reserves network. Bieszczady National Park, along with the San Valley Landscape Park, Cisna-Wetlina Landscape Park and neighboring protected areas located in Slovakia and Ukraine together constitute the East Carpathian Biosphere Reserve. The Tatra Transboundary Biosphere Reserve comprises national parks located in the Polish and Slovak Tatra Mts., while the Babia Góra Biosphere Reserve functions only on the Polish side of the border and comprises the Babia Góra National Park. In Poland, the total area included within the biosphere reserves is 1,461 km^2 (approximately 8 % of the Polish Carpathian region).

Landscape parks located in the Polish Carpathians (along with nature reserves located within their boundaries) cover 4,545 km^2, 23.1 % of the mountain area and 31.6 % of the total protected area in this region. These are protected areas under an

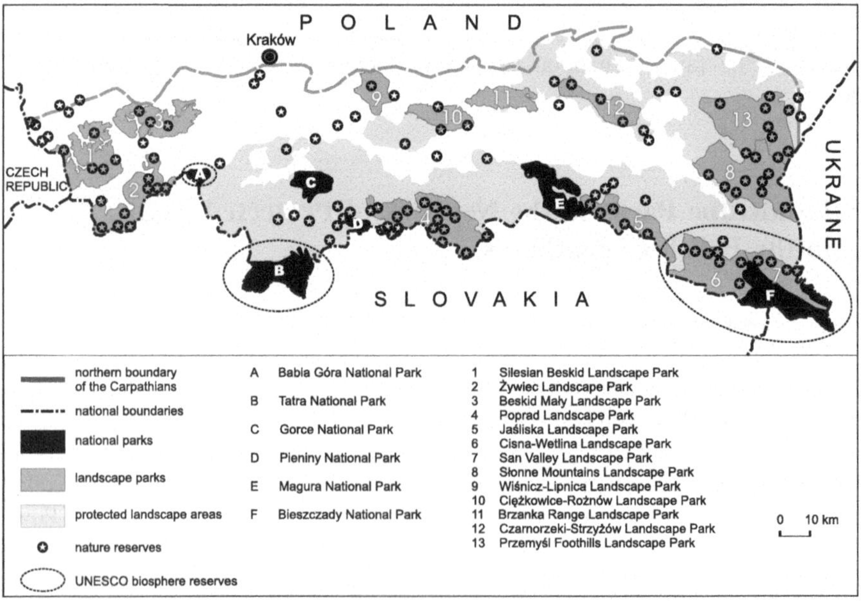

Fig. 1 Protected areas in Polish Carpathians

Table 2 Protected areas in the Polish Carpathians

Protected areas	Number	Area [sq. km]	Proportion in the Polish Carpathians [%]
National parks	6	825.80	4.21
Nature reserves	116	103.11	0.53
Landscape parks	13	4,506.30[a]	23.00
Protected landscape areas	9	7,346.80[a]	37.48
Total	144	12,782.01	65.22

[a] Area of landscape parks and protected landscape excluding nature reserve areas located within their boundaries

Source data of the Polish Central Statistical Office (GUS) and Regional Directorates for Environmental Protection, 2010

intermediate protection regime, classified as category V by the IUCN. According to the Polish Act on Environmental Protection (2004), *landscape parks cover the areas protected for their natural, historic and cultural assets, as well as scenic assets, for the purpose of preservation and popularisation of these assets under the conditions of sustainable development*. The idea of establishing landscape parks in Poland originated in the 1950s in a group of Kraków architects. While the concept has evolved, tourism has remained an important function of such parks.

In the landscape parks of the Carpathians, a plethora of factors determining the development of tourism are visible. The parks differ primarily with regard to the character and attractiveness of tourist values, their location in relation to centers of tourist flow emission, leading forms of tourism, as well as the level of the development of tourist function and its significance for the functional structure of municipalities. Some of them cover areas characterized by outstanding assets, frequently visited by tourists since as early as the nineteenth century.

Landscape park areas with the most long-standing traditions of tourism development are those of the Silesian Beskid Landscape Park and the Poprad Landscape Park. These parks are also characterized by the highest level of development and number of tourist visits. The Poprad Landscape Park, which covers the area of the Beskid Sądecki, is characterised by the highest accommodation capacity. Within its boundaries, 400 accommodation facilities may accommodate 17,300 people in total (32 people/km^2). In addition, a dense network of tourist trails and downhill skiing stations may be found in this park. The high level of tourist development stems from the fact that large tourist and health resorts, such as Krynica, Muszyna, Piwniczna and Żegiestów, have been included within the park's boundaries. This seems to be a notable exception as other intensely developed areas in the Carpathian landscape parks have not been included under protection.

This is exemplified by the Silesian Beskid Landscape Park, located on the periphery of the Bielsko-Biała industrial district, in close proximity to the Upper Silesian conurbation. This park was formed in the most intensely built up area of the Beskidy Mts., in a relatively highly populated area (337 people/km^2)

characterized by well developed tourism. Nearly 900 accommodation facilities, offering approximately 35,000 beds (36 beds/km^2) are found in municipalities administratively associated with the park. In the Silesian Beskid, however, developed areas (centers of settlements) have not been included in the boundaries of the landscape park.

A similar situation occurs in the neighboring landscape parks of the Beskid Mały and Beskid Żywiecki Mts. The parks comprise mostly forest and meadow areas, while accommodation facilities found within their boundaries are mainly holiday farms and mountain shelters. On the other hand, there is a highly developed auxiliary infrastructure within the parks' boundaries, represented by the network of tourist trails. In the Silesian Beskid Landscape Park, the density of hiking trails is 1,186 m/km^2, and trails used for other than hiking forms of tourism have also been signposted. The Silesian Beskid is also the most intensely developed mountain range in the Beskids for ski tourism, with a considerable number of ski lifts, cable cars and downhill skiing routes located within the boundaries of the landscape park.

Landscape parks located in the Bieszczady Mts. (San Valley Landscape Park and Cisna-Wetlina Landscape Park) encircle the northern and western parts of the Bieszczady National Park, which is visited annually by approximately 300,000 tourists. The tourist infrastructure of these parks is far less developed than that of the Silesian Beskid or Poprad landscape parks (there are approximately 3,000 beds in both parks in total). However, due to forced re-settlement after World War II, these are the least populated regions of the country, with a population density of a few people/km^2 (within the boundaries of San Valley Landscape Park it drops to less than 4 people/km^2), while more than a dozen villages in the area are completely abandoned. In this case, the number of beds per resident is higher than in most intensely developed, yet also relatively densely populated parks (34 beds/100 inhabitants in the Cisna-Wetlina Landscape Park and 76 beds in the San Valley Landscape Park). Such high values indirectly show the great significance of tourism in this region.

The other landscape parks of the Carpathians have a poorly developed tourist infrastructure. Nonetheless, a fundamental difference between these parks should be emphasized. Jaśliska, the Słonne Mountains and Przemyśl Foothills landscape parks were formed in areas depopulated after World War II, as were parks of the Bieszczady Wysokie. These parks cover weakly populated and highly forested areas that are rarely visited by tourists. On the other hand, other landscape parks located in the Carpathian Foothills are characterized by a high share of agriculture and are visited and used quite extensively. The areas of these landscape parks do not exhibit outstanding tourist features, yet their close proximity to cities (e.g. Kraków, Tarnów, Rzeszów, Krosno) makes them popular one-day excursion destinations. This form of tourism, however, offers relatively little to the region in terms of potential economic benefits.

4 Sustainable Tourism Development in Carpathian Landscape Parks

The far-reaching differences among landscape parks, visible both in their environmental and cultural value, socio-economic potential and tourist functions present in their areas, hinder the establishment of universal tourism development guidelines for all these areas. The situation is further complicated by considerable disproportions in tourism development of particular parks. In addition, attention should be paid to the parks' surroundings, since, as demonstrated by the example of the Silesian Beskid Landscape Park, there can be highly developed major tourist resorts, including city resorts, in the immediate neighborhood of the park's boundaries.

Three zones, characterized by different manners and ranges of tourist utilization (Table 1), can be demarcated within the boundaries of landscape parks and their vicinity. Tourism development in each zone would proceed according to the principles of sustainable development, depending on local conditions. In the most environmentally valuable places it would be driven by the priority of environmental protection, whereas in the most intensely economically utilized areas it would prioritize socio-economic matters (Table 3).

With the theoretical model of tourism development in landscape parks outlined, the question of who is to apply the model in practice now arises. Should sustainable development be controlled by a superior authority or should it be a result of a grassroots initiative? Assuming that sustainable development (also in the tourism sector) is a compromise that maintains harmony between socio-economic needs and the requirements of environmental protection in a particular region, dialogue between all parties involved, conscious and active participation of residents and planning are all of paramount importance.

In case of the studied landscape parks, the fundamental obstacle to the implementation of sustainable development is the considerable number of administrative bodies managing their areas and the lack of both a platform for cooperation and institutions to initiate cooperation and coordination. Landscape parks administrative bodies are not capable of performing such a function as they do not have appropriate decision-making powers, and have only limited human and financial resources at their disposal, as Polish landscape parks employ far fewer staff than national parks. Directors of landscape parks are only to a minor extent able to influence trends in tourism development within park boundaries. The activity of administrative bodies is mainly limited to promotional and educational activities as well as initiating activities that benefit sustainable tourism, whereas decision-making which could substantially influence development does not fall within their jurisdiction (Zawilińska 2010b). It should also be noted that in Poland a management system has been adopted in which majority of parks do not have independent administrative bodies, but create a group of parks with a joint administration. Thus, a given administrative body usually has under its control a number of protected areas, which form a spatially non-contiguous area. While park

Table 3 The application of sustainable tourism development in landscape parks

The area in a landscape park	Position of tourism	Zone characteristics
Nature reserves	"Neotenous tourism"	Areas of particular environmental value; tourist traffic either not developed or very low Tourism development subordinate to the goals of environmental protection. Areas excluded from tourist use or accessible solely for eco-tourism. Lack of tourist infrastructure or infrastructure limited to hiking trails or educational trails
All areas except built-up areas and the areas of concentrated mass tourism	"Environment-led tourism"	Extensive tourist land use. Promotion of ecological forms of tourism supporting the preservation of natural resources. Development of eco-tourism and educational projects. Reduction of tourism forms which considerably interfere with natural environment and cause changes to the landscape (e.g. construction of second homes, extension of skiing stations). Tourism-related land development should include primarily tourist trails and related infrastructure as well as mountain shelters
Non-developed areas which are places of tourism concentration		Natural environment under considerable pressure from a high number of tourists. It is necessary to eliminate the forms of tourism which endanger the environment (e.g. motorcycle tourism), and to reduce the negative environmental impacts, e.g. by channeling the tourist flow (e.g. construction of footbridges, railings, viewing platforms)
Built-up areas of towns and villages		Promotion of tourism that supports the preservation of the local cultural heritage, e.g. cognitive tourism and agro-tourism. Adaptation of existing facilities (e.g. old wooden cottages) to serve as tourist facilities, limitations of construction and development, particularly with reference to holiday buildings
Major tourist resorts	"Product-led tourism"	Intensively developed or urbanized areas, where economy is dominated by tourism. Attempts at preserving landscape patterns, support of pro-ecological initiatives, improving the quality of tourist services, formation of new tourist products using local potential

Source Zawilińska (2010b)

administrative bodies are able to initiate cooperation that favors sustainable tourism development, they do not have means to coordinate it effectively.

In Poland, local governments act as the legitimate administrator of an area and have wide decision-making competence. However, local governments are not capable of managing the development of tourism in entire parks as only their fragments lie within the boundaries of respective administrative units. The boundaries of landscape parks have been demarcated according to environmental criteria, and therefore do not correspond to administrative divisions. This leads to the fragmentation of park areas among a number of administrative units. In this context, the most fragmented Carpathian landscape parks are the Beskid Mały Landscape Park (15 communes) and the Silesian Beskid Landscape Park (14 communes). It frequently happens that an area protected as a landscape park constitutes only a small part of an administrative unit, often in a marginal position. As a result, several communes that are administratively associated with landscape parks consider problems of tourism development and park territories as insignificant.

Where there is considerable administrative fragmentation of parks and negligible decision-making power held by their administrators, it is essential to form a platform for cooperation in order to create a consistent development policy in the parks. Gołembski and Nawrot (2002) believe the function of such a platform should be performed by intercommunal associations, incorporating communes administratively associated with particular parks. Regrettably, such a solution has not been applied in the Carpathian region. Although 82 % of communes administratively associated with landscape parks belong to intercommunal associations, their territorial range does not overlap with park areas. In addition, only half of these administrative units undertake activities towards tourism development as part of their operations in an association (Zawilińska 2010b).

The lack of cooperation and comprehensive planning of tourism development over entire landscape park area creates a highly unfavorable situation. Local governments limit their operations to the boundaries of particular administrative units. Furthermore, they have a short-term view of tourism development limited to the duration of their mandate. Such a situation occurs not only in Poland and its Carpathian communes, but is also evident in other countries (Dodds and Butler 2010).

Establishing local tourist organizations could provide a wide spectrum of cooperation opportunities at the local level. According to guidelines (Cele i zadania regionalnych i lokalnych organizacji turystycznych 1999), such organizations constitute a forum for cooperation between local authorities, local representatives of the tourism sector, and other parties expressing interest in the development of tourism. 11 such organizations operate in the Polish Carpathians; unfortunately, the range of their operations does not overlap with landscape park areas. Furthermore, local authorities are hardly interested in undertaking joint activities in co-operation with local tourist organizations. Moreover, there are only few examples of cooperation between local tourist organizations and administration of protected areas. In addition, local tourist organizations primarily focus their attention on the development of tourist products and promotions, and do not shape regional tourism policy (Zawilińska 2010a).

While the lack of a formal platform for cooperation is a fundamental barrier to coherent programs of sustainable tourism development of landscape parks, it does not necessarily hinder individual cooperation between neighboring administrative units, local governments and administrative bodies of landscape parks, as well as other stakeholders. 71 % of communes administratively associated with landscape parks in the Carpathians have undertaken cooperation in tourism management with neighboring municipalities, 45 % have cooperated with administrative bodies of landscape parks and 17 % have cooperated with other organizations associated with tourism (Zawilińska 2007, 2010b). Regrettably, such cooperation is hardly stable and is not associated with coherent tourism development policy, typically, it is associated with specific tasks (e.g. demarcation of tourist trails, publishing promotional materials.

When analyzing problems of sustainable tourism development in Carpathian landscape parks, the lack of essential planning documents ultimately stands out. According to the Act on Environmental Protection dated 16 April 2004, each landscape park should have an environmental protection plan that defines spatial, quantitative and temporal range of park accessibility, preferred forms of tourism, and principles and directions of spatial development. These provisions then regulate principles of planning documents of administrative units. Unfortunately, only 3 out of 13 Carpathian landscape parks had a valid environmental protection plan in 2010. Separate plans of tourism development for entire landscape parks and local administrative units are also uncommon. The lack of tourism development plans in communes administratively associated with landscape parks is somewhat surprising, since tourism development is laid out in development strategies of almost all communes comprising landscape parks; moreover, majority of them declared tourism a preferential development option.

Attention should also be drawn to the discrepancy between guidelines that appear in scientific works and Polish planning practice. Numerous authors stress the significance of integration of various activities at the local level and the contribution of a wide circle of stakeholders to the processes of planning and tourism development in a given protected area (Philips 2002, 2003). Philips (2002) brings forward the fundamental role of including local communities in the planning process of development of IUCN V category protected area as early as possible, even before the demarcation of its boundaries. However, a tradition of such cooperation does not exist in Poland, with local communities scarcely participating in the planning process of local development. Intense involvement of local residents in the planning process of tourism development in landscape parks and their surroundings would increase the chance to meet their needs and expectations, thus mitigating existing and potential conflicts.

Another problem related to the development of tourism in Carpathian protected areas is the lack of staff in local governments skilled in dealing with tourism development. Only 63 % of communes administratively associated with Carpathian landscape parks have a tourism department; of those, 61 % are managed by a single person who combines tourism-related responsibilities with other tasks. Only 28 % of staff responsible for tourism issues have appropriate qualifications in tourism (Zawilińska 2010b). Considering the significance of tourism in the economy of most communes, the provisions laid out in local development strategies and considerable interest of the authorities and residents in the development of tourism, these values are much below expectations.

There are a number of conflicts related to the development of tourism in Carpathian landscape parks, and several weaknesses listed above add another dimension to their complexity. However, the growing awareness of the need to guide local development by the principles of sustainable tourism is evident among the employees of local governments. In this way, in spite of numerous problems hindering the coordination of tourism management, changes that can be observed within the landscape parks usually tend towards sustainable development (Zawilińska 2007).

5 Conclusions

As areas of high tourist values and a relatively mild protection regime, Carpathian landscape parks are naturally qualified for tourism development. Tourism industry should contribute to the socio-economic growth of these areas and the preservation of their natural and cultural assets while also satisfying the needs of individual tourists. Unfortunately, achieving such harmony in practice does not come without effort. The development of sustainable tourism faces numerous obstacles in Carpathian landscape parks. To a large extent, these stem from the lack of comprehensive planning and a lack of coordination between neighboring communes, landscape park administrative bodies and other stakeholders involved in the tourism development in landscape park regions. As a result, tourism develops in a relatively random manner, which does not always follow theoretical principles of protected areas management and does not allow local potential to be fully realized.

Changing this unfavorable situation will require a holistic view of the park as a system connected with its social and economic surroundings, and the elaboration of a consistent approach for the development of tourism, as well as the introduction of a common, planned policy in all areas of the park.

The implementation of optimal solutions and the mitigation of existing and future conflicts related to tourism will require the broad cooperation of landscape park administrative bodies, local governments, local business representatives, NGO's and other parties interested in the development of tourism. The cooperation should cover both the planning and implementation of tourism.

The general principles of sustainable tourism development should be applied to all areas of parks, while detailed solutions can, depending on local conditions, consider to a varying degree the environmental and socio-economical aspects. In developed tourist resorts, sustainable development should seek to ensure consistency in tourist demand and to maintain the economic importance of tourism while at the same time mitigating the negative impacts of tourism on the natural environment. In the most valuable natural areas, sustainable development implies above all the conservation of nature.

References

Agenda 21 for the Travel and Tourism Industry (1997) Towards environmentally sustainable development. WTO, WTTC, Earth Council, London

Baranowska-Janota M (1989) Basic principles of tourism development in protected mountain nature areas. In: Warszyńska J (ed) Carpathian tourist conference. Jagiellonian University, Kraków

Baranowska–Janota M, Korzeniak G (1991) Zasady turystycznego użytkowania i ochrony parków krajobrazowych. Instytut Gospodarki Przestrzennej i Komunalnej Oddział w Krakowie, Kraków

Bätzing W (1988) Ökologische und ökonomische probleme alpiner touristenzentren. Das Beispiel Gastein (Hohe Tauern/Österreich). In: Haimayer P (ed) Probleme des ländlichen Raumes im Hochgebirge, Innsbr Geogr Stud 16

Borrini-Feyerabend G (1999) Collaborative management of protected areas. In: Stolton S, Dudley N (eds) Partnerships form protection, new strategies for planning and management for protected areas. WWF and IUCN, Earthscan, London

Borrini-Feyerabend G, Kothari A, Oviedo G (2004) Indigenous and local communities and protected areas. Towards equity and enhanced conservation. Guidance on policy and practice for co-managed protected areas and community conserved areas. IUCN, Gland and Cambridge

Ceballos-Lascurain H (1996) Tourism, ecotourism and protected areas. IV World Congress on National parks and protected areas. IUCN Protected Areas Programme, Gland

Cele i zadania regionalnych i lokalnych organizacji turystycznych (1999) Polska Organizacja Turystyczna, Warszawa

Concise Statistical Yearbook of Poland (2010) Central Statistical Office, Warsaw

Dodds R, Butler R (2010) Barriers to implementing sustainable tourism policy in mass tourism destinations. Tourismos Int Multi J Tour 5(1):35–53

Durydiwka M, Kowalczyk A, Kulczyk S (2010) Definicja i zakres pojęcia "turystyka zrównoważona". In: Kowalczyk A (ed) Turystyka zrównoważona. Wydawnictwo Naukowe PWN, Warszawa

Eagles P, McCool S, Haynes C (2002) Sustainable tourism in protected areas: guidelines for planning and management. IUCN, Gland and Cambridge

Faracik R, Kurek W, Mika M, Pawlusiński R (2009) Tourism in the Polish Carpathian Mountains in light of contemporary directions in tourism development. In: Domański B, Kurek W (eds) Gospodarka i Przestrzeń. Instytut Geografii i Gospodarki Przestrzennej Uniwersytetu Jagiellońskiego, Kraków

Gołembski G, Nawrot Ł (2002) Rozwój turystyki zrównoważonej w parkach krajobrazowych: nowe trendy w Unii Europejskiej. In: Zimniewicz K (ed) Zarządzanie parkiem krajobrazowym w warunkach zrównoważonego rozwoju. Akademia Ekonomiczna w Poznaniu, Poznań

Groch J (1997) Uzdrowiska i lecznictwo uzdrowiskowe w Polsce w okresie transformacji systemu ekonomicznego państwa (na przykładzie uzdrowisk karpackich). In: Domański B, Jackowski A (eds) Geografia, Człowiek, Gospodarka. Instytut Geografii Uniwersytetu Jagiellońskiego, Kraków

Hunter C (1997) Sustainable tourism as an adaptive paradigm. Ann Tour Res 24(4):850–867

Hunter C, Green M (1995) Tourism and the environment. A sustainable relationship? Routledge, London

Jackowski A (1981) Typologia funkcjonalna miejscowości turystycznych (na przykładzie województwa nowosądeckiego). Rozprawy habilitacyjne UJ 53, Kraków
Jafari J (2000) Encyclopedia of Tourism. Routledge, London and New York
Jagusiewicz A (1979) System parków krajobrazowych w Polsce. Instytut Urbanistyki i Planowania Przestrzennego Politechniki Warszawskiej, PWN, Warszawa: Łódź
Jodłowski M (2004) Development of new branches of climbing and its impact on the natural environment of the Tatra Mountains. Fiz Geogr Geomorfol (Kiev) 46:229–235
Jodłowski M, Wójcik P (2007) Zagrożenia środowiska przyrodniczego w rejonach wspinaczkowych Tatr Reglowych. In: Pociask-Karteczka J, Matuszyk A, Skawiński P (eds) Stan i perspektywy rozwoju turystyki w Tatrzańskim Parku Narodowym. Academy of Physical Education, Krakow Tatra National Park, Cracow-Zakopane
Jones P, Burgess J (2005) Building partnership capacity for the collaborative management of marine protected areas in the UK: a preliminary analysis. J Environ Manage 77:227–243
Khan A (1996) Tourism and a European strategy or the alpine environment. In: Cater E, Lowman G (eds) Ecotourism. A sustainable opinion? Wiley, Chichester
Kistowski M (2004) Wybrane aspekty zarządzania ochroną przyrody w parkach krajobrazowych. Uniwersytet Gdański, Gdański Bogucki Wydawnictwo Naukowe, Gdańsk-Poznań
Kurek W (1990) Wpływ turystyki na przemiany społeczno-ekonomiczne obszarów wiejskich polskich Karpat. Rozprawy habilitacyjne 194. Uniwersytet Jagielloński, Kraków
Kurek W (2004) Turystyka na obszarach górskich Europy. Instytut Geografii i Gospodarki Przestrzennej Uniwersytetu Jagiellońskiego, Kraków
Kurek W (2008) Tourism in Polish Carpathians in the political transition time. In: Wyrzykowski J (ed) Conditions of the foreign tourist development in Central and Eastern Europe, tourism in geographical environment 10. University of Wrocław, Wrocław
Mika M (2004) Turystyka a przemiany środowiska przyrodniczego Beskidu Śląskiego. Instytut Geografii i Gospodarki Przestrzennej Uniwersytetu Jagiellońskiego, Kraków
Mika M (2008) Tourism on the Polish-Czech-Slovak borderland in the light of contemporary development trends. In: Wyrzykowski J (ed) Conditions of the foreign tourist development in Central and Eastern Europe, tourism in geographical environment 10. University of Wrocław, Wrocław
Mika M, Pawlusiński R (2002) Współpraca samorządów lokalnych w dziedzinie turystyki. Zesz Nauk WSBiF w Bielsku-Białej, Prace z zakresu ekonomii 1
Mika M, Pawlusiński R (2006) Some problems of the development of spa & wellness services in the Polish Carpathian Mountains. In: Wyrzykowski J (ed) Conditions of the foreign tourist development in Central and Eastern Europe, contemporary models of Spa tourism 9. University of Wrocław, Wrocław
Philips A (2003) Turning ideas on their head. A new paradigm for protected areas. George Wright Forum 20(2):8–32. http://www.georgewright.org/202phillips.pdf. Accessed 18 Octo 2010
Phillips A (2002) Management guidelines for IUCN category V protected areas: protected landscapes/seascapes. IUCN, Gland and Cambridge
Pitrus E (2006) Uzdrowiska karpackie w dobie transformacji. Turyzm, vol 16(1)
Polish Act on Environmental Protection (2004) Dziennik Ustaw No. 92, item 880
Ptaszycka-Jackowska D (2000) Współpraca parków krajobrazowych z samorządami lokalnymi w zakresie zagospodarowania przestrzennego. Biul Parków Kraj Wielkop 6:8
Ptaszycka-Jackowska D, Baranowska-Janota M (1996) Przyrodnicze obszary chronione. Możliwości użytkowania. Instytut Gospodarki Przestrzennej i Komunalnej, Warszawa
Ptaszycka-Jackowska D, Baranowska-Janota M (2003) Tourism within the Polish and Slovak transfrontier region. In: Kurek W (ed) Issues of tourism and health resort management. Institute of Geography and Spatial Management, Jagiellonian University, Cracow
Szpara K (2010) Agroturystyka w Karpatach Polskich. Pr Geogr 125. Instytut Geografii i Gospodarki Przestrzennej Uniwersytetu Jagiellońskiego, Kraków
Wachter D (1993) Winter sports and their impact on the environment: the case of Switzerland, tourism in mountain regions and the role of regional nature parks. Council of Europe, Sofia

Warszyńska J (1989) Eine modellmethode zur Bewertung der Umwelt für Erholungszwecke (am Beispiel der Vorstadzone Krakaus). Boch Geogr Arb 51

Wpływ narciarstwa i turystyki pieszej na przyrodę Masywu Pilska (1995) In: Łajczak A, Michalik S, Witkowski Z (eds) Instytut Ochrony Przyrody PAN, St Nat 41

Zawilińska B (2007) Możliwości rozwoju turystyki w parkach krajobrazowych Karpat Polskich w świetle idei zrównoważonego rozwoju. Institute of Geography and Spatial Management, Jagiellonian University, Kraków PhD dissertation

Zawilińska B (2010a) Działalność lokalnych organizacji turystycznych w Karpatach Polskich. Zesz Nauk Uniw Ekon w Krakowie 842:103–119

Zawilińska B (2010b) Możliwości rozwoju turystyki w parkach krajobrazowych Karpat Polskich w świetle idei zrównoważonego rozwoju. Uniwersytet Ekonomiczny w Krakowie, Kraków

Factors Influencing Willingness to Recommend a Visit to Water-Enhanced Tourist Attractions in Central-European Mountainous and Submontane Landscapes

Josef Navrátil and Kamil Pícha

Abstract The aim of the chapter is to reveal the predictors of likelihood of recommendation for a visit to water-enhanced tourist attractions in Central-European mountainous and submontane landscapes. Pull motivation, push motivation, on-site experience and perception of site quality were chosen as potential factors and multiple linear regression was used as the method. Pleasant 'natural' environment, heritage and culture, and accessibility were found to be factors of pull motivation; social-gathering and escape as factors of push motivation; pleasure and arousal as factors of experience; and rarity, irregularity and novelty as factors of perceived quality. The forward selection of multiple linear regressions found the pleasure experience as the main predictor for willingness to recommend. Others are: pull motivation regarding types of attractions and site quality regarding rarity and novelty.

1 Introduction

Waters play an important role in every mountainous landscape (Godde 1998). They co-create the mountainous landscape's character (Gabr 2004). They also serve as a refuge for those biotopes being endangered by human activity (Chytrý et al. 2001) and for species of organism related to those biotopes as well (Kučera 2005). These reasons help to make specific forms of water occurrences in mountainous landscapes a desired destination for tourists. Opening up accessibility to some parts of mountainous areas is, however, quite regularly in violation of the interests of nature and of landscape protection. Understandably, as the tourist load begins to manifest itself, the more environmental effect it has on those most vulnerable places in which aquatic environments and wetlands unambiguously lie (Christ et al. 2003).

J. Navrátil (✉) · K. Pícha
Faculty of Economics, University of South Bohemia in České Budějovice, Studentská 13, České Budějovice 370 05, Czech Republic
e-mail: josefnav@gmail.com

J. Kozak et al. (eds.), *The Carpathians: Integrating Nature and Society Towards Sustainability*, Environmental Science and Engineering,
DOI: 10.1007/978-3-642-12725-0_33, © Springer-Verlag Berlin Heidelberg 2013

Water, besides water-based activities, plays the role of an aesthetical background for any of outdoor tourism activities. Its calmative and relaxing effect on a human being (Říha 1987) is an important component of the quality of many outdoor recreational activities (Kakoyannis and Stankey 2002). This chapter is devoted fully to this fundamental importance of water in the landscape for tourism.

Likelihood of recommendation given by willingness to recommend represents, besides the intentions to revisit a destination, a main component of loyalty to a given destination (Oppermann 2000). Generally, it is a matter of the further intended action being influenced by the visit to the destination (Chen and Tsai 2007), which is manifested, above all, by word-of-mouth recommendations (Hui et al. 2007). As loyalty becomes a fundamental strategic component for organizations (Chi and Qu 2008) and recommendations to other people (via word-of-mouth) are one of the most often sought after types of information for people interested in travelling (Hui et al. 2007), a high attention is given to this phenomenon in the matter of destinations as well (for comprehensive literature review please see Simpson and Siguaw 2008). Word-of-mouth is also one of the important factors in image perceptions of a destination (Beerli and Martin 2004).

Non-experimental causal studies show us, that there are several main causal paths among factors of destination loyalty: image→ satisfaction→ loyalty (Chi and Qu 2008); image→ quality→ satisfaction→ loyalty (Bigné et al. 2001); quality→ value; satisfaction→ loyalty (Petrick 2004), which are given together as image→ quality→ value→ satisfaction→ loyalty (Chen and Tsai 2007), and pull motivation→ satisfaction (including value)→ loyalty (Yoon and Uysal 2005). The reverse causal path from satisfaction to perceived value was tested as significant by Duman and Mattila (2005). All these results are modified by recent research of Assaker et al. (2010), whose model suggests that the impact of factors on loyalty is modified with time.

Thus, there are many factors influencing destination loyalty via satisfaction. Yoon and Uysal (2005) summarize that, only the combination of the two main approaches to the study of the destination loyalty concept—behavioural (based on sequence purchase, proportion of patronage, or probability of purchase) and attitudinal (based on consumer brand preference or intention to buy)—may lead to the understanding of this concept and, thus, it is necessary to consider satisfaction and motivation constructs simultaneously. Rate of satisfaction (Kozak and Rimmington 2000) and particularly the overall satisfaction (Chi and Qu 2008) are considered to be the main factors of the rate of destination loyalty. These are then a function of a large number of mutually co-acting factors, particularly perceived value, perceived quality, destination image, and push and pull motivations. These factors have been well studied before (e.g. Baker and Crompton 2000; Petrick 2002, 2004; Lee et al. 2005; Yoon and Uysal 2005).

However, there is one more important factor which modified satisfaction with a visit to an attraction—that of on-site experience (Vittersø et al. 2000) as a way of satisfying a wide range of personal needs (Cohen 1979; Chhetri et al. 2004), which is, however, dynamic across time during the visit (Borrie and Roggenbuck 2001).

Experience of certain places visited means that "an individual... enters into a relationship with their surroundings" (den Breejen 2007, p. 1419), that results in different feelings that an individual experiences in different places (Chhetri et al. 2004). Leisure experiences are often operationalised through questions asking about emotion or mood (Lee and Shafer 2002) and "positive emotions are crucial to overall experience quality" (Farber and Hall 2007, p. 249). Pleasure has also been confirmed as a main factor of satisfaction and loyalty (Bigné et al. 2005).

Therefore, the aim of our chapter is to study the impact of all those factors together to reveal the impact of push motivation, pull motivation, on-site experience and perceived quality, on tourism site loyalty and as the case we have chosen water-enhanced tourist attractions in Central-European mountainous and submontane landscape.

2 Study Area

The defined objective is related to the area of South of Bohemia (the Czech Republic). The South Bohemian Region was the second most favorite destination in Czech domestic tourism in 2008, as 2.2 million tourism trips were made into this region (CzechTourism 2009, Table 12). The selected area comprises most of two tourism marketing regions (as defined by the national agency, Czech Tourism); South Bohemia and Sumava Mountains), that are the most attractive domestic destinations in the Czech Republic (Novotný 2004, p. 23).

The area comprises five bordering geomorphologic units: Sumava mountains, Sumava foothills, Novohradske Hory mountains, Novohradske Podhuri foothills and the basin of Trebon.

Based on the previous experience of authors and an extensive field survey undertaken in spring 2009 within these five geomorphologic units, several types of water-related tourist attractions were identified (sites on marked tourist paths): mountain glacier lakes, springs, water-falls, stony rivers in deep valleys, rivers in flat broad mountain valleys, canals, ponds, peat bogs, water closely linked with a historical monument, high situated point with a view on a water-course in deep timbered valleys, points with a wide view on a dominant water level.

3 Data and Methods

3.1 Data Collection and Sample

Data were collected via written questionnaires completed on-site at selected tourist attractions throughout the study area.

Selection of particular sites for questioning was done by the help of a database containing potential sites. This database was created on the grounds of a field survey and following registration of discovered sites in GIS JANITOR J/2 software (Pala 2008). With regard to the predominance of cycle tourism in all areas the database does not contain all locations but only those where there is a high probability of cycle tourists making a stop. This means a place, where a basic tourist infrastructure is available—a guidepost to tourist paths or a signboard (any type) or equipment for having a rest (benches, station roofs, etc.). Each place was linked with a type of attraction and an approximate number of visitors during one day in the summer tourist season (based on the consultation with representatives of local and regional tourism organizations). We have automatically discarded those locations with a visit rate inferior to the units in which daily average is lower than one visitor. With regard to the number of potential locations for questioning and to the validity of data a minimal number of locations to be surveyed were set at two. The selection of locations where the survey would be realized was random: (1) if a type of location was represented in the database by less than four records, two locations were selected; (2) if a type of location was represented in the database by four and more records, three locations were selected. Among the types cited in the part "Study area", the type of 'water springs' was, thus, not selected to be surveyed and analysed within this chapter.

The questionnaire survey was carried out during the summer season in 2009 (from June to September) by ten trained students at 26 localities identified as described above. A pilot research on a thirty-member sample preceded the survey in May 2009. On the basis of its results, the final version of the query tool was prepared (Robinson 1998). We used systematic sampling for selection of participants as there is no possibility to undertake such a survey with random sampling. To reduce this problem, the survey was done during both weekends and work days (Petrick et al. 2001). In case of low visitation (units of visitors per day) the researchers attempted to contact every possible visitor (Farber and Hall 2007). In case of medium visitation (tens of visitors per day) the researchers attempted to contact every fifth visitor and in the case of high visitation (hundreds of visitors per day) every tenth visitor was approached and asked if he or she wished to participate (Navrátil et al. 2010). The aim was to obtain a target of 64 fully completed questionnaires in each locality that results in the 1664 respondents in the sample.

3.2 Measures

The questionnaire was developed in order to examine causal structure (Pringle 1981) among push motivation, pull motivation, experience, perceived quality, perceived value of visit, satisfaction with visit, and tourism site loyalty. As the straight ties (in sense of causality) in the path, perceived value→ satisfaction→ loyalty have been previously confirmed (Yoon and Uysal 2005; Chen and Tsai 2007) and as multiple linear regression is adopted here, only push motivation, pull motivation, experience and quality as factors of tourism site loyalty were examined.

Motivation. The motivation construct used in this study is based on push and pull motivations as presented by Yoon and Uysal (2005). Partial push and pull motives were adapted from studies of Yoon and Uysal (2005) and Petrick et al. (2001). Some motives were removed and some added to address reasons specifically related to outdoor recreation in Central-European conditions (as in Ballantyne et al. 2008). Thus motivation to a visit was measured by a five-point Likert-like scale with "1 = Not at all important" and "5 = Very important" anchoring the scale. Respondents were asked to rate the importance of 16 push motives and 15 pull motives as reasons for visiting a particular site.

The employment of the concept of push and pull motivations allowed us to assess, also, other aspects of satisfaction (Gnoth 1997). Push motivation "related to internal or emotional aspects" (Yoon and Uysal 2005, p. 46) are viewed as "consumer dispositions" (Goossens 2000). Pull motivations "are connected to external, situational, or cognitive aspects" (Yoon and Uysal 2005, p. 46), which could be stated as "marketing stimuli" (Goosesns 2000) and, perhaps, as a substantial part of destination image.

Experience. As already mentioned, the main components of experience are emotions or moods (Farber and Hall 2007). For measurement of emotions experienced in particular locations the leading environmental psychology approach was used—the Mehrabbian-Russell model (Donovan and Rossiter 1982). The reason for this is, above all, its basis in the Stimulus-Organism-Response paradigm; its orientation on behavioural consequences "approach or avoidance" (Donovan and Rossiter 1982), and its usefulness in tourism issues (Jang and Namkung 2009) and also in outdoor tourism studies (Chhetri et al. 2004). It is quite usual that the original Mehrabian-Russell Pleasure-Arousal-Dominance scale is modified in various ways for purposes of concrete studies and researches, e.g. omitting the dominance factor (Donovan et al. 1994), unipolar approach (Jang and Namkung 2009), substitution of mood items (Chhetri et al. 2004) or selection of items (Sparks 2007). However, we have used the original 18 pleasure-arousal-dominance measures measured on semantic differential seven-point scales because of their complexity and the possibility to use them in different environments (Navrátil et al. 2011).

Quality. We understand quality as "a measure of the provider's performance" (Petrick 2004, p. 399). As our research is oriented on the visit rate to 'natural' sites, the "provider's performance" is represented by the environment itself of a place visited. Therefore, we have used, instead of the SERVQUAL scale usual in tourism (Parasuraman et al. 1991), the Mehrabian-Russell general measure of information rate. The original 14 seven-point scales of semantic differential were employed (Donovan and Rossiter 1982).

Loyalty. As a measure of destination loyalty we used willingness to recommend, which was identified by asking the question: Will you suggest this site to your friends/relatives as a destination to visit? (Yoon and Uysal 2005) This was measured on a five-point Likert scale (1 = Definitely not, 5 = Definitely yes).

3.3 Data Analysis

Factors (indicators) of each multi-item construct were first identified by explorative factor analysis (EFA) of each measurement tool. In all cases, the principal components analysis method was employed and only the factors with an eigenvalue greater than 1 were assessed, and the results were varimax rotated (Robinson 1998). All indicators revealed by EFA were evaluated by confirmatory factor analysis (CFA) for each construct separately, with all items allowed to be intercorrelated freely (Yoon and Uysal 2005). Only items with a factor loading greater than 0.5 per factor were included (Nusair and Hua 2010). Indicators of each construct were modified until the critical values of selected fit indices (Table 1) were reached (Yoon and Uysal 2005). Then, for each indicator of each construct, the composite mean was calculated, i.e. average value for indicator (CFA factor) from values of items loaded significantly on this factor (Chen and Tsai 2007). Mean value from all questionnaires was considered for each item as its importance (used in Tables 2 and 3).

These factors were considered as independent variables influencing willingness to recommend (dependent variable). The selection of factors of willingness to recommend was decided by means of the multiple linear regression (Nusair and Hua 2010) using the forward selection method for independent variables. First run of forward selection was performed. Then the data were purged from outliers. Consequently, the process of forward selection was repeated. The model was assessed based on the partial regression graphs and partial residual graphs and the method was assessed by means of F-test of importance of a regression model (Meloun and Militký 2006).

All computations were performed using STATISTICA 8.0 software package.

Table 1 Goodness of fit indices for separated evaluation of willingness to recommend factors (n = 1,664)

	Chi square/d.f.	RMSEA	GFI	CFI	NNFI
Pull motivation	4.53	0.047	0.988	0.976	0.961
Push motivation	7.39	0.061	0.984	0.952	0.922
On-site experience	3.04	0.035	0.988	0.987	0.981
Perceived quality	6.74	0.059	0.983	0.953	0.922

d.f. = degrees of freedom; RMSEA = Steiger-Lind root mean square error of approximation; GFI = Jöreskog-Sörbom goodness of fit index; CFI = Bentler comparative fit index; NNFI = Bentler-Bonett non-normed fit index

4 Results and Discussion

4.1 Indicators of Constructs

The scales measuring pull motivation, push motivation, and experience, revealed an acceptable reliability value for Cronbach's alpha (0.762, 0.726, 0.790). The value of reliability of the measurement tool is lower (0.693); however, it is very close to the acceptable value of 0.7, so it is possible to use (Yoon and Uysal 2005).

4.1.1 Motivation

On the basis of EFA, four factors of pull motives explaining 51.1 % of the total variability were identified (Table 2): pleasant 'natural' environment, heritage and culture, accessibility, and closeness. Indicators of pull motives were evaluated using fit indices of CFA. One indicator (closeness) and one item in the accessibility indicators were removed to obtain acceptable values of fit indicators.

Table 2 The results of EFA for pull motivations

Pull motivations (item mean ± standard deviation)	Pleasant 'natural' environment	Heritage, culture	Accessibility	Closeness
Location is situated in an interesting landscape. (4.22 ± 1.00)	*0.73*	–	–	–
Environment is pleasant here. (4.28 ± 0.93)	*0.70*	–	–	–
It is quiet here. (4.06 ± 1.13)	*0.68*	–	–	–
Location is culturally/artistically interesting. (2.87 ± 1.38)	–	*0.87*	–	–
Location is related to an interesting history. (3.25 ± 1.42)	–	*0.83*	–	
It is on the way that we have planned. (3.74 ± 1.36)	–	–	*0.66*	–
Location is accessible. (3.61 ± 1.28)		–	*0.63*	–
Information is provided in this location (by means of a nature trail, information board or a guide). (3.39 ± 1.40)	–	–	*0.60*	–
It is fun here. (2.84 ± 1.32)	–	–	0.56	–
It is quite close to our accommodation/ home. (2.73 ± 1.43)	–	–	–	0.87
Eigenvalue	3.88	1.52	1.23	1.03
% of total variability	25.88	10.11	8.19	6.88
Cronbach's alpha EFA factors	0.68	0.73	0.61	–
Cronbach's alpha CFA factors	0.68	0.73	0.61	–

Response format: 1 = Not at all important, 5 = Very important. Only items with varimax rotated factor loadings greater than 0.5 are shown. Items in italics were confirmed by CFA

Table 3 The results of EFA for push motivations

Push motivations (item mean ± standard deviation)	Social gathering	Escape	Self-reflection	New knowledge and experience	Relaxation
Be with friends. (3.26 ± 1.55)	*0.82*	–	–	–	–
Talk with friends during the journey about experience. (3.30 ± 1.43)	*0.73*	–	–	–	–
Enjoy. (3.25 ± 1.32)	*0.55*	–	–	–	–
Free ourselves of a stereotypical sort of day-to-day life and job. (4.13 ± 1.11)	–	*0.73*	–	–	–
Visit interesting places. (4.19 ± 1.02)	–	*0.65*	–	–	–
Change environment. (3.84 ± 1.23)	–	*0.65*	–	–	–
Relax through a physical recreational activity. (3.71 ± 1.30)	–	*0.51*	–	–	–
Reflection on site about the "good old times". (2.37 ± 1.47)	–	–	0.79	–	–
Possibility to be really myself. (2.86 ± 1.34)	–	–	0.62	–	–
Gain new knowledge. (3.06 ± 1.30)	–	–	–	0.76	–
Get to know new locations. (3.97 ± 1.28)	–	–	–	0.59	–
Experience an adventure. (3.18 ± 1.35)	–	–	–	0.58	–
Do nothing, just relax. (3.00 ± 1.38)	–	–	–	–	0.81
Eigenvalue	3.41	1.87	1.38	1.23	1.03
% of total variability	21.30	11.70	8.63	7.67	6.44
Cronbach's alpha EFA factors	0.66	0.63	0.38	0.58	–
Cronbach's alpha CFA factors	0.66	0.63	–	–	–

Response format: 1 = Not at all important, 5 = Very important. Only items with varimax rotated factor loadings greater than 0.5 are shown. Items in italics were confirmed by CFA

'Pleasant environment' and 'interesting landscapes' were the items of highest importance. On the other hand, closeness, fun, and cultural significance were the items with the lowest importance (Table 2).

Five indicators of push motivation to visit water-enhanced tourist attractions were revealed: social gathering, escape, self-reflection, new knowledge and experience, and relaxation. These five factors explain 55.7 % of variability of the dataset (Table 3). Three indicators were removed (self-reflection, new knowledge and experience and relaxation), after evaluating the indicators structure with CFA.

To visit interesting places and the escape from day-to-day life were the items of highest importance. To visit a place unknown for friends or relatives was the item with the significantly lowest importance (Table 3).

The discovered motives for a visit are analogical to the previous findings concerning motivation to the participation in outdoor recreation.

Pull motivations have many things in common with the choice of destination (Goossens 2000) since they represent the specific attractions of the destination (Dann 1981). Factors of the pull motivations thus reflect, quite often, specifics of the core tourism resources and attractions (Ritchie and Crouch 2003). Our study has identified, as a first dominant factor, the pleasant 'natural' environment and as a second factor that of heritage and culture. Two other factors have a generally geographical substance (Haggett 2001); those items connected with accessibility and closeness (distance). These factors draw our attention to the fact that a tourism attraction is not independent but that its attractiveness is also governed by its spatial situation and, therefore, by the related topics that are to be dealt with by destination marketing (Ritchie and Crouch 2003).

As concerns the push motives, it is, first of all, the social-gathering that was identified as a motive explaining maximum of variability among partial motives for visitors as it was, also, in research undertaken for Allegheny National Forest in Pennsylvania (Graefe et al. 2000). Social togetherness, social contact or social gathering belong generally to the main push motives of participation in travel and tourism (Crompton 1979) as it was confirmed in analogical studies of outdoor recreation for Delaware State Park visitors (Confer et al. 1996), for botanical gardens visitors (Ballantyne et al. 2007) as well as for other outdoor activities, such as sport fishing (Arlinghaus and Mehner 2003; Navrátil et al. 2009) and golf sites (Petrick et al. 2001). Escape and relaxation belongs, as well, among generally considered push motives to the participation in travel and tourism (Crompton 1979; Dann 1981) as well as new knowledge and experience (Ryan and Glendon 1998; Greafe et al. 2000; Yoon and Uysal 2005; Ballantyne et al. 2008). "Travel has always offered a unique opportunity for self-discovery" (Pruitt and LaFont 1995) and self-reflection is also one of the push motives known from previous studies (Dann 1981). In our case, this factor could be considered as a strongly introvert version of self-fulfilment (Hsu et al. 2007).

4.1.2 Experience

Structure of feelings being experienced in particular locations by visitors, manifests a conformity with the Mehrabian-Russell model of emotional component of the attitude (Mehrabian and Russell 1974). This could be stated based on the factor analysis that has confirmed the original Pleasure-Arousal-Dominance dimensions of this model explaining 49.7 % of the total variability of the population (Table 4). However, the percentage of the explained variability is lower than in the case of analogical studies (Donovan and Rossiter 1982; Jang and Namkung 2009). CFA confirmed presence of all three dimensions and only two items within the pleasure indicator and three items within the arousal indicator were removed.

The highest reliability was obtained, similarly to other studies, for the dimension of pleasure (Donovan and Rossiter 1982; Donovan et al. 1994). The dimension of dominance is comparable with the results of previous studies (Donovan and Rossiter 1982). Arousal dimension has lower value of reliability (Donovan and

Table 4 The results of EFA for on-site experience

On-site experience (item mean ± standard deviation)	Pleasure	Arousal	Dominance
Depressed-contented (6.13 ± 1.13)	*0.82*	–	–
Unhappy-happy (6.04 ± 1.15)	*0.81*	–	–
Unsatisfied-satisfied (5.82 ± 1.23)	*0.76*	–	–
Annoyed-pleased (6.02 ± 1.15)	*0.80*	–	–
Bored-relaxed (6.06 ± 1.14)	*0.72*	–	–
Restricted-free (5.88 ± 1.28)	0.70	–	–
Despairing-hopeful (5.34 ± 1.23)	0.56	–	–
Relaxed-stimulated (3.87 ± 2.07)	–	0.66	–
Calm-excited (3.6 ± 1.95)	–	*0.76*	–
Dull-jittery (4.41 ± 1.16)	–	0.56	–
Unaroused-aroused (4.93 ± 1.42)	–	0.58	–
Sluggish-frenzied (4.41 ± 1.31)	–	*0.63*	–
Controlled-controlling (3.92 ± 1.56)	–	–	*0.82*
Submissive-dominant (3.95 ± 1.54)	–	–	*0.73*
Influenced-influential (3.71 ± 1.60)	–	–	*0.79*
Eigenvalue	4.99	2.35	1.61
% of total variability	27.74	13.07	8.92
Cronbach's alpha EFA factors	0.87	0.67	0.70
Cronbach's alpha CFA factors	0.87	0.50	0.70

Response format: seven-point scale, 1 = I definitely experience the mood on left side, 4 = I can not decide, 7 = I definitely experience the mood on right side. Only items with varimax rotated factor loadings greater than 0.5 are shown. Items in italics were confirmed by CFA

Rossiter 1982; Donovan et al. 1994), which is, also the case in other studies (discussed by Bigné et al. 2005, p. 838), however it was included into regression analysis, as the goodness of fit indices for evaluation of experience as factor of willingness to recommend were excellent (Table 1).

4.1.3 Perceived Quality

As it was already mentioned, in our conception we understand quality as a quality of the environment that is given by its descriptive characteristics. The employed scale is pertinent to achieve the defined objective as only one characteristic of Mehrabian-Russell's general measure of information rate was not loaded with the value of at least 0.5 on any factor. In our sample, four factors of environment perception (Table 5) were revealed, that are quite consistent with the three dimensions discussed by Mehrabian and Russell (1974).

The first indicator was labelled as commonplaceness, because items such as usual-surprising, common-rare or similar-contrasting were strongly loaded on this factor. This factor represents a combination of the factors of 'novelty' and 'complexity' that are expected by the model (Donovan and Rossiter 1982). This factor symbolises the specificity of the rarity in the supply of travel and tourism.

Table 5 The results of EFA for quality perception

Quality perception (item mean ± standard deviation)	Common-placeness	Regularity	Novelty	Density
Surprising-usual (3.31 ± 1.79)	*0.76*	–	–	–
Rare-common (2.87 ± 1.63)	*0.73*	–	–	–
Heterogeneous-homogeneous (3.07 ± 1.58)	*0.60*	–	–	–
Varied-redundant (2.63 ± 1.54)	*0.66*	–	–	–
Contrasting-similar (3.48 ± 1.69)	0.53	–	–	–
Complex-simple (4.01 ± 1.67)	0.53	–	–	–
Asymetrical-symetrical (3.97 ± 1.69)	–	0.55	–	–
Random-patterned (4.99 ± 1.66)	–	0.81	–	–
Intermittent-continuous (4.66 ± 1.57)	–	0.81	–	–
Novel-familiar (3.63 ± 2.25)	–	–	0.78	–
Immediate-distant (3.99 ± 1.93)	–	–	−0.77	–
Dense-sparse (3.73 ± 1.32)	–	–		0.74
Crowded-uncrowded (3.97 ± 1.67)	–	–		0.70
Eigenvalue	3.19	1.72	1.35	1.01
% of total variability	22.79	12.29	9.66	7.22
Cronbach's alpha EFA factors	0.74	0.59	0.48	0.29
Cronbach's alpha CFA factors	0.74	0.65	0.48	–

Response format: seven-point scale, 1 = The adjective on the left definitely describes this site, 4 = I can not decide, 7 = The adjective on the left definitely describes this site. Only items with varimax rotated factor loadings greater than 0.5 are shown. Items in italics were confirmed by CFA

The second indicator is pure complexity dimension—perceived regularity of environment. The third indicator is pure novelty, labelled as novelty. The fourth factor is spatial dimension in perceived quality of environment, labelled as density. So, indicators reflect specifics of perceived quality of environment by tourists, and besides novelty there exists also the factor of rarity and the main factor of spatiality, which is density (Sheldon and Var 1984). However, as indicated by the low values of Cronbach's alpha, after CFA evaluation the density indicator was removed, as well as two items within the commonplaceness factor and one within the regularity factor.

4.2 Model of Word-of-Mouth Predictors

The forward selection method of the multiple linear regression via software STATISTICA 8.0 was used to determine the model of word-of-mouth predictors among visitors to water-enhanced tourist attractions.

Absolute values of all correlations of independent variables were inferior to 0.5 excepting the correlation between pull motivation to visit pleasant 'natural' environment and push motivation labelled as 'Escape'. The model includes only those predictors whose parameters beta were significant on the calculated

Table 6 Overall goodness of fit—results of multiple linear regression

	Sums of squares	d.f.	Mean squares	F	p-value
Regression	179.9999	5	35.99999	106.8825	0.00
Residual	531.1624	1,577	0.33682		
Total	711.1623				

significance level $p < 0.01$ based on the t test of the statistical significance of the parameter. Based on the discovery of the existence of outliers these were eliminated and the process of forward selection was repeated. The resulting model is significant (Table 6), therefore the dependent variable word-of-mouth and independent variables pleasure, heritage and culture, commonplaceness, pleasant 'natural' environment, and novelty are in linear dependence. These five measured independent variables explain 25 % of variability of willingness to recommend the locality as a good place to visit (adjusted $R^2 = 0.251$, standard error of estimate = 0.580). The degree of recommendation of a location to visit is positively dependent particularly on the degree of experience of pleasure from the visit to the location (Table 7). The importance of other predictors is markedly lower. The increasing degree of the recommendation of a location is explainable by the increase of: the importance of getting to learn the history, perception of a location to be more rare, emphasis on the possibility to visit pleasant 'natural' environment and perception of a locality as a new one.

The multiple linear regression model confirmed that the experience of pleasure has the strongest influence on destination loyalty (Bigné et al. 2005; Sparks 2007). Besides the pleasure-seeking, another important factor is manifested in the willingness to recommend novelty-seeking (Duman and Mattila 2005), as both rarity and novelty were identified by the model as being significant as well. However, it is not only novelty itself, but rarity, also, that has a special importance in tourism (McKercher and du Cros 2002), because it "contributes to the attraction" (Orams 2002, p. 286).

The perception of the quality of supply, represented by the perception of the structure of a location, is related also to the importance of a concrete offer manifested in pull motives to a visit (Yoon and Uysal 2005).

Table 7 Model of word-of-mouth predictors

	b	S.E. of b	t (1577)	p-value
Intercept	4.125110	0.124850	33.04042	0.000000
Pleasure	0.252691	0.018496	13.66204	0.000000
Heritage, culture	0.094622	0.012083	7.83102	0.000000
Commonplaceness	−0.073900	0.013831	−5.34319	0.000000
Pleasant 'natural' environment	0.068741	0.021458	3.20352	0.001385
Novelty	0.024716	0.008777	2.81609	0.004922

Results of a multiple linear regression as described in Table 6

5 Conclusions

This study tried to reveal the predictors of likelihood of recommendation for a visit to water-enhanced tourist attractions in Central-European mountainous and sub-montane landscapes. Pull motivation, push motivation, on-site experience and perception of site quality were chosen as potential factors.

Pleasant 'natural' environment, heritage and culture, and accessibility were found to be factors of pull motivation (especially pleasant environment' was one of the items of highest importance); social gathering and escape as factors of push motivation; pleasure, dominance and arousal as factors of experience; and commonplaceness, regularity and novelty as factors of perceived quality. The forward selection of multiple linear regressions found the pleasure experience as the main predictor for willingness to recommend. This is conformable to the position of Goosens (Goossens 2000), who accentuated pleasure to be very important motivation as well as a factor of destination loyalty. He observed that a number of authors consider such "experiential aspects" to play an important role in consumer choice behavior. Others are: pull motivation regarding types of attractions and site quality regarding commonplaceness and novelty.

It can be concluded that willingness to recommend a site to visit, in terms of being a substantial part of destination loyalty, is influenced by pull motivations, perceived quality and by on-site experience. Our potential determinants explain one quarter of variance, that is comparable e.g. to study of Yoon and Uysal (2005). This state could be improved by enlarging this model with variables regarding expectation (Hsu et al. 2010) or place attachment (Gu and Ryan 2008).

These findings suggest that enhancing positive experiences of visitors would be the best way to attract more new visitors. However, all other factors are very closely tied to geographical and environmental roots; so, attracting more visitors must be in harmony with the capacity of the site. If not, those roots would be undermined and the core resource of tourism, as well as wildlife and heritage, would be lost (Williams 1998; Vereczi 2002).

In any case managers of tourism destination should observe very carefully these motives as they are vital for the overall tourists' satisfaction and are considered to be the main factors of the rate of destination loyalty (Kozak and Rimmington 2000; Chi and Qu 2008).

Acknowledgments We are grateful to all attraction visitors who participate in this survey. We also thank local and regional tourism authorities for cooperation as well as we thank the students for field collecting of questionnaires. Authors also thank Elizabeth George for language revision and two anonymous reviewers for many helpful comments on previous version on manuscript. The whole field survey, participation on Forum Carpaticum 2010 as well as preparation of this chapter was supported from the Czech Science Foundation—GACR 403/09/P053 'The typology of tourist's attitudes towards environment, the case of waters in landscape.' The chapter is a part of European Social Fund and Ministry of Education, Youth and Sports of the Czech Republic Operational Programme Education for Competitiveness—CZ.1.07/2.2.00/07.0178 'Studies of rural development economics at the University of South Bohemia in České Budějovice.'

References

Arlinghaus R, Mehner T (2003) Socio-economic characterisation of specialised common carp (*Cyprinus carpio* L.) anglers in Germany, and implications for inland fisheries management and eutrophication control. Fish Res 61:19–33

Assaker G, Vinzi VE, O'Connor P (2010) Examining the effect of novelty seeking, satisfaction, and destination image on tourists' return pattern: a two factor, non-linear latent growth model. Tour Manag. doi:10.1016/j.tourman.2010.08.004

Baker DA, Crompton JL (2000) Quality, satisfaction and behavioral intentions. Ann Tour Res 27:785–804

Ballantyne R, Packer J, Hughes K (2008) Environmental awareness, interests and motives of botanic gardens visitors: implications for interpretive practice. Tour Manag 29:439–444

Beerli A, Martin JD (2004) Factors influencing destination image. Ann Tour Res 31:657–681

Bigné JE, Sanchez MI, Sanchez J (2001) Tourism image, evaluation variable and after purchase behaviour: inter-relationship. Tour Manag 22:607–616

Bigné JE, Andreu L, Gnoth J (2005) The theme park experience: an analysis of pleasure, arousal and satisfaction. Tour Manag 26:833–844

Borrie WT, Roggenbuck JW (2001) The dynamic, emergent, and multi-phasic nature of on-site wilderness experiences. J Leis Res 33:202–228

Chen Ch, Tsai D (2007) How destination image and evaluative factors affect behavioral intentions? Tour Manag 28:1115–1122

Chhetri P, Arrowsmith C, Jackson M (2004) Determining hiking experiences in nature-based tourist destinations. Tour Manag 25:31–43

Chi ChG-Q, Qu H (2008) Examining the structural relationships of destination image, tourist satisfaction and destination loyalty: an integrated approach. Tour Manag 29:624–636

Christ C, Hillel O, Matus S, Sweeting J (2003) Tourism and biodiversity. Mapping tourism's global footprint, Washington and Paris Conservation International and UNEP

Chytrý M, Kučera T, Koči M (eds) (2001) Katalog biotopů České republiky. AOPK CR, Praha

Cohen E (1979) A phenomenology of tourist experiences. J Br Sociol Assoc 13:179–201

Confer JJ, Vogelsong HG, Graefe AR, Solan DS (1996) Relationships between motivations and recreation activity preferences among delaware statte park visitors: an exploratory analysis. In: Kuentzel W (ed) Proceedings of the 1996 northeastern recreation research symposium. http://www.fs.fed.us/ne/newtown_square/publications/technical_reports/pdfs/scanned/gtr232b.pdf. Accessed 24 October 2010

Crompton JL (1979) Motivation of pleasure vacation. Ann Tour Res 6:408–424

CzechTourism (2009) Domácí a výjezdový cestovní ruch za rok 2008. http://www.czechtourism.cz/files/statistiky/aktualni_data/27_10_09_domaci_vyjezd_cr_2008.pdf. Accessed 24 October 2010

Dann GMS (1981) Tourist motivation: an appraisal. Ann Tour Res 8:187–219

den Breejen L (2007) The experiences of long distance walking: a case study of the West Highland way in Scotland. Tour Manag 28:1417–1427

Donovan RJ, Rossiter JR (1982) Store atmosphere: an environmental psychology approach. J Retail 58:34–57

Donovan RJ, Rossiter JR, Marcoolyn G, Nesdale A (1994) Store atmosphere and purchasing behavior. J Retail 70:283–294

Duman T, Mattila AS (2005) The role of affective factors on perceived cruise vacation value. Tour Manag 26:311–323

Farber ME, Hall TE (2007) Emotion and environment: visitors' extraordinary experiences along the Dalton highway in Alaska. J Leis Res 39:248–270

Gabr HS (2004) Perception of urban waterfront aesthetics along the Nile in Cairo. Egypt Coast Manag 32:155–171

Gnoth J (1997) Tourism motivation and expectation formation. Ann Tour Res 24:283–304

Godde P (1998) Community-based mountain tourism: Practices for linking conservation with enterprise. http://www.mtnforum.org/rs/econfreports/Community-BasedMountainTourism.pdf. Accessed 24 October 2010

Goossens C (2000) Tourism information and pleasure motivation. Ann Tour Res 27:301–321

Graefe AR, Thapa B, Confer JJ, Absher JD (2000) Relationships between trip motivations and selected variables among Allegheny National Forest visitors. In: Cole DN, McCool SF, Borrie WT, O'Loughlin J (eds) Wilderness science in a time of change conference—Volume 4: wilderness visitors, experiences, and visitor management; 1999 May 23-27; Missoula, MT. 1st edn. U.S. Department of Agriculture, Forest Service, Rocky Mountain Research Station, Odgen

Gu H, Ryan C (2008) Place attachment, identity and community impacts of tourism—the case of a Beijing hutong. Tour Manag 29:637–647

Haggett P (2001) Geography: a global synthesis. Pearson Education, Harlow

Hsu CHC, Cai LA, Wong KKF (2007) A model of senior tourism motivations—Anecdotes from Beijing and Shanghai. Tour Manag 28:1262–1273

Hsu CHC, Cai LA, Li M (2010) Expectation, motivation, and attitude: a tourist behavioral model. J Travel Res 43:282–296

Hui TK, Wan D, Ho A (2007) Tourists' satisfaction, recommendation and revisiting Singapore. Tour Manag 28:965–975

Jang SCh(S), Namkung Y (2009) Perceived quality, emotions, and behavioral intentions: application of an extended Mehrabian–Russell model to restaurants. J Bus Res 62:451–460

Kakoyannis Ch, Stankey GH (2002) Assessing and evaluating recreational uses of water resources: implications for an integrated management framework. USDA, Portland

Kozak M, Rimmington M (2000) Tourist satisfaction with Mallorca, Spain, as an off-season holiday destination. J Travel Res 38:260–269

Kučera T (ed) (2005) Red book on habitats of the Czech Republic. 1st edn. http://www.usbe.cas.cz/cervenakniha/eng. Accessed 24 October 2010

Lee B, Shafer CS (2002) The dynamic nature of leisure experience: an application of affect control theory. J Leis Res 34:290–310

Lee C, Lee Y, Lee B (2005) Korea's destination image formed by the 2002 world cup. Ann Tour Res 32:839–858

McKercher B, du Cros H (2002) Cultural tourism: The partnership between tourism and cultural heritage management. Haworth Press, New York

Mehrabian A, Russell JA (1974) An approach to environmental psychology. MIT Press, Cambridge

Meloun M, Militký J (2006) Kompendium statistického zpracování dat. Academia, Praha

Navrátil J, Martinát S, Kallabová E (2009) Framework for utilizing angling as a tourism development tool in rural areas. Agric Econ—Zemed Ekon 55:508–518

Navrátil J, Pícha K, Hřebcová J (2010) The importance of historical monuments for domestic tourists: The case of South-western Bohemia (Czech Republic). Morav Geogr Rep 18:45–61

Navrátil J, Pícha K, Rajchard J, Navrátilová J (2011) Impact of visit on visitors' perceptions of the environments of nature-based tourism sites. Tour 59:7–23

Novotný O (2004) Motivace k návštěvě turistických regionů v České republice. Stem/Mark, Praha. http://www.czechtourism.cz/files/statistiky/motivaceregiony.pdf. Accessed 4 February 2009

Nusair K, Hua N (2010) Comparative assessment of structural equation modeling and multiple regression research methodologies: e-comemerce context. Tour Manag 31:314–324

Oppermann M (2000) Tourism destination loyalty. J Travel Res 39:78–84

Orams MB (2002) Feeding wildlife as a tourism attraction: a review of issues and impacts. Tour Manag 23:281–293

Pala P (2008) Manuál k aplikaci JanMap v.2.4.7. http://janitor.cenia.cz/www/public/manual/janmap/index.html. Accessed 27 April 2009

Parasuraman A, Berry LL, Zeithaml VA (1991) Refinement and reassessment of the SERVQUAL scale. J Retail 67:420–450

Petrick JF (2002) Development of multimensional scale for measuring the perceived value of a service. J Travel Res 34:119–134
Petrick JF (2004) The roles of quality, perceived value and satisfaction in predicting cruise passengers' behavioral intentions. J Travel Res 42:397–407
Petrick JF, Backmann SJ, Bixler R, Norman WC (2001) Analysis of golfer motivations and constraints by experience use history. J Leis Res 33:56–70
Pringle DG (1981) Causal modelling: the simon-blalock approach. Goe Abstracts, Norwich
Pruitt D, LaFont S (1995) For love and money romance tourism in Jamaica. Ann Tour Res 22:422–440
Říha J (1987) Voda a společnost. SNTL, Praha
Ritchie JRB, Crouch GI (2003) The competitive destination: a sustainable tourism perspective. CABI Publishing, Oxon
Robinson GM (1998) Methods and techniques in human geography. Wiley, Chichester
Ryan Ch, Glendon I (1998) Application of leisure motivation scale to tourism. Ann Tour Res 25:169–184
Sheldon PJ, Var T (1984) Resident attitudes to tourism in North Wales. Tour Manag 5:40–47
Simpson PM, Siguaw JA (2008) Destination word of mouth: the role of traveller type, residents, and identity salience. J Travel Res 47:167–182
Sparks B (2007) Planning a wine tourism vacation? factors that help to predict tourist behavioural intentions. Tour Manag 28:1180–1192
Vereczi G (2002) Sustainable development of tourism: principles and practice. In: di Castri F, Balaji V (eds) Tourism, biodiversity and information, 1st edn. Backhuys Publishers, Leiden
Vittersø J, Vorkinn M, Vistad OI, Vaagland J (2000) Tourist experiences and attractions. Ann Tour Res 27:432–450
Williams S (1998) Tourism geography. Routledge, London and New York
Yoon Y, Uysal M (2005) An examination of the effects of motivation and satisfaction on destination loyalty: a structural model. Tour Manag 26:45–56

The Transnational Hiking Trail *via carpatica*: A Concept for the Sustainable Development of the Carpathians

Joachim Jaudas

Abstract The *via carpatica* will one day connect seven European countries, and it will be a model for sustainable mountain tourism. It creates new jobs, promotes an ecologically compatible landscape and infrastructure development, and enables comprehensive possibilities for arranging leisure time activities, international encounters, natural experiences and the maintenance and further enhancement of the regional cultural heritage.

1 The Vision of a *via carpatica*

A traversal on the transnational hiking trail of the Carpathian bend, *via carpatica*, is certainly a vision today. Although single sections of the Carpathian mountain bend are more or less well developed, a continuous route with paths and huts along the mountain bend is still lacking. For this reason, the creation of a transnational hiking trail, which runs through seven countries of this European mountain region, is of far reaching importance. The idea behind *via carpatica*, consists of developing and promoting this European mountain region with a sustainable social and ecologically compatible tourist concept, the core of which is hiking tourism.

The *via carpatica* according to the vision will one day connect seven European countries and it could be a general model for sustainable mountain tourism: It will create new jobs, promote an ecologically compatible landscape and infrastructure development and enable comprehensive possibilities for arranging leisure time activities, international encounters, natural experiences and the maintenance and further enhancement of the regional cultural heritage.

Our investigations in the Carpathian project developed within the INTERREG IIB/CADSES programme show that there is a strong need for further development

J. Jaudas (✉)
Institut für Sozialwissenschaftliche Forschung - ISF München, Jakob-Klar-Str. 9, D-80796 Munich, Germany
e-mail: jaudas@gmx.net

J. Kozak et al. (eds.), *The Carpathians: Integrating Nature and Society Towards Sustainability*, Environmental Science and Engineering,
DOI: 10.1007/978-3-642-12725-0_34,

in sustainable tourism, in particular for hiking tourism in the individual countries of the Carpathian bend. Sustainable hiking tourism may strengthen the local added value and create a balance to the loss of traditional economy (VASICA 2009). We therefore will present some of the expected effects of the sustainable development of the tourism sector. In conclusion, we want to outline the first steps for implementing the concept of *via carpatica.*

2 Results and Consequences of the Development of the Transnational Hiking Trail *via carpatica*

Looking at the results of the inventory of tourism infrastructure and main local stakeholders in the Romanian Carpathians, carried out by ISF München in 2006, numerous positive developments will result in the Carpathian region with the development of *via carpatica* on the basis of a sustainable tourism concept (VASICA 2009). Hereby growth potentials will be promoted as a result of hiking tourism, e.g. new economic and social perspectives will be created which will prevent further migration from the mountain areas. If this is successful, then mountain tourism will make an important political contribution by preventing forced migration throughout Europe with all its related negative consequences for the migrants themselves as well as for the receiving countries to which the migrants flee. This could lead to a model of transnational regional development in which a steady balance would be created between globalisation (here understood as an external use of the resources of mountain regions, among other things, by tourism) and the economic self-sufficiency. Along with it there would be space for independent regional development and identities.

2.1 Development of Economy and Employment

Traditional economic sectors in the mountain regions, particularly agriculture (including sheep farming) are frequently no longer able to compete, despite of their high quality. As long as there are no new work perspectives or ways to earn an income, the migration of the residents from the Carpathians is the consequence. Therefore new qualification concepts should be developed as well as employment fields.

By the development of *via carpatica* many new economical stimuli with the corresponding effects on the employment will be generated, especially in services: maintenance of paths; management of huts; accommodation and stores in the valleys; public traffic; information, advice and support for hikers and mountaineers. By this new work and training perspectives will be created for the local population. They completely replace the precarious fields of work, improving and stabilising the economical base of the mountain regions.

In consequence of this development, the local and regional economy would be strengthened without making it dependent of the financial and decision-making power of big investors. It is based on available structures and increases these as required; it enables enterprises (e.g. inns) to be again profitable which otherwise would have no future without this development. It also enables new usage of abandoned facilities (such as work or forest paths). The necessary investments could be arranged in a way that avoids foreign dependencies.

The added value and the effects on the employment will work locally and inure to the benefit of the local people. Certainly this will contribute to reduce the unintended migration flows.

2.2 Development of the Region and the Regional Culture

A sustainable form of tourism leads to a revaluation of the rural space. Not only new employment opportunities in the service sector will be created which are directly attributed to tourism, but also other activities. The development of *via carpatica* requires the ideas and imagination of the regional stakeholders, as well as design and development. Moreover, the residents themselves should "reexplore" their region before they make it approachable for strangers. The "rediscovery" of the natural, cultural, economic and social resources of one's own region strengthens stakeholders' regional identity and increases their independent ability to act. In this context the danger of transfiguration of the regional history is not to underestimate (Bätzing 1991).

Thus the *via carpatica* is not to be measured just in terms of economical success, but also on the ability of maintaining this success and reactivating the local culture in a comprehensive sense.

2.3 Networking

Until now in the Carpathian region there have hardly been any cross-national approaches towards regional development. One reason for this is attributed to the isolation of the countries of this region, which emerged from the political situation in the past, but still continues with the political upheaval since 1990.

The Carpathian Convention is a first and politically powerful attempt to create the framework for a close co-operation among all the Carpathian countries. With this convention the agreement under international law depicts an important basis for supranational activities for the future.

With a transnational hiking trail through the Carpathians international networks can already be developed and implemented. The planning and carrying out of a European project for the creation of *via carpatica* would mark the beginning of necessary networking among the project partners involved and also signify the co-operative efforts of the local and regional stakeholders.

3 Proceeding in the Construction of a *via carpatica*

At the moment there are no solid transnational structures for the installation of a *via carpatica*, although there are stakeholders on a national level, who work for the development of a *via carpatica* in their region. There is also a spontaneous international coordination. Therefore the planning of the building of a *via carpatica* is still programmatic, such as to transfer the general principles of a sustainable development and the experiences with other mountain long distance trails like the French GTA (Grande Traversée des Alpes) and the Italian *gta* (grande traversata delle alpi).

With the construction of the *via carpatica*, the principle should be that the paths and huts as well as the partners in the regions concerned should be completely reliant on the structures available, which should be further developed (bottom-up-approach). Thus the *via carpatica* will emerge step by step, depending on the stage of development of current structures in the individual regions. Numerous sections will be created with individual institutions responsible (NGOs, communes, regional administrations). This will also lead to an independent character of individual sections (variety in unity).

If these goals, results and impacts were achieved, mountain tourists one day would be able to hike on marked paths over the ridges and crests of the Carpathians, from the Danube back to the Danube. They would find a hut in the evening, where they could spend the night and get something to eat, and infrastructure in the valleys that provides an easier access to the mountains. They would find route descriptions and get information on the Internet. And all this will be made possible by the efforts made by the resident population, which in turn will also help to secure jobs. This will be an important contribution to the stabilisation of this beautiful European mountain region.

References

Bätzing W (1991) Die Alpen. Geschichte und Zukunft einer europäischen Kulturlandschaft. Beck, München

VASICA (2009) Visions and Strategies in the Carpathian Area. Interim Secretariat of the Carpathian Convention. UNEP, Vienna

Spatial Patterns of Second Homes Development in the Polish Carpathians

Mirosław Mika

Abstract The Carpathian Mountains are among the most important recreational regions in Poland. Since the early 1990s, this area experiences an intensive development of second homes, which contributes to lasting functional changes and urbanization of rural areas. This phenomenon is affected by a complex set of social, economic and legal factors that are related to the process of political and economic transformation of the country. The owners of second homes are primarily residents of towns in the Carpathian foreland areas. This study presents the conditions, the main directions and spatial differentiation of second homes development in the Polish Carpathians. Particular attention was paid to specific location of these objects, which are constructed in areas with high landscape values, and which tend to be isolated from permanent housing. The study shows also consequences and conflicts arising from the rapid pace of development of this settlement form.

1 Introduction

Since the 1950s, the phenomenon of second homes has been the subject of numerous detailed and comprehensive studies in Europe (Coppock 1977; Haldrup 2004; Dijst et al. 2005; Gallent et al. 2005; Marjavaara 2007, Módenes and López-Colás 2007; Tress 2007; Arnstberg and Bergstrom 2008; Barke 2008; Pitkänen 2008; Hidle et al. 2010; Norris and Winston 2010), North America (Meyer-Arendt 2001; Cho et al. 2003; Kuentzel and Ramaswamy 2005) and other regions of the world (Hall and Müller 2004; Hoogendoorn et al. 2005, 2009; Moss 2006). There is a rapid development of second homes in all countries and regions, where intensive processes of urbanization and disurbanization take place. In the studies,

M. Mika (✉)
Institute of Geography and Spatial Management, Jagiellonian University, Gronostajowa 7, 30-387 Kraków, Poland
e-mail: miroslaw.mika@uj.edu.pl

J. Kozak et al. (eds.), *The Carpathians: Integrating Nature and Society Towards Sustainability*, Environmental Science and Engineering,
DOI: 10.1007/978-3-642-12725-0_35, © Springer-Verlag Berlin Heidelberg 2013

different issues are being addressed, such as factors of second homes development and its impacts on natural and social environment, economic and spatial consequences, or their role in the processes of urbanization of rural areas and suburbanization.

In Poland, second homes are not a new phenomenon. But their actual development has begun only in the 1990s, together with political and socio-economical changes and the introduction of free land trade. Currently, second homes are created in all areas with attractive landscape: in the mountains, at the coast and in Lakelands, they are also one of the most important factors of urbanization of rural areas (Kowalczyk 1994; Dziegieć 1995).

The Carpathian Mountains belong to main tourist regions in Poland (Warszyńska 1995; Faracik et al. 2009). This area is socially, culturally and economically diverse. Compared to the Southern Carpathians (Romania) and Eastern Carpathians (Ukraine), it is strongly transformed by economic activity and settlement development (Kurek 1996, Kurek 2004; Górz 2003). Currently, the Polish Carpathians are subject to further functional transformation due to the gradual decline of agriculture, which together with forestry have been the main economic activities there throughout the twentieth century (Ptaszycka-Jackowska 2007). Since the 1990s, cities and medium-sized towns located in the foothills of the Carpathians have played a growing role in this process, both in terms of core-periphery relations and meeting the needs for tourism and recreation of their inhabitants. Since that time, the rapidly growing settlements of second homes have been one of the major "change conveyors".

The purpose of this study is to present the main directions and regional differences of development of second homes in the Polish Carpathians, and to indicate location models of such objects. The presented paper synthesizes results of second homes research in the Polish Carpathians, including the author's detailed research in the Silesian Beskid Mountains (Mika 1997, 2001, 2004; Mika and Faracik 2008), in the Gorce Mountains by Petko (2011), and in the Beskid Wyspowy Mountains by Faracik (2006) and Gaś (1997), studies on the development of tourism in rural (Warszyńska 1995; Górz 2003; Ptaszycka-Jackowska and Baranowska-Janota 2003; Kurek 2004; Faracik 2006; Szpara 2006; Ptaszycka-Jackowska 2007; Bielak 2009) and protected areas and their surroundings (Faracik et al. 2008; Zawilińska 2010), as well as research of the historical perspective (Groch 1977; Rogalewska 1980; Hornung 1986; Jackowski 1989). Conditions for development of second homes are presented in the context of changing social, economic, legal and political situation in Poland in the 1980s and 1990s, based on research conducted in suburban areas of big agglomerations of Warszawa (Kowalczyk 1994), Kraków (Faracik 2006) and Łódź (Matczak 1986; Dziegieć 1995; Wiluś 1997; Wojciechowska 1998; Włodarczyk 1999; Szkup 2003).

In this study, all residential buildings, privately owned and located in places different than the place of permanent residence of the owners are considered as second homes. These include the year-round buildings, seasonal wooden cottages, and permanent caravans.

2 History of Second Homes in the Polish Carpathians

There is a long tradition of building houses for recreational purposes in the Polish Carpathians. The origins of this phenomenon are associated with the beginnings of tourism development in the nineteenth century, when the region was under the Austro-Hungarian rule. The first properties were located by a small group of aristocrats and wealthy entrepreneurs in towns with health resort function, such as: Krynica, Wisła and Zakopane. At the turn of the twentieth century, other forms of hospitality begun to develop, such as guest houses and bed & breakfasts. The first second homes, along with health resorts, became the nucleus of tourism development that took place in the Carpathians in the 1920s and 1930s (Jackowski 1989; Kowalczyk 1994). In the 1930s, recreational houses began to appear also in the vicinity of industrial centers located at the feet of the western part of the Beskid Mountains, which were Bielsko and Biała (Mika 2004). They mostly belonged to German and Jewish entrepreneurs from the textile industry. Before the World War II, second homes belonged to the elite and the properties took the form of non-urban mansions.

The phenomenon of second homes in Poland took on a common character only in the late 1960s and early 1970s, at the time of socialist economy and intense urbanization and industrialization of the country (Rogalewska 1980; Kowalczyk 1994). Due to the rapid increase in urban population, the social demand for various forms of recreation was growing. As traveling abroad was very limited by political impediments, Polish tourism between the 1950s and 1980s was dominated by a social model of organized domestic tourism, based on holiday resorts belonging to state companies. Second homes, known then as "summer houses", were complementary facilities and they were supposed to accommodate short-term recreation of inhabitants of industrial regions. There were formally designated areas in suburban zones of large cities and in other easily accessible rural regions, where city dwellers were able to lease and later buy plots for recreational purposes. These areas were released on lands that were not valuable for agriculture. The recreational plots had a small area, usually a few acres, which is why second homes built on them formed dense clusters of settlements (Rogalewska 1980). These objects were normally used only during summer season, most were made out of wood and were not equipped with water and sewage infrastructure.

Another though less frequent, way to acquire a second home in the past political system was to adapt inherited farms or abandoned rural buildings for recreational purposes. In the postwar period, during the intensive industrialization of the 1960s and 1970s, many Carpathian villages became strongly depopulated as a result of rural population immigrating to cities in search of work (Górka 1995). In addition, in the 1970s and 1980s, traditional pastoralism began to fade. Some of deserted farm houses and outbuildings, as well as shepherd's huts, were leased or purchased for recreational purposes. The scale of this process in the Carpathians was not as great as in the north-eastern Poland (Masuria) or in the Czech Republic, where, after the World War II, there were many buildings abandoned by the German population, which were then adapted for second homes (Kowalczyk 1994; Vagner and Fialova 2004). In the Polish Carpathians, adapted facilities accounted for only a few percent of all houses used for private recreational purposes (Rogalewska 1980).

In the 1980s, despite the deep economic crisis, second homes were developing at a rapid pace, and spatial planning institutions were not able to fully control this process (Kowalczyk 1994). Recreational cottages were built without legal permits, on private farmlands, not in specially designated areas. In Poland, where the post-war collectivization of agricultural land was only partial, the agriculture in mountainous areas was dominated by small private farms. Although in the socialist times there was no official free trade of land, the ambiguous law regulations made it possible to buy or lease land from farmers. This led to a chaotic and uncoordinated spread of second homes, which was repeatedly criticized in scientific literature, due to the excessive concentration, architectural forms, and the impact on the environment (Groch 1977; Hornung 1986).

The first owners or lessees of recreational land and second homes in the 1960s and 1970s were representatives of the political and military elite, managers of industrial plants and people with privileges, such as former soldiers of World War II (veterans). In the 1980s, the structure of the owners began to change, and people from other social groups started to play an increasingly important role—people like doctors, researchers and university teachers and staff of industrial plants. An important social factor in the development of second homes in the 1970s and 1980s was the development of individual automotive transportation (Szkup 2003).

Before 1989, the phenomenon of second homes in Poland was not as widespread as in other socialist countries, for instant the former Czechoslovakia, where individual recreational buildings were the main social model of domestic tourism (Turnock 1989; Havrlant 2004a; Vagner and Fialova 2004). The last published comprehensive data of 1980 indicated that there were 44,000 second homes in Poland. For comparison, in the 1980s, there were approximately 500,000 second homes registered in Czechoslovakia (Fialova et al. 2009). The Polish Carpathians had a low share of second homes, in 1980 there were only about 2,500–3,000 of them in the entire region (Rogalewska 1980). There were many more objects in suburban areas of Warszawa and Łódź (Matczak 1986; Kowalczyk 1994; Wiluś 1997; Wojciechowska 1998; Włodarczyk 1999; Szkup 2003). In the Carpathian region, recreational cottages were built in some villages in the Silesian Beskid Mts., the Żywiecki Beskid Mts., around the Lake Żywieckie and in the northern part of the Wyspowy Beskid Mts, that is in mountainous areas close to major urban centers. Some villages in the Podhale region, in the vicinity of the Tatra Mts., were also considered attractive location for recreational houses (Rogalewska 1980).

3 Present Development Factors

The political changes that took place in Poland after 1989 substantially changed the conditions for the development of settlements. A rapid growth in number of second homes in the Polish Carpathians started in the early 1990s, accompanied by numerous economic, social and environmental processes. The most important systemic changes that have affected the dynamics and nature of this phenomenon in the 1990s was the introduction of free trade of land, local government reform, which gave most powers to local authorities and the introduction of new regulations regarding land use planning and new building control (Mika and Faracik 2008).

The recent development of second homes in Poland is affected by a complex set of factors, resulting from both the demand and therefore needs and living conditions of the urban population, as well as from supply of land for development, which is the effect of the changing social and economic situation of the rural population (Table 1). The most important factors are:

- demographic and spatial development of cities, increase in the number of cars and growing difficulty of urban life,
- increase in wealth of the population, especially urban dwellers, who benefited from the positive effects of economic transition,
- mental fatigue of the economically active part of society and its growing needs for contact with nature,
- pursuance of social prestige given by own recreation house outside the city,
- family ties linking urban dwellers and rural population,
- financial needs of rural population (Mika and Faracik 2008).

Cities, especially agglomerations, have become in the 1990s and 2000s the cores of economic growth in Poland (Parysek 2005), so they have been attracting people who search for work and therefore growing in terms of population and area. According to 2009 data, the urban population accounts for 61.3 % (23.3 million) of all Polish citizens. However, 39 large cities with more than 100,000 people are

Table 1 Main factors for the second homes development in the Polish Carpathians after 1990

Supply factors	Demand factors
High attractiveness of Carpathians for tourism and recreation	Demographic and spatial development of cities
Close location of the urban centers and highly industrialized Upper Silesia region	Growing difficulty of urban life and growing needs for contact with nature
Negative changes in economic situation of the rural population after 1990 and growing financial needs	Increase in wealth of the urban population
Introduction of free trade in land	Social prestige given by own recreation house
Liberal law concerning settlement development	Rural origins of some of the urban population
Improving of the road system	Way of capital investment

populated by the total of 11.01 million, which represents 28.9 % of the Polish population. In 1990 an automotive boom has started, which has increased since then the negative effects of living in an urban environment and deepened the need for a weekend "escape" from the city.

Residents of large cities were the first to experience the positive results of economic transformation (Parysek 2005). Some groups began to obtain much higher incomes than the rest of society, through different ways, such as purchasing shares in privatized companies, finding better paid jobs and running their own businesses. Between 1995 and 2009 the average wage has increased significantly—by 365 %. This income growth applies mainly to the residents of Warszawa, where the GDP per capita is 3 times higher than the national average, Poznań (2 times higher) and Kraków (1.5 times higher). The purchase of land and a house in rural area came to be seen, in some social groups, as a form of capital investment and a sign of social prestige.

One of serious social consequences of transition that is relatively rarely emphasized, is the mental fatigue (often exhaustion) of economically active population—the result of irregular working hours and emotional overexertion. It causes a strong need for contact with nature. It is not only a significant factor of second homes development but also one of the major motives of traveling of Polish tourists.

Rural origins of the urban population and their family ties with villagers play also an important role in the development of second homes. A large part of the current urban population in Poland has emigrated from the countryside in search of work in the 1960s and 1970s, or are descendants of those immigrants (Górka 1995; Soja 2008). Not only do they maintain family ties with those living in villages, but have also inherited land in rural areas.

The phenomenon of second homes in the Polish Carpathians would not have taken such a mass character if not for financial needs of people living in this region. Residents of rural areas in Poland bore a significant social cost of economic transition in the 1990s. This problem was particularly noticeable in the Carpathians. Many families lost their economic livelihood. One way to raise funds for current needs was the sale of some land, often at a reduced price. Attractively priced land quickly found buyers, even in places difficult to access by car. Frequently, local population had put strong pressure on local authorities and spatial planning departments to designate more agricultural land for development. Such cases occurred in villages of the western part of the Beskidy Mts. (Silesian Beskid, Żywiecki Beskid), which were affected by high unemployment in the mid-1990s due to the closure of coal mines in Upper Silesia (Mika 2001). At the turn of the twenty-first century, this practice was limited by the development of small businesses and the service sector, including tourism.

4 Main Directions of Spatial Development

Currently second homes in Poland are not subject to registration, not even in recreational areas designated in spatial planning documents. Therefore, it is impossible to give exact numbers of these objects. According to previous studies on the functional transformation of tourist destinations in the Polish Carpathians (Mika 2004; Faracik 2006; Szpara 2006; Bielak 2009; Zawilińska 2010), it can be estimated that in the first years of the twenty-first century, there were about 15,000 individual recreational houses. In the period 2000–2010, the Carpathian region was under high development pressure, so it can be assumed that the number of second homes currently exceeds 20,000.

Second homes in the Carpathian Mountains are not distributed evenly (Fig. 1). Differentiating factors for their occurrence are:

- accessibility from urban centers and course of major transportation routes,
- location of important tourist centers, including health resorts,
- occurrence of water reservoirs on Carpathian rivers.

The regions with the highest concentration of second homes are: the Silesian Beskid Mts. (about 4,000 objects), the Wyspowy Beskid Mts. (about 4,000 objects), the Żywiecki Beskid Mts., the Gorce Mts. and their surroundings (about 1,400 objects), as well as the Sądecki Beskid Mts. A large concentration of such objects can be found around artificial water reservoirs—Lake Rożnowskie, Lake Żywieckie, Lake Dobczyckie. An important area of development of second homes is also the valley of the Raba river, particularly small towns Myślenice and Pcim located within the mountainous part of the suburban area of Kraków, where there

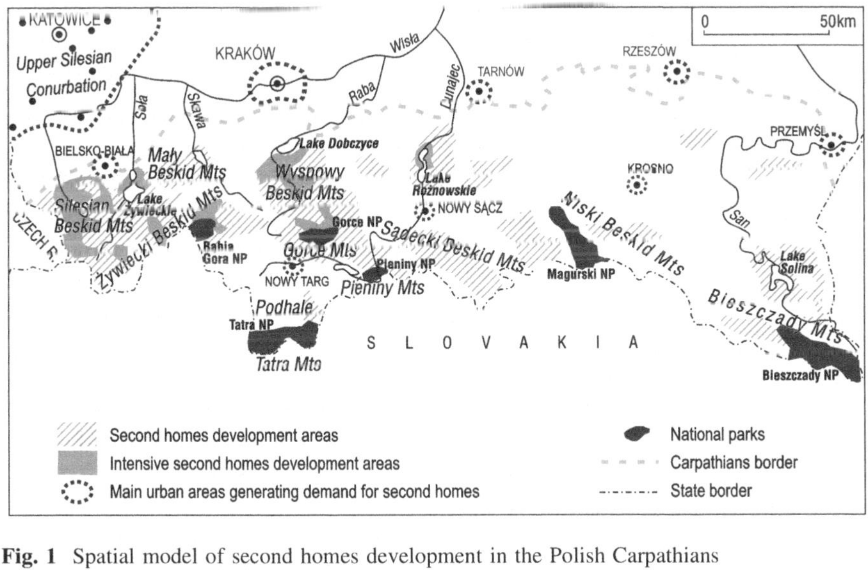

Fig. 1 Spatial model of second homes development in the Polish Carpathians

are approximately 2,500 of these objects (Faracik 2006). In the past few years, there has been a growing interest in purchase of land in the Bieszczady Mts., especially in areas around Lake Solina, in the Niski Beskid Mts. and in foothills areas near Rzeszów and Przemyśl (Fig. 1). Very attractive areas for second home location are lands in the vicinity of national parks, such as the Babia Góra National Park (Faracik et al. 2008) and the Gorce National Park (Petko 2011), where some buildings were created close to their borders.

Second homes are built both in villages and in small towns of the Carpathian region. Their number in different localities varies from a few to over a thousand, for example villages Brenna in the Silesian Beskid Mts. (1,200 second homes) and Gródek nad Dunajcem near Lake Rożnowskie (1,000). In some localities the number of these objects is equal to or even exceeds the number of permanent homes (e.g. Brenna village).

The owners of second homes in the Polish Carpathians are primarily residents of urban agglomerations located in the foothills of the Carpathians: Katowice conurbation of Upper Silesia (3.4 million inhabitants) and metropolitan area of Kraków with its satellite cities (1.45 million), as well as larger cities: Bielsko-Biała (0.18 million), Tarnów (0.12 million) and Rzeszów (0.17 million). The largest investment pressure comes from a highly industrialized area of Upper Silesia, where the average population density is about 1,900 people per km^2. The inhabitants of this region own second homes mainly in the western part of the Beskidy Mts. (Silesian Beskid, Żywiecki Beskid, Lake Żywieckie) due to their proximity and short travel time (up to 1.5 h). Similarly, the inhabitants of Kraków agglomeration locate their recreational houses in nearby mountainous areas and in the valleys of the Raba and Dunajec rivers. Due to the nationwide status of the Carpathian Mountains as a tourist destination, there are also second homes in this area owned by people from other parts of the country, particularly from central Poland (e.g. Warszawa). In the last few years, there is a growing demand for second homes generated by residents of cities located within the Carpathian Mountains, such as: Nowy Sącz, Nowy Targ, Krosno and Żywiec.

The spatial extent of recreational housing development in the Carpathians is still growing. In the 1990s, the greatest rate of growth was reported in renowned tourist resorts, such as Szczyrk, Wisła, Ustroń and in villages with recreational traditions (Mika 2004; Kurek 2008). But then, in years 1995–2005, prices of development land significantly increased in these localities. The increase was 5–7-fold, and in places up to 15-fold. For example, in 1995, 1 m^2 of land for a second home in the western part of the Beskidy Mts. (e.g. Brenna, Wisła, the area around Lake Żywieckie) could be bought for a price from 12 to 15 Polish złoty (about 4–5 USD). Now, land in these areas is rarely sold for less than 100 Polish złoty (about 33 USD) per 1 m^2, and in the most attractive locations over 250 Polish złoty (about 83 USD). Due to a gradual saturation of the Polish Carpathians with second homes and increase in land prices, potential investors are constantly looking for new, poorly developed and attractive locations. The condition of local roads has also improved which enabled an easier penetration of the mountain area, and thus extended potential areas for development of second homes.

5 Location Models of Second Homes

Permanent settlement in the Polish part of the Carpathians is highly fragmented due to historical and social circumstances of its development (Górka 1995). Characteristic elements of the Carpathian villages are numerous scattered hamlets and isolated farms in remote and high parts of the valleys and in forest clearings. Dispersion of farms coincides with the fragmentation of agricultural lands, as well as a significant share of private forests. Location of second homes corresponds with landscape fragmentation, which makes their dispersion even more of an issue.

The current law on the development of settlement gives local authorities several measures to decide about excluding land with poor soils from agricultural use and to supply it for development. Soils in mountainous areas have a low productivity value, which is why, at the request of their owners, agricultural lands may be made eligible for parceling and later sold as building plots. So today the only de facto barrier to the development of settlements in the Carpathian Mountains, besides the areas of national parks, are the boundaries of forest properties.

Liberal law and dispersion of settlements result in a relatively high degree of freedom for potential investors to choose the location place. This is the reason why contemporary characteristics of locations of second homes is largely a reflection of social needs of their owners and clearly refers to the location model of tourism and recreation by Christaller (1933)—the peripheral areas theory.

Since the 1990s, second homes are usually erected outside the permanent housing areas, in places that offer tranquility and close contact with nature. The special character of second homes location can be indicated by the fact that many buildings are located in places where permanent houses would not have been built, such as on steep slopes, very close to water bodies, in areas difficult to reach, and in a considerable distance from residential areas. Based on previous studies (Gaś 1997; Mika 2004; Faracik 2006; Petko 2011) typical locations of second homes in the Polish Carpathians can be indicated (Fig. 2):

- away from permanent settlement, on slopes, in the immediate vicinity of the forest (A),
- alongside valleys and along roads on slopes (B),
- in permanent forest clearings (C),
- in the immediate vicinity of streams and rivers (D),
- directly at water reservoirs (E) and on slopes overlooking water bodies (F),
- within permanent settlement zones (G) or in hamlets (H),
- between villages or between the village and its hamlets (I),
- at the edge of permanent settlement zones (J),
- in remote places at the end of valleys (L) and in high altitude mountain areas (K).

The most attractive location places of second homes offer scenic views, proximity to forest, lakes and streams (types: A, B, C, D, E, F). For example, in the Silesian Beskid Mts. the vast majority of objects (96.6 %) are situated on slopes

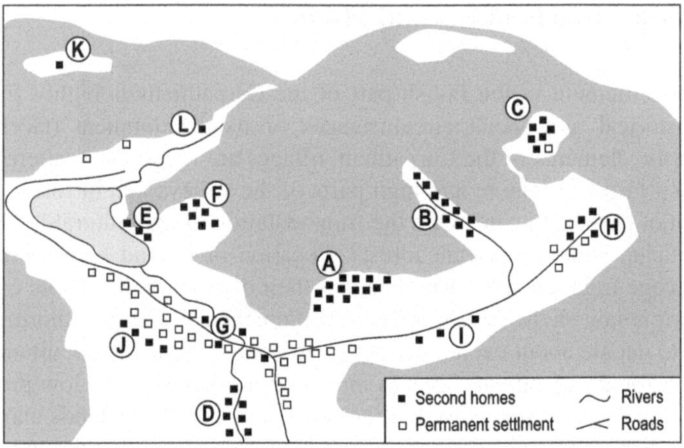

Fig. 2 Second homes location types in the Polish Carpathians (*explanation in the text*)

and up to 15.5 % of them lie within 100 m from the forest (Mika 2004). Second homes are also built on slopes with a significant inclination (40°). Figure 3 is the illustration of these trends in Brenna in the Silesian Beskid Mts., one of the most urbanized villages in the Polish Carpathians, a result of second homes development (1,200 objects). There are also cases when second homes are present within national parks borders, for example in the Gorce National Park. This is not the result of illegal building but the effect of adaptation of former shepherd huts for recreation (K). In the western part of the Beskidy Mts. forest clearings, once only with isolated farms (C), are being also rapidly built upon.

Despite the fact that second homes are no longer built in specially designated areas, they still have a tendency to be concentrated in some areas, and in many cases they form clusters in the landscape, isolated from permanent settlements. This situation is the result of several overlapping reasons, which are:

- selling a few or several contiguous parcels by the owner, who acts upon the principle of maximizing profits with minimum loss of his land,
- buying adjoining parcels by their prospective owners because of attractive landscape, proximity to forest and water bodies, but also in order to reduce the cost of construction, to have common access roads, common water intake and access to electricity sources. Sometimes the reasons for this situation are social, such as the formation of the neighboring communities of specific social groups (e.g. artists and scientists),
- settlement concentration as a result of fragmentation of recreational land—when owners of second homes resell parts of their land (Mika 2004; Mika and Faracik 2008).

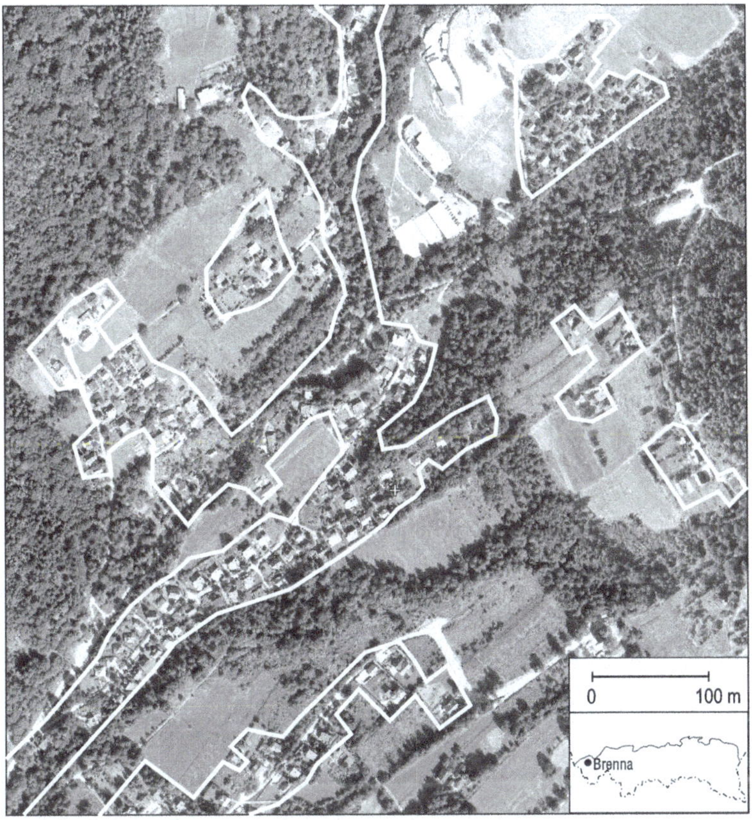

Fig. 3 Directions of second homes development in the Jatny valley, Brenna in the Silesian Beskid Mts. The second homes are marked with the *white outline* and represent the location types "B" and "C" shown in Fig. 2

6 Impacts and Conflicts

One of the most visible consequences of the development of second homes in the Polish Carpathians is the urbanization of rural areas. Since 1980s, the rapid and uncontrolled development of this kind of settlement has caused substantial changes in land use. In some areas, "recreational agglomerations" arose in the landscape, with tens or even hundreds of buildings. Creation of second homes has caused a change in spatial and morphological patterns of settlement from rural to urban (density of architecture, the layout of buildings and a dense network of access roads). As the result, the mountain scenery and countryside landscape became urbanized and in some places heavily urbanized. Most second homes built in the past several years are year-round buildings and are erected on permanent foundation. Based on detailed research in the Silesian Beskid Mts. (Mika 2004), it has been estimated that the erection of one such house caused permanent transformation of 385 m^2 of biologically active land with high recreational values. The trend towards dispersion of second homes is stimulating the development of permanent settlements, which creates a synergistic effect of urbanization.

The development of second homes strongly impacts the quality of the natural environment and its functioning. Households situated in high parts of the valleys or in a proximity to water reservoirs cause bacteriological pollution of water bodies and, locally, of water sources. Due to financial shortcomings of local authorities, the rapid development of recreational housing estates is not accompanied by the development of water and sewage infrastructure. Although modern buildings are already equipped with household sewage treatment, there is still a lot of waste mismanagement in case of houses created in the past.

The issues of architectural aesthetics and the impact of dense development on the harmony and spatial arrangement of a mountain landscape are also significant. Regulations regarding the form of buildings were introduced only in the construction law of 1994, which indicated that the objects created in mountainous regions should refer to the traditional character of settlements, i.e. should have at least a gabled roof (Mika 2004). Other regulations were also introduced, such as those regarding the distance that must be maintained between the building and the plot border, so that the distance between houses can not now be less than 8 m. Second homes built before those regulations often do not meet those architectural requirements and form clusters of chaotic and unsightly buildings where the distance between houses is sometimes less than 4 m (Fig. 3).

Second homes from the 1970s and 1980s currently undergo serious qualitative transformations. Originally wooden objects are in many cases rebuilt in brick, and in this process, their usable area is greatly increased. This process is particularly spectacular in the suburban area of Kraków, where, in the 1990s, in recreation areas, and even in allotment gardens, brick second houses have been illegally built, which eventually became the first houses. The lack of legal regulations on the use of recreational areas at this time was the reason, why areas with the originally recreational purpose have been built upon with large permanent housing estates. It proves how strongly the phenomenon of second homes is associated with the processes of suburbanization in the vicinity of large cities (Faracik 2006).

The second homes stampede is considered a source of environmental and functional conflicts. This phenomenon is negatively evaluated by those involved in efforts to protect the environment and by foresters, who emphasize the excessive and chaotic pace of urbanization, the rapid loss of biologically active areas and the impacts it has on fauna and flora of forests (Faracik et al. 2008). There is a good reason to fear that the urbanization of the mountains may lead to a critical decrease of their value for tourism. Tourism is now an important factor supporting the economic development of some of the Carpathian villages, and the loss of space reduces their attractiveness for different forms of close to nature tourism. This conflict also has a social dimension because commercial mass tourism causes noise, and one of the main reasons for building second homes is to satisfy the need for peace and quiet. Both phenomena (mass tourism and second homes) overlap and intermingle in all tourist resorts in the Carpathians, which leads to a clear functional conflicts, and in time could weaken their competitiveness (Kurek and Mika 2008; Mika 2010), and exacerbate social conflicts.

7 Conclusions

The development of second homes in the Polish Carpathians is different than the one in the mountains of Czech Republic and Slovakia, where large-scale objects have been created at the time of socialism (1960–1980s). These objects currently are in the process of modernization and functional transformation (Havrlant 2004b, 2007; Fialova et al. 2009). In Poland, a boom in second homes development took place only after 1989, as a result of economic and social changes.

Second homes are nowadays an important factor in functional transformation of rural space in the Polish Carpathians. As a result of growing urban development and increased affluence of Polish society, it is constantly evolving. Being a dynamic form of the settlement, it contributes to permanent development and urbanization of attractive tourist areas in the mountains. Since 1990s, the development of second homes in the Carpathians is mainly regulated by market mechanisms of real estate, and in terms of natural conditions, limited by boundaries of national parks and forests. Construction of second homes raises much controversy because of the intensity and their location, which are, in many cases, incompatible with the needs of nature conservation and the need to protect the space for the further development of tourism function.

The presented image of spatial development of second homes in the Polish Carpathians has some characteristic features, which are:

- dynamism and spontaneity,
- strong spatial concentration, especially in areas of high natural and landscape values,
- tendency to spatial isolation from permanent settlement,
- significant consumption of space,
- the permanent nature of buildings.

The effects of rapid development of second homes include also a number of transformations in the social and economic situation in rural areas. These changes however, make up a distinct range of problems, such as the relationships between immigrant population and permanent residents, city residents moving permanently to villages, the impact of second homes on the stimulation of rural enterprise, spread of innovation, or rental of second homes for commercial purposes, which causes them to become competition for the villagers providing tourism services, such as renting rooms to tourists or running tourist farms (agritourism).

The spread of second homes is one of the symptoms of the suburbanization process of cities located in the Carpathian foreland areas. Gradually, these objects cease the recreation function, becoming the first homes. This process has clearly started already the end of the 1990s in the Silesian Beskid Mts. and the northern part of the Beskid Wyspowy Mts. Furthermore, it is not confined to areas close to the conurbation of Upper Silesia and to Kraków agglomeration, reaching ever further into the Carpathians as a result of improved accessibility.

The presented models of the location of second homes in the Polish Carpathians confirm development trends, as well as the motives of city residents for choosing a place to build their second homes, reported in Western European countries. The specific Polish character of this phenomenon is emphasized by the fact, that in the 1980s and 1990s it was not controlled by the institutions responsible for development and spatial planning, taking on a very spontaneous and chaotic form. In some localities it has led to their excessive urbanization, to decline in natural values and to an irreversible loss of traditional agricultural functions.

References

Arnstberg KO, Bergstrom I (2008) No place like second home: weekends, holidays, retirement and urban sprawl. In: Couch Ch, Petschel-Held G, Leontidou L (eds) Urban sprawl in Europe: landscape, land-use change and policy. Wiley-Blackwell, Oxford

Barke M (2008) Second homes in the Spanish housing market: one market or two? J Hous Built Environ 23(4):277–295

Bielak B (2009) Funkcja turystyczna zbiorników zaporowych w otoczenia w Karpatach Polskich. Instytut Geografii i Gospodarki Przestrzennej UJ, Kraków

Cho S-H, Newman DH, Wear DN (2003) Impacts of second home development on housing prices in the southern Appalachian Highlands. Rev Urban Reg Dev Stud 15(3):208–225

Christaller W (1933) Die zentralen Orte in Süddeutschland. Gustav Fischer, Jena

Coppock JT (1977) Second homes: curse or blessing? Oxford Press, Oxford

Dijst M, Lanzendorf M, Barendregt A, Smit L (2005) Second homes in Germany and the Netherlands: ownership and travel impact explained. Tijdschr Econ Soc Geogr 96(2):139–152

Dziegieć E (1995) Urbanizacja turystyczna terenów wiejskich w Polsce. Turyzm 5(1):5–56

Faracik R (2006) Turystyka w strefie podmiejskiej Krakowa. Instytut Geografii i Gospodarki Przestrzennej UJ, Kraków

Faracik R, Kurek W, Mika M, Pawlusiński R (2009) Turystyka w Karpatach Polskich w świetle współczesnych kierunków rozwoju. In: Domański B, Kurek W (eds) Gospodarka i przestrzeń. Instytut Geografii i Gospodarki Przestrzennej UJ, Kraków

Faracik R, Mika M, Pawlusiński R (2008) Plan rozwoju i zarządzania turystyką w regionie Babiej Góry. Instytut Geografii i Gospodarki Przestrzennej UJ, Kraków

Fialova D, Kadlecowa V, Nozickova V, Vagner J (2009) Changes in second housing in the integrating Europe. In: Wilk W (ed) Global changes: their regional and local aspects. Uniwersytet Warszawski, Warszawa

Gallent N, Mace A, Tewdwr-Jones M (2005) Second homes: European perspectives and UK policies. Ashgate Publ, Aldershot

Gaś M (1997) Drugie domy jako element zagospodarowania turystycznego wsi w strefie podmiejskiej Krakowa. Folia Geogr Ser Geogr-Oecon 17:29–30

Górka Z (1995) Osadnictwo. In: Warszyńska J (ed) Karpaty Polskie: Przyroda, człowiek i jego działalność. Uniwersytet Jagielloński, Kraków

Górz B (2003) Społeczeństwo i gospodarka Podhala w okresie transformacji. Wyd. Nauk Akademii Pedagogicznej w Krakowie, Kraków

Groch J (1977) Letnie domki na Zarabiu jako przykład budownictwa rekreacyjnego dla mieszkańców Krakowa. Folia Geogr Ser Geogr-Oecon 10:17–26

Haldrup M (2004) Laid-back mobilities: second-home holidays in time and space. Tour Geogr 6(4):434–454
Hall CM, Müller DK (eds) (2004) Tourism, mobility and second homes: between elite landscape and common ground. Channel View Publications, Clevedon
Havrlant J (2004a) Vyvoj druheho bydleni v Moravskosleskych Beskydech. In: Vagner J, Fialova D (eds) Regionalni diferenciace druheho bydleni v Cesku. Univer. Karlova v Praze, Prague
Havrlant J (2004b) Changes in the use of second housing facilities in the Moravian-Silesian Beskids Mts. Morav Geogr Rep 12(2):24–34
Havrlant J (2007) Współczesne trendy w rozwoju funkcji turystyczno-rekreacyjnej Beskidów Czeskich. In: Kurek W, Mika M (ed) Studia nad turystyką. Tradycje, stan obecny i perspektywy badawcze. Instytut Geografii i Gospodarki Przestrzennej UJ, Kraków
Hidle K, Ellingsen W, Cruickshank J (2010) Political conceptions of second home mobility. Sociol Rural 50(2):139–155
Hoogendoorn G, Mellett R, Visser G (2005) Second homes tourism in Africa. Reflections on the South African experience. Urban Forum 16(2–3):112–154
Hoogendoorn G, Visser G, Marais L (2009) Changing countrysides, changing villages: Second homes in Rhodes, South Africa. S Afr Geogr J 91(2):75–83
Hornung M (1986) Główne problemy przestrzenne budownictwa letniskowego na terenach górskich województwa bielskiego. Zesz Nauk AWF w Krakowie 37:56–80
Jackowski A (1989) History of tourism in Polish Carpathians. Zesz Nauk UJ Pr Geogr 80:75–83
Kowalczyk A (1994) Geograficzno społeczne problemy zjawiska drugich domów. Uniwersytet Warszawski, Warszawa
Kuentzel WF, Ramaswamy VM (2005) Tourism and amenity migration: a longitudinal analysis. Ann Tour Res 32(2):419–438
Kurek W (1996) Agriculture versus tourism in rural areas of the Polish Carpathians. Geojournal 38(2):191–196
Kurek W (2004) Turystyka na obszarach górskich Europy. Instytut Geografii i Gospodarki Przestrzennej UJ, Kraków
Kurek W (2008) Tourism in Polish Carpathians in the political transition time. In: Wyrzykowski J (ed) Conditions of the foreign tourist development in central and Eastern Europe: tourism in geographical environment. Uniwersytet Wrocławski, Wrocław
Kurek W, Mika M (2008) Miejscowości turystyczne w dobie konkurencji (na przykładzie Polskich Karpat. In: Gołembski G (ed) Turystyka jako czynnik wzrostu konkurencyjności regionów w dobie globalizacji. Wyd. AE Poznań, Poznań
Marjavaara R (2007) The displacement myth: second home tourism in the Stockholm Archipelago. Tour Geogr 9(3):296–317
Matczak A (1986) Budownictwo letniskowe w strefie podmiejskiej Łodzi. Acta Univ Lodz Folia Geogr 7:37–165
Meyer-Arendt KJ (2001) Recreational development and shoreline modification along the north coast of Yucatán, Mexico. Tour Geogr 3(1):87–104
Mika M (1997) Drugie domy w Beskidzie Śląskim. Turyzm 7(1):25–42
Mika M (2001) Współczesne przemiany i zagrożenia środowiska przyrodniczego Beskidu Śląskiego związane z turystyką. In: German K, Balon J (eds) Przemiany środowiska przyrodniczego i jego funkcjonowanie. Problemy Ekologii Krajobrazu, Kraków
Mika M (2004) Turystyka a przemiany środowiska przyrodniczego Beskidu Śląskiego. Instytut Geografii i Gospodarki Przestrzennej UJ, Kraków
Mika M (2010) Regional and cross-border competition as a development factor in tourist reception areas in the Western Carpathians. In: Richard D (ed) Development durable des territories: de la mobilisation des acteurs aux demarches pearticipatives. Universitarires Blaise Pascal, Clermont-Ferrand
Mika M, Faracik R (2008) Second homes as a factor of the transformation of rural areas in the Polish Carpathians. Folia Geogr Acta Fac Stud Humanit Nat Univ Prešoviensis 12:245–255
Moss LAG (ed) (2006) The amenity migrants: seeking and sustaining mountains and their culture. CAB Int, Wallingford

Módenes J-A, López-Colás J (2007) Second homes and compact cities in Spain: two elements of the same system? Tijdschr Econ Soc Geogr 98(3):325–335
Norris M, Winston N (2010) Second-home owners: escaping, investing or retiring? Tour Geogr 12(4):546–567
Parysek J (2005) Miasta polskie na przełomie XX i XXI w. Rozwój i przekształcenia strukturalne. Bogucki Wyd Nauk, Poznań
Petko J (2011) Drugie domy jako element zagospodarowania turystycznego Gorców, Dissertation. Instytut Geografii i Gospodarki Przestrzennej UJ, Kraków
Pitkänen K (2008) Second-home landscape: the meaning(s) of landscape for second-home tourism in Finnish Lakeland. Tour Geogr 10(2):169–192
Ptaszycka-Jackowska D (2007) Gospodarowanie przestrzenią turystyczną w Polskich Karpatach. Pr Geogr IGiGP UJ 117:99–112
Ptaszycka-Jackowska D, Baranowska-Janota M (2003) Tourism within the Polish and Slovak transfrontier region. Pr Geogr IGiGP UJ 111:43–60
Rogalewska B (1980) O strukturze przestrzennej budownictwa letniskowego w Polsce. Przegl Geogr 3:575–582
Soja M (2008) Cykle rozwoju ludności Karpat Polskich w XIX i XX wieku. Instytut Geografii i Gospodarki Przestrzennej UJ, Kraków
Szkup R (2003) Kształtowanie podmiejskiej przestrzeni wypoczynkowej. Przykład zachodniego sektora strefy podmiejskiej Łodzi. Wyd. Uniwersytetu Łódzkiego, Łódź
Szpara K (2006) Turystyka wiejska w Bieszczadach Wysokich. Dissertation, Instytut Geografii i Gospodarki Przestrzennej UJ, Kraków
Tress G (2007) Seasonality of second-home use in Denmark. Landscape Ser 1(7):151–179
Turnock D (1989) The human geography of Eastern Europe. Routledge, London
Vagner J, Fialova D (2004) Regionalni diferenciace druheho bydleni v Cesku. Univer Karlova v Praze, Prague
Warszyńska J (ed) (1995) Karpaty Polskie Przyroda, człowiek i jego działalność. Uniwersytet Jagielloński, Kraków
Wiluś R (1997) Rozwój funkcji turystycznej w dolinie rzeki Warty na odcinku od Działoszyna do Uniejowa. Łódzkie Tow, Nauk, Łódź
Włodarczyk B (1999) Przemiany form aktywności turystycznej. Przykład krawędzi Wyżyny Łódzkiej. Łódzkie Tow, Nauk, Łódź
Wojciechowska J (1998) Kolonizacja turystyczna terenów nadpilicznych. Łódzkie Tow Nauk, Łódź
Zawilińska B (2010) Możliwości rozwoju turystyki w parkach krajobrazowych Karpat Polskich w świetle idei zrównoważonego rozwoju. Uniwersytet Ekonomiczny, Kraków

Large Dominant Enterprises Versus Development of Rural Mountainous Areas: The Case of the Polish Carpathian Communes

Magdalena Dej and Grzegorz Micek

Abstract This chapter focuses on selected aspects of local impact of dominant enterprises in nonmetropolitan rural areas in the Polish Carpathian Mountains (Carpathians). It has been shown that companies localized in the Carpathians and possessing a dominant share in local labor markets do not possess such positions in the sphere of their local impact. It has also been argued that Carpathian communes where large enterprises are present differ significantly from each other from the point of view of their levels of development. The research has confirmed that dominant enterprises do not exert a significant local impact, as they are focused on taking advantage of a cheap, local labor force. The influence of a large company depends on the specifics of its activity and the features of the community.

1 Introduction

The agricultural- and tourism-oriented economic specialization of mountainous rural areas may suggest that the prevailing majority of them should represent a balanced dimensional structure of enterprises and a lack of companies dominating the local economies. However, as it has already been shown at the first stage of research, a significant number of large companies operate in rural areas. Hence, by analogy to similar research conducted in small and medium urban centers (Sobala-Gwosdz 2000; Domański and Gwosdz 2005), our study analyzes the influence of large companies on their local environment, communities, and economy in mountainous areas.

It is broadly assumed that the multifunctional character of a specified area may act as a distinguisher of its success (Heffner 2007; Bański 2008). However, a

M. Dej · G. Micek (✉)
Institute of Geography and Spatial Management, Jagiellonian University, Gronostajowa 7, 30-387 Kraków, Poland
e-mail: grzegorz.micek@uj.edu.pl

J. Kozak et al. (eds.), *The Carpathians: Integrating Nature and Society Towards Sustainability*, Environmental Science and Engineering,
DOI: 10.1007/978-3-642-12725-0_36, © Springer-Verlag Berlin Heidelberg 2013

contradictory question can be asked: to what extent does the mono-functional character of a stated community, expressed indirectly by the domination of a single company, influence local development? The purpose of this study was to determine the level and scope of the local influence of large enterprises in nonmetropolitan rural areas in the Polish Carpathian Mountains. The analysis of the influence was done in two stages. First, we assessed how the communes where dominant enterprises are present differ from other rural areas of the Carpathians. In the next stage, based on case studies of three communes, the perceived local impact of selected companies was analyzed.

1.1 Previous Research on Entrepreneurship in Rural Areas

Deliberations on the development of rural areas presently represent one of the main trends of the geography of rural areas and are inseparably linked to the issue of multifunctional development. In such development, the role of the major stimulus is assigned to local entrepreneurship, expressed by the performance of small or medium-sized companies. The role of large companies active in rural areas is practically ignored.

Simultaneously, we want to emphasize that deliberations on entrepreneurship in rural areas are not a completely new trend. In global literature, the first studies dealing with the problem of entrepreneurship in rural areas are represented by works concerning industrial counter-urbanization (Berry 1976; Keeble 1989) and nonmetropolitan industrialization (Lonsdale and Seyler 1979). Many chapters have also been devoted to the analysis of city-country (urban-rural) shifts, primarily regarded as kinds of a manufacturing shift (Keeble 1980, 1984; Fothergill et al. 1985; Keeble and Tyler 1995). The discussion on a city-countryside shift is also present in the Polish literature (Grzeszczak 1998), but to a limited degree.

A significant set of works connected with the analyzed issue are those that describe the regions of success and the problem regions in Polish rural areas, such as primarily the research devoted to the identification and typology of rural problem areas (Kulikowski 1992; Bański 1999; Rosner 1999, 2002). Research on the activation of rural areas (Kłodziński 2002, 2006) also corresponds to this problem.

In this chapter, particular emphasis shall be placed on the multifunctional development of rural areas as discussed in the studies mentioned above. This development is mostly regarded as an activation of rural areas via stimulating small enterprises. This problem has been undertaken very frequently in recent times (Heffner 2000; Stasiak 2000). It is characteristic that even these works that are aimed at analyzing rural entrepreneurship focus almost exclusively on the sector of small and medium enterprises (Hałasiewicz and Kaleta 2000; Kłodziński 2000; Kłodziński and Fedyszak-Radziejowska 2002).

1.2 The Forms of Local Impact of Large Companies

The local influence of big companies in Poland has, so far, belonged to the domain of economies of small and middle-sized towns. These centers, previously dependent on huge factories (Domański 1992, 1997) and now facing the threat of losing markets, have encountered the necessity of restructuring employment. Some of the Polish small and medium-sized towns (e.g. Mielec) have been successful with the problem of a bankrupt factory by diverting new investments and urgently introducing proinvestment instruments (Domański and Gwosdz 2005). Most of the mono-functional urban centers, however, have closed or are still passing through a huge crisis, as the restructuring actions undertaken have resulted in failure (Gorzelak et al. 2007).

The local impact of these large companies is a problem that shall be considered from multiple points of view in this chapter. This influence can potentially interfere with several spheres, including economic as well as social and cultural ones; it can also influence local spatial management.

When analyzing the influence of dominant enterprises on the economic sphere, one usually emphasizes their performance in local labour markets. This dimension has been thoroughly studied by Stryjakiewicz (2004) in the case of the operations of a foreign pharmaceutical enterprise company (GlaxoSmithKline) in Poznań. On one hand, dominant companies offer significant numbers of jobs, thus helping to improve the financial standing of the inhabitants, reducing unemployment (also hidden unemployment), and supporting the labor activation of local populations. On the other hand, due to little competition from other companies, they can easily set the rules, which usually means dictating that the employees have relatively poor conditions of employment. From an economic point of view, we shall also discuss the influence of dominant companies on local entrepreneurship. The influence on the local economy can be stimulated by the existing factory, due to the multiplier effect, resulting from increased supply and consumer demand. On the other hand, in the case of a large company that does not cooperate with local suppliers and is based only on the local labor force, this influence can result in restricting the entrepreneurship. Establishing a large company contributes also to the revenues of the community which can be defined as an indisputably positive influence, including revenues derived from real estate taxes and a local share in personal (PIT) and corporate income tax (CIT).

Of course, there are also other areas in which the companies influence the local milieu, such as the social and cultural dimensions (e.g., the influence of the factory on migration tendencies and population structures, education aspirations of the youth, local lifestyle, social activity or mutual relations of the inhabitants) or spatial management (e.g. the influence of the company on land use, quality and quantity of the elements of technical infrastructure, changes in the physiognomy of the surroundings). There is a scarcity of chapters that use multidimensional local impact analysis to investigate a single plant. However, some research has already been carried out on the influence of a group of factories (Domański 2001; Domański and Gwosdz 2005). The relationships of foreign investors with local authorities have been studied by Dziemianowicz (1999, 2007).

1.3 Terminology of the Research

The northern border of the Carpathians has been defined by Balon et al. (1995). Other borders of the study area follow the Poland–Slovakia and Poland–Ukraine borders. The term "nonmetropolitan rural communes" refers to all communes with administrative status of the rural entities, which have not been located within the borders of the Kraków Metropolitan Area or within a distance of less than 8 km from other cities of more than 50,000 inhabitants. Subsequently, 123 communities located in the Polish Carpathians have been identified, 22 of which have possessed dominant enterprises in their areas (Fig. 1).

A dominant enterprise has been defined as a company characterized by the employment of at least 100 persons by the end of 2008. Simultaneously, it has been assumed that a dominant enterprise in a given commune must employ at least 1.5 times more people than the next largest company. Applying more radical boundary conditions (e.g. prevailing over five times in the area of employment) would result in limiting the set of communes with dominant enterprises to only a few (precisely: two) administrative entities in the area of the Polish Carpathians.

1.4 Dominant Enterprises in the Polish Carpathians

The Polish Carpathians have always been regarded as a poorly developed region in comparison to other parts of the country (Adamus and Luchter 1995). After a period of rapid growth, based on the exploitation and processing of crude oil at the

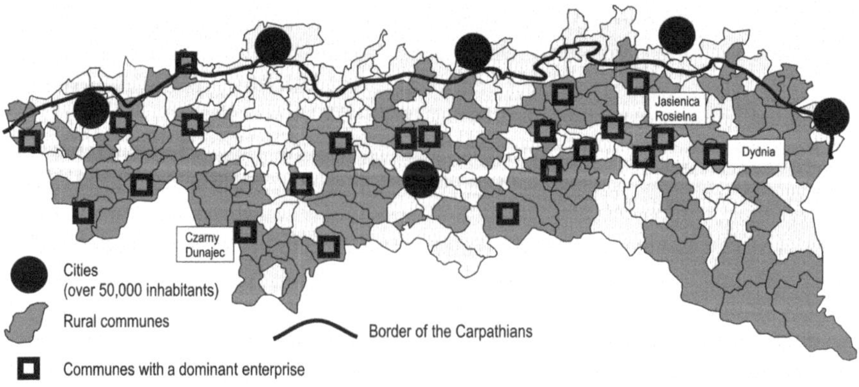

Fig. 1 Rural communes in the Polish Carpathian Mountains

turn of the nineteenth and twentieth centuries, a period of stagnation occurred after World War II. During the economy of the socialist period (1945–1989), the Carpathians became a region almost forsaken by central government administration and, hence, strongly underinvested. New, large-scale industrial investments omitted the Carpathians both because of a concentration of efforts in other regions of the country (which occurred in the latter years of the period) and due to protection of natural areas. Only some mass commuting from rural areas to towns and the development of tourist services in some rural areas was observed. It is surprising that the period of economic transformation at the turn of the twentieth and twenty-first centuries also did not bring about any spectacular investments in the Carpathians. An exemplification of a lack of interest by external investors may be due to the fact that between 1995 and 2008, none of the foreign enterprises invested more than 2 million USD in the Małopolskie part of the Polish Carpathians (based on Domański et al. 2009). Nonetheless, there are now 22 enterprises dominating local markets in the Polish Carpathians, and the prevailing majority of them are local companies; over three-fourths of them are owned by private Polish entrepreneurs. Over half of the dominant companies were established after 1989, while only eight during the time period of the prior economy. Hence, it can be clearly stated that these are relatively young companies that dominate local markets. Among the 22 companies analyzed, 12 are in manufacturing, 5 of which represent the food industry. There are not many dominant service companies (5 altogether).

2 Research Methods

To determine the scale of influence of large companies on the local communities it was necessary to first carry out a statistical comparative analysis. This allowed us to compare the areas of domination of large companies with other rural, nonmetropolitan areas from the point of view of various indicators of local development (Table 1). Both the level and the dynamics of development have been analyzed.

The selection of variables has resulted from prior determination of the dimensions of local development. The following kinds of variables have been studied: demography (population growth), infrastructure (represented by the percentage of inhabitants with access to water supply and sewage systems), socio-economic circumstances (entrepreneurship level, percentage of commuters, unemployment), standard of living (communes' revenues from PIT and CIT, new housing) and local government performance (communes' own revenues, gaining European Union (EU) funds). When selecting the variables, three coefficients applied from Bański (2008) were used to describe the success of rural areas: the communes' population (representing the demographic potential of the area), the number of businesses registered within the rural private sector, and the communes' own revenues.

Table 1 Index of local development (LDI) and its variables in two groups of communes in the Polish Carpathians

Variable	Communes without a dominant enterprise	Communes with a dominant enterprise	Significance (Levene's test for equality of variances)	Significance (*t* test equality of means)	Years
Number of inhabitants	7,782	9,559	0.76	0.04	2008
Unemployment rate (%)	7.82	7.89	0.51	0.95	2008
Number of incoming commuters per 1,000 inhabitants	18.02	34.00	0.00	0.00	2006
Number of private companies per 1,000 inhabitants	57.22	57.58	0.66	0.94	2008
Amount of a commune's own revenues per 1 inhabitant	714.30	809.07	0.69	0.26	2008
Amount of CIT and PIT[a] settled per 1 inhabitant	266.31	294.73	0.62	0.28	2008
Amount of EU funds (investment funds) acquired per 1 inhabitant	118.60	244.87	0.02	0.07	2006–2008
Number of constructed buildings per 1,000 inhabitants	3.53	3.35	0.42	0.69	2008
Percentage of households with access to water pipelines	35.42	36.10	0.13	0.90	2008
Percentage of households with access to sewage system	23.67	27.23	1.00	0.47	2008
LDI	0.05	0.21	0.88	0.08	–

[a] corporate (CIT) and personal income tax (PIT)

Based on variables listed in Tables 1 and 2, synthetic indices of local development (LDI) and growth (LGI) have been constructed using the sum of standardized values. In order to calculate LGI for each commune, growth variables were used for the period between 2003 and 2008. LDI or LGI equals zero when the commune is on average standing. Positive values of LDI and LGI mean that the commune performed better than average. The next step was to conduct Levene's test for equality of variances for each variable. Consequently, by way of applying the test for equality of the means for two independent samples, the significance of the difference between medians for two types of administrative entities (communes with or without dominant enterprises) were tested.

Three nonmetropolitan Carpathian communes (Czarny Dunajec, Dydnia, and Jasienica Rosielna, see Fig. 1) were then studied in detail to illustrate the direct influence of large companies on their local milieu. At the beginning of 2010,

Table 2 Index of local growth (LGI) and its variables in two groups of communes (2003–2008) in the Polish Carpathians

Variable	Communes without a dominant enterprise	Communes with a dominant enterprise	Significance (Levene's test for equality of variances)	Significance (*t* test equality of means)
Change of number of inhabitants (2003 = 100)	101.08	101.57	0.86	0.37
Change of unemployment rate (in % points) (2003 = 100)	−6.86	−6.75	0.73	0.85
Change of number of private companies per 1,000 inhabitants (2003 = 100)	104.31	106.51	0.08	0.34
Change of amount of commune's own revenues per 1 inhabitant (2003 = 100)	198.18	196.77	0.90	0.90
Change of amount of CIT and PIT[a] settled per 1 inhabitant (2003 = 100)	289.99	276.34	0.53	0.32
Change of number of constructed buildings per 1,000 inhabitants (2003 = 100)	173.25	131.58	0.02	0.18
Change of number of households with access to water pipelines	390.27	351.68	0.11	0.72
Change of number of households with access to sewage system	1,851.41	2,275.41	0.32	0.26
LGI	0.00	−0.01	0.09	0.85

[a] corporate (CIT) and personal income tax (PIT)

survey research was conducted in selected communes among the inhabitants, and in-depth inquiries were carried out with the managers of the companies. Questionnaires were distributed among the inhabitants via primary and secondary schools situated in specified communes. Out of 1,200 questionnaires distributed, 517 filled-in questionnaires were returned. The sample of inhabitants in all communes represented a similar structure of education and age; only in Jasienica Rosielna was the percentage of persons with secondary education extraordinarily high (46 %). Employees of dominant companies comprised between 12 % (Czarny Dunajec) and 37 % (Dydnia) of the respondents.

3 Results of the Comparative Statistical Analysis

Communes where dominant enterprises are present usually possess larger populations than those that do not host such companies. They also represent a significantly higher number of incoming commuters (labor immigration) than other communes (Table 1). Positive deviations, however with low statistical significance have only been observed in the area of acquired EU funds and, finally, LDI (Fig. 2). Two groups of communes did not differ significantly with respect to other variables.

For communes possessing dominant enterprises, the LDI value varied significantly ranging from −0.55 to 2.14; in 10 among 22 communes, negative LDI values were registered.

Variables used for constructing LGI for the years 2003–2008 do not present statistically significant differences between both types of communes researched (Table 2). Hence, it can be assumed that the dynamic of development of communes analyzed did not depend on the presence of dominant enterprises during that period.

Thus, although communes with large enterprises represented a slightly higher level of local development, most of the observed differences were statistically insignificant (during the statistical analysis, it was found that neither the year of establishment nor the provenience of investor, local or external, influenced the level of the communes' development). Some of communes have been developing faster, while others slower. To determine that this was due not to the presence of an enterprise itself but rather to other detailed factors that affected the local milieu, it was necessary to conduct case studies.

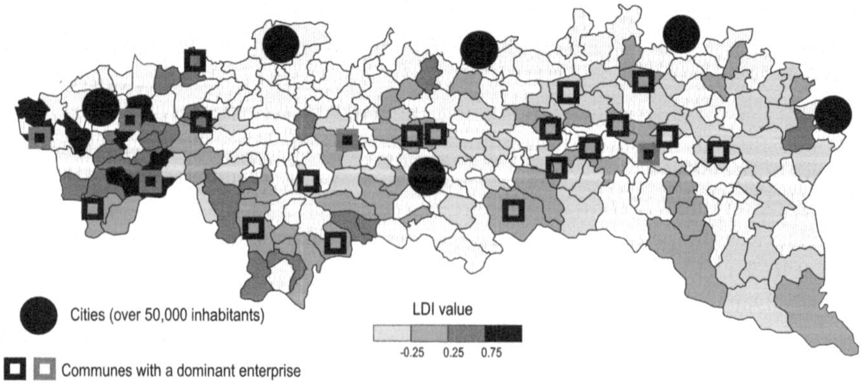

Fig. 2 LDI in the Polish Carpathians

4 Case Studies of Selected Communes

Less-developed (Dydnia, Jasienica Rosielna) and medium-developed (Czarny Dunajec) communes were chosen for further research, assuming that the influence of companies on their local milieus might be small. The selected communes have shown little activity; for instance, none of them acquired EU support for investments in the years 2006–2008, while the communes' own revenues (including tax incomes) were very low.

The influence of a dominant enterprise on its local milieu can become visible in many areas. This is confirmed by the opinions of the inhabitants in the communes where selected dominant companies are located. The chosen plants differ in the sphere of their activity (production of clothes, milk processing, and production of furniture) and the time of their establishment, while the numbers of employees are similar, comprising 200–245 persons (Table 3).

The inhabitants of communes selected for the research were asked to assess how the commune would have developed without a dominant enterprise and to indicate how, in their opinion, the presence of such a plant influences the local milieu. The positive influence of the presence of dominant companies on the communes' development was perceived more frequently in all three communes. The percentage of opinions stating that the commune would have developed either better or much better without the dominant enterprise was marginal in each case, never exceeding a few percentage of the responses (Fig. 3). Opinions expressed by inhabitants of the Czarny Dunajec and Dydnia communes were similar, with a high percentage of responses *I do not know* and those stating a lack of connection between the situation of the commune and the presence of the plant. In both of these communes mentioned, the percentage of respondents who perceived a positive influence of these companies on local development (responses stating that the commune would develop either worse or much worse without the dominant enterprise) is lower than in the case of Jasienica Rosielna. An interesting feature in the case of Jasienica Rosielna is the lowest percentage of respondents who were not able to indicate a link between the local development and the presence of the plant, as well as those, who perceived such a presence negatively.

Interesting conclusions can be derived from the opinions of the inhabitants on the types of influence of the dominant enterprises on the local milieu. Within the whole spectrum of responses, those suggesting a positive influence prevail, usually due to the fact that a large company contributes to a decreasing unemployment rate by offering jobs. In the case of each commune, such opinions comprised approximately 50 % of all indicated types of influence on the local milieu. Simultaneously, it should also be emphasized that the analyzed areas significantly differ in their unemployment rates. GUS (Central Statistical Office) data show that in 2009 the percentage of unemployed registered in the group of the population within working age was exceptionally high in the Dydnia commune, comprising 20.5 %, significantly exceeding the mean for Brzozów County (16.6 %) and for all rural communes of the Podkarpackie province (11.6 %). The situation in the

Table 3 Basic information about selected communes in the Polish Carpathians

Commune	Jasienica Rosielna	Czarny Dunajec	Dydnia
Local development index (LDI)	−0.48	−0.02	−0.55
Local growth index (LGI) (2003–2008)	−0.29	−0.29	−0.29
Key problems in the development of communes	Weak water supply system Poor utilization of EU financial means High unemployment Large emigration of urban centers	Low individual revenues and tax incomes Low entrepreneurship Huge labor emigration	Low tax incomes Poor utilization of EU financial means Huge emigration to urban centers Poor accessibility
Company			
	Okręgowa Spółdzielnia Mleczarska Jasienica Rosielna (Dairy Cooperative)	Gamatex	Company X[a]
Area of activity	Milk processing	Production of clothes	Production of furniture elements
Year of establishment	1927	1975	1991
Employment (2008)	220	200	245

[a] The company refused to disclose its name

Source Hoppenstedt Bonnier Information Polska database

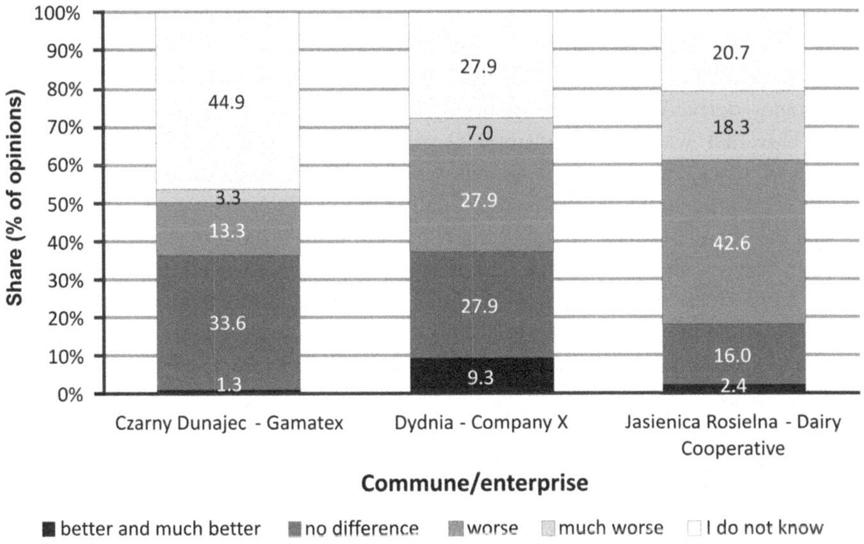

Fig. 3 Inhabitants' opinions about how the communes in the Polish Carpathians would have developed without a dominant enterprise

Jasienica Rosielna commune was slightly better (16.4 %). On the other hand, a low level of unemployment was characteristic for Czarny Dunajec, with the value of this rate comprising only 3.6 %. It was less than in the whole Nowy Targ County (5.7 %), and even less than in the case of all rural communes of the Małopolskie province (6.8 %).

Only a small percentage of inhabitants of the researched communes stated that financial advantages in the form of wages offered by the plants were important (about 3–4 % of all indicated types of influence in the case of each commune). Advantages in the form of increasing the communes' prestige, which made them more well-known and recognizable in the region or even in the country due to the presence of a dominant enterprise, were indicated in all communes. Simultaneously, in each commune there were some inhabitants who perceived a negative influence of dominant companies on the local natural environment. Moreover, in Jasienica Rosielna and Czarny Dunajec, a positive influence on the communes' budgets (due to tax incomes) was mentioned, as well as accelerating the local economy by increasing the demand.

Some types of influence have only rarely been indicated by inhabitants of a communes. Hence, some respondents from Jasienica Rosielna placed stress on advantages in the form of good dairy products available in local shops and on the financial advantages of local suppliers (farmers producing milk). As derived from information received in the Dairy Cooperative, the company buys milk from about 250–300 farmers living in 30 communes of the Małopolskie and Podkarpackie provinces, from whom about 50 are inhabitants of Brzozów County. The relatively high price of purchased milk (on average 1 Polish Zloty, PLN/liter) resulted in paying about 1.7 million PLN (about 393,000 EUR) to inhabitants of this county in 2009, approximately 36,000 PLN (about 8,310 EUR) per supplier. On the other hand, respondents from the Czarny Dunajec commune have perceived some advantages derived from the support of Gamatex in professional training of its staff. This fact was also confirmed by the company owner. Immediately after 3 years of work, employees of Gamatex will undergo an exam organized by the Craftsmen Guild to achieve a formal qualification.

The employees surveyed in the Dydnia commune did not represent such a positive opinion about the dominant enterprise. It was mentioned that the company negatively influences its local surroundings, provoking conflicts among inhabitants (some time ago the company entered into a dispute with the Commune Office with respect to spatial management of areas surrounding the plant). Several persons also mentioned that the company exploits the employees, offering poor employment conditions.

Finally, a high percentage of responses was found stating that the company does not influence its local milieu in any way, or responses of *I am not able to answer* in relation to Gamatex. These two kinds of opinions comprise a total of 40 % of all responses given by inhabitants of this commune. This means that the company does not play a significant role in the local scene and is not perceived as a serious factor in local development. This response is more understandable because most of the employees are persons living outside the community (in less developed regions), which is linked to a high level of local entrepreneurship and low unemployment rate both in the commune itself and the Nowy Targ County as a whole. As mentioned by the owner of Gamatex, the Czarny Dunajec commune is a difficult area for employers, due to well-developed tourism, resulting in high local entrepreneurship; moreover, the problems in acquiring a labor force become deepened due to popular labor migrations abroad.

The question of companies' influence on local communes and counties was also asked of the entrepreneurs themselves. Gamatex has mostly emphasized the advantage of providing jobs, while the companies from Jasienica Rosielna and Dydnia, budget revenues from taxes. In Jasienica Rosielna, the director of the Dairy Cooperative emphasized the significance of offering jobs close to employees' homes. In this commune, public transportation is limited to PKS (national bus carrier), and private bus companies are not present, which seriously restricts the opportunities of inhabitants to take up work outside the commune.

Offering jobs and reducing unemployment is the most frequently perceived advantage of the presence of large companies. Hence, it was important to look at the inhabitants' opinions about entrepreneurships as employers. Among the three companies analyzed, the best opinion was obtained by the Dairy Cooperative in Jasienica Rosielna: 44 % of all answers indicated this plant to be either a good or a very good employer (Fig. 4). For the other two employers, this rate was much lower (27.8 % for Gamatex and only 9.3 % for the company from Dydnia). Simultaneously, the companies from Czarny Dunajec and Jasienica Rosielna received a very small percentage of negative opinions (*rather bad and very bad employer*), at a level of 4–5 %, contrary to the company from Dydnia (25.6 %).

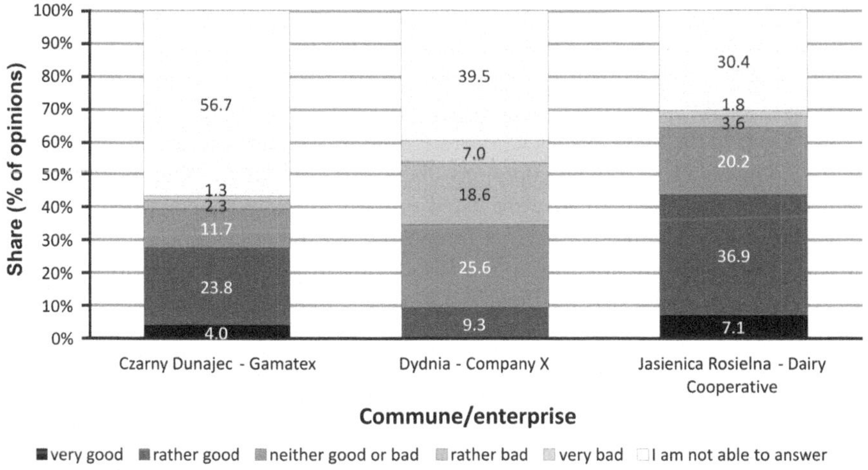

Fig. 4 Inhabitants' opinions of companies' influence on the local communities in the Polish Carpathians

The very high percentage of responses *I am not able to answer* in Czarny Dunajec confirms that this company is poorly present in the consciousness of the inhabitants of Czarny Dunajec.

The surveyed inhabitants of the communes were also asked to justify the opinions about entrepreneurships as employers. In the case of Jasienica Rosielna, the company enjoyed a good opinion and they mostly emphasized good employment conditions, expressed as on-time salaries and taxes and fees paid against the actual revenues of the employees. An even larger group of respondents (21.2 % of opinions) perceived the enterprise positively as creating jobs (Fig. 5).

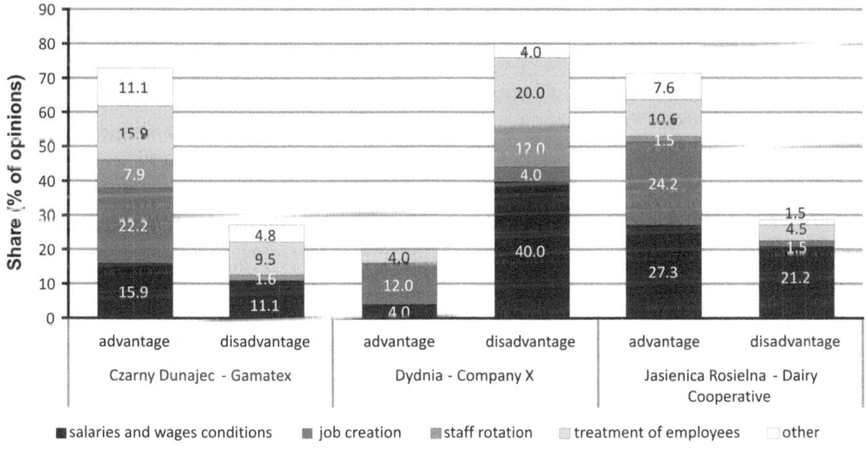

Fig. 5 Inhabitants' opinions about entrepreneurships as employers in communes in the Polish Carpathians

Taking care of the employees and being well treated by the management was also mentioned (10.6 % of the answers). Single opinions, emphasizing good wages, stable work conditions in full-time employment, good management, and a well-developed social sphere, were also expressed. On the other hand, the worst part of employment in the OSM (Dairy Cooperative) in Jasienica Rosielna was considered the low wages (21.2 % of the opinions about this company). Other negative features of working in this plant were listed only by single respondents.

In the case of Gamatex, the most frequently mentioned advantage was the fact that the company creates jobs (22.2 %), followed by the proper treatment of employees and taking care of them (15.9 % of opinions), as well as stable work conditions (7.9 %). Negative opinions were listed rarely; they included most frequently low wages (11.1 %) and bad treatment of employees (9.5 %).

The third enterprise analyzed, a company from the Dydnia commune in Brzozów County, was assessed much more severely by its inhabitants. In this case, negative aspects of working for this company were listed more frequently than positive ones. Respondents most frequently mentioned low wages (40.0 %), poor treatment of employees (20.0 %), and large staff rotation, resulting in a lack of stability in employment (12.0 %). Among the listed advantages, creating jobs was listed (12.0 %), while other aspects (proper treatment of employees and high wages) were only mentioned by single individuals.

The opinions of the inhabitants of the three communes indicate important differences in their social perception, resulting from the uneven employment conditions offered by dominant enterprises. The only common denominator of the formulated opinions is the emphasis on low wages as the key disadvantage of working for these enterprises. The inhabitants' opinions in this area are fully justified. For instance in the Dairy Cooperative in Jasienica Rosielna the 2009 average monthly gross salary (excluding management) was 1,829 PLN (approximately 422 EUR). At the same time, the average monthly gross salary in Brzozów County was 2,503 PLN (578 EUR), which was one of the lower county averages in Poland. In comparison, the 2009 average for the whole country was 3,315 PLN (766 EUR).

Large companies in the Carpathian rural communes offer jobs for locals; however, these jobs can hardly be found attractive. In Brzozów County (Dydnia and Jasienica Rosielna communes), the dominant position of these companies, resulting from a shortage of alternatives, allows them to a great extent to dictate employment conditions. Simultaneously, they do not provide an appealing offer for people living outside the communes, and employees are mostly recruited from the closest surroundings of the plant in the local commune or county. In Czarny Dunajec, on the other hand, the relatively low attractiveness of jobs in the dominant enterprise does not motivate the locals to seek employment in this company; most of the employees come from the less-developed region of Orawa.

According to the respondents, in each of the communes the presence of a dominant company limits job migrations of the local population (30–40 % of responses). At the same time, in the opinions of those surveyed, these companies do not influence the career plans or the education profiles of local youth.

5 Conclusions

The statistical analysis has shown that the Polish Carpathian communes in which dominant enterprises are present are strongly differentiated in their levels of development. They represent slightly higher values of the synthetic development index than other communes; nonetheless, it cannot be stated that these enterprises have an important or direct local impact. It is hard to find the dominant enterprises in Carpathian rural communes acting as the engines of local economies; they are focused rather on taking advantage of a local cheap labor force. Due to their profiles of activity, these enterprises do not require a qualified labor force. Hence, they do not attract highly educated management from outside or create incentives for the local youth to take up education aimed at gaining jobs in these companies in the future. In general, these companies do not offer a wide range of opportunities for the development of employees' skills.

The influence of a large company depends on the specific character of its activity. Certain kinds of production (i.e. manufacturing a large number of food products) require a base of local raw materials (as in the case of the Dairy Cooperative in Jasienica Rosielna). The scale and scope of the influence depend also on local economic conditions. In areas characterized by a high local entrepreneurship level, the dominant company is poorly perceived by the locals (as in the case of Czarny Dunajec), because the inhabitants may find other employment opportunities, such as in tourism. In this way, a company active in such areas offers more attractive conditions that would encourage the people to seek employment there. In an area of low entrepreneurship level, a large company can dictate its conditions yet still find people prepared to work (Dydnia, Jasienica Rosielna). If the local economy and labor market is more difficult due to a high unemployment rate and low level of economic development, then worse conditions are offered by dominant enterprises to their employees.

Acknowledgments The analyses conducted in the present chapter are elements of broader research financed by a grant from the Polish Ministry of Science and Higher Education, no. K/PBP/000/291, *Mechanizmy i skutki oddziaływania dużych przedsiębiorstw w pozametropolitalnych obszarach wiejskich Polski* ("Mechanisms and effects of influence of large enterprises in nonmetropolitan rural areas in Poland").

References

Adamus J, Luchter L (1995) Przemysł. In: Warszyńska J (ed) Karpaty Polskie. Przyroda, człowiek i jego działalność. Wydawnictwo UJ, Kraków

Balon J, German K, Kozak J, Malara H, Widacki W, Ziaja W (1995) Regiony fizycznogeograficzne. In: Warszyńska J (ed) Karpaty Polskie. Przyroda, człowiek i jego działalność. Wydawnictwo UJ, Kraków

Bański J (1999) Obszary problemowe w rolnictwie Polski. Prace Geograficzne no 172, IGiPZ PAN, Wydawnictwo Continuo, Wrocław

Bański J (2008) Wiejskie obszary sukcesu. Rural Stud 14
Berry BJL (1976) Urbanization and counterurbanization. Sage Publications, Beverly Hills—London
Domanski B (1992) Social control over the milltown: Industrial paternalism under socialism and capitalism. Tijdschr Econ Soc Geogr 83(5):353–360
Domanski B (1997) Industrial control over the socialist town: benevolence or exploitation. Praeger, London
Domański B (2001) Kapitał zagraniczny w przemyśle Polski: prawidłowości rozmieszczenia, uwarunkowania i skutki. IGiGP UJ, Kraków
Domański B, Gwosdz K (eds) (2005) Dziesięć lat doświadczeń pierwszej polskiej specjalnej strefy ekonomicznej. Mielec 1995–2005. Instytut Geografii i Gospodarki Przestrzennej UJ, Kraków
Domański B, Dej M, Działek J, Górecki J, Guzik R, Gwosdz K, Kocaj A, Micek G (2009) Inwestycje zagraniczne w Małopolsce 2008. Urząd Marszałkowski Województwa Małopolskiego, Kraków
Dziemianowicz W (ed) (1999) Inwestycje zagraniczne w gminie. Jak przyciągnąć, obsłużyć i zatrzymać inwestora. PAIiIZ i Fundusz Współpracy, Warszawa
Dziemianowicz W (2007) Konkurencyjność gmin w kontekście relacji władze lokalne—inwestorzy zagraniczni. Wydawnictwo Uniwersytetu Warszawskiego, Warszawa
Heffner K (2007) Rozwój społeczno-gospodarczy obszarów wiejskich. Definicje – uwarunkowania – zależności – czynniki – skutki. Badania zróżnicowania rozwoju obszarów wiejskich. In: Rosner A (ed) Zróżnicowanie poziomu rozwoju społeczno-gospodarczego obszarów wiejskich a zróżnicowanie dynamiki przemian. IRWiR PAN, Warszawa
Fothergill S, Gudgin G, Kitson M, Monk S (1985) Rural industrialization: trends and causes. In: Haeley MJ, Ilbery BW (eds) The industrialization of the countryside, pp 147–159
Gorzelak G, Jałowiecki B, Kozak M, Smętkowski M (2007) Władze miast wobec zagrożenia degradacją społeczno-gospodarczą w warunkach transformacji systemowej. Stud Reg Lokal 3(29):5–31
Grzeszczak J (1998) Przesunięcie miasto – wieś w przemyśle krajów Unii Europejskiej. IGiPZ PAN, Warszawa
Hałasiewicz A, Kaleta A (2000) Przedsiębiorczość na obszarach wiejskich. FAPA, Warszawa
Heffner K (2000) Kreacja ośrodków wielofunkcyjnego rozwoju na obszarach wiejskich. Wieś i Rolnictwo 2, Warszawa
Keeble D (1980) Industrial decline, regional policy and the urban-rural manufacturing shift in the United Kingdom. Environ Plan A 12(8):945–962
Keeble D (1984) The urban-rural manufacturing shift. Geogr 69(303):163–166
Keeble D (1989) The dynamics of European industrial counter urbanization in the 1980s: corporate restructuring or indigenous growth? Geogr J 155(1):70–74
Keeble D, Tyler P (1995) Enterprising behaviour and the urban-rural shift. Urban Stud 32(6):975–997
Kłodziński M (ed) (2000) Rozwój przedsiębiorczości wiejskiej w warunkach integracji z UE. IRWiR PAN, Warszawa
Kłodziński M (2002) Aktywizacja gospodarcza obszarów wiejskich. IRWiR PAN, Warszawa
Kłodziński M, Fedyszak-Radziejowska B (eds) (2002) Przedsiębiorczość wiejska w Polsce i w krajach UE. IRWiR PAN, Warszawa
Kłodziński M (2006) Aktywizacja społeczno-gospodarcza gmin wiejskich i małych miast. IRWR PAN, Warszawa
Kulikowski R (1992) Obszary problemowe rolnictwa w Polsce. Biul KPZK PAN 158:23–40
Lonsdale RE, Seyler HL (eds) (1979) Non-metropolitan industrialization. Winston and Sons, New York
Rosner A (ed) (1999) Typologia wiejskich obszarów problemowych. IRWiR PAN, Warszawa
Rosner A (ed) (2002) Wiejskie obszary kumulacji barier rozwojowych. IRWiR PAN, Warszawa

Sobala-Gwosdz A (2000) The influence of large manufacturing firms on the local development of Jarosław in the 1990s. In: Marszał T (ed) Local economy and urban development in Poland. Uniwersytet Łódzki, Łódź, 62–69

Stasiak A (ed) (2000) Możliwości wielofunkcyjnego rozwoju wsi polskiej w kontekście integracji z UE. Stud KPZK PAN 110

Stryjakiewicz T (ed) (2004) Wpływ inwestorów zagranicznych na rozwój regionalny i lokalny na przykładzie GlaxoSmithKline Pharmaceuticals S.A. w Poznaniu. Bogucki Wydawnictwo Naukowe, Poznań

Cities and Towns in the Polish Carpathians: Opportunities and Limitations of Urban Development

Joanna Więcław-Michniewska

Abstract The development of a region tends to be strictly related to the socio-economic status and competitiveness of its towns and cities. The objective of this research was to identify the competitiveness of Polish Carpathian towns, focusing on essentially geographical factors, such as topography, and various socioeconomic factors that were combined into the final emulation indicator and provided a synthetic picture of opportunities and limitations of urban development. The results of the study show that the natural environment has exerted a major influence on the potential for urban expansion in the Polish Carpathian Mountains. The physical environment, while limiting spatial expansion, has been, however, an asset in the mountains and its skillful use could improve urban competitiveness. The use of natural assets ought to be linked with care for the environment in the spirit of sustainable development.

1 Introduction

Mountainous areas have typically offered more difficult conditions for human settlement than lowlands. A harsher climate and less fertile soils hamper agriculture; land morphology has not been conducive to communication and transport networks, while trade and commerce have tended to be less dynamic. An analysis of changes in mountainous areas during the twentieth century helped to identify regions overly exposed to levels of anthropogenic pressure harmful to the environment, as well as regions with poor economic development, which were often arenas of political conflict (Borsdorf and Braun 2008).

J. Więcław-Michniewska (✉)
Institute of Geography and Spatial Management, Jagiellonian University, Gronostajowa 7, 30-387 Kraków, Poland
e-mail: j.wieclaw-michniewska@uj.edu.pl

J. Kozak et al. (eds.), *The Carpathians: Integrating Nature and Society Towards Sustainability*, Environmental Science and Engineering,
DOI: 10.1007/978-3-642-12725-0_37, © Springer-Verlag Berlin Heidelberg 2013

Historically, towns acted as foci for socioeconomic processes, and competitiveness was inherent in their evolution, especially in terms of long-term development strategies and the improvement of local living standards. When considering the development of socioeconomic processes in a mountainous area, it becomes apparent that towns served as centers of economic growth with a visible impact on their hinterland and its prosperity (Boesch 2006; Illés 2008). The role of urban areas may increase as a result of rapid modernization of economic processes, especially with increasing mobility of the means of production. Environmental conditions are still important, especially in areas with sparser urban networks, such as in mountains. The towns and cities involved in this study differed in terms of the nature of their relationship with the surrounding mountains, but, in contrast with non-mountain areas in their respective countries, there were very few Carpathian areas with the degree of urbanization that would reach the level of the Petroşani Depression in Romania (Fourny 2000).

It is worth noting that mountain towns, regardless of the specific mountain range or even country, show remarkable similarities. These similarities come from the strong presence of mountains in the culture and the daily life of local communities. This is why mountain towns stand out by building their identities on values related to mountains, since settlements were the places where the cultural features of mountain communities were expressed to their fullest extent (Fourny 1999; Hoyaux 2000; Debarbieux and Rudaz 2008). The strength of the common mountain values is reflected in numerous associations of mountain towns and nongovernmental organizations (NGOs) representing private citizens. Examples include The Association of Alpine States (ARGE ALP), an association active in six alpine countries that pursues cross-border cooperation to solve common socioeconomic problems and to raise environmental awareness, and the Rhodopes Mountains Association in Bulgaria, aiming at the development of the economic potential of the Rhodope region and improving local government performance (Kiryakow 2003). Regional sustainable development, which includes the contribution and input of towns, has been an important research topic (Borsdorf and Braun 2008). This coincides with the fact that solutions adopted to address socioeconomic development in various mountain areas primarily involve natural resources; land-use change; support for the services market, especially in tourism; preservation of cultural values, including mountain architecture; and cross-border cooperation (Pogačnik 2003; Björnsen Gurung et al. 2009).

2 Cities and Towns in the Polish Carpathians: Setting a Context

The development of a region tends to be strictly related to the socioeconomic status and competitiveness of its towns and cities. Many factors (for example, urban policy, local entrepreneurship or political and legal frameworks at the different levels) may have different impacts on different towns, depending on the resources at their disposal, whether natural, cultural, economic, or social. The Carpathian Mountains are among the least urbanized regions in Poland. In 2009, their urbanization index was just below 40 % (compared to 61 % for Poland): the towns selected for the study accounted for just 4 % of the total urban population of the country and less than 7 % of the overall number of towns (Bank Danych Regionalnych 2009). Towns of the Polish Carpathian Mountains vary widely in terms of their population potential and socioeconomic condition, as well as their potential to improve these aspects (Górka 2004; Illés 2008; Więcław-Michniewska 2011).

The Carpathian settlement network is characterized by unevenness in terms of both town sizes and their distribution in the region. Bielsko-Biała is the largest city with 175,400 inhabitants, followed by Nowy Sącz (84,600) and Krosno (47,500). However, small towns dominate the region with more than half of them below 10,000 inhabitants, of which nearly half are below 5,000. Medium-sized towns (i.e., with populations of more than 25,000) can be found in the western part of the area (Żywiec, Cieszyn), in the south (Nowy Targ, Zakopane), and in the north (Bochnia), with the remaining three (Gorlice, Jasło, Sanok) forming (together with the larger Krosno) a string of "powiat" capitals (middle tier of administrative division) in the east. During the last decade, there was a drop in urban population in the Polish Carpathian Mountains, especially in larger towns (as a result of progressive suburbanization), which was consistent with patterns observed in the Alps and in the whole Carpathian region (Borsdorf 2006; Illés 2008).

Overall, the density of the urban network decreases southward, a direction associated with an increasing altitude and increasingly mountainous land relief. Urbanization decreases also noticeably from west to east. Certain eastern powiats (Brzozowski, Leski, Bieszczadzki) have only one town each (the powiat capital), which contrasts with larger numbers of towns in the west, such as in the Cieszyński Powiat (five towns). This disparity in the level of urbanization is most likely related to the recent history of city rights than to natural conditions. Indeed, among nearly 50 former towns that lost their formal city rights during the twentieth century, most are found in the eastern part of the area (Fig. 1). Many of them managed to retain their urban character and also often their urban functions, which has allowed them to reclaim their city rights, a trend particularly noticeable in the first decade of this century (Fig. 1).

Mountainous regions, such as the Western Carpathian Mountains, constitute extremely valuable natural areas (Denisiuk 1995; Turnock 2002). The Polish part of the Carpathian Mountains is characterized by a high degree of biodiversity, a wealth of plants and wildlife, and one of Europe's largest forest resources. The mountain range also constitutes an important headwater area that influences hydrology and climate over large areas (Framework Convention on the Protection and Sustainable Development of the Carpathians 2003). The beauty and diversity of the mountain and foothill landscapes also play a key role, especially in terms of economic development. Perception of the Carpathian landscape, viewed as a set of natural components in their natural state or affected by human activity, involves

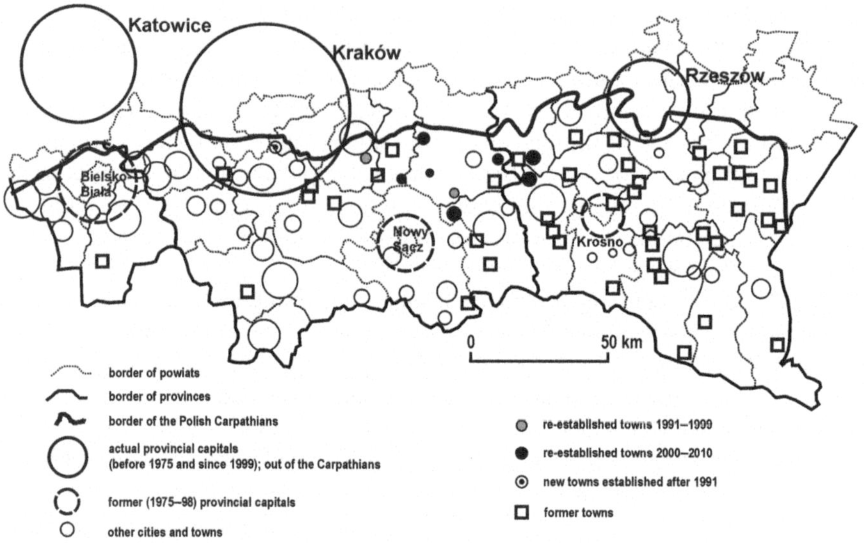

Fig. 1 Urban network in the Polish Carpathians

ethnographic and cultural elements that are highly appreciated by the public. For this reason skillful utilization of natural and cultural assets could offer an opportunity for growth and development in many Carpathian towns.

Urban competitiveness is often evaluated from the perspective of regional development. In addition, it is often seen in the context of the economic, social, and cultural dimensions of its growth potential (Gaczek and Rykiel 2000; Markowski 2005; Parysek and Mierzejewska 2009a). For this reason, to determine the competitiveness various authors use different approaches and widely ranging indicators, depending on the adopted methodology and objectives (Harańczyk 2002; Dziemianowicz 2005; Słodczyk and Szafranek 2008; Parysek and Mierzejewska 2009b). The purpose of this study is to identify the competitiveness of Carpathian towns with respect to geographical factors, such as topography, and various socioeconomic factors that determine a town's capability to achieve generally accepted improvement in living and working standards. This aim was approached by defining a complex emulation indicator using a range of environmental and socioeconomic variables.

3 Methods

To balance socioeconomic and environmental factors, the composite emulation indicator was built of five synthetic factors (topographic, environmental, anthropogenic pressure, socioeconomic, and functional), each of which was itself composed of several variables (Table 1). The selection of variables and factors was

Table 1 Synthetic factors and basic variables forming the emulation indicator

Synthetic factor	Basic variable	Unit (year)
Environmental	Public green area (generally accessible)	ha per 1,000 population (2009)
	Agricultural land area	ha per 1,000 population (2005)
	Forest land	ha per 1,000 population (2005)
	Mean altitude	m a.s.l. (2009)
	Range of altitude within town	m (2009)
Anthropogenic pressure	Untreated wastewater	Dam^3 per 1,000 population
	Mixed waste	t per capita (2009)
	Average annual concentration of SO_2	$\mu m/m^3$ (2008)
	Average annual concentration of NO_2	$\mu m/m^3$ (2008)
	Average annual concentration of dust	$\mu m/m^3$ (2008)
Socioeconomic	Natural increase	per 1,000 population (2009)
	Net internal migration employed person	per 1,000 population (2009)
	Employed person	Percent of the total population (2008)
	Nonworking age population per 100 persons of working age	Persons (2009)
	Distance from the nearest upper rank urban center	km (2009)
	Economic entities	per 1,000 population (2009)
	Tourists accommodated	Persons per capita (2008)
	Usable floor space of dwellings	m^2 per capita (2008)
	Dwellings completed	Numbers (2008)
	Libraries	per 1,000 population (2009)
	Sewage network	km per km^2 (2009)
	Investment outlays on protection of air and climate	% of total expenditures (2008)
	Investment outlays on protection of waters and wastewater management	% of total expenditures (2008)
	Investment outlays on waste management and protection of surface	% of total expenditures (2008)
Functional	Employed person by NACE sections	
	Industry	% of total employment (2003)
	Services	% of total employment (2003)
	Enterprises by NACE sections	
	Industry	per 1,000 population (2009)
	Services	per 1,000 population (2009)

(continued)

Table 1 (continued)

Synthetic factor	Basic variable	Unit (year)
Topographic	Site type by relief	
	Narrow valley	-
	Slope or hilltop	-
	Broad valley or small basin	-
	Vast inner mountain basin or foothill margin	-

determined by the availability, resolution, and reliability of current statistical data. Thirty measures obtained from the Regional Data Bank of the Central Statistical Office and supplemental variables (Stan środowiska w województwie podkarpackim w latach 2000–2007 2008; Bank Danych Regionalnych 2009; Raport o stanie środowiska w województwie małopolskim w 2008 Roku 2009; Stan środowiska w województwie śląskim w 2008) were used to build four out of five factors, while topographic maps were used to build the topographic factor. In a third step a set of basic variables was selected through elimination of redundant data, while trying to maintain the correct proportion between variables for each factor. This study, therefore, has used only selected variables characterizing aspects deemed important in order to estimate the competitiveness of the selected sample of towns. Four factors (environmental, anthropogenic pressure, socioeconomic, and functional) were computed with Perkal's method, with the topographic factor evaluated on a basis of maps. Each factor was then divided into either four or five classes. To obtain the synthetic emulation indicator, a ranking method was applied by scoring each class, which helped to combine variables expressed in various scales and factors with different numbers of classes (Runge 2006).

Towns and cities were selected for this study on the basis of their location within the Polish Carpathian Mountains in accordance with the classification by Kondracki (1978). Since drawing precise regional boundaries tends to be a difficult task (Kondracki 1978; Balon et al. 1995), towns located at the foot of the Carpathian Mountains were included in the study if a clear majority of their territory was within the Carpathians. As of January 1, 2010, 61 towns and cities defined in this way were located within the Polish Carpathian Mountains, of which 56 were included in the study. Some towns and cities located at the foot of the Carpathians such as Kraków and Kańczuga, were excluded, although other authors included these within the Carpathian urban network (Górka 1995; Illés and Zoltán 2007). Three towns that regained their town rights in 2009 and 2010, including Bobowa, Brzostek, and Kołaczyce, were also excluded for the lack of relevant statistical data. It must be noted that the selected towns and cities do not form a uniform settlement system, as a result of their different histories and their current position in three different provinces: Silesia, Małopolska, and Podkarpacie. The latter is significant, since each province has a different potential for supporting infrastructure development of its towns and specific regional development policies

(Strategia rozwoju województwa małopolskiego 2000; Małopolska 2015. Strategia rozwoju województwa podkarpackiego na lata 2007–2013, 2006; Strategia rozwoju województwa podkarpackiego na lata 2007–2020, 2006).

4 Results

The considerable diversity of the land relief is among the main factors shaping the natural environment of the Polish Carpathian Mountains. For this reason the Carpathian towns selected for this study varied in their location with respect to landforms. The topography determined their potential for growth not only spatially, but also economically and financially, such as due to the suitability of land for specific types of buildings and related cost involved in their erection. A similar type of observation was made by Fourny (2000) who considered locations and functions of alpine towns. The spatial arrangement of towns can either favor or hinder urban growth, especially in terms of land available for development.

In this study we identified the extent of favorable locations of towns in various topographical situations. It was assumed that spatial expansion of towns was the easiest in large intermountain basins or in smaller basins at the confluence of two or three valleys, between hills. Less favorable locations included narrow valleys between mountain ranges or below mountain passes. Most towns did not have the best conditions for spatial growth, and nearly half of them belonged to either unfavorable or somewhat unfavorable location types (Fig. 2). Towns with the most

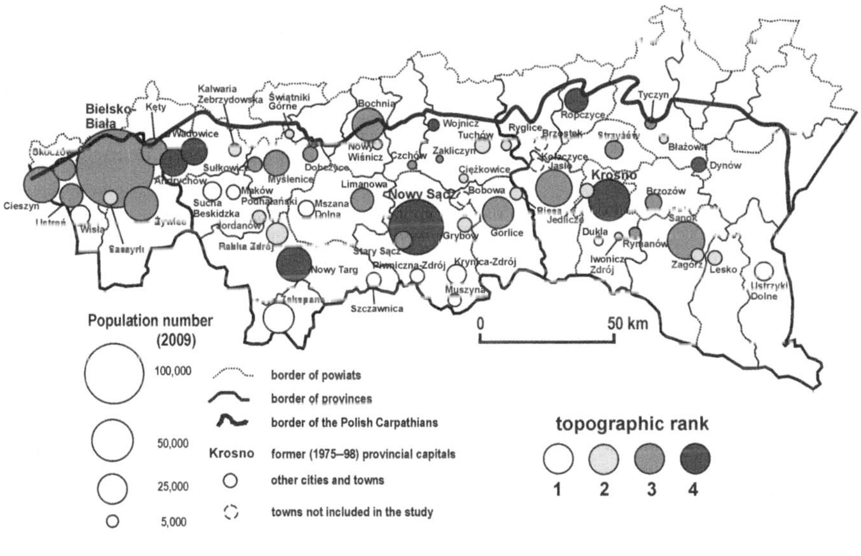

Fig. 2 Topographic factor for towns in the Polish Carpathians. Legend: *1* unfavorable, *2* rather unfavorable, *3* somewhat favorable, *4* favorable

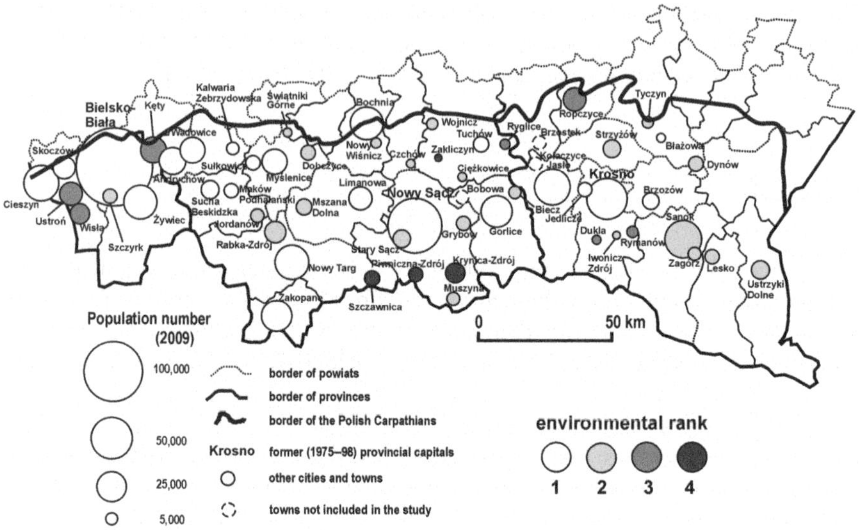

Fig. 3 Environmental factor for the towns in the Polish Carpathians. Legend: *1* unfavorable, *2* rather unfavorable, *3* favorable, *4* very favorable

favorable location were found in the northern section of the Polish Carpathian Mountains, close to the boundary of the region, along a line stretching west to east from Andrychów to Wadowice, Wojnicz, and Dynów. There were three larger towns, including Krosno, Nowy Sącz, and Nowy Targ, located centrally in wide mountain basins and identified to have favorable conditions for spatial expansion. The least favorable situation was found in four mountain spas, including Szczawnica, Piwniczna-Zdrój, Muszyna, and Krynica-Zdrój, as well as in the towns of Dukla, Ustrzyki Dolne, Maków Podhalański, and in Zakopane, the latter wedged between the Gubałówka ridge and the boundary of the Tatra National Park.

The study estimated the influence of the environment on the urban potential for growth (environmental factor) by taking into account the elevation range of the developed land and the proportion of undeveloped land (municipal green areas, agricultural land, and forests) in the total town area (Fig. 3). Elevation was used to account for the sensitivity of the natural environment to anthropogenic pressure, known to increase both upwards and deeper into the mountains. Most of the towns are located in the lower-foothill zone of the Polish Carpathian Mountains, between 200 and 300 m a.s.l. Some mountain spas (Ustroń, Krynica, Szczawnica) and ski resorts (Szczyrk, Wisła) are higher up at 500–600 m a.s.l. Zakopane, the highest town in Poland, has a town center at altitudes of 800–900 m a.s.l. Other variables of the environmental factor were selected so as to account for the presence of undeveloped land within towns, on the one hand, and for the impact of the environment on the quality of life on the other. The latter has been studied with a variety of methods (Borsdorf 1999a,b) and was mentioned by several authors in terms of the importance of building an attractive living environment contributing

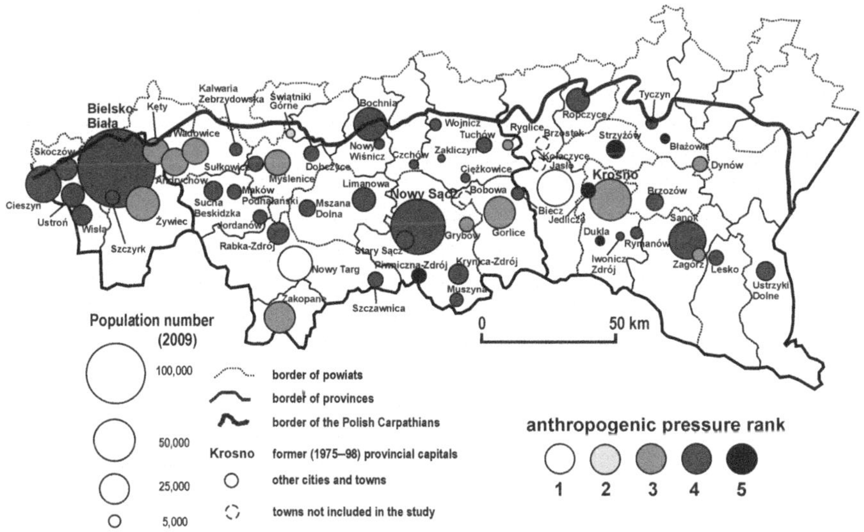

Fig. 4 Anthropogenic pressure in the towns of the Polish Carpathians. Legend: *1* very high, *2* high, *3* medium, *4* low, *5* very low

to the competitiveness of towns and regions (McCann 2007). The highest value of the environmental factor was found in the southern part of the area (particularly in the mountain spas of Szczawnica, Muszyna, and Piwniczna-Zdrój), followed by favorable conditions of towns in the west (Ustroń and Wisła), in eastern spas (Iwonicz and Rymanów), and in several foothill towns (Zakliczyn, Kęty, Ropczyce, and Ryglice). Overall, favorable environmental conditions, although to a varying degree, were found in nearly 20 % of all towns, while approximately 40 % of them had unfavourable environmental conditions (Fig. 3).

The study of the relationship between the town and its natural environment was completed by classifying towns according to their anthropogenic pressure. Potential pollution (degradation) of the environment was identified as an adverse factor for potential urban growth. Generally, the condition of the natural environment tended to be average: approximately 40 % of the towns were classified in either the satisfactory or average groups and another 40 % in the good condition category (Fig. 4). No apparent relationship was identified between environmental conditions and location, although towns in the eastern part of the Polish Carpathian Mountains were found to have slightly better conditions than other towns in the study area. Among the towns, there were only small differences in air quality; however, considerable differences in the amount of untreated wastewater were found. The most degraded environment overall was found in Nowy Targ, Jasło, and Świątniki Górne, which coincided with their highest untreated wastewater values. High dust and gas pollution due to the use of traditional heating fuels was particularly acute in the town of Nowy Targ, but it was present also in other Carpathian towns. Existing air quality protection programs (in Nowy Targ and in the Tatra powiat)

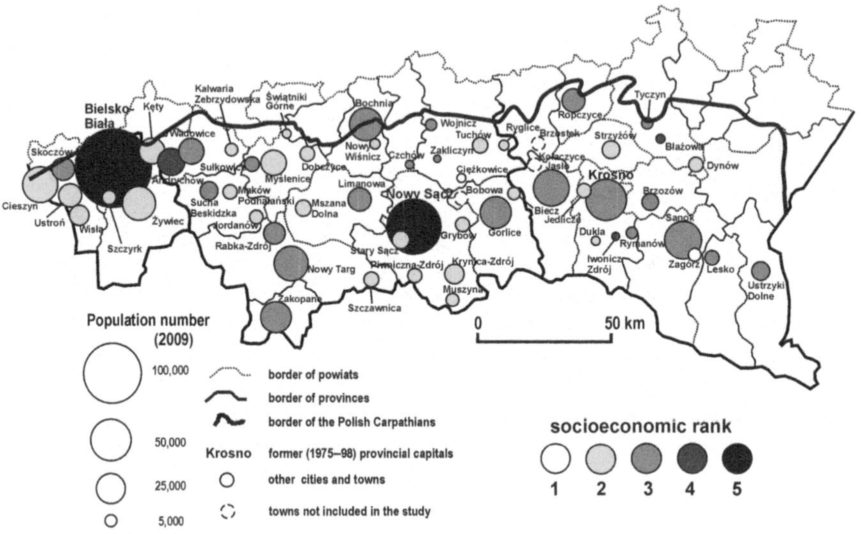

Fig. 5 Socioeconomic factor for the towns in the Polish Carpathians. Legend: *1* very poor, *2* poor, *3* medium, *4* good, *5* very good

have failed to reach the expected lower pollution levels (Raport o stanie środowiska w województwie małopolskim w 2008 roku 2009). At the other end of the scale there were towns with very good environmental conditions, including Jedlicze, Strzyżów, Dukla, and Piwniczna-Zdrój.

A number of social and economic variables were used to create the socioeconomic and functional indicators, which together reflect the degree of socioeconomic development of the towns. As for the functional indicator (map not presented here), towns of various sizes formed a cluster with a high proportion of services in the economy in the northwestern part of the area (from Bielsko-Biała to Dobczyce). A similar degree of development was observed in towns with strong tourist and spa functions (Zakopane, Krynica and Muszyna), while in the eastern part of the area no town stood out in particular. Looking at the socioeconomic factor, the general pattern of the Carpathian towns is one of relatively low and uniform socioeconomic development. Indeed, 87 % of the towns were rated either category 2 (poor, 48 %) or 3 (medium, 39 %). Out of the remaining 13 %, two towns, Nowy Sącz and Bielsko-Biała (former provincial capitals during 1975–1998), stood out as the highest category ranks, followed by three other towns in the "good" development category. Two small towns, Ciężkowice and Zagórz, were labelled with the lowest "very poor" category (Fig. 5).

The final emulation indicator provided a synthesis of all five factors. The top category included two large towns (Nowy Sącz and Bielsko-Biała), a smaller town (Wadowice) between them, along with Ropczyce in the eastern foothills area (Fig. 6). Approximately 15 % of the towns were highly competitive, and these were mostly medium-sized towns with a relatively harmonious degree of

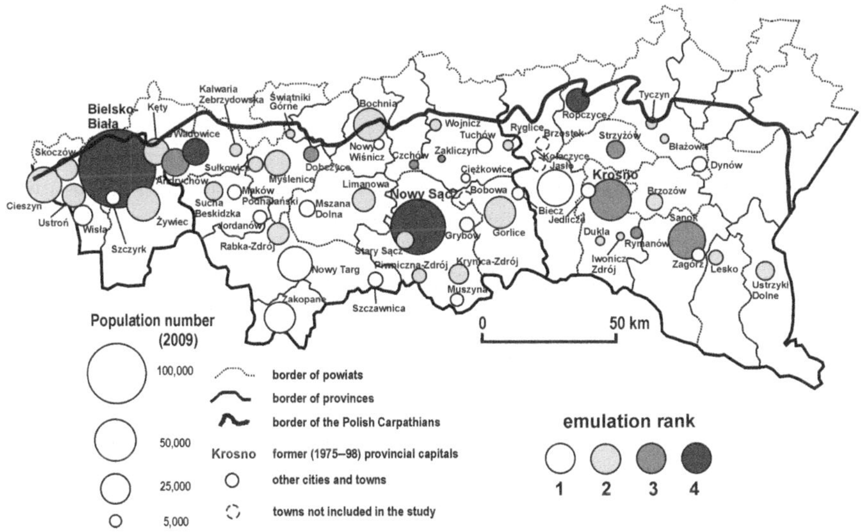

Fig. 6 Emulation indicator for towns in the Polish Carpathians. Legend: *1* low, *2* medium, *3* high, *4* very high

functional-economic level of development, typically performing higher administrative functions (seat of powiat authorities). At the other end of the spectrum there were towns with low ranks in one or more factors: unfavorable topographic conditions (e.g., Mszana Dolna and Wisła), poor environmental conditions (Nowy Targ and Jasło), poor socioeconomic development (Dynów and Maków Podhalański) or low functional rank (Szczawnica and Tuchów), that were not offset by other components of the emulation indicator. Nearly half of the towns have fallen into category 2 (medium-level emulation indicator). An analysis of the relationship between the functional, the socioeconomic and the anthropogenic pressure factors showed that Carpathian towns and cities displayed a relatively weak level of economic development, although they also did not exert an excessively negative impact on the environment.

It is difficult to find any meaningful spatial differentiation in the level of competitiveness between the western and eastern part the Polish Carpathian Mountains. In the future, however, this division may become more apparent. The eastern part of the area has a less developed settlement network and a lower population density overall (due to historical reasons) and is bordered by less developed northeastern Slovak peripheries and by the Ukraine, which has become less accessible since 2007 due to Schengen regulations. Contrary, the western part of the area borders the highly developed region of the Czech Silesia and tourist regions of the Slovak Tatra Mountains.

5 Conclusion

Current trends in economic and social development suggest that towns will remain the centers organizing the socioeconomic environment well into the twenty-first century. At the global scale the dominant role will be played by a small number of large cities; but at a smaller scale, such as local and regional, even small towns may influence the development, especially in mountainous areas (Knox 2002; Boesch 2006). In the Polish Carpathian Mountains the natural environment has significantly influenced the potential for urban expansion. Environmental conditions and anthropogenic pressures were found particularly strong when comparing results of this study with other studies on the socioeconomic dimension of Polish urban competitiveness (Dziemianowicz 2005). Topography, which limits urban expansion, when combined with the skillful use of environmental assets may improve urban competitiveness. The towns of Cieszyn and Zakopane are good examples here, as these towns topped an investment appeal category ranking (Dziemianowicz 2005), but were ranked low in the overall emulation indicator (e.g., Zakopane ranked low because of its very low topographical rank). Typical barriers to settlement, such as steep slopes, high altitude, and harsh climate, became a springboard for the economic growth of established mountain-based industries, such as tourism and winter sports (Więcław-Michniewska 2011). The use of natural assets, however, should be linked with care for the environment for the purpose of sustainable development. Indeed, climate change has already been forcing winter sports resorts to develop their summer season potential (Bourdeau 2008). A potential opportunity for the growth of the competitiveness of mountain towns can be found in the modernization of economic systems at a global scale and in the development of new technologies allowing remote access over a wider area. Towns can also take the opportunity of their attractive location in the mountains and provide adequate services to develop themselves into attractive places to live (Borsdorf 1999a; Papa et al. 2006; Mainet 2007). There are, however, open questions about the scale to which this could realistically improve a town's potential and boost its future growth opportunities.

References

Balon J, German J, Kozak J, Malara J, Widacki J, Ziaja W (1995) Regiony fizycznogeograficzne. In: Warszyńska J (ed) Karpaty Polskie. Przyroda, człowiek i jego działalność, Uniwersytet Jagielloński, Kraków

Bank Danych Regionalnych (2009) GUS. http://www.stat.gov.pl/bdl/app/strona.html?p_name=indeks. Accessed 15 September 2009

Björnsen Gurung A, Bokwa A, Chełmicki W, Elbakidze M, Hirschmugl M, Hostert P, Ibisch P, Kozak J, Kuemmerle T, Matei E, Ostapowicz K, Pociask-Karteczka J, Schmidt L, van der Linden S, Zebisch M (2009) Global change research in the Carpathian mountain region. Mt Res Dev 29(3):282–288. doi:10.1659/mrd.1105

Boesch M (2006) Analysing the influence of cultural differences on regional development in the Alps. DIAMONT data infrastructure for the Alps/Mountain orientated network technology work package 5: final report. www.alpine-space.org/uploads/media/DIAMONT_WP5_report_final.pdf. Accessed 15 September 2009

Borsdorf A (1999a) Quality of life in Alpine towns—with examples from Innsbruck and Bregenz. Rev Geogr Alp 87(1):163–169. doi:10.3406/rga.1999.2931

Borsdorf A (1999b) La qualité de vie dans les villes Alpines. Le cas d'Innsbruck. Rev Geogr Alp 87(4):81–91. doi:10.3406/rga.1999.2969

Borsdorf A (2006) Land-Stadt Entwicklung in den Alpen. Dorf oder Metropolis? In: Psenner R, Lackner R (eds) Die Alpen im Jahr 2020. Alpine space: man and environment 1, Innsbruck. http://www.uibk.ac.at/alpinerraum/publications/vol1/08_borsdorf.pdf. Accessed 15 September 2009

Borsdorf A, Braun V (2008) The European and global dimension of mountain research—an overview. J Alp Res 4:117–129. http://rga.revues.org/index630.html. Accessed 15 September 2009

Bourdeau P (2008) Les défis environnementaux et culturels des stations de montagne. Une approche à partir du cas français. Téoros Rev Rech Tour 27(2):23–30

Debarbieux B, Rudaz G (2008) Linking mountain identities throughout the world: the experience of Swiss communities. Cult Geogr 15:497–517. doi:10.1177/1474474008094619

Denisiuk Z (1995) Ochrona przyrody i krajobrazu. In: Warszyńska J (ed) Karpaty Polskie. Przyroda, człowiek i jego działalność. Uniwersytet Jagielloński, Kraków

Dziemianowicz W (2005) Ranking atrakcyjności inwestycyjnej miast Polski—refleksje po czterech edycjach badań. Pr St Geogr 35:109–127

Fourny MCh (1999) Affirmation identitaire et politiques territoriales des villes Alpines. Rev Geogr Alp 87(1):171–180. doi:10.3406/rga.1999.2932

Fourny MCh (2000) L'identité Alpine: un enjeu géopolitique pour les villes. Histoire des alpes-storia delle Alpi–Geschichte der Alpen. 2000/2005 Ville et montagne. http://www.arc.usi.ch/en/index/aisa/ris_ist_labi_aisa_rivista.htm. Accessed 15 September 2009

Framework convention on the protection and sustainable development of the Carpathians (2003) http://www.carpathianconvention.org/NR/rdonlyres/65B750E0-30A7-4AE4-BE8A-45EF13D989AA/0/CarpathianConvention.pdf. Accessed 15 September 2009

Gaczek WM, Rykiel Z (2000) Konkurencyjność regionów a ich adaptacyjność. In: Ciok S, Ilnicki D (eds) Przekształcenia regionalnych struktur funkcjonalno przestrzennych. Uniwersytet Wrocławski, Wrocław

Górka Z (1995) Osadnictwo. In: Warszyńska J (ed) Karpaty Polskie. Przyroda, człowiek i jego działalność. Uniwersytet Jagielloński, Kraków

Górka Z (2004) Les villes des Carpates Polonaises a l'époque de la transition socio-economique de la Pologne. Zesz Nauk UJ Pr Geogr 113:145–152

Harańczyk A (2002) Uwarunkowania konkurencyjności miast w Polsce. In: Słodczyk J (ed) Przemiany bazy ekonomicznej i struktury przestrzennej miast. Uniwersytet Opolski, Opole

Hoyaux AF (2000) Habiter la Ville et la Montagne: Essai de Géographie Phénoménologique sur les relations des habitants au Lieu, à l'Espace et au Territoire (Exemple de Grenoble et Chambéry). Dissertation, Université Joseph-Fourier–Grenoble I. http://tel.archives-ouvertes.fr/docs/00/04/74/98/PDF/tel-00007867.pdf. Accessed 15 September 2009

Illés I (2008) Visions and Strategies in the Carpathian Area (VASICA). Centre for Regional Studies of Hungarian Academy of Sciences, Transdanubian Research Institute. http://w3.dti.rkk.hu/kiadv/discussion/discussionspec07.pdf. Accessed 15 September 2009

Illés I, Zoltán G (eds) (2007) Socio-economic analysis of the Carpathian area. Centre for Regional Studies of Hungarian Academy of Sciences, Transdanubian Research Institute. http://www.carpathianconvention.org/NR/rdonlyres/61729C4D-4624-4A41-898F-50D00FFF4F0B/0/SocioeconomicanalysisoftheCarpathianArea.pdf. Accessed 15 September 2009

Kiryakow D (2003) The participation of the population. In: Spatial planning for the sustainable development of particular types of European areas: mountains, coastal zones, rural zones, flood-plains and alluvial valleys. Council of Europe Publishing, Strasbourg

Knox PL (2002) World cities and the organization of global space. In: Johnston RJ, Taylor PJ, Watts MJ (eds) Geographies of global change, 2nd edn. Blackwell, Oxford

Kondracki J (1978) Geografia fizyczna Polski. PWN, Warszawa

Mainet H (2007) L'attractivité territoriale dans les projets de développement des petites villes: approche géographique, Proc XLIIIe Colloque de l'ASRDLF (Association de Science Régionale de Langue Française), Les dynamiques territoriales: débats et enjeux des différentes approches disciplinaires, Grenoble et Chambéry, 11–12 et 13 juillet 2007. http://edytem.univ-savoie.fr/d/asrdlf2007/pub/resumes/textes/Mainet.pdf. Accessed 15 September 2009

Małopolska (2015) Strategia rozwoju województwa małopolskiego na lata 2007–2013 (2006) Zarząd Województwa Małopolskiego. http://www.wrotamalopolski.pl/root_BIP/BIP_w_Malopolsce/root_UM/podmiotowe/Strategia/dokument/SRWM+2015/default.htm. Accessed 15 September 2009

Markowski T (2005) Przedmiotowa i podmiotowa konkurencyjność regionów. Biul KPZK PAN 219:24–37

McCann EJ (2007) Inequality and politics in the creative city-region: questions of livability and state strategy. Int J Urban Reg Res 31(1):188–196. doi:10.1111/j.1468-2427.2007.00713

Papa F, Collet L, Landel PA (2006) TIC et développement durable : du maillage à la valorisation des territoires de montagne. Rev Geogr Alp 94(3):89–97. doi:10.3406/rga.2006.2409

Parysek JJ, Mierzejewska L (2009a) Problemy funkcjonowania i rozwoju miast polskich z perspektywy 2009 r. Rozw Reg Polit Reg 6:9–25

Parysek JJ, Mierzejewska L (2009b) Wewnątrzregionalna konkurencyjność i komplementarność obszarów. Rozw Reg Polit Reg 6:101–123

Pogačnik A (2003) Instruments of regional planning for mountain and coastal zones. In: Spatial planning for the sustainable development of particular types of European areas: mountains, coastal zones, rural zones, flood-plains and alluvial valleys. Council of Europe Publishing, Strasbourg

Raport o stanie środowiska w województwie małopolskim w 2008 Roku (2009) WIOś w Krakowie, Kraków. http://www.krakow.pios.gov.pl/publikacje/raporty/raport08/index.htm. Accessed 15 September 2009

Runge J (2006) Metody badań w geografii społeczno-ekonomicznej—elementy metodologii, wybrane narzędzia badawcze. Wydawnictwo UŚ, Katowice

Słodczyk J, Szafranek E (eds) (2008) Mechanizmy i uwarunkowania budowania konkurencyjności miast. Uniwersytet Opolski, Opole

Stan środowiska w województwie podkarpackim w latach (2000–2007) (2008) WIOś, Rzeszów. http://www.wios.rzeszow.pl/pl/1,60,170,367/2/raport_za_lata_2000__2007.html. Accessed 15 September 2009

Stan środowiska w województwie śląskim w (2008) http://www.katowice.pios.gov.pl/index.php?tekst=monitoring/raporty/2008/i. Accessed 15 September 2009

Strategia rozwoju województwa małopolskiego (2000) Urząd Marszałkowski Województwa Małopolskiego http://www.malopolskie.pl/strategia/pliki/strategiamalopolskiverpolska.pdf. Accessed 15 September 2009

Strategia rozwoju województwa podkarpackiego na lata 2007–2020 (2006) Zarząd Województwa Podkarpackiego. http://www.si.podkarpackie.pl/pliki/2008/strat_woj.pdf. Accessed 15 September 2009

Turnock D (2002) Ecoregion-based conservation in the Carpathians and the land-use implications. Land Use Policy 19(1):47–63. doi:10.1016/S0264-8377(01)00039-4

Więcław-Michniewska J (2011) Wybrane warunki rozwoju funkcji turystycznej miast Karpat Polskich w świetle badań osadniczych. Zesz Nauk UJ Pr Geogr 125:179–196

Part IV
Methods

Data, Techniques, and Methods in the Carpathian Research

Katarzyna Ostapowicz

The Carpathian Mountains represent a complex system with various factors influencing synergies between environmental and human subsystems of the entire region. Understanding these natural and social processes and phenomena requires access to and use of high quality and accurate data, and sophisticated technologies and methods. In particular, development of geographic information science and technology (GIS&T) over the last three decades (Goodchild 2010) has played a significant role in advancing the availability of geospatial data and development of digital geodatabases. Abundant satellite data, including Landsat data from the National Aeronautics and Space Administration (NASA) or ENVISAT data from the European Space Agency (ESA) archives, are good examples. At the same time, technological advances in remote sensing (RS) and geographic information systems (GIS) have developed multiple and integrative spatial–temporal approaches to assess, analyze, and model natural or social processes, addressing vulnerability, resilience, and sustainability in unique mountain systems such as the Carpathians.

While parts I–III of this book have introduced various aspects of the Carpathian environmental and human subsystems, part IV is composed of a set of eleven chapters is dedicated to methodological issues that frame Carpathian studies and often lead to specific and innovative research approaches focused on mountainous areas. The methodological aspects of these studies are mostly related to two components: (1) data collection and preprocessing, and (2) techniques and methods to assess, analyze, and model phenomena or the processes in question. As the data collected are mostly spatial, these issues are strongly connected with the use of GIS&T.

The first six chapters of this section focus on the methodology of data collection (from historical maps to field-collected data) and preprocessing. In the first chapter of this section, availability and content of hydrological maps for the Polish part of

K. Ostapowicz (✉)
Institute of Geography and Spatial Management, Jagiellonian University, Gronostajowa 7, 30-387 Kraków, Poland
e-mail: katarzyna.ostapowicz@uj.edu.pl

J. Kozak et al. (eds.), *The Carpathians: Integrating Nature and Society Towards Sustainability*, Environmental Science and Engineering,
DOI: 10.1007/978-3-642-12725-0_38,

the Carpathians is discussed. The next five chapters focus on the application and practical use of terrestrial photography and Light Detection and Ranging (LiDAR) and computer tomography in various fields, like tourism, natural hazards assessment, or forestry. The last five chapters of the section mostly discuss the methods used for analysis of data, in addition to issues of data collection, integration, and standarization. In chapter seven, methods to assess land cover changes with the optical RS data from Moderate Resolution Imaging Spectroradiometer (MODIS) in the pan-Carpathian scale is proposed. The next two chapters describe approaches to estimate spatial landscape patterns, while the last two deal with connectivity assessment of protected areas.

Acquisition and access to accurate geodata are fundamental elements of environmental and human studies, which are often difficult to carry out, particularly when working with historical data. Siwek et al. address this important issue in the context of hydrographical maps of the Polish part of the Carpathians. They review availability of these maps since 1785 with a focus on the Hydrographic Map of Poland (Mapa Hydrograficzna Polski). This map, at a scale of 1:50,000, has been published since the early 1960s and is currently connected with two European Union directives: the Water Framework Directive (European Commission 2000) and the Infrastructure for Spatial Information in the European Community (INSPIRE) Directive (European Commission 2012). The authors point out that in recent years substantial progress in the digital mapping of hydrographic features in Poland has occurred; however, there are still areas with poor map coverage. In the chapter they also underline advantages of hydrographic maps in various fields of research and practice, such as water resource management, land development, spatial management, flood protection, and environmental conservation, since these provide valuable long-term data. Siwek et al. also discuss a situation specific to mountainous areas such as the Carpathians in which more water features are contained per unit area than lowlands, and thus these areas require a different approach for the generalization of map content.

The need to use specific approaches in mountainous regions is also emphasized by Kolecka in "True-3D Imaging of Mountainous Regions: Case Study from the Polish Tatra Mountains" but in the context of terrain relief and its visualization. She introduces an alternative to the classical, two-dimensional (2D) method that may allow a more complex, three-dimensional (3D) visualization of maps for mountain environments. For selected areas in the Tatra Mountains she used a lenticular foil technology that generates 3D analogue maps and images of tourist and climbing routes using ortophotomaps and digital elevation models (DEMs). In this way climbing routes can be overlaid on a 3D model presenting either the whole mountain ridge, one valley, or even a particular rock wall, with interesting and convincing visualization effects. Kolecka also underlines the practical value of these maps in regional promotion that could help preserve the natural and cultural heritage for future generations.

The issue of data availability (including historical information) and also data and modeling accuracy is discussed in two chapters by Chrustek et al., who discuss snow avalanche hazard mapping. In the first study they focus on new, promising possibilities of receiving very accurate information about avalanches from a single (historical) terrestrial photograph to update an avalanche cadastre. In general, the idea is based on creating a georeferenced orthoimage from a terrestrial photograph by means of photogrammetry and computer vision, and then creating on-screen vectorization of the avalanche extents. In "Obtaining Snow Avalanche Information by Means of Terrestrial Photogrammetry: Evaluation of a New Approach", they investigate the influence of input data quality on estimating potential avalanche release areas. The authors compare different types of DEMs, placing emphasis on very high spatial resolution information generated from Airborne Laser Scanning (ALS) and Terrestrial Laser Scanning (TLS). Their study confirm that not only quality but also spatial resolution of DEMs (related to the level of generalization) influence the accuracy of release area estimation and hence various snow avalanche parameters calculated by dynamic models.

The question of how to acquire more detailed and accurate data is the focus of a study by Wężyk et al. They elaborate upon a semiautomatic method of the ALS point-cloud data processing for revising the digital forest map of the Tatra National Park, Poland, and updating attributes stored in the descriptive forest database. The study demonstrates that a relatively new technology like LiDAR collects very detailed information about forests, even in inaccessible mountain areas like extremely steep slopes or isolated ledges. Another type of remote measurements of forest stands is presented by Niemtur et al. who analyze fir stands infested by fungal pathogens in different parts of the Polish Carpathians by using computer tomography. The advantage of this method is that it is not invasive (measurements are carried out without tree cutting) and increases high accuracy of collected data.

These chapters all confirm that in local-scale studies very detailed, high accuracy data are required. However, because research scale determines data generalization, regional research is typically based on data that have a relatively high generalization level. Therefore, information obtained from these data provides mostly an overview of trends in natural and social processes, although sometimes locations of hotspots important for the studied phenomena can be identified as well. For instance, in "Assessment of Land Cover Changes in the Carpathian Mountains with MODIS Data", Jaśkowiec investigates land cover changes in the pan-Carpathian scale using medium spatial resolution RS data (MODIS) and one of the MODIS derived products: the Normalized Difference Vegetation Index (NDVI) dataset. She demonstrates that these data determine with good accuracy general trends in contemporary land cover changes in the Carpathian region, while simultaneously showing the need to combine MODIS data with satellite imagery of higher spatial resolution, such as Landsat data, in order to analyze land cover changes in more detail and from a long-term perspective.

In studies by Smaliychuk and Kruhlov and also by Mkrtchian research are conducted on a local scale and focus on landscape spatial pattern assessments that are connected to, and of key importance for, nature conservation. In both chapters different sources of geodata, RS, and GIS methods have been introduced, and the value of their use is discussed and emphasized. Smaliychuk and Kruhlov analyze forest cover changes between the 1980s and the 2000s with respect to the natural landscape structure and proximity to roads and settlements in two municipalities in the Ukrainian Carpathians: Stara Sil' and Boberka. As a source of information they use topographic maps and Google Earth, which are analyzed with apply simple GIS methods like vectorization and map algebra. The important intermediate product of the study is a geodata set of natural geo ecosystems (GES). The results show an increase in forest cover due to forest succession on former agricultural land to be a common trend. Mkrtchian investigates possibilities and advantages of using habitat quality and diversity measures in the design of ecological networks in the Ukrainian Carpathians. He showes that habitat diversity (assess using NDVI derived from Landsat data) may be approximated by the measures related to topographic variance, and correlates with ecological and socioeconomic factors, thereby determining land suitability classes for conservation purposes.

The last two chapters of Part IV assess nature conservation from a slightly different angle, focusing on methodological issues related to the assessment of landscape connectivity and development of ecological networks. In addition, like the other studies, these chapters also discuss the use of different geodata and GIS methods. Bianchin and Neubert analyze structural connectivity of the existing network of protected areas using a nearest-neighbor analysis. The area under investigation stretches from the Baltic Sea to Ukraine and to the Adriatic Sea, covering large parts of Central and Eastern Europe. An important part of this study are storing and integration of different types of national, as well as European and world, data sets of protected areas. Integration of national data into a coherent European data set is always a crucial and difficult issue in regional studies. In contrast, Deodatus et al. look more closely at methodological aspects of the creation of functional and consolidated ecological corridors in the Carpathians. They carry out a pilot study at two locations in Ukraine, proposing delimitation of corridors connecting the protected areas in Ukraine with those in Romania and Poland. The methodology is based on landscape ecological modeling, using the habitat requirements of brown bear, European bison, lynx, and wildcat to locate the most suitable corridor areas. Manageable corridors were created by identifying interconnected land management units with a minimal number of obstacles for wildlife and conflicts with land use, and forming the shortest possible connections. It is noteworthy that corridors and their management plans were developed in consultation with practitioners: users and owners of the land.

The eleven chapters of Part IV of this book present only a small proportion of the contributions to the twelve thematic sessions of the 1st Forum Carpaticum that addressed various methodological issues related to data, techniques, and methods. The clear outcome from the meeting was that a complex and complete description of the Carpathian Mountains system and its changes, in the pan-Carpathian scale, is only possible with the wide use of new technologies and tools for data acquisition, processing, analysis and modeling, and with a coherent environmental information system using all aspects of the geospatial technologies (Kozak et al. 2011). The first attempts to build a pan-Carpathian database, such as the Carpathian geoportal developed under the INTERREG CADSES framework (Carpathian Project 2009), were important, yet could be treated only as the preliminary steps toward this goal, and therefore were not sufficient for the scientific community and various stakeholders. However, a much more elaborated information system for the Carpathians seems to be a logical consequence of the implementation of the INSPIRE directive (European Commission 2012) and the huge inflow of remotely sensed data. The full use of the current and future capabilities of the geospatial technologies by researchers and practitioners requires better access to, and knowledge about, the existing or new data sets, and further development of methodologies that analyze and model the environmental and human subsystems of the Carpathian Mountains. Several studies on land cover change detection (with a strong focus on innovative RS technologies) and its causes and consequences (e.g., Kuemmerle et al. 2011; Griffiths et al. 2012), as well as studies on natural conservation that use sophisticated GIS technologies and methods (e.g., Kuemmerle et al. 2010; Ziółkowska et al. 2012) exemplify these recommended research trends. Past and present Carpathian studies, including these introduced in this book, demonstrate that the current level of knowledge about the Carpathian Mountains cannot be achieved without proper and wide use of GIS&T.

References

Carpathian Project (2009) VASICA. Visions and Strategies in the Carpathian area. Carpathian Project, Vienna. http://www.carpathianproject.eu/portal/. Accessed 30 April 2012

European Commission (2000) Directive 2000/60/EC of the European Parliament and of the Council of 23 October 2000 establishing a framework for Community action in the field of water policy. Official J European Comm L 327(43):1–72

European Commission (2012) INSPIRE Infrastructure for Spatial Information in the European Community. http://inspire.jrc.ec.europa.eu. Accessed 17 May 2012

Goodchild MF (2010) Twenty years of progress: GIScience in 2010. J Spatial Inf Sci 1:3–20

Griffiths P, Kuemmerle T, Kennedy RE, Abrudan IV, Knorn J, Hostert P (2012) Using annual time-series of Landsat images to assess the effects of forest restitution in post-socialist Ro-mania. Remote Sens Enviro 118:199–214

Kozak J, Björnsen Gurung A, Ostapowicz K (2011) Research agenda for the Carpathians 2010–2015. Integrating nature and society towards sustainability. Institute of Geography and Spatial management, Jagiellonian University, Kraków

Kuemmerle T, Olofsson P, Chaskovskyy O, Baumann M, Ostapowicz K, Woodcock CE, Houghton RA, Hostert P, Keeton WS, Radeloff VC (2011) Post-Soviet farmland abandonment, forest recovery, and carbon sequestration in western Ukraine. Glob Chang Biol 17:1335–1349

Kuemmerle T, Perzanowski K, Chaskovskyy O, Ostapowicz K, Halada L, Bashta AT, Kruhlov I, Hostert P, Waller DM, Radeloff VC (2010) European bison habitat suitability in the Carpathian mountains. Biol Conserv 143(4):908–916

Ziółkowska E, Ostapowicz K, Kuemmerle T, Perzanowski K, Radeloff VC, Kozak J (2012) Po-tential habitat connectivity of European bison (Bison bonasus) in the Carpathians. Biol Conserv 146(1):188–196

Hydrographic Maps as Sources of Information About the Polish Carpathians

Janusz Siwek, Wojciech Chełmicki and Maria Baścik

Abstract The chapter is a review of both historical and modern hydrographic maps of the Polish Carpathians. The emphasis of the review is on the Hydrographic Map of Poland (Mapa Hydrograficzna Polski), at a scale of 1:50,000, which has been published since the early 1960s. Special attention is paid to the hydrological characteristics of mountainous areas such as the Carpathians, which contain more water features per unit area than lowlands. Given this diverse water distribution, a different approach is needed for the generalization of map contents, especially with respect to the shape of the river network and groundwater outflows as springs, bog springs, and seeps. Hydrographic maps are sources of information that are useful for different purposes, such as water resource management, land development, spatial management, flood protection, and environmental conservation. The recording of hydrological phenomena allows for an assessment of changes in the aquatic environment over the long-term.

1 Introduction

Mountains are headwater areas for most large rivers. The supply of water across neighboring lowlands depends on water influx from mountains. The same is true of the Polish Carpathians, which serve as water towers for the large water system of the Vistula River. The spatial distribution of a variety of water features in the

Wojciech Chełmicki – deceased in October 2011

J. Siwek (✉) · W. Chełmicki · M. Baścik
Institute of Geography and Spatial Management, Jagiellonian University,
Gronostajowa 7, 30-387 Kraków, Poland
e-mail: j.siwek@uj.edu.pl

J. Kozak et al. (eds.), *The Carpathians: Integrating Nature and Society Towards Sustainability*, Environmental Science and Engineering,
DOI: 10.1007/978-3-642-12725-0_39, © Springer-Verlag Berlin Heidelberg 2013

Carpathian Mountains results from a diverse combination of geology, landforms, climate, soil types, and plant cover.

Economic activity plays an increasing role in the creation and use of water resources. Water resources are also affected by changes in land use related to economic activity. The diversity of hydrological phenomena reflects different conditions for water resource renewal. A hydrographic map is a perfect means of summarizing water phenomena and the variety of economic uses of water resources. The purpose of this chapter is to review and assess the usefulness of hydrographic maps of the Polish Carpathians, with an emphasis on the Hydrographic Map of Poland (*Mapa Hydrograficzna Polski*), at a scale of 1:50,000, which has been published since the 1960s.

2 Historical Outline of the Development of Hydrographic Maps of the Polish Carpathians

Jan Długosz, a Polish historian and chronicle writer who lived in the fifteenth century, described rivers in detail in a chronicle titled Chronicles of the Glorious Kingdom of Poland (*Annales seu cronicae incliti Regni Poloniae*), and specifically in the part titled Chorographía of the Kingdom of Poland (*Chorographia Regni Poloniae*). His description of rivers was very detailed for his era. It was a pioneering work, accurately describing the river network of the fifteenth century Poland. Particularly important at the time was information about waterways (transportation, floating timber), which were generally well depicted on historical maps.

One of the first hydrographic maps of Poland was a map by Karol de Perthées, a Polish cartographer of French origin. He created the map in 1785 and called it Hydrographic Map of Poland (*Carte hydrographique de la Pologne*) at a scale of about 1:1,350,000. His map was published in Paris in 1809. It depicted the location of about 3,100 rivers and 150 lakes (Graf et al. 2008). The map also covered the northern part of the Carpathian Mountains.

The first thematic map of the Tatra Mountains was published in 1796. It was printed from engraved copperplates. The map's author was a professor of natural history at the University of Lwów (Lemberg)-Baltazar Hacquet. The river network is marked on the map, but only large rivers have proper names. Another map, the Kingdom of Galicia and Lodomeria (scale: 1:288,000) (Fig. 1), was published in 1824 and it was the most accurate map of the region at the time (Liesganig 1824). This map was prepared and published by Józef Liesganig who was a mathematician in the town of Košice (now in Slovakia). He prepared the map based on a military survey carried out by Austrian topographers in the years 1763–1785 (Baścik 2006). It was called "Josephinian" because it was made for the purpose of completing information for a land cadastre ordered by the Austrian Emperor Joseph II of the House of Habsburg. One more map worth mentioning is the first

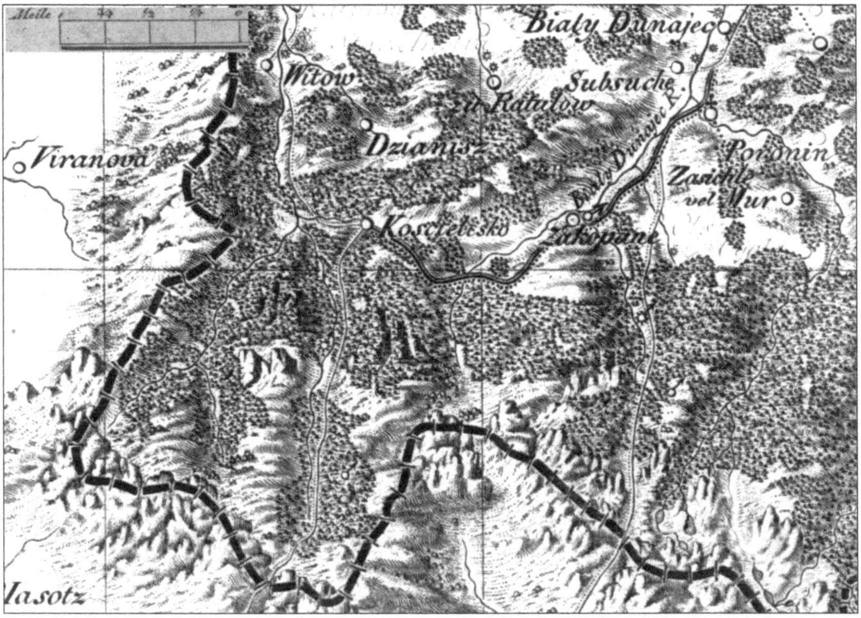

Fig. 1 Fragment of map of the Kingdom of Galicia and Lodomeria (scale: 1:288,000) by J. Liesganig (1824) showing headwater area of the Dunajec River in the Western Tatra Mountains. (Courtesy of Wiesław Siarzewski)

detailed map of the Tatra Mountains by Albrecht Sydow (scale: 1:200,000), which was published in 1830. As discussed by Szaflarski (1972), the river network on the map was clearly different from the river network known today, which is due to the lack of good triangulation methods at the time.

Wincenty Pol was the founder of the Department of Geography at the Jagiellonian University. His work was a special contribution to the field of the hydrography of the Carpathians. The manuscript of his monograph titled The Northern Side of the Carpathians (*Północne stoki Karpat*) along with accompanying maps is dated 1844. His most significant work was titled Hydrography (*Hydrografia*) (Pol 1851). It contained a detailed description of the river network supplemented with information on the location of their respective springs. Though the original hydrographic maps of Poland created by Pol were lost, they were later reconstructed by cartographers. Poland's river network was later redrawn based on the descriptions made by Pol (Niemcówna 1923).

An extensive Hydrography of the Polish Lands (Hydrografia ziem polskich) was published by Ludomir Sawicki (1912), who was the father of modern geography at the Jagiellonian University. His work included a hydrographic map of Poland with the river network, watershed boundaries, and wetlands.

Lake cartography of the Tatra Mountains also has a long history. Before World War I, bathymetric maps of the 8 largest lakes of the Tatra Mountains were drawn by Eugeniusz K. Dziewulski (Trafas 1996). In the years 1908–1910, measurements of lakes were performed by Ludomir Sawicki (1912). As a result, the Atlas of Tatra Mountains Lakes (*Atlas jezior tatrzańskich*) (Sawicki 1929) was compiled and contained topographic maps of 15 lakes (scale: 1:2,000). This was the first work providing so much detail on the bathymetry of the lakes. This project was subsequently continued and contributions were made by W. Ormicki, Z. Korosadowicz, J. Szaflarski, and J. Młodziejewski. J. Szaflarski drew topographic bathymetric maps of Tatra lakes at a scale of 1:500 with 5 m interval isobaths. Altogether, 36 lakes were measured by 1935. The mapping process led to the creation of a three-volume Atlas of Tatra Mountains Lakes (Szaflarski 1933, 1935; Gajda et al. 1936).

Hydrographic maps published before World War II focused on river networks, selected springs, lakes, and wetlands. In a sense, they presented limited information. Prewar maps lacked information on hydrological characteristics such as differences in the magnitude of stream flow and river runoff.

An increase in the number of gauging stations in the 1950s and 1960s and a corresponding increase in the amount of hydrological information available allowed for the determination of more reliable values of specific elements of the water balance. This led to the creation of the first runoff maps. Dynowska (1971) produced a map of river regime types for Poland. The map designated areas in the Carpathians with the same percentage of groundwater, pluvial, and nival recharge.

In most of the Polish Carpathians, except the Tatra Mountains and the Podhale foothills recharge from the surface (pluvial and nival) is the dominant factor. In the Tatra Mountains and the Podhale foothills recharge from the surface and underground recharge are equally important. The Map of River Runoff Regimes (*Mapa reżimu odpływu rzecznego*) created by Dynowska was also published many years later in edited form as part of the Atlas of the Republic of Poland (*Atlas Rzeczypospolitej Polskiej*) (Dynowska 1994a). The western part of the Polish Carpathians was determined to be an area dominated by pluvial and nival recharge, while the area to the east of the Dunajec River is an area dominated by nival and pluvial recharge. The Bieszczady mountain range is an area of strong nival recharge. It was determined that variability in diurnal discharge in major Carpathian rivers is high or very high.

In 1974, the Hydrological Map of the World was published, where the map of Poland served as an example at a scale of 1:2,500,000 (Dynowska et al. 1974, 1976).

The value of specific runoff for the Carpathian foothills (Pogórze Karpackie) was determined to be 5–10 $dm^3\ s^{-1}\ km^{-2}$, while for the Beskidy range, it was 10–20 $dm^3\ s^{-1}\ km^{-2}$, and in the rest of the area it was 20–50 $dm^3\ s^{-1}\ km^{-2}$. The runoff coefficient (precipitation/runoff) for the Carpathian foothills was determined to be 0.3–0.5, and over 0.5 for the rest of the Carpathians. Hydrographic maps, similar to those created in 1976, were published in atlases of the provinces of Kraków (Dobija et al. 1979), Bielsko (Dobija et al. 1981), and Tarnów (Dobija et al. 1988). The maps contained graphs of the average monthly coefficient of discharge for major Carpathian rivers (discharge for specific months with respect to the average yearly discharge).

The Hydrological Atlas of Poland (*Atlas Hydrologiczny Polski*) (Stachy 1987) contained an array of maps of river runoff, a map of flooding periods, a map of groundwater runoff, and a map of the raw water balance, as well as other maps. Especially worth mentioning are maps of snow cover in the Carpathians over the 6-month winter period. Some of the maps were modified and later published in the Atlas of the Republic of Poland. Examples include maps of average (Stachy and Biernat 1994) and maximum specific runoff (Fal and Punzet 1994). The latter characteristic was estimated for the Tatra Mountains to be over 4,000 $dm^3\ s^{-1}\ km^{-2}$.

In 1995, an extensive geographic monograph on the Polish Carpathians was published (Warszyńska 1995). One chapter was dedicated specifically to hydrology (Dynowska 1995) and contains isoline maps featuring mean, maximum, and minimum specific runoff values at 50 % probability of occurrence. It also includes a map of summer low-flow characteristics based on the work of Tlałka (1982).

In addition to the maps mentioned above, the first cartographic summary of the regional distribution of springs in Poland was published by Dynowska (1994b) as part of the Atlas of Natural Resources and Threats to the Geographic Environment of Poland (*Atlas zasobów walorów i zagrożeń środowiska geograficznego Polski*). Springs in the Carpathian foothills are predominantly very-low-discharge fissure springs in a porous aquifer. The Beskidy Mountains and the Spisko-Gubałowskie foothills are areas with a large number of low-discharge fissure flysch-type springs as well as seepage springs in weathering rock cover. The Orawsko-Nowotarska basin is an area of very few, very-low-discharge springs located in sand and weathering rock cover. Generally, the Tatra Mountains are an area featuring mostly low-discharge springs in crystalline rocks, limestone, dolomite, and in sandstone. An exception to the rule are high-discharge karst springs in the Western Tatra Mountains. The atlas also contains a map of mineral waters (Płochniewski and Turek 1994).

All the maps mentioned above are general reference maps at a scale of 1:1,000,000 or less. Detailed maps constitute a separate category. The Hydrographic Map of Poland (*Mapa Hydrograficzna Polski*) at a scale of 1:50,000 is an example of a detailed map.

3 Hydrographic Map of Poland at a Scale of 1:50,000

The idea to produce a detailed hydrographic map of Poland at a scale of 1:50,000 was first conceived in the 1950s. The map was to include various categories of water features along with elements of the economic use of water resources.

In 1954, the Instruction on the Development of the Detailed Hydrographic Map (*Instrukcja do opracowania szczegółowej mapy hydrograficznej*) was published (Klimaszewski et al. 1954, 1956) followed by updated editions (Celmer et al. 1958, 1959). The instructions became the basis for the first detailed hydrographic map of the Polish part of the Tatra Mountains. As a result of these efforts, hydrographic maps of the Western Tatras (Wit and Ziemońska 1960) and the High Tatras (Wit-Jóźwik 1974) were created. The maps, along with an extensive commentary, are the most detailed depiction of water networks on the northern slopes of the Tatras to date.

A later edition that appeared as a chart called Hydrography in the Atlas of Tatra National Park (Atlas Tatrzańskiego Parku Narodowego) (1985) was identical to the one originally printed. The 2006 edition (map sheets: Zakopane and Murzasichle) is a slightly generalized, updated version of the original map. The updates primarily concern water management facilities and the current state of water quality, as well as changes resulting from the introduction of GIS-3 Technical Guidelines (*Wytyczne techniczne GIS-3*) (Główny Geodeta Kraju 2005). At present, in addition to its graphical form, the map is available in a digital format and accessible to the general public at the website of the Polish Head Office of Geodesy and Cartography (www.geoportal.gov.pl). The georeferenced map may also be obtained via the Web Map Service (WMS), thus making it possible to use it directly in GIS (geographic information system) applications.

The first map sheets covering areas outside of the Tatra Mountains were created in the 1980s. Map sheets at a scale of 1:50,000 covered the Western Carpathians and were made using the "1965" reference system. Today, the area covered is located in the Śląskie Province.

Later editions of hydrographic maps presented a slightly different picture because of changes of the national reference systems in Poland. In 2006, map sheets developed via standard GIS-3 methods covered the Western Carpathians in the Śląskie and Małopolskie Provinces (Fig. 2). No map sheet has been published thus far that would cover the current Podkarpackie Province.

The hydrographic maps of Poland at a scale of 1:50,000 are substantially correlated with the Water Framework Directive (2000/60/EC) and the directive establishing an Infrastructure for Spatial Information in the European Community (INSPIRE) (2007/2/EC). The maps cover most of the hydrographic elements listed in the Directives. In recent years, there has been a substantial progress in the

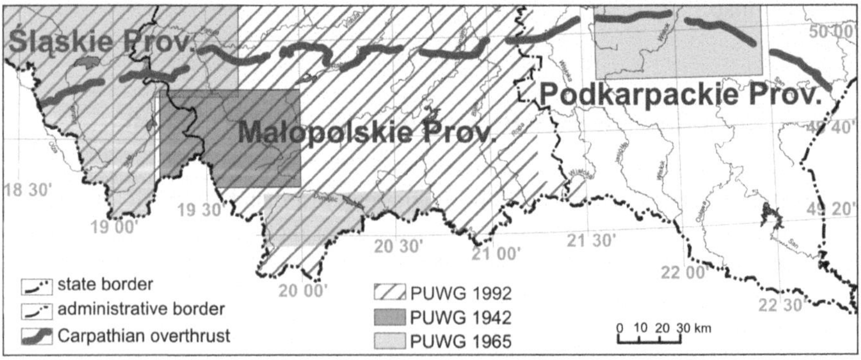

Fig. 2 Hydrographic map coverage of the Polish Carpathians in 2010 and the spatial reference systems (PUWG) used

digital mapping of hydrographic features in Poland; however, there are still areas with poor map coverage, especially in the northeastern part of the country.

4 Map Contents

The content of the latest edition of the Hydrographic Map of Poland is shown using over 70 symbols and includes the following: topographic drainage divides, closed drainage basins, surface waters, groundwater outflows, groundwater in the first horizon with the water table represented using hydro-isobaths, ground permeability, water features and water management facilities, artificial drainage areas, anthropogenic terrain modifications, surface water pollution, precipitation measurement stations, water-level measurement stations for surface water and groundwater, and stream discharge measurement stations. Detailed information about the symbols used on the Hydrographic Map of Poland 1:50,000 can be found in the GIS-3 Technical Guidelines (Główny Geodeta Kraju 2005).

Comments printed on the reverse side of the map address the following issues: natural characteristics of the area, geological structure and lithology, topographic drainage divides, precipitation, surface waters, hydrological characteristics, groundwater, characteristics of the period when hydrographic analysis was performed, surface water quality, and changes in water circulation.

Among the most important changes with respect to the first map sheets produced are new information on the depth of groundwater in the first horizon (using hydro-isobaths) and new information on the lithology and permeability of surface formations represented with a colored field in the background of the map. Drainage divide symbols and symbols for selected categories of features have also changed since the first edition (Fig. 3).

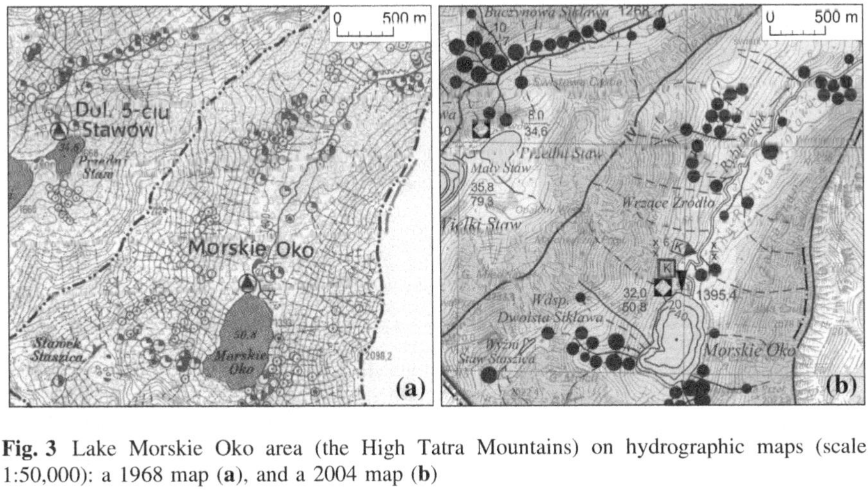

Fig. 3 Lake Morskie Oko area (the High Tatra Mountains) on hydrographic maps (scale: 1:50,000): a 1968 map (**a**), and a 2004 map (**b**)

4.1 Assessment of the Map's Credibility

The geographic environment of Poland is characterized by a wide range of water circulation conditions, intensities and spatial densities of water features, as well as different degrees of hydrotechnical engineering. Mountainous areas such as the Carpathians are characterized by a much more dense river network and a greater density of groundwater outflows than the rest of the country. Local geological issues, especially tectonic activity in flysch areas, have created a unique layout of the first groundwater table. This leads to problems associated with map interpretation.

The first problem concerns differences in hydrological identification and consequently accuracy in mapping. Most areas in the Beskidy Mountains were not mapped in great detail or were mapped as part of master theses in the 1960s and 1970s. Advanced tools such as global positioning systems (GPS) were not available at the time. The quality of base maps varied, as did mapping accuracy. Fieldwork was conducted during different time periods and under different meteorological conditions. For that very reason, an accurate transfer of handwritten data onto maps was either not possible or created reliability problems (Fig. 4).

This is also a cause of differences in map content between certain areas. This is especially true of point data such as low-discharge springs typical of the flysch part of the Carpathians. Sometimes striking differences between neighboring areas do not result from real differences in the terrain but reflect a different degree of recognition of its features. To the extent that groundwater outflow map data may be considered reliable, the incomplete nature of the mapping process for the Carpathian region as a whole makes broader comparative analysis impossible

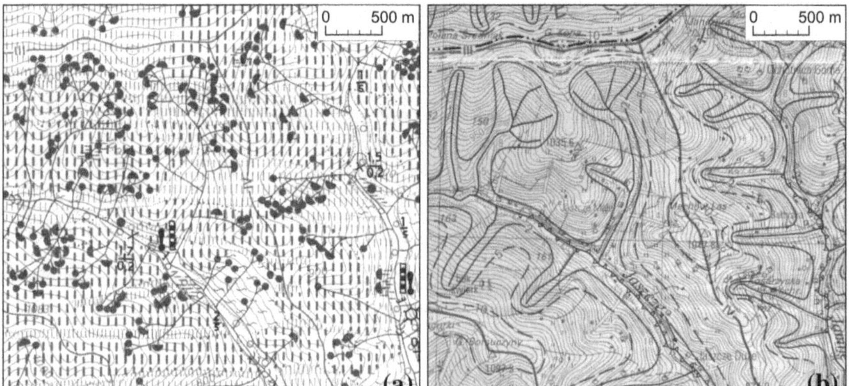

Fig. 4 The catchment of Jaszcze (the Beskidy Mountains). None of the numerous small springs presented on the hydrographic test map from 1960 (**a**), were placed on the hydrographic map from 2002 (**b**) (both fragments have the same spatial extent). The reason was that the first mapping (**a**) was based on detailed fieldwork and covered only a small area while the second map (**b**) is a part of map covering the whole Poland, and is based on less detailed fieldwork

(e.g., number of springs in each drainage basin). In many cases, spring density calculated using different sheets of the Hydrographic Map may lead to incorrect conclusions.

Even when terrain analysis is done properly, the results may not always be easily transferrable onto a map at a scale of 1:50,000, where a broad range of information will be generalized in the map creation process. The procedure for hydrographic map generalization, although standardized for all of Poland, requires modifications for mountainous areas, especially the Carpathians. Such modifications take into account the unique nature of the mountain environment and the topical purpose of the map. The generalization problem is especially noticeable for the generalization of a river network (Fig. 5) or numerous low-discharge springs. River network characteristics are directly based on cartographic materials at a scale of 1:50,000. Meanwhile, the actual river network, represented at a larger, more detailed scale of 1:10,000 possesses a much more complex structure than might be inferred by looking at a topographic map.

Another problematic issue are hydro-isobaths under conditions of discontinuity of the first groundwater table. This is especially true of flysch-type bedrock strongly affected by tectonic movements and local lithology. As much as the course of hydro-isobaths in river valleys does not evoke much doubt, in case of mountain ranges this information is not entirely reliable and may be considered as purely approximate in nature. These and other problems not mentioned here do not apply to most lowland and highland areas where the spatial density of hydrographic features does not require a radical selection of information for the purpose of making a map at a scale of 1:50,000 and geological structure does not create interpretation problems (e.g., groundwater table).

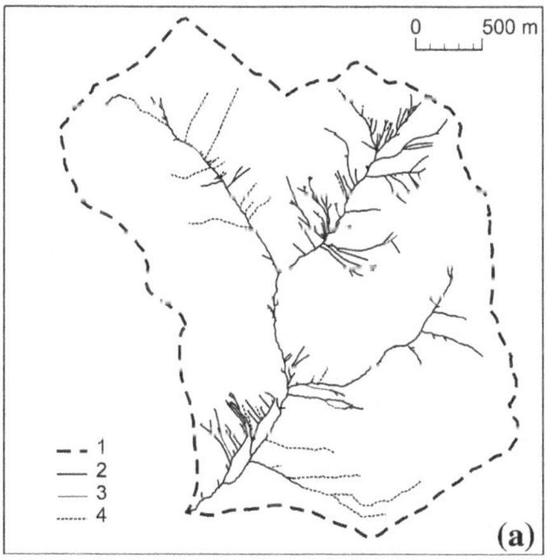

Fig. 5 Comparison of the river networks mapped based on field measurements (Siwek et al. 2009) (**a**), and presented on the topographic map of Poland 1:10,000 (**b**) *1* watersheds, *2* permanent streams, *3* streams in debris, *4* dry valleys

5 Practical Applications of the Map

Large-scale hydrographic maps are useful sources of information for water resource management, drinking water protection zoning, protection plans for natural reserves (national parks and landscape parks), environmental conservation, environmental impact assessments, water use permit studies, residential land management in municipalities, evaluation of waste hazards, flood protection studies, and many others purposes. The following Polish institutions frequently use hydrographic maps: the Institute of Urban Development, Regional Water Management Boards, National and Regional Inspectorates of Environmental Protection, General and Regional Directorates of Environmental Conservation, and local governments, as well as institutions managing national parks and other protected areas.

The detailed recording of hydrological features allows one to assess changes in the water environment over the long term. Changes in the location of springs and the magnitude of groundwater discharge are of special significance. As the "pressure" of tourist activity in mountain national parks increases, the documentation value of hydrographic maps becomes undeniable, especially in longer periods. Finally, there exists an urgent need to update hydrographic maps by extending them to cover the eastern parts of the Polish Carpathians. There is also a need to update hydrographic map content for areas where the terrain has not yet been fully documented in detail.

References

Baścik M (ed) (2006) Dolina Kościeliska. Śladami badaczy, artystów i wędrowców, 2nd edn. TPN, Zakopane

Celmer T, Klimaszewski M, Pietkiewicz S, Stephan W, Werner-Więckowska H, Wilgat T, Wit K (1958) Instrukcja do zdjęcia hydrograficznego Polski. Dok Geogr 3

Celmer T, Klimaszewski M, Pietkiewicz S, Stephan W, Werner-Więckowska H, Wilgat T, Wit K (1959) Instrukcja do zdjęcia hydrograficznego Polski. Dok Geogr 4

Dobija A, Dynowska I, Tlałka A (1979) Hydrologia. In: Trafas K (ed) Atlas miejskiego województwa krakowskiego. Komisja Nauk Geograficznych PAN, Urząd Wojewódzki w Krakowie, Kraków

Dobija A, Dynowska I, Tlałka A (1981) Hydrologia. In:Trafas K (ed) Atlas województwa bielskiego. Komisja Nauk Geograficznych PAN, Urząd Wojewódzki w Bielsku-Białej, Kraków

Dobija A, Dynowska I, Tlałka A (1988) Hydrologia. In: Trafas K (ed) Atlas województwa tarnowskiego. Komisja Nauk Geograficznych PAN, Urząd Wojewódzki w Tarnowie, Kraków

Dynowska I (1971) Typy reżimów rzecznych w Polsce. Zesz Nauk UJ Pr Geogr 28

Dynowska I (1994a) Reżim odpływu rzecznego. In: Atlas Rzeczypospolitej Polskiej. IGiPZ PAN, Warszawa

Dynowska I (1994b) Źródła. In: Atlas zasobów walorów i zagrożeń środowiska geograficznego Polski. IGiPZ PAN, Warszawa

Dynowska I (1995) Wody. In: Warszyńska J (ed) (1995) Karpaty Polskie. Przyroda, człowiek i jego działalność. Wydawnictwo UJ, Kraków

Dynowska I, Dobija A, Tlałka A, Trafas K (1974) Projekt mapy hydrologicznej świata w skali 1:2500000 na przykładzie Polski. Pol Przegl Kartogr 6(2):61–65 (+map)

Dynowska I, Dobija A, Tlałka A, Trafas K (1976) Hydrological map of the world, scale 1:2500000. Geogr Pol 33:67–71

Fal B, Punzet J (1994) Maksymalny odpływ jednostkowy. In: Atlas Rzeczypospolitej Polskiej. IGiPZ PAN, Warszawa

Gajda R, Ormicki W, Szaflarski J (1936) Atlas jezior tatrzańskich III. Zbiorowe Prace Nauke Koła Geografów Uczn UJ 3A, Kraków

Główny Geodeta Kraju (2005) Wytyczne techniczne GIS-3. Mapa hydrograficzna Polski, skala 1:50 000 w formie analogowej i numerycznej. GUGIK

Graf R, Kaniecki A, Medyńska-Gulij B (2008) Dawne mapy jako źródło informacji o wodach śródlądowych i stopniu ich antropogenicznych przeobrażeń. Bad Fizjogr Pol Zach Ser A Geogr Fiz 59:11–27

Klimaszewski M, Pietkiewicz S, Więckowska H, Wit K (1954) Instrukcja do opracowania szczegółowej mapy hydrograficznej Polski. Biul Geogr 7

Klimaszewski M (1956) The detailed hydrographical map of Poland. Przegl Geogr 28(suppl):41–47

Niemcówna S (1923) Wincenty Pol jako geograf. Orbis, Kraków

Liesganig J (1824) Koenigreich Galizien und Lodomerien herausgegeben im Jahre 1790. scale 1:288 000. Wien

Płochniewski Z, Turck S (1994) Wody mineralne. In: Atlas zasobów walorów i zagrożeń środowiska geograficznego Polski. IGiPZ PAN, Warszawa

Pol W (1851) Hydrografja. Kraków

Sawicki L (1912) Hydrografia ziem polskich. In: Encyklopedia Polska. 1 Geografia fizyczna ziem polskich. Charakterystyka fizyczna ludności. AU, Kraków

Sawicki L (1929) Atlas jezior tatrzańskich. Pr Kom Geogr PAU 1, Kraków

Siwek J, Kołodziej A, Laszczak E, Mocior E, Plenzler J, Płaczkowska E, Rozmus M, Rzonca B, Ścisłowicz B, Wójcik S, Ziółkowski L (2009) Geologiczne i geomorfologiczne uwarunkowania wykształcenia sieci hydrograficznej w zlewni górnej Wołosatki (Bieszczady Wysokie). Kwart AGH Geol 35(2):249–261

Stachy J (ed) (1987) Atlas Hydrologiczny Polski t 1. Wydawnictwa Geologiczne, Warszawa

Stachy J, Biernat B (1994) Średni niski odpływ jednostkowy. In: Atlas Rzeczypospolitej Polskiej. IGiPZ PAN, Warszawa

Szaflarski J (1933) Atlas jezior tatrzańskich Tab I-VI. Zbior Pr Nauk Koła Geogr Uczn UJ 3A, Kraków

Szaflarski J (1935) Atlas jezior tatrzańskich II Tab I-V. Zbior Pr Nauk Koła Geogr Uczn UJ 3A, Kraków

Szaflarski J (1972) Poznanie Tatr. Szkice z rozwoju wiedzy o Tatrach do połowy XIX wieku, Warszawa

Tlałka A (1982) Przestrzenne zróżnicowanie niżówek letnich w dorzeczu górnej Wisły. Rozpr Habil UJ 63.1–119

Trafas K (1996) Od mapy Spisza F. Czakiego do mapy satelitarnej Tatr. In: Kotarba A (ed) Przyroda Tatrzańskiego Parku Narodowego a człowiek, 1. Nauki o Ziemi. TPN, Polskie Towarzystwo Przyjaciół Nauk o Ziemi, Kraków–Zakopane

Warszyńska J (ed.) (1995) Karpaty Polskie. Przyroda, człowiek i jego działalność. Wydawnictwo UJ, Kraków

Wit K, Ziemońska Z (1960) Hydrografia Tatr Zachodnich. Objaśnienia do mapy hydrograficznej Tatry Zachodnie 1:50 000. Instytut Geografii PAN, Kraków

Wit-Jóźwik K (1974) Hydrografia Tatr Wysokich. Objaśnienia do mapy hydrograficznej Tatry Wysokie 1:50 000. Dok Geogr 5

True-3D Imaging of Mountainous Regions: Case Study from the Polish Tatra Mountains

Natalia Kolecka

Abstract 3-dimensional (3D) maps, showing the whole mountain ridge, one valley or even a particular rock wall with the climbing route may increasingly offer an alternative to classical maps using traditional 2-dimensional presentation of relief. Geographic information systems (GIS) software enables creating 3D views on the screen, however it is not a sufficient solution as the map are used mostly in the field. In this work a lenticular foil technology is utilized to produce 3D analogue maps and images of tourist and climbing routs in the Polish part of the Tatra Mountains that can be perceived with an unaided eye. The advantages of the approach are producing more realistic and visually interesting illustrations of high mountain sites that may help to promote the region among visitors.

1 Introduction

Since the beginning of the nineteenth century the mapping campaigns in the mountain areas were supported by professional surveying technologies, that enabled creation of relatively accurate maps in medium scales. Yet the relief representation remained a very problematic aspect (Imhof 1982). Hill profiles and hachures were complemented by different methods, such as contour lines, shading, rock and cliff drawing or hypsometric tints. Development of photogrammetric and satellite techniques had a revolutionary influence on updating the topographic information and creating orthophotomaps. Nowadays, the geographic information systems (GIS) and computer graphics enable to manipulate and create an abstract representation of various areas (Haeberling and Hurni 2001). The most popular is a

N. Kolecka (✉)
Institute of Geography and Spatial Management, Jagiellonian University, Gronostajowa 7, 30-387 Kraków, Poland
e-mail: nkolecka@gis.geo.uj.edu.pl

J. Kozak et al. (eds.), *The Carpathians: Integrating Nature and Society Towards Sustainability*, Environmental Science and Engineering,
DOI: 10.1007/978-3-642-12725-0_40, © Springer-Verlag Berlin Heidelberg 2013

3-dimensional (3D) on-screen visualization—both a static and a dynamic one. The geoinformation offers many possibilities to explore various visualization strategies. In tourism domain, the main purpose is to promote unique landscapes, natural environment or cultural heritage, and give an overview of the place we are going to visit. It should be easily understandable by everyone (Almer et al. 2004).

More than 50 % of map users are not able to spontaneously interpret height information from the contour lines (Buchroithner et al. 2004). For that reason some effort should be made to enable the map user to perceive relief without an additional equipment, such as anaglyph or polarization glasses. For a long time the technology of raise-relief maps has been known. Solid 3-dimensional models of towns, mountain ranges or other landscapes can be frequently met in museums, schools, mountain lodges, etc. They are made of plastic or other materials similar to plastic cast, fully colored and textured, sometimes very detailed masterpieces produced by enthusiasts (Fig. 1). They seem to be rather a kind of art than automatic production.

Another solution to the problem of 3D perceptions is the autostereoscopy, especially the lenticular method, that can be distributed to the broad community as hard copies. It was applied for the first time in 3D terrain modeling at the Dresden University of Technology, Institute for Cartography (Buchroithner et al. 2003, 2004; Buchroithner and Waelder 2003). Nowadays, the technology is used by lenticular companies, that produce small and large format prints, but they mainly concentrate on entertainment applications, billboards, gadgets for children, etc. Some of them, however, produce also maps of countries, continents or large mountain ranges. In the USA the lenticular technology was used by Ian White to produce 3D map of New York, Seattle and Washington (Wagman 2005), including skyscrapers, underground network and street view.

The aim of this study is to present the lenticular method and perspectives of the technology in mapping and visualization of the mountainous areas. The following sections contain a description of the case study area in the Polish Tatra Mountains, an overview of principles of the lenticular technology and the process of the

Fig. 1 Solid map of the Tatra Mountains (permission from Jerzy Zasadni, geologist and researcher at the University of Science and Technology, Kraków, Poland)

lenticular images production. Finally, the results obtained for the case study area are presented followed by short conclusions.

2 Study Area

As the test study area surroundings of the Gąsienicowa Valley and a legendary alpine tourist route "Orla Perć" were chosen. It is one of the most popular places in the Polish part of the Tatra Mountains. Differences in altitude reach 800 m, landscape changes from forests, through alpine grasslands, right up to rock walls. Broad valleys neighbor steep rock cliffs.

The Tatra Mountains are the highest mountain range in the Carpathians with an area of 785 km^2, the major part (610 km^2) of which lies in Slovakia. The north-western peak of Rysy (2499 m) is the highest Polish mountain. An attractive alpine landscape and easy access make the Tatra Mountains an extremely popular tourists destination.

The history of cartographic representation of this region is very long. The first attempt to depict the Tatra Mountains was made by an anonymous artist at the turn of the fourteenth and fifteenth centuries. It is a fresco, painted in the parish church in Poprad, Slovakia. More professional panoramas originate from the eighteenth century, and with the technique development their quality was growing (Lassota 2007). Nowadays they are used by several publishing companies in mass tourist "map" production. They also provide climbers with route-schemas, that undoubtedly are less commercial. According to climbers opinions on specialized guidebooks, maps and schemas, expressed on many websites or blogs, there is a lack of accurate and pictorial materials on the Tatra Mountains. The recent publication of a so called *topo* is, however, a great breakthrough in this field. The *topos* (MasterTopo 2010) are big size posters, consisting of detailed color photography representation of particular rock cliffs, and lines showing exact location of climbing routes with difficulty-classification and specific names (Fig. 2). It is likely that the lenticular technology would improve the relief interpretation, especially in the context of difficulties with the depth perception (Buchroithner et al. 2004).

3 The Lenticular Method

3.1 The Lenticular Technology

The ability to perceive the world in 3D is called the depth perception (Goldstein 2002). Looking with two eyes provides our brain with two images of reality, taken from slightly different angles. They are then combined into one 3-dimensional

Fig. 2 A part of MasterTopo poster with climbing routes on Kazalnica at the Morskie Oko lake (permission from MasterTopo 2010)

space, where distance to each object is estimated by the brain by means of "visual triangulation". In order to perceive space in a flat image, we should recreate the two pictures representing different perspectives of the same scene with slight relative displacements of objects (disparities) (Fig. 3) and provide each eye with a left and a right image. Then the brain composes them to create the illusion of depth. Traditionally, stereoscopy may support this process. However, separation of input images is achieved with an additional equipment, such as anaglyph or polarization glasses and stereoscopes. An interesting alternative to traditional stereoscopy is a technique called autostereoscopy, that enables displaying stereo-images without any headgear. The most popular solutions use parallax barrier and lenticular lens (Roberts 2003).

A lenticular image is created out of the input images (more than 2), that are cut into thin stripes and then interlaced. The key problem—separation of the sub images—is realized using a transparent plastic foil, covered with parallel micro-lenses lined up in columns. Views are arranged under lenticules so that each eye is projected a different view (Fig. 4). The brain then processes these views to a single coherent 3D image. The lenticular print can simultaneously provide a 3D view to many observers, unlike other methods that demand additional equipment, which is usually provided only to one observer, e.g. stereoscope or polarization glasses

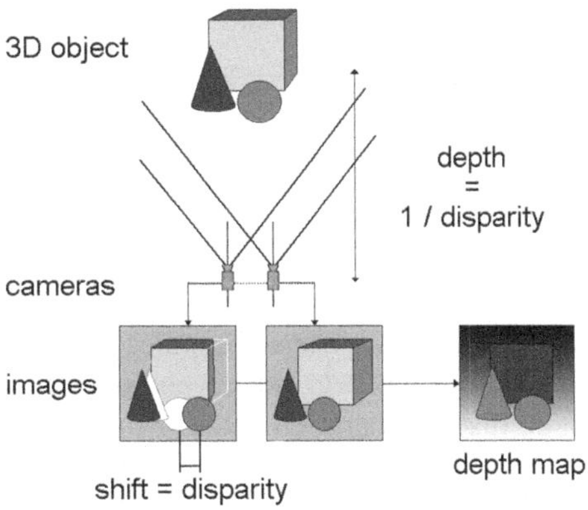

Fig. 3 Principle of stereovision—disparity and depth (*source* Henkel 2010, modified)

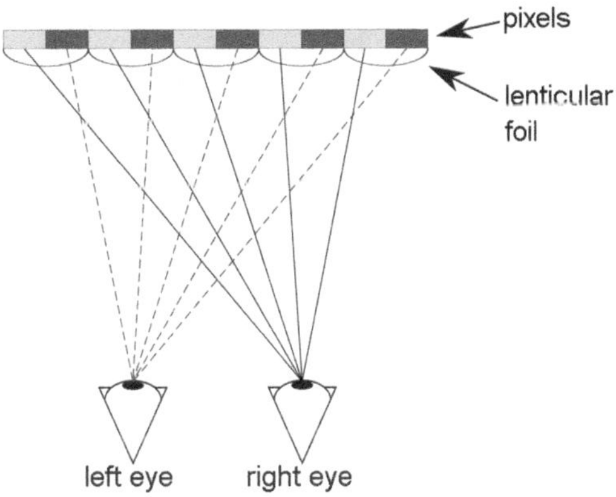

Fig. 4 Principle of the lenticular method (*source* 3D Forums 2010)

(Buchroithner et al. 2004). The 3D impression is present as long as one stereopair of images can be seen (Park et al. 1995; StereoGraphics 1997; Buchroithner et al. 2004).

To achieve good results in lenticular printing, a suitable type of lens has to be chosen. The most important parameter of the lenticular foil is the number of lenses per inch (LPI), called lens density. It ranges from 10 LPI to 200 LPI and is responsible for the resolution of the lenticular image, number of sub images that can be interlaced under lens and viewing distance. The low number of sub images influences leaps between them. The following parameters: lens radius, thickness of the foil, viewing aperture and refractive index are matched to the particular image and effect that should be achieved. For example, the thicker the foil, the bigger lens radius and the smaller viewing angle. Thick lens with narrow viewing angle are the optimal choice for 3D viewing. Increase of the lens thickness decreases resolution (Park et al. 1995; StereoGraphics 1997; Buchroithner et al. 2004). The dependencies between various parameters are presented in Table 1.

The lenticular images' production starts with creation of sub images using a depth map. It is followed by image interlacing and printing; the interlaced image is

Table 1 Dependencies between lenticular foil resolution, thickness and perception distance (*source* Lentimax 2010)

Resolution [LPI]	20	40	60	75	100
Thickness [mm]	3.3–3.8	0.9–1.8	0.8	0.65	0.4
Perception distance [m]	8–1.5	4–1	2–0.3	1.5–0.2	1–0.15

produced out of subimages. There are several ways how to obtain such images (Park et al. 1995; StereoGraphics 1997; Buchroithner et al. 2004). For example, if working in the field, stereo photographs of the object can be taken keeping constant distance between places from which the adjacent photographs are taken and using rather parallel than convergent camera orientation. Similar method can be used while working with a digital 3D model, by defining proper camera positions and orientation. But operating on the geographic data demands quite a different approach, that utilizes Digital Elevation Model (DEM) in a raster format to create illusion of space. The process of taking photo of the object from several stations can be imitated by generating the subimages out of one 2D image and a depth map.

The depth map is an array displayed as a grayscale image. The horizontal coordinates correspond to rows and columns, and Z-values (32 bits, float) are stored in the array's elements instead of intensity information. It can also show the distance from each visible point of original object to the viewer or the camera (Marshall 1994). The rule of thumb is to assign white color to pixels closest to the viewer, and black to the most distant ones. In practice the Z-values are often normalized to the range of 0-255 (8 bits, integer) to create an image that can be input to a graphical software.

The task is straightforward if the final lenticular image presents an aerial perspective. However the terrestrial perspective demands quite different processing of the input DEM. The depth map must be created for the particular camera position by computing spatial distance from each terrain point to the camera. The distance value is then stored in a new map, having the same extent and resolution as the input DEM.

The created depth map and the source 2D image serve as input for the software that out of them generates several stereo views (sub images). This views are used to prepare lenticular images.

In the interlacing process the input subimages are cut into very narrow strips and arranged to match the lenticular lens. Under each lens one stripe from each subimage must be printed in a correct order (Fig. 5). There are many freeware (LIC 2010) and commercial software packages for interlacing, and the process is usually fast.

Before the interlaced image is printed, the printer must be calibrated. It is a process, when the printer is aligned to specific lens by iteratively performing the pitch test. The print is very carefully glued to the lenticular foil, to preserve correspondence between image stripes and lens. Any disturbance can damage the 3D effect.

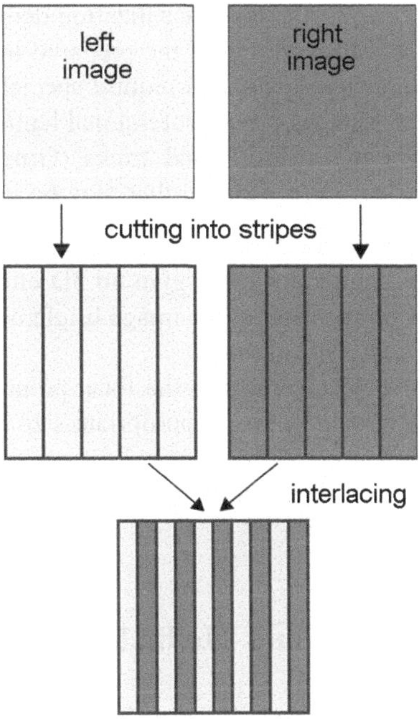

Fig. 5 Example of creating an interlaced image out of two subimages (Buchroithner et al. 2004)

3.2 Layout and Design

At the beginning of the work a concept of how and what an image will present must be clear. It is connected with the input data—image, DEM, vector information, lettering, as well as with layout. Some information, like rivers, roads and paths in a vector format may be draped on the ground. Other, like text, can hover above the ground, as not to occlude the background image. The chosen font should be sans, as low resolution can make fine details like serifs disappear. Minimum text size is also important, as it is determined by the lenticular foil resolution. Placement of the information with reference to the relief determines further processing. The layers that are to be draped on the relief should be merged with the input image, to be taken into account while producing sub images. The hovering layers should be added to the sub images. Table 2 presents the general aspects of the information placement.

Table 2 Composition of the image and additional information into 3D model (Buchroithner et al. 2004, modified)

Data type	Placement
Image data	Directly on the relief
Roads, rivers, boundaries, tourist routes	Directly on the relief—merged with image data
Signatures	A bit above the relief—placed manually
Lettering, geographic names	Considerably above the relief

Some types of visualization demand to include information that will not appear on a background of the prepared image, but next to it. This is a case of maps and their legends which require special treatment (Buchroithner et al. 2003, 2004).

Making use of professional lenticular companies experience, it is worth to know about some tips and tricks (Gruendemann and Habermann 2009; Outeraspect 2010; Tracer 2010), that may be useful for beginners:

- the picture should be rather bright,
- solid color areas give no 3D effect, so each object should be textured,
- in the front of the image bright and vivid colors should be used, and cold ones in the background,
- objects closest to the focal point should be the sharpest in the lenticular image,
- text must have appropriate size to be readable (e.g. for 75 LPI—at least 10 pt. font),
- sans fonts are better than serif.

4 Data and Methods

To prepare stereomates for lenticular images of the test area the following input data, received from the Tatra National Park GIS Section, were used:

- DEM with the horizontal resolution of 0.75 m,
- orthophotomaps in RGB, resolution of 0.75 m,
- vector layer with tourist routes, including information about the route color.

The mountainous area has been presented in two ways:

- from the birds-eye view,
- from the terrestrial perspective.

Two methods were used to create the depth map. First, the depth map was created directly out of the DEM in orthoprojection, displayed in grayscale, where the lowest parts of terrain are black and peaks are white (Fig. 6a). In the other approach, the depth map was computed for the particular camera position. The coordinates of the camera (in the global coordinate system) served as an input to the self-written Python script, computing spatial distance from each terrain point to the camera, and saving the data to the new geotiff file (Fig. 6b). The camera coordinates were also used for composing the terrestrial perspective view in ESRI ArcScene software (Fig. 6c).

It was decided, that only the input image (orthophotomap) and the vector layer containing tourist routes will appear in the final composition. The geographic data needed also additional information, such as signatures of major peaks and passes, their names and altitude. No map legend was planned, as the only available size was 10 cm by 15 cm.

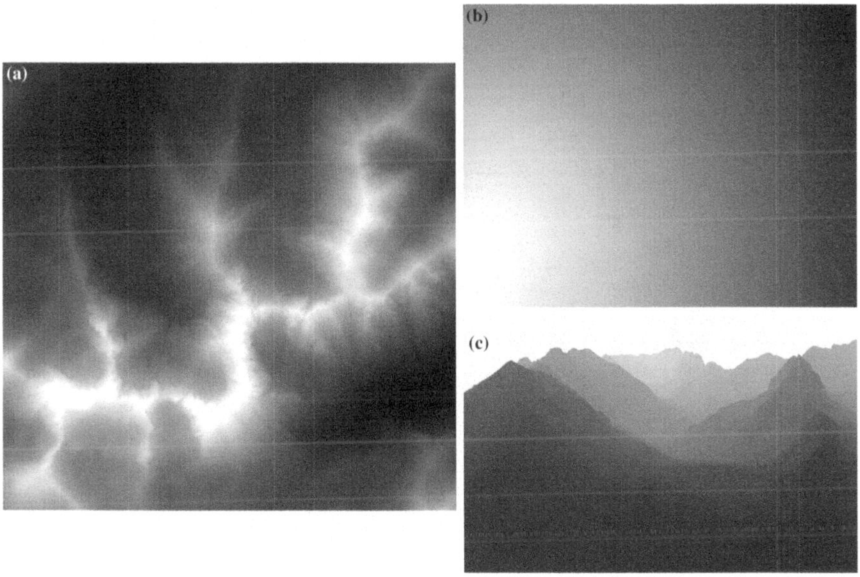

Fig. 6 Depth maps for the test area: aerial perspective (**a**), terrestrial perspective—in orthogonal projection (**b**), brightness decreases with distance to the camera; in camera coordinate system (**c**)

The data pre-processing started with the integration of the background image with tourist routes, as they should be placed on the terrain. This task was made in a graphics software.

The terrestrial perspective view was set in such a way as to attractively present the Hala Gąsienicowa surroundings (Fig. 7). On the base of the depth maps and the corresponding images (orthophotomaps) the sub images were created. To achieve good results number of sub images equaled 10 in both cases (Fig. 7).

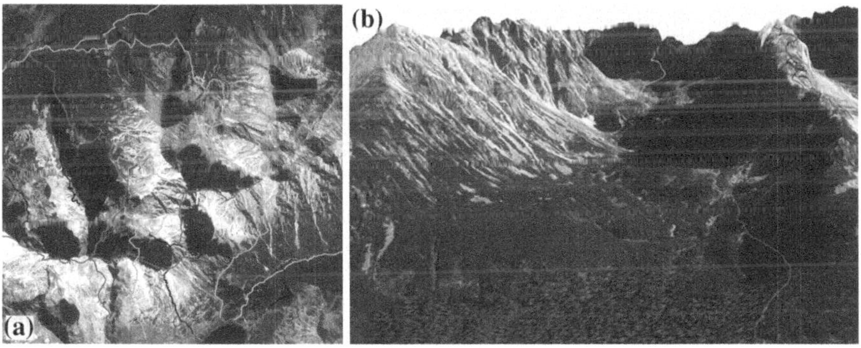

Fig. 7 Exemplary sub images for the test area: aerial perspective (a); terrestrial perspective (b), before adding labels

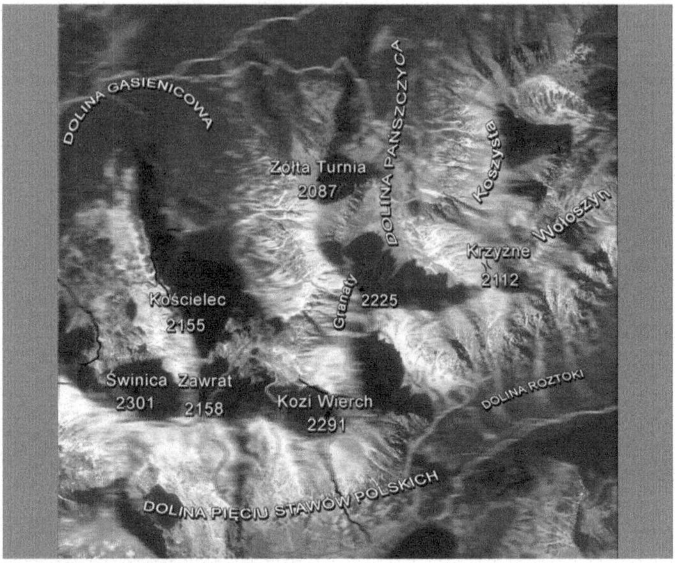

Fig. 8 Image for the Polish part of the Tatra Mountains in aerial perspective ready to print

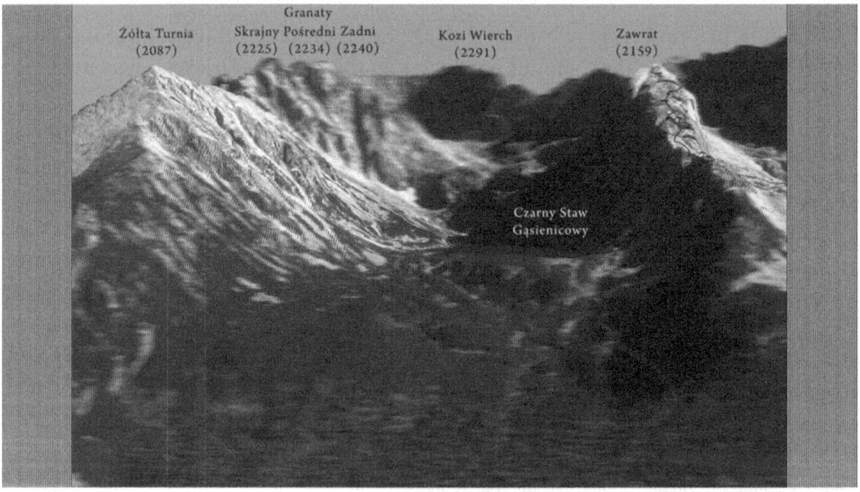

Fig. 9 Image for the Polish part of the Tatra Mountains in terrestrial perspective ready to print

The information that was intended to hover over the terrain was added afterwards in the same place in each sub image. This part of work was done in a Gimp software—a freely distributed GNU Image Manipulation Program. Two ready-to-print interlaced images, suitable for 300 dpi printer resolution and lenticular foil of 100 LPI were created out of the two sets of sub images. As the zero-parallax plane (no disparity between objects in the sub images) was set in the middle of the depth ranges, the corresponding fragments are the sharpest and the other ones seem a bit blurred. They were printed and integrated with the lenticular foil very carefully to avoid disturbances that could damage the 3D effect. The output sizes of the aerial and terrestrial images were 10 × 10 and 10 × 15 cm, respectively.

5 Results and Discussion

Two lenticular images in a postcard size presenting aerial and terrestrial perspective were produced (Figs. 8, 9). The images can be successfully viewed with unaided eyes. Depth impression is very strong and there are no leaps between the subimages. On the terrestrial view the steep parts of the relief are a bit blurred because pixels of the input orthophotomap representing cliffs were stretched to cover vertical surfaces in the terrestrial perspective. The text information is readable and does not occlude images content, because it seems to hover over the terrain (Fig. 8).

Recent development of the commercial and free software opens up access to the lenticular technology to wider community, not limited to professionals. However, even if the software enables creation of sub images and interlacing, the general concept, design and layout need to be done manually, often with a huge workload. Unlike large wall maps (Buchroithner et al. 2004), the postcards were designed as insets to guide-books.

These postcards can be viewed auto-stereoscopically with unaided eyes. Neither glasses, nor special illumination is necessary and 3D impression is given to many people simultaneously, what is a great advantage in comparison to other techniques, such as holography, stereoscopic or anaglyph viewing. As Buchroithner et al. (2004) stated, this kind of spatial visualization yields great benefits "for all individuals who have difficulties to spontaneously extract relief information from conventional maps". The second advantage of the lenticular technology is that they are easily portable. According to what the inventor of the 3D urban maps Ian White said (Wagman 2005), "people don't always want to sit down with a computer to look at geographic information. The high-tech adaptation of a century-old novelty card printing technique may provide one answer".

6 Conclusions

The high quality of the samples, their portability and no need to use any additional equipment are undisputable advantages. Contrary to anaglyph technology, lenticular images are colorful. The technology has a great potential for guide books. Lenticular insets could be added instead of flat maps or panoramas to give visitors a better overview of a place they are going to visit. Prints in larger format may offer an interesting alternative to wall maps and can be displayed in places frequently visited by tourists. A field of potential interest are lenticular guide maps for climbers, showing details of routes, difficulties and dangers of the wall. Most of all, development and popularization of the lenticular technology is important for the people having problems with relief interpretation out of traditional maps.

The results achieved may be treated as an exemplary work, encouraging broad community in Poland to make use of the lenticular technology in modern cartography, with applications in tourism or spatial planning.

References

3D Forums (2010) http://www.3d-forums.com/. Accessed 06 Oct 2010

Almer A, Schnabel T, Schardt M, Stelzl H (2004) Real-time visualization of geoinformation focusing on tourism applications. ISPRS Int Arch Photogramm Remote Sens Spat Inf 36

Buchroithner M, Waelder O (2003) True-3D presentations of the mars surface. In: ICA/ACI (ed) In: Proceedings of the 21st international Cartographic conference, Durban

Buchroithner MF, Waelder O, Habermann K, Koenig B (2003) Generating a true-3D image map of high relief terrain using lenticular foil. In: ICA/ACI (ed) In: Proceedings of the 21st international Cartographic conference, Durban

Buchroithner M, Habermann K, Gruendemann T (2004) True 3D visualization of mountainous terrain by means of lenticular foil technology. In: 4th ICA Mountain Cartography Workshop, Vall de Nuria. http://www.mountaincartography.org/publications/papers/papers_nuria_04/buchroithner.pdf. Accessed: 10 Oct 2010

Goldstein EB (2002) Sensation and perception, 6th edn. Wadsworth, Pacific Grove, CA

Gruendemann T, Habermann K (2009) Autostereoscopic maps on the basis of lenticular foils. In: Buchroithner M (ed) ICA SPIE Europe symposium, True-3D in Cartography, Dresden

Haeberling C, Hurni L (2001) Mountain cartography: revival of a classic domain. ISPRS J Photogramm Remote Sens 57:134–158

Henkel R (2010) Stereo disparities and depth. http://homepages.inf.ed.ac.uk/rbf/CVonline/. Accessed 7 Oct 2010

Imhof E (1982) Cartographic relief presentation. Walter de Gruyter, Berlin, reissued by ESRI Press, Redlands 2007

Lassota L (2007) Panoramy tatrzańskie. W górach 3(13). http://www.wgorach.com/ ?id = 55274 &location = f&msg = 1&lang_id = PL. Accessed 10 Oct 2010

Lentimax (2010) Technical notes. http://www.lentimax.eu. Accessed 5 Oct 2010

LIC (2010) Lenticular image creator by Jason Bennett. http://www.conceptjewel.com/ lic/index.html. Accessed 5 Oct 2010

Marshall AD (1994) Vision systems. http://www.cs.cf.ac.uk/Dave/Vision_lecture/ Vision_-lecture_caller.html. Accessed 5 Oct 2010

MasterTopo (2010) Topo for climbers. http://mastertopo.pl/. Accessed 25 Oct 2010

Outeraspect (2010) A company with a strong focus on lenticular technology. http://www.outeraspect.com/index.php. Accessed 5 Oct 2010

Park D, Kim K, Lee C, Son J, Lee Y (1995) Lenticular stereoscopic imaging and displaying techniques with no special glasses, vol 3. In: Proceedings of the 1995 IEEE international conference on image processing (ICIP-95), Washington, DC, USA, p. 3137. October 1995

Roberts DE (2003) History of lenticular and related autostereoscopic methods. Leap Technologies, Hillsboro

StereoGraphics (1997) Developers' handbook. Background on creating images for crystaleyes® and SimulEyes®. http://www.cs.unc.edu/Research/stc/index.html. Accessed 28 Oct 2010

Tracer (2010) Lenticular printing company. http://www.tracer1.com/. Accessed 5 Oct 2010
Wagman (2005) Urban mapping gives a vintage printing technique a high-tech revival. http://www.directionsmag.com/. Accessed 28 Oct 2010

Obtaining Snow Avalanche Information by Means of Terrestrial Photogrammetry—Evaluation of a New Approach

Paweł Chrustek, Natalia Kolecka and Yves Bühler

Abstract Recent snow avalanche hazard mapping tools and procedures offer methods to improve the accuracy and reliability of risk and hazard localization. The validation of numerical mass movement models mainly depends on recorded historical avalanche data sets such as avalanche outlines and release volumes. These data sets are often unavailable or of an unknown accuracy. Avalanche characteristics such as release area, flow height and flow path, runout distance and total amount of released snow mass are essential parameters for proper calibration and evaluation of numerical simulation tools. Incorrectly calibrated models can influence decisions-making which directly affects human safety. The acquisition of high quality data regarding observed avalanche events is often hindered by the high risk permanently present in avalanche terrain. This chapter describes a method based on photogrammetry and computer vision that allows using a single terrestrial photograph with unknown exterior and interior orientation parameters to accurately map avalanche outlines and release areas. We evaluate this method by comparing its results with GPS measurements made in the field and discuss the optimization of measurement efficiency, costs and human safety.

P. Chrustek (✉) · N. Kolecka
Institute of Geography and Spatial Management, Jagiellonian University, Gronostajowa 7 30387 Kraków, Poland
e-mail: p.chrustek@annapasek.org

P. Chrustek
Anna Pasek Foundation, Małobądzka 101 42500 Będzin, Poland

Y. Bühler
WSL Institute for Snow and Avalanche Research SLF, Flüelastr. 11 7260 Davos, Switzerland

J. Kozak et al. (eds.), *The Carpathians: Integrating Nature and Society Towards Sustainability*, Environmental Science and Engineering,
DOI: 10.1007/978-3-642-12725-0_41,

1 Introduction

Each year snow avalanches cause a great number of accidents in mountain areas. This force of nature brings not only casualties, but also significant forest and infrastructure damage. Within the 2009/2010 winter season, a total of 237 people were killed by snow avalanches across Europe and North America (International Commission for Alpine Rescue 2010). Rescue statistics show that the common cause of avalanche accidents is a difficulty in spatial risk factor evaluation. This problem can be related to the lack of knowledge and experience, as well as objective causes (e.g. weather conditions), which may affect both amateurs and specialists. For this reason, it is very important to supplement snow avalanche education with the knowledge from the field of geographic information system (GIS) and remote sensing of environment that provide methods and tools for improved spatial risk and hazard location (Chrustek 2009).

Avalanche hazard mapping tools and procedures developed by avalanche specialists from all over the world offer an increasing number of methods for providing more accurate risk and hazard localization. Numerical avalanche dynamics models like RAMMS (Christen et al. 2010), SAMOS (Sampl and Zwinger 2004) or ELBA+ (Volk and Kleemayr 1999, Sauermoser and Illmer 2002) coupled with GIS have become an essential part of snow engineering and hazard mapping studies (Christen et al. 2010).

Unfortunately, a constant fundamental problem in the science of snow mass movements is to record and document occurring events. Post-event mapping is helpful to determine release areas, volumes of the released masses, runout distances and inundation areas. This kind of data is necessary for developing and evaluating new mitigation methods and tools. Avalanche mapping is also important for risk assessment verification and accident documentation. Beside conventional avalanche data (altitude, aspect, size, etc.) a general outline marked on maps should be a part of each avalanche survey (Meister and Jeller 2009). Obtaining more precise spatial variability of roughness parameters (e.g. type of land cover) is also a crucial research step for avalanche specialists, especially in the context of avalanche dynamic analyses. Different land cover types influence the friction parameters which are input parameters for dynamic models (Gruber 2001).

On a global scale, the documentation of avalanches remains sparse and therefore incomplete, and their accuracy is unknown. Currently, detection and mapping of observed avalanches relies mainly on isolated observations acquired by individual experts under field conditions. Quite often, only avalanches causing accidents or resulting in heavy damages are mapped (Bühler et al. 2009).

At certain locations, such as Davos in Switzerland, long-term records of well documented events exist. However, because of the changing climate and missing records, historical data may not show the complete picture of the current threats.

A common lack in high quality data of localized snow avalanche releases and depositions is more often caused by the high risk permanently present within avalanche areas (especially in the release zone), rather than limited availability of measurement devices. This is the reason that the most popular method used by experts is manual mapping based on remote observations. However, according to opinions of avalanche experts this method requires high skill levels, and very often leads to numerous discrepancies between field measurements and drawn extents (Meister and Jeller 2009). Traditional hand-held Global Navigation Satellite System (GNSS) measurements allow cheap, easy and accurate mapping of avalanches, but they are time consuming and often of restricted applicability due to avalanche danger. The quality of these measurements strictly depends on the available satellite signal which may be significantly reduced in the complex terrain (Chrustek et al. 2010).

Remote and high resolution measurements are possible through a combination of Light Detection And Ranging (LiDAR) and GNSS technology (e.g. Deems and Painter 2006; Jörg et al. 2006; Prokop et al. 2008; Vallet 2008). Optical scanners provide powerful technologies for safer and quicker measurements. Terrestrial Laser Scanning (TLS) can also be used for other applications in snow avalanche research (e.g. spatial snow-depth distribution monitoring and evaluating physical snowdrift or snowpack models; Prokop et al. 2008). Some technical parameters of TLS systems create, however, some important limitations in using this technology, e.g., the limited scope of the laser and heavy weight of the equipment. It means that TLS is usually restricted to use in easily accessible terrains and needs to be operated by a specialized team (Prokop et al. 2008; Prokop 2009).

Airborne Laser Scanning (ALS) provides similar high quality data, but is more effective for large area mapping. Unfortunately, it is also a considerably more time consuming solution (Deems and Painter 2006).

A new approach to obtain accurate post-event avalanche information that covers large areas and applies airborne optical remote sensing techniques, is discussed by Bühler et al. (2009). This innovative method, thanks to the advanced digital camera sensor, allows obtaining information about avalanche deposit of large, medium and small avalanches, even within shadowed areas.

The most important disadvantage of advanced LiDAR and digital imaging techniques is their cost and limited repetition rate. For many years, obtaining this kind of data has been impossible for many operational and research budgets. Application of these technologies also requires the involvement of large amounts of time for detailed measurement planning, and qualified staff that may operate system and process data (Deems and Painter 2006; Prokop 2009).

Single terrestrial photographs can also be used as a source of valuable geographical information. As registration perpendicular to the surface is the most effective, high-oblique terrestrial images seem to be a powerful supplement to aerial imagery for the mountainous areas (Buchroithner 2002). Recently, Aschenwald et al. (2001) and Corripio (2004) presented an approach to incorporate a single terrestrial photograph into geographical analysis. Their methods, however, employed photographs taken from known or measured locations, and this subsequently limits potential data sources to a new or well-documented set of photographs.

To georeference terrestrial photographs, one needs to know the function that creates a relation between an image pixel and a point in the global 3-dimensional (3D) coordinate system. It incorporates a perspective projection of the visible part of a digital elevation model (DEM) on the photograph, to create so called virtual photograph of the DEM, and scaling according to the photograph resolution. The algorithm presented in Aschenwald et al. (2001) aims at georectifying high-oblique terrestrial images to obtain information about landscape of mountainous areas using transformation derived from classical photogrammetry theory (Slama et al. 1980). Corripio (2004) follows computer vision theory and uses terrestrial photographs to assess snow surface albedo. In both cases, required input data are: camera and target global coordinates, elevation and visibility information, Ground Control Points (GCPs), photograph, camera focal length, and additionally (Corripio 2004) photograph resolution.

Terrain coordinates of a camera and a target (the aim of the camera) are needed to determine the angular elements of the camera orientation. It is very difficult and error-prone to find appropriate ground points collinear with the center of the photograph, e.g. using the existing orthophotomap (Aschenwald et al. 2001). In this case, however, they are approximate and additional corrections need to be included. Corripio (2004) proposed a correction due to the roll of the camera around the viewing direction axis and suggests “manual photograph orientation by trial and error”, based on visual inspection of GCPs measured in the photo and global coordinates projected on the photo. Aschenwald et al. (2001) suggested the method which allows to locate target coordinates based on creating a grid around the input target and checking each cell to optimize the target coordinates. In this case also GCPs were useful to investigate errors in input information by mean deviation of the projected points from their known position. Since this input seems to be crucial for achieving good results, it would be highly recommended to find another solution to this problem.

An interesting approach in the avalanche outline mapping context was presented by Meister and Jeller (2009). They used a digital “Atlas of Switzerland 2” (2004) and digital terrestrial pictures containing Global Positioning System (GPS) coordination and azimuth parameters and based on the digital DEM panorama and tools for adjusting digital pictures, avalanche outlines were drawn on screen.

Our aim was to develop an innovative method which allows to map avalanche extent using a single terrestrial photograph with unknown exterior and interior orientation parameters. Our idea is based on creating an orthoimage from the terrestrial photograph by means of photogrammetric and computer vision rules, and subsequent visual interpretation.

2 Methods

2.1 Theoretical Background

The proposed methodology consists of three principal steps. First, the photograph needs to be oriented in a global coordinate system to obtain camera position and to perform function mapping of 3D points into a 2-dimensional (2D) image. Next, the viewshed analysis is performed to determine parts of terrain that are not visible from the particular camera position. In the third step, the visible DEM points are projected into the photograph by means of the mapping function, to obtain color information out of the photo. Preliminary work that needs to be done includes preparation of the input data (digital images, directly from the digital cameras or scanned copy of analog photographs, and DEMs in a raster format), and identification and measurement of GCPs (both the image and ground coordinates). The final output file is a georeferenced dataset.

To orientate a terrestrial photograph, the Direct Linear Transformation (DLT) method is applied (Abdel-Aziz and Karara 1971; Luhmann et al. 2006; Kraus 2007). The DLT establishes the relationship between the 2D image coordinates and the 3D object coordinates, using projective transformation rules and GCPs. The control points must have image and global coordinates measured to compute the camera orientation parameters. A minimum of 6 points are necessary to solve the DLT; however, to cope with images from non-calibrated camera or scanned old photographs, more GCPs are necessary to obtain accurate results. The GCPs should also be evenly distributed. In an ideal case, the image should be evenly covered with points, which ensures strong bundle geometry. Global coordinates can be surveyed in the field, most likely with GNSS device, or determined by examining existing aerial orthophotomaps and elevation models. As GCPs evident and easy to identify features should be chosen, e.g. peaks, rock walls, stones, single shrubs or trees. This principle, however, has some constraints. It is not easy to find good and trustworthy GCPs using terrestrial images and existing orthophotomaps, as they both present different terrain representation recorded in different seasons. The steeper slopes, the greater differences are observed. The problem with locating GCPs is also linked to old photographs which are likely to present former landscape features, e.g. vegetation, which no longer exist. Nevertheless in many cases the task is feasible, especially when the photograph gives a wider context, not limited to the nearest surroundings of the investigated object. Another possibility is to survey control points in order to obtain more accurate results.

Eleven DLT parameters can be computed from linear equations, so no approximate values of the unknowns are required (Luhmann et al. 2006; Kraus 2007):

$$x = \frac{L_1X + L_2Y + L_3Z + L}{L_9X + L_{10}Y + L_{11}Z + 1}$$
$$y = \frac{L_5X + L_6Y + L_7Z + L}{L_9X + L_{10}Y + L_{11}Z + 1} \quad (1)$$

where $L_1 \ldots L_{11}$ are DLT parameters; x, y are image coordinates, and X, Y, Z are object coordinates in the ground coordinate system.

From $L_1 \ldots L_{11}$, the coordinates of the projection center: X0, Y0, Z0 (camera location) are computed (Fig. 1), as they are necessary to determine visible and hidden parts of terrain (Luhmann et al. 2006).

Knowing the coordinates of the projection center, the invisible parts of terrain should be eliminated in order to avoid projecting hidden points into the photograph. This goal is achieved by applying the viewshed analysis to the raster DEM. The approach introduced by Wang et al. (2000) is utilized, which uses the concept of reference planes. The results were confronted with standard GIS performance by means of the two viewshed raster subtractions. It was found that the differences did not exceed 2 % of the input DEM area. Small discrepancies could have resulted from the different algorithms.

A traditional orthoimage is created when relief displacements are removed from the original image (Kraus 1992; Novak 1992; Mikhail et al. 2001; Okeke 2001; Kraus 2007). It is done using DEM, whereby the interior and exterior parameters of the image are used. There are two ways to conduct such rectifications: forward and backward projection, the latter was used in this work (Fig. 2). Here, for each pixel of the new orthoimage, the ground planimetric (X, Y) coordinates are obtained from the ground coordinates and pixel size of the DEM. Also the Z-coordinates are extracted from the DEM. Each pixel is projected back to the image space using the DLT Eq. (1), and then attributed with the image luminance value, obtained by resampling.

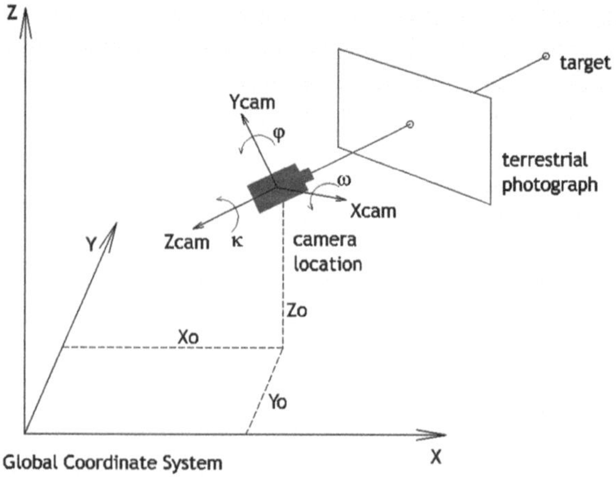

Fig. 1 Camera position and viewing angles in terrestrial photogrammetry (based on Luhmann et al. 2006, modified)

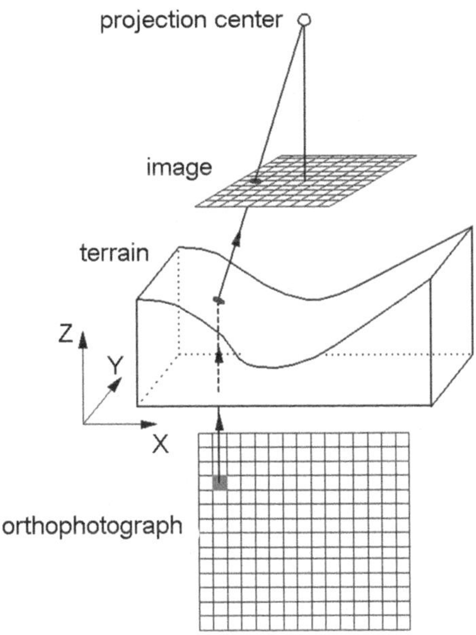

Fig. 2 Backward projection during orthoimage creation in the aerial case (based on Luhmann et al. 2006, modified)

However, in the terrestrial case, where the surface occlusions occur, only the visible set of points can be transformed into the image space. They are selected by multiplying input DEM by the visibility map. Output from this procedure is the orthoimage, saved as a TIFF file and georeferencing parameters stored in a TFW world file.

To assess the quality of the received results, accuracy of the GCPs ground coordinates are determined, and information regarding whether they were obtained by survey or by measurement on the orthophotomap and DEM was recorded. The essential step for DLT quality assessment was the comparison of GCPs coordinates measured in the photograph with GCPs global coordinates projected onto the photo. The residuals (optical error) were computed as follows:

$$
\begin{aligned}
v_x &= x - \frac{L_1X + L_2Y + L_3Z + L_4}{L_9X + L_{10}Y + L_{11}Z + 1} \\
v_y &= y - \frac{L_5X + L_6Y + L_7Z + L_8}{L_9X + L_{10}Y + L_{11}Z + 1} \\
v_{xy} &= \frac{1}{n}\sum_{i=1}^{n} \sqrt{v_x^2 + v_y^2}
\end{aligned}
\tag{2}
$$

where v_x, v_y are differences between measured and computed image coordinates of the control points by x and y, n is the number of points and v_{xy} is a total optical error.

The output orthoimage should be visually investigated through checking the general appearance of the image and comparing it with existing data, e.g. traditional (aerial) orthophotomap. Usually it can be seen if such an image contributes some new, additional information, or if it rather contains smudges or other artifacts. Finally, computed orthoimages become a base layer for manual or automatic vectorization process. Correct interpretation of the processed image strictly depends on the operator experience, the selected map scale and assumed automatic classification method.

2.2 Experimental Work

The study areas were located in the Polish Tatra Mountains and the eastern part of the Swiss Alps around Davos. To test the approach three different locations were chosen (Fig. 3):

1. Dorfberg Mountain (DB)—a 2592 × 3872 pixels digital photograph, taken with non-calibrated digital SLR Sony A100 camera equipped with 10.8 millions pixels CCD-matrix and a zoom lens 17–50 mm. RGB colors show south-eastern exposed slopes of the Dorfberg Mountain and surroundings (picture dated 9th February 2010). The photograph recorded a few small and medium size snow avalanches, the biggest one in the central part of the picture (approximately 220 m long) was also measured by GPS Trimble GeoXH device (with decimeter accuracy). Metadata are not available from the EXIF file.

Fig. 3 The test images with marked GCPs: DB (*upper left*), GC (*upper right*) and PG (*lower panel*)

2. Goryczkowa Czuba (GC)—a 1013 × 661 pixel copy of the scanned analog photograph, RGB colors show Czuba Goryczkowa Mountain and avalanche rescue action that took place on 11th January 1985. Part of medium size avalanche (approximately 600 m long) visible on the picture was released by a tourist. It killed one person and caused injuries to another.
3. Pośredni Goryczkowy (PG)—a 2592 × 3872 pixels digital photograph, taken in summer with non-calibrated digital SLR Nikon D80 camera equipped with 10.8 millions pixels CCD-matrix and a 50 mm lens. RGB colors show eastern slopes of the Pośredni Goryczkowy Mountain. Metadata are available from the EXIF file. This photograph will be used for verification of the method.

DEMs for the GC and PG test sites were available as the TIN (Triangulated Irregular Network) model based on contour lines with 5 m intervals digitized from topographic maps 1:10,000), mass points and hardlines. It was converted to the raster format with spatial resolution of 1 m. DEM for the DB site, with 2 m resolution, was produced from aerial images.

GCPs were measured in ArcGIS software, on a basis of orthophotomaps and DEM. Though the orthophotomaps resolution was 0.25 m (RGB) for the GC i PG and 0.5 m (RGB) for the DB test sites, finding GCPs was not trivial, because highly-oblique and vertical perspectives gave quite a different insight into the investigated terrain. The best points to use were main peaks, corners of rock walls, single stones and trees, avalanche fences or clumps of dwarf mountain pine. We attempted to distribute them evenly, especially to surround or to locate some points inside the area of interest. 9 (for the DB and GC) and 21 GCPs (for the PG) were found resulting in various optical errors (Table 1).

Using GCPs coordinates, DLT, exterior and interior camera orientation parameters were derived. The statistical parameters for the input datasets were then computed: minimum, maximum and average distance from the projection center to GCPs and ground pixel size (Table 2).

On the basis of the camera location derived from DLT parameters, for each of the datasets the visibility map was created. The visibility map was then used to remove hidden terrain points from further computations.

Finally, orthoimages were computed and became base layers for vectorization process. Avalanches outlines and selected land cover features were manually vectorized on screen using GIS software.

Table 1 Comparison of GCPs coordinates measured in the photograph and GCPs global coordinates projected into the photo

Test Sites	DB		PG		GC	
Coordinates	x	y	x	y	x	y
Mean residual [pixels]	−2.57	0.44	−0.16	0.17	−1.53	0.80
Standard deviation [pixels]	4.18	3.30	10.08	5.82	2.99	7.04
Optical error by x and y [pixels]	4.70	3.30	9.51	5.49	3.29	6.91
Total optical error [pixels]	5.66		7.65		10.98	

Table 2 Object distance and ground pixel size for the input datasets

		DB	PG	GC
Object distance [m]	Min	1,490	550	10
	Max	1,700	870	500
	Mean	1,595	710	255
Ground pixel size [m]	Min	0.32	0.07	0.05
	Max	0.37	0.10	2.41
	Mean	0.35	0.09	1.20

Evaluation of the process was performed mainly based on the visual examination and the standard procedure of DLT quality assessment, e.g. comparison of the measured image and global GCPs coordinates projected back into the photo (see Table 2).

The result from the DB test site was also compared with field GPS data collected after the avalanche flow (Fig. 4c). Quantitative evaluation of the results from the PG test site was additionally supported by the comparison with the traditional ortophotomap. For this reason, some dwarf mountain pine patches were vectorized. Both vectorization on the traditional orthophotomap and terrestrial orthoimage were made manually on screen, using the same map scale 1:500.

3 Results

Despite the low contrast of the snow surface, both release and deposit zones are visible in the processed orthoimage from the DB test site (Fig. 4d). Our tests showed that more detailed shape of the avalanche outline may be obtained when the outline is marked on the input photograph. The vectorization process on the orthoimage becomes then much easier. Figure 4c shows the comparison between vectorized and field measured data. This example strongly demonstrates that high accuracy avalanche outline mapping based on traditional terrestrial photograph is possible. The measured vertical differences between the outlines do not exceed 15 m, though vertical distance between camera and the avalanche exceeded 1,500 m. Therefore, the most important advantage of this method is that it works very well without accessing dangerous areas. It also saves time, cost and effort.

Very promising results were obtained when analyzing the orthoimage processed from the old scanned photograph in the CG test site (Fig. 5). This example showed that processing of an old photograph with limited identification of reliable locations of the GCPs is possible (most of characteristic terrain features were completely covered by the snow). Because of the lower resolution of the source photograph, the vectorization of the visible avalanche outline was also much easier when the outline was marked on the source image (Fig. 5b). The avalanche boundaries visible on the orthoimage processed from the raw picture are not so

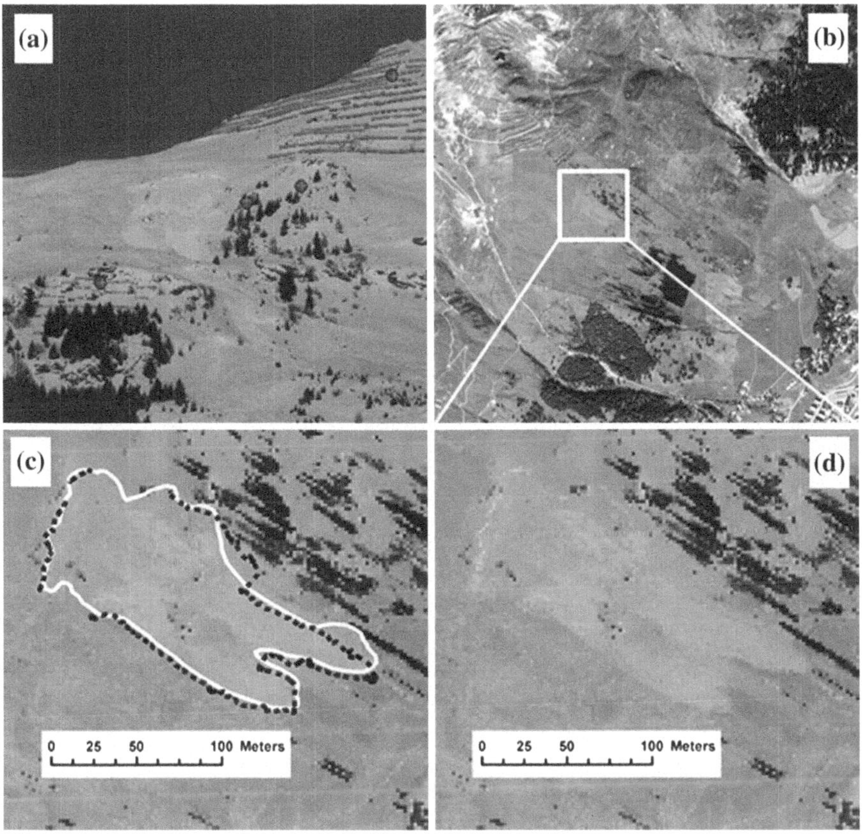

Fig. 4 DB site photo (**a**), orthoimage on the traditional orthophotomap background (**b**, **d**) vectorized avalanche outline (*white*) compared with field GPS measurements (*dashed black outline*) (**c**)

clear (Fig. 5c), and the whole outline of the avalanche was not visible As the documentation prepared after the rescue action contains a detailed description, it can help to identify precise location of the runout limits and draw them on the map in the next step.

In the PG site, terrain details visible in the summer photo were easier to identify. Hence, significant visual differences between traditional (aerial) and terrestrial orthoimage were not noticed (Fig. 6b, d). The area of test dwarf pine patch are very similar in both cases (3,550 m^2 for the traditional orthophotomap, and 3,750 m^2 for the orthoimage,) with the common part equal to 3,110 m^2 (over 80 % of the dwarf pine patch area). Reasons for the discrepancies could be vegetation changes during two years (orthophotomap was created in 2009) or errors generated by the operator during the vectorization process.

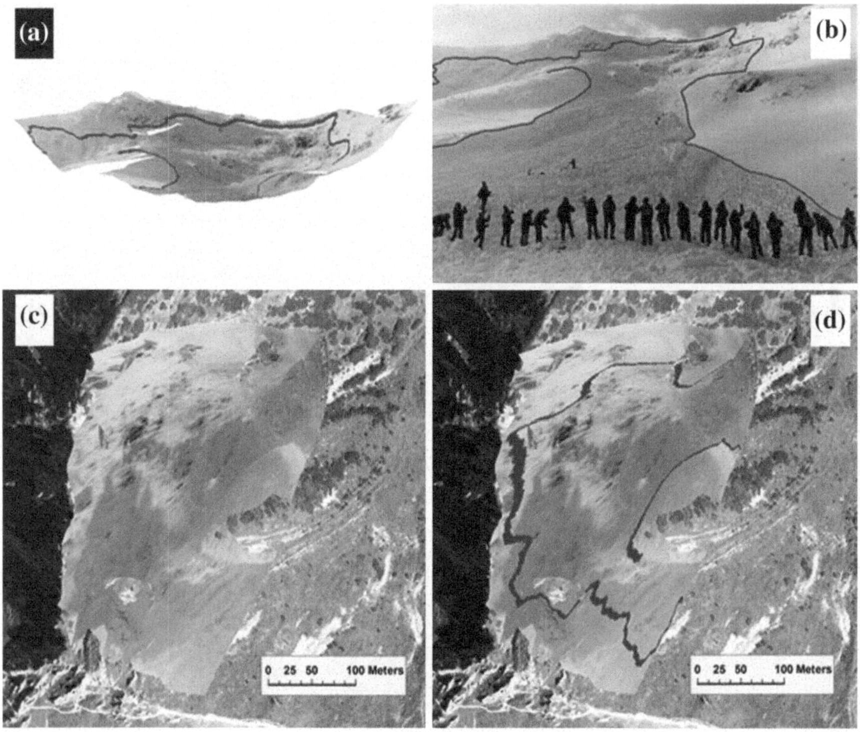

Fig. 5 CG site 3D visualization of the draped orthoimage on the DEM (**a**), original photo (**b**), orthoimage on the traditional orthophotomap background (**c**), and orthoimage with avalanche outline (*continuous line*) drawn earlier on the source image (**d**)

4 Discussion and Conclusions

Our chapter presents the method that minimizes necessary input information and replaces approximate data with mathematical calculations, based on the DLT method that has several advantages as compared to earlier works by Aschenwald et al. (2001) and Corripio (2004) (Table 3). First, mathematical computations seem to be more reliable than error-prone visual estimation. Another DLT advantage is that it works with terrestrial, oblique and aerial photographs, making this method multi-purpose. Mathematical formulae also allow to compute errors of camera orientation and assess its influence on further processing of the photograph.

Similarly to Aschenwald et al. (2001) and Corripio (2004) in the proposed method GCPs need to be collected. This is the most time-consuming and problematic part of the georectification process. What affects accuracy of the GCPs measurement are photograph and orthophotomap resolution and DEM accuracy or—in case of a field survey—GPS accuracy.

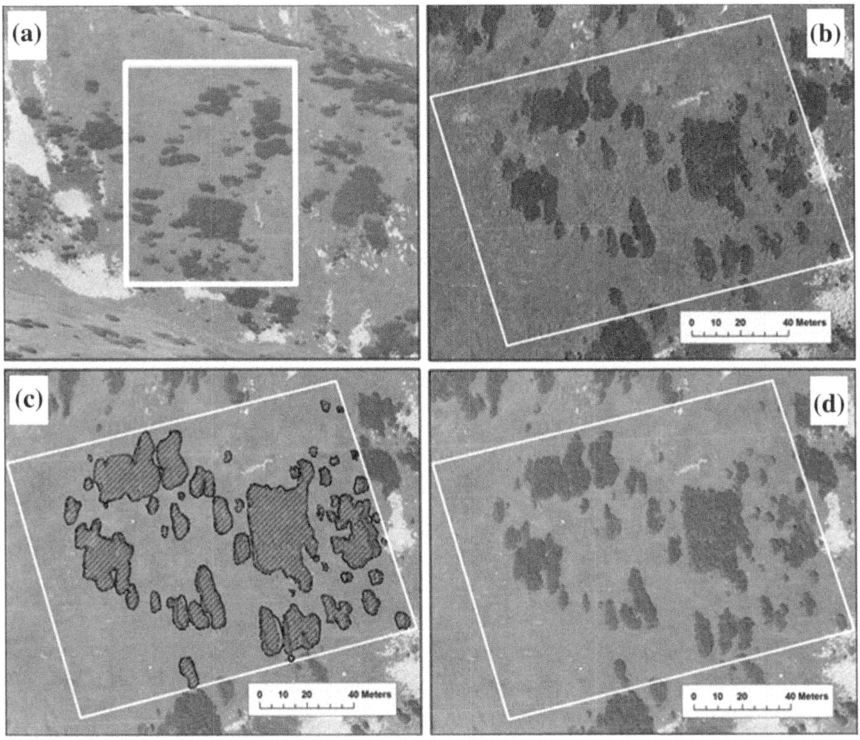

Fig. 6 Pośredni Goryczkowy site photo (**a**), the traditional orthophotomap (**b**), compared to the orthoimage (**d**, 0.25 m resolution), and the comparison between vectorized dwarf mountain pines polygons (**c**): *solid line*—the traditional orthophotomap; *dashed-dot line*—the orthoimage; white hatched areas show the intersection of both layers

It is difficult to find reliable GCPs on old photographs, whether in the orthophotomap or in the field, due to changes of landscape features and snow cover present in winter images. The problem applies also to very steep slopes that look different from the terrestrial and aerial perspectives or can be hardly accessible for surveys. Nevertheless, as the DLT method needs only a minimum of six GCPs, so it is highly probable that they can be found, as proven in the GC test site.

In comparison to Aschenwald et al. (2001) and Corripio (2004), an additional improvement of the method is the implementation of the viewshed analysis into the final procedure that has not been proposed until now. The algorithm of Wang et al. (2000) was successfully programmed and tested to assess its quality, with the results almost identical to viewshed procedures widely used in proprietary GIS software.

The positional accuracy and quality of the final result largely depends on the input DEM quality. As elevation models of high mountain areas produced from aerial photographs have major inaccuracies in a very steep or shaded terrain, these errors will be propagated into the derived data (Foote and Huebner 1995; Krupnik 2003).

Table 3 Comparison of the method proposed in this chapter with methods proposed by Aschenwald et al. (2001) and Corripio (2004)

Methods	Aschenwald et al. 2001	Corripio 2004	The method proposed in this chapter
Theory	Photogrammetry (space ren)	Computer vision	Photogrammetry (direct linear transformation)
Input data	Camera and target coordinates DEM—list of XYZ-points in text file Visibility map—(0/1) text file—in the same order as DEM points GCPs—minimum number 15 Photograph Camera focal length	Camera and target coordinates DEM—list of XYZ-points in text file Visibility map—(0/1) text file—in the same order as DEM point GCPs—minimum number not defined Photograph camera focal length	DEM—list of XYZ-points in text file Visibility map—(0/1) text file—in the same order as DEM points GCPs—minimum number 6 Photograph
Obtaining camera and target coordinates	Fixed camera position Image center from diagonals of the photo Camera and target coordinates—from orthophotomap, by approximation	Fixed camera position Target by approximation	Omitted
GCPs utilization	For quality assessment For camera position optimization—iterative checking of different positions in 1 m grid around the approximate coordinates	For quality control In the photo distance between each GCP measured in the photo and back-projected GCP is checked Trial and error camera and target optimization	For photo orientation with DLT For quality assessment
Additional corrections	Image size reduced to a smaller extent	Roll of the camera around the viewing direction axis Earth curvature and light refraction—included but can be neglected due to marginal influence on the results	Any photo/image can be processed Rolls and tilts—included in DLT

Analyzed examples show that the processing of terrestrial images from different sources and different times, including old analog photographs, is possible. With a good performance of the algorithm, high-resolution orthoimages can be easily obtained, allowing visual interpretation in a standard GIS software. Also for shaded slopes, results may provide as good radiometric information as RGB aerial photographs. The unquestionable advantage of such data is their low price. They can also be obtained almost anytime, independently of the season or equipment. Bad weather conditions can be an obstacle, but it is a common feature of many data-collecting techniques, e.g. laser scanning. Imaging at night is impossible as well. A drawback of terrestrial orthoimage is that some parts of the terrain may not be visible, yet it could be easily compensated with multiple terrestrial photographs.

The potential range of applications is wider than avalanche mapping and may be used to gather information on other natural hazards in difficult terrain such as debris flows, landslides and rock falls. As it is possible to use the method for historical photographs, interesting studies on environmental changes can be conducted as well, especially combined with repeat-photography (Kaim and Kolecka 2010).

Future work will concentrate on introducing space resection based on approximate elements derived from DLT method—to obtain more accurate camera orientation parameters, introducing extension limits of the output orthoimage to avoid projecting unwanted points of the photograph, and building a graphical user interface which will allow easy application of the procedure.

Acknowledgments We would like to express our gratitude to the Foundation for Polish Science for financial support of Paweł Chrustek. Performing the analyses was possible thanks to the VENTURES program organized by the Foundation of Polish Science and co-funded by the European Regional Development Fund under the Operational Program Innovative Economy 2007-2013. Natalia Kolecka is a grant holder of "Doctus" Programme. We also would like to express gratitude to our colleagues Marek Świerk from the Anna Pasek Foundation in Poland, for assistance in collecting field data, Wojciech Bartkowski and Jakub Radliński from the Volunteer Mountain Rescue Service (*Górskie Ochotnicze Pogotowie Ratunkowe*, GOPR), for assistance in collecting data and materials concerning historical avalanches in the Tatra Mountains.

References

Abdel-Aziz YI, Karara HM (1971) Direct linear transformation from comparator coordinates into object space coordinates in close-range photogrammetry. In: Proceedings of the ASP/UI Symposium on Close-Range Photogrammetry. American Society of Photogrammetry, Falls Church, VA

Aschenwald J, Leichter K, Tasser E, Tappeiner U (2001) Spatio-temporal landscape analysis in mountainous terrain by means of small format photography: a methodological approach. IEEE Trans Geosci Remote Sens 39:885–893

Atlas of Switzerland 2 (2004) [DVD or 2 CD-ROMs]. Swiss Federal Office of Topography, Wabern

Buchroithner M (2002) Creating the virtual Eiger North Face. ISPRS J Photogramm Remote Sens 57:114–125

Bühler Y, Hüni A, Christen M, Meister R, Kellenberger T (2009) Automated detection and mapping of avalanche deposits using airborne optical remote sensing data. Cold Reg Sci Technol 57:99–106

Christen M, Kowalski J, Bartelt P (2010) RAMMS: Numerical simulation of dense snow avalanches in three-dimensional terrain. Cold Reg Sci Technol 63:1–14

Chrustek P (2009) Promotion of secure mountain exploration by the Anna Pasek Foundation. In: Schweizer J, Van Herwijnen A (eds) International Snow Science Workshop. 27 September to 2 October 2009, Davos, Switzerland. Proceedings Swiss Federal Institute for Forest, Snow and Landscape Research, Birmensdorf

Chrustek P, Biskupič M, Kolecka N (2010) Comparison of different methods for obtaining snow avalanche data. In: Ostapowicz K, Kozak J (eds) Conference Proceedings of the 1st Forum Carpaticum, Integrating Nature and Society Towards Sustainability, Institute of Geography and Spatial Management, Jagiellonian University, Kraków

Corripio JG (2004) Snow surface albedo estimation using terrestrial photography. Int J Remote Sens 25:5705–5729

Deems J, Painter T (2006) Lidar measurement of snow depth: Accuracy and error sources. In: Proceedings of the International Snow Science Workshop, 1–6 October Telluride, CO, USA

Foote KE, Huebner DJ (1995) Error, Accuracy and Precision. http://www.colorado.edu/geography. Accessed 28 October 2010

Gruber U (2001) Using GIS for Avalanche Hazard Mapping in Switzerland. In: Proceedings of the 2001 ESRI International User Conference, July 9-13, 2001, San Diego

International Commission for Alpine Rescue (2010) Statistic—People rescued from snow avalanches—2009/2010. http://www.ikar-cisa.org. Accessed 28 October 2010

Jörg P, Fromm R, Sailer R, Schaffhauser A (2006) Measuring snow depth with Terrestrial Laser ranging system. Measuring snow depth with a terrestrial laser ranging system. In: Proceedings of the International Snow Science Workshop, 1–6 October 2006, Telluride, CO, USA

Kaim D, Kolecka N (2010) Zmiany pokrycia terenu w Tatrach Polskich na podstawie powtórzonej fotografii naziemnej. Pr Geogr 123:31–45

Kraus K (1992) Photogrammetry Fundamentals and Processes. Dummler Verlag, Bonn

Kraus K (2007) Photogrammetry: geometry from images and laser scans. Walter de Gruyter, Berlin

Krupnik A (2003) Accuracy prediction for ortho-image generation. Photogramm Rec 18(101):41–58

Luhmann T, Robson S, Kyle S, Harley I (2006) Close Range Photogrammetry. Principles, Methods and Applications. Whittles Publishing, Scotland

Meister R, Jeller P (2009) Avalanche outline mapping with a digital GPS camera. In: Schweizer J, Van Herwijnen A (eds) International Snow Science Workshop. 27 September to 2 October 2009, Davos, Switzerland. Proceedings Swiss Federal Institute for Forest, Snow and Landscape Research, Birmensdorf

Mikhail EM, Bethel JS, McGlone JC (2001) Introduction to modern photogrammetry. Wiley, New York

Novak K (1992) Rectification of digital imagery. Photogramm Eng Remote Sens 58:339–344

Okeke FI (2001) Review of Digital Image Orthorectification Techniques. http://www.gisdevelopment.net. Accessed 28 October 2010

Prokop A (2009) Terrestrial laser scanning for snow depth observations: An update on technical developments and applications. In: Schweizer J, Van Herwijnen A (eds) International snow science workshop. 27 September to 2 October 2009, Davos, Switzerland. Proceedings Swiss Federal Institute for Forest, Snow and Landscape Research, Birmensdorf

Prokop A, Schirmer M, Rub M, Lehning M, Stocker M (2008) A comparison of measurement methods: terrestrial laser scanning, tachymetry and snow probing for the determination of the spatial snow-depth distribution on slopes. Ann Glaciol 49:210–216

Sampl P, Zwinger T (2004) Avalanche simulation with SAMOS. Ann Glaciol 38:393–398

Sauermoser S, Illmer D (2002) The use of different avalanche calculation models practical experiences. In: International congress INTERPRAEVENT 2, Matsumoto, Japan

Slama CC, Theurer C, Henriksen SW (1980) Manual of Photogrammetry. American Society of Photogrammetry, Falls Church, VA

Vallet J (2008) High Precision LiDAR Mapping for Complex Mountain Topography. In: Hurni L, Kriz K (eds) Proceedings of the 6th ICA Mountain Cartography Workshop 11–15 February 2008, Zurich, Switzerland

Volk G, Kleemayr K (1999) ELBA—Ein GIS-gekoppeltes Lawinensimulationsmodell. Anwendungen und Perspektiven. Österr Z Vermess Geoinf 2 + 3:84–92

Wang J, Robinson GJ, White K (2000) Generating viewsheds without using sightlines. Photogramm Eng Remote Sens 66:87–90

Using High Resolution LiDAR Data for Snow Avalanche Hazard Mapping

Paweł Chrustek, Piotr Wężyk, Natalia Kolecka, Marek Biskupič, Yves Bühler and Marc Christen

Abstract Each year in the Carpathian Mountains and the Sudety Mountains snow avalanches cause a great number of accidents. Avalanches also threaten buildings and affect the environment. The latest studies in Poland aim to implement advanced snow avalanche hazard mapping procedures, which would allow the creation of complex cartographic products for the location of avalanche hazard areas. These preliminary studies showed that results of these procedures strongly depend on the quality of the input digital surface data. The main goal of this study is to investigate this problem in detail through comparison of different types of Digital Elevation Models (DEMs), putting stress on high resolution DEMs generated from airborne and terrestrial laser scanning, in the context of estimating potential avalanche release areas and making run-out calculations. Test sites in the Tatra Mountains in the Carpathians and in the Karkonosze Mountains in the Sudety Mountains were selected for this study. The analysis was performed using Swiss Rapid Mass Movements (RAMMS) model and modified script on delineation automated release area. The study recognized that not only quality but also

P. Chrustek (✉) · N. Kolecka
Institute of Geography and Spatial Management, Jagiellonian University, Gronostajowa 7, 30-387 Kraków, Poland
e-mail: p.chrustek@annapasek.org

P. Chrustek · P. Wężyk · M. Biskupič
Anna Pasek Foundation, Małobądzka 101 42-500 Będzin, Poland

P. Wężyk
Agricultural University in Kraków, Faculty of Forestry, Al. 29 Listopada 46, 31-425 Kraków, Poland

M. Biskupič
Charles University in Prague, Faculty of Science, Albertov 6, 12843 Praha, Czech Republic

Y. Bühler · M. Christen
WSL Institute for Snow and Avalanche Research SLF, Flüelastr. 11, 7260 Davos, Switzerland

J. Kozak et al. (eds.), *The Carpathians: Integrating Nature and Society Towards Sustainability*, Environmental Science and Engineering,
DOI: 10.1007/978-3-642-12725-0_42,

resolution of a digital surface models influence the accuracy of release area and volume estimation, calculated topography parameters, location of avalanche track and other parameters calculated by dynamic models.

1 Introduction

Avalanche hazard mapping is a set of procedures used by land planning authorities as a tool to prevent settlements, roads and railways being constructed in areas that are endangered by avalanches (Gruber 2001). It has proven to be one of the most economic effective hazard mitigation measures in Switzerland (Gruber and Margreth 2001). The goal of the procedure is to estimate the areas exposed to the avalanche hazard and related risk.

Hazard mapping started early as first large and extreme avalanches threatened human settlements, caused life losses and serious damages. In the early attempts run-outs were simply plotted on topographic maps and so called avalanche cadastres were created. With the development of knowledge about avalanche flow and rheology more sophisticated approaches were used. Nowadays in the Alps many guidelines for avalanche hazard zoning have been established, most of them based on avalanche dynamics simulations (Jamieson 2008). These numerical simulations coupled with tools for delineation of release areas, historical avalanches and snow depth records are undoubtedly the crucial part of avalanche hazard mapping (Maggioni 2005).

Generating potential release areas is the first and crucial step in the avalanche hazard mapping process. It determines their location and helps to calculate release volumes used as input parameters for further dynamics calculations. A method for automated delineation of snow avalanche release area was described by Maggioni (2005). The procedure based on geographic information systems (GIS) classifies release areas based on vegetation and such topographic parameters as inclination, planar curvature, altitude generated from digital elevation model (DEM). The method finds widespread application in mountainous regions where historical avalanche events have been poorly documented (or no documentation exists at all) (Maggioni 2005). However, it has a tendency for some generalization, mainly because the results are used for analyzing extremely large avalanches. Developing a reliable automated release method and adapting it to smaller avalanches, however, requires extensive testing with various numerical models. The impact of initial conditions (release location, dimension and volume) on model results (run out distance and flow) must be extensively tested in various regions.

In the course of several decades the estimation of avalanche run-outs has been the scope of scientific investigations in both Europe and North America. To predict avalanche speed and run-out, dynamic models use physical laws (conservation of momentum, conservation of mass). Numerical modeling requires data on initial avalanche conditions (dimensions of release zones, snow cover entrainment and friction parameters). The complexity of terrain makes avalanche flow simulations a very demanding and challenging task. Many simplifications and assumptions have to be implemented.

First attempts to simulate the avalanche flow were done in the former Soviet Union in Tbilisi. Dry friction and force increasing linearly with speed were introduced as frictional forces (Salm 2004). According to Salm (2004), the implementation of the Coulomb law of friction in avalanche dynamics is one of the most ingenious theories in this field. In 1955, in a chapter *Über die Zerstörungskraft von Lawinen* (On the destructive force of avalanches) dealing with dense snow avalanches Voellmy assumed that their flow was similar to fluids and proposed to include in the avalanche model two parameters: Coulomb friction and turbulence. Salm and others adapted the model to better fit observed run-outs and included the back pressure due to the deceleration in the run-out zone (Salm et al. 1990). This model is well known as the Voellmy-Salm model and allows to estimate the flow depth at a given flow width. It has been used widely across Europe to design avalanche hazard maps.

For many years various dynamic models have been developed (e.g. Perla et al. 1980; Sampl and Zwinger 2004) but only two of them: AVAL-1D and RAMMS (Rapid Mass Movements) (Christen et al. 2002, 2010a) were released as a commercially available solution by the WSL Institute for Snow and Avalanche Research SLF in Davos, Switzerland. Using them, however, requires experience in avalanche science and GIS technology. A DEM is the basic input for all these models; its quality directly affects the result of calculations.

Technological developments have also facilitated quick and more efficient acquisition methods of detailed geodata. The LiDAR (Light Detection And Ranging) has been known from the last decade as a rapid, accurate and adaptable method for 3-dimensional (3D) surveying of the Earth surface and profiling the atmosphere via either satellite laser scanning (SLS), airborne laser scanning (ALS) or terrestrial laser scanning (TLS). All these technologies deliver information which can be integrated with other sensors, like airborne digital cameras (e.g. CIR orthophotomaps), hyperspectral linear scanners or thermal imaging cameras (Wężyk 2006).

Several advantages of using high resolution LiDAR data for better understanding natural processes in complex terrain and obtaining snow and avalanche data were found (e.g. Vallet et al. 2000; Deems and Painter 2006; Jörg et al. 2006, Prokop et al. 2008; Vallet 2008), however, detailed studies about its influence on hazard mapping process and procedures (including dynamics calculations) have not been published yet. Introduction to this kind of studies, but mainly in the context of estimating release area delimitation, was presented by Chrustek and Wężyk (2009) and McCollister and Comey (2009).

Mountains in Poland (above 500 m a.s.l.) cover only 3.1% of the total country area, so avalanches are a less serious problem than in the Alpine regions. This does not make it less important. Each year Polish mountains witness a few fatal accidents caused by avalanches. The greatest tragedy took place on the 28th of March, 1968, when the avalanche in the Karkonosze Mountains area killed nineteen people. Moreover, snow avalanches in Poland bring significant damages to forested areas. During winter seasons of 2008/2009 and 2009/2010 many new avalanche paths in the Tatra Mountains were activated and caused many fatalities and infrastructure damages. It confirmed the importance of implementation of avalanche mapping procedures for this region (Chrustek and Biskupič 2010).

Considering current technological progress e.g. global positioning system (GPS) receivers integrated with handheld computers or mobile GIS software implemented into cell phones that are brought into general use in mountain exploring, it seems that avalanche mapping showing a range of avalanche risk/hazard areas might be very useful in future in safety context. Not only for planners, engineers or administrators of recreational areas, but also for mountain rescuers, specialists during their field work and tourists performing a wide range of winter mountain activities.

However, preliminary studies in Poland (Chrustek 2005, 2009) recognized that results of hazard mapping procedures strongly depend on the quality of the input digital surface data. DEM quality and its resolution influences determining release areas, release volumes, simulated avalanche flow and consequently precision of the hazard zoning, especially when analyzing smaller avalanches in a more complex terrain. This impact is not obvious thus detailed analysis must be performed, before avalanche hazard mapping procedures can be fully implemented in the region of the Sudety Mountains and the Carpathians.

The chapter is part of this research and its main goal is to compare different types of DEMs, in particular high resolution DEMs generated from ALS and TLS data, in the context of estimating potential avalanche release areas and making run-out calculations, that are fundamental steps of hazard zonation. The analysis was performed using the Swiss RAMMS model and modified script for release area delineation.

1.1 Test Areas

Goryczkowy test site (surroundings of the Kasprowy Wierch peak) in the Polish Tatra Mountains was chosen as a main test site (Fig. 1). The region has a high frequency of snow avalanches due to the presence of long and steep slopes. On the other hand, it has been a very popular tourist destination for a long time, and has highly developed tourist infrastructure (mountain hotels, ski lifts and ski routes). Therefore, the highest number of incidents related to avalanches is recorded in this region. Part of the analysis was made in the Mały Staw test site located in the Karkonosze Mountains, Poland (the Western Sudety Mountains) (Fig. 1). Here frequent avalanche accidents are caused mainly by small and medium avalanches in a more complex terrain.

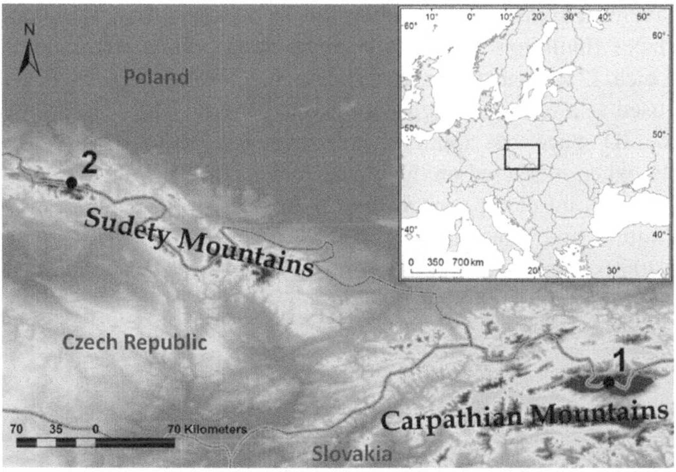

Fig. 1 Location of the test sites. (*1*) the Goryczkowy test site in the Polish Tatra Mountains (Western Central Carpathians). (*2*) the Mały Staw test site located in the Karkonosze Mountains, Poland (the western Sudety)

2 Materials

As already indicated, DEM is a fundamental dataset for avalanche hazard mapping. Higher-resolution DEMs are required for channeled and inhomogeneous terrain, especially for small avalanche events (volumes < 25,000 m3). Large, extreme events, travelling at high speed, appear not to react to small scale terrain features, suggesting that some simulations can be performed on low-resolution DEMs and yield realistic results. High-resolution DEMs seem to be crucial for small and frequent avalanches. In the same way wet snow avalanches, travelling at a lower speed than dry snow avalanches, may require a higher DEM resolution than dry snow avalanches (Christen et al. 2010a). The most suitable values of the DEM resolution in step are not standardized thus for the comparison we used three types of surface data which were finally resampled to spatial resolutions of 1, 5, 10 and 25 m.

2.1 ALS Data

The ALS system used in August 2007 over the Tatra Mountains was based on two LMS-Q560 (full waveform) scanners (Riegl) mounted on a special platform under the DA42 airplane. Those two scanners (forward and backward looking) allowed obtaining very dense point cloud, even up to 40 laser beams on 1 m^2 of the ground. Dedicated RiscanPro Software (Riegl) allowed generating the first (FE for digital surface model, DSM, generation) and the last echo (LE for DEM generation) from the full waveform signal. The matching of 33 separate scans was conducted on planar surfaces (buildings' roofs with minimum 6 points) measured with dGPS receiver (Leica 1230) and tachimeter (Leica 407 power). The EUPOS-ASG network was used as a reference signal for geoprocessing. The accuracy of the point cloud in the 3D space was approximately 0.06, 0.02 and 0.01 m in X, Y and Z, respectively, when measured on the planar roof surfaces (Wężyk et al. 2008). The whole matched point cloud was cut into 500 × 500 m tiles for the purpose of a ground filtration and generation of the DEM (1 × 1 m GRID) based on the Axelsson (2000) algorithm with the Terrasolid software. A similar approach was used to create the DEM for Mały Staw test site.

2.2 TLS Data

The Laser Profile Measuring System LPM-321 (Riegl), used by authors in the Polish Tatra Mountains in July 2009, allowed long range 3D profiling up to 6,000 m with the high accuracy. The distance meter comprises of the state-of-the-art digital signal processing and echo waveform analysis, enabling precise distance measurements even under reduced visibility conditions. The scanner can detect up to 3 target distances per measurement. The combination with mounted high resolution digital camera calibrated and accurately orientated makes a hybrid sensor system, which allows to obtain colored point clouds. The scan range of the LPM-321 is −20 to 130 ° vertically, and 360 ° horizontally. The accuracy of the measurements is 25 mm (Riegl 2010).

Because of the large size of the Goryczkowy test site, it was divided into three parts to make scanning more efficient. TLS data pre-processing consisted of point cloud denoising and decimation, removing redundant and isolated points, meshing and mesh cleaning (VRMesh 2010). Three point clouds had to be merged and registered in the global coordinate system. The procedure comprised of an approximate manual registration of input point clouds, followed by automatic alignment based on least square matching (Shan and Toth 2008; Heritage and Large 2009). Positioning in the global coordinate system was achieved by means of automatic alignment of the TLS points to the existing photogrammetric model. There were more than 1 million points in the merged cloud, with average spacing of 0.803 m.

One of the characteristic features of the TLS technology are occlusions, that cause some gaps within the data. In the Goryczkowy dataset two significant holes appeared. Therefore, the primary TLS model was updated with the altitude information obtained from the photogrammetric model (Fig. 2). TLS data for Mały Staw test site were not available.

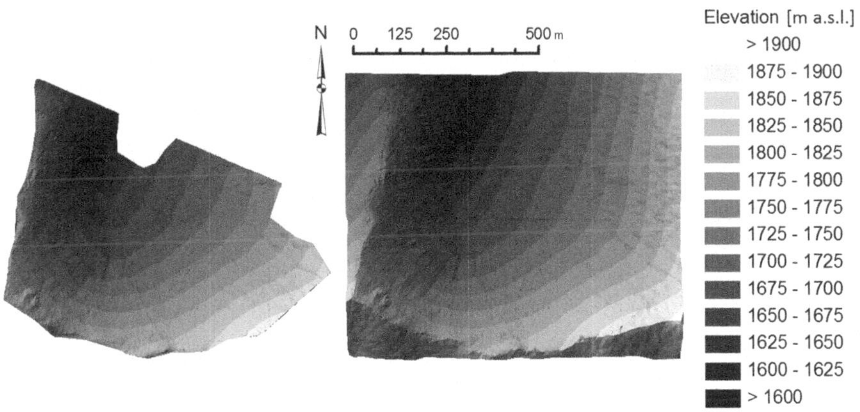

Fig. 2 TIN of the Goryczkowy test site produced out of the TLS point clouds only (*left*) and updated with the altitude data from the photogrammetric model (*right*)

2.3 Topo Data

Topo data were derived from the TIN model based on contour lines with 5 m intervals (digitized from topographic maps 1:10,000), mass points and structure lines obtained from aerial images processing. This type of data covers large areas in Polish mountainous regions and is the fundamental part of public administration GIS databases at the moment.

3 Methodology

Comparative analyses were performed based on two distinct avalanche hazard mapping procedures: generating potential release areas (PRAs) and avalanche dynamic calculations (including potential extreme event calculations based on the generated PRAs and "back calculations" based on the recorded historical avalanche). Each procedure is presented in detail below.

3.1 Generating Potential Release Areas

Automatic procedure for release area delineation proposed by Gruber et al. (2002) and Maggioni (2005) was used for the analysis at the Goryczkowy test site. This method is not directly adaptable to such areas as the Polish mountains (Chrustek 2005) thus some steps were changed due to morphological differences between mountain ranges in Poland and Switzerland. Upgrade procedure was written by the authors using the Python Script and implemented in the ESRI ArcInfo ver. 9.3.

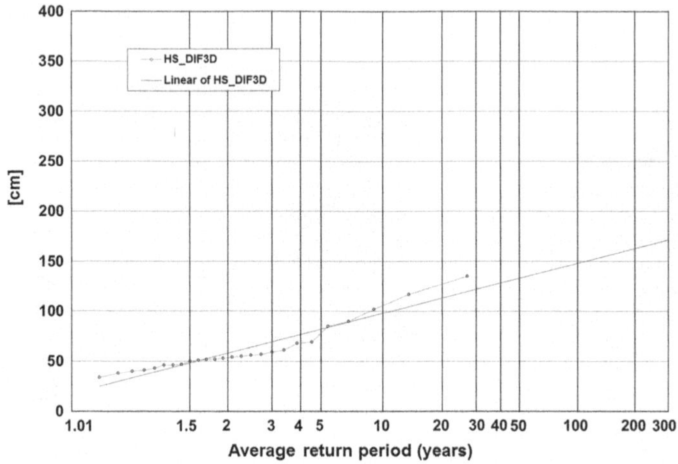

Fig. 3 Data and Gumbel statistics extrapolation of the maximum annual value of HS_DIF3D of the Kasprowy Wierch High Mountain Observatory (1,987 m a.s.l., 26 winters)

Besides 1 m resolution, spatial resolutions of the DEMs used in this analysis are the most widely used spatial resolutions for generating release areas (e.g. Gruber 2001; Gruber et al. 2002; Gruber and Bartelt 2007; Pagliardi et al. 2009). Generating PRAs from ALS and TLS LiDAR with DEM spatial resolution higher than 1 m is possible but it does not seem to be justified, as such a precise terrain differentiation disappears under snow cover, which causes natural process of "smoothing" the surface.

All of the release areas in the ESRI shape file format derived from the automatic procedure and their base DEM models were loaded into Swiss RAMMS model (Christen et al. 2010a) and then release parameters like areas, mean angle, mean altitude, estimated release volumes or mass were calculated. Estimated release height value was calculated according to the Swiss guidelines for avalanche hazard mapping (Salm et al. 1990). This procedure assumes that the maximum yearly increase of snow cover within three days (HS_DIF3D) is representative for the fracture depth of the avalanche. For hazard mapping purposes, the yearly maxima of HS_DIF3D were extrapolated using Gumbel extreme-values statistics (Reiss and Thomas 1997) and the data from the Kasprowy Wierch High Mountain Observatory which is situated about 1 km away from the Goryczkowy test site (Fig. 3).

The calculated value was corrected using the cosine of slope angle, elevation difference value and snow drift statistics. Finally, the value of 113 cm for 100 years return period was derived.

3.2 Avalanche Dynamics Calculations

RAMMS model solves the depth-averaged equations governing avalanche flow with accurate second order solution schemes. RAMMS deals with the avalanche flow using two different approaches: the standard Voellmy-Salm approach or random kinetic energy model (RKE). The Voellmy-Salm model is implemented in RAMMS through its AVAL-1D code. Coupled with an easy to use interface it is an invaluable tool for avalanche engineers dealing with hazard zoning and mitigation (Christen et al. 2010a,b).

Three model inputs must be specified to perform a numerical calculation: (1) a digital elevation model (DEM) (2) release zone area and fracture height and (3) model friction parameters.

3.2.1 Calculating Potential Extreme Avalanche Event

The goal of this comparison test was to recognize the influence of the different types of DEMs (spatial resolutions with 1 and 25 m) and relation between avalanche volumes on the calculation results.

PRAs in the ESRI shapefile format generated from different ALS and Topo DEMs for the Goryczkowa test site were used in the dynamics calculations using RAMMS model. TLS data was not used in this analysis because of the limited spatial extent of the dataset.

Generated PRAs extents, their topographic parameters and predicted extreme fracture height in 100 year return period were assumed as input parameters for modelling extreme run-outs. The input variable friction parameters were calculated using automatic procedure implemented in the RAMMS model. The procedure classifies terrain features like slope angle, planar curvature and altitude into categories such as open slope or flat, terrain or channelled or gully and forested or non-forested areas (Christen et al. 2010a). Some of the default parameters (like avalanche volume which influences calculated friction parameters) were slightly modified based on the calibrating analysis for this region. Assumed snow density was constant and equal to 300 kg/m^3. In each case the 5 m spatial resolution was used for the calculation.

3.2.2 Back Calculation Based on the Recorded Historical Avalanche

The final step of the analysis was an evaluation how the type and resolution of the DEMs influences dynamics analysis results, with particular emphasis on small avalanches in a more complex terrain. For this study a documented avalanche event from Mały Staw test site in the Karkonosze Mountains was chosen. On 26 January 2003 it killed one climber and caused serious injuries to two another. The avalanche with estimated release volume about 550 m^3 and 300 m length was

Fig. 4 Documented avalanche track from 26 January 2003 (the Mały Staw test site)

released close to the ridge and then flowed down through the rocky couloir (Fig. 4). Following parameters were used for the release area in further dynamics calculation: 2D area (1,215 m^2), mean angle (39.9 °), mean altitude (1,349.1 m a.s.l.), release height (35 cm), estimated avalanche volume (555 m^3), snow density (300 kg/m^3) and release mass (166.5 tons).

Input friction parameters were calculated in the same way as in the previous analysis. ALS and Topo data were used, with two different resolutions (1 and 25 m). Measured release height (35 cm) and release area (drawn on the map based on the accident documentation) were used for the calculation. In each case 1 m resolution was used for the calculation.

4 Results and Discussion

4.1 Results of Generating PRAs

Results of all calculations are presented in Table 1. When comparing automatically generated values for such topographic parameters as mean angle and mean altitude, there are no noticeable quantitative differences.

The biggest differences are observed for area, volume and mass parameters. The differences were caused both by the types of input DEMs and their spatial resolutions. Calculated values for ALS and TLS data related to the area are very similar but the values for Topo DEMs are always smaller than the other. For example, if an assumption is made that area values for ALS data are 100 %, then percentage differences are between 0.2–2.2% (when comparing ALS and TLS

Table 1 Automated generated PRAs for Goryczkowy test site with release parameters calculated in RAMMS model

Type of PRA	2D Area (m^2)	Mean angle (degree)	Mean altitude (m a.s.l.)	Estimated volume (m^3)	Mass (t)
ALS 1 m	83,500	34.5	1,744.4	1,14,511	34,353.3
TLS 1 m	82,300	34.5	1,743.1	1,12,811	33,843.3
Topo 1 m	79,200	34.6	1,744.5	1,08,779	32,633.7
ALS 5 m	84,000	34.4	1,744.8	1,15,005	34,501.5
TLS 5 m	84,200	34.3	1,744.5	1,15,158	34,547.4
Topo 5 m	79,200	34.5	1,745.2	1,08,620	32,586.0
ALS 10 m	84,700	34.1	1,744.6	115,644	34,693.2
TLS 10 m	83,100	34.1	1,742.4	1,13,416	34,024.8
Topo 10 m	80,100	34.3	1,743.1	1,09,564	32,869.2
ALS 25 m	79,000	33.2	1,734.3	1,06,650	31,995.0
TLS 25 m	77,300	33.7	1,737.6	1,05,018	31,505.4
Topo 25 m	76,500	34.1	1,737.2	1,04,426	31,327.8

results minimum value refers to the 5 m and maximum value to the 25 m spatial resolution of DEM) and 3.2–5.7% (when comparing ALS and Topo results – minimum value refers to 25 m and maximum to 5 m spatial resolution of DEM).

Degrading spatial resolution of the DEMs causes reduction of differences between the area, volume and mass values. On the other hand calculated values decrease pixel size with the increasing spatial resolution of the DEMs. Degrading resolution of LiDAR (ALS and TLS) DEMs from 1 to 25 m causes bigger differences in calculated release values (areas, volumes, mass) than of Topo DEMs. (e.g. for the release mass 6.86 % for ALS, 6.91 % for TSL and 4 % for TOPO).

Recognized maximum differences appear to be quite large, especially when it is assumed that these values describe release parameters for a single extreme avalanche.

4.2 *Results of Avalanche Dynamics Calculations*

4.2.1 Results of Calculations of the Potential Extreme Avalanche Event

For various input DEM resolutions maximum velocity, flow height and pressure were calculated (Table 2).

The quantitative differences between output parameters calculated for different DEMs with different resolution do not seem to be significant but more discrepancies were noticed when analyzing their spatial variations (Fig. 5).

Analyzed examples showed that ALS models allow to predict avalanche flow process more precisely (even after reducing the model resolution) than Topo models, including also such terrain as the surrounding of the Goryczkowy test site where topographic surface is not very complex.

Table 2 Selected output parameters for 100 year return period avalanche calculated using RAMMS model (for Goryczkowy test site)

Type of the input dataset used in calculation	Max velocity (m/s)	Max flow height (m)	Max pressure (kPa)
ALS 1 m	30.67	16.71	282.16
Topo 1 m	30.01	14.50	270.13
ALS 25 m	30.21	15.55	273.87
Topo 25 m	29.62	14.52	263.13

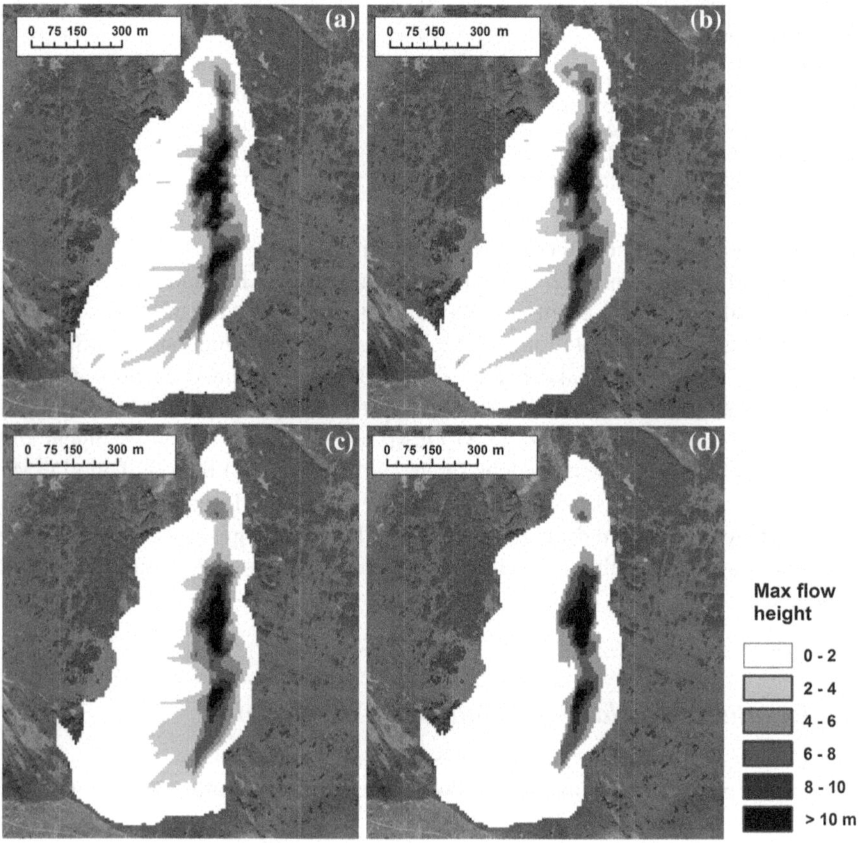

Fig. 5 Maximum flow height of predicted avalanche in the Goryczkowy test site, calculated in the RAMMS model: **a** 1 m ALS dataset, **b** 1 m Topo dataset, **c** 25 m ALS dataset, **d** 25 m Topo dataset

The influence of various DEM types and different resolutions on the maximum distance was also investigated. When comparing 1 m ALS and 1 m Topo data differences between calculated distances were about 25 m (distance for Topo dataset was greater, Fig. 5a, b). The same comparison for 25 m resolution datasets showed that the difference was much bigger – about 40 m (distance for ALS dataset was greater, Fig. 5b, c). The biggest difference was obtained after changing resolution in ALS dataset. Difference between maximum flow distance for 1 and 25 m was about 50 m (distance for ALS 25 m was greater, Fig. 5a, c). The same comparison for Topo dataset showed difference that was about 30 m (distance for Topo 1 m was greater, Fig. 5b, d).

Based on the results presented in Table 2 it can be stated that parameter differences between calculated PRAs are noticeable during dynamics calculation as well. This test showed that differences were bigger when spatial resolution were changed for ALS data. These discrepancies for Topo data were less noticeable. However, direct relation between estimated volume and maximum distance calculated by the model was not investigated (when comparing results from different resolutions). Interesting results were obtained when analyzing ALS data. Despite of much lower estimated volume (by 5.4%, 2,358.3 t less), calculated distance for 25 m dataset was 40 m longer than for 1 m dataset. It means that smoothing the

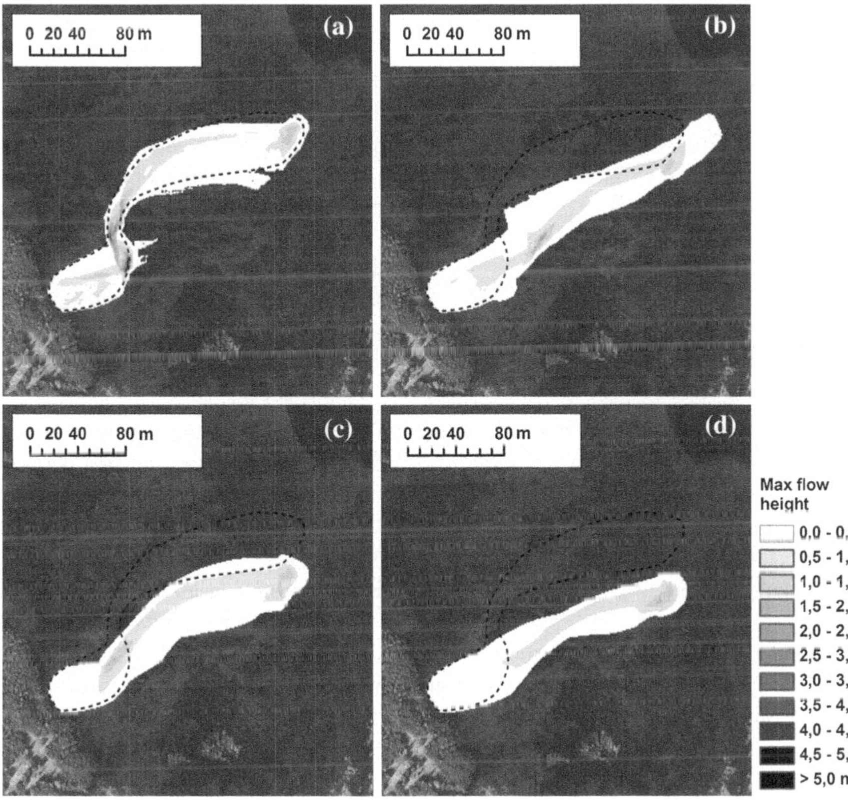

Fig. 6 Maximum avalanche flow height in the Mały Staw test site calculated in the RAMMS model: **a** 1 m ALS dataset, **b** 1 m Topo dataset, **c** 25 m ALS dataset, **d** 25 m Topo dataset; dashed line shows extents of the documented avalanche

Table 3 Selected output parameters for simulation of the documented avalanche calculated using the Swiss model RAMMS (for Mały Staw Lake test site)

Type of the input dataset used in calculation	Max velocity (m/s)	Max flow height (m)	Max pressure (kPa)
ALS 1 m	20.67	5.18	128.20
Topo 1 m	24.41	2.23	178.79
ALS 25 m	21.25	1.63	135.42
Topo 25 m	19.78	1.56	117.40

topography while decreasing the spatial resolution, strongly influences calculation results. This influence is more significant when analyzing LiDAR data.

4.2.2 Results of the Back Calculations

In this example, quantitative differences between calculated parameters are significant (Table 3) but the biggest discrepancies were observed when analyzing tracks of the avalanche flows (Fig. 6). High similarity of the simulated track with observed track was obtained only from high resolution ALS data (1 m resolution). ALS data resampled to smaller 25 m resolution lost important surface details, however, its simulated avalanche profile is more similar to the documented avalanche than the one obtained from 1 m Topo model. Results from Topo models are completely unsatisfactory (Fig. 6) as calculated avalanche flows were completely different from the observed one.

5 Conclusions

Presented tests proved that many variables influence hazard mapping results. One of the most important factors is a quality of DEMs. It influences the precision of the release area estimation, calculated topography parameters, calculated release volume, location of avalanche track and another parameters calculated by dynamic models. DEMs generated from LiDAR data (ALS or TLS) introduced new quality for avalanche modeling but influence of this improvement on all hazard mapping procedures must be extensively tested. When comparing automatically generated PRAs from Topo and LiDAR models (for common spatial resolutions of 1, 5, 10 and 25 m), it was observed that reduction of the spatial resolution causes reduction of PRA areas. It affected also directly the calculated release volume. Based on the presented test, differences between results obtained from 1 and 25 m resolution could achieve up to 7%.

Using DEMs with different resolutions only slightly affected calculation of such topographic parameters as: mean angle and mean altitude. There were no noticeable quantitative differences between these parameters computed for various

DEMs, which seems very important as they directly affect calculations of an avalanche release volume.

Presented example for the Mały Staw test site showed that only high resolution DEMs obtained from LiDAR data were able to simulate proper avalanche flow for small events which in case of the Carpathian Mountains and the Sudety Mountains kill more people every year than large catastrophic events with large return periods. For extreme events, the relation between DEM spatial resolution and avalanche model accuracy is not so obvious. When analyzing the extreme scenario for the Goryczkowy test site, significant influence of DEM spatial resolution on the final results was noticed. The resolution changes affect calculation results more significantly, when basing on LiDAR data. This relation was noticed both while analyzing release areas and performing dynamic calculations. It may prove quite important for avalanche specialists who deal with hazard mapping, because different analysis steps require different spatial resolutions of DEMs. However, it is still uncertain how differences between calculated avalanche volume, type of input DEM and its resolution can influence simulated avalanche flow and calculated avalanche maximum distance. To better understand of this problem, more tests in different regions have to be performed.

While the high resolution terrain data is still available for selected regions only, the future studies should also contain an analysis based on free of charge elevation data which is acquired by a satellite-borne sensors (like SRTM or ASTER). This data are easily downloadable and cover all mountainous regions on Earth. Spatial resolution of these data is not higher than 25 m, therefore they have a limited applicability in studies of small avalanches in a more complex terrain. However, they may be useful for large scale avalanche hazard mapping, especially for the region without access to any other digital elevation data.

Acknowledgments We would like to express our gratitude to the Foundation for Polish Science for financial support of Paweł Chrustek. Performing analyses was possible mainly thanks to the VENTURES program organized by the Foundation of Polish Science and co-funded by the European Regional Development Fund under the Operational Program Innovative Economy 2007–2013. We would also like to thank to the Anna Pasek Foundation for additional financial support. Natalia Kolecka is a grant holder of "Doctus" Programme. We would like to also thank the Karkonosze National Park, ProGea Consulting Company and Kraków Branch of the Institute of Meteorology and Water Management (IMGW) for deriving the Geodata (ALS, GIS, Meteo data), Andrzej Brzeziński and Jakub Radliński from the Mountain Rescue Services (GOPR), for assistance in collecting data and materials concerning avalanches in the Karkonosze Mountains and Marek Świerk from the Anna Pasek Foundation in Poland, for assistance in collecting field data.

References

Axelsson P (2000) DEM generation from laser scanner data using adaptive TIN models. Int Arch Photogram Remote Sens 33/4B:203–210

Christen M, Bartelt P, Gruber U (2002) AVAL-1D: an avalanche dynamics program for the practice. Interpraevent 2002. Matsumoto - Japan. Congress publication 2:715–725

Christen M, Kowalski J, Bartelt P (2010a) RAMMS: Numerical simulation of dense snow avalanches in three-dimensional terrain. Cold Reg Sci Technol 63:1–14

Christen M, Bartelt P, Kowalski J (2010b) Back calculation of the In den Arelen avalanche with RAMMS: interpretation of model results. Ann Glaciol 51(54):161–168

Chrustek P (2005) Zastosowanie GIS w typologii obszarów potencjalnego występowania lawin śnieżnych na przykładzie rejonu Kasprowego Wierchu w Tatrach. MSc Thesis. Jagiellonian University, Kraków

Chrustek P, Wężyk P (2009) Using high resolution LiDAR data to estimate potential avalanche release areas on the example of polish mountain regions. In: Schweizer J, Van Herwijnen A (eds) International Snow Science Workshop. 27 Sep to 2 Oct 2009, Davos, Switzerland. Proceedings. Swiss Federal Institute for Forest, Snow and Landscape Research Birmensdorf

Chrustek P, Biskupič M (2010) Extreme snow avalanche events in Tatra Mountains. In: Ostapowicz K, Kozak J (eds) Conference proceedings of the 1st forum carpaticum, integrating nature and society towards sustainability, Cracow, Institute of Geography and Spatial Management, Jagiellonian University, Kraków

Deems J, Painter T (2006) Lidar measurement of snow depth: Accuracy and error sources. In: proceedings of the international snow science workshop, 1–6 October 2006, Telluride, CO, USA

Gruber U (2001) Using GIS for avalanche hazard mapping in Switzerland. In: proceedings of the 2001 ESRI international user conference, July 9–13, 2001, San Diego, USA

Gruber U, Bartelt P (2007) Snow avalanche hazard modelling of large areas using shallow water numerical methods and GIS. Environ Model Softw 22(10):1472–1481

Gruber U, Maggioni M, Stoffel A (2002) Definition and characterization of potential avalanche release areas. In: Proceedings of the 22nd ESRI international user conference, October 15, 2002, San Diego, USA

Gruber U, Margreth S (2001) Winter 1999: a valuable test of the avalanche-hazard mapping procedure in Switzerland. Ann Glaciol 32:328–332

Heritage GL, Large ARG (2009) Laser scanning for the environmental sciences. Blackwell Publishing ltd, West Sussex

Jamieson B, Margreth S, Jones A (2008) Application and limitations of dynamic models for snow avalanche hazard mapping. In: Proceedings of the ISSW 2008, Whistler, Canada

Jörg P, Fromm R, Sailer R, Schaffhauser A (2006) Measuring snow depth with a terrestrial laser ranging system. In: proceedings of the international snow science workshop, 1–6 October 2006, Telluride, CO, USA

Maggioni M (2005): Avalanche release areas and their influence on uncertainty in avalanche hazard mapping. PhD Thesis, Department of Geography, University of Zurich UZH, Switzerland

Maggioni M, Gruber U (2003) The influence of topographic parameters on avalanche release dimension and frequency. Cold Reg Sci Technol 37:407–419

McCollister CH, Comey R (2009) Using LiDAR (light distancing and ranging) data to more accurately describe avalanche terrain. In: Schweizer J, Van Herwijnen A (eds) international snow science workshop. 27 Sep to 2 Oct 2009, Davos, Switzerland. Proceedings. Swiss federal institute for forest, snow and landscape research, birmensdorf

Pagliardi M, Barbolini M, Corradeghini P, Ferro F (2009) Identification of areas potentially affected by extreme snow avalanche combining expert rules, flow-routing algorithms and statistical analysis. In: Schweizer J, Van Herwijnen A (eds) International snow science workshop. 27 Sep to 2 Oct 2009, Davos, Switzerland. Proceedings. Swiss Federal Institute for Forest, Snow and Landscape Research, Birmensdorf

Perla R, Cheng TT, McClung D (1980) A two-parameter model of snow-avalanche motion. J Glaciol 26(94):197–207

Prokop A, Schirmer M, Rub M, Lehning M, Stocker M (2008) A comparison of measurement methods: terrestrial laser scanning, tachymetry and snow probing for the determination of the spatial snow-depth distribution on slopes. Ann Glaciol 49:210–216

Reiss R, Thomas M (1997) Statistical analysis of extreme values. Birkhäuser Verlag, Basel

Riegl (2010) Data sheet of the LPM-321 http://www.riegl.com/uploads/tx_pxpriegldownloads/10_DataSheet_LPM-321_18-03-2010.pdf. Accessed 20 October 2010

Salm B (2004) A short and personal history of snow avalanche dynamics. Cold Reg Sci Technol 39(2–3):83–92

Salm A, Burkard A, Gubler HU (1990) Berechnung von Fliesslawinen eine Anleitung für Praktiker mit Beispielen. Eidgenössisches Institut für Schnee und Lawinenforschung Technical Report 47, Davos

Sampl P, Zwinger T (2004) Avalanche simulation with SAMOS. Ann Glaciol 38:393–398

Sauermoser S, Illmer D (2002) The use of different avalanche calculation models practical experiences. In: International congress Interpraevent, Matsumoto, Japan

Shan J, Toth CK (ed) (2008) Topographic laser ranging and scanning. Principles and Processing, Taylor & Francis Group, Boca Raton

Vallet J, Skaloud J, Koelbl O, Merminod B (2000) Development of a helicopter-based integrated system for avalanche mapping and hazard management. Int Arch Photogramm Remote Sens Spatial Inf Sci 33:565–572

Vallet J (2008) High precision LiDAR mapping for complex mountain topography. In: Hurni L, Kriz K (eds) Proceedings of the 6th ICA mountain cartography workshop 11–15 February 2008, Zurich, Switzerland

Voellmy A (1955) Ueber die Zerstörungskraft von Lawinen. Schweiz Bauztg 73:159–285

VRMesh (2010) http://www.vrmesh.com/. Accessed 14 October 2010

Wężyk P, Borowiec N, Szombara S, Wańczyk R (2008) Generation of digital surface and terrain models in Tatra based on point cloud from airbone laser scanning (ALS). Archive of photogrammetry, Cartography and remote sensing. ISBN 978-83-61576-08-2, Szczecin, 18b, 651–661

Wężyk P (2006) Introduction to the laser scanning technology in forestry. Annals of Geomatics 4:119–132

Use of Airborne Laser Scanning Data for a Revision and Update of a Digital Forest Map and its Descriptive Database: A Case Study from the Tatra National Park

Piotr Wężyk, Marta Szostak and Piotr Tompalski

Abstract The aim of this study was to elaborate the semiautomatic method of airborne laser scanning (ALS) point-cloud data processing for revising a digital forest map in the Tatra National Park, Poland, and updating attributes stored in its descriptive forest database. We proved that elements such as gaps, clearings, biogroups, dead trees, wind-damaged trees, and areas of low canopy closure could be detected with high accuracy, based on this light detection and ranging technology. By using this method and geographic information system techniques, we were able to update selected attributes of the descriptive database, such as forest stand height, which were collected during earlier forest inventories. Moreover, parameters of some parts of forest stands, such as those growing on extremely steep slopes or isolated ledges that are hard to determine using traditional methods, are easily measured with ALS technology. The difference in stand height using the ALS method and that measured by the traditional forest inventory was equal to 1.58 m (2.87 m for absolute differences).

1 Introduction

Protected forest areas are one of the most important targets of the European Union (EU) directives and conventions. Some of these regulations concern saving habitats and biodiversity of bird species (92/43/EEC), while others allow for public access to spatial information (such as INSPIRE). For many years the Polish national parks have been trying to establish the standards and structure of geographic information systems (GIS), which would help to manage the protected areas. For more than 10 years the Polish State Forests National Forest Holding, which administrates approximately 25 % of the area of Poland, has been

P. Wężyk (✉) · M. Szostak · P. Tompalski
Department of Forest Ecology, Agricultural University of Kraków,
29 Listopada 46 31-425 Kraków, Poland
e-mail: rlwezyk@cyf-kr.edu.pl

J. Kozak et al. (eds.), *The Carpathians: Integrating Nature and Society Towards Sustainability*, Environmental Science and Engineering,
DOI: 10.1007/978-3-642-12725-0_43,

introducing a standard—the State Forests Informatics System (SFIS) *System Informatyczny Lasów Państwowych* (SILP)—which is a descriptive database only. At the same time, another standardized GIS product, called the Digital Forest Map (DFM) *Leśna Mapa Numeryczna* (LMN) was introduced in 420 forest districts. The DFM is a geometrical database that contains polygons (borders of forest compartments with a unique ID), lines (e.g. roads), and points. Together, SFIS and DFM, connected to each other, create a topologically correct, geometric layer with many attributes from the forest inventory, and can be managed with the use of GIS software. Due to the direct connection of the Polish State Forests and national parks, and the current necessity of information exchange between those institutions, some national parks have considered using a method similar to the SFIS database standards that also include the geometric structure of DFM. The Tatra National Park, one of the first users of the geoinformation technology, implemented those two products (SFIS and DFM) in 2008. Those using the National Park protection plans are accustomed to work with these standards. On the other hand, it is obvious that this solution cannot describe the entire richness of information concerning the protected areas, but it is a good starting point for proper modifications. Also, areas within the European Ecological Network Natura 2000 are subjects of such inventories.

During forest inventory measurements, aerial photos and orthophotomaps are often acquired to illustrate the changes that have taken place in the forest (fires, clear-cuts, avalanches, etc.). Unfortunately, very often the geometrical quality of DFM based on orthophotomaps is poor, especially in mountainous areas, because of central projection. The shifting (error) of the tree crown location, which often is the natural indicator for compartment borders, could reach many meters, depending on the distance from the main point. This problem occurs especially for tall trees growing on steep slopes (Wężyk and Mansberger 1997). The solution for this problem could be other remote sensing technologies like airborne laser scanning (ALS), which can provide accurate data for large areas that are difficult to access. The ALS, one of the types of light detection and ranging (LiDAR) technologies, is a modern remote sensing technology used for collecting 3D information, which provides precise information about terrain elevation and the structure of vegetation (Hyyppa and Inkinen 1999; Persson et al. 2002; Maltamo et al. 2004; Holmgren and Jonsson 2004; Hyyppä et al. 2004; McGaughey et al. 2004; Andersen et al. 2006). The filtration and classification of point clouds received from ALS enables a creation of three types of models (raster layers): a digital terrain model (DTM), digital surface model (DSM), or normalized digital surface model (nDSM), representing terrain and objects on it (Axelsson 2000; Wężyk et al. 2008a; Tompalski et al. 2009).

Currently, ALS is applied in many fields related to forestry, such as forest management, ecological assessment of biomass, or carbon sequestration (Wężyk et al. 2008b; Goluch et al. 2009). In the case of forest management, this technology could be used to measure many forest taxation parameters, especially the height of trees, which is an important parameter playing a crucial role in the determination of forest stand volume. With the traditional procedures the height of a given tree is measured by its age, health status, and site fertility; however, these data are simultaneously affected by management operations and random events occurring in the stand. Moreover, measuring of all forest tree heights was previously impossible to perform, as the process would have been too time-consuming. In addition, the error made by an operator during traditional height measurements with the use of a hypsometer often exceeds 5 %, and is caused by the subjectivity in pointing at the tree top or when the tree is leaned. Unfortunately, although some pilot ALS projects (e.g. Wężyk and Solecki 2008; Wężyk et al. 2008a, b) have been implemented regarding its practical usage for the purpose of forest inventory, there is still a lack of legal regulations and internal executive instructions that would allow the implementation of this technology in Poland on a large scale.

The aim of our study was to scrutinize the use of the automatic processing of ALS point-cloud data for the purpose of revising the DFM (such as geometric errors or lack of information) and updating the attributes stored in the descriptive database. We determined if it is possible to use ALS to map and revise geometrical objects, such as openings, gaps, clear-cuts, biogroups, dead trees, areas of low canopy closure, and the borders of forest compartments. Moreover, we ran spatial analyses for additional raster and vector layers based on nDSM, or even raw ALS point-cloud data, which could give us important feedback in cases where an update of geometry is necessary and the topology of the DFM should be rebuilt. An additional goal was to elaborate the method of automated stand height verification based on the obtained ALS data.

2 Study Area

The test area was located in the central part of the Tatra National Park (TNP) 21,164 ha, Fig. 1), which was established in 1954 for the protection of the highest part of the Polish Carpathians. It consisted of 270 selected compartments with a total area equal to 1,398.66 ha, and elevation ranging from 1,057 to 2,169 m. Norway spruce (*Picea abies*) was the dominant species, covering 52 % (157 compartments, 730.24 ha) of the study area. Dwarf mountain pine (*Pinus mugo*) covered 44.86 % of study area (101 compartments, 627.40 ha), and the remaining 3 % consisted of other species, such as European beech (*Fagus silvatica*), European mountain ash (*Sorbus aucuparia*), sycamore maple (*Acer pseudoplatanus*), European larch (*Larix decidua*), and willows (*Salix* sp.).

The age diversity of the Norway spruce stands in the study area was large and had its roots in the history of the Tatra forest (Table 1). The youngest forest compartments (<20 years; class I) were the minority (0.4 % of the total area). Age classes II–IX (20–180 years) covered together approximately 83.5 % (where each age class covered more than 6 % of total Norway spruce area). The oldest stands (>180 years) constituted about 16.1 % of the total spruce stands area. The largest and most numerous class was the age class III (41–60 years), covering more than 16 % of the Norway spruce area.

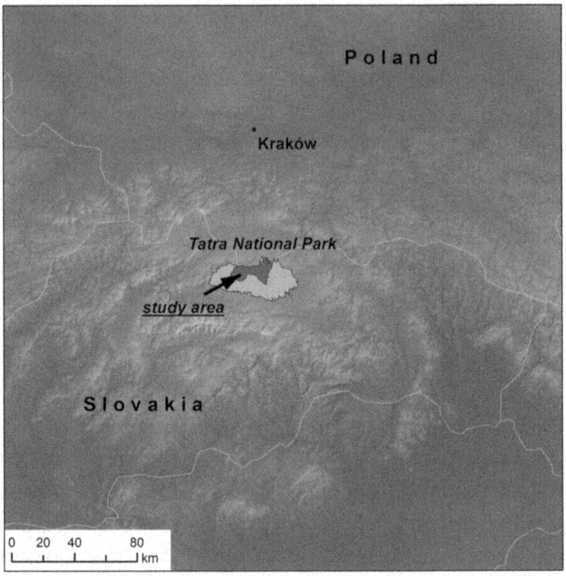

Fig. 1 Location of the study area

Table 1 Age and area structure of the Norway spruce stand in the study area, Tatra National Park, Poland

Age class (20 years)	Area (ha)	Area (%)	Number of compartments
I (0–20)	3.15	0.4	3
II (21–40)	97.72	13.4	25
III (41–60)	118.65	16.2	27
IV (61–80)	64.81	8.9	15
V (81–100)	82.78	11.3	22
VI (101–120)	44.28	6.1	14
VII (121–140)	91.18	12.5	13
VIII (141–160)	62.76	8.6	15
IX (161–180)	47.36	6.5	11
X (181–200)	35.91	4.9	5
XI (201–220)	10.11	1.4	2
XII (221–240)	44.90	6.1	4
XIII (241–260)	26.63	3.6	1
All	730.24	100.0	157

3 Data and Methods

3.1 ALS Point-Cloud Data

The LiDAR system used in the project was based on two simultaneously working LMS-Q560 full waveform scanners mounted under a DA42 airplane (Diamond Airborne Sensing). These two Riegl scanners (looking forward and backward) made it possible to obtain a very dense ALS point cloud, up to 40 pts/m^2.

RiscanPro Software (Riegl) was used to select the first echo (FE for DSM generation) and last echo (LE for DTM) from the laser wave. The matching of the 33 separate scans was performed on planar roof surfaces measured with a tachymeter (Leica 407 power) and dGPS receivers (Leica 1230). The local EUPOS-ASG (Poland) and SKPOS (Slovakia) permanent global positioning system (GPS) base stations were used for post-processing, and the accuracy of the obtained ALS point-cloud data was approximately: 0.06 m X, 0.02 m Y, and 0.01 m Z (Wężyk et al. 2008a). Subsequently, the whole matched point-cloud data was split into 500 × 500 m tiles, and Terra solid software was used for the DTM generation with 1 m spatial resolution, based on the approach of Axelsson (2000).

3.2 Processing of ALS Data

Further analysis was done in several steps. A DSM with spatial resolution of 1.0 m was generated using FUSION software (McGaughey 2007). Both DTM and DSM models were subtracted in order to create a nDSM, which represented the tree heights (HALS). This raster layer was then a further input for object based image analysis (OBIA), based on the segmentation and classification (de Kok and Wężyk 2008) within Trimble eCognition Developer software (Trimble 2010).

In order to automatically recognize gaps, the nDSM was classified into two categories: below and above 1.0 m (normalized, relative height). The supervised, semi-automated classification was performed with eCognition Developer 8 and consisted of a number of additional steps that have been required for filtering and refining the classified objects with OBIA specific algorithms based on the shape of objects and properties of their neighbors (de Kok and Wężyk 2008). The workflow of the analysis consisted of:

1. full waveform analysis, extracting echo number (data vendors, defined by end user),
2. storing point-cloud data in LAS ASPRS format (XYZ, echo number),
3. creating point-cloud tiles (500 × 500 m), batch processing,
4. filtering (low points, ghost points), running macros,
5. ground classification (Axelsson 2000),
6. generation of DTM from class "ground",
7. creating DSM (maximum point height in raster cell),
8. point-cloud normalization (relative height),
9. joining geometrical and descriptive datasets,
10. creating grid layer representing compartment height (based on the descriptive database), converting vector to raster,
11. shrinking compartment polygons using 5 m buffer,
12. creating an ALS subset for each inside part of compartment (clipping),
13. calculating the stand height as ALS 95th percentile (HALS95) for individual compartments,

14. height thresholding on nDSM layer (h > 1 m),
15. searching for gaps (>0.02 ha) and refining their shape (generalization, pixel-based object resizing, image object fusion),
16. calculating gap and compartment statistics (e.g. mean slope),
17. merging descriptive information (database relations),
18. statistical analysis and reports.

This analysis resulted in a forest stand mask (high tree vegetation) and areas with "no tree/low dwarf pine" canopy cover. According to the Polish forest guidelines (IUL 2003), areas smaller than 0.02 ha are not classified as gaps and therefore were not included in the vegetation mask. The occurrence of the gaps was assessed for different stands (data from DFM and the descriptive database).

It was also checked whether the gaps occurrence is correlated with terrain characteristics (elevation and slope degree). These analyses were done only for Norway spruce stands, which was the dominating species in the test area.

The reference forest compartment borders were included in the geometrical database (digital forest map), which was delivered by the GIS Division of the Tatra National Park. These GIS layers were updated in 2008 during a forest inventory performed with the use of satellite (IKONOS-2) and airborne orthophotomaps. Since the geometrical accuracy of the compartment borders was not sufficient for the ALS analysis, a buffer (value −5 m) was applied on all polygons. This ensured that only an inside part of each compartment was subject to processing.

Having obtained the revised vectors of the compartment borders, the tree heights were updated in the descriptive forest database, which was one of the main goals of the study. The descriptive database was directly joined with the DFM through a unique compartment ID (forest_address). The mean height (HALS95) of the compartment was determined as the 95th the percentile of the ALS point cloud within the borders of each forest stand (Wężyk and Solecki 2008; Kane et al. 2010). The grid layer HALS95 was generated based on this statistical analysis.

Using the descriptive forest database (SFIS) based on the forest inventory data from 2008, a raster representing the compartment height (HSFIS) was generated as well. These rasters (HALS95, HSFIS) were compared to each other using map algebra in ArcGIS. Two new fields in the descriptive forest database were created and filled with the following attributes: stand height based on ALS data (HALS95) and the difference between ALS and traditional inventory data (HDiff = HALS95 − HSFIS). We also compared these differences with types of stands and age from the DFM and descriptive database by using the nonparametrical Wilcoxon test.

4 Results

As a result of semiautomatic and supervised nDSM processing, 1,412 forest gaps were detected with a total area equal to 193.25 ha (13.8 % of the whole test area; mean gap area 0.13 ha; Fig. 2, Table 2). We noticed that in the areas dominated by

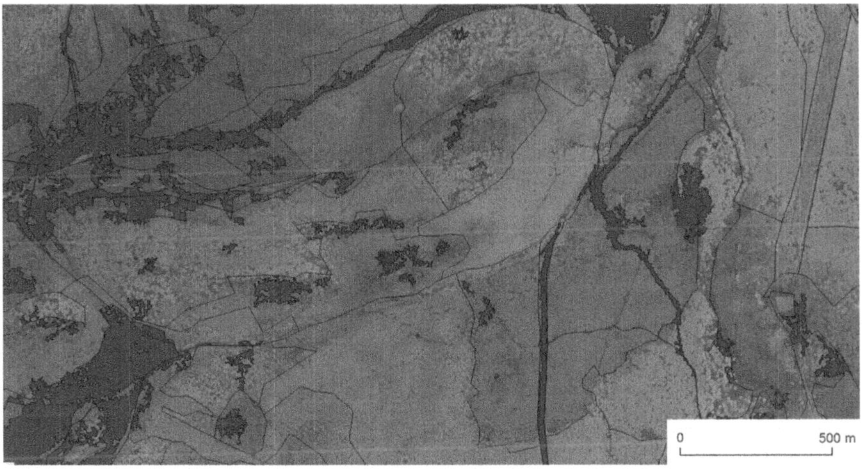

Fig. 2 Forest gaps (*dark*) with an area of over 200 m^2 in the Tatra National Park, Poland

Table 2 Results of semiautomatic gap detection using the nDSM analysis

Tree species	Number of analyzed compartments	Total area (ha)	Number of detected gaps	Gap area (ha)	Gap share (%)	Average gap size (ha)
Dwarf mountain pine	101	627.40	848	151.87	24.2	0.18
Norway spruce	157	730.24	543	39.29	5.4	0.07
Other species	12	41.01	21	2.10	0.17	0.09
Total	270	1,398.66	1,412	193.25	13.8	0.14

Table 3 Gaps in the Norway spruce stands using the nDSM analysis

Norway spruce age classes	Number of gaps	Gap area (ha)	Gap share (%)
I (0–20)	8	0.54	17.2
II (21–40)	67	7.14	7.3
III (41–60)	55	4.08	3.4
IV (61–80)	9	0.61	0.9
V (81–100)	44	4.26	5.2
VI (101–120)	43	3.92	8.8
VII (121–140)	64	4.32	4.7
VIII (141–160)	37	2.34	3.7
IX (161–180)	54	3.13	6.6
X (181–200)	13	0.59	1.7
XI (201–220)	14	0.57	5.6
XII (221–240)	86	5.74	12.8
XIII (241–260)	49	2.04	7.7
Total	543	39.29	5.4

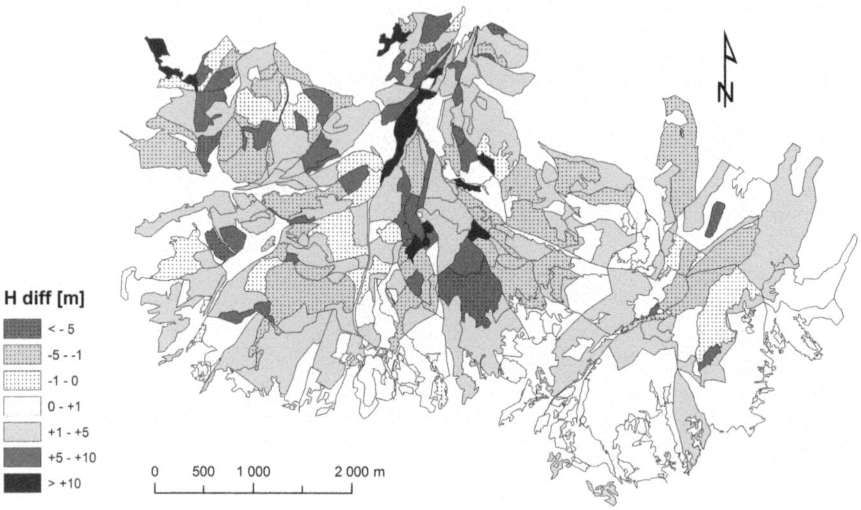

Fig. 3 Height difference (HDiff) between ALS-based and traditional field measurements

the dwarf pine, the detected gaps occupied 24.2 % of the study area (848 gaps with a total area equal to 151.87 ha, mean area 0.18 ha). In spruce stands, detected gaps covered around 5.5 % of total spruce stand area (543 gaps, total area equal to 39.29 ha).

Analysis for Norway spruce stands, divided into age classes (Table 3) demonstrates that the total area of the gaps was highest (7.14 ha, 7.3 %) in the stands of age class II. However, the highest contribution to the gap area was made by age classes I and XII (17.2 and 12.8 %, respectively). It was difficult to find any tendency in the number of gaps, their size, and spatial distribution regarding the age of the stands because there are many biotic and abiotic factors that influence the forest ecosystems that can form a gap and determine its shape.

The number of gaps and gap area were weakly correlated with mean elevation (0.29 and 0.29, respectively) and forest stand age of the compartment (0.32 and 0.21, respectively). Correlations between the number of gaps and gap area, and mean slope and aspect were insignificant.

The nonparametrical Wilcoxon test showed that for the most numerous stand types (the Norway spruce and dwarf mountain pine), as well as for the whole area differences in stand height (HDiff) were in most cases significant ($p < 0.05$). The differences for spruce stands in the most numerous age classes (II–XI) were highly significant for younger stands (<60 years), and either highly significant or significant with probability between $0.01 < p < 0.05$ for stands older than 60 years.

The mean HDiff was equal to 1.58 m (2.87 m for absolute differences; Figs. 3, 4, 5). After introducing the compartment area as a weight to normalize the calculation of the differences, the results changed slightly to 1.07 m (2.42 m for absolute differences). This indicates that the tree height values in the SFIS descriptive database are slightly underestimated compared to the ALS data.

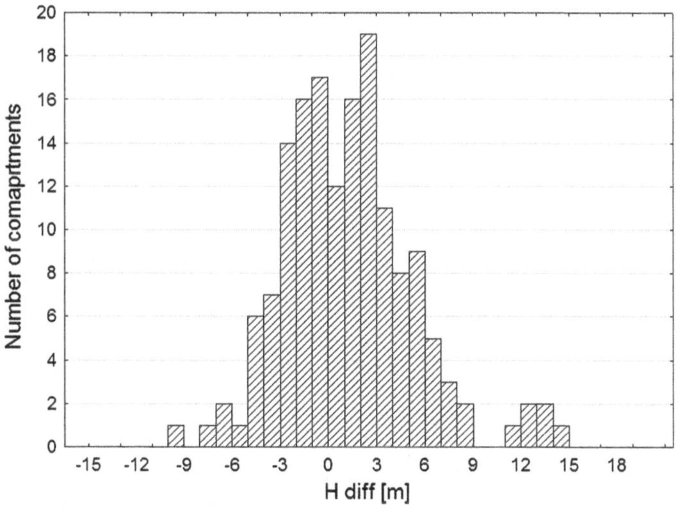

Fig. 4 Height difference (HDiff) histogram for the analyzed Norway spruce stands

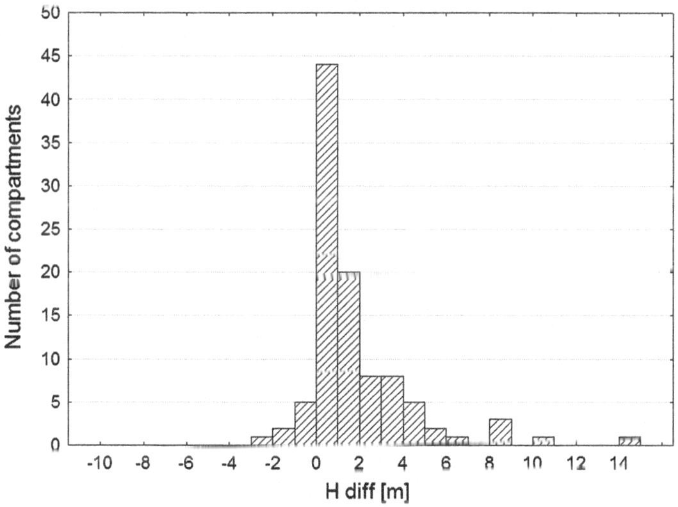

Fig. 5 Height difference (HDiff) histogram for the analyzed dwarf mountain pine patches

In the case of Norway spruce mean HDiff was equal to 1.27 m (3.29 m for absolute differences, Table 4). Weighted with the stands area, the mean differences were equal only to 0.49 (2.92 m for absolute differences). For the dwarf mountain pine, the mean difference was equal to 1.81 m (1.95 m for absolute differences). When weighted with the stand area, the value changed to 1.54 m (1.61 m for absolute differences).

Table 4 Differences between tree heights derived from the ALS data and SFIS reference values

Tree species	Age class	HDiff		Area weighted HDiff	
		Mean (m)	Absolute value (m)	Mean (m)	Absolute value (m)
European beech		3.87	7.06	3.99	7.03
European mountain ash		6.88	6.88	6.81	6.81
Sycamore maple		7.54	7.54	7.54	7.54
European larch		1.95	1.95	1.97	1.97
Willow sp.		−0.47	0.47	−0.47	0.47
Dwarf mountain pine		1.81	1.95	1.54	1.61
Norway spruce		1.27	3.29	0.49	2.92
	I	10.69	10.69	11.60	11.60
	II	5.81	5.81	5.00	5.00
	III	3.16	3.64	3.46	4.33
	IV	1.85	2.38	1.35	1.88
	V	0.32	2.52	−0.30	2.92
	VI	−1.23	2.08	−2.05	2.54
	VII	−2.22	2.33	−2.66	2.73
	VIII	−2.02	3.31	−1.51	2.52
	IX	−0.71	1.71	0.21	1.86
	X	−0.16	1.16	−0.59	0.67
	XI	−2.32	2.32	−2.21	2.21
	XII	−2.20	2.20	−3.18	3.18
	XIII	−0.99	0.99	−0.99	0.99
All tree species		1.58	2.87	1.07	2.42

The differences in stand heights were higher for deciduous stands than for coniferous stands, but due to their small total area (about 2 % of study area), their influence on the results was rather limited.

Differences in various Norway spruce stand age classes (Table 4) show that for age classes <100 years, the SFIS reference height values are lower than those obtained from ALS data. The mean differences turned out to be highest for age class I (10.69 m), and were decreasing with the age of the stand. In older stands (>100 years) HDiff values were negative, that is tree heights were lower for the ALS-based method than for traditional measurements.

5 Discussion and Conclusions

The results have confirmed the usefulness of ALS for an update of forest GIS databases, both descriptive (the height attribute) and geometrical (forest compartment border correction, and gap detection). The use of a supervised semi-automatic analysis based on the nDSM model can accelerate and replace expensive traditional forest surveys. Moreover, it can also provide accurate gap borders and distribution. However, no internal regulations describing the defining of the location of gaps or biogroups in the Polish DFM specification currently exist.

The lack of a radial shift on the nDSM model, in comparison to orthophotos usually generated on DTM models, makes ALS technology more appropriate for the update of geometrical layers. The shift of treetops and crown edges on airborne orthophotomaps can reach a few or even tens of meters (depending on radial distance from the main point of the image), and in such situations it cannot be treated as a reference image for the DFM update. Automatic height analysis based on ALS data gives the opportunity for an objective choice when setting inventory plot location, as it is a very rare situation that the compartment is homogenous in height, age, or species composition. One of the main problems of forest inventory, with respect to the stand height determination, is a height measurement, since a small sample of trees measured in the field with a hypsometer during a forest inventory may not be a good representation of the whole stand. The analysis of ALS point-cloud data for the whole compartment is possible and statistically more reliable than the traditionally measured height of a few trees in a stand (e.g., Abraham and Adolt 2006; Hollaus et al. 2006; Wężyk et al. 2008b; Wężyk et al. 2010). The results clearly show that the highest differences occur for the youngest stands, which can be explained by their very fast growth and variations of tree heights. Moreover, the possibility of correct measurements within young stands is limited, due to the dense canopy and a lack of tree-top visibility.

The analysis of ALS data can also result in the extraction of another important forest parameter that is currently assessed very subjectively: canopy cover. The integration of descriptive information referring to the vertical structure of a stand taken from the database with ALS analysis offers new solutions in the future. As the next step of ALS data processing it would be possible to create maps representing the spatial distribution of wood volume. ALS data can also be used to calculate the length of the crown or to analyze tree height variability within one compartment, and to detect and measure tree groups (e.g. above the tree line), gaps, and dead tree groups, which are very important for monitoring a sustainable forest ecosystem (Næsset and Okland 2002; Yu et al.2004; Wężyk et al. 2008b; Hollaus et al. 2009).

Finally, generation of highly accurate DTMs and DSMs that is possible using full waveform, dense ALS data (Wężyk et al. 2008a) and systematic monitoring of forest stands based on ALS data supported with digital photogrammetry, assures keeping the descriptive and geometric forest GIS databases updated, without the need for forest inventory field measurements, which are both time- and cost-consuming. Mountainous regions can especially benefit from this technology, as it can be utilized in many aspects of sustainable forest management and environment protection.

References

Abraham J, Adolt R (2006) Stand height estimations using aerial images and laser scanning data. EARSeL. In: Proceedings of the workshop on 3D remote sensing in forestry, Austria

Andersen HE, Reutebuch SE, Mcgaughey RJ (2006) A rigorous assessment of tree height measurements obtained using airborne lidar and conventional field methods. Can J Remote Sens 32(5):355–366

Axelsson P (2000) DEM generation from laser scanner data using adaptive TIN models. Int Arch Photogramm Remote Sens 33/4B:203–210

de Kok R, Wężyk P (2008) Principles of full autonomy in image interpretation. The basic architectural design for a sequential process with image objects. In: Blaschke T, Lang S, Hay GJ (eds) Object-based image analysis. series: lecture notes in geoinformation and cartography. Springer, Berlin

Goluch P, Borkowski A, Jozkow G, Tymkow P, Mokwa M (2009) Application of digital terrain model generated from airborne laser scanning data in hydrodynamic modelling. St Geotech Mech 31(3):61–72

Hollaus M, Dorigo W, Wagner W, Schadauer K, Höfle B, Maier B (2009) Operational wide-area stem volume estimation based on airborne laser scanning and national forest inventory data. Int J Remote Sens 30(19):5159–5175

Hollaus M, Wagner W, Eberhöfer C, Karel W (2006) Accuracy of large-scale canopy heights derived from LiDAR data under operational constraints in a complex alpine environment. ISPRS J Photogramm Remote Sens 60(5):323–338

Holmgren J, Jonsson T (2004) Large scale airborne laser scanning of forest resources in Sweden. Int Arch Photogramm Remote Sens Spat Inf Sci 36(8/W2):157–160

Hyyppä J, Inkinen M (1999) Detecting and estimating attributes for single trees using laser scanner. Photogramm J Finl 16(2):27–42

Hyyppä J, Hyyppä H, Litkey P, Yu X, Haggrén H, Rönnholm P, Pyysalo U, Pitkanen J, Maltamo M (2004) Algorithms and methods of airborne laser-scanning for forest measurements. Int Arch Photogramm Remote Sens Spat Inf Sci 36(8/W2):82–89

IUL (2003) Forest management guide. Part 3. Polish State Forest Holding, Warsaw

Kane R, McGaughey RJ, Bakker JD, Gersonde RF, Lutz JA, Franklin JF (2010) Comparisons between field- and LiDAR-based measures of stand structural complexity. Can J For Res 40(4):761–773

Maltamo M, Mustonen K, Hyyppä J, Pitkänen J, Yu X (2004) The accuracy of estimating individual tree variables with airborne laser scanning in boreal nature reserve. Can J For Res 34(9):1791–1801

McGaughey RJ (2007) Fusion/ldv: Software for lidar data analysis and visualization. Software manual USDA Forest Service, Pacific Northwest Research Station University of Washington, Seattle

McGaughey R, Carson W, Reutebuch S, Andersen H-E (2004) Direct measurement of individual tree characteristics from LIDAR data. In: Proceedings of the annual ASPRS conference, Denver, 23–28 May 2004. American Society of Photogrammetry and Remote Sensing (ASPRS), Bethesda

Næsset E, Okland T (2002) Estimating tree height and tree crown properties using airborne scanning laser in a boreal nature reserve. Remote Sens Environ 79:105–115

Persson A, Holmgren J, Söderman U (2002) Detecting and measuring individual trees using an airborne laser scanner. Photogramm Eng Remote Sens 68(9):925–932

Tompalski P, Wężyk P, de Kok R, Kukawski M (2009) Determining the number of trees using airborne laser scanning and true orthoimagery. Ann Geomat 7(2):133–141

Trimble (2010) Cognition developer 8.0.2 reference book. Trimble Documentation, Munich

Wężyk P, Borowiec N, Szombara S, Wańczyk R (2008a) Generation of digital surface and terrain models of the Tatras Mountains based on airborne laser scanning (ALS) point cloud. Arch Fotogram Kartogr Teledetekcji 18:651–661

Wężyk P, Mansberger R (1997) Przykład wykorzystania ortofotografii cyfrowej i systemu GIS w leśnictwie. Arch Fotogram Kartogr Teledetekcji 6:133–151

Wężyk P, Solecki K (2008) Airborne laser scanning (ALS): based determination of the Chojna forest district tree stand heights. Arch Fotogram Kartogr Teledetekcji 18:663–672

Wężyk P, Szostak M, Tompalski P (2010) Aktualizacja baz danych SILP oraz LMN w oparciu o dane lotniczego skaningu laserowego. Arch Fotogram Kartogr Teledetekcji 21:437–446

Wężyk P, Tompalski P, Szostak M, Glista M, Pierzchalski M (2008b) Describing the selected canopy layer parameters of the Scots pine stands using ALS data. In: Hill RA, Rosette J, Suarez J (eds) Proceedings of SilviLaser 2008: 8th international conference on LiDAR applications in forest assessment and inventory, September 2008. Edinburgh, UK

Yu X, Hyyppä J, Hyyppä H, Maltamo M (2004) Effects of flight altitude on tree height estimation using airborne laser-scanning. Int Arch Photogramm Remote Sens Spat Inf Sci 36(8/W2):96–101

Computer Tomography in Wood-Decay Assessment of Silver Fir (*Abies Alba* Mill.) Stands in the Polish Part of the Carpathians

Stanisław Niemtur, Elżbieta Chomicz and Mariusz Kapsa

Abstract Silver fir (*Abies alba* Mill.) is one of the major tree species (with beech and spruce) in the Polish Carpathians which during the last twenty years significantly improved its vitality. In our study we analyzed the degree of fir stands infestation by fungal pathogens using computer tomography. The test stands at age from 60 to 200 years were located in four different parts of the Polish Carpathians. Our results confirmed large share of fir trees in selected stands with wood decay that was proportionally increasing with stand age. It was also found that the percentage of silver fir with damaged wood by fungal pathogens was similar, and in some stands even bigger than of the Norway spruce, which is generally considered to be susceptible to damage by fungi. Among inspected silver fir trees in seed stands we found also some healthy individuals despite their age above 150 years. We suggest further investigations to determine potential differences between silver fir and Norway spruce in damages caused by fungi in Carpathian forests.

1 Introduction

Silver fir (*Abies alba* Mill.) is one of the major tree species in the Polish part of the Carpathians (in the Carpathian Natural Forest Region; CNFR), being a natural component of mixed stands with beech and spruce, mostly in the lower mountain zone. Together with Norway spruce this tree species is considered as an ecological stabilizer of the Carpathian forests, due to greater resistance to wind, insect pests and fungal pathogens. Because of low requirements for light, silver fir is a perfect species to form multigenerational stands with diversified structure that could increase forest biodiversity.

S. Niemtur (✉) · E. Chomicz · M. Kapsa
Department of Forestry Management in Mountain Regions, Forest Research Institute, Fredry 39 30-605 Kraków, Poland
e-mail: S.Niemtur@ibles.waw.pl

J. Kozak et al. (eds.), *The Carpathians: Integrating Nature and Society Towards Sustainability*, Environmental Science and Engineering,
DOI: 10.1007/978-3-642-12725-0_44,

Knowledge about development of fir stands is important because area of fir stands in the Polish Carpathians will most likely increase in consequence of revitalization of existing fir stands. A relatively good regeneration of silver fir enables a wider usage of this species in silviculture and restoring fir in its natural range and sites, allowing a conversion of stands with species composition not suitable to site conditions, and preventing calamities and decline of spruce stands.

Present occurrence of fir stands in the Polish Carpathians is determined by the past direction of forest management, especially the excessive planting of spruce. It is also a consequence of large-fir stands regression caused mainly by air pollution which was observed not only in the Carpathians but also in other parts of Europe between 1970s and 1990s. Currently, after a period of the reduction of fir stands area (from 28.3 in 1967 to 23 % in 1990) that favored spruce and beech, silver fir recently comes again into prominence in revitalization process (Fig. 1; Niemtur 2007; Niemtur et al. 2009).

Increase of the silver fir cover is connected also with the large-scale dieback of spruce stands in the Beskidy Mountains and conversion of artificial spruce monocultures to mixed stands with high share of fir. One of the most important reasons for the reduction of spruce stands vitality in this region are fungi diseases. The most common cause of root rot are fungi from the genus *Armillaria* and *Heterobasidion*, considered to be the most economically important root pathogen of coniferous trees (Rykowski and Sierota 1984).

In general, the area of stands with fungal diseases has persisted in Poland for many years at a high level. According to recent evaluations it amounts to 2,000–3,000 km^2 (Lech 2003; Małecka 2010). Especially vulnerable are the Carpathian stands, including native stands in the national parks (Niemtur and Chomicz 2008), e.g. silver fir which are one of the most productive European woody species in mountain ecosystems, with high requirements for air moisture.

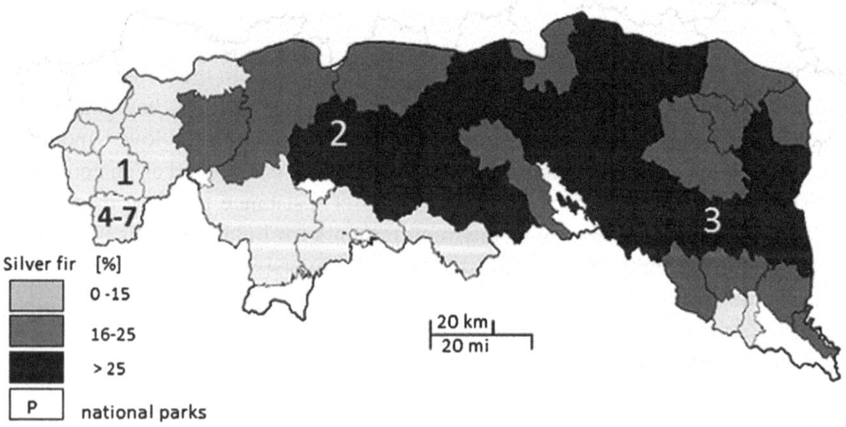

Fig. 1 Spatial distribution of fir stands in the Carpathian Natural Forest Region, CNFR. 1–7: forest districts with experimental plots (see Table 1 for description)

Exception to this is a relatively xerophilic relict in an alpine ecotype in Wallis canton in the south-western Switzerland (Bernicchia et al. 2007). A usually high requirement of fir for air moisture favors spread of fungal diseases. There are numerous fungi able to cause the rot, but contrary to spruce, information about size of these damages in fir stands is missing, particularly in the stands of fir trees selected for seed harvest (seed stands) with long felling age.

Current mass planting of fir in the Carpathians forests and promotion of natural regeneration of this species may also raise concerns whether the historical model of the Carpathian forests with "extensive spruce management" will not be replaced by fir or if there are no factors that in future may weaken vitality of fir stands on a mass scale, similarly as contemporary spruce stands. Therefore we decided to carry out investigations of the fungal pathogens threat and the occurrence of butt rot of fir trees.

The main objective of our study was to determine a degree of infestation of fir trees by fungal pathogens in various locations in the Polish Carpathians. Additional objectives were to estimate how an age of fir trees affects a degree of infestation by fungi, and to compare spruce and fir trunk damages in the same site conditions in a mixed stand.

2 Methods

Investigations were carried out in four fir stands with selected seed-trees, located in different parts of the CNFR. In one location (plots 4–7, forest district Ujsoły) the fir stands were in four different ages (Table 1, Fig. 1).

Investigated fir stands were situated in the lower mountain belt (from 610 m in the Bieszczady Mountains to 850 m in the Silesian Beskid). In each stand 30 seed-trees with similar diameter and height have been selected and a wood rot presence at the bottom of the trunk was studied. Similar rules were applied in selecting 10 fir trees in stands of different ages (Ujsoły District, stands 4–7). The wood rot presence was controlled at the bottom of the tree trunk. To compare the results with spruce stand condition we measured also 30 spruce trees in Węgierska Górka Forest District.

Table 1 Location of experimental plots in fir stands

No	Region	Forest districts	Age	Altitude (m)
1	Silesian Beskid	Węgierska Górka, Radziechowy	140	833
2	Beskid Wyspowy	Limanowa, Lubogoszcz	140	772
3	Bieszczady	Baligród, Czarne	140	610
4	Silesian Beskid	Ujsoły, Nickulina	60	720
5		Ujsoły, Nickulina	70	690
6		Ujsoły, Kiczora	120	710
7		Ujsoły, Sól	200	650

We used the Picus Sonic tomograph to estimate wood damages inside the trunk, without cutting or boring the selected trees. Measurement level was determined at 10 cm from the surface of soil from the side of the slope. We used 8–10 measuring points for each tree, depending on the size of the circumference (Fig. 2) according to the tomograph manual (Göcke and Rust 2007).

In each measuring point an electrode was hammered until contact with wood, and connected with a magnetic sensor. The impulse sound was induced at successive electrodes connected with the magnetic sensor by 3-fold tap of the Metal Hammer Lite version. The shape of the trunk cross section was generated by measuring an appropriate distance between the measurement points with the Picus Calliper (Argus Electronic 2009). After sending the information from the sensors to a laptop, cross-sectional tomography trunk was generated by the Picus PC Program Version Q71.6.

Wood on a cross section was assigned by tomograph to one of three categories: solid (wood without any changes), damaged (structure of the wood completely destroyed) and other (wood with early signs of pathogen infection). Percentage of each category of wood was generated in the form of tomogram's legend (Fig. 2).

3 Results

Our results showed that in the Silesian Beskid (the Węgierska Górka Forest District) all tested firs were infested by fungi (Fig. 3), similar situation (close to 90 % of trees; plots 2 and 3) was observed in Limanowa (Beskid Wyspowy) and Baligród (Bieszczady). Average size of damaged wood zone on a cross section was

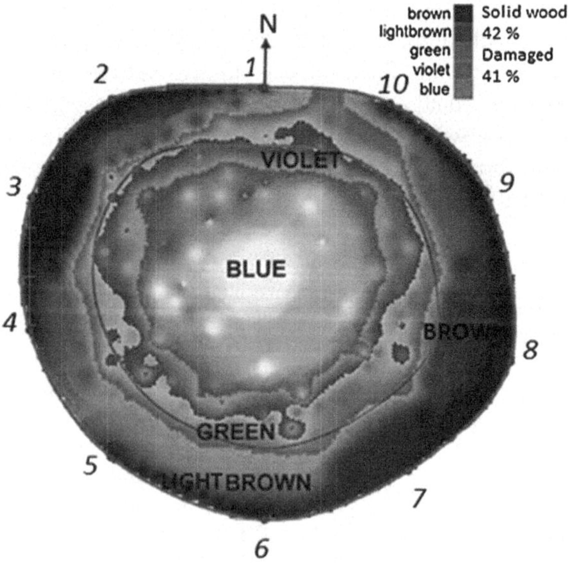

Fig. 2 Tomogram with 10 measuring points and damages presented by zones. *Blue* and *violet*: area with completely destroyed structure of the wood; *brown*: wood without any changes (solid wood); *green*: initial phase

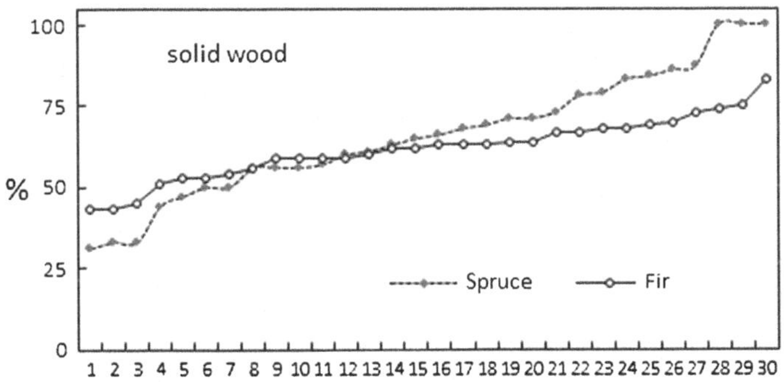

Fig. 3 Percentage of solid wood of 30 spruces and firs on experimental plot 1 in the Węgierska Górka forest district

also the highest in the Węgierska Górka district, amounting to over 22 % of the cross-section area, whereas it was 14 % in Limanowa and 13 % in Baligród (Table 2).

In case of diagnosed spruce trees selected in spruce–fir mixed stands, percentage of spruce trunks with decay was even less than in case of fir trees, amounting to 90 % (3 trees without damages; Figs. 3, 4).

Among the investigated firs in the Beskid Wyspowy (Limanowa) we found four trees without any damage. In case of several fir trees, butt rot appeared to be triggered by mechanical damage leading to side infection (Fig. 5).

Also in Baligród among 30 fir trees we found tomograms of four trees without any visible sign of fungal infection (Fig. 6).

Table 2 The proportion of damaged wood in a cross-section area of studied firs and spruces

Forest districts	Number of trees	Proportion of damaged wood (%)			
		Average	Minimum	Maximum	Variation coefficient
1. Węgierska Górka/ spruce/	30	34.1	0	69	56.0
Węgierska Górka/fir/	30	38.4	17	57	24.4
2. Limanowa	30	27.5	0	65	60.9
3. Baligród	30	24.8	0	63	65.9
4. Nickulina, age: 60 years	10	9.8	0	33	117.1
5. Nickulina, age: 70 years	10	13.5	0	35	93.1
6. Kiczora, age: 120 years	10	21.5	0	36	56.6
7. Sól, age: 200 years	10	53.1	16	83	32.9

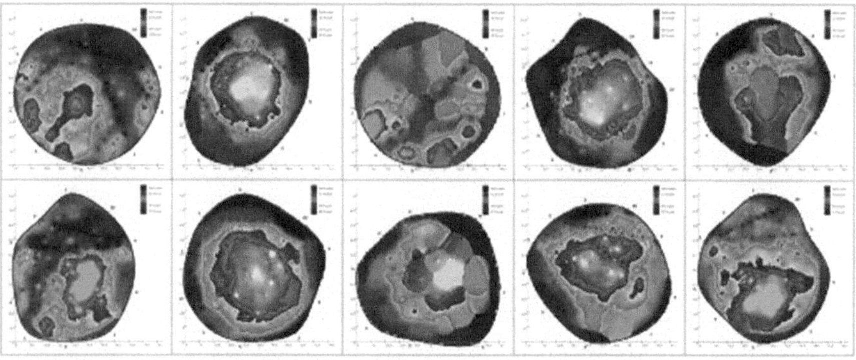

Fig. 4 Tomograms of tested firs chosen from 30 trees infected by fungal pathogens on experimental plot 1 in the Węgierska Górka forest District

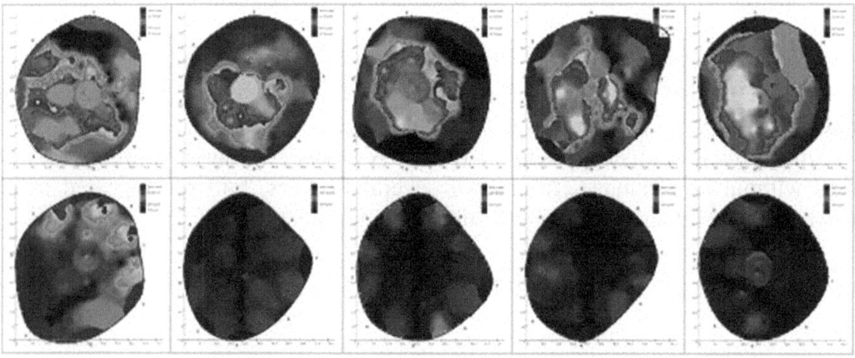

Fig. 5 Tomograms chosen from 30 tested firs on experimental plot 2 in the Limanowa Forest District

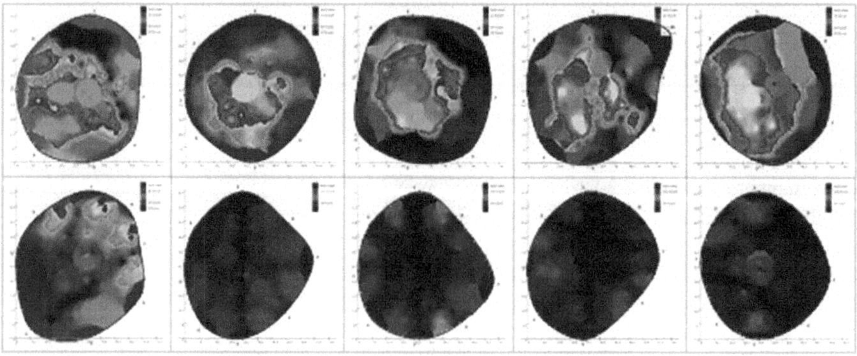

Fig. 6 Tomograms chosen from 30 tested firs on experimental plot 3 in the Baligród forest district

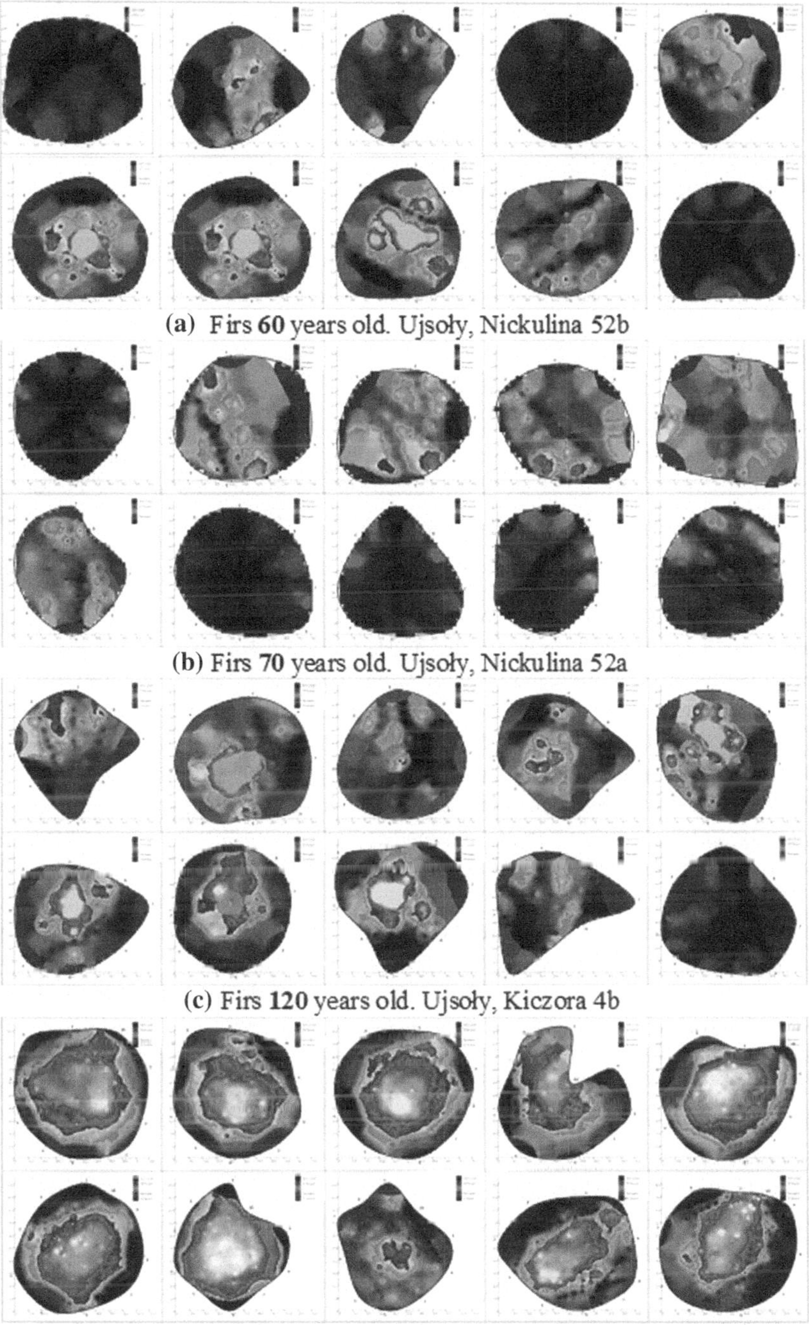

Fig. 7 Tomograms of fir trees of different ages: A—60 years; B—70 years; C—120 years; D—200 years

We also noticed the close relationship between age of fir trees and the degree of wood destruction by fungi pathogens (Fig. 7, Table 2). Wood damage was found even in 60-year old fir stands (Fig. 7).

4 Discussion and Conclusions

Our investigations confirmed that the sonic tomograph is a very useful uninvasive method which allows to analyze condition of trees. The results showed a large share of fir trees with high percentage of the wood with structure completely destroyed at the bottom of a trunk by fungi pathogens. In addition, examinations carried out in fir seed stands confirm that wood damage of fir trees by fungi was similar or higher than that of older spruce trees (Fig. 3).

It should be noticed that we found also healthy individuals among inspected fir trees which could be mother trees for future regenerations and stands. To identify such trees for forest tree breeding program a wider usage of the computer tomography method is necessary.

Further research is required to determine potential differences in damage character between silver fir and Norway spruce, especially because these species, together with European beech, are the most important forest trees in Carpathians, often forming mixed stands. Frequent wood damage caused by fungi found in trunks of fir from various parts of the Polish Carpathians justify further health condition observations of this species.

In contrast to spruce, fir is planted on increasingly larger areas. Due to a good quality of natural fir regenerations observed for several years, its economic and environmental significance in the Carpathian forests will grow. In addition, the analysis of changes in the range of the climate vertical belts (Hess 1965) in the Carpathian forests confirms the need to take into account in forest management a natural tendency to reduce share of spruce in favor of fir even in the upper mountain zone (Niemtur and Pierzgalski 2012).

In general, an increase of the probability of wood decay with tree age is well known (Bernadzki 2003) and intensification of wood decay occurrence in relation to stand age was widely described for spruce stands (Norokorpi 1979; Krzan 1988; Lech 2003; Barszcz 2004), also on the basis of our previous research with the Picus Sonic tomograph (Niemtur et al. 2008). Therefore similarly as for spruce, it is necessary to reconsider the cutting ages of fir stands as this species is no longer an endangered one, and for over twenty years shows clear symptoms of revitalization. In this context, it must be remembered that likewise the threat of fungal pathogens, the quality of natural regeneration in fir stands can be increased with the age of mother trees. This relation was confirmed long time ago by the results of the experiment carried out in 1914 in Tharandt (Miś et al. 2008).

Acknowledgments The authors wish to thank mgr inż. Marek Pierzchała from Forest Research Institute in Kraków for his great help in field work.

References

Argus Electronic Gmbh (2009) Description of tree disease detection tool of the company. Informative materials of Argus Electronic. Rostock. http://www.argus-electronic.de. Accessed 30 September 2012

Barszcz A (2004) The dependency of the quality and size structure and of the value of timber on the origin and age of stands. Electronic J Pol Agr Univ 7(2). http://www.ejpau.media.pl/volume7/issue2/forestry/art-08.html. Accessed 31 October 2012

Bernadzki E (2003) Struktura wieku i zagrożenie zgnilizną drewna starych drzewostanów sosnowych. Sylwan 5:3–12

Bernicchia A, Savino E, Pérez Gorjón S (2007) Aphyllophoraceous wood-inhabiting fungi on Abies alba in Italy. Mycotaxon 100:185–188

Göcke L, Rust S (2007) Picus Sonic Tomograph. Manual, Rostock

Hess M (1965) Piętra klimatyczne w polskich Karpatach zachodnich. Zesz Nauk UJ Pr Geogr 11:1–255

Krzan Z (1988) Badania nad zgniliznami odziomkowymi świerka Picea abies (L.)Karst. w Tatrzańskim Parku Narodowym. Ochr Przyr 46:113–120

Lech P (2003) Zagrożenie drzewostanów świerkowych w Polsce przez patogeny korzeni w świetle wyników monitoringu fitopatologicznego lasów gospodarczych. Sesja Naukowa "Drzewostany świerkowe: stan, problemy, perspektywy rozwojowe". PTL, Ustroń-Jaszowiec

Małecka M (2010) Grzybowe choroby infekcyjne w ubiegłym roku. Las Polski 6:26–27

Miś R, Kukuła J, Węgiel A (2008) Effect of the age of the maternal pine Pinus sylvestris L. on the growth and development of progeny stands. Acta Sci Pol Silvarum Colendarum Ratio et Industria Lignaria 7(3):23–35

Niemtur S (2007) Carpathians forests: actual state and an attempt to a forecast. Probl Zagosp Ziem Górsk 54:31–44

Niemtur S, Chomicz E (2008) Computer tomography of Norway spruces (Picea abies(l.) Karst.) in threatened stands with the various protective status. Probl Zagosp Ziem Górsk 55:99–112

Niemtur S, Sierota Z, Chomicz E, Żółciak A (2008) Występowanie, konsekwencje gospodarcze zgnilizny czerwonej i możliwości jej ograniczania w drzewostanach świerkowych Krainy Karpackiej. Forest Research Institute (IBL) Final Report BLP 295. IBL, Kraków

Niemtur S, Chomicz E, Kapsa M, Pierzchała M, Poznański R, Zięba S (2009) Wzorcowe modele lasów karpackich, uwarunkowania ich rozwoju i zasady prowadzenia. Forest Research Institute (IBL) Final Report BLP 325. IBL, Kraków

Niemtur S, Pierzgalski E (2012) Effects of forest ecosystems changes on hydrological processes in Western Carpathians. In: Krecek J, Haigh MJ, Hofer T, Kubin E (eds) Management of mountain watersheds. Springer, Berlin

Norokorpi Y (1979) Old Norway spruce stands, amount of decay and decay causing microbes in Northern Finland. Comm Inst For Fenniae 97(6):1–77

Rykowski K, Sierota Z (1984) Aspekt ekonomiczny występowania huby korzeni w drzewostanach sosnowych na gruntach porolnych. Sylwan 1:11–21

Assessment of Land Cover Changes in the Carpathian Mountains with MODIS Data

Barbara Jaśkowiec

Abstract The study aimed to assess general trends of contemporary land cover change in the Carpathian Mountains. Land cover change was calculated using a series of ten MODIS NDVI data sets covering the second half of July for the years 2000–2009. Two standard change detection methods were used: NDVI differencing as well as trend analysis. Changes recorded between 2001 and 2009 were not widespread. Almost 92 % of the total Carpathians area remained unaltered, and NDVI values decreased on average 0.2 % annually. The negative trend dominated in the mountains, although it varied with elevation.

1 Introduction

The term *change* is usually defined as a transformation of Earth surface, which can be seen in satellite monitoring as a spectral or spatial movement of analyzed entity over time (Coppin et al. 2004; Mandrone et al. 2004). Land changes strongly affect functioning of the Earth geosystems, especially in regional and global perspectives. They influence e.g., biotic diversity, contribute to local, regional and global climate changes. Moreover, they are basic source of soil degradation, and affect the ecosystem services which are essential from a human perspective (Lambin et al. 2001). Land cover change is usually defined either as *conversion* or *modification* (Lambin et al. 2001; Coppin et al. 2004). Land cover conversion is characterized by a complete replacement of one cover type by another (e.g., meadows replaced by forests), whereas land cover modification is defined as more elusive alterations which influence the character of land cover without changing its type. In general, modifications of land cover are much more widespread (Ramankutty et al. 2006). For both types of land cover changes, various approaches can be used to reveal

B. Jaśkowiec (✉)
Institute of Geography and Spatial Management, Jagiellonian University,
Gronostajowa 7 30-387 Kraków, Poland
e-mail: bjaskowiec@gis.geo.uj.edu.pl

J. Kozak et al. (eds.), *The Carpathians: Integrating Nature and Society Towards Sustainability*, Environmental Science and Engineering,
DOI: 10.1007/978-3-642-12725-0_45, © Springer-Verlag Berlin Heidelberg 2013

these alterations. They can be divided into two groups of methods: bi-temporal change detection on the basis of the snap-shot model or trend analysis of time series (Coppin et al. 2004).

From many different techniques of bi-temporal digital change detection image differencing is one the most frequently used (e.g., Lyon et al. 1998; Hayes and Sader 2001; Coppin et al. 2002; Lafortezza et al. 2004; Pu et al. 2008). It employs original or transformed imagery to differentiate from subsequent images and to map change or no-change pixels on a basis of calculated difference. Use of indices (e.g., vegetation indices) allows to eliminate variations in reflectance caused by various spatial resolution, band ranges, atmosphere conditions or others.

Studies with trend analysis (temporal trajectory analysis) are not so common like more conventional methods of digital change detection (Coppin et al. 2002). They are usually applied for products with high temporal resolution (e.g., Advanced Very High Resolution Radiometer (AVHRR), SPOT-VEGETATION, MODIS). This type of analysis demands comparison of development curves (profiles) of various relevant indices. Change detection based on such profiles turned out to be appropriate for studies of variability of the Earth surface attributes in a regional scale (Coppin et al. 2004). In addition, this approach allows to detect more subtle changes, to characterize trend and direction of alterations and to predict future changes (Coppin et al. 2002).

Central and Eastern Europe experienced drastic changes after fall of the Iron Curtain in 1989 that have triggered extensive land cover and land use change. The Carpathian Mountains, ecologically rather homogeneous and the least disturbed range in Europe, are densely cut by political borders (Kuemmerle et al. 2006) what allows to monitor differences in land cover change dynamics between particular countries. Various studies at diverse spatial scales have been carried out in different Carpathian regions (Kozak 2005; Kuemmerle et al. 2005, 2006, 2009a, b, c, 2011; Kozak et al. 2007; Mihai et al. 2007; Knorn et al. 2009; Main-Knorn et al. 2009; Müller et al. 2009; Baumann et al. 2011; Griffiths et al. 2012a; Knorn et al. 2012). The main process recognized by these investigations was cropland abandonment that occurred at unprecedented rates especially in the western Ukraine (Baumann et al. 2011) and also in the southern Romania (Argeş County; Kuemmerle et al. 2009b, c; Müller et al. 2009). Another studies revealed slight increase of forest cover in the Polish Western Carpathians (Kozak 2003), forest succession as an affect of marginal parts arable land abandonment in the Western Ukraine (Kuemmerle et al. 2009a) or in the border triangle of Poland, Czech Republic and Slovakia in the Western Carpathians (Main-Knorn et al. 2009). At the same time, in the interior of the Ukrainian Carpathians (Kuemmerle et al. 2009a), in the Romanian Carpathians (Griffiths et al. 2012a; Knorn et al. 2012), as well as in the Polish part of the study area located in the border region of Poland, Czech Republic and Slovakia (Main-Knorn et al. 2009) forest disturbances were noticed (decrease of forest cover was noticed).

Most of these studies were carried out in a regional scale and were using Landsat images for multiple dates. Thus until recently a detailed spatial land cover change data at the pan-Carpathian level were missing. Lately, first results of land cover change mapping (a new approach to consistent mapping of land cover change) across the entire Carpathians based on Landsat images has been presented by Griffiths et al. (2012b).

Apart from satellite-derived data information on land cover change can be obtained from the CORINE Land Cover 2000–2006 Changes database. However, these data do not take into consideration the territory of Ukraine. What is more, there are some differences between particular countries. According to this database, only ~1,692 km^2 have changed, what is less than 1 % of the Carpathians area. More than 64 % of all noticed alterations concerned forests converted into transitional woodland-shrub. This fact could indicate decrease of vegetation. On the other hand, some alterations (like turning transitional woodland-shrub or pastures into forests) represent increase of vegetation cover. Hence, satellite-based vegetation indices may reveal spatial patterns and annual rates of change.

In the Carpathians little attention has been paid until now to the use of satellite-derived products characterized by high temporal resolution and medium spatial resolution, ensuring large area coverage and possibility to trace annual dynamics of land cover. Such products are provided by e.g., Moderate Resolution Imaging Spectroradiometer (MODIS) sensor. Some of them, like MODIS NDVI (Normalized Difference Vegetation Index) product, may be successfully applied to study land cover changes (Carraro et al. 2008). Therefore, the purpose of this study is to determine general directions of contemporary land cover change in the Carpathian Mountains based on MODIS NDVI data.

2 Study Area

The Carpathian Mountains are one of the biggest European mountain ranges, with an arc stretching across the territory of eight countries: Austria, Czech Republic, Slovakia, Poland, Ukraine, Romania, Hungary and Serbia (Fig. 1). Elevation ranges from around 100 m in foothills to 2,655 m in the Gerlach Peak (High Tatras, Slovakia). In this study the entire Carpathians area, in total ~237,000 km^2, was analyzed. The generalized boundaries of such area were adapted from Kondracki (1989).

The Carpathian Mountains are a vital watershed for Central Europe. This region is a headwater area of the major tributaries of four European catchments: Danube, Dniester, Vistula and Oder. The climate of the entire Carpathians is temperate continental with continentality increasing from west to east. This region is very important at the pan-European level from a viewpoint of biodiversity (Turnock 2002; Kuemmerle et al. 2009c). This region includes the largest semi-natural old-growth forests in the Central and the Western Europe. Moreover, they provide the habitat for many endemic species, as well as Europe's largest wolf, brown bear and bison populations.

The Carpathian landscape is characterized by fragmented land use pattern with relatively large forest patches interlaced with small patches of other land use types such as croplands, grasslands, pastures and settlements (Kuemmerle et al. 2006;

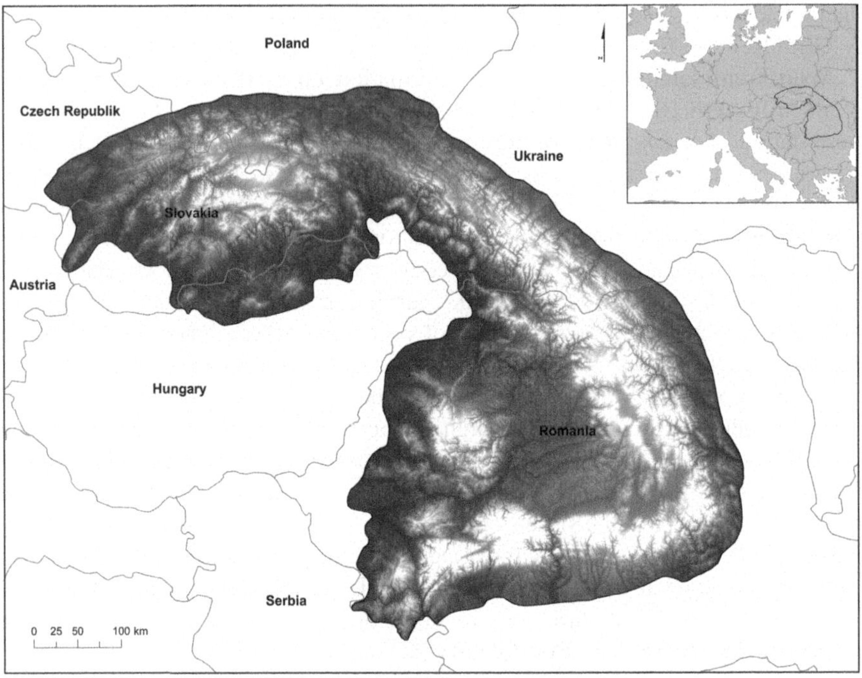

Fig. 1 Study area

Kozak et al. 2008). In general, forests cover about 50 % of the whole area of the Carpathians, and even up to 80–90 % between 1,000 and 1,500 m (Kozak et al. 2008) while e.g., alpine areas are relatively isolated and cover only 2 % of the region. The state of current land cover pattern is mostly the consequence of land use legacies from Habsburg empire and socialist times (Turnock 2002). However, after the political breakdown in 1989, the land use system has experienced rapid changes (e.g., land reforms; Turnock 2002; Hostert et al. 2008; Kozak 2009). Furthermore, a new impulse of change has been released after the accession of Central European countries to the European Union (EU) in 2004 and 2007, however it has not affected the territory of Ukraine. In 1990s some areas in the Carpathian region experienced abandonment of farmland (about 15–20 % of farmland) or agricultural parcelization (Hostert et al. 2008). Forest cover decreased due to illegal logging (especially in Ukraine). Forest disturbance rates generally increased in all Carpathian countries; they were higher in Slovakia and Ukraine than elsewhere. Later on, the introduction of EU policies, predominantly the Common Agricultural Policy (CAP), has brought some farmland back to production. Moreover, some areas have been reforested. On the other hand, forest cover has been reduced due to legal logging or urban sprawl.

3 Data

In this study, detection of land cover changes relies on data acquired by MODIS, an instrument on board of Terra and Aqua satellites from National Aeronautics and Space Administration (NASA). These two instruments image the entire Earth's surface every one to two days (MODISWeb 2010). They gather data in 36 spectral bands, at varying spatial resolutions (250, 500 or 1,000 m).

Data acquired by the MODIS sensors were used to generate products at different pre-processing stages. Analysis in this study was based on the Vegetation Indices (MOD13Q1) product. This product provides "consistent spatial and temporal comparisons of vegetation conditions" (LP DAAC 2010) every 16 days, at a nominal spatial resolution of 250 m (Huete et al. 1999). It consists of twelve layers (Scientific Data Sets; Table 1), from which two has been chosen: NDVI and pixel reliability.

Radiometric measurements of vegetation not only mark its existence, but also provide information on both size and condition of vegetation within a pixel (Huete et al. 1999), that is why they are applied to distinguish different types of vegetation and vegetated areas from artificial surfaces (Lee et al. 2004). Hence, vegetation indices have been often used in detection of land cover change (both conversion and modification) in a regional scale (e.g., Lyon et al. 1998; Casacchia et al. 2002; Taddei and Puzzolo 2004; Lafortezza et al. 2004; Mandrone et al. 2004; Yagüe and García 2004; Schweitzer et al. 2005; Stellmes et al. 2005; Jarocińska and Zagajewski 2008; Pu et al. 2008). Their development is based on differential absorption, transmittance and reflectance energy by vegetation in some range of electromagnetic spectrum (Lyon et al. 1998). It has been proven that the ratio of near-infrared band and red band is significantly correlated with amount of biomass (Lyon et al. 1998). The most widely used vegetation index is NDVI described by Eq. (1), where NIR and R are respectively the near-infrared and the red spectral bands:

Table 1 Scientific data sets of MOD13Q1

Scientific Data Sets	Units	Minimum value	Maximum value	Scale factor
NDVI	*NDVI*	−2,000	10,000	0.0001
EVI	EVI	−2,000	10,000	0.0001
VI Quality	Bits	0	65,534	–
Red reflectance	Reflectance	0	10,000	0.0001
NIR reflectance	Reflectance	0	10,000	0.0001
Blue reflectance	Reflectance	0	10,000	0.0001
MIR reflectance	Reflectance	0	10,000	0.0001
View zenith angle	Degree	−9,000	9,000	0.01
Sun zenith angle	Degree	−9,000	9,000	0.01
Relative azimuth angle	Degree	−3,600	3,600	0.1
Composite day of the year	Julian day of year	1	366	–
Pixel reliability	*Rank*	0	3	–

Source After LP DAAC 2010, modified. Data sets used in this study shown in italics

$$NDVI = \frac{NIR - R}{NIR + R} \tag{1}$$

By design, NDVI value varies between $<-1; +1>$, depending on both quantity and condition of vegetation (Lee et al. 2004). Areas containing a dense vegetation canopy (including forest) will tend to higher NDVI values (than e.g., dry or scarcely vegetated areas) due to their relatively high reflectance in the near-infrared and low reflectance in the visible range (Mycke-Dominko and Slinkina 2004). The average NDVI values for vegetation vary from 0.4 to 0.8 (Jarocińska and Zagajewski 2008), while water bodies and bare areas are characterized by negative or close to 0 values (Lee et al. 2004).

MODIS NDVI is computed based on atmospherically corrected reflectance of MODIS bands 1 (R; 620–679 nm) and 2 (NIR; 841–876 nm) (Eq. 1; Huete et al. 1999). MODIS NDVI algorithm works on a pixel basis (Huete et al. 1999; Solano et al. 2010). A single value per pixel is computed on a basis of all observations within a 16-day period. When all 16-day observations are collected, a filter based on quality, clouds, viewing geometry and solar zenith angle is applied. Then, data are processed by means of one of two techniques depending on the quantity and quality of available pixels. MODIS NDVI values are within the range from −0.2 to 1. Non-standard value (−0.2 instead of −1) as a lower NDVI limit could be related to use of Terra satellite for land observation (for land objects a lower limit of NDVI is about 0). In practice, MODIS NDVI values are calibrated to the range from −2,000 to 10,000 (Table 1).

The analysis used a series of ten MODIS NDVI data sets covering the second half of July (to capture a peak of vegetation greenness) from 2000 to 2009. Data were gained from the Land Processes Distributed Active Archive Center (LP DAAC) server for two scenes: h19v03 and h19v04.

In addition, the Shuttle Radar Topography Mission Digital Elevation Model (SRTM DEM) was used as ancillary data for summarizing land cover change. The spatial resolution of SRTM DEM was 100 m.

4 Methods

First, a quality of MODIS NDVI data was assessed on the basis of pixel reliability layer. Unsatisfactory pixels (with values −1, 2 and 3 denoting respectively: no data, snow/ice and clouds) were eliminated together with 1 pixel-wide buffer. Data for the year 2000 were removed from the further analysis due to their bad quality across most of the Carpathian area.

During NDVI differencing a methodology proposed by Pu et al. (2008) has been used. They assumed that both the increase and decrease part of NDVI difference have a normal distribution which determines the mean and standard deviation for each part of the histogram (μ_H, σ_H for the increase and μ_L, σ_L for the decrease). To identify the occurrence of changes two thresholds determined separately based on the mean and standard deviation should be established (Lafortezza et al. 2004; Pu et al. 2008). Using Pu et al. (2008) approach, the NDVI difference images were thus divided into three classes:

- ΔNDVI < $(\mu_L + \sigma_L)$ → decrease of vegetation,
- $(\mu_L + \sigma_L) \geq \Delta$NDVI $\geq (\mu_H\text{-}\sigma_H)$ → no change,
- ΔNDVI > $(\mu_H\text{-}\sigma_H)$ → increase of vegetation.

Trend analysis was performed for time series from 2001 to 2009 using linear regression method, in two scales: for the whole Carpathians and with respect to elevation classes (elevation ranges: below 100, 100–200, 200–300, 300–500, 500–1,000, 1,000–1,500, 1,500–2,000, above 2,000 m). To describe changes in the analyzed period three widely used indices were chosen: trend coefficient, individual chain index and average rate of change.

Trend coefficient (b_1) expresses the average rate of change of a given feature in analysed time t (Starzyńska 2002). It is computed according to the Eq. (2), where $\hat{y}$ means theoretical value of trend function and b_0 is Y-intercept.

$$\hat{y} = b_0 + b_1 t \quad t \in \langle 1, n \rangle \tag{2}$$

Individual chain indices ($i_{yt/t\text{-}1}$) describe the dynamics of change (Starzyńska 2002). They are defined as a ratio of value of given phenomena in the time t and the time preceding $t-1$ (Eq. 3).

$$i_{y_{t/t-1}} = \frac{y_t}{y_{t-1}} 100 \quad t \in \langle 1, n \rangle \tag{3}$$

The average rate of change (G) is determined as an unweighted geometric mean of individual chain indices (Eq. 4; Starzyńska 2002). It enables to assess changes of analyzed phenomena in the whole observed period.

$$G = \sqrt[n]{\prod_{t-1}^{n} i_{y_{t/t-1}}} \quad t \in \langle 1, n \rangle \tag{4}$$

5 Results

The NDVI differencing image 2001–2009 covers ~235,300 km^2, 1 % of the study area was excluded because of low data quality. Changes recorded between 2001 and 2009 were not much widespread and covered only 8 % of the analyzed area. As for observed changes, decreases in the NDVI values were predominant (Fig. 2). Areas with drops of NDVI values covered about 5 % (~12,300 km^2) of the mountains, while territories where increase of NDVI was noticed occupied only less than 3 % (~6,600 km^2).

Negative NDVI differences were registered in the western part of the Apuseni Mountains (especially in the Criş Mountains), central (Transylvanian Plain) and north-western part of the Transylvanian Plateau (Someş Plateau), or in the Northern Mountain (Cserhát, Mátra or Bükk Mountains) situated within the Inner

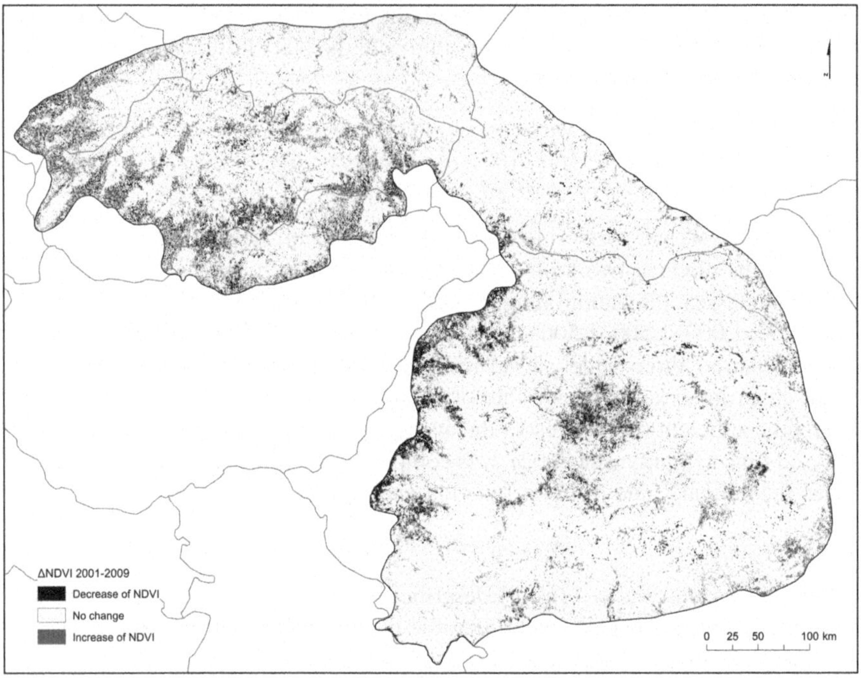

Fig. 2 Changes of NDVI between 2001 and 2009

Western Carpathians (Fig. 2). Regions where the increase of NDVI was found are more dispersed and cover smaller areas. The most pronounced areas with positive NDVI changes were located in the south-eastern part of the Moldavian-Muntenian Carpathians (Outer Eastern Carpathians) and in the western part of the Western Carpathians (Fig. 2).

Areas without significant changes in NDVI (NDVI difference value close to 0) were noticed in the south-western part of the Southern Carpathians (within the Retezat-Godeanu Mountain group as well as Banat Mountains; Fig. 2).

The NDVI values varied from 0.75 to 0.80 between 2001 and 2009. Nevertheless, differences between particular years were very slight. The most visible change (a drop) occurred in the year 2007. Between 2001 and 2009 the negative trend in NDVI dominated in the Carpathians. Trend coefficient calculated for the whole area was approximately −0.0021.

The average rate of change was also negative. Between 2001 and 2009 NDVI value decreased on average 0.2 % annually. However, the change rates were not stable in the analyzed period. During the first three seasons the change rate was close to −1 %. After that, rapid changes occurred, although only for two periods the average rate of change was positive: 2004/05 (2.4 %) and 2007/08 (5.3 %). The last analyzed season (2008/09) did not experience any significant alterations in context of trend dynamics.

Table 2 Trend coefficient and the average rate of change for elevation classes

Elevation class [m a.s.l.]	Trend coefficient	Average rate of change [%]
100<	−0.0030	−0.83
<100–200)	−0.0046	−0.86
<200–300)	−0.0042	−0.37
<300–500)	−0.0039	−0.24
<500–1,000)	−0.0012	−0.03
<1,000–1,500)	0.0028	−0.04
<1,500–2,000)	0.0059	−0.11
2,000≤	0.0058	−0.01

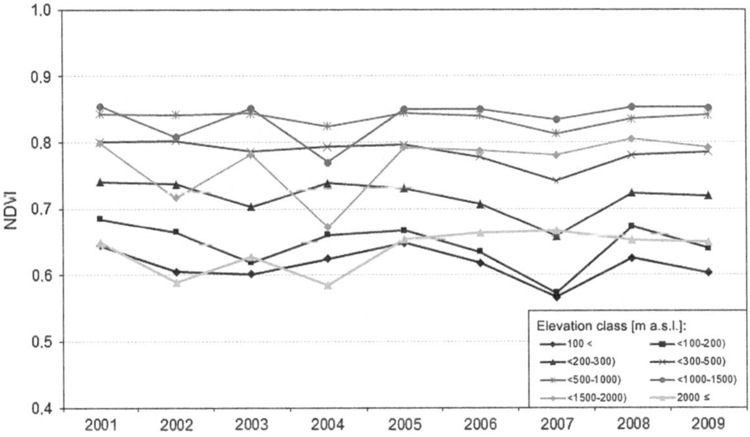

Fig. 3 Trend trajectories for elevation classes

Looking into particular elevation classes, trend coefficient belonged to the range <−0.0046; 0.0059> (Table 2). Trend trajectories were diverse: rapid drops appeared either in 2004 or 2007 (Fig. 3).

The average rate of change was negative for all elevation classes (Table 2). The largest decreases were recorded in regions below 200 m a.s.l (Fig. 4). Areas located between 500 and 1,500 m a.s.l. or above 2,000 m a.s.l. experienced the slightest alterations of vegetation.

6 Discussion and Conclusions

Observed vegetation changes did not occur equally across the Carpathians. In general, the most noticeable and intensive alterations were noticed within the lower, usually more populated, parts of the mountains, especially below 200 m a.s.l. This fact could be connected with changes in either agricultural land structure or crop structure within arable lands. Spectral response varies depending either on

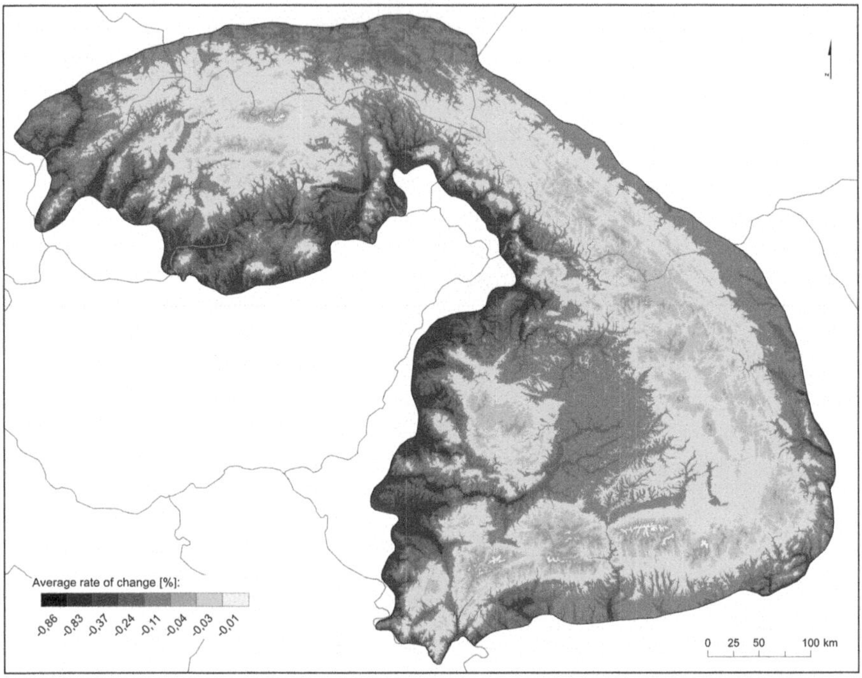

Fig. 4 The average rate of change for elevation classes

type of agricultural land use (meadow, pasture or arable land), brightness coefficient of different crops, their phenological stage, or on moisture conditions (Ciołkosz et al. 1999). The elevated areas experienced relatively the smallest modifications. The low rate of change recorded in the highest part of the Carpathians could be connected with minor human pressure on these areas.

Marked NDVI changes can be associated with some processes of land cover change. Decrease of vegetation greenness could refer to e.g., forest cover alterations that resulted from forest degradation and legal/illegal logging, or transformation of semi natural areas into artificial surfaces (urban sprawl). On the other hand, increase of greenness might be connected with e.g., growth of forest cover areas, or cropland abandonment and in consequence natural succession or afforestation.

Some NDVI changes observed in the Western Ukraine Carpathians are related to land abandonment. Baumann et al. (2011) also noticed effects of widespread farmland abandonment between 1986/1989 and 2006/2008, especially in the plains and in the Carpathian foothills, although, they were much more widespread that changes revealed based on MODIS data. Reasons for such differences may include the following: higher spatial resolution of Landsat images used, or another time period, as most of these changes occurred mainly after the transition period (in the 1990s). Vegetation changes, observed in many regions of the Carpathians, based on MODIS NDVI data, occurred also in forest areas and corresponded with other results. For example, in the Ukrainian Carpathians changes in forest area coincided with forest disturbances revealed by Kuemmerle et al. (2009a) that occurred between 2000 and 2007. Similar results were received for the protected areas in the northern part of the Romanian Carpathians. Knorn et al. (2012) noticed well visible disturbances for two periods: 2002–2006 and 2006–2009 for the Romanian Carpathians, especially in the eastern part of Maramures, western part of Rodna, and western part of Calimani.

Changes revealed in NDVI differencing were also much more widespread than those found in CORINE Land Cover 2000–2006 Changes database. According to the CORINE database, majority of noticed alterations occurred in the Western Carpathians [e.g., well visible effects of forest windbreaks (Vel'ká Kalamita) on the southern slopes of the Tatra Mountains] or higher parts of the analyzed region. Such a difference in relation to changes identified in this study may be linked to the fact that majority of changes recorded on the basis of the NDVI differencing could be connected with changes in agricultural areas. They are related to a spectral change within a given land cover type, therefore they might be considered mostly as a modification of land cover, while CORINE Land Cover 2000–2006 Changes database contains typically land cover conversions.

Although such analysis has been proven to be suitable for land cover change research, it has some limitations.

First of all, results are affected by coarse spatial resolution of MODIS data (250 m). The study of Price (2003) shows that in some areas MODIS data can produce virtually the same results as Landsat ETM+ data. Nevertheless, in the Carpathians many agricultural fields or forest clear cuts are much smaller than minimal mapping unit of such analysis. The heterogeneity of Carpathian ecosystems requires high resolution data in order to resolve potential uncertainties and verify the results.

Selection of a single NDVI image from particular years could also influence the results due to the possibility of change in phenological seasons in individual years.

Another point is a delimitation of threshold values in NDVI differencing method. Using a traditional approach (the thresholds specified by 1 SD from the mean or with some modification to the SD; Pu et al. 2008) would extend areas with noticeable changes of NDVI values.

In such analysis, apart from interannual fluctuation of NDVI, there are also observed interannual changes. It would be beneficial to take into account in future analysis the trajectory of NDVI changes within one year. Above all, it might help to eliminate the seasonality effect. Another possibility is to consider various ancillary data (e.g., meteorological data relating to precipitation, drought periods or length of snow cover period) to explain the interannual fluctuations of NDVI.

Moreover, as an effect of data availability, time span used in this study is quite short to characterize long-term trends in land cover change. Extension of analyzed period could help to define the character of registered changes in an unequivocal way, as well as comparison results of this study with classification of land cover types.

This study showed that it is possible to map land cover change for large area using MODIS NDVI data, however it requires further improvements to separate land cover conversion from modification. In the future, the analysis can be extended in several ways. First, the noise elimination may help to exclude effects of modification. In determining phenological stages, reduction of noise or fitting a model to data observed is necessary (Sakamoto et al. 2005), similarly as the analysis of change trajectories (temporal trajectory profiles) for particular years. Next, a pixel-based classification created on the basis of change trajectories may improve tracking of land cover change in the whole time span. Finally, the results of analysis should be verify based on the high spatial resolution data.

References

Baumann M, Kuemmerle T, Elbakidze M, Ozdogan M, Radeloff VC, Keuler NS, Prishchepov AV, Kruhlov I, Hostert P (2011) Patterns and drivers of post-socialist farmland abandonment in Western Ukraine. Land Use Policy 28:552–562

Bhattacharya A, Srivastawa PK, Pal SK (2005) Change detection analysis of eastern coastal zone of India using satellite remote sensing data and GIS. In: Erasmi S, Cyffka B, Kappas M (eds) Remote sensing and GIS for environmental studies: application in geography, göttingen geographische abhandlungen 113. Verlag Erich Goltze GmbH&Co, Göttingen

Carraro H, Goncalves P, Caetano M (2008) Contribution of multispectral and multitemporal information from MODIS images to land cover classification. Remote Sens Environ 112:986–997

Casacchia R, Salvatori R, Grignetti A, Mandrone S, Capraro A, Bogliolo MP (2002) Land use changes in southern Lazio, Italy, by satellite data multitemporal analysis. In: Bruzzone L, Smits P (eds) Analysis of multi-temporal remote sensing images, series in remote sensing 2. World Scientific, New Jersey

Ciołkosz A, Miszalski J, Olędzki JR (1999) Interpretacja zdjęć lotniczych. Wydawnictwo Naukowe PWN, Warszawa

Coppin P, Lambin E, Jonckheere I, Muys B (2002) Digital change detection methods in natural ecosystem monitoring: a review. In: Bruzzone L, Smits P (eds) Analysis of multi-temporal remote sensing images, series in remote sensing 2. World Scientific, New Jersey

Coppin P, Jonckheere I, Nackaerts K, Muys B, Lambin E (2004) Digital change detection methods in ecosystem monitoring: a review. Int J Remote Sens 25:1565–1596

Griffiths P, Kuemmerle T, Kennedy RE, Abrudan IV, Knorn J, Hostert P (2012a) Using annual time-series of Landsat images to assess the effects of forest restitution in post-socialist Romania. Remote Sens Environ 118:199–214

Griffiths P, Kuemmerle T, Knorn J, Hostert P (2012b) Pixel-based compositing approaches for fine-scale land cover change mapping across the entire Carpathian ecoregion—first results. In: Boltižiar M (eds) The conference abstracts of the 2nd Forum Carpaticum 2012: From Data to Knowledge-from Knowlegde to Action

Hayes DJ, Sader SA (2001) Comparison of change-detection techniques for monitoring tropical forest clearing and vegetation regrowth in a time series. Photogramm Eng Remote Sens 67:1067–1075

Hostert P, Kuemmerle T, Radeloff VC, Müller D (2008) Post socialist land-use and land-cover change in the Carpathian Mountains. IHDP update 2:70–73

Huete A, Justice C, van Leeuwen W (1999) MODIS vegetation index (MOD13) Algorithm Theoretical Basis Document v3. http://modis.gsfc.nasa.gov/data/atbd/atbd_mod13.pdf. Accessed 10 October 2010

Jarocińska A, Zagajewski B (2008) Analiza kondycji roślinności zlewni Bystrzanki, Konferencja Jubileuszowa – 40-lecie Stacji Naukowo-Badawczej IGiPZ PAN w Szymbarku, XVIII Ogólnopolskie Sympozjum Zintegrowanego Monitoringu Środowiska Przyrodniczego. http://www.igipz.pan.pl/krakow/szymbark/pdfy/Jarocinska_Zagajewski.pdf. Accessed 24 October 2010

Knorn J, Rabe A, Radeloff VC, Kuemmerle T, Kozak J, Hostert P (2009) Land cover mapping of large areas using chain classification of neighboring Landsat satellite images. Remote Sens Environ 113:957–964

Knorn J, Kuemmerle T, Radeloff VC, Szabo A, Mindrescu M, Keeton WS, Abrudan I, Griffiths P, Gancz V, Hostert P (2012) Forest restitution and protected area effectiveness in post-socialist Romania. Biol Conserv 146:204–212

Kondracki J (1989) Karpaty. Wydawnictwa Szkolne i Pedagogiczne, Warszawa

Kozak J (2003) Forest cover changes in the Western Carpathians over the past 180 years: a case study from the Orawa Region in Poland. Mt Res Dev 23:369–375

Kozak J (2005) Zmiany powierzchni lasów w Karpatach Polskich na tle innych gór świata. Wydawnictwo Uniwersytetu Jagiellońskiego, Kraków

Kozak J (2009) Land use change in the Northern Carpathians. In: global change and sustainable development in mountain regions, Jandl R, Borsdorf A, van Miegroet H, Lackner R, Psenner R (eds) Alpine Space—Man and Environment 7. Innsbruck University Press, Innsbruck

Kozak J, Estreguil C, Ostapowicz K (2008) European forest cover mapping with high resolution satellite data: the Carpathians case study. Int J Appl Earth Obs Geoinf 10:44–55

Kozak J, Estreguil C, Vogt P (2007) Forest cover and pattern changes in the Carpathians over last decades. Eur J Forest Res 126:77–90

Kuemmerle T, Chaskovskyy O, Knorn J, Radeloff VC, Kruhlov I, Keeton WS, Hostert P (2009a) Forest cover change and illegal logging in the Ukrainian Carpathians in the transition period from 1988 to 2007. Remote Sens Environ 113:1194–1207

Kuemmerle T, Kozak J, Radeloff VC, Hostert P (2009b) Differences in forest disturbance among land ownership types in Poland during and after socialism. J Land Use Sci 4(1):73–83

Kuemmerle T, Müller D, Griffiths P, Rusu M (2009c) Land use change in Southern Romania after the collapse of socialism. Reg Environ Change 9:1–12

Kuemmerle T, Olofsson P, Chaskovskyy O, Baumann M, Ostapowicz K, Woodcock CE, Houghton RA, Hostert P, Keeton WS, Radeloff VC (2011) Post-Soviet farmland abandonment, forest recovery, and carbon sequestration in western Ukraine. Glob Change Biol 17(3):1335–1349

Kuemmerle T, Radeloff VC, Perzanowski K, Hostert P (2006) Cross-border comparison of land cover and landscape pattern in Eastern Europe using a hybrid classification technique. Remote Sens Environ 103:449–464

Kuemmerle T, Hostert P, Damm A, Radeloff V (2005) Cross-border monitoring of landscape dynamics in Eastern Europe. In: Erasmi S, Cyffka B, Kappas M (eds) Remote senings and GIS for environmental studies: application in geography, Göttingen Geographische Abhandlungen 113. Verlag Erich Goltze GmbH&Co, Göttingen

Lafortezza R, Iasillo D, Lovreglio R (2004) Multitemporal assessment of land cover transitions in burnt forest areas using Landsat TM/ETM images. In: Smits P, Bruzzone L (eds) Analysis of multi-temporal remote sensing images, series in remote sensing 3. World Scientific, New Jersey

Lambin EF, Turner BL, Geist HJ, Agbola SB, Angelsen A, Bruce JW, Coomes OT, Dirzo R, Fischer G, Folke C, George PS, Homewood K, Imbernon J, Leemans R, Li X, Moran EF, Mortimore M, Ramakrishnen PS, Richards JF, Skånes H, Steffen W, Dione GD, Svedin U, Veldkamp TA, Vogel C, Xu J (2001) The causes of land-use and land-cover change: moving beyond the myths. Glob Environ Change 11:261–269

Lee SH, Roh DK, Kim CM, Cho HK (2004) Change detection of forest burnt area using multitemporal Landsat TM and ETM + data. In: Smits P, Bruzzone L (eds) Analysis of multitemporal remote sensing images, series in remote sensing 3. World Scientific, New Jersey

LP DAAC (2010) https://lpdaac.usgs.gov. Accessed 24 October 2010

Lyon JG, Yuan D, Lunetta RS, Elvidge CD (1998) A change detection experiment using vegetation indices. Photogramm Eng Remote Sens 64:143–150

Main-Knorn M, Hostert P, Kozak J, Kuemmerle T (2009) How pollution legacies and land use histories shape post-communist forest cover trends in the Western Carpathians. For Ecol Manage 258:60–70

Mandrone S, Casacchia R, Grignetti A, Salvatori R, Paterno C (2004) Multitemporal analysis of land use in northern Lazio, Italy. In: Smits P, Bruzzone L (eds) Analysis of multi-temporal remote sensing images, series in remote sensing 3. World Scientific, New Jersey

Mihai B, Savulescu I, Sandric I (2007) Change detection analysis (1986–2002) of vegetation cover in Romania: a study of alpine, subalpine, and forest landscapes in the Iezer Mountains, Southern Carpathians. Mt Res Dev 27:250–258

MODIS Web (2010) http://modis.gsfc.nasa.gov. Accessed 24 October 2010

Müller D, Kuemmerle T, Rusu M, Griffiths P (2009) Lost in transition: determinants of post-socialist cropland abandonment in Romania. J Land Use Sci 4:109–129

Mycke-Dominko M, Slinkina O (2004) Ocena podatności lasu na pożar na podstawie obrazów satelitarnych NOAA-AVHRR. http://www.sgp.geodezja.org.pl/ptfit/wydawnictwa/bialobrzegi/Bialobrzegi2004/37-mycke-dominkopopr.doc. Accessed 24 October 2010

Price JC (2003) Comparing MODIS and ETM + data for regional and global land classification. Remote Sens Environ 86:491–499

Pu R, Gong P, Tian Y, Miao X, Carruthers RI, Anderson GL (2008) Using classification and NDVI differencing methods for monitoring sparse vegetation coverage: a case study of salt cedar in Nevada, USA. Int J Remote Sens 29:3987–4011

Ramankutty N, Graumlich L, Achard F, Alves D, Chhabra A, DeFries RS, Foley JA, Geist H, Houghton RA, Goldewijk KK, Lambin EF, Millington A, Rasmussen K, Reid RS, Turner BL II (2006) Global land-cover change: recent progress, remaining challenges. In: Lambin EF, Geist HJ (eds) Land-use and land-cover change: local processes and global impacts. Springer, Berlin

Sakamoto T, Yokozawa M, Toritami H, Shibayama M, Ishitsuka N, Ohno H (2005) A crop phenology detection method using time-series MODIS data. Remote Sens Environ 96:366–374

Schweitzer C, Rücker GR, Conrad C, Bendix J, Strunz G (2005) Knowledge-based land use classification combining expert knowledge, GIS, multi-temporal Landsat 7 ETM + and MODIS time series data in Khorezm, Uzbekistan. In: Erasmi S, Cyffka B, Kappas M (eds) Remote sening and GIS for environmental studies: application in geography, Göttingen Geographische Abhandlungen 113. Verlag Erich Goltze GmbH&Co, Göttingen

Solano R, Didan K, Jacobson A, Huete A (2010) MODIS Vegetation Indices (MOD13) C5 User's guide. terrestrial biophysics and remote sensing lab, The University of Arizona. http://tbrs.arizona.edu/project/MODIS/MOD13.C5-UsersGuide-HTML-v1.00/index.html. Accessed 24 October 2010

Starzyńska W (2002) Statystyka praktyczna. Wydawnictwo Naukowe PWN, Warszawa

Stellmes M, Sommer S, Hill J (2005) Regional vegetation monitoring based on NOAA AVHRR time series in the Mediterranean area. In: Erasmi S, Cyffka B, Kappas M (eds) Remote sensing and GIS for environmental studies: application in geography, göttingen geographische abhandlungen 113. Verlag Erich Goltze GmbH&Co, Göttingen

Taddei R, Puzzolo V (2004) Phyto-climatic classification of the vegetation cover in eastern Sicily using historic series of multiseasonal Landsat MSS images. In: Smits P, Bruzzone L (eds) Analysis of multi-temporal remote sensing images, series in remote sensing 3. World Scientific, New Jersey

Turnock D (2002) Ecoregion-based conservation in the Carpathians and the land-use implications. Land Use Policy 19:47–63

Yagüe J, García P (2004) Approaching CORINE land cover over Castilla and León (central Spain) with a multitemporal NOAA-AVHRR NDVI MCV series. In: Smits P, Bruzzone L (eds) Analysis of multi-temporal remote sensing images, series in remote sensing 3. World Scientific, New Jersey

Using Habitat Quality and Diversity Measures to Assess Conservation Priorities for Sites in the Ukrainian Carpathians

Alex Mkrtchian

Abstract The chapter investigates the possibilities and advantages of using habitat quality and diversity measures in the design of ecological networks in the Ukrainian Carpathians. It was shown that habitat diversity could be indicated by the measures of topographic variance. The latter correlate well with some characteristics connected with ecological as well as socioeconomic factors determining land suitability for conservation. The received measure of topographic variation correlates significantly with the abundance of rare and endangered plant species entered into the Ukrainian Red Book. The correlation of this abundance with the Normalized Difference Vegetation Index (NDVI) values is not as pronounced and is highly dependent on local conditions. An attempt was made to arrive at an integral measure of the appropriateness for assigning protected status, using easily obtained remote sensing and elevation data.

1 Introduction

The Carpathians are one of the largest mountain ranges in Europe, providing an essential habitat and refuge for many endangered species of plants and animals and Europe's largest area of virgin forests, and they constitute a major ecological, economic, cultural, recreational, and living environment in the heart of Europe, which is shared by numerous peoples and countries.

In the Ukraine, the Carpathians are the largest and highest mountain range, hosting a large part of the country's biodiversity, as well as providing indispensable ecosystem services and being an important area for local and regional tourism and recreation. Thus, the Carpathians are the priority area for conservation activities in Ukraine.

A. Mkrtchian (✉)
Ivan Franko National University of Lviv, Doroshenka st. 41, Lviv 79000, Ukraine
e-mail: alemkrt@gmail.com

J. Kozak et al. (eds.), *The Carpathians: Integrating Nature and Society Towards Sustainability*, Environmental Science and Engineering,
DOI: 10.1007/978-3-642-12725-0_46,

Establishing a large ecological network and assigning protected status to substantial areas is a resource-consuming enterprise, involving large management costs and even more indirect costs of land use restrictions. Thus, the placement, zoning, and management of protected areas should be established upon sound scientific research regarding the aims of conservation and the best ways to achieve them. The conservation aims can be broadly divided into three groups: the conservation of rare, endangered, or valuable species (the realm of conservation biology); the conservation of valuable ecosystems functions (e.g., soil and water protection, and carbon retention capacity); and the protection of scenic landscapes and natural monuments for their aesthetic qualities and attractiveness for tourists and local people. Careful planning of the placement, zoning, and management of protected areas requires that reliable data on the occurrence and distribution of biodiversity data, as well as some other ecosystem properties determining the suitability of areas for conservation, are available. Because protected areas are characterized by restrictions placed on land utilization and certain kinds of economic and social activities that can burden society, the placement of such areas should be based on sound methodology, minimizing such a burden, while maximizing the aims of conservation. It is recognized, for example, that cheaper or less socially disruptive reserve networks are more likely to be implemented.

The Ukrainian Carpathians currently host a rather dense network of protected areas with various protection regimes, including the Carpathian Biosphere Reserve, the Gorgany Natural Reserve, several national nature parks and regional landscape parks, and scores of small, protected areas of local significance. The current state of the Ukrainian Carpathians protected areas network and its legal and organizational basis and functions have been briefly summarized by Deodatus and Protsenko (2010, Appendices 2, 3).

It is widely recognized that the present structure and management regime of this network is far from optimal and should be modified according to modern standards, approaches, and practices to be effectively integrated into the European ecological network (EECONET). For the last dozen years, several relevant legal acts have been passed, including two laws "On the State Program for the Development of the National Ecological Network in Ukraine for 2000–2015" (2000) and "On the Ecological Network of Ukraine" (2004). The topic became the subject of a number of scientific studies and projects, and a number of proposals, recommendations, and suggested schemes have been put forward (Sheliag-Sosonko 1999; Brusak et al. 2006; Deodatus and Protsenko 2010).

All of the proposed projects and schemes are heuristic and tentative, because an "ideal" scheme should take into consideration the exhaustive inventories of species habitats and ecological characteristics, as well as valuable ecosystems functions, natural monuments, and scenic landscapes. This can be a practically unfeasible task. However, methods could be developed that render more objectivity by explicitly accounting for the factors related to conservation aims. Remote sensed imagery could be used as a data source for the task, considering its accessibility and low costs, compared to field surveys. However, as species and processes can rarely be directly seen and measured, remote sensing is usually used to derive some quantitative measures related to ecological characteristics meaningful to conservation.

The aims of this study concern the investigation of the possibilities of using two quantitative measures as indicators of proper conservation targets. The first measure is the Normalized Difference Vegetation Index (NDVI), easily determined and mapped using spectral reflectance measurements acquired in the red and near-infrared regions (Rouse et al. 1973). Values of NDVI could correlate with the Leaf Area Index, plant productivity, biomass, chlorophyll concentration in leaves, fractional vegetation cover, and some other ecosystem properties that together indicate the degree of presence and the current state of natural vegetation of the area (Glenn et al. 2008). Thus, NDVI could be regarded as indicative of the overall "naturalness" of the site conditions determined by the ecosystem's status, primary production, the proportion of area covered by natural vegetation, etc.

The second measure pertains to the local habitat diversity, mostly determined by abiotic factors like soils, microclimate, hydrologic conditions, etc. The plant and animal habitats are characterized by the sets of indirect, direct, and resource environmental gradients (Austin and Smith 1989). This means that the increased diversity of abiotic factors in certain area leads to the increased amount of potential habitats, thereby contributing to its biodiversity. In areas of complex topography, topographic variables make efficient predictors of habitat characteristics, given the high spatial accuracy of modern digital elevation models (DEMs) (Guisan and Zimmermann 2000).

In this study, an attempt is made to derive these measures from freely available digital data and to test their suitability as indicators of conservation priority by comparing them to the data on rare and endangered species habitats. Some suggestions are made for the improvement of the existing protected areas network using these measures.

2 Study Area

The study area is located in the Ukrainian Carpathians together with parts of the adjacent foothills and plains. The Ukrainian Carpathians can be roughly divided into three parallel, relatively narrow belts extending from northwest to southeast. The northern macroslope of the range makes up the spruce belt, where secondary spruce forest stands dominate with admixtures of beech and fir. The highest axial part of the range consists of connected or isolated massifs, in their highest parts (starting from 1,500–1,600 m) covered with subalpine vegetation. The southern macroslope makes up the beech belt, where beech forests dominate. In the northern and southern foothills of the range, oak forests dominate with an admixture of beech and hornbeam, while Precarpathian and Transcarpathian plains are mostly arable lands with patches of broad-leaved forests. The detailed description of the natural conditions of the region can be found in Herenchuk (1968), which also contains its regionalization scheme cited below.

3 Materials and Methods

The source of elevation data was the processed SRTM DEM (Shuttle Radar Topography Mission Digital Elevation Model) version 4.1 (Jarvis et al. 2008), available online from the CGIAR Consortium for Spatial Information (http://srtm.csi.cgiar.org). These data with 3-arc second resolution are in decimal degrees based on World Geodetic System (WGS) 84. For the calculation of NDVI, the Landsat 7 ETM+ multiband image was used (NASA Landsat Program 2009) from the online USGS archive (http://glovis.usgs.gov). The data were acquired on October 13, 2009, and are available in the Universal Transverse Mercator (UTM) geographic coordinate system (zone 35). The multiband image consists of 8 band layers, of which bands 3 and 4 (both having 30 m resolution) were utilized in the study.

The SRTM DEM data were projected to the UTM coordinate system and resampled to 90 m spatial resolution. The Landsat layers were resampled to 90 m resolution, for better compatibility with the DEM and to facilitate further calculations.

The resampled DEM was used to obtain the three measures of habitat diversity:

1. the variance of elevation;
2. the variance of slope values;
3. The relief horizontal dissection and aquatic habitat richness measured by a local density of stream network.

Variances of elevation and slope values were measured in local windows by taking variance measures among values in local neighborhoods (e.g., 3×3 square around a central pixel). The local variance of elevation indicates mainly the diversity of microclimates (temperature, humidity, and precipitation values) and insolation regimes, while the variance in slope values indirectly indicates the complexity and diversity of geological structures and processes with diversity in soil edaphic and hydrologic properties. The third measure, the stream network density, was derived using the method described by Montgomery and Foufoula-Georgiou (1993), wherein the variable flow-accumulation area thresholds are calculated using slope value raised to a power.

The evaluation of these measures should take into account different scale domains in the landscape. Currently, the scaling problem is one of the most important in ecology (Alpin 2007). It is known from landscape and ecological studies that the reaction of different species to environmental conditions reveals itself on certain spatial scales, which could be designated as the spatial scales of certain species. Likewise, processes shaping the landscapes reveal themselves in different spatial scales; therefore, landscape management should comprise different scales, each one pertinent to a certain set of species (McGarrigal and Marks 1994). Wiens (1989) shows examples when changes in scale affect the size and even the sign of habitat relationships and interspecific associations, concluding that scaling issues are fundamental to all ecological investigations, as they are in other sciences (Wiens 1989). Processes determining and influencing most ecological factors also often show scale-dependency (Clark 1985; Levin 1992; Blöschl 1995). Margalef (1968) introduced the notion of a diversity spectrum, in which biological diversity is examined as a function of scale. These studies show that the values of criteria for determining the appropriateness of sites for assigning protected status should be estimated not only on a scale dictated by the size of the pixels of the raster image but also on a range of scales embracing the most important ecological processes and the living space of species selected as conservation targets.

This was achieved by the calculation of measures in a series of local neighborhoods with different window sizes, each being a triple of its predecessor: 270 × 270 m, 810 × 810 m, 2,430 × 2,430 m, and 7,290 × 7,290 m. While the sizes were arbitrarily chosen, the spectrum is meant to encompass the processes pertaining to the movement of the large part of vertebrate species. Then, the averages among all scale levels were used as values to be entered into the model.

To make some assessment of the validity of these criteria, correlational analysis was carried out using data on geographic distribution of the diversity of a set of plant species. While biodiversity is a rather complicated concept, the species count is often regarded as its surrogate (Purvis and Hector 2000). The hypothetical set of target species contained the vascular plant species included in the Ukrainian Red Book, the official registry of the rare and endangered species composed by the Ministry for Environmental Protection of the Ukraine, as prescribed and regulated by Ukrainian law (Ministry for Environmental Protection of Ukraine 1996). Most of the species in the Ukrainian Red Book are also listed in the International Union for Conservation of Nature (IUCN) Red List of Threatened Species.

In whole, 92 species of the division *Angiospermae*, together with the 14 species of the group of lower vascular plants (divisions of *Pinophyta*, *Lycopodiophyta*, and *Pteridophyta*), were selected for analysis (106 species in total). These were plant species characteristic of the Ukrainian Carpathians, whose habitats are described in the Ukrainian Red Book in sufficient detail to be mapped. Then, the digital maps were produced with the delineations of the habitats of each of these species.

It should be noted that the habitat descriptions used were often pretty rough, so the produced habitat map is not assumed to be of high spatial accuracy. But in the absence of the detailed spatial species inventories, this approach is regarded to be the only feasible one and suitable enough to obtain the relative measures of biodiversity. After creating the digital habitat map layer, the habitats of different species were overlaid to obtain a map layer of species richness, which depicts the amount of species present in a certain place (Fig. 1). It shows that the highest values are characteristic of subalpine meadows on the highest areas of the region, while the beech belt of Transcarpathia has by far larger rare and endangered plant species richness than the northern spruce belt, foothills, and adjacent plains.

The correlation coefficients were calculated between species richness values, measures of habitat diversity and NDVI values determined in neighborhoods of

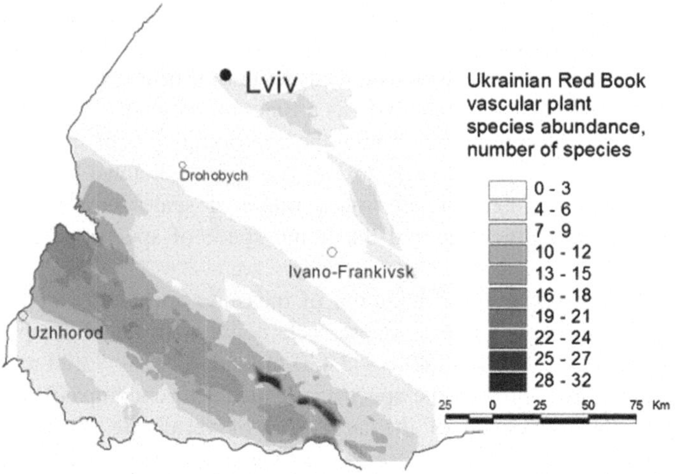

Fig. 1 The amount present of rare and endangered vascular plant species in the Ukrainian Carpathians as listed in the Ukrainian red book

different sizes. While the reliabilities of these coefficients were hard to strictly measure due to the presence of spatial autocorrelations (which complicates greatly the estimation of the degrees of freedom), the presence of the large and consistently repeated correlation coefficients suggests a strong possibility that an accurate relationship exists between values.

Lastly, an attempt was made to arrive at an integral measure of appropriateness to assigning protected status. To this aim, NDVI values over all the scale levels were standardized (subtracting the minimal value and then dividing by the maximum one) to 0–1 scale, and then the average values were calculated. The latter mean of the NDVI was averaged among scale spectrum. Similarly, the three measures of habitat diversity (the variances of elevation and slope values, and the stream network density), each averaged over four scale levels, were standardized to 0–1 scale and then summed to arrive at the integral measure of habitat diversity—the terrain variability index. At the last step, the averaged NDVI measures were multiplied by the integral measures of habitat diversity, to obtain the final assessment.

Finally, the correlation coefficients were calculated for the measures of terrain diversity and NDVI, and calculated for different neighborhood windows (corresponding to different spatial scales) and the observed rare and endangered vascular plant species count.

For the preparation and analysis of the data, several geographic information system (GIS) software products were used, including commercial ArcView and ArcGIS software packages by ESRI, and free and open ILWIS GIS software.

4 Results

Figure 2 illustrates the distribution of NDVI values calculated from the Landsat images, while Fig. 3 shows the terrain variability index arrived at by combining the values of the variance of elevation, the variance of slope values, and the local stream network density. These can be visually compared to Fig. 1, depicting the spatial density of habitats of the rare and endangered plant species.

As can be seen from Table 1, all correlations are positive, yet the correlations between species count and terrain diversity are substantially higher than the correlations between species count and NDVI values. All correlations appear to consistently increase with the spatial scale. The spatial pattern of correlations was also visually analyzed, showing strong negative correlations of species count and NDVI index in the highest mountain ranges with subalpine meadows, and characterized by low NDVI values except for the largest density of Red Book species. The highest positive correlations were characteristic of beech belt, while in spruce belt the correlations were insignificant. While the overall correlation is small, it is positive and steadily rises with the increase in spatial window, from 0.01 for 90 m to almost 0.1 for 7.29 km.

The correlation with terrain diversity turned out to be much higher: from 0.6 for the 90 m neighborhood to almost 0.73 for the 7.29 km neighborhood. It proves the hypothesis that terrain variability is a good indicator of the diversity of plant habitats. The spatial distribution of correlation values (Fig. 4) shows that the largest correlations are characteristic of highest axial massifs with the presence of subalpine associations; high correlation is also found in beech belt and in dissected Precarpathian plains, while the spruce belt again shows insignificant correlations.

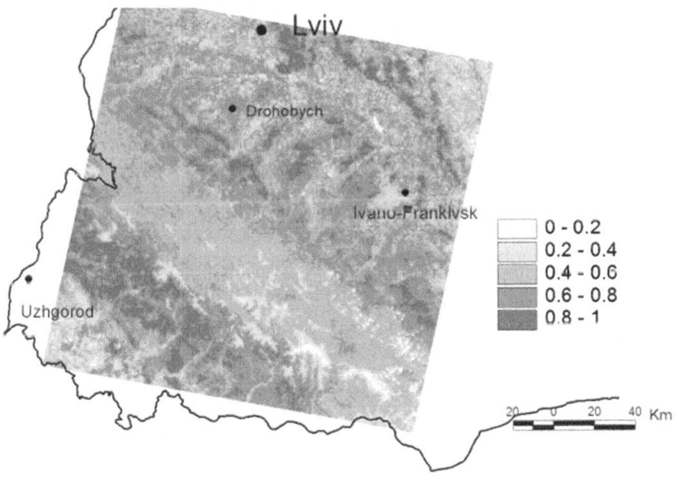

Fig. 2 The distribution of scale-averaged NDVI values in the Ukrainian Carpathians

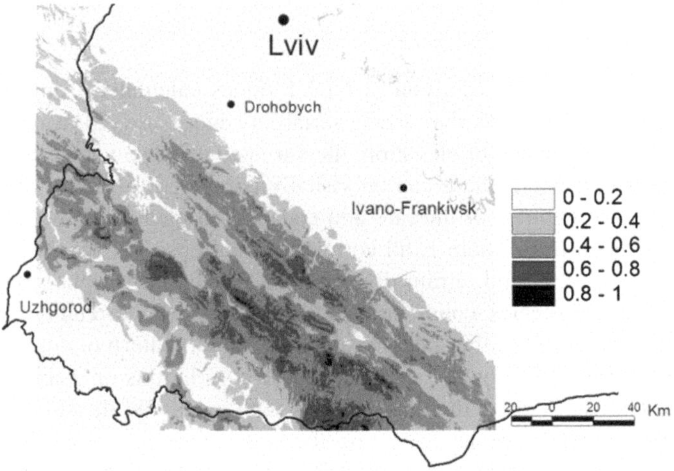

Fig. 3 The terrain variability (*TV*) index in the Ukrainian Carpathians

Table 1 The correlation coefficients between rare and endangered plant species count that were present and the NDVI and terrain diversity values in the Ukrainian Carpathians

Neighborhood (m)	NDVI	Terrain diversity
90	0.0187	–
270	0.0306	0.6032
810	0.0372	0.6154
2,430	0.0539	0.6790
7,290	0.0979	0.7275

It should be noted that the increases in correlation coefficients with the increases in the size of the averaging window could be an artifact, due to the fact that the plant habitats used to calculate the amount of species present (Fig. 1) were delineated with relatively low spatial precision.

The final stage of the research consisted of the combination of information on terrain diversity and NDVI and the creation of the map layer showing the integral values of appropriateness of sites for assigning a conservation status. The appropriateness is higher in mountains than on adjacent plains, while the highest values of the appropriateness index are observed on the south macroslope of the Polonyna range, characterized by the highly dissected relief with a big variance of altitudes and covered by rich beech and mixed forests. The second region considered among the most appropriate is the Vyhorlat-Hutyn low-mountain range made up of volcanic rocks. The third most appropriate region is the northern macroslope of the Skybovi Carpathians, made up of flysch rocks and covered with mixed forests.

The sites with the highest values of the appropriateness index lie mostly on both slopes of the transversal valley in the southwestern part of the Ukrainian Carpathians, known as the Perechyn-Lypha natural region (Fig. 5). Of all the administrative regions of Ukraine, the Transcarpathian region has by far the largest area of the sites most appropriate for assigning a protected status.

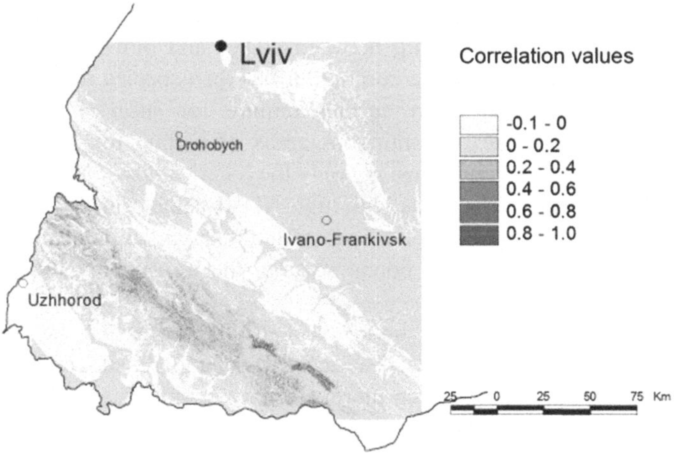

Fig. 4 Correlations between the amount of rare and endangered vascular plant species (Fig. 1) and the terrain variability (Fig. 3) in the Ukrainian Carpathians

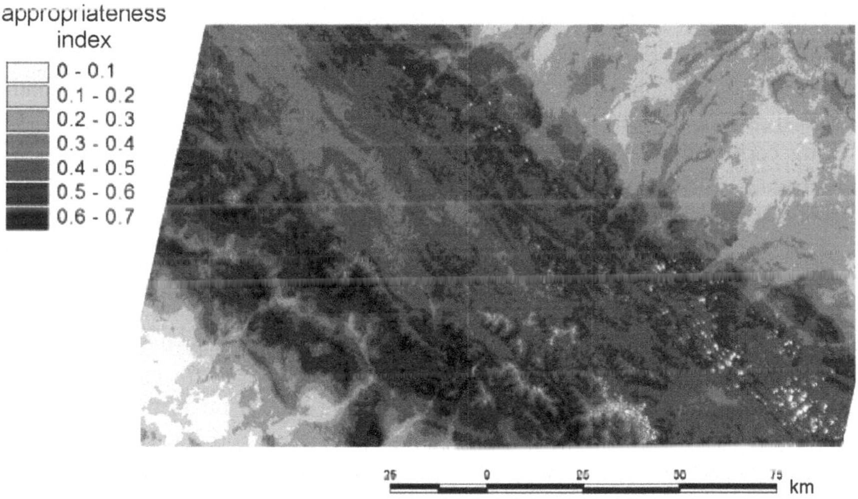

Fig. 5 The most appropriate sites for nature-conservation activities (shown in *black*) in the Ukrainian Carpathians

5 Discussion

There are a number of studies of relations between NDVI values and various ecosystem parameters and properties. Bawa et al. (2002) studying in the Western Ghats, India, found a positive correlation between mean NDVI and tree species diversity for different vegetation types, concluding that satellite imagery can identify broad patterns of tree species diversity in tropical forests. Oindo et al. (2000), examining the relationship between NDVI and bird species richness in Kenya, revealed a strong positive correlation between species richness and maximum average NDVI. As many species require for successive survival and reproduction the existence of continuous areas of certain minimal extent with favorable environmental conditions, not only the local values of NDVI but also the presence of sufficiently large areas with high NDVI values can be of importance. As Danell et al. (1996) showed in their research, the larger and more homogeneous the area of forest coverage in a boreal section, the more mammalian herbivore species coexist there.

To explain such findings, Skidmore et al. (2003) suggested the productivity hypothesis, predicting that when resources are abundant and reliable, species become more specialized, allowing more species per unit area.

NDVI can be a viable indicator for the assessment of the appropriateness for assigning protected status for the sites, due to the fact that much of the animals prefer undisturbed areas of natural vegetation with sufficient extent, low disturbance factors like noise, and sufficiently high primary ecosystem production, which all are pretty well correlated with NDVI.

NDVI value can also relate to the ability of an ecosystem to fix carbon, thus contributing to carbon's withdrawal from the atmosphere and alleviating its harmful influence on the global climate. Prince (1991), Running et al. (2004) showed that NDVI has been related to net primary production at broad spatial scales. Frank and Karn (2003) concluded that NDVI has the potential for predicting carbon flux in semiarid grasslands and shrublands. Glenn et al. (2008) compared NDVI with another image-derived indices in their ability to indicate carbon flux values in various ecosystems.

There are also a number of studies proving the relationships between the measures of terrain variability and the measures of biodiversity. By analyzing floristic data Hofer et al. (2008) discovered that the standard deviation of altitude explained a high proportion of the variation in γ diversity—the total species richness over an area (linear regression model, R2 = 0.63)—concluding that terrain variability at a landscape scale has strong effects on niche or microsite diversity and is an appropriate estimator of relative species richness in landscapes that are topographically heterogeneous and gradient dominated. Tang et al. (2006) used the elevation range (i.e., the difference between the highest and the lowest elevation) as a surrogate of habitat heterogeneity and found positive relationships between elevation range and the densities of plant genera and mammal species.

Habitat heterogeneity has long been shown to influence species diversity of birds (MacArthur 1964; Pianka 1966; Karr and Roth 1971). Richerson and Lum (1980) showed that in California topographic heterogeneity measured as a standard deviation of elevations in an area explains 19 % of the total flora diversity, ranking as the second highest contribution among all environment predictors after the rainfall values. Thus, they claimed that topographic heterogeneity has a strong effect on patterns of the total flora and most subdivisions. Owen (1989) found a significant relationship between variance in elevation and species richness for reptilian taxons.

These findings can be explained by the heterogeneity hypothesis stating that diverse ecosystems support richer assemblages of biological species compared with simple ecosystems (Podolsky 1994).

It should be noted that in case of future environmental changes and disturbances, like widely expected ubiquitous climate changes, areas with diverse abiotic environments could provide more shelter and refugia for species challenged by unfavorable conditions. For instance, in mountainous areas species facing a warmer and drier climate than that to which they have evolved could just shift their habitats several hundred meters upwards, while on plains they would have to migrate hundreds and thousands kilometers for a relatively short time span, which can be highly problematic.

Areas with a high level of terrain dissection are often at the same time harder for infrastructure development, construction activities, and less suitable for intensive agriculture. This makes the withdrawal of these areas from most economic activities for conservation purposes to be less burdensome for society.

The combining of these two criteria allowed for an assessment of the appropriateness of sites for assigning a conservation status, based on the values of NDVI and terrain variability integrated among several scales (see Fig. 5). This integral assessment can in turn be compared with the existing network of protected areas, to identify potential gaps in the latter. It can be inferred that, while most protected areas are situated in the northeastern part of the Ukrainian Carpathians, the ecosystems of its southwestern part (the Transcarpathian region) are greatly underrepresented in the existing network.

The proposed methodology, however, should be regarded as tentative, as it can be improved by elaborating the criteria of habitat diversity and richness, as well as the rules of their aggregation, to better meet the aims of nature conservation. More reliable estimates of the relations between habitat properties and species richness can be obtained by utilizing detailed field inventories of species, which is very labor- and time-consuming, yet an important task.

6 Conclusions

The assessment and mapping of the appropriateness for assigning protected status can be improved by making use of some easily mapped indicators of ecosystem properties defining their suitability for conservation. The terrain variability index calculated in a GIS from freely available DEMs were shown to correlate significantly with the diversity of rare and endangered vascular plant species. This correlation, however, differs spatially and with the scale of analysis. NDVI, calculated from multispectral spatial imagery, showed a lower correlation with rare plant species occurrences, yet it can also be indicative for conservation suitability assessment due to its connections with potential animal habitats, carbon-fixing capacity, and socioeconomic considerations involving the costs of conservation.

It should be noted that the beech forest belt shows significantly higher correlation values between rare plant species occurrences, topographical variability and NDVI indices, while in the spruce belt where most forests are secondary and are often in discord with the abiotic environment conditions, this correlation is nearly absent.

The output of this proposed methodology for the assessment and mapping of the appropriateness for assigning protected status can be used by conservation planners to help improve the existing protected areas network, inform the selection of new conservation areas for minimal cost, and facilitate the exploration of trade-offs between conservation and socioeconomic objectives.

References

Alpin P (2007) On scales and dynamics of observing the environment. Int J Remote Sens 27(11):2123–2140

Austin MP, Smith TM (1989) A new model for the continuum concept. Vegetatio 83:35–47

Bawa K, Rose J, Ganeshaiah KN, Barve N, Kiran MC, Umashaanker R (2002) Assessing biodiversity from space: an example from the Western Ghats. India Conserv Ecol 6(2):7

Blöschl G (1995) Scale issues in hydrological modelling: a review. Hydrol Process 9(3–4):251–290

Brusak V, Bezushko A, Voznyi Y, Zinko Y, Feldbaba-Klushyna L, Masikevych Y, Matveyev S, Movchan Y, Popovych S, Prychodko M (2006) Skhema ecomerezhi Ukrajinskyh Karpat (natsionalnyj riven'). Living Ukraine 9(10):8–9

Clark WC (1985) Scales of climate impacts. Clim Chang 7:5–27

Danell K, Lundberg P, Niemela P (1996) Species richness in mammalian herbivores: patterns in the boreal zone. Ecography 19(4):404–409

Deodatus FD, Protsenko L (eds) (2010) Creation of ecological corridors in Ukraine. A manual on stakeholder involvement and landscape-ecological modelling to connect protected areas, based on a pilot in the Carpathians. State Agency for Protected Areas of the Ministry of Environmental Protection of Ukraine, Altenburg and Wymenga Ecological Consultants, InterEcoCentre, Kyiv

Frank A, Karn J (2003) Vegetation indices, CO2 flux, and biomass for Northern Plains grasslands. J Range Manag 56:382–387

Glenn EP, Huete AR, Nagler PL, Nelson SG (2008) Relationship between remotely-sensed vegetation indices, canopy attributes and plant physiological processes: what vegetation indices can and cannot tell us about the landscape. Sens 8(4):2136–2160

Guisan A, Zimmermann NE (2000) Predictive habitat distribution models in ecology. Ecol Model 135:147–186

Herenchuk K (ed) (1968) Pryroda Ukrajinskyh Karpat. Lviv University, Lviv

Hofer G, Wagner H, Herzog F, Edwards PG (2008) Effects of topographic variability on the scaling of plant species richness in gradient dominated landscapes. Ecography 31(1):131–139

Jarvis A, Reuter HI, Nelson A, Guevara E (2008) Hole-filled seamless SRTM data V4. International centre for tropical agriculture (CIAT). http://srtm.csi.cgiar.org. Accessed 31 Oct 2010

Karr JR, Roth RR (1971) Vegetation structure and avian diversity in several new world areas. Am Nat 105:423–435

Levin SA (1992) The problem of pattern and scale in ecology. Ecol 73:1943–1983
MacArthur RH (1964) Environmental factors affecting bird species diversity. Am Nat 98:387–397
Margalef R (1968) Perspectives in ecological theory. Chicago University Press, Chicago
McGarigal K, Marks B (1994) Fragstats: spatial pattern analysis program for quantifying landscape structure. Reference manual, Corvallis
Ministry for environmental protection of Ukraine (1996) Ukrainian red book, 2nd edn. http://mail.menr.gov.ua/publ/redbook/redbook.php?lang=lat. Accessed 31 Oct 2010
Montgomery D, Foufoula-Georgiou E (1993) Channel network source representation for digital elevation models. Water Resour Res 29(12):3925–3934
NASA Landsat Program (2009) Landsat 7 ETM+ scene L71185026_02620070522, Geo-cover. USGS, Sioux Falls
Oindo BO, de By RA, Skidmore AK (2000) Interannual variability of NDVI and bird species diversity in Kenya. Int J Appl Earth Obs Geoinf 2(3–4):172–180
Owen JG (1989) Patterns of herpetofaumal species richness: relation to temperature, precipitation, and variance in elevation. J Biogeogr 16(2):141–150
Pianka ER (1966) Latitudinal gradients in species diversity: a review of concepts. Am Nat 100:33–46
Podolsky R (1994) Ecological hot spots. A method for estimating biodiversity directly from digital Earth imagery. Earth Obs Mag 30–36
Prince SD (1991) Satellite remote sensing of primary production. Int J Remote Sens 12:1301–1311
Purvis A, Hector A (2000) Getting the measure of biodiversity. Nature 405:212–219
Richerson PJ, Lum K (1980) Patterns of plant species diversity in California: relation to weather and topography. Am Nat 116(14):504–536
Rouse JW, Haas RH, Schell JA, Deering DW (1973) Monitoring vegetation systems in the Great Plains with ERTS. In: Third ERTS Symposium, 10–14 December 1973, Greenbelt, MD. NASA, Washington, DC. SP-351(I), pp 309–317
Running S, Nemani R, Heinsch F, Zhao M, Reeves M, Hashimoto H (2004) A continuous satellite-derived measure of global primary production. Biosci 54:547–560
Sheliag-Sosonko Y (1999) Rozbudova ecomerezhi Ukrajiny. Phytosociocenter, Kyiv
Skidmore AK, Oindo BO, Said MY (2003) Biodiversity assessment by remote sensing. In: Proceedings of the 30th international symposium on remote sensing of the environment: information for risk management and sustainable development, Nov 10–14, 2003, Honolulu, Hawaii. International Center for Remote Sensing of Environment, Tucson, AZ
Tang Z, Wang Z, Zheng C, Fang J (2006) Biodiversity in China's mountains. Front Ecol Environ 4(7):347 352
Wiens JA (1989) Spatial scaling in ecology. Funct Ecol 3(4):385–397

Recent Forest Cover Change in Low Mountain Landscapes of Lviv Oblast in the Ukrainian Carpathians

Anatoliy Smaliychuk and Ivan Kruhlov

Abstract This chapter analyzes forest cover change (FCC) with respect to the natural landscape structure and proximity to roads and settlements in two municipalities: Stara Sil' and Boberka in the Ukrainian Carpathians. A natural landscape is perceived as a mosaic of natural geoecosystems (GES), whose pattern is defined by landforms, which also reflect distribution of other hydro climatic and biotic components, including the potential natural vegetation (PNV). The forest cover of the 1980s was digitized from topographic maps, while its recent change was manually detected using high-resolution images from the Google Earth. The FCC data were overlaid with the natural GES data for further analysis. Proximity to roads and settlements was included into the analysis as an additional economic parameter. The important intermediate result of the study is a geodataset of natural GES that includes 35 types with attributes on topographic location, average altitude, mean slope, soil, and dominant PNV sub-formation. We distinguished five types of FCC. The total area with the FCC is estimated as 79.4 ha (2.5 %) for Stara Sil' and 715.9 ha (10.5 %) for Boberka. A common trend is the increase of the forested area due to forest succession on former agricultural land.

1 Introduction

Conservation and sustainable use of landscape and biotic diversity are stipulated by international (e.g. UNEP 2003) and national legislation (Verkhovna Rada 2009). Successful implementation of these decisions requires, among other

---

An erratum to this chapter is available at 10.1007/978-3-642-12725-0_50

A. Smaliychuk (✉) · I. Kruhlov
Faculty of Geography, Ivan Franko National University of Lviv,
Doroshenka St. 41 79000 Lviv, Ukraine
e-mail: zljukalviv@gmail.com

J. Kozak et al. (eds.), *The Carpathians: Integrating Nature and Society Towards Sustainability*, Environmental Science and Engineering,
DOI: 10.1007/978-3-642-12725-0_47,

preconditions, accurate and current geodata about the structure and recent dynamics of the landscapes.

Two past decades have marked significant socioeconomic transformations in the Carpathian countries, including Ukraine. Drastic changes have occurred particularly in land ownership, land use, and as a consequence, in land cover (e.g. Kuemmerle et al. 2006, 2007, 2008, 2009; Ressl et al. 2008). These dynamics are caused by political and economic factors, which have a nationwide, and even international, impact. However, it is assumed that the effects of these factors manifest themselves somewhat differently, depending on local natural and socioeconomic conditions. For instance, the Ukrainian Carpathians embrace low- and medium-mountain areas with different landform types, assemblages of altitudinal bioclimatic belts, and thus with different patterns of natural and cultural geoecosystems (Herenchuk 1968; Holubets et al. 1988; Kruhlov 2008; Kruhlov et al. 2008). Recent forest cover change (FCC) studies in the Ukrainian Carpathians, which are based on medium-resolution Landsat TM/ETM+ remote sensing data, reveal some peculiarities and trends in land cover dynamics in the northwestern part (Kuemmerle et al. 2006, 2007, 2008) and in the whole region (Kuemmerle et al. 2009).

The goal of this study is to examine how natural landscape and economic location influence FCC on the local level. We intended to use large-scale maps and fine-resolution forest cover data. To reach this goal, detailed FCC and natural geoecosystem (GES) maps have to be created and the FCC pattern has to be analyzed over the background of the natural GES mosaic. This was accomplished by using two model municipalities located in the flysch low mountains with significant portions of agricultural land.

The two municipalities, Stara Sil' and Boberka, belong to Lviv Oblast (Fig. 1). Stara Sil' has an area of 3,208 ha, population of 2,119 inhabitants (The State Statistics Committee of Ukraine 2002), and is located at the edge of the Carpathians, where the Marginal Beskydy low-mountain meso-ecoregion borders with the upland of Upper Dniester Fore-Carpathians (Kruhlov 2008; Kruhlov et al. 2008). The elevation varies from 304 to 651 m. The natural vegetation is represented by broadleaved forests with domination of pedunculate oak (*Quercus robus*) and European beech (*Fagus sylvatica*), often mixed with hornbeam (*Carpinus betulus*) and silver fir (*Abies alba*), on brownish podzolic pseudogleyic and brown mountain soils (Kruhlov et al. 2008). Boberka has an area of 6,819 ha, a population of 1,654 inhabitants (The State Statistics Committee of Ukraine 2002), and belongs to the Sian-Stryi Verkhovyna meso-ecoregion of elevated low mountains (Kruhlov 2008) with the elevation range from 536 to 826 m occupied by natural mixed beech–silver fir and beech–Norway spruce (*Picea abies*) forests on brown mountain soils (Kruhlov et al. 2008).

Both areas have a long history of agricultural development: low elevation and gentle slopes afforded vast expansion of grasslands and cultivated fields around the villages. Stara Sil' is proximate to the raion (district) administrative town Staryi Sambir at 7 km. It was a major regional centre of salt production until the end of the Nineteenth century. At the end of the World War II, a fairly large collective farm was created here with the share of arable land of about 50 %. It collapsed in

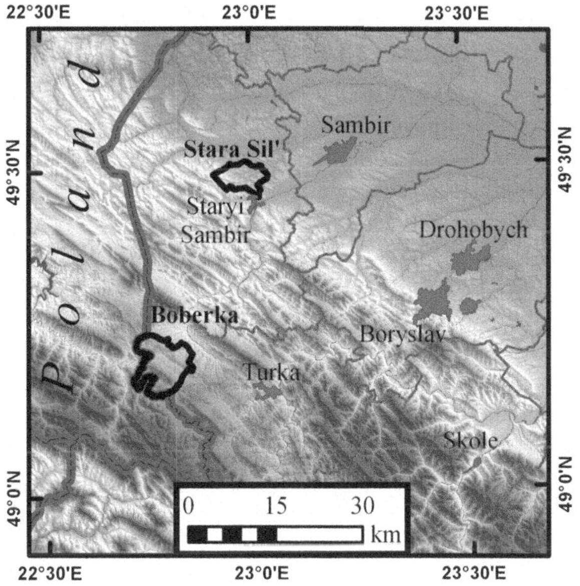

Fig. 1 Location of the study areas, Stara Sil' and Boberka, in the Lviv Oblast of the Ukranian Carpathians

2000, and agricultural land is now partly abandoned and partly used individually by local people as a household-scale natural economy (Anonymous 2008). The local population was partly resettled from here after the World War II. The area is known for the preserved ethnical architecture, clothes, and customs. The traditional economies were forestry and agriculture, and in some places the land is still traditionally cultivated on artificially terraced slopes. The municipality is included into the Nadsianskyi Regional Landscape Park, a part of the East Carpathian Biosphere Reserve, and is considered attractive for the development of ecotourism and traditional agriculture (Maryskevych and Niewiadomski 2005).

2 Materials and Methods

In this study, a natural landscape is interpreted as a natural morphogenic geoecosystem: a geospatial model of an area, in which landform elements (georelief) determine differentiation of local hydro climatic and natural (potential) biotic features (Kruhlov 2006), namely potential natural vegetation (PNV) (e.g. Tüxen 1956). Thus, the landforms determine the spatial pattern of the GES. The FCC pattern is analyzed over the background of the natural GES mosaic, thus providing insight about the importance of natural factors for forest cover dynamics. Proximity to roads and settlements was also chosen as an important economic factor of FCC (e.g. Kuemmerle et al. 2009). A geographic information system (GIS) was used as the main tool of the analysis. The general scheme of the data processing is presented in Fig. 2.

The materials used in the study included:

- topographic maps from the USSR period-1980s at a scale of 1:50,000 and 1950s at a scale of 1:25,000 (only for Stara Sil') (Anonymous 2006)
- satellite images (2005–2008) with spatial resolution at about 2.5 m available from Google Earth (earth.google.com),
- Shuttle Radar Topography Mission Digital Elevation Model (SRTM DEM), 3-arc-second (Jarvis et al. 2008),
- cadastral scheme at a scale of 1:50,000 with borders of the model communities (The State Committee of Ukraine on Land Resources 1993),
- literature on climate, soil, and vegetation of the study areas (Herenchuk 1968; Holubets et al. 1988; Kruhlov et al. 2008),
- field material on the interrelations of landforms, soil, and vegetation collected within 72 plots used as additional source for the description of the GES.

All analogue map data were scanned and transformed into a single coordinate system: UTM Zone 34 N. Boundaries of the municipalities were vectorized from rather imprecise cadastral schemes and corrected using high-resolution satellite

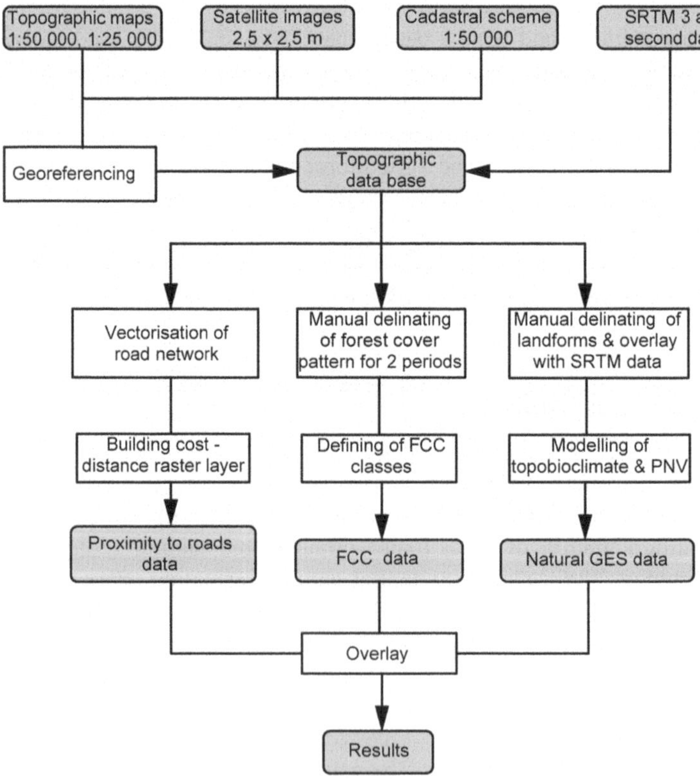

Fig. 2 Study workflow and data processing

images and large-scale topographic maps, as the boundaries mostly run along visible landscape features (roads, watersheds, streams, or forestry blocks).

The landforms were manually delineated from scanned 1:50,000 topographic maps and characterized by average slope, aspect, and mean elevation (representing altitudinal elevation belts) using statistical overlays with SRTM 3-arc-second data. Since large-scale vegetation and soil data are not available, the PNV and the soil were established as a function of morphometry, surface drainage and elevation within the landforms using information from the literature and the field observations. The forest cover pattern of the 1980s was digitized from the topographic maps, while its recent changes were manually detected using high-resolution images available in Google Earth to detect forest or non-forest (agriculture) transformations. Then, the FCC data were overlaid with the natural GES data for further analysis.

The road network and boundaries of settlements were vectorized from the satellite images. Proximity to roads (including paved, forested, and field roads) and settlements were included into the analysis as a proxy variable describing the economic accessibility, since it is assumed that the FCC may occur more intensively in more remote areas (Kuemmerle et al. 2009). We calculated also a cost-distance surface from the roads and settlement edges using slope gradients reclassified and assigned as cost values from 1 to 4.

3 Results

The important intermediate result of the study is a geodataset of natural GES that includes 35 types with attributes on topographic location, average altitude, mean slope gradients, type of bioclimatic altitudinal belts, soil, and dominant PNV subformation (according to Kruhlov et al. 2008; Table 1, Fig. 3).

We distinguished five types of FCC. Four of them can be aggregated in a group with secondary succession changes at the areas, which were formerly extensively used by agriculture. These types included areas where grassland was replaced by deciduous forest, mixed forest, coniferous forest, or shrubland. The fifth type of forest cover changes shows the areas where former forest or shrubland had been replaced by grassland. The last FCC type is the result of logging.

The total area with FCC for Stara Sil' is estimated at 79.4 ha (2.5 % of overall area), while for Boberka, 715.9 ha (10.5 %). In Stara Sil' replacement of grassland with broadleaves forest dominates. However, in Boberka, the occurrence of coniferous forest is more frequent (Table 2, Fig. 4).
Average elevations were calculated for each FCC type using the SRTM DEM. There were significant differences between the two study areas (Fig. 5): whereas in Boberka mean elevation of the whole municipality and the area with FCC were the same (644 m a.s.l.), in Stara Sil' these values were equal to 379 and 415 m a.s.l. respectively. Slope classes were also calculated from the SRTM DEM and overlaid with the FCC areas (Fig. 6).

Table 1 Natural GES types in Stara sil' and Boberka

GES types	Area, ha		Topographic location	Bioclimatic belt	Soil	PNV type
	Stara Sil'	Boberka				
A1a	152.2	–	Alluvial floodplains and terraces	Warm, moderately warm	Luvisol haplic	*Alneta glutinosae*
A1b	–	536.9		Moderately cool, cool	Fluvisol umbric and Leptosol umbric	*Alneta incanae*
A2a	110.1	–	V-shape alluvial valleys, gullies	Warm, moderately warm	Luvisol haplic	*Alneta glutinosae*
A2b	1.5	418.0		Moderately cool, cool	Fluvisol umbric and Leptosol umbric	*Alneta incanae*
A3a	29.1	–	Ravines	Warm, moderately warm	–	*Fraxineto-Querceta*
A3b	–	85.9		Moderately cool, cool	–	*Fraxineto-Fageta*
B1a	291.5	–	Watershed surfaces and gentle slopes (0 – 6°)	Warm	Albeluvisol stagnic	*Carpineto-Querceta*
B1b	233.7	–		Moderately warm		*Abieto-Querceta*
B1c	0.9	976.7		Moderately cool, cool	Cambisol dystric	*Fageto-Abieta*
B2a	139.7	–	Watershed surfaces (>6°)	Warm, moderately warm	Albeluvisol stagnic	*Fageto-Querceta*
B2b	14.5	585.7		Moderately cool, cool	Cambisol dystric	*Abieto-Fageta*
C1a	15.8	–	Flat surfaces (0 – 2°)	Warm, moderately warm	Albeluvisol stagnic	*Fraxineto-Querceta*
C1b	–	9.4		Moderately cool, cool	Cambisol dystric	*Fraxineto-Fageta*
C2a	134.2	–	Concave gentle slopes (3 – 5°)	Warm, moderately warm	Albeluvisol stagnic	*Fraxineto-Querceta*
C2b	–	290.8		Moderately cool, cool	Cambisol dystric	*Fraxineto-Fageta*

(continued)

Table 1 (continued)

GES types	Area, ha		Topographic location	Bioclimatic belt	Soil	PNV type
	Stara Sil'	Boberka				
C3a	170.0	–	Concave moderate slopes (6 – 12°)	Warm, moderately warm	Albeluvisol stagnic	*Fraxineto-Querceta*
C3b	5.0	533.2		Moderately cool, cool	Cambisol dystric	*Fraxineto-Fageta*
C4c	9.8	59.8	Concave moderate-to-steep slopes (13 – 20°)	Moderately cool, cool	Cambisol dystric	*Fageto-Abieta*
C5c	–	13.3	Concave steep slopes (>20°)	Moderately cool, cool	Cambisol dystric	*Fageto-Abieta*
D1a	77.5	–	Flat surfaces (0 – 2°)	Warm	Albeluvisol stagnic	*Carpineto-Querceta*
D1b	3.0	–		Moderately warm		*Abieto-Querceta*
D1c	–	39.2		Moderately cool, cool	Cambisol dystric	*Fageto-Abieta*
D2a	596.4	–	Straight and convex gentle slopes (3 – 5°)	Warm	Albeluvisol stagnic	*Carpineto-Querceta*
D2b	267.7	–		Moderately warm		*Abieto-Querceta*
D2c	–	920.0		Moderately cool, cool	Cambisol dystric	*Fageto-Abieta*
D3a	834.1	–	Straight and convex moderate slopes (6 – 12°)	Warm, moderately warm	Albeluvisol stagnic	*Fageto-Querceta*
D3b	71.6	1,989.8		Moderately cool, cool	Cambisol dystric	*Abieto-Fageta*
D4b	51.4	343.8	Straight and convex moderate-to-steep (13 – 20°)	Moderately cool, cool	Cambisol dystric	*Abieto-Fageta*
D5b	–	15.2	Straight and convex steep slopes (>20°)	Moderately cool, cool	Cambisol dystric	*Abieto-Fageta*

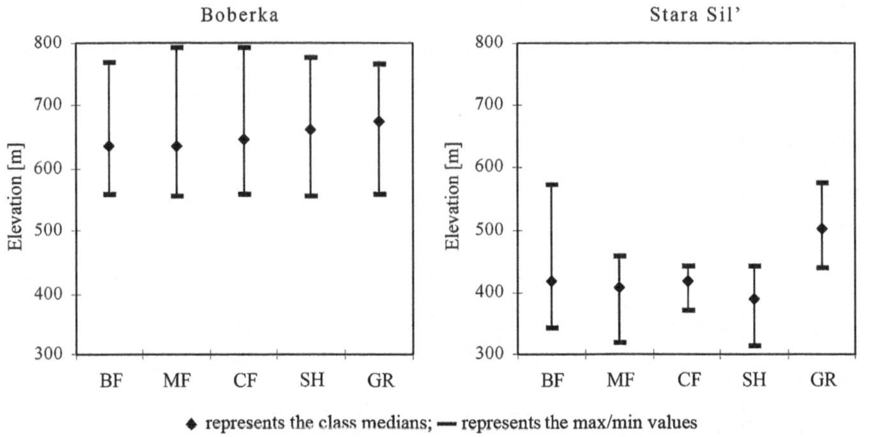

Fig. 3 Fragment of the 1:50,000 map of geoecosystems: **a** Boberka, **b** Stara Sil' (explanations to geoecosystems types in Table 1)

Table 2 Distribution of forest cover change types within study communities. Replacement of grassland with deciduous forest (*GR–DF*); mixed forest (*GR–MF*); coniferous forest (*GR–CF*); shrubland (*GR–SH*). Replacement of forest/shrubland with grassland (*F/Sh–GR*)

Types of forest cover change	GR–DF ha/ %	GR–MF ha/ %	GR–CF ha/ %	GR–SH ha/ %	F/Sh–GR ha/ %
Stara Sil'	32.4/40.8	19.1/24.0	19/23.9	6.1/7.7	2.8/3.5
Boberka	210.5/29.4	163.2/22.8	229.1/32.0	76.6/10.7	36.5/5.1

The FCC data were overlaid with the natural GES data to analyze the dependence of the change on the PNV types (Fig. 7). In Stara Sil', the majority of FCC could be found in *Fageto-Querceta* (over 50 %) and *Abieto-Querceta* (about 10 %) PNV types. However, in Boberka, FCC mainly occurred in the GES with *Abieto-Fageta* (over 40 %), *Alneta incanae* and *Fageto-Abieta* (about 10 % each) PNV types.

Proximity to roads and settlement edges was calculated and results were aggregated into seven classes for further interpretation. These data were also overlaid with the FCC data (Fig. 8). Proximity to roads and settlements is worse in Boberka due to its cul-de-sac location.

4 Discussion

Farmland abandonment and natural reforestation are the main traits of the land cover change in the Ukrainian Carpathians (Kuemmerle et al. 2007, 2008, 2009). Our study reveals the same situation in both of the municipalities. There are some differences in FCC distribution in the two municipalities, which is mainly caused by the variety of GES and PNV types, which is higher in Stara Sil'.

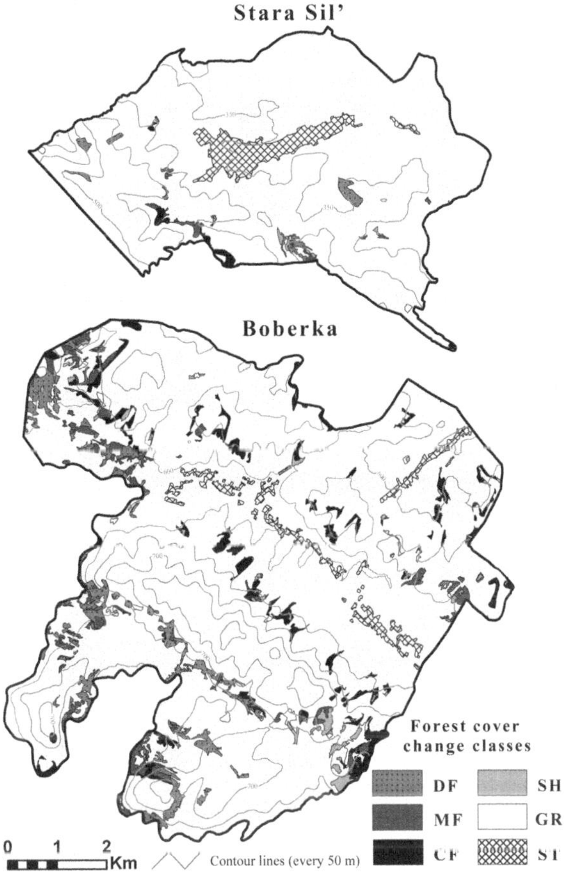

Fig. 4 Forest cover change. Replacement of former land cover types with deciduous forest (*DF*); mixed forest (*MF*); coniferous forest (*CF*); shrubland (*SH*); grassland (*GR*); settlements (*ST*)

In Stara Sil', which is located at the border of mountains and plains, FCC (predominantly Type 5: replacement of forest by grassland) occurred more on mountain ridges with higher elevations than on the plains with lower elevation, which corresponds well with the results of previous studies (Kuemmerle et al. 2007).

FCC more often occurs on moderate and steep slopes than on gentle ones, because gentle slopes (up to 6°) are predominantly occupied by nonforest land cover types. The moderate slopes (6–12°) are primarily impacted by FCC: forest succession on former grassland is most explicitly manifested here. This particular case differs from previous studies on FCC in the Ukrainian Carpathians, which revealed that FCC more often occurs on gentle slopes and valley bottoms that are predominately used in agriculture (Kuemmerle et al. 2008).

Our study also shows that FCC occurs with more or less the same intensity in all GES/PNV types. Absence of FCC in the GES with *Carpineto-Querceta* (Stara Sil') can be explained by a very insignificant portion of actual forest cover: these

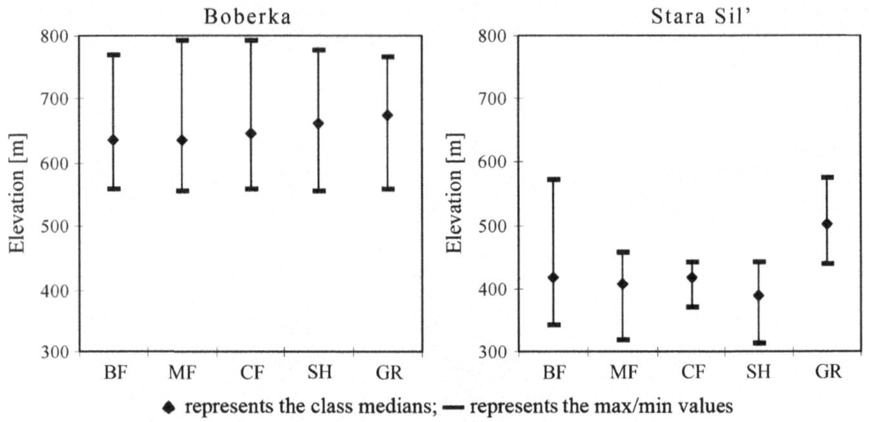

Fig. 5 Distribution of elevation for each forest cover change class. Replacement of former land cover types with deciduous forest (*DF*); mixed forest (*MF*); coniferous forest (*CF*); shrubland (*SH*); grassland (*GR*)

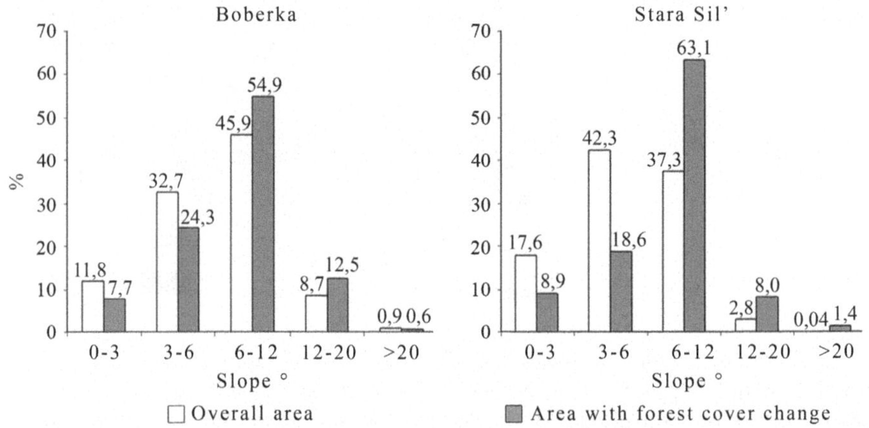

Fig. 6 Distribution of forest cover change areas according to slope classes

are predominantly gentle watershed surfaces occupied by permanent grassland and agricultural fields.

There is no significant difference in the proximity of the FCC patches to roads and settlements in the two municipalities, and this differs from the study results of Kuemmerle et al. (2009), which showed that FCC more often occurred on remote and nonaccessible places. FCC often occurs on the areas remote from settlements (but not from roads), at forest edges, or inside the forest. In Boberka, FCC patches are concentrated at the abandoned places, where villages existed before World War II: the shrub encroachment on grasslands and arable land had started here in the middle of the Twentieth century (e.g. Angelstam et al. 2003).

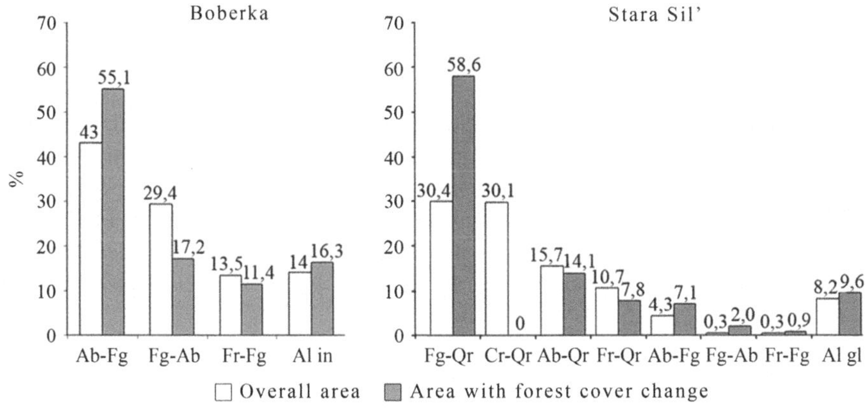

Fig. 7 Distribution of forest cover change areas according to PNV types. *Fg-Qr Fageto-Querceta*; *Cr-Qr Carpineto-Querceta*; *Ab-Qr Abieto-Querceta*; *Fr-Qr Fraxineto-Querceta*; *Ab-Fg Abieto-Fageta*; *Fg-Ab Fageto-Abieta*; *Fr-Fg Fraxineto-Fageta*; *Al gl Alneta glutinosae*; *Al in Alneta incanae*

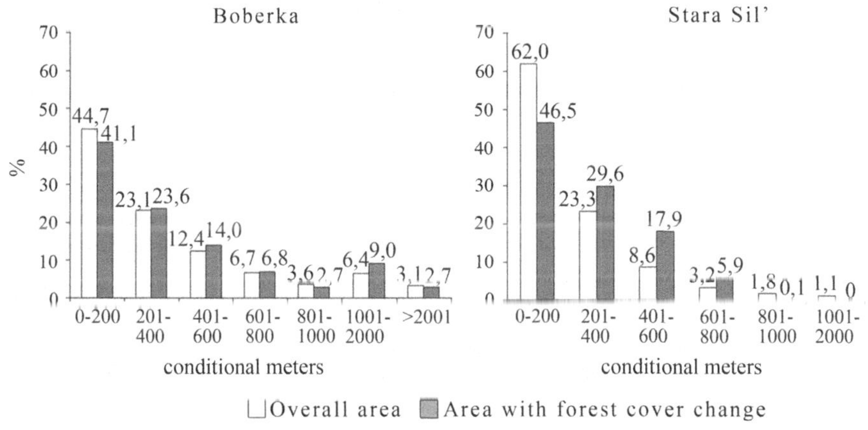

Fig. 8 Distribution of forest cover change according to proximity to roads and settlements

There is an increase of deciduous forest areas in Stara Sil', while coniferous forest expanded in Boberka. This can be partly explained by the differences in elevations and climatic conditions: the cooler climate in Boberka better supports regeneration of coniferous species. However, field checks showed that the dominant species of the regenerating patches is Norway spruce: this finding does not agree with the PNV types of the sites, and it has a distinct cultural origin. It could be explained by self-seeding of spruce from the adjacent areas of spruce plantations. Also, this can lead to the instability and future dieback of these spontaneous secondary spruce stands due to unsuitable site conditions.

5 Conclusions

The low mountain landscapes in Lviv Oblast experienced an increase of forested area because of the abandonment of agricultural lands. During the last decades the annual increase of forest areas was estimated at 0.14 and 0.35 % for Stara Sil' and Boberka, respectively. In most cases FCC correlates with the GES pattern and proximity to field and forest roads and settlement edges. In addition, FCC more often occurs on moderate slopes (in two municipalities) and higher elevation (in Stara Sil'). Higher reforestation rates in Boberka are most likely caused by the remote economic location of the municipality.

References

Angelstam P, Boresjö-Bronge L, Mikusiński G, Sporrong U, Wästfelt A (2003) Assessing village authenticity with satellite images: a method to identify intact cultural landscapes in Europe. Ambio 32(8):594–604

Anonymous (2006) Topographicheskie karty 1:50 000. http://maps.vlasenko.net/soviet-military-topographic-map/map50k. Accessed 15 November 2007

Anonymous (2008) Mista i sela Ukrainy. Lvivschyna. http://who-is-who.com.ua/bookmaket/mistalvov2008/7/19/1.html. Accessed 9 October 2008

Herenchuk KI (1968) Pryroda Ukrayinskykh Karpat. Vydavnytstvo Lvivskoho Universytetu, Lviv

Holubets MA, Honchar MT, Komendar VI, Kucheryavyi VA, Odynak YP (eds) (1988) Ukrainskiye Karpaty. Priroda, Naukova Dumka

Jarvis A, Reuter HI, Nelson A, Guevara E (2008) Hole-filled SRTM for the globe Version 4, available from the CGIAR-CSI SRTM 90 m Database. http://srtm.csi.cgiar.org. Accessed 4 March 2009

Kruhlov I (2006) Landshaft yak geoecosystema. Visnyk Lvivskoho Univ Ser Geogr 33:186–193

Kruhlov I (2008) Delimitatsiya, metryzatsiya ta klasyfikatsiya morfogennykh ekoregioniv Ukrayinskukh Karpat. Ukr Geogr Zhurnal 3:59–68

Kruhlov I, Mukha B, Senchyna B (2008) Natural geoecosystems. In: Roth M, Nobis R, Stetsuk V, Kruhlov I (eds) Transformation processes in the western Ukraine. Concepts for sustainable land use, Weissensee Verlag

Kuemmerle T, Chaskovskyy O, Knorn J, Radeloff VC, Kruhlov I, Keeton WS, Hostert P (2009) Forest covers change and illegal logging in the Ukrainian Carpathians in the transition period from 1988 to 2007. Remote Sens Environ 113:1194–1207

Kuemmerle T, Hostert P, Radeloff VC, Perzanowski K, Kruhlov I (2007) Post-socialist forest disturbance in the Carpathian border region of Poland, Slovakia, and Ukraine. Ecol Appl 17:1279–1295

Kuemmerle T, Hostert P, Radeloff VC, van der Linden S, Perzanowski K, Kruhlov I (2008) Cross-border comparison of post-socialist farmland abandonment in the Carpathians. Ecosyst 11:614–628

Kuemmerle T, Radeloff VC, Perzanowski K, Hostert P (2006) Cross-border comparison of land cover and landscape pattern in Eastern Europe using a hybrid classification technique. Remote Sens Environ 103:449–464

Maryskevych O, Niewiadomski Z (2005) Regionalnyi landshaftnyi park Nadsyanskyi. www.bieszczady.pl/download/pdf/broszura_NRPK_v.UA.pdf. Accessed 27 September 2008

Ressl R, Endlicher W, Lehmann S, Baruth B, Bosch B, Rosenthal G (2008) Current land use structure of upper Dniester basin and recent changes in the model communities. In: Roth M, Nobis R, Stetsuk V, Kruhlov I (eds) Transformation processes in the Western Ukraine. Concepts for sustainable land use, Weissensee Verlag

The State Committee of Ukraine on Land Resources (1993) Skhema zemlekorystuvannya Stararosambirs'kogo i Turkivskoho rayoniv masshtabu 1:50 000. Lvivskyi filial instytutu zemleustroyu, Lviv

The State Statistics Committee of Ukraine (2002) Demografichnyi shchorichnyk za 2001 rik. Kyiv

Tüxen R (1956) Die heutige potentielle natürliche vegetation als gegenstand der vegetations-kartierung. Angew Pflanzensoz 13:5–42

UNEP (2003) Framework Convention on the Protection and Sustainable Development of the Carpathians. http://www.carpathianconvention.org/index.htm. Accessed 19 April 2009

Verkhovna Rada (2009) Zakon Ukrayny pro landshafty (proekt). http://gska2.rada.gov.ua/pls/zweb_n/webproc4_1?id=&pf3511=36258. Accessed 27 December 2009

Connectivity of Protected Areas in Central European Border Regions

Sylvi Bianchin and Marco Neubert

Abstract Within the scope of the project TransEcoNet (Transnational Ecological Networks in Central Europe), the transnational ecologically important areas in border regions within Central and Eastern Europe were analyzed. In this study, an ecological network is to be understood as the existing network of protected areas. The area under investigation stretches from the Baltic Sea to the Ukraine and to the Adriatic Sea, covering the NUTS 3 (Nomenclature of Territorial Units for Statistics, level 3) regions adjoining selected inner Central European national boundaries. The analyses are based on spatial data and are performed in a Geographical Information System. After collecting and harmonizing national as well as European data sets of protected areas, all areas were classified according to the international standard provided by the categories of International Union for Conservation of Nature and reclassified as core and non-core areas according to their protection status and size. The connectivity of the network was estimated using the results of a nearest neighbor analysis. In general, the results showed a good level of coverage by nature protection sites as well as a good connectivity between them within the area under investigation. Thus, other border areas besides the Green Belt, the former Iron Curtain, also form important ecological networks and provide important ecological functions and services for society.

S. Bianchin · M. Neubert (✉)
Leibniz Institute of Ecological Urban and Regional Development,
Weberplatz 1 01217 Dresden, Germany
e-mail: m.neubert@ioer.de

S. Bianchin
e-mail: sylvi.bianchin@tu-dresden.de

J. Kozak et al. (eds.), *The Carpathians: Integrating Nature and Society Towards Sustainability*, Environmental Science and Engineering,
DOI: 10.1007/978-3-642-12725-0_48,

1 Introduction

Protected areas such as national parks, nature parks and biosphere reserves are often isolated "islands" for protecting the world's biodiversity. They are separated by weakly protected and unprotected landscapes, traffic corridors as well as settlements. It is often the case that animal and plant species dispose of less space for migration, dispersion and reproduction than necessary. To preserve both natural and cultural heritage in the long run, the TransEcoNet project is thus striving for a better connection of protected landscapes with those that are weakly protected and unprotected across national borders.

The protected area network is part of the ecological network, which represents a major contribution to protecting, maintaining and enhancing biodiversity (Jones-Walters 2007). Ecological networks are a coherent system of natural and/or semi-natural landscape elements that are configured and managed with the objective of maintaining or, as the case may be, of restoring ecological functions as a means of conserving biodiversity. They also provide appropriate opportunities for the sustainable use of natural resources (Bennett 2004). Ecological networks are, therefore, a tool for supporting the maintenance, restoration and reestablishment of functional ecological relations between different elements of a landscape (Finck and Riecken 2001; Riecken et al. 2004). Actions taken to maintain or to restore these interactions include: conserving a representative array of habitats, allowing species populations access to a sufficient surface area, allowing seasonal migration, permitting genetic exchange between different local populations, allowing local populations to move away from degrading habitats and securing the integrity of vital environmental processes (Bruszik at al. 2006). By focusing on the ecological interactions across landscapes, ecological networks explicitly include relations between semi-natural and natural areas and cultivated areas and, therefore, they identify appropriate opportunities within the landscape matrix for the sustainable use of natural resources: agriculture, forestry, fishing, human settlements, recreation, etc. (Finck and Riecken 2001; Bennett 2004; Riecken et al. 2004).

The European Community Directive 92/43/EEC (the Habitats Directive) regulates the development of the European ecological network Natura 2000. Article 10 of this framework states that the ecological coherence of the Natura 2000 network should be improved and that the management of features of the landscape, which are of major importance for wild fauna and flora, should be encouraged. One suggested assessment criterion for selecting sites eligible for identification as sites of community importance and designation as special areas of conservation (Annex III of the Habitats Directive, stage 2) should be the geographical situation of the site in relation to migration routes of species and whether it belongs to a continuous ecosystem situated on both sides of one or more internal community frontiers. This criterion clearly shows the importance of the transboundary coherence of ecological networks.

Furthermore, the Convention on Biological Diversity (CBD) recognizes the importance of linkages and outlined actions in the Programme of Work on

Protected Areas (PoWPA). One of the goals is that by 2015, all protected areas and protected area systems are integrated into the wider land- and seascape, by applying the ecosystem approach and taking into account connectivity as well as the concept of ecological networks. Unfortunately cross-border cooperation in this field is not yet well developed. Most plans are being or have been developed either at the regional or sub-national level (Mackovčin 2000; Jongman et al. 2004; Fuchs et al. 2010; Leibenath et al. 2010).

The first objective of our study was to develop a common and harmonized transboundary database of the existing protected areas in Central and Eastern Europe (programme area of the Central Europe Territorial Cooperation Programme). The second objective was to analyze the transboundary connectivity of the protected areas across borders using nearest neighbor analysis.

2 Study Area

The project's study area was situated within or rather between the wide-ranging ecological networks of the Alps, the Carpathians and the Green Belt. The investigation area (Fig. 1) consists of NUTS 3 regions that adjoin the borders between Germany, the Czech Republic, Poland, Slovakia, Ukraine, Austria, Hungary, Slovenia, Italy and Croatia. The eastern part of the study area overlaps the Carpathian Ecoregion (17.6 %).

3 Methodology

Since the database serves as the basis for our analysis, an attempt was made to collect the national databases on protected areas from the relevant countries. If they were not available, we used the database from the European Union. In the case of the Ukraine, the World Database on Protected Areas (WDPA) was considered (Table 1).

During the data preparation process, we combined the national, European and WDPA datasets and harmonized them along the borders using geographical information systems (GIS) methods mainly due to projection problems.

3.1 Inventory of Protected Areas and its Core Areas

The inventory of protected areas provides the base data for further analyses within the project. For the preparation of this database for the entire investigation area, the categories of the International Union for Conservation of Nature (IUCN) according to Dudley (2008) were assigned to every protected area (Table 1).

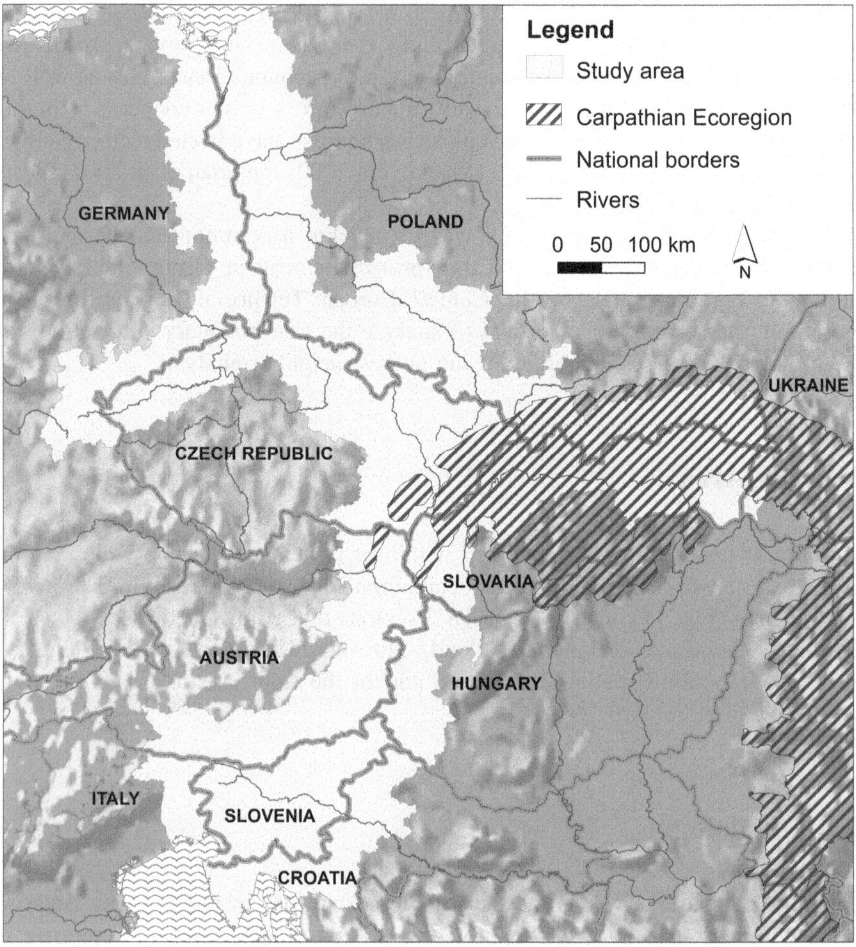

Fig. 1 Location of the study area and the Carpathian Ecoregion

Natura 2000 sites are currently not included in the IUCN categories. From a nature conservation point of view, they were valued between IUCN category IV (habitat/species management area) and V (protected landscape/seascape). That means that they have a higher value than protected landscapes but a lower one than nature reserves. If a Natura 2000 area is protected fully or partly by an IUCN status of IV or better at the same time these higher existing categories have been used instead.

For areas with a very low protection status, for example, the landscape protected areas in Poland and the areas of ecological importance in Slovenia, we created a new project-related category and did not assign the IUCN category. Unfortunately, country and even regional specific conservation issues pertaining to certain protected areas could not be taken into account during the labeling

Table 1 Overview of the databases and sources used (data status 2009) (changed according to Neubert et al. 2010)

Country	IUCN I (a, b)	IUCN II	IUCN III	IUCN IV	Natura 2000	IUCN V	Other national protection status
Austria	–	National park[b]	Natural monument[b]	Nature conservation area[b]	SPA/SCI[b]	Landscape conservation area/nature park/protected part of landscape[b]	Additional areas[b]
Croatia	Strict reserve[a]	National park[a]	Natural monument[a]	Special reserve/natural monument[a]	–	Protected landscape/nature park[a]	Regional park/horticultural monument/park forest[a]
Czech Republic	Strict reserve/wilderness area[a]	National park[a]	Natural monument[a]	–	SPA/SCI[a]	Landscape protected area[a]	Additional areas[a]
Germany	–	National park[a]	–	Biosphere reserve/nature protection area[a]	SPA/SCI[a]	Nature park/landscape protected area[a]	–
Hungary	–	National park[a]	–	Nature protection area[a]	SPA/SCI[a]	Nature park[a]	–
Italy	Strict nature reserve[b]	National park[b]	–	Nature conservation area[b]	SPA/SCI[b]	–	–

(continued)

Table 1 (continued)

Country	IUCN I (a, b)	IUCN II	IUCN III	IUCN IV	Natura 2000	IUCN V	Other national protection status
Poland	–	National park[a]	–	Nature reserves[a]	SPA/SCI[b]	Landscape parks[a]	Protected landscape[a]
Ukraine	–	–	–	Nature conservation area[c]	–	–	–
Slovakia	Wilderness area[b]	National park[b]	Natural monument[b]	Habitat/ species management area[b]	SPA/SCI[b]	Protected landscape/ seascape[b]	Additional areas[b]
Slovenia	Wilderness area[b]	National park[b]	Natural monument[b]	Habitat/species management area[b]	SPA/SCI[b]	Protected landscape/ seascape[b]	Ecological important areas[a]

[a] National dataset
[b] EU dataset
[c] WDPA dataset
SPA special protected area
SCI sites of community importance

procedure because of the size of the investigation area and the amount of data being analyzed.

The definition of core and non-core areas which we analyzed in our study was based on the inventory of protected areas and their classification where core areas are defined as:

- IUCN categories I (Ia and Ib) and II, these areas are already defined as core areas by the IUCN,
- IUCN categories III, IV and Natura 2000 sites with an area larger than 10 km^2.

All the remaining areas are defined as non-core areas. Areas with IUCN categories III, IV and Natura 2000 sites which are less than 10 km^2 were considered as being too small for our analysis. The IUCN category V and areas of national protection status were considered as being too weak in terms of their protection status to be considered as core areas.

3.2 Connectivity Assessment

Based on the information from the database and the definition of the core areas we estimated the connectivity of protected areas which is defined as a measure of the ability of organisms to move among separated patches of suitable habitat (Hilty et al. 2006).

We analyzed the connectivity of the protected areas using the common nearest neighbor method performed with the software vLATE (a free ArcGIS extension). This method was used because it is relatively simple and provides an overview of how the protected areas are connected among each other for a large area. The nearest neighbor analysis comes from landscape metrics and measures the distance from the edge of each protected area to the edge of the nearest protected area. The land use between the protected areas is not considered (Lang and Tiede 2003). From the information of the edge to edge distance of the protected areas the connectivity can be derived. A small distance to the nearest neighbor means the protected area is "well connected" and a large distance means it is "poorly connected".

Other methods such as a cost surface analysis (e.g. Nikolakaki 2004; Gurrutxaga et al. 2010) or an analysis based on graph theory (e.g. Goetz et al. 2009; Galpern et al. 2011) would have required a large amount of detailed data and would have been too complex for the large study area.

In general our analysis was performed in two steps:

1. assessment of the protected areas according to the core area definition and identification of the type of nearest neighbor (core area or non-core area) using the nearest neighbor index. As an additional result, stepping stones (between 500 and 5,000 m in terms of distance) and satellites (above 5,000 m in terms of

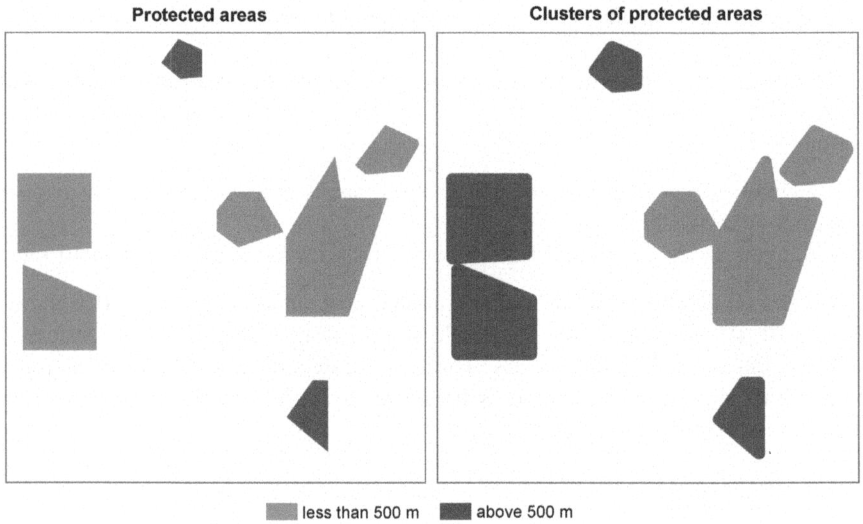

Fig. 2 Differences between the two nearest neighbor analyses: step 1 (*left*) and step 2 (*right*)

distance) were classified according to the calculated distance between the protected areas (Fig. 2).

2. analysis of the protected areas with a buffer (250 m) to cluster the nearest neighbor which are within a 500 m distance using ArcGIS and, subsequently, an additional vLATE analysis of these clusters.

With the nearest neighbor index we also obtained information about the relationship between two objects (here between two protected areas) (Fig. 2, left). Therefore, a buffer (250 m) was used to include more than only one neighbor within a 500 m distance. The results were clustered protected areas which are less than 500 m apart from each other. After creating these clusters, a second nearest neighbor analysis using vLATE was performed to identify the distances between these clusters (Fig. 2, right). The distances were selected according to Planungsbüro für angewandten Naturschutz GmbH (2006) and Bastian and Schreiber (1999).

Step 1 allowed to identify single isolated protected areas and step 2 to detect isolated groups of protected areas which are less than 500 m apart from each other (clusters). The advantage of step 2 was that the significance of objects having only a single neighbor (as is the case with the two on the left side in the figures) but which are distant to others can be minimized.

3.3 The Transboundary Connectivity Assessment

Finally, we evaluated the transboundary connectivity based on the spatial analysis of the protection status defined by the IUCN categories on either side of national borders. The following four different match categories were distinguished:

1. corresponding protection status (same protection status on both sides of the border),
2. differing protection status (different protection status on both sides of the border),
3. unilateral protection status (protection only on one side of the border),
4. no protection status (on either side of the border).

4 Results

4.1 Protected Areas and its Core Areas

The statistical analysis of the protected areas and their protection status according to IUCN categories shows that nearly 65 % of the study area has no protection status. The Natura 2000 sites cover 17.3 % of the whole study area, whereas only 2.8 % are either strict reserves, wilderness areas or national parks (IUCN category I and II). In the part of the study area which covers the Carpathian Ecoregion only nearly 50 % of the whole area is protected. The highest protection category (IUCN category I and II) covers only 8 % of the area (Fig. 3).

The results of the core area analysis are illustrated in Fig. 4. Nearly 60 % of the protected area network in the study area is defined as core area. In the Carpathian Ecoregion, over 67 % of the protected areas are defined as core areas with a very high percentage of areas with a high protection status (IUCN category I and II).

4.2 Connectivity Assessment Using Nearest Neighbor Analysis

The nearest neighbor analysis showed that most protected areas in the study area are well connected with each other (Fig. 5).

Nearly 95 % of the protected areas in the study area and 96 % of the Carpathian Ecoregion are less than 500 m away from other protected areas. Under 0.5 % of the areas in both regions are satellites with a distance above 5,000 m to the next protected area.

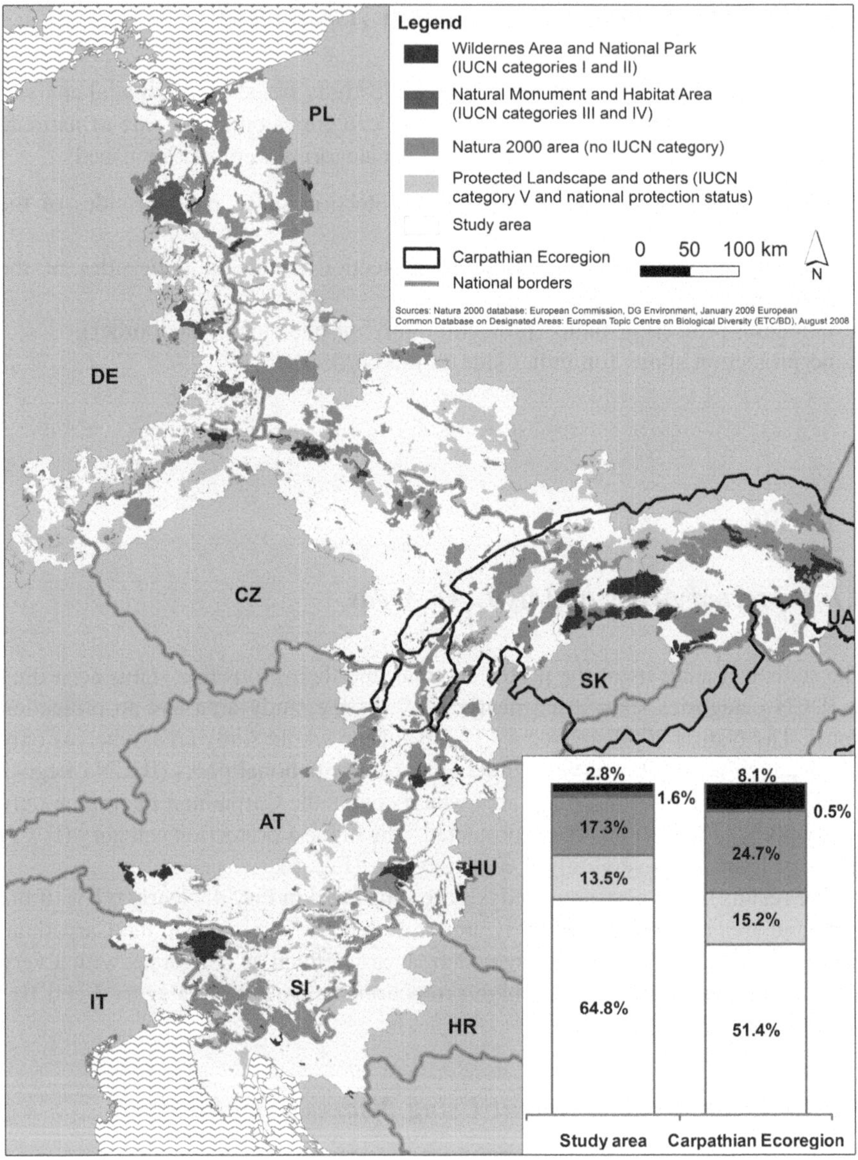

Fig. 3 The inventory of the protected areas in the study area and the overlapping part of the Carpathian Ecoregion

The results from the nearest neighbor analysis step 1 (classification of the nearest neighbor by type) reveal that core areas rarely have other core areas as a nearest neighbor (Fig. 6). For the most part, core areas are located closely to non-

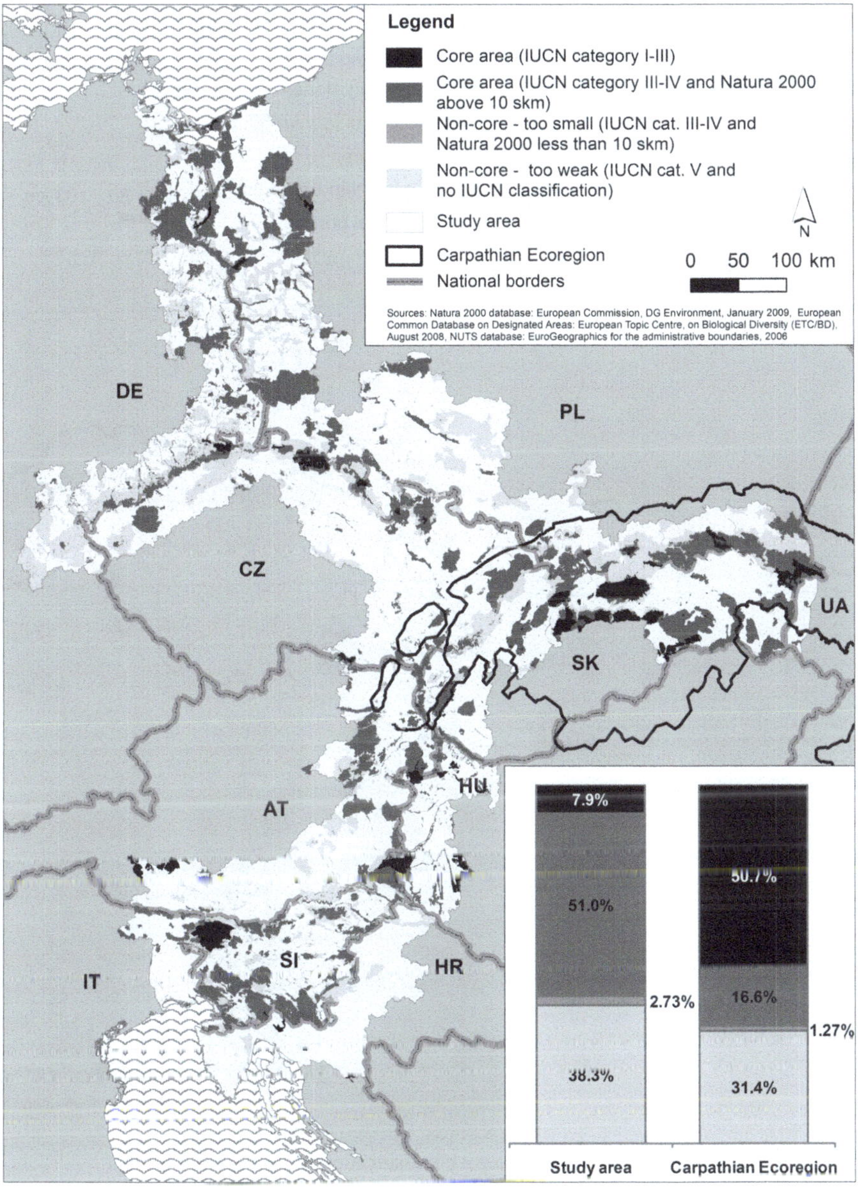

Fig. 4 Core areas in the study area and the overlapping part of the Carpathian Ecoregion

core areas. Around 5 % of the existing protected areas have no other protected area in the vicinity of 500 m.

The results of the statistical analysis of the nearest neighbor by distance and protection status (Fig. 7) show that the highest IUCN categories (I, Ia and Ib) are

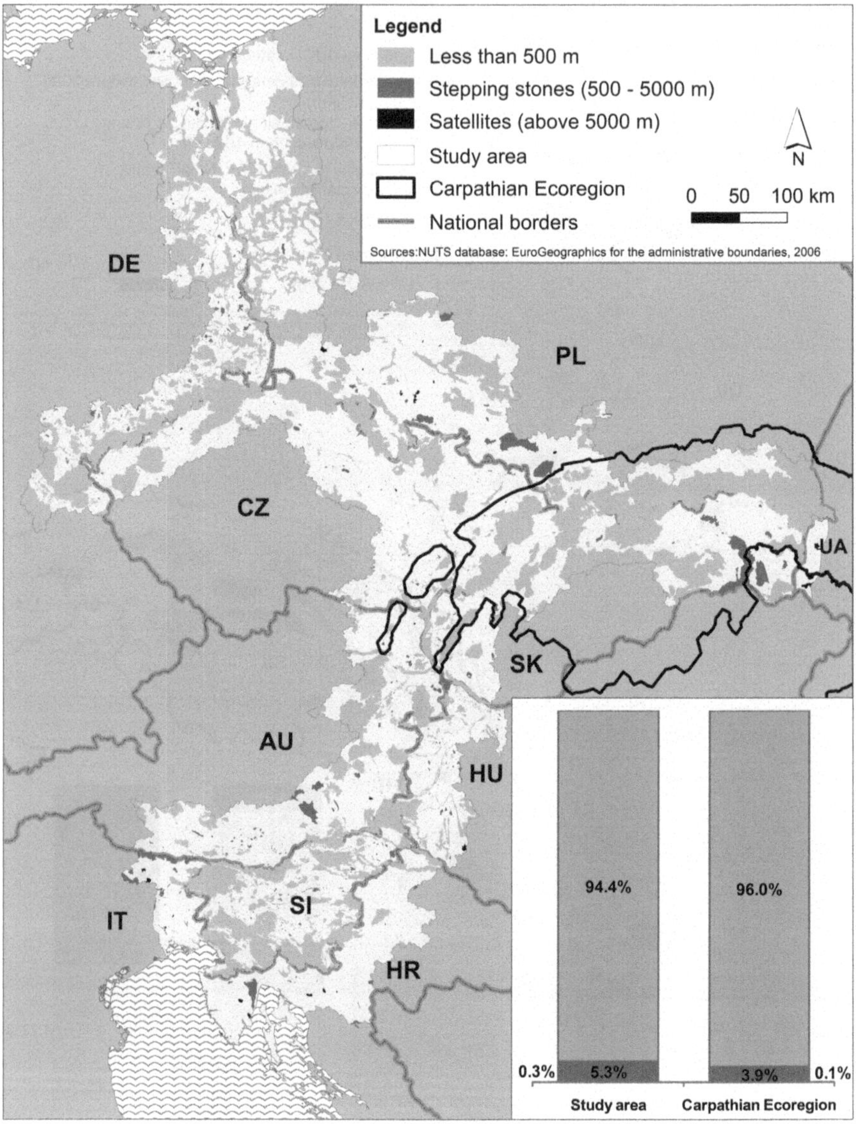

Fig. 5 The nearest neighbor analysis (step 1): classification of the nearest neighbor by distance for the study area and the overlapping part of the Carpathian Ecoregion

not as well connected as protected areas with the IUCN category II. Therefore, it can be concluded that the protected areas with an IUCN category I, Ia and Ib are more or less small and isolated. The Natura 2000 areas should be well connected (Article 10 of the European Community Directive 92/43/EEC), which currently

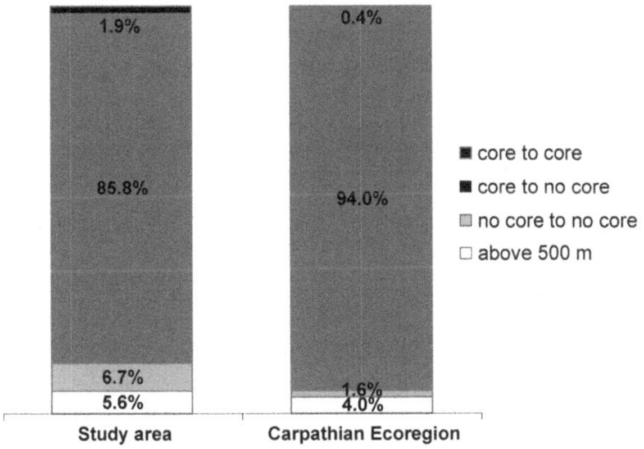

Fig. 6 The nearest neighbor by type for the study area and the overlapping part of the Carpathian Ecoregion

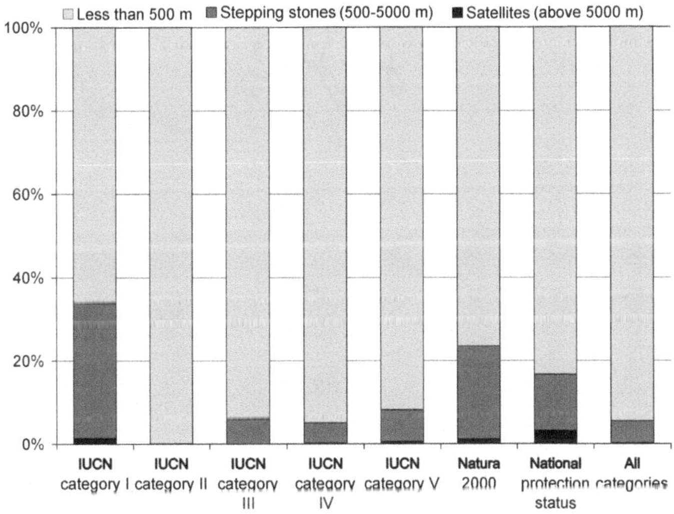

Fig. 7 The nearest neighbor by distance and protection status for the study area

has not yet been achieved. Only 5 % of the existing protected areas in the study area have no other protected area in the vicinity of 500 m.

The results of step 2 (Fig. 8) showed that especially in southwest Poland, the southern part of Austria and the eastern part of Slovenia well connected protected areas according to step 1 (less than 500 m) were reclassified as stepping stones after the buffering operation (500–5,000 m).

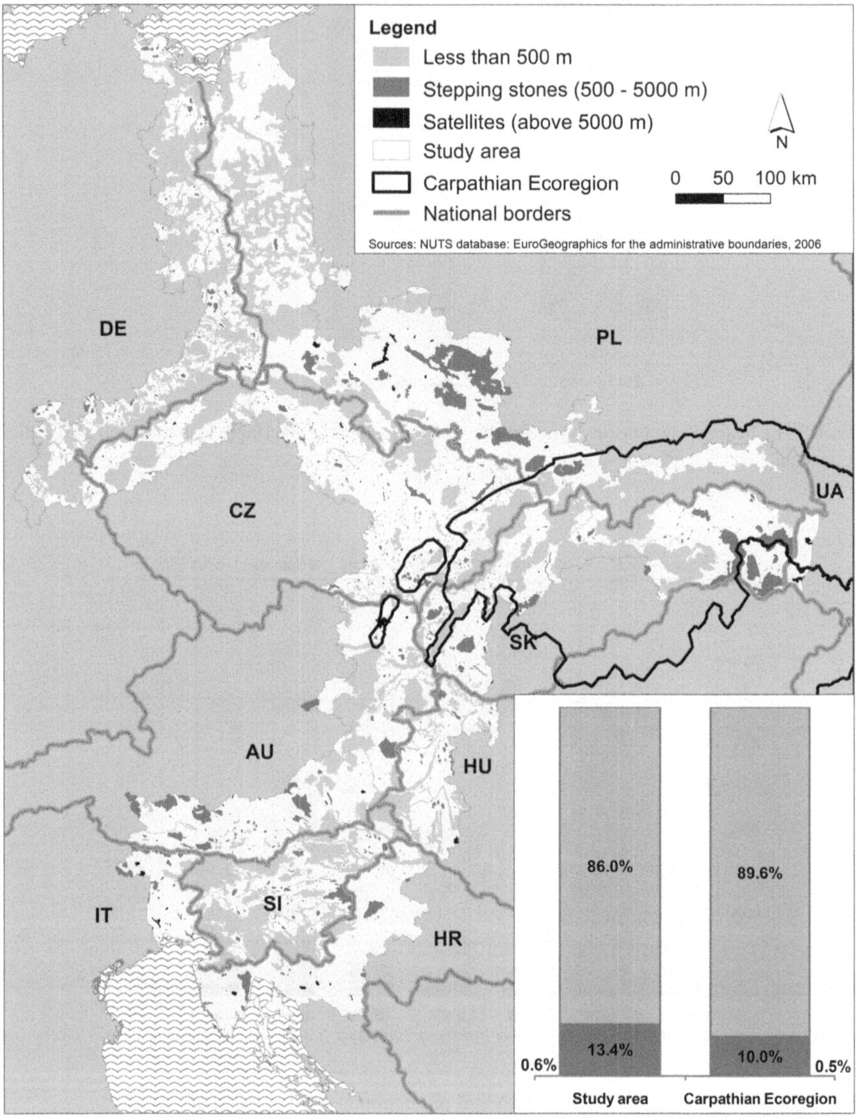

Fig. 8 The nearest neighbor analysis (step 2): classification of the nearest neighbor by distance for the study area and the overlapping part of the Carpathian Ecoregion

4.3 Transboundary Connectivity

The outputs of the analysis of the transboundary connectivity give an idea of the situation of the connectivity of protected area network across borders (Fig. 9). Overall 16 % of the protected areas in the study area and 23 % in the part of the

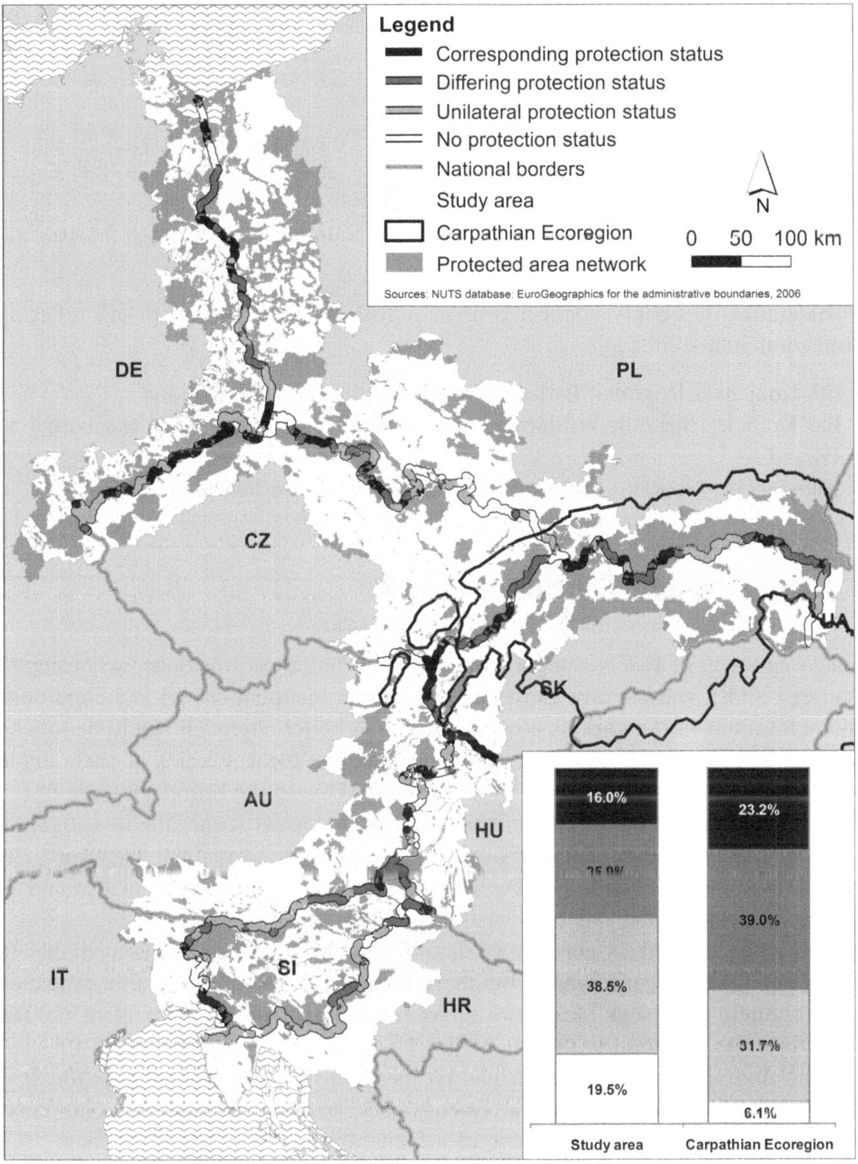

Fig. 9 The transboundary connectivity of the protected areas across borders for the study area and the overlapping part of the Carpathian Ecoregion

Carpathian Ecoregion are transnational, they have the same protection status on both sides of the border. Good examples of corresponding protection status across borders are:

- the Saxon and Bohemian Switzerland National Park between Germany and the Czech Republic,
- the Karkonosze/Krkonoše National Park between Poland and the Czech Republic,
- the Tatra Mountains National Park between Poland and Slovakia,
- the Natura 2000 site of the Morava-Dyje floodplain between the Czech Republic, Slovakia and Austria,
- the Neusiedler See-Seewinkel Fertö-Hanság National Park between Austria and Hungary.

Examples of weakly connected areas across borders (unilateral and differing protection status) are:

- the Kozjanski Regional Park in Slovenia at the border to Croatia,
- the Králický Sněžník Wilderness area in the Czech Republic at the border to Poland,
- Babiogórski National Park in Poland at the border to Slovakia.

5 Discussion and Conclusions

The assignment of IUCN categories to the protected areas was done, according to Dudley (2008), without any knowledge of specific national or regional conditions and regulations of the various protected areas. A better approach would be to take a closer look at the specific management plans and local policies of each single protected area and categorize them accordingly. Due to the European scale of the analysis and the corresponding number of protected areas within the investigation area (19,256), this approach was practically impossible to realize. Therefore, the results represent an initial overview of the situation in the investigation area and an additional detailed analysis on a smaller scale should be conducted.

The areas defined as core areas should represent the best areas available to ensure biodiversity conservation but there is little evidence that existing protected areas actually represent ideal core areas for biodiversity (Boitani et al. 2007). Therefore, the assumption that the defined core areas are the best areas for supporting biodiversity conservation has not been proven in terms of functionality. Within another part of the TransEcoNet project, selected core and non-core areas were further analyzed in the course of a landscape functionality assessment.

The step 1 of the nearest neighbor analysis only takes into consideration the nearest protected area. However, two protected areas can be classified as being well connected (below 250 m), even if they are only connected to each other and all other protected areas are far away. With the second step, an attempt was made to overcome this problem with the help of a previously used clustering procedure. However, the problem was not solved satisfactorily. It needs to be further analyzed and perhaps another approach including a buffer procedure will obtain better results.

Besides the above mentioned weaknesses of this approach, one of its strengths is the development of a harmonized transboundary database which was used as the basis for further research (Neubert et al. 2010, 2011). Furthermore, the results of the analysis offer an overview of the connectivity of the protected area network across borders in Central European countries on a large scale. The cross border cooperation and data exchange between the ten countries involved is one of the achievements of the project. This provides the basis for further analyses and the development of common plans and strategies for a better connection between protected, weakly protected and unprotected landscapes across national borders.

The CBD programme of work on protected areas emphasizes the importance of establishing protected areas in a mosaic of land and water habitats to facilitate maintenance of ecological processes. Specific activities of this programme refer to "linking habitats", such as buffer zones around protected areas, biological corridors and ecological stepping stones. Another aim is to establish and strengthen regional networks, transboundary protected areas and collaboration between neighboring protected areas across national boundaries (Bennett and Mulongoy 2006).

Protected areas play a crucial role in maintaining the ecological integrity of ecological networks, because they often form all or a large part of the core areas. In some regions protected area networks may themselves function as ecological networks (Bennett and Mulongoy 2006). Thus, protected areas form the backbone of the ecological network. The presented results of the inventory of the protected area network illustrate the current state with a focus on transboundary connectivity.

In general, our analyses indicate the weaknesses of the current situation as well as the potentials for further development of the network of protected areas. To face the still existing challenges concerning the implementation of joint transnational strategies like CBD or Natura 2000, it needs to be strengthened and supported by cross-border cooperation.

Acknowledgments We would like to thank all project partners of the TransEcoNet project involved in the analysis for their contributions and useful comments: A. Hahn (University of Dresden), Ch. Renetzeder (University of Vienna), W. Lazowski (Austrian League for Nature Conservation Burgenland), G. Király (University of West Hungary), B. Barboric (Geodetic Institute of Slovenia), H. Skokanova (The Silva Tarouca Research Institute for Landscape and Ornament Gardening), D. Ribeiro (University of Nova Gorica).

This project is implemented through the CENTRAL EUROPE Program co-financed by the ERDF (project no. 1CE061P3).

References

Bastian O, Schreiber KF (1999) Analyse und ökologische Bewertung der Landschaft. 2. überarbeitete Auflage. Spektrum Akademischer Verlag GmbH, Heidelberg

Bennett G (2004) Integrating biodiversity conservation and sustainable use: lessons learned from ecological networks. IUCN, Gland

Bennett G, Mulongoy KJ (2006) Review of Experience with Ecological Networks, Corridors and Buffer Zone. Secretariat of the convention on biological diversity. Technical Series No. 23, Montreal

Boitani L, Falcucci A, Maiorano L, Rondinini C (2007) Ecological networks as conceptual frameworks or operational tools in conservation. Conserv Biol 21(6):1414–1422

Bruszik A, Rientjes S, Delbaere B, van Uden G, Richard D, Terry A, Bonin M (2006) Assessment of the state of affairs concerning the Pan-European ecological network (Final draft—31 Aug 2006)

Dudley N (2008) Guidelines for applying protected area management categories. IUCN, Gland

Finck P, Riecken U (2001) Nationaler Biotopverbund aus bundessicht. Flächenpool-Lösungen—ein Fortschritt für den Vollzug der Eingriffsregelung? Tagungsband zur Oppenheimer Arbeitstagung, 2. Teil: Planung vernetzter Biotopsysteme—Umsetzung und Konsequenzen. Tagungsband Oppenheim

Fuchs D, Hänel K, Lipski A, Reich M, Finck P, Riecken, U (2010) Länderübergreifender Biotopverbund in Deutschland Grundlagen und Fachkonzept. Naturschutz und biologische Vielfalt Heft 96, Bundesamt für Naturschutz

Galpern P, Manseau M, Fall A (2011) Patch-based graphs of landscape connectivity: a guide to construction, analysis and application for conservation. Biol Conserv 144:44–55

Goetz SJ, Jantz P, Jantz CA (2009) Connectivity of core habitat in the Northeastern United States: parks and protected areas in a landscape context. Remote Sens Environ 113:1421–1429

Gurrutxaga M, Lozano PJ, del Barrio G (2010) GIS-based approach for incorporating the connectivity of ecological networks into regional planning. J Nat Conserv 18:318–326

Hilty JA, Lidicker Jr WZ, Merenlender AM (2006) Corridor ecology: the science and practice of linking landscapes for biodiversity conservation. Island Press, Washington

Jones-Walters L (2007) Pan-European ecological network. J Nat Conserv 15:262–264

Jongman RHG, Külvik M, Kristiansen I (2004) European ecological networks and greenways. Landsc Urban Plan 68:305–319

Lang S, Tiede D (2003) V-LATE extension für ArcGIS—vektorbasiertes tool zur quantitativen landschaftsstrukturanalyse. ESRI Anwenderkonferenz. CDROM, Innsbruck

Leibenath M, Blum A, Stutzriemer S (2010) Transboundary cooperation in establishing ecological networks: the case of Germany's external borders. Landsc Urban Plan 94:84–93

Mackovčin P (2000) A multi-level ecological network in the Czech Republic: implementating the territorial system of ecological stability. GeoJ 51(3):211–220

Neubert M, Bianchin S, Syrbe R, Stöcker C (2010) Report—inventory of ecological networks. http://www.central2013.eu/fileadmin/user_upload/Downloads/outputlib/inventory_networks.pdf. Accessed 20 Aug 2012

Neubert M, Bianchin S, Syrbe R, Stöcker C (2011) Report—gaps in ecological networks. http://www.central2013.eu/fileadmin/user_upload/Downloads/outputlib/gaps_in_eclogical_networks.pdf. Accessed 20 Aug 2012

Nikolakaki P (2004) A GIS site-selection process for habitat creation: estimating connectivity of habitat patches. Landscape Urban Plann 68:77–94

Planungsbüro für angewandten Naturschutz GmbH (2006) Übersicht zur Abschätzung von maximalen Entfernungen zwischen Biotoptypen für Tierpopulationen in Bayern. Stand: Dezember 2006. http://www.pan-gmbh.com/content/dload/TabEntfernungen.pdf. Accessed 20 Aug 2012

Riecken U, Ullrich K, Finck P (2004) Biotopverbund. In: Konold W, Böcker R., Hampicke U (eds) Handbuch Naturschutz und Landschaftspflege: Kompendium zu Schutz und Entwicklung von Lebensräumen und Landschaften, 13. Erg. Lfg. 9/04. ecomed, Landsberg

Creation of Ecological Corridors in the Ukrainian Carpathians

Floris Deodatus, Ivan Kruhlov, Leonid Protsenko, Andriy-Taras Bashta, Vitaliy Korzhyk, Stefan Tatuh, Mykola Bilokon, Mykhaylo Shkitak, Iaroslav Movchan, Sebastian Catanoiu, Razvan Deju and Kajetan Perzanowski

F. Deodatus (✉)
Altenburg & Wymenga Ecological Consultants, Box 32 9269ZR Feanwâlden, The Netherlands
e-mail: fdeodatus@gmail.com

I. Kruhlov
Faculty of Geography, Ivan Franko National University, vul. Doroshenka 41, Lviv 79000, Ukraine

L. Protsenko
InterEcoCentre, r.65, 31/33 Kudriavskaya str, Kyiv 04053, Ukraine

A.-T. Bashta
Institute of the Ecology of the Carpathians, National Academy of Sciences of Ukraine, vul. Kozelnytska 4, Lviv 79026, Ukraine

V. Korzhyk
Vyzhnytsky NNP, 61, Golovna str, v, Beregomet, Vyzhnytskyi rayon, Chernivetska Oblast 59233, Ukraine

S. Tatuh · M. Shkitak
Administration for Environment and Natural Resources in Lviv Oblast, vul. Stryiska 98, Lviv 79026, Ukraine

M. Bilokon
State Department of Environment Protection in Chernivetskiy Oblast, 35, Mayakovskogo str, Chernivtsi 58003, Ukraine

I. Movchan
National Aviation University, 1, Kosmonavta Komarova prospect, Kiev 03680, Ukraine

S. Catanoiu · R. Deju
Vanatori Neamt Natural Park, Zimbrului no.2 617500 Agaipa, Romania

K. Perzanowski
Carpathian Wildlife Research Station, MIZ PAS Ogrodowa 10 38-700 Ustrzyki Dolne, Poland

K. Perzanowski
Catholic University of Lublin, Konstantynow 1H 20-708 Lublin, Poland

J. Kozak et al. (eds.), *The Carpathians: Integrating Nature and Society Towards Sustainability*, Environmental Science and Engineering,
DOI: 10.1007/978-3-642-12725-0_49, © Springer-Verlag Berlin Heidelberg 2013

Abstract In order to develop a methodology for the creation of functional and consolidated ecological corridors for the Carpathians, a pilot study has been conducted at two locations in Ukraine creating corridors connecting Ukrainian protected areas with protected areas in Romania and Poland. The methodology was based on landscape ecological modelling, using the habitat requirements of brown bear, European bison, lynx and wildcat to locate the most suitable corridor areas. Manageable corridors were created by identifying interconnected land management units with a minimum of obstacles for wildlife and conflicts with land use, and forming the shortest possible connection. The location of the corridors and their management plans were developed in consultation with the users and owners of the land. Approval and inclusion of the corridors in the spatial planning system was achieved following a model elaborated after analysis of the Ukrainian institutional and regulatory framework related to ecological network development.

1 Introduction

Several biological theories such as the theory of island biogeography and the meta-population theory deal with the limitations in space and time of animal populations (MacArthur and Wilson 1967; Hanski 1998, 2005). These theories support the generally accepted opinion that the survival of large mammals depends on large land areas with appropriate habitat providing food and other functions during all seasons for a population big enough to maintain sufficient genetic variability to cope with environmental changes, disease and inbreeding.

The Carpathians are one of the largest continoues natural areas in Europe with a high biological diversity represented by the last large European populations of large mammals such as brown bear (*Ursus arctos*), lynx (*Lynx lynx*), and wolf (*Canis lupus*). For Europe, the Carpathians do not only form an important reservoir of these species, but they also play a role connecting wildlife areas in Eastern, Western and Southern Europe. During the last century significant changes occurred in the Carpathians regarding land use and land cover. The most striking trends in the Carpathian landscape since the 1990s are:

- privatisation and fragmentation of land,
- farmland abandonment,
- encroachment of farmland and pastures by forests,
- developing road infrastructure and urbanisation,
- unsustainable development of tourism and recreational facilities.

As a result, the Carpathians tend to turn into a fragmented landscape of isolated forest blocks with little possibilities for animals to move from one to another. To cope with current and fragmentation and future habitat loss, most Carpathian countries have established a framework for the development of an ecological network including legislation, spatial planning and policy targets (Jongman and

Kamphorst 2002; Bennett and Mulongoy 2006; van Maanen et al. 2006; Simeonova et al. 2009). Generally this has led to the consolidation of protected area systems established mainly in marginal areas with a low human population density. However, since complexes of protected areas are usually separated by zones with high human influence, such as agriculture, settlements and infrastructure, the connectivity between these protected areas has hardly been improved. In many cases fragmentation continues due to expanding traffic infrastructure, tourism facilities, settlements and other development.

The objective of our study is to develop a methodology for the creation of ecological corridors connecting (protected) core areas within the Carpathian ecological network, considering landscape ecology, land use, ownership and legislation. The specific purpose of this work is to help closing the gap between the possibilities offered by advanced scientific modelling techniques available today for corridor mapping (van Maanen et al. 2006; Beier et al. 2007, 2008) and the realities and requirements related to the creation and consolidation of ecological corridors in the actual land management system (Bennett and Mulongoy 2006; Lombard et al. 2010; Mackey and Watson 2010). The study is carried out as a pilot study in the Ukrainian Carpathians, to allow realistic investigation and testing of procedures required to deal with legislation, stakeholders and other aspects of the land use system. The general methodology developed is meant to be applicable as a model for corridor development in the Carpathians.

In most Carpathian countries, frameworks for ecological network development have been realized following initially the Landscape Stabilisation Approach and since 1995 the Pan-European Biological and Landscape Diversity Strategy (Bennett and Mulongoy 2006). Ecological networks have been designed in all countries, based on Natura 2000 in EU countries. Most of these frameworks include provisions for the development of connections between core areas for conservation. In most countries specific legislation addressing connectivity has been adopted, but this has not been effective everywhere (Jongman and Kamphorst 2002; Bennett and Mulongoy 2006; Jędrzejewski et al. 2009). Corridor development is most advanced in Czech Republic, Slovakia and Hungary, where ecological corridors connecting protected areas have been created and are functional. In the other countries corridors have only been established on pilot basis, but in many cases, corridors remain features on maps waiting for implementation. This can be explained by the lack of experience of those responsible for the development of ecological networks in dealing with land use issues outside protected areas, and the lack of understanding of how to deal with administrative mechanisms to match land use with conservation (Nassauer and Opdam 2008; van der Windt and Swart 2008).

The planning of ecological corridors usually involves GIS (Geographical Information System) modeling techniques (Marulli and Mallarch 2005; Hepcan et al. 2009; Gurrutxaga et al. 2010; Roy et al. 2010), which are predominantly focused on the identification of the ecological linkages in a landscape through a least-cost approach, relying on the suitability assessment of landscapes for selected species based on the parameters provided by species experts (Schadt et al. 2002;

Beier et al. 2008). Modeling results can be validated and optimized using the reliability theory (Jordan 2000), uncertainty analysis (Schadt et al. 2002; Beier et al. 2009), network flows (Phillips et al. 2008), graph theory (Urban et al. 2009; Rayfield et al. 2010), or Markov chains (Billionnet 2010). Special software has been designed in several occasions to assist the GIS modeling of the ecological corridors (e.g. Majka et al. 2007).

2 Study Area

The Ukrainian part of the Carpathian Ecoregion covers approximately 21,000 km^2 (Kruhlov 2008), which is about 10 % of the whole ecoregion area. The Ukrainian Carpathians are crucial for the ecological connectivity as they occupy the narrow part of the mountain arc linking its massive northern and southern sections, thus forming a bottleneck for animal migration within the Carpathians, as well as between the Southern, Western, and Eastern Europe (Webster et al. 2001). This part of the Carpathians consists of a series of parallel low and medium mountain ridges (usually up to 1,500 m a.s.l.) stretching in the NW–SE direction and predominantly formed by flysch. They are mainly covered with beech and spruce forests; however, several elevated ridge tops (higher than 1,500 m) have subalpine and alpine vegetation (Herenchuk 1968). Human settlements are rather densely dispersed and they are represented mainly by medium and large villages (1–3 thousands of inhabitants) located in river valleys. The economy of the region is mainly determined by forestry, recreation, and nature conservation (Kubijovyc 1984; Burdusel et al. 2006; Anon. 2007). Forest fragmentation is limited, but forests are disturbed by clear-cuts in widely-spread cultural spruce stands (Kuemmerle et al. 2006). The area of these clear-cuts is gradually increasing as well as abandoned agricultural land is gradually changing into forest (Kuemmerle et al. 2008). Forests are state-owned, while agricultural land (mostly grassland) is predominantly private.

Terrestrial mammals occurring in the Ukrainian Carpathians include rare species such as brown bear (*Ursus arctos*), lynx (*Lynx lynx*), wildcat (*Felis sylvestris*), European mink (*Mustela lutreola*), otter (*Lutra lutra*) and European bison (*Bison bonasus*). There are also populations of red deer (*Cervus elaphus*), roe deer (*Capreolus capreolus*), wolf (*Canis lupus*), fox (*Vulpes vulpes*), pine marten (*Martes martes*), ermine (*Mustela erminea*), Carpathian squirrel (*Sciurus vulgaris carpathicus*) and common dormouse (*Muscardinus avellanarius*). Species of amphibians and reptiles include Carpathian newt (*Lissotriton montadoni*), Alpine newt (*Mesotriton alpestris*), yellow-bellied toad (*Bombina variegata*), spotted salamander (*Salamandra salamandra*), Aesculapian snake (*Zamenis longissimus*), and smooth snake (*Coronella austriaca*). During the last 20 years populations of large mammals decreased due to human influence such as poaching and other disturbances. Particularly, numbers of red deer and roe deer have dropped drastically, and elk has almost disappeared from the Ukrainian Carpathians (Domnich

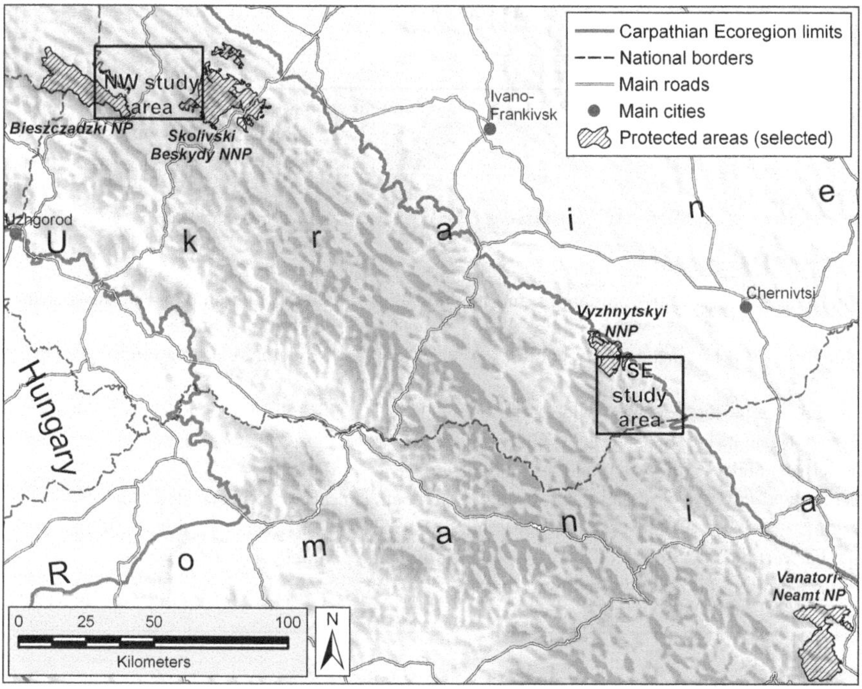

Fig. 1 The Ukrainian Carpathians showing the pilot areas and the related protected areas

et al. 2009). Downward trends of carnivores are also observed but accurate data are lacking (Nowell and Jackson 1996; Servheen et al. 1998; Anon. 2004; Bashta and Potish 2005, 2007).

The pilot study has been carried out at two locations in the north-west and the south-east of the Ukrainian Carpathians (Fig. 1), establishing local level ecological corridors between selected protected areas in Ukraine, Poland and Romania. The location in the north-west is the area between Skolivski Beskydy National Nature Park in Ukraine and Bieszczadzki National Park in Poland, while in the south-east the area between Vyzhnytskyi National Nature Park and the border with Romania is covered in order to create the Ukrainian part of a corridor towards Vanatori-Neamt Natural Park. Characteristics of the landscape of the two locations are presented in Table 1.

3 Methods

In this study, ecological networks and particularly ecological corridor areas are considered as natural, socio-economic and legal entities, since they exist in the landscape inhabited, transformed and managed by humans. Corridor modelling has

Table 1 Landscape features of the pilot study areas (Kuemmerle et al. 2006, 2007, 2008, 2009; Hostert et al. 2008; Kruhlov 2008)

Landscape feature	North-west area	South-east area
Meso-ecoregions	Sian-Stryi Verkhovyna and Internal Beskydy	Bukovyna internal mountains and Pokuttia-Bukovyna external mountains
Rocks	Flysch	Flysch
Elevation average and range (m a.s.l.)	750 (580–1,100)	830 (560–1,200)
Landform	Low mountains and medium mountains	Medium mountains and dissected low mountains
Climate	Moderately cool	Moderately cool and moderately warm
Natural vegetation	Beech–spruce forests	Beech–spruce and beech forests
Human population	~12,000 (17 villages)	~2,000 (2 villages)
Dominant land use	Forestry and agriculture	Forestry
Actual land cover	Grassland, forest and settlement	Forest and grassland and settlement
Disturbances	Forest clear-cutting, hunting	Forest clear-cutting, hunting
Land cover change trend	Forest encroachment on grassland, moderate expansion of built-up areas	Stable

been based on the habitat requirements of selected species, referred to as model species hereafter[1], which are considered to represent the overall habitat requirements of all species to be facilitated by the corridor. In the modelling process, available corridor area is identified for each of these model species in a stepwise process, subsequently eliminating land of low connectivity due to physical barriers, habitat suitability, human disturbance, future development and unwillingness of land users and land owners to contribute to the corridor protection (Fig. 2). Hence, manageable corridors are obtained by completing the following steps:

1. identification and delineation of possible corridors, using habitat suitability criteria determined for selected model species,
2. tailoring the corridor boundaries to the extent of existing administrative, ownership, and management units in order to create manageable areas to realise and maintain connectivity,
3. determining and implementing required protection measures and arrangements with regard to legislation and management.

Following earlier corridor modelling practice in other areas (e.g. van Maanen et al. 2006; Beier et al. 2007, 2008) four model species were identified: brown bear, European bison (further referred to as bison), lynx, and wildcat. These species have rather rigid and distinctive habitat requirements, together covering

[1] The terms "umbrella species" and "focal species" are sometimes also used in this context (van Maanen et al. 2006; Beier et al. 2008).

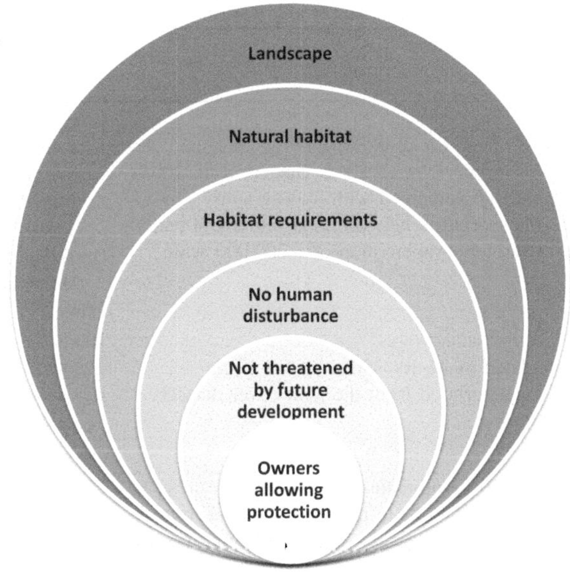

Fig. 2 Stepwise approach for corridor modeling, subsequently eliminating land of low connectivity due to physical barriers, habitat suitability, human disturbance, future development and unwillingness of land users and land owners

the requirements of all large terrestrial mammal species of the Carpathians and therefore they can be regarded as "umbrella" for those species. For each of the model species an ecological profile has been prepared describing habitat requirements based on expert knowledge as well as publications on habitat utilization (Slobodian 1988, 1993; Turianyn 1988; Nowel and Jackson 1996; Servheem et al. 1998; Anon. 2004; Bashta 2004; Pucek et al. 2004; Bashta and Potish 2005; Ray et al. 2005; van Maanen et al. 2006; Krasinska and Krasinski 2007; Klar et al. 2008; Kuijper et al. 2009). Habitat suitability and resistance values for landscape features were represented as a separate raster geo-dataset in a GIS. The habitat suitability values were established by experts for each model species and assigned to the respective geo-datasets using a standardized scale from 1 to 100 with 0 as a restrictive value (Table 2).

The raster datasets on habitat suitability determined by the species experts were additively overlaid to establish one overall habitat suitability data layer for each of the model species. Assuming that the focal species have different tolerance to human presence, the additive overlay was adjusted for this feature as follows. Bison was considered as the most tolerant to human disturbance and therefore its human proximity factor received the weight of 1.0, for lynx this factor was set to 3.0, while for the bear and the wildcat it was estimated at 2.0. The additive overlay was adjusted with the weight factor of 1.0 for all other the landscape features.

Subsequently, possible corridors were manually drafted based on habitat suitability maps for each species, and these "species corridors" were merged into a single "robust" corridor (van Maanen et al. 2006; Nassauer and Opdam 2008). The course of these preliminary corridors has been evaluated by the species experts to identify so-called "bottleneck areas" which are relatively narrow

Table 2 Landscape features and their suitability values for model species

Landscape feature	Category	Suitability values (1–100)			
		Bear	Bison	Lynx	Wildcat
Land cover types derived from Landsat TM/ETM + images (Kuemmerle et al. 2006; Hostert et al. 2008) and supplemented with hydrography, transportation network, and settlement pattern from topographic map of 1:200,000 scale	Coniferous forest	100	70	100	50
	Deciduous and mixed forest	100	100	100	100
	Grassland and shrub land	20	50	10	20
	Other	0	0	0	0
Forest/grassland ratio calculated in % for 250 m radius circle neighborhood from the land cover dataset (above)	100/0 %	100	75	100	75
	75/25 %	75	100	50	100
	50/50 %	50	75	10	75
	25/75 %	25	50	0	25
	0/100 %	0	10	0	0
Altitudinal bioclimatic belts (elevation intervals in m a.s.l.) stratified from the SRTM 3-arc-second digital elevation model (Jarvis et al. 2008)	0–350	50	100	50	100
	350–700	80	100	80	100
	700–1,100	100	50	100	50
	1,100–1,300	100	30	100	30
	1,300–1,500	80	10	80	10
	1,500–1,800	30	0	30	0
	>1,800	10	0	10	0
Terrain roughness (m) calculated as an elevation magnitude within a 250 m radius circle from the SRTM data	0–50	50	100	50	100
	50–100	100	50	100	80
	100–200	100	30	100	50
	>200	100	10	100	30
Human proximity is calculated using a cost-distance function, where distance is estimated to settlements and roads, and cost is defined by the terrain's slope. The proximity values are standardized to a 1–100 scale					

corridor parts crossing agricultural and settled areas. Field checks of these "bottleneck areas" have been made to determine the least problematic passages for the animals. As a result, the best corridor option for all four model species has been determined based on habitat suitability and the most passable bottleneck areas.

In order to function and to be able to cope with land use changes and development projects in the future, ecological corridors need to meet the following conditions (Jongman and Kamphorst 2002; Bennett and Mulongoy 2006; Mackey and Watson 2010):

- be composed of manageable units,
- of which boundaries follow administrative, natural or landownership boundaries,
- accepted and respected by all stakeholders,
- recognized and respected by spatial planning authorities,
- recognized and approved by local and higher administration,

- managed according to agreed and effective management arrangements.

To comply with these criteria, the "raw" corridors as they were modelled above are fitted in the existing land use system by identifying adjacent manageable land units (e.g. forest blocks) within the area covered by the corridor area based on habitat suitability and the best bottlenecks, forming a chain between the protected areas to be connected. The selection of these areas to determine the final corridor boundaries is achieved through direct consultations with land owners and land users with the aim to agree on a final course of the corridor using boundaries of municipalities and forest management units, taken from administrative and forestry maps. Areas where no agreement on compatible land use could be reached or where future incompatible developments were unavoidable could be excluded from the corridor following the consultations.

The Ukrainian laws supporting the development of the Ukrainian ecological network are the Laws "On the State Programme of Ukraine's National Ecological Network Development for 2000–2015" (2000) (Government of Ukraine 2000) and "On the Ecological Network of Ukraine" (2004) (Government of Ukraine 2004). These laws define the different elements of the ecological network and ensure the Government's support on the development of the ecological network (Brusak et al. 2006). To facilitate the implementation of these laws, the Ministry of Environmental Protection of Ukraine has issued the directive "Methodological Recommendations for the development of regional and local Econet schemes" (Order 13/11/2009 No. 604) (Government of Ukraine 2009). When corridors are endorsed by the relevant authorities (particularly the Land Resources Department and Regional Council), incompatible land use can be sanctioned and future projects have to take into account the integrity of the corridors. In order to determine the procedure to be followed for the creation of ecological corridors, which are recognized by the law and authorities, the sections of these Methodological Recommendations relevant for corridor development were extracted (Deodatus and Protsenko 2010) and combined with the modelling methodology for corridors applied (van Maanen et al. 2006). The different steps were further elaborated in consultation with government staff during the implementation of the procedure.

4 Results

The main result of this study is a methodology for the creation of ecological corridors developed in the context of landscape, ecology, land use and legislation, based on a pilot connecting protected areas in two different locations.

Habitat suitability maps prepared for different model species showed very similar suitability patterns in each pilot. Habitat suitability scores were generally higher for bison and lower for lynx, which is most probably related to human proximity. Nevertheless, merging "species corridors" into one "robust" corridor was straight forward in both areas.

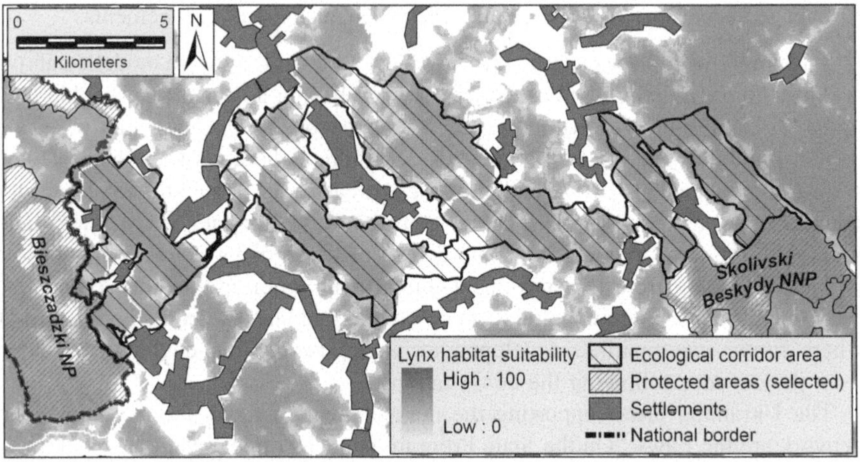

Fig. 3 The approved north-eastern corridor area related to the habitat suitability for lynx and the final corridor geometry tailored to land management units

The main barriers for connectivity were formed by residential areas mainly located in river valleys and to a lesser extent by agricultural areas, particularly when tree cover was poor. The potential corridor area in the south-eastern pilot is mainly covered by forest and has limited residential and agricultural areas. Only one “bottleneck area” with low connectivity due to a road and farmland was found here. The potential corridor area in the north-western pilot however, with a higher human occupation counted nine of such bottleneck areas. Field checks proved easy accessibility for animals of the bottleneck in the south-eastern pilot, as the hay-fields in the valley bottoms were only partly fenced, and patches of trees and shrubs provided cover for eventually crossing animals. Local people witnessed that the place is frequently crossed by deer and boars. In the north-western pilot four of the nine bottlenecks were hardly accessible for the animals, due to the high occurrence of agricultural land, residential area, fences and roads. Therefore the course of the final corridor here has been determined by the location of the other five “bottlenecks” which appeared to provide sufficient conditions for animal passage (Fig. 3). Consultations with land users and the local administration resulted in both pilots in agreement on the final corridor boundaries aligned with existing management units such as forestry management units and municipality borders.

After agreement on the location of the ecological corridors, approval has been received from local authorities and relevant technical authorities, including the spatial planning authorities. According to the Ukrainian regulations, a file has been elaborated for this purpose according to the specifications presented in the Directive “Methodological Recommendations for the development of regional and local Econet schemes” (2009) of the Ministry of Environmental Protection of Ukraine. Through extraction of relevant sections from this directive, a model has

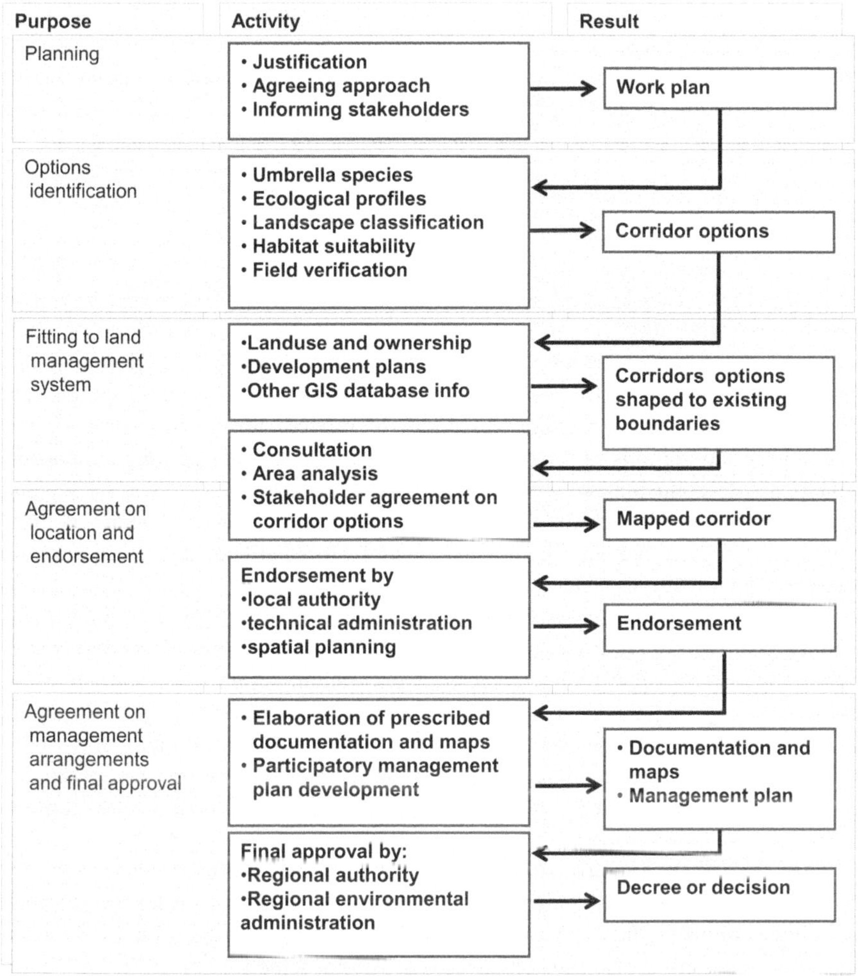

Fig. 4 The model for ecological corridor creation elaborated in the Ukrainian Carpathians, including corridor location identification, corridor management and inclusion in the regulatory land management system

been derived adapted to the application of ecological corridor creation, which resulted in the draft directive "Methodological Recommendations for Ecological Corridor Scheme Development". The file compiled according to this directive (the Ecological Corridor Scheme) includes a number of maps with prescribed formats as well as text sections presenting justification, general environmental context information, stakeholders, and a management plan of the proposed corridor. This file played a central role in the process leading to endorsement and to the inclusion of the ecological corridor into the spatial planning system. In the final stage of the procedure, corridor maps, corridor description and management plans have been approved by the relevant level state authorities ensuring their management on the long term. An outline of the entire procedure developed (Fig. 4) disposed of country-specific terminology and procedures can be considered as a guideline for corridor development in Carpathian countries.

5 Discussion

Several authors (Hanski 2002; O'Donell 2007; Nassauer and Opdam 2008; van der Windt and Swart 2008) claim close collaboration between scientists and government officers to be instrumental for effective ecological network development, as it provides opportunities for synergy and appropriate policy development and implementation. In the corridor development process presented in this chapter researchers and administrators fulfilled an indispensable role in ecological corridor development, contributing specific and highly complementary knowledge and experience. Since the use of GIS is increasingly important in spatial planning, administrators can benefit on the one hand from qualified staff from research institutes and from new technology. On the other hand, collaboration makes research staff more familiar with tuning their work to administrative requirements, leading to more effective use of scientific results. The process of corridor development is also an opportunity to create broad public support for ecological corridors by using major events for radio or television broadcastings. Governmental as well as non-governmental stakeholders developed commitment and understanding with regard to biodiversity conservation and the creation of an ecological network, while they were intensively involved in the corridor development process.

This study has shown that pilot projects are very useful to understand the constraints and gaps in the current framework for ecological corridor development and help to improve the methodologies used. Generally GIS has been a very useful and time-efficient tool that helped to focus the selection process to identify appropriate corridor area. Moreover GIS is a very flexible tool for the composition of cheap high-quality maps and on-screen support to decision processes through desktop comparison of corridor options. At the same time, the use of GIS has its limitations. It helps ordering information, but this process should be controlled and interpreted critically by the users, by comparing GIS output with field information. The finalisation of the corridors requires therefore "handwork", by visual map interpretation. The identification of "bottleneck areas" for field verification proved to be a useful approach, which, on the one hand, ensured realistic results of the corridor mapping and, on the other hand, helped to optimize time and resources required for field work.

To ensure the perpetual functioning of ecological corridors, agreements need to be reached on their management (Jędrzejewski et al. 2009; Mackey and Watson 2010). This can be done in a management plan specifying crucial elements such as stakeholders, responsibilities, measures and timing. Management plans are often based on a zoning system and it is usually convenient to match management zone boundaries as much as possible with land ownership and land use, to minimize the arrangements to be made with stakeholders. In western Carpathian countries (Hungary, Czech Republic and Slovakia) management plans or other arrangements are included in the respective legislation on ecological corridor development (Jongman and Kamphorst 2002). In Ukraine, however, both laws on ecological network development refer but indirectly to the use of management plans (Deodatus and Protsenko 2010). Therefore, the inclusion of an ecological corridor management plan has been proposed as compulsory in the Ecological Corridor Scheme in the "Methodological Recommendations for Ecological Corridor Scheme Development". The major challenge to make corridors work is the establishment of management arrangements accepted by all stakeholders (Bennett and Mulongoy 2006; Chettri et al. 2007; Lombard et al. 2010) and their enforcement. The final challenge is to adapt arrangements and supporting documentation to the requirements of the spatial planning authorities. Their acceptance and integration of the designed corridors into spatial management plans can assure their existence and functioning in the future, making them more resistant to threats such as infrastructure and other development plans (Nassauer and Opdam 2008).

Connectivity issues are usually (at least technically) more easily dealt with in agricultural areas as agriculture is not necessarily conflicting with connectivity (Lombard et al. 2010). Agriculture in most of the Carpathians has mainly an extensive character which is potentially compatible with wildlife presence. However, it may cause sometimes conflicts between wildlife and land users (poaching, crop damage, cattle predation). Handling this type of conflicts as well as the consolidation of extensive farming can be supported through an adequate High Nature Value farming policy in wildlife areas such as corridors (Hoogeveen et al. 2002; Andrews and Rebane 2005). A very useful tool reducing wildlife-human conflicts in this case and creating sometimes also opportunities for farmers is the land-swap instrument, which involves the swap of nature areas (parts of forests or protected areas) for agricultural land located in strategic parts of corridors. If crossings between ecological corridors and transport infrastructure cannot be avoided, constructions are recommended to enable connectivity such as wildlife overpasses and underpasses (Iuell et al. 2003; Jędrzejewski et al. 2009).

6 Conclusions

A pilot project to develop ecological corridors from paper plans to land allocated to connect biodiversity core areas agreed by all stakeholders provides a setting to develop and test a model for ecological corridor creation including lessons learned for wider application. Crucial for the functionality of corridors with regard to species migration are the ecological characteristics of the species selected to determine the habitat suitability criteria of the model. In this Ukrainian study, the establishment of two corridors has been realized as a result of combining the spatial modelling of corridors based on ecology and landscape with the administrative process that leads to inclusion of these corridors in the governmental land management system. By doing so, contributions of scientist and government authorities in the process are better geared to their purpose. The engagement of all stakeholders optimized the design and support for the establishment of the corridors. GIS maps provided effective support in this process, giving stakeholders accurate information on the location of optimal habitats of biological species and barriers. Field verification for the validation of this information and its interpretation proved to be essential. The actual choice of corridor location and boundaries should be realized through a dialogue of relevant authorities, land owners and land users to establish acceptance and future support. Using "bottleneck areas" as a key for corridor identification contributed very much to the efficiency of this process by focusing attention of consultations and analysis on these areas, enabling the exclusion of unsuitable corridor parts at an early stage. GIS maps indicating habitat suitability were very effective to locate these areas. Established corridors only make sense when their management is covered by agreements among stakeholders. The elaboration of management plans is therefore part of the model for corridor creation, and alignment of corridors with existing land management unit boundaries such as forest management units and municipality borders turned out to be instrumental.

Acknowledgments We are grateful to the valuable contributions of the following colleagues to this paper: Akos Gabor Ugron, Anatoliy Deyneka, Bohdan Prots, Borys Bagley, Dries Kuiper, Eddy Wymenga, Edith Oudt, Guus Schutjes, Hans Kampf, Harald Egerer, Harmanna Groothof, Henk Zingstra, Hieke van den Akker, Igor Ivanenko, Jana Urbancikova, Jan Kadlecik, Jan Seffer, Joep van den Vlasakker, Maaike Krol, Meeuwes Brouwer, Mike Baltzer, Mykhaylo Kokhanets, Mykhaylo Oprysko, Oksana Maryskevych, Oleg Kohan, Patrick Hostert, Tobias Kuemmerle, Ton Verhagen, Vasyl Pryndak, Victor Melnichuk, Volodymyr Domashlinets, Włodzimierz Jędrzejewski, Yuriy Lylo and Yuriy Zhebchuk, and Zbigniew Niewiadomski. Funding was provided by the Netherlands International Biodiversity Policy Programme BBI-MATRA of the Netherlands Ministry of Agriculture, Nature and Food Quality and the Ministry of Foreign Affairs, and the work was carried out with the Ministry of Environmental Protection of Ukraine.

References

Andrews J, Rebane M (2005) Farming and wildlife: a practical management handbook. USPB Conservation Management Series, Kiev

Anon. (2004) Status and trends for large carnivores in Europe. LCIE report for the UNEP-WCMC project biodiversity trends and threats in Europe. Large carnivore initiative for Europe, Kraków

Anon (2007) Carpathians environment outlook 2007. UNEP, Geneve

Bashta AT (2004) O restytucji, rozmieszczeniu i liczebności żubra Bison bonasus w Beskidach Ukraińskich (Karpaty Wschodnie). Rocz Bieszcz 12:253–260

Bashta AT, Potish L (2005) Population status and distribution of the wildcat Felis silvestris (Felidae, Carnivora) in Transcarpathia (Ukraine). Nat Conserv 61(4):53–59

Bashta AT, Potish L (2007) Mammals of the Transcarpathian region (Ukraine). NASU, Lviv

Beier P, Majka DR, Jenness J (2007) Conceptual steps for designing wildlife corridors. http://www.corridordesign.org. Accessed 14 Oct 2010

Beier P, Majka DR, Newell S, Garding E (2008) Best management practices for wildlife corridors. Northern Arizona University, Flagstaff

Beier P, Majka DR, Newell SL (2009) Uncertainty analysis of least-cost modeling for designing wildlife linkages. Ecol Appl 19:2067–2077

Bennett G, Mulongoy KJ, (2006) Review of Experience with Ecological Networks, Corridors and Buffer Zones. CBD Technical Series 23, http://www.cbd.int/ts/. Accessed 30 June 2012

Billionnet A (2010) Integer programming for optimizing habitat network permeability. Manag Environ Qual Intern J 21:570–588

Brusak V, Bezusko A, Voznyi Y, Zinko Y, Feldbaba-Klushyna L, Masikevych Y, Matveyev S, Movchan Y, Hjhjvych S, Pruhodko M (2006) Ecological network scheme of the Ukrainian Carpathians. Living Ukr 9–10:8–9

Burdusel E, Kanianska R, Maryskevych O (2006) Policy consultation on sustainable agriculture and rural development in the Carpathians. UNEP-Vienna ISCC, Vienna

Chettri N, Sharma E, Shakya B, Bajracharya B (2007) Developing forested conservation corridors in the Kangchenjunga Landscape, Eastern Himalaya. Mt Res Dev 27(3):211–214

Deodatus FD, Protsenko L (eds) (2010) Creation of ecological corridors in Ukraine. A manual on stakeholder involvement and landscape-ecological modelling to connect protected areas, based on a pilot in the Carpathians. State agency for protected areas of the ministry of environmental protection of Ukraine, Altenburg and Wymenga Ecological Consultants, InterEcoCentre, Kiev

Domnich VI, Smirnova IA, Shadura AN, Domnich AV (2009) The alteration of the number of *Cervidae* and *Canidae* under the influence of the game and poaching in Ukraine. Proceedings of the XXIX international union of game biologists congress, Moscow

Government of Ukraine (2000) Law of Ukraine "On the State Programme of Ukraine's National Ecological Network Development for 2000–2015" (2000). Government of Ukraine

Government of Ukraine (2004) Law of Ukraine "On the Ecological Network of Ukraine" (2004). Government of Ukraine

Government of Ukraine (2009) Methodological Recommendations for the development of regional and local Econet schemes (Order 13/11/2009 No. 604). Government of Ukraine

Gurrutxaga M, Lozano PJ, del Barrio P (2010) GIS-based approach for incorporating the connectivity of ecological networks into regional planning. J Nat Conserv 18:318–326

Hanski I (1998) Connecting the parameters of local extinction and metapopulation dynamics. Oikos 83(2):390–396

Hanski I (2002) In the midst of ecology, conservation, and competing interests in the society. Ann Zool Fennici 39:183–186

Hanski I (2005) Landscape fragmentation, biodiversity loss and the societal response. The longterm consequences of our use of natural resources may be surprising and unpleasant. EMBO Rep 6(5):388–392

Hepcan S, Hepcan ÇC, Bouwma IM, Jongman RHG, Özkan MB (2009) Ecological networks as a new approach for nature conservation in Turkey: a case study of Izmir province. Landsc Urban Plan 90:143–154

Herenchuk KI (ed) (1968) Pryroda ukrayinskykh karpat. Vydavnytstvo Lvivskoho Universytetu, Lviv

Hoogeveen Y, Gabrielsen P, Petersen JE (2002) High nature value farming areas. Defining the concept and developing an agri-environmental indicator. Proceedings of an expert meeting, European Environment Agency, Copenhagen, 21–22 Feb 2002

Hostert P, Chaskovskyy O, Knorn J, Kuemmerle T (2008) Ukrainian Carpathians land cover map. Geomatics Lab, Humboldt University, Berlin

Iuell B, Bekker GJ, Cuperus R, Dufek J, Fry G, Hicks C, Hlavac V, Keller V, Rosell B, Sangwine CT, Torslov N, le Wandall Maire B (eds) (2003) COST 341—Wildlife and Traffic: a European Handbook for Identifying Conflicts and Designing Solutions. European Co-operation in the Field of Scientific and Technical Research, Brussels

Jarvis A, Reuter HI, Nelson A, Guevara E (2008) Hole-Filled SRTM for the Globe Version 3. Int Center for Trop Agric, Cali, Colombia. http://srtm.csi.cigar.org. Accessed 12 May 2007
Jędrzejewski W, Nowak S, Kurek R, Myslajek RW, Stachura K, Zawadzka B, Pchalek M (2009) Animals and roads—methods of mitigating the negative impact of roads on wildlife. Mammal Research Institute, Polish Academy of Sciences, Białowieża
Jongman R, Kamphorst D (2002) Ecological corridors in land use planning and development policies. Nature and environment 125. Council of Europe Publishing, Strasbourg
Jordan F (2000) A reliability-theory approach to corridor design. Ecol Model 128:211–220
Klar N, Fernández N, Kramer-Schadt S, Hermann M, Trinzen M, Brüttner I, Niemitz C (2008) Habitat selection models for European wildcat conservation. Biol Conserv 141:308–319
Krasinska M, Krasinski Z (2007) European Bison: the nature monograph. Mammal Research Institute, Polish Academy of Sciences, Białowieża
Kruhlov I (2008) Delimitatsiya, metryzatsiya ta klasyfikatsiyaomrfogennykh ekoregioniv Ukrayinskukh Karpat. Ukr Geogr Zhurnal 3:59-68
Kubijovyc V (1984) Carpathian Mountains (Karpaty). Encyclopedia of Ukraine, vol 1. Canadian Institute of Ukrainian Studies, Edmonton
Kuemmerle T, Hostert P, Perzanowski K, Radeloff VC (2006) Cross-border comparison of land cover and landscape pattern in Eastern Europe using a hybrid classification technique. Remote Sens Environ 103:449–464
Kuemmerle T, Hostert P, Radeloff VC, Perzanowski K, Kruhlov I (2007) Post-Socialist forest disturbance in the Carpathian border region of Poland, Slovakia, and Ukraine. Ecol Appl 17:1279–1295
Kuemmerle T, Hostert P, Radeloff VC, van der Linden S, Perzanowski K, Kruhlov I (2008) Cross-border comparison of post-socialist farmland abandonment in the Carpathians. Ecosyst 11:614–628
Kuemmerle T, Chaskovskyy O, Knorn J, Radeloff VC, Kruhlov I, Keeton WS, Hostert P (2009) Forest cover change and illegal logging in the Ukrainian Carpathians in the transition period from 1988 to 2007. Remote Sens Environ 113:1194–1207
Kuijper DPJ, Cromsigt JPGM, Churski M, Adam B, Jedrzejewska W (2009) Do ungulates preferentially feed in forest gaps in European temperate forests? For Ecol Manag 258:1528–1535
Lombard AT, Cowling RM, Fabricius C (2010) Designing conservation corridors in production landscapes: assessment methods, implementation issues, and lessons learned. Ecol Soc 15(3):7
MacArthur RH, Wilson EO (1967) The theory of island biogeography. Princeton University Press, Princeton
Mackey B, Watson J (2010) Connectivity conservation and the Great Eastern Ranges corridor. Department of Environment, Climate Change and Water NSW, Sydney
Majka D, Beier P, Jenness J (2007) Corridor designer ArcGIS toolbox tutorial. http://www.corridordesign.org. Accessed 13 Oct 2010
Marulli J, Mallarch JM (2005) A GIS methodology for assessing ecological connectivity: application to the Barcelona Metropolitan area. Landsc Urban Plan 71:243–262
Nassauer JI, Opdam P (2008) Design in science: extending the landscape ecology paradigm. Landsc Ecol 23:633–644
Nowell K, Jackson P (1996) Wild cats: status survey and conservation action plan. IUCN, Gland
O'Donell F (ed) (2007) National environmental policy of Ukraine: assessment and develop-ment strategy. Ministry of Environmental Protection of Ukraine—GEF, Kiev
Phillips SJ, Williams P, Midgley G, Archer A (2008) Optimizing dispersal corridors for the Cape Prote aceae using network flow. Ecol Appl 18:1200–1211
Pucek Z, Belousova IP, Krasińska M, Krasiński ZA, Olech W (2004) Status Survey and Conservation Action Plan European Bison. IUCN, Gland
Ray JC, Redford KH, Steneck RS, Berger J (2005) Large carnivores and the conservation of biodiversity. Island Press, Washington

Rayfield B, Fortin MJ, Fall A (2010) The sensitivity of least-cost habitat graphs to relative cost surface values. Landsc Ecol 25:519–532

Roy A, Devi BSS, Debnath B, Murthy MSR (2010) Geospatial modelling for identification of potential ecological corridors in Orissa. J Indian Soc Remote Sens 38:387–399

Schadt S, Knauer F, Kaczensky P, Revilla E, Wiegand T, Trep L (2002) Rule-based assessment of suitable habitat and patch connectivity for the Eurasian lynx. Ecol Appl 12:1469–1483

Servheen C, Herrero H, Peyton B (1998) Bears: status survey and conservation action plan. IUCN, Gland

Simeonova V, Bos E, Jongman R, Zingstra H (2009) Implementation of ecological networks in different socio-economic contexts. Guiding principles based on experiences in central and Eastern Europe. Alterra rapport 1896. Alterra, Wageningen

Slobodian AA (1988) Ecology and ethology of brown bear. Knowledge about the mammal fauna of Ukraine, its rational use and conservation. Naukova dumka, Kyiv

Slobodian AA (1993) Ukraine/Bears. Brown bear, Polar bear, Asiatic black bear. Distribution, ecology, use and conservation. Nauka, Moskva

Turianyn II (1988) Cats of the Ukrainian Carpathians. Knowledge about the mammal fauna of Ukraine, its rational use and conservation. Naukova dumka, Kyiv

Urban DL, Minor ES, Treml EA, Schick RS (2009) Graph models of habitat mosaics. Ecol Lett 12:260–273

Van Der Windt HJ, Swart JAA (2008) Ecological corridors, connecting science and politics: the case of the Green River in the Netherlands. J Appl Ecol 45:124–132

Van Maanen E , Altenburg W, Klaver R, Predoiu G, Popa M, Ionescu O, Jutj R, Negus S, Ionescu G (2006) Safeguarding of the Romanian Carpathian ecological network. A vision for large carnivores and biodiversity in Eastern Europe. A&W ecological consultants, Veenwouden, The Netherlands. ICAS Wildlife Unit, Brasov, Romania

Webster R, Holt S, Avis C (eds) (2001) The status of the Carpathians. A report developed as a part of the Carpathian Ecoregion initiative. WWF, Vienna

Erratum to: Recent Forest Cover Change in Low Mountain Landscapes of Lviv Oblast in the Ukrainian Carpathians

Anatoliy Smaliychuk and Ivan Kruhlov

Erratum to:
Chapter "Recent Forest Cover Change in Low Mountain Landscapes of Lviv Oblast in the Ukrainian Carpathians" in: J. Kozak et al. (eds.), *The Carpathians: Integrating Nature and Society Towards Sustainability*, DOI 10.1007/978-3-642-12725-0_47

There is a mistake in the figures 3 and 5—one figure was put twice—Fig. 5 is correct but Fig. 3 should be replaced with the following figure.

The online version of the original chapter can be found under DOI 10.1007/978-3-642-12725-0_47

A. Smaliychuk (✉) · I. Kruhlov
Faculty of Geography, Ivan Franko National University of Lviv,
Doroshenka St. 41 79000 Lviv, Ukraine
e-mail: zljukalviv@gmail.com

J. Kozak et al. (eds.), *The Carpathians: Integrating Nature and Society Towards Sustainability*, Environmental Science and Engineering,
DOI: 10.1007/978-3-642-12725-0_50, © Springer-Verlag Berlin Heidelberg 2013

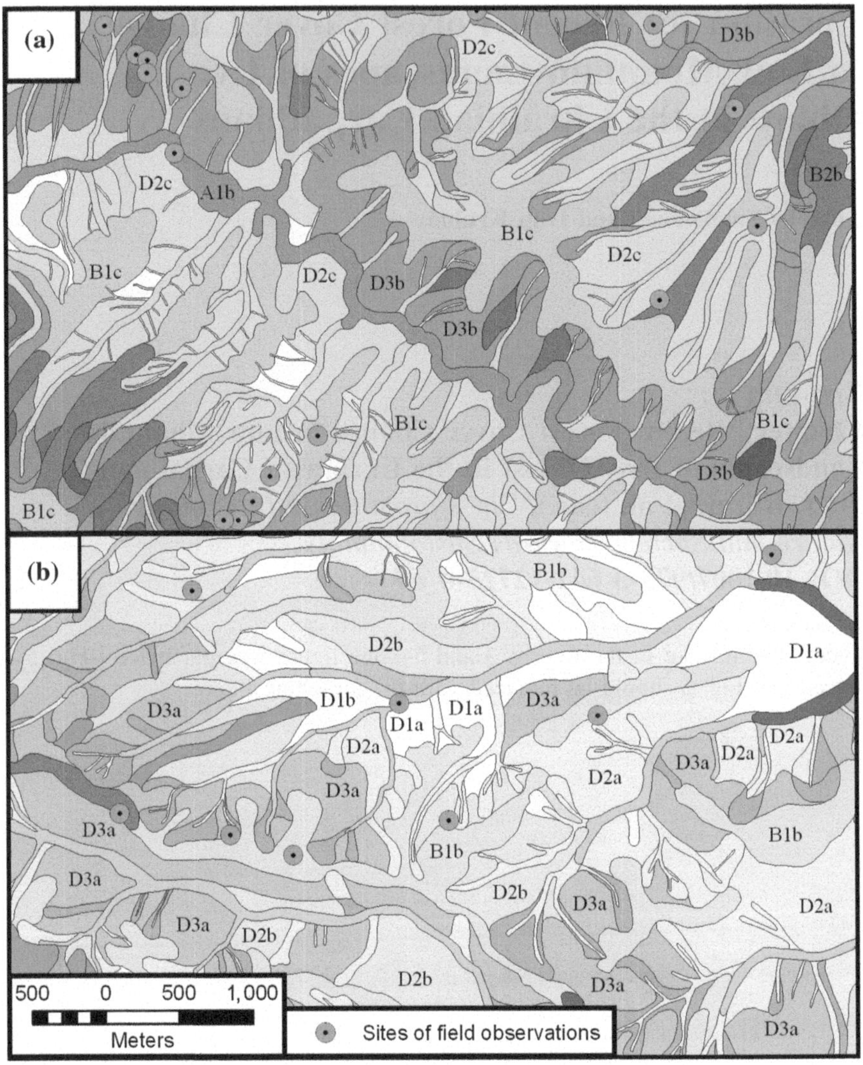

Fig. 3 Fragment of the 1:50,000 map of geoecosystems: **a** Boberka, **b** Stara Sil' (explanations to geoecosystems types in Table 1)

The manufacturer's authorised representative in the EU is Springer Nature Customer Service Centre GmbH, Europaplatz 3, 69115 Heidelberg, Germany. If you have any concerns regarding our products, please contact ProductSafety@springernature.com

Printed and bound by CPI Group (UK) Ltd, Croydon, CR0 4YY
30/09/2026
02225602-0019